Single-Variable Calculus

ROBERT A. ADAMS
Professor, Department of Mathematics
The University of British Columbia

Paul Nathan
675-0994

Single-Variable Calculus

REVISED EDITION

ADDISON-WESLEY PUBLISHERS LIMITED
Don Mills, Ontario • Reading, Massachusetts
Menlo Park, California • London
Amsterdam • Sydney

Sponsoring Editor: *Ron Doleman*
Production Editor: *Laura R. Skinger*
Text Designer: *Margaret Ong Tsao*
Cover Designer: *Gord Pronk*
Illustrator: *ARVAK*
Art Coordinator and Illustrator: *Joseph K. Vetere*
Production Manager: *Sherry Berg*

Canadian Cataloguing in Publication Data

Adams, R. A. (Robert Alexander), 1940–
Single-variable calculus

Includes index.
ISBN 0-201-11114-4

1. Calculus, Differential. 2. Calculus, Integral.
I. Title.

QA303.A32 1986 515.3 C86-094201-5

ISBN 0-201-11114-4
ABCDEFGHIJ-DO-91 90 89 88 87 86

Dedicated to the Memory of

DR. RONALD C. RIDDELL

Preface

Single-Variable Calculus provides a complete introduction to the calculus of functions of a single real variable. It is organized for a two-semester course and treats all the topics usually covered over such a period. Since there is more material in the text than can be presented in two semesters, instructors are encouraged to select from the various optional topics. The level of presentation is suitable for general science and engineering students, and it is assumed that these students have some facility in high-school algebra and have been exposed to elementary analytic geometry (equations of lines and circles). As many students are intimidated by the size of older calculus textbooks, the author has endeavored to keep this text compact. Multivariate calculus, which deserves a book of its own, has been omitted, and other topics included have been condensed with no loss of clarity. As a result, this text is not as massive or bulky in appearance as many other books available in recent years.

Principal Features

- There is an emphasis on geometry. The interpretation of the derivative as the slope of the tangent line is introduced early and reinforced in examples and exercises thereafter.
- Trigonometric functions, exponentials, and logarithms are developed before applications of differentiation so that these functions can be *freely* used in the applications.
- Precise statements are given for theorems. Proofs are given immediately if short, but are sometimes postponed to the end of the current section or to the Appendices.

- Exercises are varied in type and difficulty. Besides providing drill in the basic techniques, many exercises are intended to sharpen skills in algebraic manipulation.
- Some theoretical exercises are included to stimulate deeper understanding of the topics concerned. More difficult and/or theoretical exercises are marked with an asterisk (*).
- Some exercises are phrased in terms of differential equations and initial-value problems. These are marked with a dagger (†).
- Approximation methods are accompanied by error estimates, which are often derived, and examples and exercises that encourage the use of scientific calculators.

Three theoretical appendices are included. The first deals with mathematical induction, the second with formal proofs of limit theorems and the derivation of properties of continuous functions, and the third with the Riemann integral. This material would not be suitable for most classes but may be used selectively by an instructor to provide enrichment for individual able and motivated students.

Some material on differential equations is provided, although no special chapter is set aside for it. The equations $y' = ky$ and $y'' + k^2y = 0$ are discussed along with the functions that provide their solutions. Solutions of more general constant coefficient and Euler equations are provided via exercises. A section on linear and separable first order equations is provided in Chapter 7.

The division of material into "core" and "optional" in the chart on page ix is somewhat arbitrary. The basic concepts and techniques of differential calculus are developed in Chapters 2 through 4 and those of integral calculus in Chapters 6 and 7. All this material draws on the discussion of functions and limits given in Chapter 1. Some of the applications of differentiation in Chapter 4 and many of those of integration in Chapter 7 can be omitted without loss of coherence. Chapter 5 (on parametric and polar curves) can be omitted. If it is, Section 7.4 will also have to be omitted.

The object of the last two chapters is to obtain and apply the power series representations (Maclaurin series) for the elementary functions $\ln(1 + x)$, e^x, $\sin(x)$, $\cos(x)$, and $(1 + x)^r$. Where time permits, these can be approached via a study of numerical series (Chapter 8). As a shorter alternative, one can begin with Section 9.5 (Taylor's formula) and use this to derive the Maclaurin series of Section 9.3. The applications in Sections 9.4 and 9.5 can then be done.

Acknowledgments

Preliminary versions of this text have been used for two years (by about 2500 students per year) in first-year calculus courses at the University of British Columbia. The author is grateful to many students and colleagues there for their encouragement and useful comments and criticisms. In addition, the author acknowledges with gratitude the careful reviews and thoughtful suggestions received from reviewers elsewhere, in particular, Professors D. Gates and H. de Bryn of Vanier College, K. Taylor of the University of Saskatchewan, L. Florence of the University of Toronto, P. Kumar of Columbia University, J. Bebernes of the University of Colorado, A. Gerhard of the

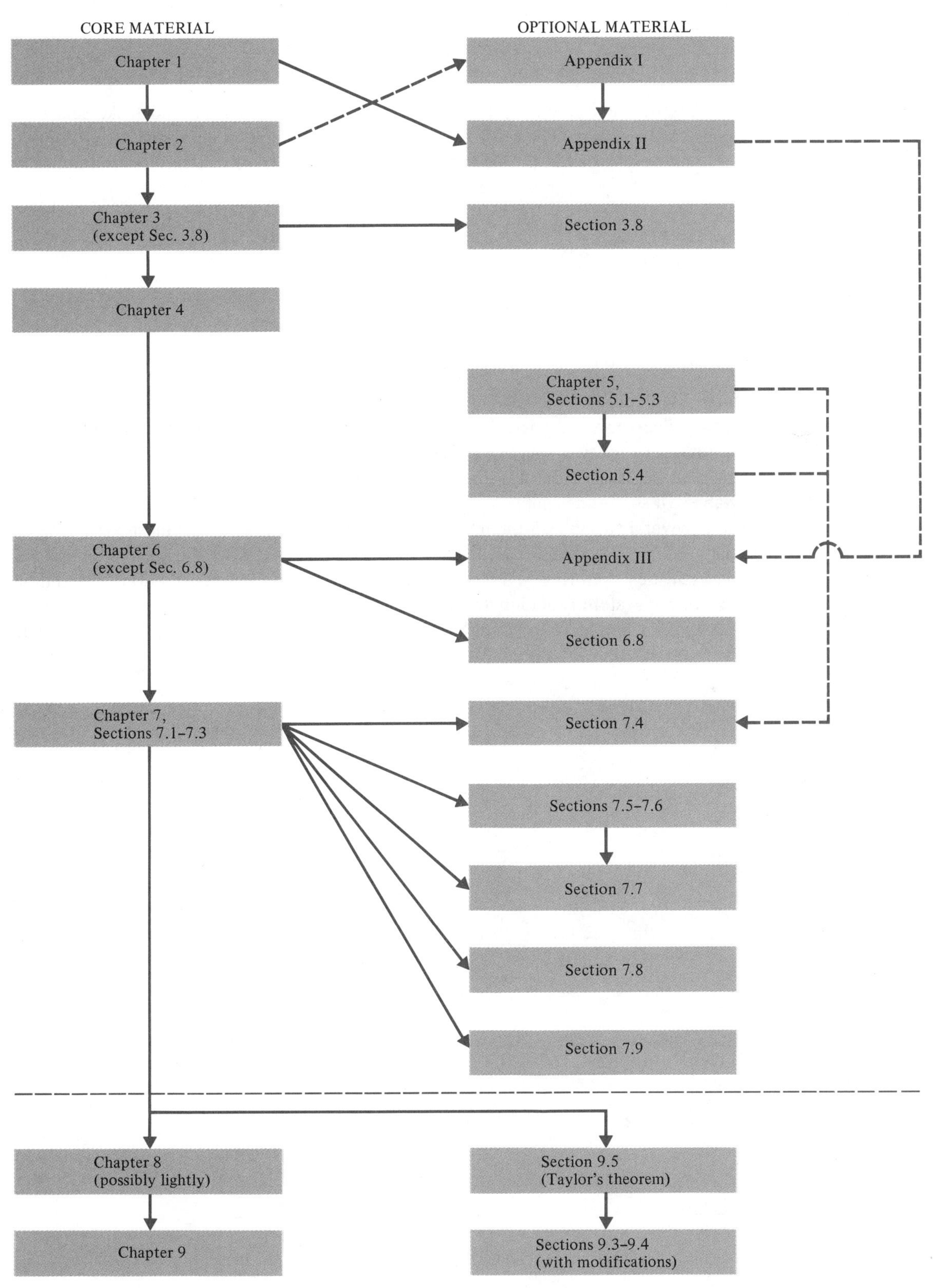
CORE MATERIAL
OPTIONAL MATERIAL
Chapter 1
Appendix I
Chapter 2
Appendix II
Chapter 3
(except Sec. 3.8)
Section 3.8
Chapter 4
Chapter 5,
Sections 5.1–5.3
Section 5.4
Chapter 6
(except Sec. 6.8)
Appendix III
Section 6.8
Chapter 7,
Sections 7.1–7.3
Section 7.4
Sections 7.5–7.6
Section 7.7
Section 7.8
Section 7.9
Chapter 8
(possibly lightly)
Section 9.5
(Taylor's theorem)
Chapter 9
Sections 9.3–9.4
(with modifications)

University of Manitoba, S. Willard of the University of Alberta, and J. A. Senez of Concordia University. The efforts of these and others have ensured that errors, omissions, and obscurities in the original versions were corrected. Additional comments and revisions from our readers are welcome.

Vancouver, Canada *R.A.*
January 1983

Preface to Revised Edition

This revised edition includes a new chapter, Chapter 10, ''Partial Differentiation.'' A complete course in real-variable calculus is normally spread over three or four semesters with the first two concerned mainly with functions of a single variable and the remainder devoted to the differentiation and integration of functions of two or more variables. It sometimes happens that a short introduction to the differentiation of functions of several variables is included towards the end of the second semester of a single-variable course. This chapter is intended for such situations. Also, the ''errors, omissions and obscurities'' referred to above have been corrected in this new edition. The author would like to gratefully acknowledge the careful review of Chapter 10 by the following: J. Wainwright, University of Waterloo, R. D. Norman, Queen's University, E. J. Barbeau, University of Toronto, and D. W. Trim, University of Manitoba.

Vancouver, Canada *R.A.*
May 1986

Contents

1 *FUNCTIONS, LIMITS, AND CONTINUITY* *1*

1.1 The Real Line and the Cartesian Plane 2
Intervals 4
The Absolute Value 6
Coordinates and Graphs 8

1.2 Functions 14
Operations on Functions 19
Odd and Even Functions 20
Inverse Functions 21

1.3 Limits 26

1.4 Extensions of the Limit Concept 33
One-Sided Limits 33
Limits at Infinity 36
Infinite Limits 38

1.5 Continuity 41

2 *DIFFERENTIATION: DEFINITION, INTERPRETATION, AND TECHNIQUES* **49**

2.1 Tangent Lines and Their Slopes 50
Normals 54

2.2 The Derivative 56
Leibniz Notation and Differentials 63

2.3 Differentiation Rules (Sums, Products, Quotients) 66
Sums and Constant Multiples 66
The Product Rule 67
The Reciprocal Rule 69
The Quotient Rule 70

2.4 The Chain Rule 72
Derivatives of Inverse Functions 75
Proof of the Chain Rule 76

2.5 Interpretations of the Derivative 79
Velocity and Acceleration 80
Marginals 82

2.6 Higher-Order Derivatives 84
Differential Equations and Initial-Value Problems 87

2.7 Implicit Differentiation 88

2.8 The Mean-Value Theorem 92
Increasing and Decreasing Functions 96
Proof of the Mean-Value Theorem 98

2.9 Antiderivatives and Indefinite Integrals 101
Review Exercises for Chapter 2 105

3 *THE ELEMENTARY TRANSCENDENTAL FUNCTIONS* **109**

3.1 The Circular (Trigonometric) Functions 110
The Addition Formulas 115
The Functions Tangent, Secant, Cotangent and Cosecant 117
Some Trigonometry 119

3.2 Derivatives of the Trigonometric Functions 123
The Projectile Problem 127
Simple Harmonic Motion 130

3.3 The Inverse Trigonometric Functions 134

3.4 The Natural Logarithm 143

3.5 The Exponential Function 149

3.6 General Exponentials and Logarithms 159
Logarithmic Differentiation 162

3.7 Growth and Decay Problems 165
Logistic Growth 169

3.8 The Hyperbolic Functions and Their Inverses 172
Review Exercises for Chapter 3 177

4 *VARIOUS APPLICATIONS OF DIFFERENTIATION* ***179***

4.1 Critical Points and Extreme Values 180
The First-Derivative Test 184

4.2 Concavity and Inflections 189
The Second-Derivative Test 193

4.3 Sketching the Graph of a Function 196

4.4 Optimization Problems 207
Suggested Procedures for Solving Optimization Problems 210

4.5 Related Rates 218
Procedures for Solving Related-Rates Problems 220

4.6 Tangent-Line Approximations 226
The Error Estimate 227
Errors in Measurement 229
Newton's Method 231

4.7 Indeterminate Forms 237
The Generalized Mean-Value Theorem 238
l'Hôpital's Rules 239

5 *CURVES IN THE PLANE* ***245***

5.1 Parametric Curves 246
General Plane Curves: Parametrizations 249
Some Interesting Parametric Curves 250

5.2 Smooth Curves 254

The Slope of a Parametric Curve 254

Sketching Parametric Curves 256

5.3 Velocity and Acceleration in the Plane 258

Plane Vectors 259

Position, Velocity, and Acceleration as Vectors 263

5.4 Polar Coordinates and Polar Curves 268

Some Polar Curves 270

Conics 272

The Slope of a Polar Curve 275

6 *INTEGRATION* ***279***

6.1 Approximating Areas Using Rectangles 280

6.2 The Definite Integral 289

Properties of the Definite Integral 294

The Mean-Value Theorem for Integrals 296

6.3 The Fundamental Theorem of Calculus 300

6.4 The Method of Substitution 307

Trigonometric Integrals 311

6.5 Inverse Substitutions 316

The Inverse Trigonometric Substitutions 317

Completing the Square 320

Other Substitutions 321

6.6 The Method of Partial Fractions 324

6.7 Integration by Parts 334

Reduction Formulas 338

Summary of Techniques of Integration 340

Review Exercises on Techniques of Integration 341

6.8 Approximate Integration 342

The Trapezoid Rule 343

Simpson's Rule 347

6.9 Improper Integrals 351

Integrals of Bounded, Piecewise Continuous Functions 356

Estimating Convergence and Divergence 357

7 *APPLICATIONS OF INTEGRATION* **361**

7.1 Areas of Plane Regions 362
Area Between Two Curves 363

7.2 Volumes 368
Solids of Revolution 372
Cylindrical Shells 375

7.3 Arc Length and Surface Area 380
Arc Length 380
Areas of Surfaces of Revolution 384

7.4 Geometric Applications for Polar and Parametric Curves 387
Regions Bounded by Polar Curves 387
Regions Bounded by Parametric Curves 390
Arc Lengths for Parametric and Polar Curves 394

7.5 Mass, Moments, and Center of Mass 398
Moments and Center of Mass 401

7.6 Centroids 406
The Pappus Theorem 409

7.7 Other Physical Applications 411
Hydrostatic Pressure 411
Work 413
Potential and Kinetic Energy 415

7.8 Probability 418
Expectation, Mean, Variance, and Standard Deviation 422
The Normal Distribution 425

7.9 First-Order Separable and Linear Differential Equations 430
Separable Equations 430
First-Order Linear Equations 434

8 *INFINITE SERIES* **439**

8.1 Sequences and Convergence 440

8.2 Infinite Series 447

8.3 Convergence Tests for Positive Series 453
The Comparison Tests 454

The Integral Test 458
The Ratio Test 460

8.4 Absolute and Conditional Convergence 463
The Alternating Series Test 465

8.5 Estimating the Sum of a Series 471
Integral Bounds 471
Geometric Bounds 475
Alternating Series Bounds 476

9 *POWER SERIES REPRESENTATIONS OF FUNCTIONS* ***479***

9.1 Power Series 480
Algebraic Operations on Power Series 482

9.2 Differentiation and Integration of Power Series 485
Proofs of Theorems 4 and 5 488

9.3 Taylor and Maclaurin Series 492
Maclaurin Series for the Elementary Functions 493
Other Maclaurin and Taylor Series 496

9.4 Applications of Taylor and Maclaurin Series 500
Approximating the Values of Functions 500
Functions Defined by Integrals 501
Indeterminate Forms 502

9.5 Taylor's Theorem 504
Applications of Taylor's Formula 507
The Binomial Series 509

10 *PARTIAL DIFFERENTIATION* ***513***

10.1 Functions of Several Variables 514
Graphical Representation 514
Limits and Continuity 517

10.2 Partial Derivatives 522
Tangent Planes 524
Higher Order Derivatives 526

10.3 The Chain Rule and Differentials 531
Homogeneous Functions 535
Higher Order Derivatives 535
Differentials and Differentiability 536

Proof of the Chain Rule 538

10.4 Gradients and Directional Derivatives 541

Directional Derivatives 542

Tangent Lines to Level Curves 545

Higher Dimensional Vectors 546

The Gradient in Higher Dimensions 546

Tangents to the Graphs of Functions 548

10.5 Extreme Values 551

Extreme Values of Functions Defined on Closed, Bounded Sets 557

Extreme Value Problems with Constraints 557

APPENDICES ***563***

I Mathematical Induction 563

II The Theoretical Foundations of Calculus 566

Limits of Functions 567

Continous Functions 573

Completeness and Sequential Limits 574

Continous Functions on a Closed, Finite Interval 575

III The Riemann Integral 578

Uniform Continuity 582

TABLES ***587***

1 Trigonometric Functions 587

2 Exponential Functions 589

3 Natural Logarithms 590

4 Powers and Roots 591

5 Table of Integrals 592

ANSWERS TO ODD-NUMBERED EXERCISES ***599***

INDEX ***640***

1 Functions, Limits, and Continuity

1.1 The Real Line and the Cartesian Plane

1.2 Functions

1.3 Limits

1.4 Extensions of the Limit Concept

1.5 Continuity

Underpinning the study of calculus are the concepts of real number, coordinate system, and function. In the first two sections of this chapter we will review these concepts and set out the terminology and symbols we will use in referring to them throughout the book. The remaining sections introduce and explore the concept of limit, an operation on functions. The use of limits distinguishes calculus from other branches of mathematics (arithmetic, algebra, geometry) you have already encountered.

1.1 THE REAL LINE AND THE CARTESIAN PLANE

Elementary calculus depends heavily on properties of **real numbers,** that is, numbers expressible in decimal form such as

$$\begin{aligned} 5 &= 5.00000\ldots \\ -\tfrac{3}{4} &= -0.750000\ldots \\ \tfrac{1}{3} &= 0.3333\ldots \\ \sqrt{2} &= 1.4142\ldots \\ \pi &= 3.14159\ldots \end{aligned}$$

We expect that as a student of calculus you already have some familiarity with the real numbers and with the Cartesian coordinate system in the plane. Both are treated only briefly here to fix the terminology.

The real numbers can be represented geometrically as points on a number line, which we will call the *real line,* represented in Fig. 1.1. The symbol $\mathbb{R}$ is used to denote either the real number system or, equivalently, the real line.

The properties of the real number system fall into three categories: algebraic properties, order properties, and completeness. The algebraic properties will already be familiar to you and we will not dwell on them here; roughly speaking, they assert that to real numbers there may be applied all the usual operations of arithmetic—addition, subtraction, multiplication, and division (except by zero), that these operations always produce real numbers, and that the real numbers satisfy all the usual laws of arithmetic.

The *order properties* refer to the order in which the numbers appear on the real line. If x lies to the left of y, then we say $x < y$ or $y > x$. Of course $x \leq y$ means that either $x < y$ or $x = y$. The order properties can be summarized as follows:

i) If $x < y$ and z is any real number, then $x + z < y + z$.

ii) If $x < y$ and $z > 0$, then $xz < yz$.

iii) If $x < y$ and $z < 0$, then $xz > yz$; in particular, $-x > -y$.

iv) If $0 < x < y$ then $0 < \dfrac{1}{y} < \dfrac{1}{x}$.

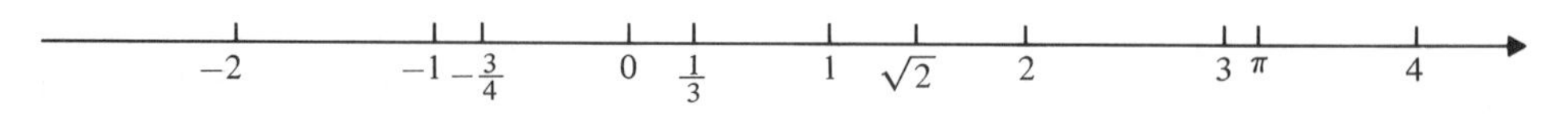

Figure 1.1

Note especially the rules for multiplying an inequality by a number. If the number is positive the inequality is preserved; if the number is negative the inequality is reversed.

The *completeness* property of the real number system is more subtle and difficult to understand. One way to state it is as follows: If A is any set of real numbers having at least one number in it, and if there exists a real number y with the property that $x \le y$ for every x in A, then there exists a *smallest* number y with the same property. Roughly speaking, this says that there can be no holes or gaps on the real line—every point corresponds to a real number. Certain important results in calculus require the completeness property for their proofs. Most of these results can be derived with no great difficulty from a few basic theorems, in particular Theorems 5 and 6 in Section 1.5. We do not prove these theorems in this chapter, but we sketch their proofs in Appendix 2, which is concerned with the theoretical foundations of calculus. The techniques for formal proofs involving limits in that appendix often are not studied in first courses in calculus but are deferred to subsequent courses in mathematical analysis. We will, however, make some direct use of completeness when we study infinite sequences and series in Chapter 8.

We distinguish three special subsets of the real numbers:

i) the natural numbers, namely, the numbers 1, 2, 3, 4, . . .
ii) the integers, namely, the numbers 0, ± 1, ± 2, ± 3, . . .
iii) the rational numbers, that is, numbers that can be expessed in the form m/n where m is an integer and n is a natural number.

The rational numbers are precisely those real numbers with decimal expansions that are either:

a) terminating, that is, ending with an infinite string of zeros, or
b) repeating, that is, ending with a string of digits that repeats over and over.

Example 1 Show that the numbers

a) $1.323232\ldots = 1.\overline{32}$ and
b) $0.3405405405\ldots = 0.3\overline{405}$

are rational numbers by expressing each as a quotient of two integers. (The bars indicate the pattern of repeating digits.)

Solution

a) Let $x = 1.323232\ldots$ Then $x - 1 = .323232\ldots$ and $100x = 132.323232\ldots = 132 + .323232\ldots = 132 + x - 1$.

Therefore, $99x = 131$ and $x = \dfrac{131}{99}$.

b) Let $y = 0.3405405405\ldots$ Then $10y = 3.405405405\ldots$ and $10y - 3 = .405405405\ldots$ Also,

$$10000y = 3405.405405405\ldots = 3405 + 10y - 3.$$

Therefore, $9990y = 3402$ and $y = 3402/9990 = 63/185$.

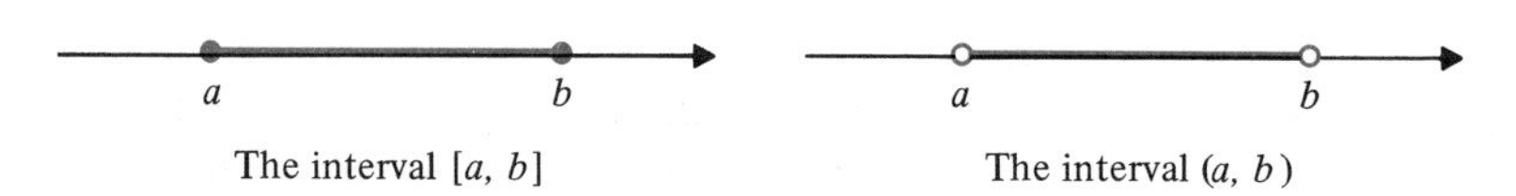

Figure 1.2

The set of rational numbers possesses all the algebraic and order properties of the real numbers, but not the completeness property. There is, for example, no rational number whose square is 2. Hence there is a "hole" on the "rational line" where $\sqrt{2}$ should be. To see this, suppose that there were a rational number $x = m/n$ such that $x^2 = 2$. We can assume that any common factors in the fraction m/n have been canceled, so that m and n are not both even integers. Since $m^2 = 2n^2$, m^2 must be even, and therefore m must also be even. (A product of odd numbers is always odd.) Since m is even we can write $m = 2k$ where k is an integer. Thus $4k^2 = 2n^2$ and $n^2 = 2k^2$; n^2 is even and so is n. We have arrived at a contradiction; we assumed m and n were not both even and then proved they were both even. Accordingly there can be no rational number whose square is 2. Because the real line has no such "holes" in it, the real line is the appropriate setting for studying limits and therefore calculus.

Intervals

A subset of the real line is called an **interval** if it contains at least two numbers and also contains all real numbers between any two of its elements. For example, the set of real numbes x such that $x > 6$ is an interval, but the set of real numbers y such that $y \neq 0$ is not an interval. (Why?)

If $-\infty < a < b < \infty$ we often refer to

i) the *closed interval* from a to b, denoted $[a, b]$, consisting of all real numbers x satisfying $a \leq x \leq b$, and

ii) the *open interval* from a to b, denoted (a, b), consisting of all real numbers x satisfying $a < x < b$. (See Fig. 1.2. Note the use of solid and open circles to indicate that the closed interval contains its endpoints while the open interval does not.)

We can extend the above notations to various half-open intervals and infinite intervals (see Fig. 1.3).

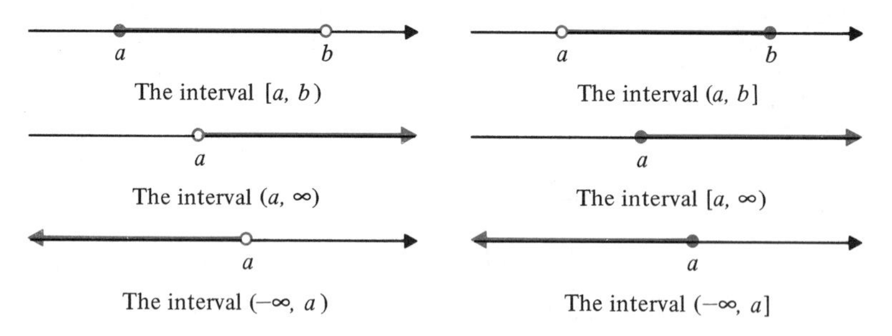

Figure 1.3

The whole real line may also be regarded as an interval: $\mathbb{R} = (-\infty, \infty)$.

Example 2 Solve the following inequalities and graph their solutions on the real line:

a) $2x - 1 > x + 3$

b) $\dfrac{x}{-3} < 2x - 1$

c) $\dfrac{2}{x - 1} \geq 5$

d) $\dfrac{3}{x - 1} \leq \dfrac{-2}{x}$

Solution

a) Adding $1 - x$ to both sides we get $2x - x > 3 + 1$, so $x > 4$. (See Fig. 1.4.)

b) Multiplying both sides by -3 and remembering to reverse the direction of the inequality (since we are multiplying by a negative number), we get $x > -6x + 3$. Thus $7x > 3$ and $x > 3/7$. (See Fig. 1.5.)

c) We would like to multiply both sides by $x - 1$, but this will necessitate reversing the inequalty if $x - 1 < 0$, so we break the problem into two cases.

Case I. $x - 1 > 0$, that is, $x > 1$. Then $2 \geq 5(x - 1) = 5x - 5$. Therefore $7 \geq 5x$ and $x \leq 7/5$. This case leads to solutions $1 < x \leq 7/5$.

Case II. $x - 1 < 0$, that is, $x < 1$. Then $2 \leq 5(x - 1)$ and $x \geq 7/5$. Since no real numbers x satisfy both $x < 1$ and $x \geq 7/5$, this case leads to no solutions.

The only solutions x of the given inequality satisfy $1 < x \leq 7/5$. (See Fig. 1.6.)

d) Note that one side or the other is not defined if $x = 1$ or $x = 0$. We would like to multiply the inequality by $(x - 1)x$ to clear it of fractions. Accordingly, we must distinguish two cases: $(x - 1)x > 0$ and $(x - 1)x < 0$.

Case I. $x > 1$ or $x < 0$. Then $(x - 1)x > 0$ and

$$3x \leq -2(x - 1)$$
$$5x \leq 2,$$

so $x \leq 2/5$. This case produces the solutions $x < 0$.

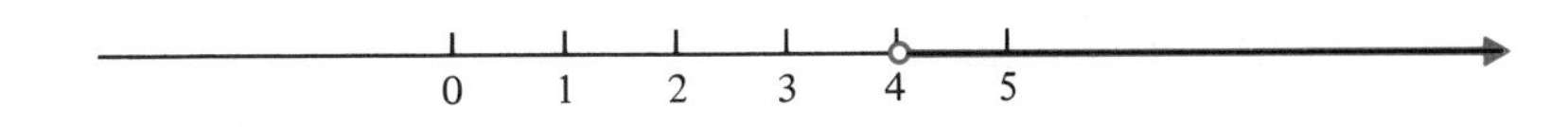

Figure 1.4

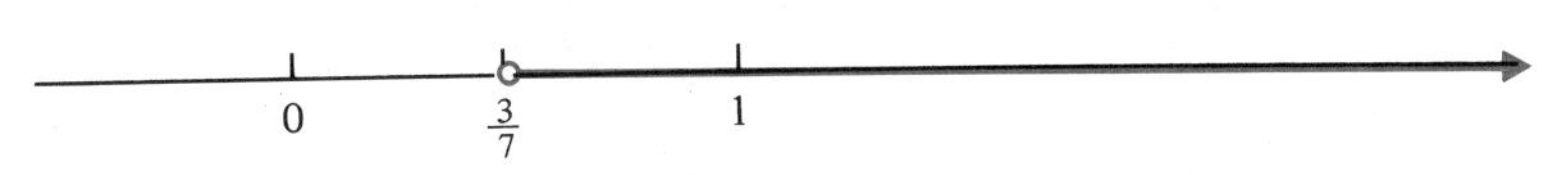

Figure 1.5

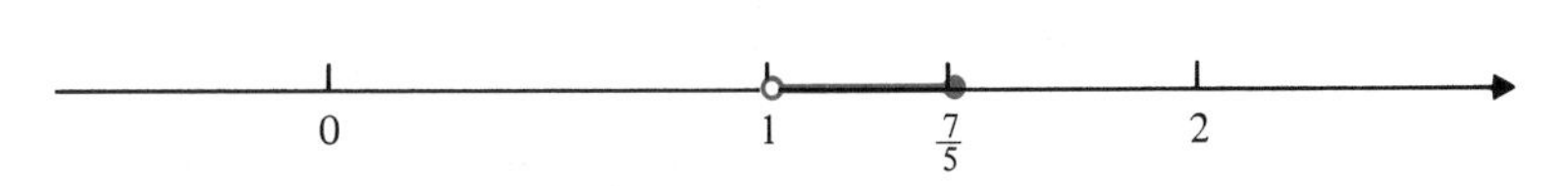

Figure 1.6

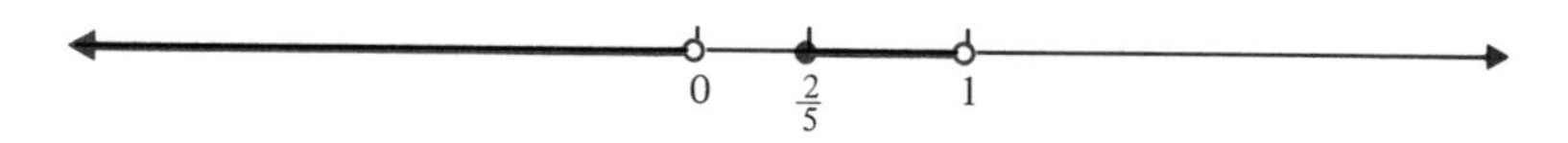

Figure 1.7

Case II. $0 < x < 1$. Then $(x - 1)x < 0$ and

$$3x \geq -2(x - 1)$$
$$5x \geq 2,$$

so $x \geq 2/5$. This case produces the solutions $2/5 \leq x < 1$.

The solution set is the union of intervals: $(-\infty, 0) \cup [2/5, 1)$, as shown in Fig. 1.7.

Note the use of the symbol $\cup$ to denote the union of intervals. A real number is in the union of two intervals if it is in either interval. Similarly, we shall use the symbol $\cap$ to denote intersection. A real number x is in the intersection $I \cap J$ of two intervals I and J if x belongs to both of the intervals I and J. For example,

$$[1, 3) \cap [2, 4] = [2, 3).$$

The Absolute Value

For any real number x, the **absolute value** of x, denoted $|x|$, is defined by

$$|x| = \begin{cases} x, & \text{if } x \geq 0 \\ -x, & \text{if } x < 0. \end{cases}$$

For example, $|2| = 2$ since $2 \geq 0$, but $|-3| = -(-3) = 3$ since $-3 < 0$. Evidently $|x| \geq 0$ for every real number x, and $|x| = 0$ only if $x = 0$.

Geometrically, $|x|$ represents the (nonnegative) distance from x to 0 on the real number line. More generally, $|x - y|$ represents the (nonnegative) distance between the points x and y on the real line, since this distance is the same as that from $x - y$ to 0 (see Fig. 1.8):

$$|x - y| = \begin{cases} x - y, & \text{if } x \geq y \\ y - x, & \text{if } x < y. \end{cases}$$

The absolute value function has the following properties: for any real numbers a and b,

$$|ab| = |a|\,|b|$$
$$|a \pm b| \leq |a| + |b| \qquad \text{(the triangle inequality).}$$

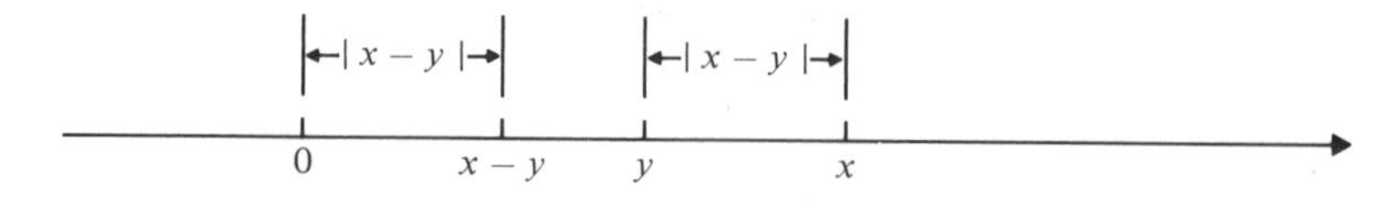

Figure 1.8

The first property is easily checked by considering the four possible cases ($a \geq 0$, $b < 0$), ($a \geq 0$, $b \geq 0$), ($a < 0$, $b < 0$), and ($a < 0$, $b \geq 0$). The second property follows from the first because $\pm 2ab \leq |2ab| = 2|a||b|$. We have

$$\begin{aligned}|a \pm b|^2 &= (a + b)^2 \\ &= a^2 \pm 2ab + b^2 \leq |a|^2 + 2|a||b| + |b|^2 \\ &= (|a| + |b|)^2,\end{aligned}$$

and taking (positive) square roots we obtain

$$|a \pm b| \leq |a| + |b|.$$

This property is called the triangle inequality because it states that the distance from a to $\pm b$ does not exceed the sum of the distances from each of these points to 0. (The length of one side of a triangle does not exceed the sum of the lengths of the other two sides.)

Example 3 Solve:

a) $|2x - 3| = 5$
b) $|x + 1| < |x - 1|$

Solution a) We can break the equation $|2x - 3| = 5$ into two cases.

Case I. $2x - 3 \geq 0$ $\left(\text{or } x \geq \frac{3}{2}\right)$. Thus $|2x - 3| = 2x - 3$ and the equation becomes $2x - 3 = 5$ and yields the solution $x = 4$.

Case II. $2x - 3 < 0$ $\left(\text{or } x < \frac{3}{2}\right)$. Then $|2x - 3| = -(2x - 3) = 3 - 2x$. In this case the equation is $3 - 2x = 5$, and the solution is $x = -1$.

There are, therefore, two solutions, $x = 4$ and $x = -1$.

An easier way to get this answer is to interpret the equation geometrically, as represented in Fig. 1.9. If $|2x - 3| = 5$, then $|2(x - 3/2)| = 5$, so $2|x - 3/2| = 5$ and $|x - 3/2| = 5/2$. This says that the distance from x to 3/2 is 5/2. There are two such numbers, x, namely, $x = 3/2 + 5/2 = 4$ and $x = 3/2 - 5/2 = -1$.

b) We could proceed as in (a) by considering all possible cases (where $x + 1$ is positive or negative and where $x - 1$ is positive or negative). However, observe that the inequality says that the distance of x from -1 is less than the distance

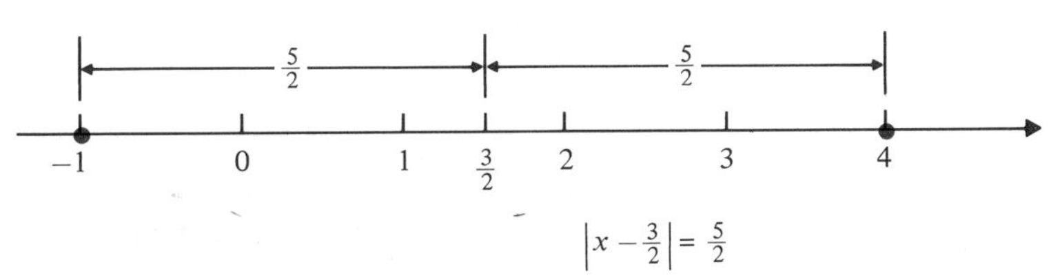

Figure 1.9

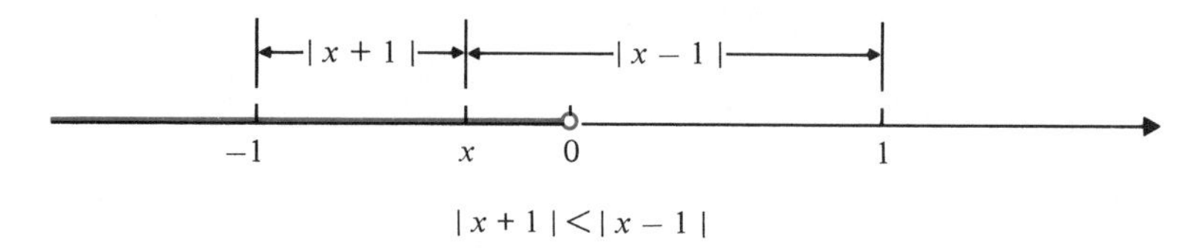

Figure 1.10

x from 1. Thus the solution consists of all real numbers x less than 0. This is shown in Fig. 1.10.

Coordinates and Graphs

The positions of all points in a plane can be measured with respect to a pair of perpendicular real lines in the plane intersecting at the 0-point of each. It is conventional to take one line horizontal and call it the x-axis, and the other line vertical and call it the y-axis. The point of intersection of the axes is called the origin, denoted 0. The position of any point P is then specified by an ordered pair (x, y) of real numbers, called the coordinates of P; x measures the displacement of P from 0 measured parallel to the x-axis, and y the displacement measured parallel to the y-axis (see Fig. 1.11). By the Pythagorean theorem the distance from O to P is

$$|OP| = \sqrt{x^2 + y^2}.$$

More generally, the distance between points $P_1 = (x_1, y_1)$ and $P_2 = (x_2, y_2)$ is $\sqrt{(x_2 - x_1)^2 + (y_2 - y_1)^2}$, as shown in Fig. 1.12.

Given any equation (or inequality) involving the variables x and y, the *graph* of the equation (or inequality) is the set of all points $P = (x, y)$ whose coordinates x and y satisfy the equation. See, for example, Figs. 1.13 and 1.14.

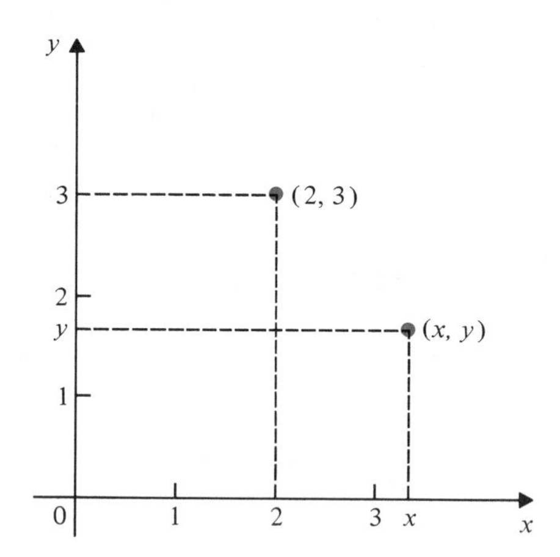

Figure 1.11

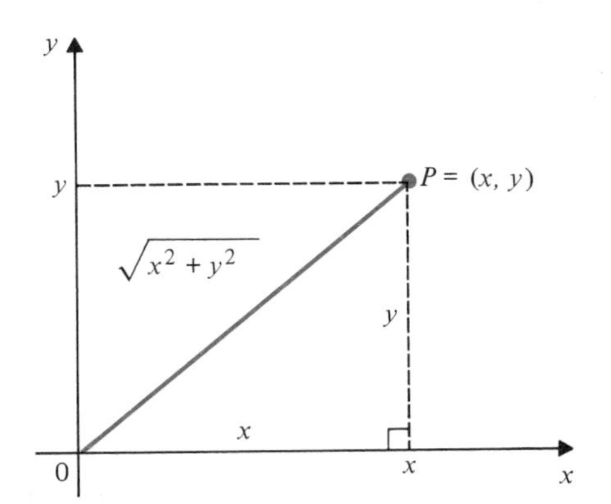

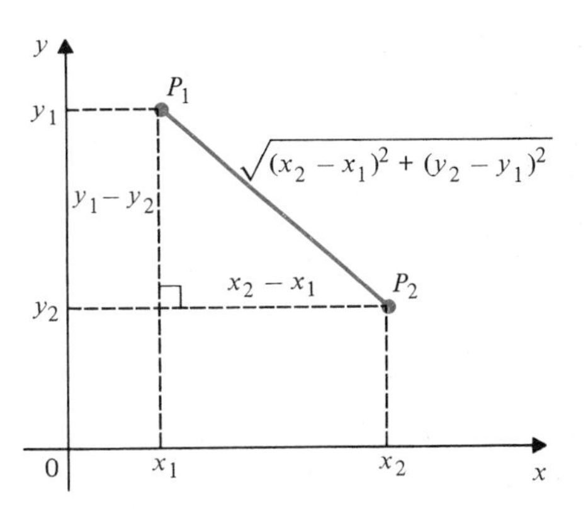

Figure 1.12

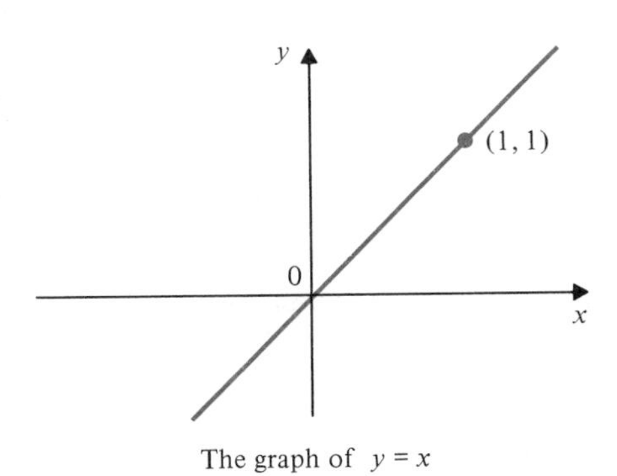

The graph of $y = x$

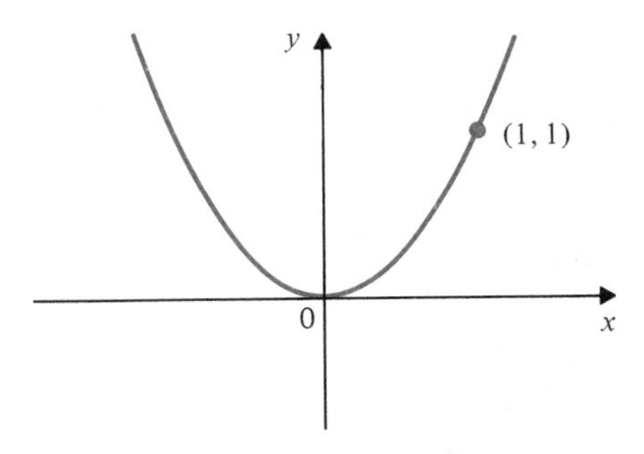

The graph of $y = x^2$

Figure 1.13

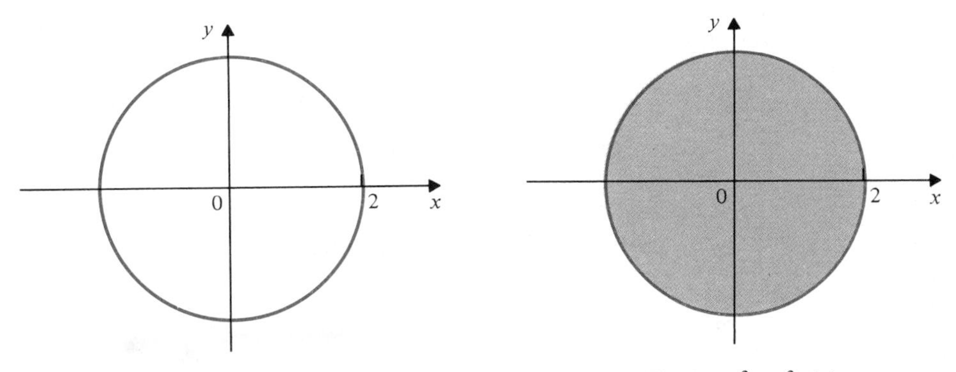

The circle $x^2 + y^2 = 4$

The disc $x^2 + y^2 \leqslant 4$

Figure 1.14

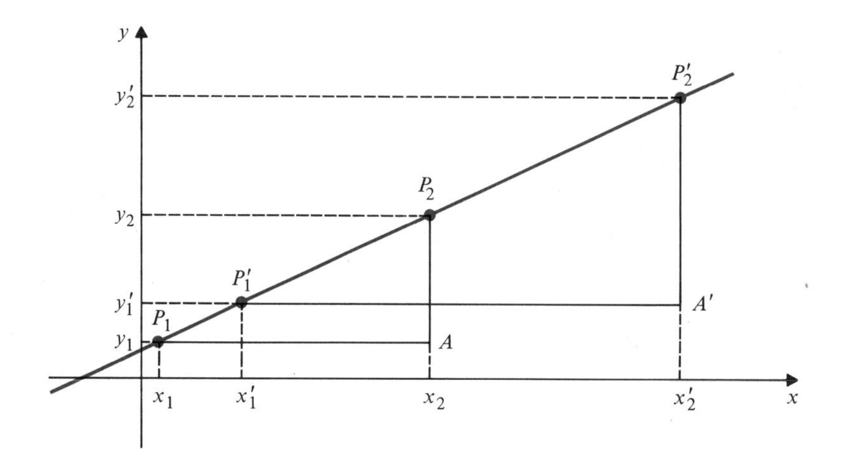

Figure 1.15

Any nonvertical straight line in the plane has the property that the ratio

$$m = \frac{y_2 - y_1}{x_2 - x_1}$$

has the same value for every choice of a pair of points $P_1 = (x_1, y_1)$ and $P_2 = (x_2, y_2)$ on the line. The value of m is the **slope** of the line (see Fig. 1.15). In Figure 1.15, triangles P_1AP_2 and $P_1'A'P_2'$ are similar. Thus,

$$\frac{y_2 - y_1}{x_2 - x_1} = \frac{y_2' - y_1'}{x_2' - x_1'} = m.$$

If $P = (x, y)$ is any point on the straight line with slope m passing through $P_1 = (x_1, y_1)$ then

$$\frac{y - y_1}{x - x_1} = m$$

so

$$y = y_1 + m(x - x_1).$$

This is called the **point-slope form** of the equation of a straight line and is illustrated in Fig. 1.16. It will be very useful when we study tangent lines in Chapter 2. A horizontal line through P_1 has slope 0 and equation $y = y_1$. A vertical line through P_1 has infinite slope, so its equation is $x = x_1$.

A linear equation is of the form

$$Ax + By + C = 0,$$

where A and B are not both zero. The graph of such an equation is always a straight line. It has infinite slope if $B = 0$ and slope $-A/B$ otherwise. Two lines are parallel if they have the same slope; they are perpendicular if the product of their slopes is -1. Some other forms of equations of straight lines are

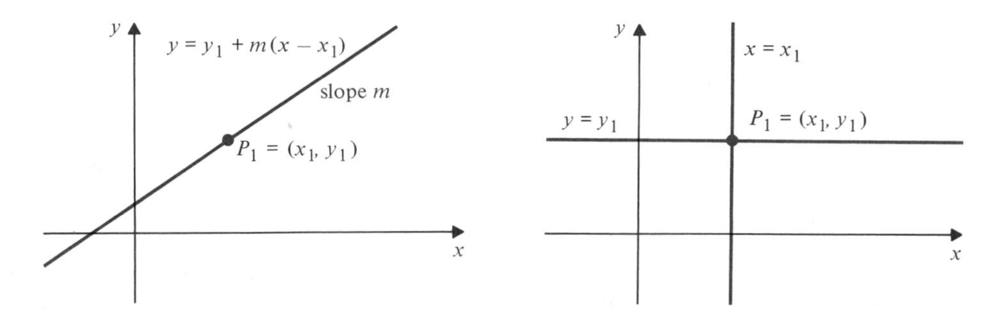

Figure 1.16

i) The slope–y-intercept form: $y = mx + b$. This is a straight line with slope m and y-intercept b, that is, it passes through the point $(0, b)$.

ii) The two-intercept form: $\frac{x}{a} + \frac{y}{b} = 1$. This line has x-intercept a and y-intercept b; that is, it passes through the points $(a, 0)$ and $(0, b)$.

Example 4 Figure 1.17 illustrates the following examples:

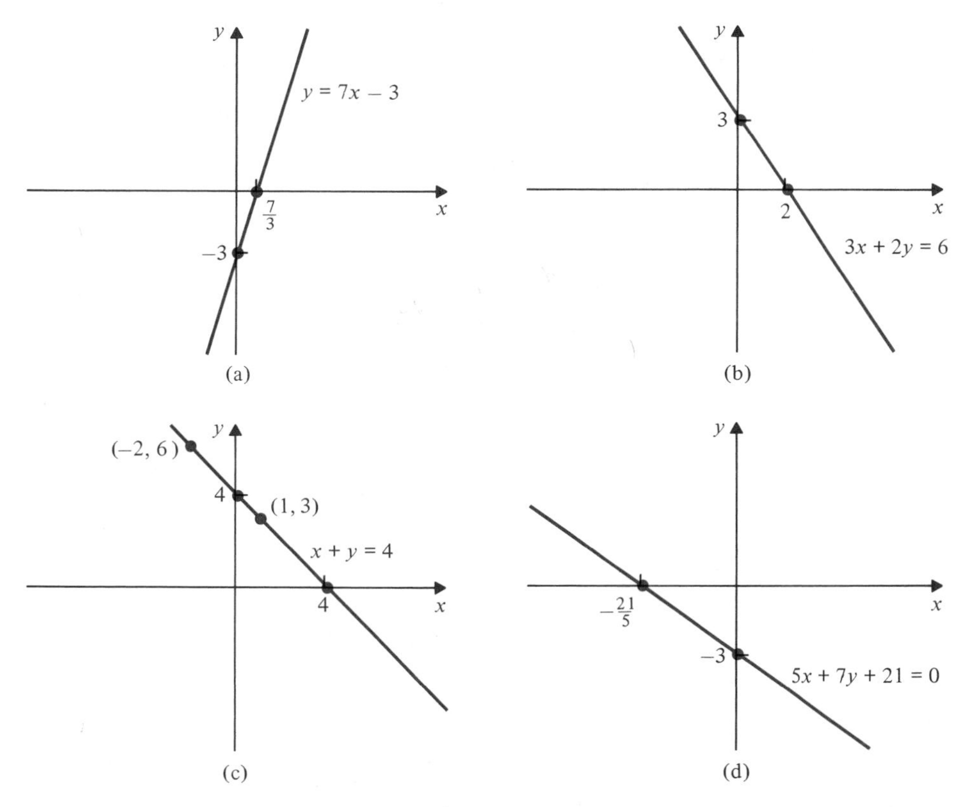

Figure 1.17

a) The straight line passing through the point $(0, -3)$ and having slope 7 has equation $y = 7x - 3$.

b) The straight line passing through the points $(2, 0)$ and $(0, 3)$ has equation $x/2 + y/3 = 1$ or $3x + 2y = 6$.

c) The straight line passing through the points $(1, 3)$ and $(-2, 6)$ has a slope of $(6 - 3)/(-2 - 1) = -1$. Hence it has equation $y - 3 = -1(x - 1)$ or $x + y = 4$.

d) The equation $5x + 7y + 21 = 0$ represents a straight line with slope $-5/7$. The y-intercept of this line is -3; the x-intercept is $-21/5$.

Example 5 Identify the graph of the equation $x^2 + y^2 - 2x + 4y = 4$.

Solution Observe that the expression $x^2 - 2x$ contains the first two terms of the perfect square $(x - 1)^2 = x^2 - 2x + 1$. Hence $x^2 - 2x = (x - 1)^2 - 1$. Similarly, $y^2 + 4y = (y + 2)^2 - 4$. Hence the given equation can be written

$$(x - 1)^2 + (y + 2)^2 = 1 + 4 + 4 = 9 = 3^2.$$

This says that the distance from the point (x, y) to the $(1, -2)$ is equal to 3. Hence the equation represents a circle of radius 3 centered at the point $(1, -2)$ (see Fig. 1.18).

Example 6 Sketch the graph of the equation $y = x^2 - 4x$.

Solution We can make a table of values to obtain the coordinates of points on the graph.

x	-2	-1	0	1	2	3	4	5	6
y	12	5	0	-3	-4	-3	0	5	12

The graph is a parabola with vertex at $(2, -4)$ and a vertical axis, as shown in Fig. 1.19. The equation may be rewritten $y = (x - 2)^2 - 4$.

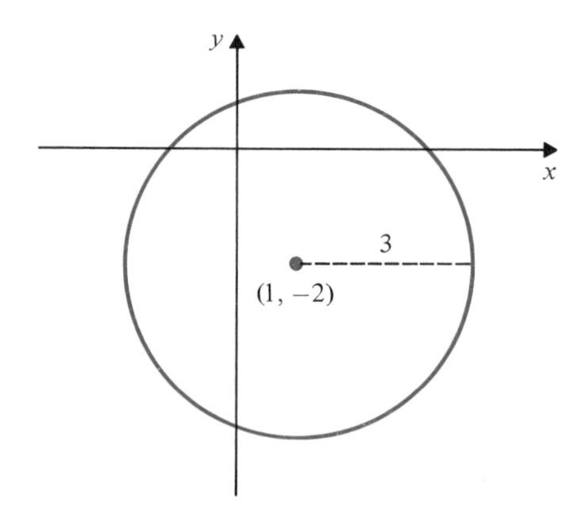

Figure 1.18

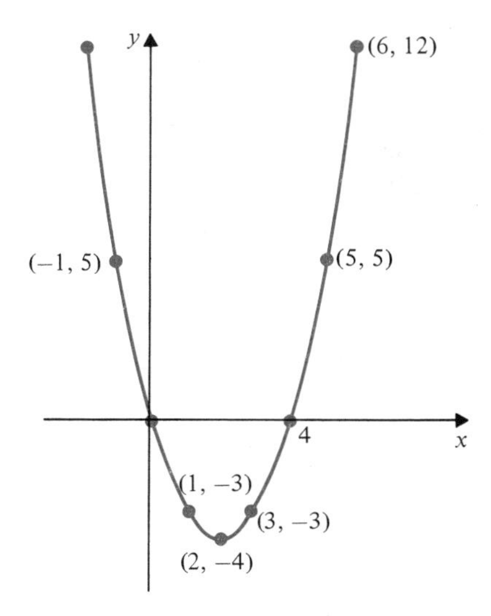

Figure 1.19

In the study of calculus we will encounter many kinds of curves and their equations. You are probably already familiar with some of them (straight lines, circles, parabolas, and perhaps ellipses and hyperbolas) but certainly not with all of them. Calculus will provide useful tools to help us sketch the graphs of new equations without having to calculate the coordinates of a great many points.

EXERCISES

1. Express the following repeating decimals as quotients of integers:

a) $9.090909\ldots = 9.\overline{09}$ **b)** $3.2101010\ldots = 3.2\overline{10}$

c) $0.00044044044\ldots = 0.00\overline{044}$ **d)** $0.285714285714\ldots = 0.\overline{285714}$

In Exercises 2–15 solve the given inequality and graph the solution set.

2. $5x - 3 \le 7 - 3x$

3. $3(2 - x) < 2(3 + x)$

4. $\dfrac{3x - 4}{2} \le \dfrac{6 - x}{4}$

5. $x^2 - 1 \ge 2x$

6. $x^3 > 4x$

7. $\dfrac{1}{2 - x} < 3$

8. $x^2 - x \le 2$

9. $\dfrac{x + 1}{x} > 2$

10. $\dfrac{x}{2} > 1 + \dfrac{4}{x}$

11. $\dfrac{3}{x - 1} < \dfrac{2}{x + 1}$

12. $|x - 3| > 1$

13. $|x + 2| \ge |x + 3|$

14. $|4 - 3x| \le 9$

15. $\left|\dfrac{x + 1}{2}\right| \ge 1$

In Exercises 16–23 find equations of the straight lines satisfying the given conditions.

16. passes through (2, 3) and has slope -4

17. passes through (5, -3) and has slope 0

18. passes through (-2, -10) and has slope 20

19. passes through (2, 1) and (4, 5)

20. passes through (0, -4) and (8, 0)

21. passes through the origin and is parallel to the line $x + 3y = 17$

22. passes through (-2, -3) and is parallel to the line $y = 4 - 2x$

23. passes through the point (1, 1) and is perpendicular to the line $y = 3x$

In Exercises 24–43 sketch the graph of the given equation or inequality. When possible, identify the graph.

24. $y = 2 + \dfrac{x - 1}{3}$

25. $\dfrac{x}{4} + y = 1$

26. $2x - 5y + 5 = 0$

27. $\dfrac{x}{2} - \dfrac{y}{3} = 2$

28. $x^2 + y^2 = 25$

29. $x^2 + y^2 = 5$

30. $(x + 1)^2 + y^2 = 25$

31. $x^2 + (y - 4)^2 = 16$

32. $x^2 + 4x + y^2 = 12$

33. $x^2 + y^2 + 2x - 2y = 2$

34. $y = 2x^2 - 1$

35. $y = x^2 + 2x + 2$

36. $x^2 - y^2 = 0$

37. $x^2 - y^2 = 1$

38. $4x^2 + 9y^2 = 36$

39. $xy = 1$

40. $2x - y \geq 1$

41. $x + y < 2$

42. $x^2 + y^2 - 6y \leq 0$

43. $y \leq 1 - x^2$

1.2 FUNCTIONS

Definition 1 A **function** f is a rule that assigns to each real number x in some set $\mathcal{D}(f)$ (called the *domain* of f) a *unique* real number $f(x)$ called the value of f at x.

Example 1 The squaring function on $\mathbb{R}$ is the function f that assigns to each real number x its square, x^2:

$$f(x) = x^2 \qquad (x \text{ in } \mathbb{R}).$$

The domain of f is the set of all real numbers: $\mathcal{D}(f) = \mathbb{R}$.

While, strictly speaking, we should call the function f, and reserve the notation $f(x)$ to denote the value of f at x, we sometimes loosely refer to the function as $f(x)$. Thus, in the example above, the squaring function is often simply called the function x^2. Another way of denoting a function is as follows:

$$f\colon x \to f(x).$$

This is read as "the function f that takes x to $f(x)$." The squaring function can then be denoted

$$f: x \rightarrow x^2$$

or, even more simply, without using any f,

$$x \rightarrow x^2.$$

Example 2 Let F be the function defined by

$$F(x) = 3(x - 1) + 5 \qquad (x \text{ in } \mathbb{R}).$$

Find the values of F at the points 0, 2, -4, $t + 2$, and $F(2)$.

Solution

$$F(0) = 3(-1) + 5 = 2$$
$$F(2) = 3(2 - 1) + 5 = 8$$
$$F(-4) = 3(-4 - 1) + 5 = -10$$
$$F(t + 2) = 3(t + 2 - 1) + 5 = 3t + 8$$
$$F(F(2)) = F(8) = 3(7) + 5 = 26$$

Definition 2 The **range** of a function f is the set of all real numbers y that are obtained as values of the function; that is, it is the set of all numbers $y = f(x)$ corresponding to all numbers x in the domain of f. The range of f is denoted $\mathcal{R}(f)$.

Definition 3 The **graph** of a function f is the graph of the equation $y = f(x)$ in the Cartesian plane.

Figure 1.20 illustrates the domain, range, and graph of the function f. Note that the domain of the function f can be represented as a set of points on the x-axis; the range is a set on the y-axis. The graph of a function is such that any vertical line through a point on the domain meets the graph at *exactly one point* (because $f(x)$ is a unique number for each x). The horizontal line through this point on the graph

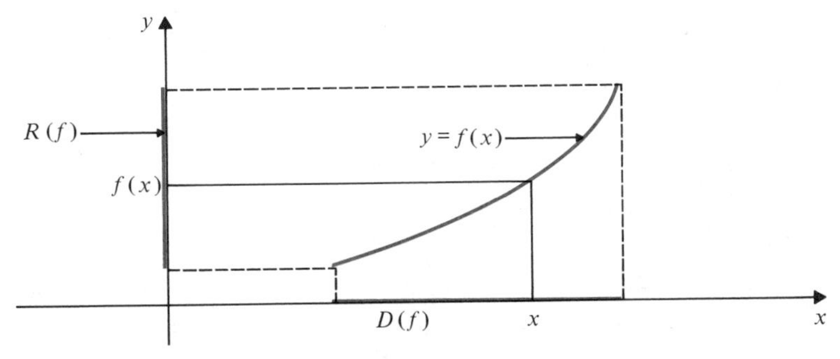

Figure 1.20

meets the y-axis at $f(x)$. Thus $f(x)$ is the vertical displacement of that point on the graph.

Remark: The domain of a function f may or may not be specified. If it is not specified, always assume it is *the largest set of real numbers for which $f(x)$ makes sense as a real number*.

Example 3 Specify the domain and range and sketch the graph of each of the following functions:

a) $f(x) = x^2$ b) $g(x) = x^2 \quad (-1 \leq x < 2)$

c) $h(x) = \sqrt{x}$ d) $k(x) = |x|$

e) $S(x) = \dfrac{1}{x}$ f) $T(x) = \begin{cases} -x \text{ if } -1 \leq x \leq 0 \\ 1 \text{ if } x > 0 \end{cases}$

Solutions See Figs. 1.21–1.26.

a)
$$\mathcal{D}(F) = \mathbb{R} = (-\infty, \infty)$$
$$\mathcal{R}(f) = [0, \infty)$$

b)
$$\mathcal{D}(g) = [-1, 2)$$
$$\mathcal{R}(g) = [0, 4)$$

c) Since squares cannot be negative.
$$\mathcal{D}(h) = [0, \infty)$$
$$\mathcal{R}(h) = [0, \infty).$$

Note that the uniqueness condition in the definition of *function* requires that there be only one value of $h(x) = \sqrt{x}$ for each value of x in $\mathcal{D}(h)$. By convention, the term *square root* and the symbol $\sqrt{}$ always refer to the *nonnegative* square root. Thus, while there are two numbers whose square is 4, only one of these is called $\sqrt{4}$, namely $\sqrt{4} = 2$. If we want to give both solutions of the equation $x^2 = a$ (a > 0) we can indicate them as $x = \sqrt{a}$ or $x = -\sqrt{a}$.

d)
$$k(x) = |x| = \begin{cases} x \text{ if } x \geq 0 \\ -x \text{ if } x < 0 \end{cases}$$
$$\mathcal{D}(k) = \mathbb{R}, \qquad \mathcal{R}(k) = [0, \infty)$$

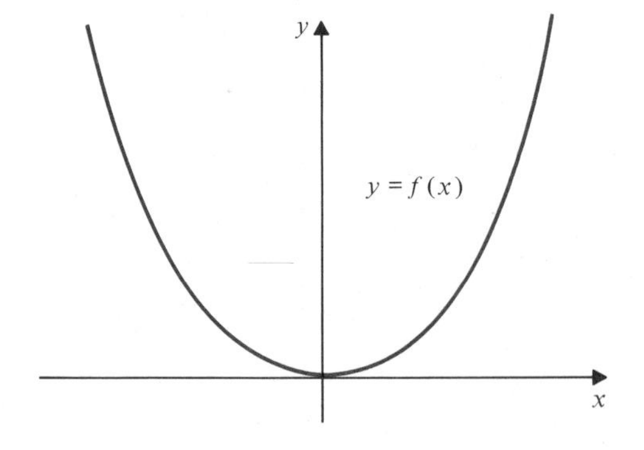

Figure 1.21

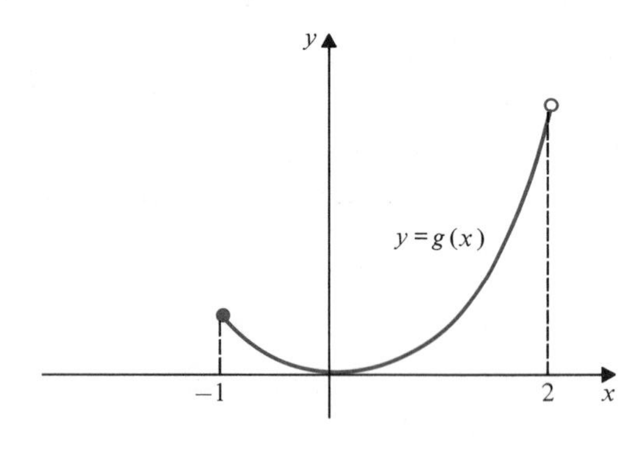

Figure 1.22

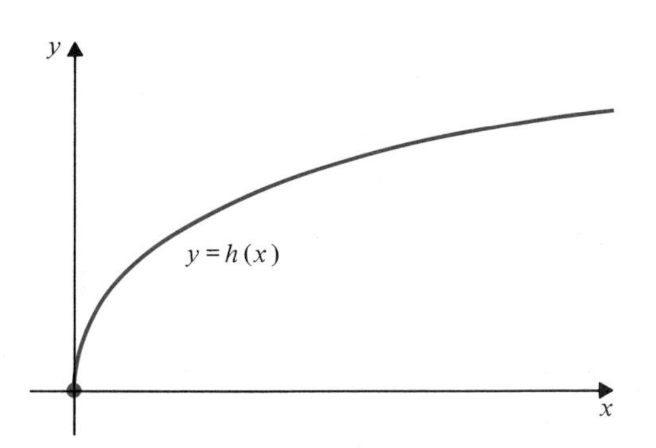

Figure 1.23

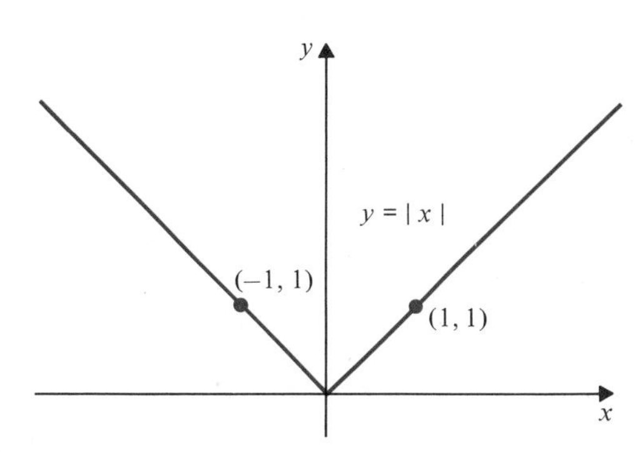

Figure 1.24

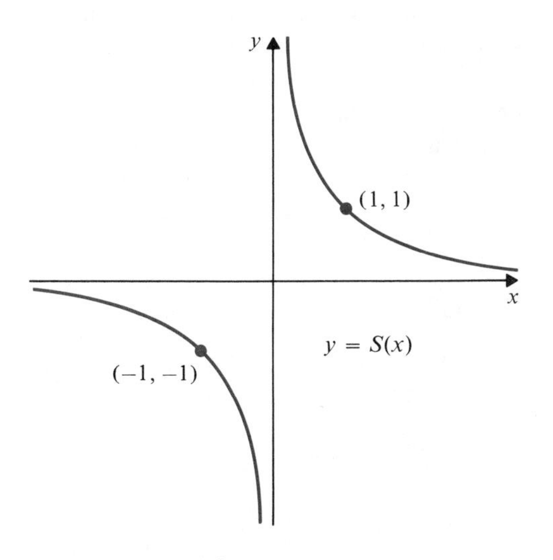

Figure 1.25

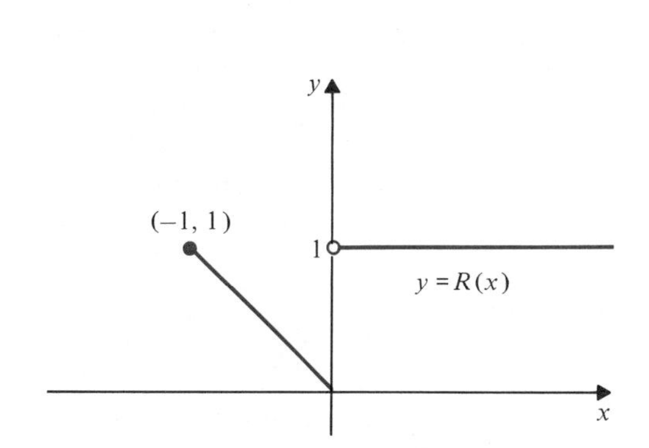

Figure 1.26

Observe that if a is any real number, then

$$\sqrt{a^2} = |a|.$$

We cannot assert that $\sqrt{a^2} = a$ unless we know beforehand that $a \geq 0$.

e) We cannot divide by 0, so

$$\begin{aligned} \mathcal{D}(S) &= (-\infty, 0) \cup (0, \infty) \\ &= \{x : x \neq 0\} \end{aligned}$$

(the set of all x such that $x \neq 0$).

$$\mathcal{R}(S) = (-\infty, 0) \cup (0, \infty).$$

f)
$$\mathcal{D}(T) = [-1, \infty)$$
$$\mathcal{R}(T) = [0, 1]$$

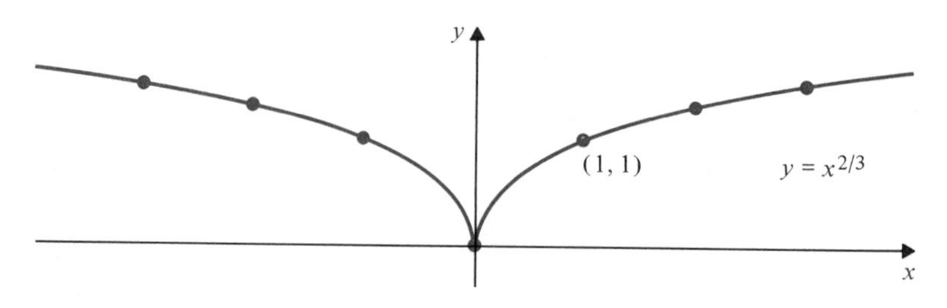

Figure 1.27

At the moment, if we wish to sketch the graph of a function f either we must identify the equation $y = f(x)$ beforehand as representing a standard curve such as a straight line or part of a circle or we must calculate a number of points and connect them by a "reasonable" curve. Thus we can recognize the graph of $f(x) = 2x - 3$ as the straight line with slope 2 and y-intercept -3, and we can recognize $g(x) = \sqrt{16 - x^2}$ as the function whose graph will be the upper semicircle of the circle with equation $x^2 + y^2 = 16$. For more complicated functions, however, we are reduced to the laborious computation of coordinates of points as illustrated in Example 6 of Section 1.1. That this computational method can have its pitfalls is illustrated by the following example.

Example 4 Use a table of values to help you sketch the graph of the function $f(x) = x^{2/3}$.

Solution Using a scientific calculator we can compute the following approximate values:

x	-3	-2	-1	0	1	2	3
$y = x^{2/3}$	2.08	1.59	1	0	1	1.59	2.08

The graph $y = x^{2/3}$ is depicted in Fig. 1.27. Note that nothing in the table of values suggests the infinitely sharp point (cusp) on the curve at the origin. Unless we calculate coordinates of many more points near $x = 0$ we would be likely to miss this very important feature of the graph and might even draw the graph passing smoothly through the origin, as shown in Fig. 1.28.

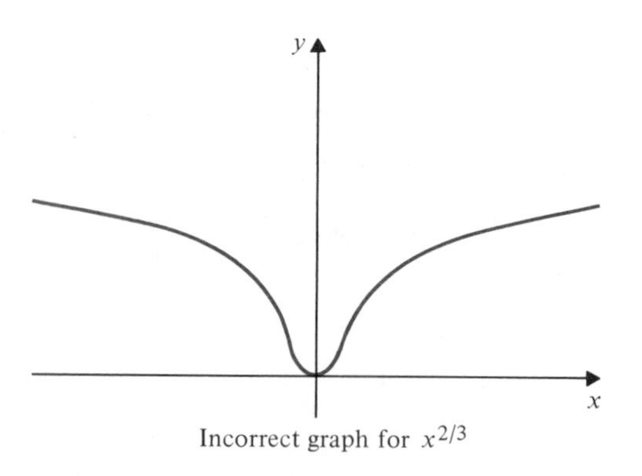

Figure 1.28

Using calculus we will be able to sketch graphs without missing such important points, and without having to calculate many coordinates.

Operations on Functions

Like numbers, functions can be added, subtracted, multiplied, and divided (except when the denominator is 0) to produce new functions. If f and g are functions, then for every x that belongs to the domains of both f and g we define:

$$\begin{aligned} f+g: &\quad x \to f(x)+g(x), \\ f-g: &\quad x \to f(x)-g(x), \\ fg: &\quad x \to f(x)g(x), \\ f/g: &\quad x \to f(x)/g(x) \qquad (\text{where } g(x) \neq 0). \end{aligned}$$

Functions can also be multiplied by constants:

$$cf: \quad x \to cf(x)$$

Example 5 If $f(x) = \sqrt{x+1}$ and $g(x) = \dfrac{x-1}{x}$, then

$$\begin{aligned} (f+g)(x) &= \sqrt{x+1} + \frac{x-1}{x}, \\ (f-g)(x) &= \sqrt{x+1} - \frac{x-1}{x}, \\ (fg)(x) &= \frac{\sqrt{x+1}(x-1)}{x}, \\ (f/g)(x) &= \frac{\sqrt{x+1}(x)}{x-1}. \end{aligned}$$

We have $\mathcal{D}(f+g) = \mathcal{D}(f-g) = \mathcal{D}(fg) = [-1, 0) \cup (0, \infty)$, and $\mathcal{D}(f/g) = [-1, 0) \cup (0, 1) \cup (1, \infty)$. Note that even though $x\sqrt{x+1}/(x-1)$ makes sense when $x = 0$, $(f/g)(x)$ is not defined at $x = 0$ because $g(0)$ is not defined.

The graph of a sum or difference of functions can be obtained by adding or subtracting the heights to the graphs of each function, as shown in Fig. 1.29.

Definition 4

> The **composition** $f \circ g$ or $f(g)$ of two functions f and g is defined by
>
> $$f \circ g(x) = f(g(x))$$
>
> for every x for which $f(g(x))$ makes sense, that is, for every x in $\mathcal{D}(g)$ such that $g(x)$ is in $\mathcal{D}(f)$.

Example 6 Given $f(x) = \sqrt{x}$ and $g(x) = x + 1$ we have the following.

$$f \circ g(x) = f(g(x)) = \sqrt{g(x)} = \sqrt{x+1}, \ \mathcal{D}(f \circ g) = [-1, \infty)$$

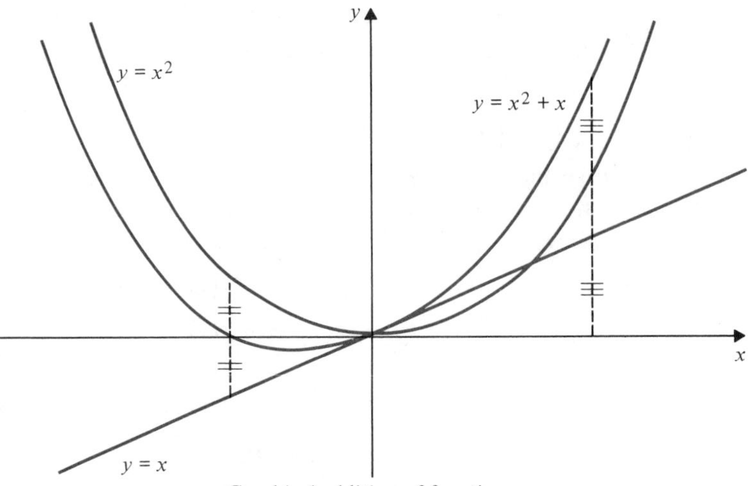

Graphical addition of functions

Figure 1.29

$$g \circ f(x) = g(f(x)) = f(x) + 1 = \sqrt{x} + 1, \mathcal{D}(g \circ f) = [0, \infty)$$

$$f \circ f(x) = f(f(x)) = \sqrt{f(x)} = \sqrt{\sqrt{x}} = x^{1/4}, \mathcal{D}(f \circ f) = [0, \infty)$$

$$g \circ g(x) = g(g(x)) = g(x) + 1 = (x + 1) + 1 = x + 2, \mathcal{D}(g \circ g) = \mathbb{R}$$

$$g \circ f \circ g(x) = g(f(g(x))) = f(g(x)) + 1 = \sqrt{x + 1} + 1, \mathcal{D}(g \circ f \circ g) = [-1, \infty)$$

Example 7 If $H(x) = \dfrac{1 - x}{1 + x}$ calculate $H \circ H(x)$ and specify its domain.

Solution

$$H \circ H(x) = H(H(x)) = \frac{1 - H(x)}{1 + H(x)}$$

$$= \frac{1 - \dfrac{1 - x}{1 + x}}{1 + \dfrac{1 - x}{1 + x}} = \frac{1 + x - 1 + x}{1 + x + 1 - x} = x,$$

provided that $x \neq -1$. Note that, although x is defined on $\mathbb{R}$, the domain of $H \circ H$ is $(-\infty, -1) \cup (-1, \infty)$. Because $H(x)$ is not defined for $x = -1$, neither is $H(H(x))$.

Odd and Even Functions

Definition 5

A function f is said to be **odd** if, whenever x belongs to $\mathcal{D}(f)$, $-$x also belongs to $\mathcal{D}(f)$ and

$$f(-x) = -f(x).$$

A function f is **even** if, whenever x belongs to $\mathcal{D}(f)$, $-x$ also belongs to $\mathcal{D}(f)$ and

$$f(-x) = f(x).$$

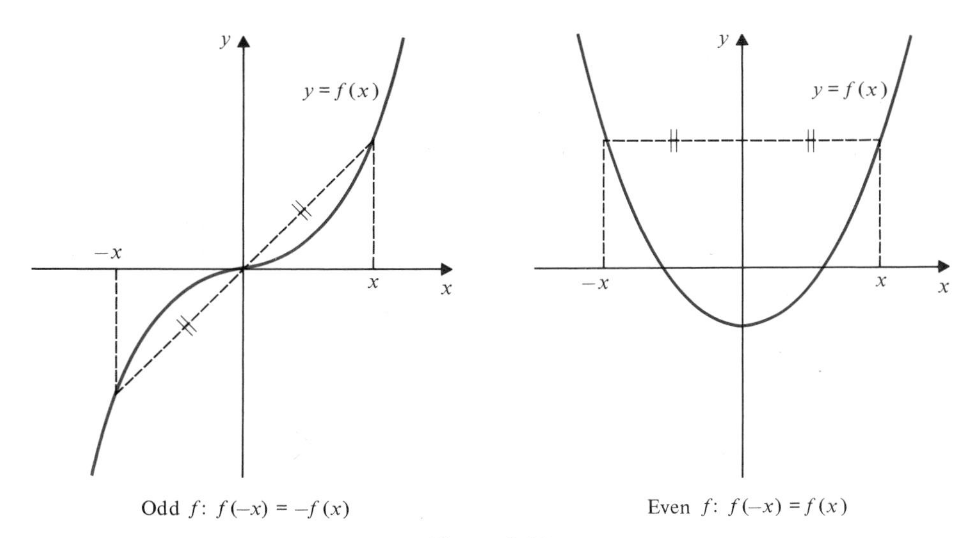

Figure 1.30

Odd powers of x (x, x^3, x^5, . . ., x^{-1}, x^{-3}, x^{-5}, . . .) are odd functions; even powers (1, x^2, x^4, x^6, . . ., x^{-2}, x^{-4}, x^{-6}, . . .) are even functions. The absolute value function $|x|$ is an even function.

The graph of an odd function is symmetric about the origin; that of an even function is symmetric about the y-axis. (See Fig. 1.30.) These symmetries are useful when we want to sketch the graphs of even and odd functions. Note also that an odd function defined at $x = 0$ must vanish there (because $f(0) = f(-0) = -f(0)$ implies $2f(0) = 0$, and so $f(0) = 0$).

Inverse Functions

Given any function f and any point x in its domain, there exists *exactly one* point y in its range such that $y = f(x)$. In graphical terms, a vertical line through x meets the graph of f exactly once, and y is the height of that point. Suppose that f is such that, given any y in its range, there is *exactly one* x in its domain such that $y = f(x)$. That is, a horizontal line through y meets the graph of f at exactly one point. Such a function f is said to be *one-to-one*.

If f is one-to-one, then each y in the range determines a unique x in the domain, so the equation $y = f(x)$ can be solved for x as a function of y; $x = f^{-1}(y)$. The function that we have denoted f^{-1} is called the *inverse of* f. (Do not confuse inverse functions with reciprocals. f^{-1} does not mean $1/f$.)

Here is a formal definition of these concepts. They are illustrated in Fig. 1.31.

Definition 6

A function f is **one-to-one** if $f(x_1) \neq f(x_2)$ whenever x_1 and x_2 belong to the domain of f and $x_1 \neq x_2$. A one-to-one function is also said to be **invertible,** and its **inverse function** f^{-1} is defined as follows: for every y in the range of f, $f^{-1}(y)$ is the unique element x in the domain of f such that $y = f(x)$.

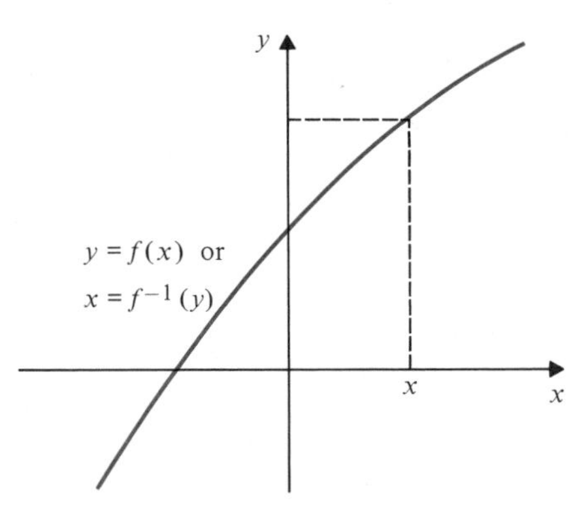

f is one-to-one and invertible.
$y = f(x)$ is equivalent to $x = f^{-1}(y)$

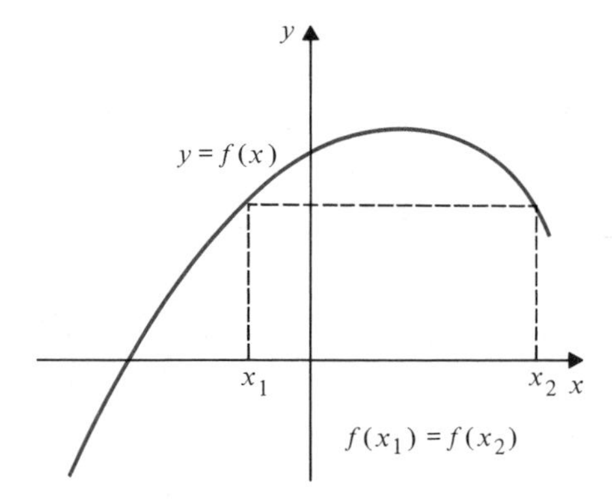

f is not one-to-one and
does not have an inverse.

Figure 1.31

Observe the following.

> i) $\mathscr{D}(f^{-1}) = \mathscr{R}(f)$.
> ii) $\mathscr{R}(f^{-1}) = \mathscr{D}(f)$.
> iii) $x = f^{-1}(y) \Leftrightarrow y = f(x)$.
> iv) f is the inverse of f^{-1}; that is, $(f^{-1})^{-1} = f$.
> v) $f(f^{-1}(y)) = y$ for every y in $\mathscr{R}(f)$.
> vi) $f^{-1}(f(x)) = x$ for every x in $\mathscr{D}(f)$.

(The symbol $\Leftrightarrow$ should be read ''is equivalent to'' or ''means the same as'' or ''if and only if.'')

If S is a set of real numbers, the **identity function** I_S on S is defined by

$$I_S(x) = x \qquad \text{for } x \text{ in } S.$$

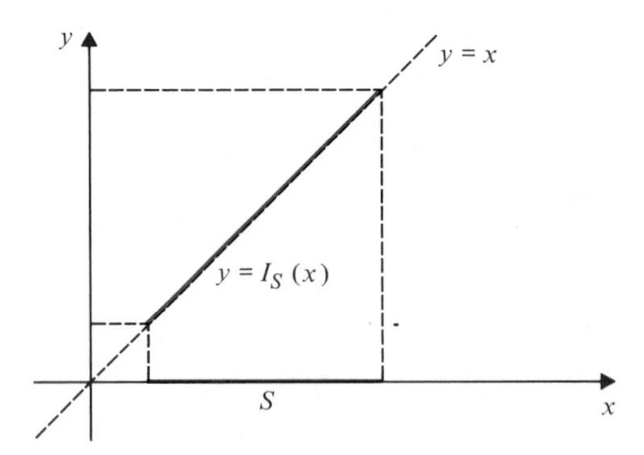

Figure 1.32

The graph of I_S is depicted in Fig. 1.32. In terms of this identity function, properties (v) and (vi) above can be restated in the form

$$f \circ f^{-1} = I_{\mathcal{R}(f)}$$

$$f^{-1} \circ f = I_{\mathcal{D}(f)}$$

As defined above, f^{-1} is expressed as a function of y rather than of x. Since we usually prefer to express functions as functions of x, we interchange x and y to get

$$y = f^{-1}(x) \Leftrightarrow x = f(y).$$

Since the points (x, y) and (y, x) in the Cartesian plane are mirror images in the line $y = x$, it follows that the curves $y = f(x)$ and $x = f(y)$ are also mirror images of one another in that line. See Fig. 1.33.

The graph of f^{-1} is the mirror image of the graph of f in the line $y = x$.

Example 8 Show that $f(x) = 3x - 5$ is invertible and find its inverse function f^{-1}.

Solution If $f(x_1) = f(x_2)$, then $3x_1 - 5 = 3x_2 - 5$, so $3(x_1 - x_2) = 0$ and $x_1 = x_2$. Therefore f is invertible. If $y = f^{-1}(x)$, then

$$x = f(y) = 3y - 5,$$

so $y = \dfrac{x + 5}{3}$. Hence $f^{-1}(x) = \dfrac{x + 5}{3}$. See Fig. 1.34.

Example 9 What is the inverse of the function $f: x \to \sqrt{x}$?

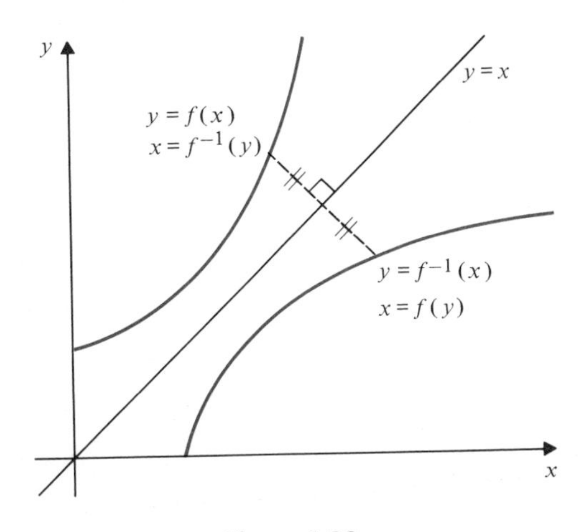

Figure 1.33

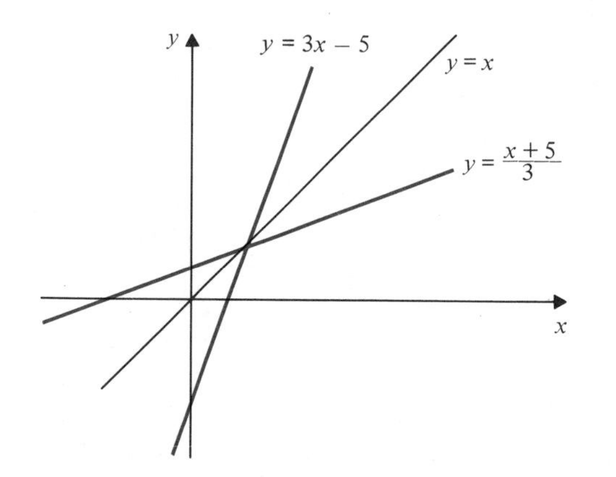

Figure 1.34

Figure 1.35

Solution If $y = f^{-1}(x)$, then $x = f(y) = \sqrt{y}$, so y must be nonnegative. Clearly, $y = x^2$. Since $\mathcal{D}(f^{-1}) = \mathcal{R}(f) = [0, \infty)$ and $\mathcal{R}(f^{-1}) = \mathcal{D}(f) = [0, \infty)$, we have

$$f^{-1}(x) = x^2 \qquad \text{for } x \geq 0,$$

as illustrated in Fig. 1.35.

Example 9 raises an interesting question. Suppose that we had started with the function $g(x) = x^2$ on $(\mathbb{R})$ and tried to invert it. We would of course have failed because g is not one-to-one and hence not invertible. (See Fig. 1.36.) In order to be able to invert a function with values x^2 we must *restrict the domain* so that the function becomes one-to-one. If we define

$$G(x) = x^2 \qquad \text{for } x \geq 0,$$

then $\mathcal{D}(G) = [0, \infty)$ and G is one-to-one and invertible. Clearly, $G^{-1}(x) = \sqrt{x}$ for $x \geq 0$. We shall make extensive use of this restriction procedure when we define the inverses of the trigonometric functions in Section 4.3.

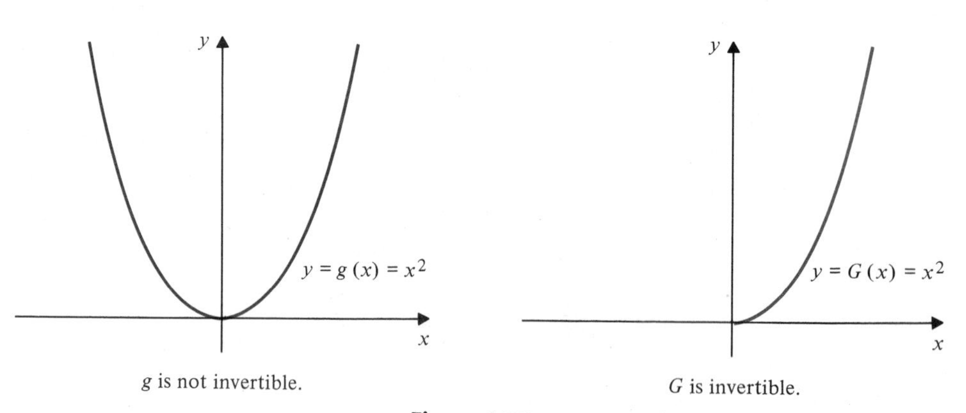

g is not invertible. G is invertible.

Figure 1.36

Definition 7 A function f is said to be **self-inverse** if $f^{-1} = f$, that is, if $f(f(x)) = x$ for every x in the domain of f.

The function $f(x) = 1/x$ is self-inverse, as is the function H in Example 7 in this section. The graph of a self-inverse function must be its own mirror image in the line $y = x$ and so must be symmetric about that line.

EXERCISES

Specify the domains and ranges of the functions in Exercises 1–6.

1. $f(x) = \dfrac{x + 1}{x}$

2. $g(x) = \dfrac{1}{x^2 + 1}$

3. $F(x) = \dfrac{x}{x^2 - x}$

4. $H(x) = \sqrt{x^2 - x}$

5. $h(x) = (x - 2)^2 - 4$

6. $G(x) = \dfrac{x - 1}{|x + 1|}$

Sketch the graphs of the functions in Exercises 7–14. Which of these functions are even? Which are odd? Which are neither even nor odd?

7. $f(x) = x^2 + 1$

8. $f(x) = x^3$

9. $f(x) = \sqrt{|x|}$

10. $g(x) = \sqrt{x^2 - 2x + 1}$

11. $F(x) = \sqrt{4 - x^2}$

12. $G(x) = -\sqrt{4 - x^2}$

13. $f(x) = |x^2 - 1|$

14. $f(x) = |x| + |x - 2|$

15. Sketch the graph of the equation $|x| + |y| = 1$.

16. Show that the inequality $|a - b| \geq \big||a| - |b|\big|$ holds for any real numbers a and b.

For the functions f and g in Exercises 17–19 calculate $f + g$, fg, f/g, g/f, $f \circ f$, $f \circ g$, $g \circ f$, and $g \circ g$. Specify the domain of each function you construct.

17. $f(x) = \dfrac{1}{x^2 - 1}$, $g(x) = \dfrac{1}{x}$

18. $f(x) = x^4$, $g(x) = \sqrt{x - 1}$

19. $f(x) = \sqrt{1 - x^2}$, $g(x) = 2 + x$

20. Find all values of the constants A and B for which the function $F(x) = Ax + B$ satisfies the following.

a) $F \circ F(x) = F(x)$ for all x

b) $F \circ F(x) = x$ for all x

For the functions $f(x)$ in Exercises 21–31 calculate the inverse function f^{-1}. Specify the domain and range of f and f^{-1}.

21. $f(x) = x - 1$

22. $f(x) = 2x - 1$

23. $f(x) = \sqrt{x - 1}$

24. $f(x) = x^3$

25. $f(x) = 1 + \sqrt[3]{x}$

26. $f(x) = \dfrac{1}{x + 1}$

27. $f(x) = x^2, \quad x < 0$

28. $f(x) = \dfrac{x^2}{x^2 + 1}, \quad x \geq 0$

29. $f(x) = \dfrac{x}{\sqrt{x^2 + 1}}$

30. $f(x) = \dfrac{x^2 - 1}{x^2 + 1}, \quad x < 0$

31. $f(x) = \begin{cases} x^2 + 1, & \text{if } x \geq 0 \\ x + 1, & \text{if } x < 0 \end{cases}$

32. For what values of the constants a, b, and c is the function $f(x) = (x - a)/(bx - c)$ self-inverse?

33. If f is an even function and g is an odd function, must any of the functions $f \circ f$, $g \circ g$, $f \circ g$, or $g \circ f$ be odd or even?

34. If f is both an even function and an odd function, show that $f(x)$ is identically zero on its domain.

35. Suppose that the domain of a function f is symmetric about the origin, that is, $-x$ belongs to $\mathcal{D}(f)$ whenever x belongs to $\mathcal{D}(f)$. Show that

a) f is the sum of an even function and an odd function. (*Hint:* Let $E(x) = [f(x) + f(-x)]/2$.

b) there is only one way of writing f as the sum of an even and an odd function. (That is, the even and odd functions in (a) are uniquely determined by f.)

1.3 LIMITS

The concept of *limit* is the cornerstone on which the development of calculus rests. Before we attempt any definition of this concept we will illustrate the idea with an example.

Example 1 Let $f(x) = \dfrac{x^2 - 1}{x - 1}$. Observe the following:

i) f is defined for all real numbers x except $x = 1$; that is,

$$\mathcal{D}(f) = (-\infty, 1) \cup (1, \infty).$$

ii) $f(x) = x + 1$ if x is in $\mathcal{D}(f)$.

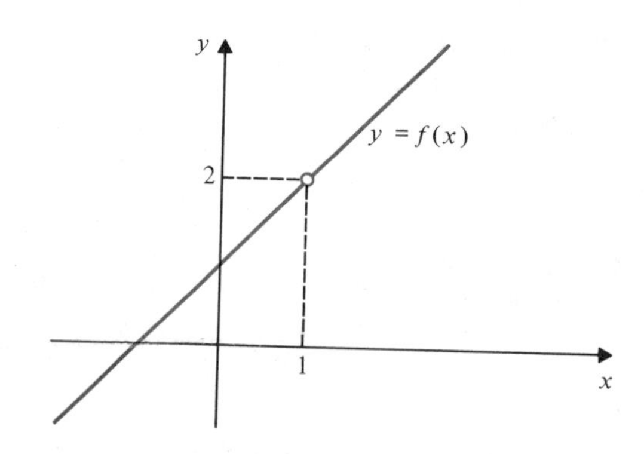

Figure 1.37

The graph of f is thus that of the straight line $y = x + 1$ with one point removed, namely (1, 2). This removed point is shown as a "hole" in the graph (see Fig. 1.37). Even though $f(1)$ is not defined, it is clear that $f(x)$ approaches as close as we like to 2 as x approaches closer and closer to 1. We say that $f(x)$ approaches the limit 2 as x approaches 1, and we write

$$\lim_{x \to 1} f(x) = \lim_{x \to 1} \frac{x^2 - 1}{x - 1} = 2.$$

Definition 8

We say that the function f approaches the **limit** L as x approaches a, and we write

$$\lim_{x \to a} f(x) = L,$$

if $f(x)$ is defined for all x nearby the point a, except possibly at a itself, and if f approaches arbitrarily close to L as x approaches a.

This is not a very precise definition; we have not said just what we mean by such vague phrases as "nearby the point," "arbitrarily close," and "approaches." The intent of the definition, however, should be clear enough so that it can be used for most of our purposes. When we wish to prove theoretical results about limits such as those given in Theorems 1 to 3 in this section, we must have a more precise definition of limit. We will not actually prove such results in this chapter (the proofs are in Appendix 2), but we include the more formal definition here anyway, so you can see how the vague phrases from the earlier definition are made precise.

Definition 8a

Suppose that there exist numbers b, a, and c with $b < a < c$ such that f is defined on the interval (b, a) and on the interval (a, c). Suppose that there exists a real number L with the following property: If ϵ is any positive number, then there exists another positive number δ depending on ϵ such that

$$|f(x) - L| < \epsilon \text{ whenever } 0 < |x - a| < \delta.$$

(That is, the distance of $f(x)$ from L is smaller than ϵ whenever the distance of x from a is smaller than δ but $x \neq a$.) (See Fig. 1.38.) Then we say that the limit of $f(x)$ as x approaches a is L:

$$\lim_{x \to a} f(x) = L.$$

Note that this definition makes no demand of f at the point $x = a$. f may or may not be defined at $x = a$, and if it is defined there, $f(a)$ may or may not be equal to L.

Definition 8, the less precise version of the definition of limit, is quite adequate for our needs in computing values of limits.

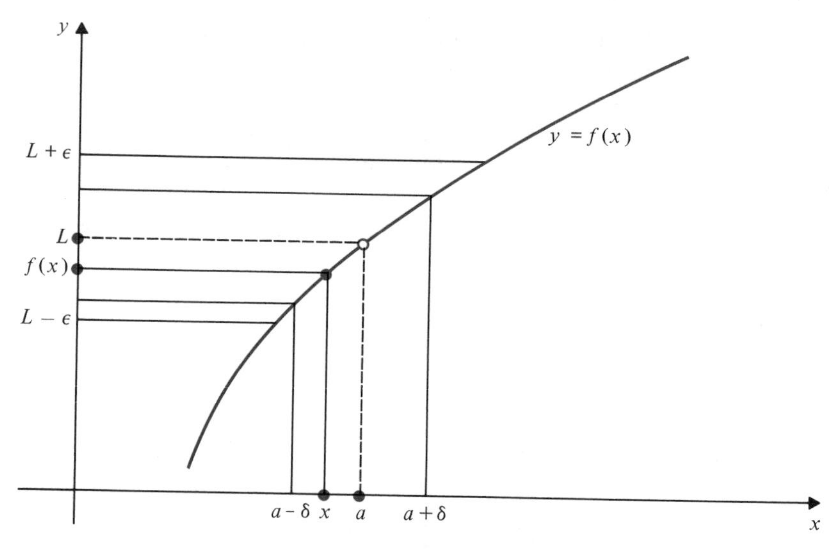

$|f(x) - L| < \epsilon$ provided $0 < |x - a| < \delta$.

Figure 1.38

Example 2 Find $\lim_{x \to 1} \dfrac{x-3}{x+7}$.

Solution As x approaches arbitrarily close to 1, the numerator $x - 3$ approaches arbitrarily close to -2, and the denominator $x + 7$ approaches arbitrarily close to 8. Hence,

$$\lim_{x \to 1} \frac{x-3}{x+7} = \frac{-2}{8} = -\frac{1}{4}.$$

Example 3 Find $\lim_{x \to 4} \dfrac{\sqrt{x} - 2}{x^2 - 5x + 4}$.

Solution If we try to proceed as in Example 2 and evaluate the limits of the numerator and denominator separately we will be led (as in Example 1) to the value 0/0, which is meaningless. In this case we must really determine whether there is a number to which $f(x) = \dfrac{\sqrt{x} - 2}{x^2 - 5x + 4}$ approaches arbitrarily close as x approaches 4 without actually letting x become equal to 4. Using a pocket calculator we can determine values for $f(x)$ for values of x close to 4:

x	f(x)
3.9	.0867525
3.99	.0836643
3.999	.0833663
3.9999	.0833367
3.99999	.0833366

x	f(x)
4.1	.0801473
4.01	.0830046
4.001	.0833003
4.0001	.0833298
4.00001	.0833300

$f(x)$ does indeed appear to be getting closer and closer to some limit as x approaches very close to 4 from either side. That is,

$$\lim_{x\to 4} f(x) \simeq 0.08333. \ . \ . $$

The limit can be determined exactly by astute factoring and cancellation, similar to the procedure we used in Example 1:

$$\frac{\sqrt{x}-2}{x^2-5x+4} = \frac{\sqrt{x}-2}{(x-4)(x-1)} = \frac{\sqrt{x}-2}{(\sqrt{x}-2)(\sqrt{x}+2)(x-1)} = \frac{1}{(\sqrt{x}+2)(x-1)}$$

if $x \neq 4$. Hence,

$$\lim_{x\to 4} \frac{\sqrt{x}-2}{x^2-5x+4} = \frac{1}{(2+2)(4-1)} = \frac{1}{12} = 0.083333. \ . \ . $$

It is because both the numerator and denominator of the original fraction contained factors that vanished at $x = 4$ (the factor $\sqrt{x} - 2$ in each case) that the original fraction gave the meaningless 0/0 when 4 was substituted for x. Proper evaluation of the limit depended on canceling out these vanishing factors first.

The principal properties of limits that we will need hereafter are summarized in Theorems 1, 2, and 3. These properties should seem intuitively clear once you understand the concept of limit. They can be rigorously proved using Definition 8a. See Appendix II for details.

Theorem 1 (*Uniqueness of Limits*) A function cannot have two different limits at the same point; if $\lim_{x\to a} f(x) = L$ and $\lim_{x\to a} f(x) = M$, then necessarily $L = M$. □

Theorem 2 If $\lim_{x\to a} f(x) = L$ and $\lim_{x\to a} g(x) = M$, then the following conclusions hold.

i) $\lim_{x\to a} (f(x) + g(x)) = L + M$ (The limit of a sum is the sum of the limits.)

ii) $\lim_{x\to a} (f(x) - g(x)) = L - M$ (The limit of a difference is the difference of the limits.)

iii) $\lim_{x\to a} (f(x)g(x)) = LM$ (The limit of a product is the product of the limits.)

iv) $\lim_{x\to a} \dfrac{f(x)}{g(x)} = \dfrac{L}{M}$, provided $M \neq 0$. (The limit of a quotient is the quotient of the limits.)

v) $\lim_{x\to a} cf(x) = cL$ for any constant c.

vi) If $f(x) \leq g(x)$ near a, then $L \leq M$. □

Theorem 3 (*The Squeeze Theorem*) Suppose that $f(x) \leq g(x) \leq h(x)$ for all x near a (except possibly at $x = a$). If

$$\lim_{x\to a} f(x) = \lim_{x\to a} h(x) = L,$$

then $\lim_{x\to a} g(x) = L$ also. □

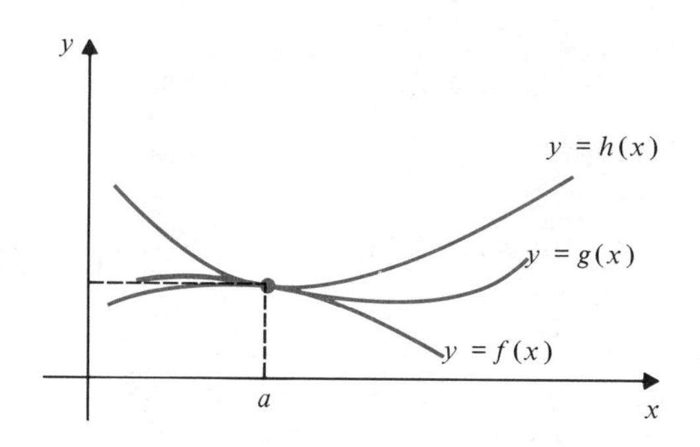

Figure 1.39

Figure 1.39 shows that the graph of g is trapped between those of f and h, which come together at $x = a$.

Example 4 Determine which of the following limits exist and evaluate those that do.

a) $\lim_{x\to 2} (x^2 + 5x)$

b) $\lim_{x\to -3} \dfrac{x^2 + x - 6}{x + 3}$

c) $\lim_{x\to 0} \dfrac{1}{x}$

d) $\lim_{x\to \pi} \dfrac{x - \pi}{x^2}$

e) $\lim_{x\to 3} \dfrac{x - 3}{\sqrt{x} - \sqrt{3}}$

f) $\lim_{x\to 0} \dfrac{x}{|x|}$

g) $\lim_{h\to 0} \dfrac{\dfrac{1}{x + h} - \dfrac{1}{x}}{h}$

h) $\lim_{x\to 1} f(x)$ where $f(x) = \begin{cases} x + 1, & \text{if } x \neq 1 \\ 4, & \text{if } x = 1 \end{cases}$

Solution

a) Evidently x^2 approaches 4, and $5x$ approaches 10 as x approaches 2. Thus $\lim_{x\to 2}(x^2 + 5x) = 4 + 10 = 14$.

b) We cannot just substitute $x = -3$ because the expression $(x^2 + x - 6)/(x + 3)$ becomes the meaningless form 0/0 in this case. This happens because $x + 3$ is a factor of both the numerator and the denominator. We have

$$\lim_{x\to -3} \frac{x^2 + x - 6}{x + 3} = \lim_{x\to -3} \frac{(x + 3)(x - 2)}{x + 3} = \lim_{x\to -3} (x - 2) = -5.$$

c) $1/x$ takes on larger and larger values as x approaches 0 (positive values if $x > 0$ and negative values if $x < 0$). Hence $\lim_{x\to 0} 1/x$ does not exist. There is no *unique* real number that $\dfrac{1}{x}$ approaches as x approaches 0.

d) In the expression $(x - \pi)/x^2$, the numerator approaches 0 but the denominator approaches π^2 as x approaches π. Hence,

$$\lim_{x\to \pi} \frac{x - \pi}{x^2} = \frac{0}{\pi^2} = 0.$$

e) The expression $(x - 3)/(\sqrt{x} - \sqrt{3})$ becomes 0/0 if we substitute $x = 3$. Evidently there is a common factor to cancel:

$$\lim_{x\to 3}\frac{x-3}{\sqrt{x}-\sqrt{3}} = \lim_{x\to 3}\frac{(\sqrt{x}-\sqrt{3})(\sqrt{x}+\sqrt{3})}{\sqrt{x}-\sqrt{3}} = \lim_{x\to 3}(\sqrt{x}+\sqrt{3}) = 2\sqrt{3}.$$

f) We have

$$\frac{x}{|x|} = \begin{cases} 1, & \text{if } x > 0 \\ -1, & \text{if } x < 0. \end{cases}$$

The function is not defined at $x = 0$. It appears that $x/|x|$ approaches different numbers (1 and -1) as x approaches 0 from the positive and negative sides. Since limits must be unique, we conclude that $\lim_{x\to 0} x/|x|$ does not exist.

The function $x/|x|$ is frequently called sgn x (that is, *signum* x) after the Latin word for sign, because its value (1 or -1) depends on the sign of x. The function sgn x will be of some use to us later. Its graph is illustrated in Fig. 1.40.

g) The expression $((1/x + h) - 1/x)/h$ is the meaningless 0/0 if $h = 0$ because both numerator and denominator have the factor h. To cancel this factor we need to do a little algebra:

$$\lim_{h\to 0}\frac{\dfrac{1}{x+h}-\dfrac{1}{x}}{h} = \lim_{h\to 0}\frac{x-(x+h)}{x(x+h)h}$$
$$= \lim_{h\to 0}\frac{-h}{x(x+h)h}$$
$$= \lim_{h\to 0}\frac{-1}{x(x+h)}$$
$$= \frac{-1}{x^2}, \quad x \neq 0.$$

h) Since $f(x) = x + 1$ for $x \neq 1$, and since $x + 1$ evidently approaches 2 as x approaches 1, we have $\lim_{x\to 1} f(x) = 2$. Note that in this case the limit is not $f(1)$. (Here $f(1) = 4$.) We say that this function f is *not continuous* at $x = 1$. See Fig. 1.41. (See Section 1.5 for a discussion of continuity.)

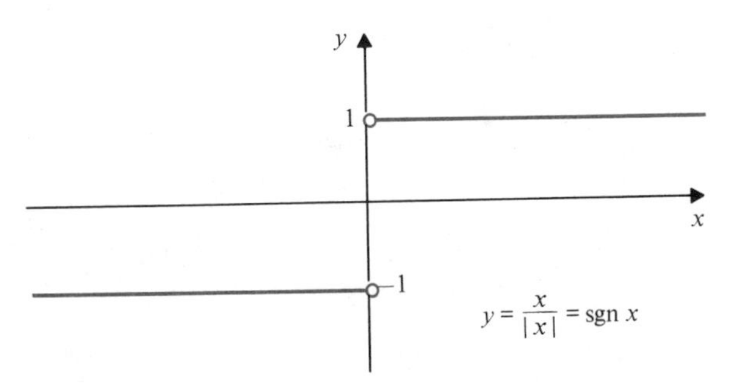

Figure 1.40

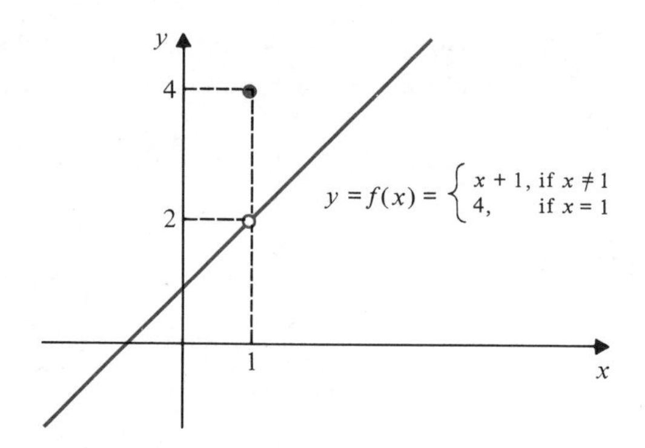

Figure 1.41

Example 5 Find $\lim_{x\to 0} f(x)$ given that $|f(x) - 2| \leq x^2$ for all x.

Solution Since $|f(x) - 2| \leq x^2$ we have $-x^2 \leq f(x) - 2 \leq x^2$. (That $f(x) - 2$ is no farther from 0 than is x^2 implies that $f(x) - 2$ lies between $-x^2$ and x^2.) Hence, $2 - x^2 \leq f(x) \leq 2 + x^2$. But

$$\lim_{x\to 0} (2 - x^2) = 2 = \lim_{x\to 0} (2 + x^2).$$

By the squeeze theorem, $\lim_{x\to 0} f(x) = 2$.

EXERCISES

In Exercises 1–30 evaluate the limit or explain why it does not exist.

1. $\lim_{x\to 4} (x^2 - 9)$
2. $\lim_{x\to -3} (x^2 - 9)$
3. $\lim_{x\to -1} (1 + x + x^2 + x^3)$
4. $\lim_{x\to 2} \dfrac{x - 1}{x^2 + 1}$
5. $\lim_{x\to 1} \dfrac{x^2 - 1}{x + 1}$
6. $\lim_{x\to -1} \dfrac{x^2 - 1}{x + 1}$
7. $\lim_{x\to 3} \dfrac{x^2 - 6x + 9}{x^2 - 9}$
8. $\lim_{x\to -2} \dfrac{x^2 + 2x}{x^2 - 4}$
9. $\lim_{x\to -2} \dfrac{x^2 + x - 2}{x + 2}$
10. $\lim_{h\to 0} \dfrac{\sqrt{4 + h} - 2}{h}$
11. $\lim_{x\to -1} \dfrac{x^3 + 1}{x + 1}$
12. $\lim_{x\to \pi} \dfrac{(x - \pi)^2}{\pi x}$
13. $\lim_{x\to 4} |x - 4|$
14. $\lim_{x\to -2} |x - 2|$
15. $\lim_{x\to 0} \dfrac{|x - 2|}{x - 2}$
16. $\lim_{x\to 2} \dfrac{|x - 2|}{x - 2}$
17. $\lim_{t\to 1} \dfrac{t^2 - 1}{t^2 - 2t + 1}$
18. $\lim_{x\to 2} \dfrac{\sqrt{4 - 4x + x^2}}{x - 2}$
19. $\lim_{x\to 1} \dfrac{x - 4\sqrt{x} + 3}{x^2 - 1}$
20. $\lim_{x\to 0} \dfrac{x^2 + 3x}{(x + 2)^2 - (x - 2)^2}$

21. $\lim_{x\to -2} \frac{x^4 - 16}{x^3 + 8}$

22. $\lim_{x\to 8} \frac{3x^{2/3} - 12}{x^{1/3} - 2}$

23. $\lim_{x\to 5} \frac{x^2 + 3x - 10}{3x^2 + 16x + 5}$

24. $\lim_{x\to -5} \frac{x^2 + 3x - 10}{3x^2 + 16x + 5}$

25. $\lim_{x\to 1} \frac{(x^2 - 1)^2}{x^3 - 2x^2 + x}$

26. $\lim_{x\to -\sqrt{2}} \frac{|x^2 - 2|}{x^2 + 2\sqrt{2}x + 2}$

27. $\lim_{x\to 2} \left(\frac{1}{x - 2} - \frac{4}{x^2 - 4}\right)$

28. $\lim_{x\to 2} \left(\frac{1}{x - 2} - \frac{1}{x^2 - 4}\right)$

29. $\lim_{x\to 0} \frac{\sqrt{2 + x^2} - \sqrt{2 - x^2}}{x^2}$

30. $\lim_{x\to 0} \frac{|3x - 1| - |3x + 1|}{x}$

31. Use a calculator to compute values for $f(x) = \frac{x^3 - x}{5x^2 - 4x - 1}$ for several values of x near 1 (say $x = 1 \pm 0.1$, 1 ± 0.01, 1 ± 0.001, and so on). Guess the value of $\lim_{x\to 1} f(x)$ and then try to verify your guess.

32. Repeat Exercise 31 for the function $f(x) = \frac{2x\sqrt{x} + x - 8\sqrt{x} - 4}{x + \sqrt{x} - 6}$ for values of x near 4.

33. Suppose that $x^4 < f(x) < x^2$ if $|x| < 1$ and $x^2 < f(x) < x^4$ if $|x| > 1$. Find

a) $\lim_{x\to -1} f(x)$

b) $\lim_{x\to 0} f(x)$

c) $\lim_{x\to 1} f(x)$

34. Suppose $|f(x)| \le g(x)$ where $\lim_{x\to a} g(x) = 0$. Find $\lim_{x\to a} f(x)$.

1.4 EXTENSIONS OF THE LIMIT CONCEPT

In this section we will extend the concept of limit to allow for three situations not covered by the definition of limit given in the previous section:

i) one-sided limits, that is, limits as x approaches a from one side only (the left or the right);

ii) limits at infinity, where x becomes arbitrarily large, positive or negative;

iii) infinite limits, which are not really limits at all but provide useful symbolism for describing the behavior of functions whose values become arbitrarily large, positive or negative.

One-Sided Limits

Let us reconsider the signum function, $\operatorname{sgn} x = x/|x|$, introduced in Example 4(f) of the previous section. We argued there that $\lim_{x\to 0} \operatorname{sgn} x$ does not exist, because sgn x approaches different values (1 or -1) as x approaches 0 from the right-hand side (through positive values) or from the left-hand side (through negative values). If we

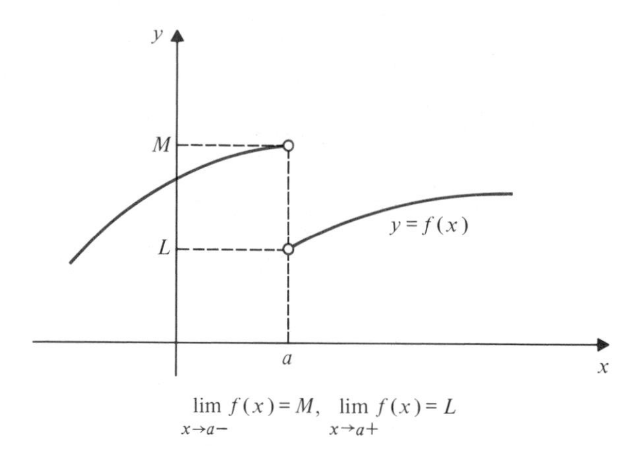

Figure 1.42

are only interested in letting x approach 0 from one side, we could argue that an appropriate limit exists; specifically,

$$\lim_{x\to 0+} \operatorname{sgn} x = 1, \qquad \lim_{x\to 0-} \operatorname{sgn} x = -1,$$

where we read the symbol $\lim_{x\to 0+}$ as "the limit as x approaches 0 from the right (or from the positive side of 0)," and the symbol $\lim_{x\to 0-}$ as "the limit as x approaches 0 from the left (or from the negative side of 0)."

Definition 9

We say that the function f has **right-hand limit** L as x approaches a, or that *the limit of $f(x)$ as x approaches a from the right is L* provided $f(x)$ is defined on the interval (a, b) for some $b > a$ and $f(x)$ approaches arbitrarily close to L as x decreases toward a. The notation for right-hand limit is

$$\lim_{x\to a+} f(x).$$

Similarly, f has **left-hand limit** M as x approaches a, or *the limit of $f(x)$ as x approaches a from the left is M* provided $f(x)$ is defined on the interval (b, a) for some $b < a$ and $f(x)$ approaches arbitrarily close to M as x increases toward a. The notation for left-hand limit is

$$\lim_{x\to a-} f(x).$$

Right-hand and left-hand limits are illustrated in Fig. 1.42.

As was the case for limits studied in the previous section, $f(a)$ need not exist for either $\lim_{x\to a-} f(x)$ or $\lim_{x\to a+} f(x)$ to exist, and if $f(a)$ does exist it need be neither of these one-sided limits.

Example 1 a) The function $f(x) = \sqrt{1 - x^2}$ is defined only on the interval $[-1, 1]$. (Why?) Evidently $f(-1) = 0$ and $f(1) = 0$. If $-1 < a < 1$ then $\lim_{x\to a} f(x) =$

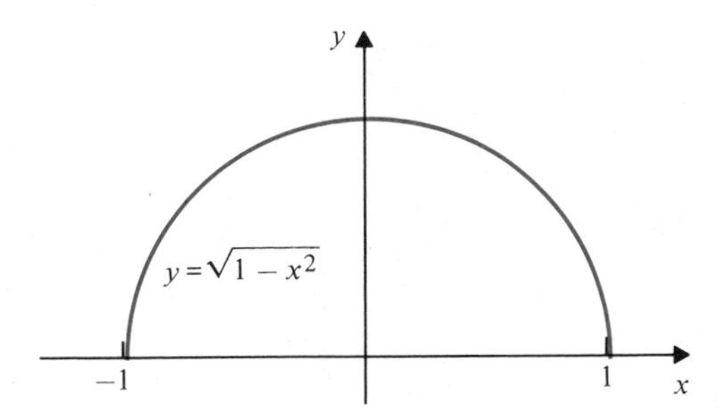

Figure 1.43

$\sqrt{1 - a^2} = f(a)$. However, f has only a right-hand limit at -1 and a left-hand limit at 1 (see Fig. 1.43.)

$$\lim_{x \to -1+} f(x) = 0, \qquad \lim_{x \to 1-} f(x) = 0.$$

b) Let $[x]$ denote the greatest integer that does not exceed x. Thus $[3] = 3$, $[2.7] = 2$, $[-1/2] = -1$, $[0] = 0$. If n is any integer, then

$$\lim_{x \to n+} [x] = n, \qquad \lim_{x \to n-} [x] = n - 1.$$

See Fig. 1.44. If a is not an integer, then $n < a < n + 1$ for some integer n, and

$$\lim_{x \to a+} [x] = \lim_{x \to a-} [x] = \lim_{x \to a} [x] = n.$$

It is always true that $\lim_{x \to a} f(x) = L$ holds if and only if $\lim_{x \to a+} f(x) = L = \lim_{x \to a-} f(x)$. If $\lim_{x \to a+} f(x)$ and $\lim_{x \to a-} f(x)$ both exist but are not equal, then $\lim_{x \to a} f(x)$ does not exist.

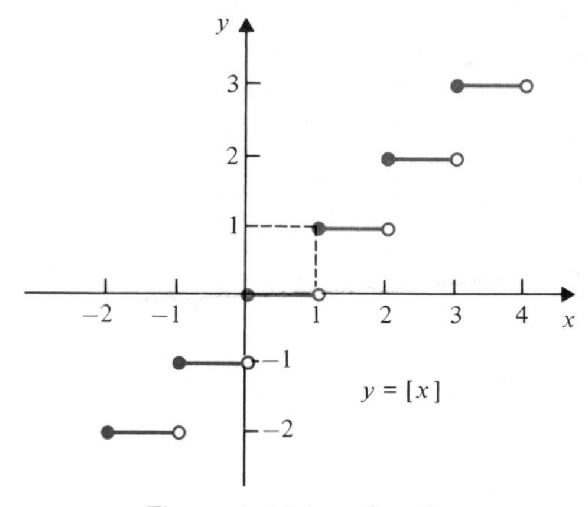

The greatest integer function

Figure 1.44

Left-hand limits and right-hand limits have all the properties possessed by ordinary (two-sided) limits as stated in Theorems 1, 2, amd 3 of Section 1.3.

Limits at Infinity

If a function f is defined on the whole real line or on a semi-infinite interval of the form (a, ∞) or $(-\infty, a)$, it is appropriate to ask how the values of $f(x)$ behave as x becomes very large, positive (approaches infinity) or negative (approaches negative infinity).

Example 2 Consider the function

$$f(x) = \frac{3x^2 + 1}{2x^2 + 5}.$$

As x becomes large (either positive or negative) both the numerator and the denominator of $f(x)$ become large. The quotient, $f(x)$, however, does not become large. To see this, divide both numerator and denominator by x^2 (the highest power of x present) and thus, rewrite $f(x)$ in the form

$$f(x) = \frac{3 + \dfrac{1}{x^2}}{2 + \dfrac{5}{x^2}}, \qquad \text{valid for } x \neq 0.$$

As x approaches infinity or negative infinity, $1/x^2$ approaches 0, so the numerator approaches 3 and the denominator approaches 2. Thus we say that

$$\lim_{x\to\infty} f(x) = \frac{3}{2}, \qquad \lim_{x\to-\infty} f(x) = \frac{3}{2}.$$

In graphical terms, as depicted in Fig. 1.45, the graph of f approaches the horizontal straight line $y = 3/2$ as x recedes far to the right or left of 0. We call the line $y = 3/2$ a (horizontal) *asymptote* of the graph of f. Asymptotes are straight lines that graphs approach as they recede very far from the origin. We will study asymptotes in some detail in Section 4.3.

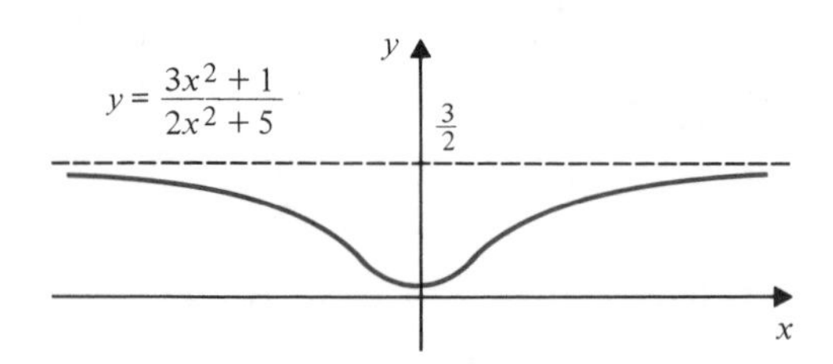

Figure 1.45

Definition 10

If $f(x)$ is defined on an interval (a, ∞) and if $f(x)$ approaches arbitrarily close to L as x recedes to arbitrarily large positive numbers, then we say that $f(x)$ tends to the limit L as x tends to infinity, and we write

$$\lim_{x\to\infty} f(x) = L.$$

Similarly, if $f(x)$ is defined on an interval $(-\infty, a)$ and if $f(x)$ approaches arbitrarily close to M as x recedes to arbitrarily large negative numbers, then we say that $f(x)$ tends to the limit M as x tends to negative infinity, and we write

$$\lim_{x\to-\infty} f(x) = M.$$

As in Definition 8, this definition contains somewhat vague terms ("arbitrarily close," "arbitrarily large") that would have to be made more precise if we wanted to use the definition to provide proofs of statements. (See Appendix II.) The version we gave in Definition 10, however, will serve our purposes. Many functions can be manipulated algebraically into a form where the limits at infinity or negative infinity can be determined by inspection. If the function is a quotient whose numerator and denominator involve sums of various powers of x, this can be achieved (as in Example 2 in this section) by dividing the numerator and denominator by the highest power of x in the denominator.

Example 3

a) $$\lim_{x\to\infty}\frac{1+x}{x-2x^3} = \lim_{x\to\infty}\frac{\dfrac{1}{x^3}+\dfrac{1}{x^2}}{\dfrac{1}{x^2}-2} = \frac{0}{-2} = 0 = \lim_{x\to-\infty}\frac{1+x}{x-2x^3}$$

b) $$\lim_{x\to\infty}\frac{\sqrt{x}(x+1)(x^2+1)}{(2\sqrt{x}+3)^2(1-\sqrt{x})^5} = \lim_{x\to\infty}\frac{1\left(1+\dfrac{1}{x}\right)\left(1+\dfrac{1}{x^2}\right)}{\left(2+\dfrac{3}{\sqrt{x}}\right)^2\left(\dfrac{1}{\sqrt{x}}-1\right)^5} = \frac{1}{4(-1)} = -\frac{1}{4}$$

The expression in (b) does not have a limit at $-\infty$. Why?

Example 4 Find

a) $\lim_{x\to\infty}(\sqrt{x^2+x}-x)$

b) $\lim_{x\to-\infty}(\sqrt{x^2+x}-x)$

Solution In (a) we are trying to find the limit of the difference of two functions each of which becomes arbitrarily large as x increases to infinity. Again an algebraic trick is needed to render the limit obvious:

$$\lim_{x\to\infty}(\sqrt{x^2+x}-x) = \lim_{x\to\infty}\frac{(\sqrt{x^2+x}-x)(\sqrt{x^2+x}+x)}{\sqrt{x^2+x}+x}$$
$$= \lim_{x\to\infty}\frac{x^2+x-x^2}{\sqrt{x^2+x}+x}$$
$$= \lim_{x\to\infty}\frac{x}{x\sqrt{1+\frac{1}{x}}+x} = \lim_{x\to\infty}\frac{1}{\sqrt{1+\frac{1}{x}}+1} = \frac{1}{2}.$$

At first glance, the situation in (b) may appear similar, but it is not. We are now dealing with a sum rather than a difference of functions with large values:

$$\lim_{x\to-\infty}(\sqrt{x^2+x}-x) = \lim_{x\to-\infty}\left(|x|\sqrt{1+\frac{1}{x}}-x\right) \qquad \text{Remember } \sqrt{x^2}=|x|.$$
$$= \lim_{x\to-\infty} -x\left(\sqrt{1+\frac{1}{x}}+1\right), \qquad \text{does not exist. The function becomes infinite.}$$

(*Note:* $|x| = -x$ for negative x.)

Infinite Limits

Consider the function $1/x^2$. Evidently $\lim_{x\to 0} 1/x^2$ does not exist; there is no real number L to which $1/x^2$ approaches arbitrarily close as x approaches 0. In fact, the values of $1/x^2$ become arbitrarily large as x draws closer and closer to zero from either side (see Fig. 1.46). It is convenient to describe this behavior by saying that $1/x^2$ approaches infinity as x approaches zero and to write

$$\lim_{x\to 0}\frac{1}{x^2} = \infty.$$

Note that in writing this we are *not* saying that $\lim_{x\to 0} 1/x^2$ *exists*. (Infinity is not a real number.) Rather we are saying that that limit *does not exist* because $1/x^2$ becomes arbitrarily large near $x = 0$.

Definition 11

> If $f(x)$ becomes arbitrarily large positive as x approaches arbitrarily close to the value a, then we say that $f(x)$ tends to infinity as x approaches a, and we write $\lim_{x\to a} f(x) = \infty$. Similarly, if $f(x)$ becomes arbitrarily large negative as x approaches a, then we write $\lim_{x\to a} f(x) = -\infty$.

The above definition can also be extended to cover such notions as $\lim_{x\to a+} f(x) = \infty$ (or $-\infty$), $\lim_{x\to a-} f(x) = \infty$ (or $-\infty$), $\lim_{x\to\infty} f(x) = \infty$ (or $-\infty$), and $\lim_{x\to-\infty} f(x) = \infty$ (or $-\infty$). Reconsidering Example 4(b) above, we can now assert that

$$\lim_{x\to-\infty}(\sqrt{x^2+x}-x) = \infty,$$

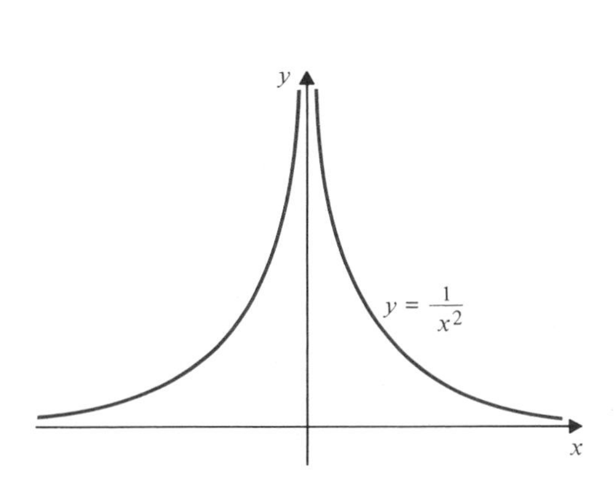

Figure 1.46

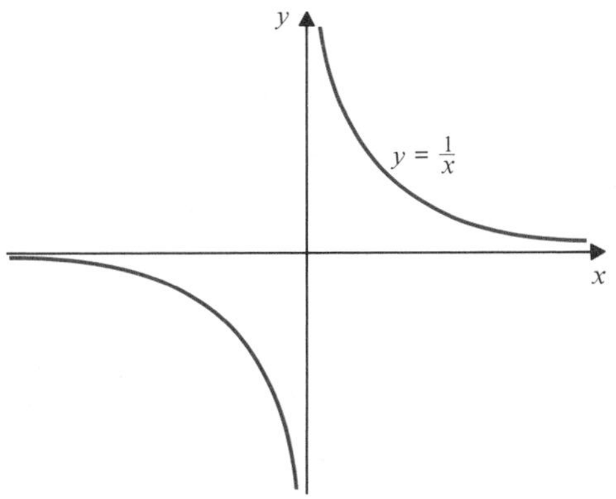

Figure 1.47

which, we stress again, in no way contradicts our earlier assertion that this limit does not exist.

Example 5 The function $1/x$ has no limit as x approaches 0, but in this case we cannot even say that $1/x$ approaches infinity (or negative infinity) because the values of $1/x$ are positive for $x > 0$ and negative for $x < 0$. We can only write $\lim_{x\to 0} 1/x$ does not exist. (See Fig. 1.47.) However, we can make assertions about one-sided limits of $1/x$ at 0:

$$\lim_{x\to 0+} \frac{1}{x} = \infty, \qquad \lim_{x\to 0-} \frac{1}{x} = -\infty.$$

Again we are not saying that either of these one-sided limits exists (as a real number), but rather that one is infinity and the other negative infinity.

Example 6 a) $\displaystyle\lim_{x\to 2} \frac{1}{|x-2|} = \infty.$

b) $\displaystyle\lim_{x\to -5} \frac{1}{5+x}$ does not exist; but $\displaystyle\lim_{x\to -5+} \frac{1}{5+x} = \infty$ (the limit as x approaches -5 from the right is infinite) and $\displaystyle\lim_{x\to -5-} \frac{1}{5+x} = -\infty$ (the limit as x approaches -5 from the left is $-\infty$).

c) $\displaystyle\lim_{x\to 2} \frac{x-1}{4-x^2}$ does not exist, $\displaystyle\lim_{x\to 2-} \frac{x-1}{4-x^2} = \infty$, $\displaystyle\lim_{x\to 2+} \frac{x-1}{4-x^2} = -\infty$; $\displaystyle\lim_{x\to -2} \frac{x-1}{4-x^2}$ does not exist, $\displaystyle\lim_{x\to -2-} \frac{x-1}{4-x^2} = \infty$, $\displaystyle\lim_{x\to -2+} \frac{x-1}{4-x^2} = -\infty$.

When a function $f(x)$ approaches infinity or negative infinity as x approaches some finite point a, then the graph of f approaches the vertical straight line $x = a$ as it recedes to infinity near a. Thus the vertical line $x = a$ is called a (vertical) asymptote of the graph of f. We persist in calling it an asymptote even if only a one-sided limit is infinite there. Thus the y-axis is a vertical asymptote of each of the curves of $y = 1/x$ and $y = 1/x^2$. The line $x = 2$ is a vertical asymptote of the graph $y = 1/|x - 2|$; the line $x = -5$ is a vertical asymptote of the graph $y = 1/(5 + x)$; both of the lines $x = 2$ and $x = -2$ are vertical asymptotes of the graph $y = (x - 1)/(4 - x^2)$.

EXERCISES

In Exercises 1–38 determine the indicated limit, if it exists. If not, determine whether the limit is infinity, negative infinity, or neither.

1. $\lim_{x\to 3} \dfrac{1}{3 - x}$

2. $\lim_{x\to 3} \dfrac{1}{(3 - x)^3}$

3. $\lim_{x\to 3} \dfrac{1}{(3 - x)^2}$

4. $\lim_{x\to 3} \dfrac{x^3}{(3 - x)^2}$

5. $\lim_{x\to 3-} \dfrac{1}{3 - x}$

6. $\lim_{x\to 3+} \dfrac{1}{3 - x}$

7. $\lim_{x\to -5/2} \dfrac{2x + 5}{5x + 2}$

8. $\lim_{x\to a+} \dfrac{x - a}{x^2 - a^2}$

9. $\lim_{x\to a} \dfrac{|x - a|}{x^2 - a^2}$

10. $\lim_{x\to a-} \dfrac{|x - a|}{x^2 - a^2}$

11. $\lim_{x\to 2+} \sqrt{2 - x}$

12. $\lim_{x\to 2-} \sqrt{2 - x}$

13. $\lim_{x\to 1-} \dfrac{2x + 3}{x^2 + x - 2}$

14. $\lim_{x\to 1+} \dfrac{2x + 3}{x^2 + x - 2}$

15. $\lim_{x\to 2+} \dfrac{x}{(2 - x)^3}$

16. $\lim_{x\to 1-} \sqrt{1 - x^2}$

17. $\lim_{x\to 1+} \dfrac{1}{|x - 1|}$

18. $\lim_{x\to 1-} \dfrac{1}{|x - 1|}$

19. $\lim_{x\to -2-} \dfrac{x^2 - 4}{|x + 2|}$

20. $\lim_{x\to -2+} \dfrac{x^2 - 4}{|x + 2|}$

21. $\lim_{x\to -3+} \dfrac{x + 3}{|x + 3|}$

22. $\lim_{x\to -3-} \dfrac{x + 3}{|x + 3|}$

23. $\lim_{x\to -3} \dfrac{x + 3}{|x + 3|}$

24. $\lim_{x\to 0} \dfrac{4 - 5x}{x^4}$

25. $\lim_{x\to 2} \dfrac{x - 3}{x^2 - 4x + 4}$

26. $\lim_{x\to 1+} \dfrac{\sqrt{x^2 - x}}{x - 1}$

27. $\lim_{x\to\infty} \dfrac{3x^3 - 5x^2 + 7}{8 + 2x - 5x^3}$

28. $\lim_{x\to -\infty} \dfrac{x^2 - 2}{x - x^2}$

29. $\lim_{x\to\infty} \dfrac{x\sqrt{x + 1}\,(1 - \sqrt{2x + 3})}{7 - 6x + 4x^2}$

30. $\lim_{x\to\infty} \dfrac{x^2 + 3}{x^3 + 2}$

31. $\lim_{x\to\infty} \dfrac{x + x^3 + x^5}{1 + x^2 + x^3}$

32. $\lim_{x\to -\infty} \dfrac{Ax^2 + Bx + C}{Dx^2 + Ex + F}$

33. $\lim_{x\to\infty} \frac{3x + 2\sqrt{x}}{1 - x}$

34. $\lim_{x\to-\infty} \frac{x}{\sqrt{x^2 + 1}}$

35. $\lim_{x\to\infty} \left(\frac{x}{x + 1} - \frac{x}{x - 1}\right)$

36. $\lim_{x\to-\infty} (\sqrt{x^2 + 2x} - \sqrt{x^2 - 2x})$

37. $\lim_{x\to-\infty} (\sqrt{x^2 + 2x} + x)$

38. $\lim_{x\to-\infty} \frac{2x - 5}{|3x + 2|}$

39. Specify any horizontal or vertical asymptotes of the graphs of the functions in Exercises 1–38.

1.5 CONTINUITY

Definition 12

A function f is said to be **continuous at the point *a*** if

$$\lim_{x\to a} f(x) = f(a).$$

According to this definition, three conditions must be satisfied if f is to be continuous at a:

i) $f(x)$ must be defined nearby and at $x = a$,
ii) $\lim_{x\to a} f(x)$ must exist, and
iii) the limit in (ii) must be equal to $f(a)$.

In graphical terms, f is continuous at $x = a$ if its graph extends some distance to the left and right of the point $(a, f(a))$ and has no break in it at that point. In Fig. 1.48 f is continuous at $x = a$ because its graph has no break there. It is *discontinuous* at b and c because $f(x)$ does not have a limit as x approaches either of these points. (It has left-hand and right-hand limits, but they are not equal.) It is discontinuous at $x = d$ because $\lim_{x\to d} f(x)$, although it exists, is not equal to $f(d)$.

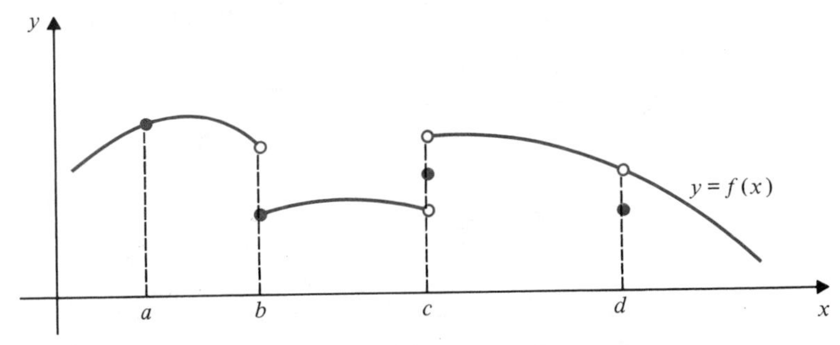

f is continuous at a but not at $b, c,$ or d.

Figure 1.48

We say that f is **continuous on the right** at $x = a$ if $\lim_{x \to a+} f(x) = f(a)$ and that it is **continuous on the left** if $\lim_{x \to a-} f(x) = f(a)$. Evidently, if f is continuous on the right and on the left at $x = a$ then it is continuous at $x = a$. In Fig. 1.48 f is continuous on the right and left at $x = a$ and continuous on the right at $x = b$. It is not continuous on either side at c or d.

A function is **continuous on the interval *I*** if it is continuous at every point of I. If the interval is closed, the function need only be continuous on the right at the left endpoint and continuous on the left at the right endpoint (as well as continuous at every interior point).

Example 1

a) $f(x) = 3x - 2$ on $[-1, 2]$. f is continuous on $[-1, 2]$. See Fig. 1.49.

b) $f(x) = 1/x$ on $[-1, 0) \cup (0, 1]$. f is continuous where defined.

c) $H(x) = \begin{cases} 1, & \text{if } x > 0 \\ 0, & \text{if } x \leq 0 \end{cases}$

H is continuous everywhere except at $x = 0$, where it is continuous on the left but not on the right (see Fig. 1.50).

d) The greatest integer function $[x]$ of Example 1(b) of Section 1.4 is continuous except at the integers. At each integer the function is continuous on the right but discontinuous on the left:

$$\lim_{x \to n+} [x] = n = [n], \qquad \lim_{x \to n-} [x] = n - 1 \neq [n].$$

Most of the functions encountered in elementary calculus are continuous wherever they are defined. Some properties of continuous functions are collected in the following theorem. The proof involves using the definition of continuity and appropriate properties of limits given in Theorem 2 in Section 1.3. (See Appendix II for further discussion.)

Theorem 4

a) If the functions f and g are continuous at $x = a$, then the functions $f + g$, $f - g$, and fg are also continuous at $x = a$. If $g(a) \neq 0$, then f/g is continuous at $x = a$.

b) If f is continuous at L and if $\lim_{x \to a} g(x) = L$, then

$$\lim_{x \to a} f(g(x)) = f(L) = f(\lim_{x \to a} g(x)).$$

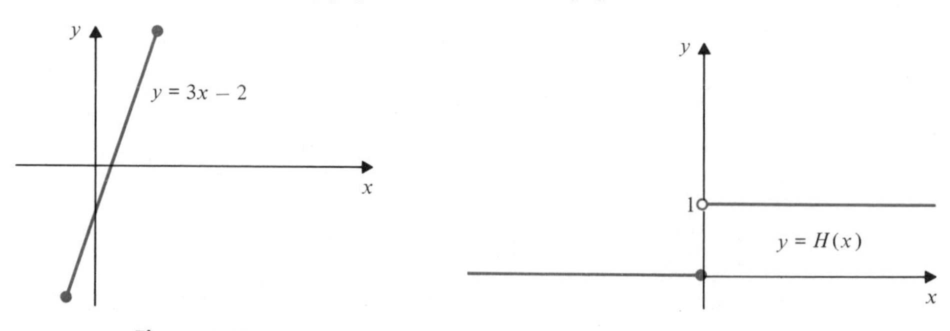

Figure 1.49

Figure 1.50

In particular, if g is continuous at a (so that $L = g(a)$), then

$$\lim_{x \to a} f(g(x)) = f(g(a)),$$

that is, the composition $f \circ g$ is continuous at a.

c) The function $f(x) = C$ (constant) is continuous on the whole real line.

d) If r is a rational number, then the function $f(x) = x^r$ is continuous wherever it is defined. In particular, if n is a positive integer, then x^n is continuous on the whole real line. □

Roughly speaking, parts (a) and (b) of Theorem 4 assure us that sums, differences, products, quotients, and compositions of continuous functions are continuous wherever they are defined. Parts (c) and (d) provide us with some specific examples of continuous functions from which we may determine others by taking such combinations.

A **polynomial** is a function $P(x)$ that is the sum of a finite number of terms each of which is a constant multiple of a nonnegative integer power of x; that is, $P(x)$ is of the form

$$P(x) = a_0 + a_1x + a_2x^2 + a_3x^3 + \cdots + a_nx^n,$$

where $a_0, a_1, \ldots, a_n$ are constants and $a_n \neq 0$. This polynomial is said to be of degree n. For instance, $5x^2 - 3x + 1$ is a polynomial of degree 2.

Theorem 4 assures us that every polynomial is continuous everywhere on the real line. (Why?) If $P(x)$ and $Q(x)$ are two polynomials, the fraction $P(x)/Q(x)$ is called a **rational function.** By Theorem 4, every rational function is continuous on the real line except at points x where $Q(x) = 0$.

Example 2

a) The function $5 - 2x^3 + (3/4)\,x^7$ is a polynomial, and so it is continuous on the real line $\mathbb{R}$.

b) The function $f(x) = (x^2 - x)/(x^2 + 3)$ is a rational function whose denominator does not vanish for any x. Thus $f(x)$ is continuous on $\mathbb{R}$.

c) The function $x/(x^2 - 4)$ is a rational function whose denominator vanishes at $x = 2$ and $x = -2$. It is continuous on $\mathbb{R}$ except at these two points.

d) The function $\sqrt{x}$ is defined on $[0, \infty)$. It is continuous on that interval (by Theorem 4(d)); in particular, it is continuous only on the right at $x = 0$.

e) The function $f(x) = \sqrt{x^2 - 1}$ has domain $(-\infty, -1] \cup [1, \infty)$. Since $x^2 - 1$ is continuous everywhere on $\mathbb{R}$, $f(x)$ is continuous on its domain, by Theorem 4(b).

f) The function $1/(\sqrt{x + 2})$ has domain $(-2, \infty)$ and is continuous there since $x + 2$ is continuous everywhere on $\mathbb{R}$.

The following two theorems contain very important and useful results about continuous functions. They are more subtle than the results quoted in Theorem 4; the proofs (see Appendix II) require a careful study of the implications of the completeness property of the real numbers. Both results should seem obviously true on an intuitive level.

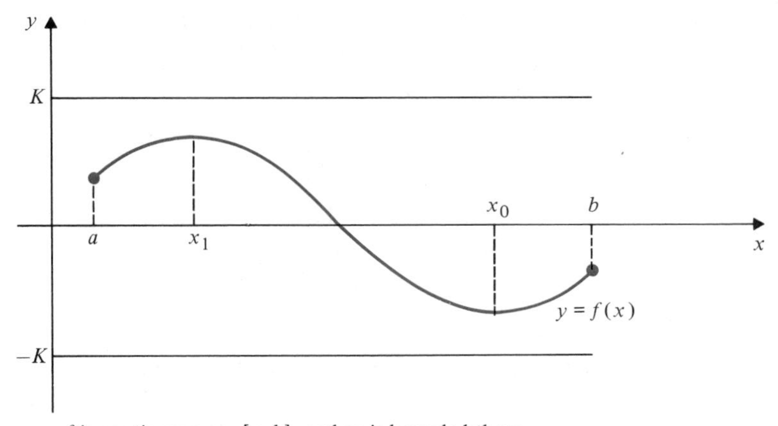

f is continuous on $[a, b]$ and so is bounded there.
f assumes a maximum value at the point x_1 and a minimum value at x_0.

Figure 1.51

Theorem 5 If f is continuous on the closed, finite interval $[a, b]$, then

i) there is a positive real number K such that

$$|f(x)| \leq K$$

on $[a, b]$, and

ii) there are points x_0 and x_1 in $[a, b]$ such that for every x in $[a, b]$,

$$f(x_0) \leq f(x) \leq f(x_1). \square$$

Conclusion (i) asserts that $f(x)$ is bounded on $[a, b]$; that is, it cannot take on arbitrarily large positive or negative values. Conclusion (ii) asserts that f takes on maximum and minimum values at points of $[a, b]$. These results are illustrated in Fig. 1.51.

Theorem 5 will be especially useful when we study optimization problems in Section 4.4. Such problems require us to find maximum or minimum values of functions defined on intervals and arise frequently in applications of mathematics.

Example 3 What is the area of the largest rectangular plot that can be fenced on all sides using 200 meters of fence?

Solution If the length and width of the plot are x and y (meters), respectively, then the perimeter is $2x + 2y$ (see Fig. 1.52). Since we have only 200 meters of fence, we want $2x + 2y = 200$, so $y = 100 - x$. The area of the plot can be expressed as a function of x alone:

$$A(x) = xy = x(100 - x) = 100x - x^2.$$

Evidently x and y must both be nonnegative, so $0 \leq x \leq 100$. We are looking for the maximum value of $A(x)$ on the interval $[0, 100]$. Theorem 5 assures us that such a maximum value exists. Since $A(x) > 0$ if $0 < x < 100$ and $A(0) = 0$ and

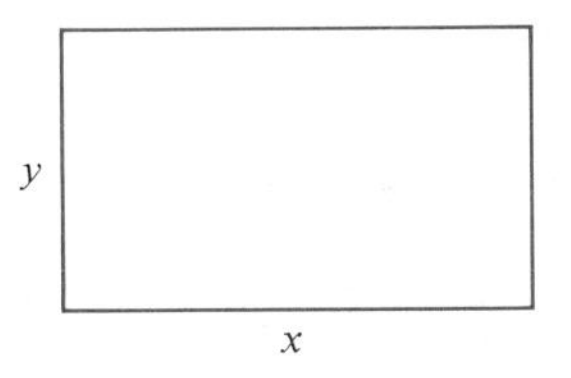

Figure 1.52

$A(100) = 0$, the maximum must occur at a point x in the open interval $(0, 100)$. Calculus will provide us with useful tools for finding points where such maximum (or minimum) values occur. In this case, however, we can find the maximum value without using any calculus; we need only complete the square in the quadratic polynomial $A(x)$:

$$\begin{aligned} A(x) &= 2500 - (2500 - 100x + x^2) \\ &= 2500 - (50 - x)^2. \end{aligned}$$

Clearly $A(x) \leq 2500$ for every x, and $A(50) = 2500$. Thus the largest rectangle with perimeter 200 meters has area 2500 m^2. It is a square of side 50 m.

Theorem 5 may fail if any of its conditions are not satisfied. That is, a function $f(x)$ may fail to be bounded, or may fail to have a maximum or minimum value, if its domain is not a finite interval, or not closed, or if f fails to be continuous anywhere in the domain.

Example 4 a) If $f(x) = x$ on $\mathbb{R}$, then f is not bounded, and so it does not have a maximum or minimum value.

b) If $f(x) = x$ on $(1, 2)$, then f, though bounded, has no maximum or minimum value. Indeed, its range is also the open interval $(1, 2)$, which has no greatest or least element.

c) Let $f(x) = \begin{cases} \dfrac{1}{x}, & \text{if } -1 \leq x \leq 1 \text{ and } x \neq 0 \\ 0, & \text{if } x = 0. \end{cases}$

Then f is defined on the closed, finite interval $[-1, 1]$ but it is not continuous at the point $x = 0$ in that interval. The function is not bounded on $[-1, 1]$ and does not have a maximum or minimum value there. See Fig. 1.53.

Theorem 6 (*The Intermediate-Value Theorem*) If f is continuous on $[a, b]$ and if s is a real number lying between $f(a)$ and $f(b)$, then there exists at least one real number c between a and b such that $f(c) = s$. □

Intuitively, we can see that since the graph of f has no breaks between a and b and since it starts out at height $f(a)$ and ends up at height $f(b)$, it must somewhere cross the horizontal line at height s between $f(a)$ and $f(b)$. This is depicted in Fig. 1.54.

Theorem 6 is often used to show that certain equations have solutions.

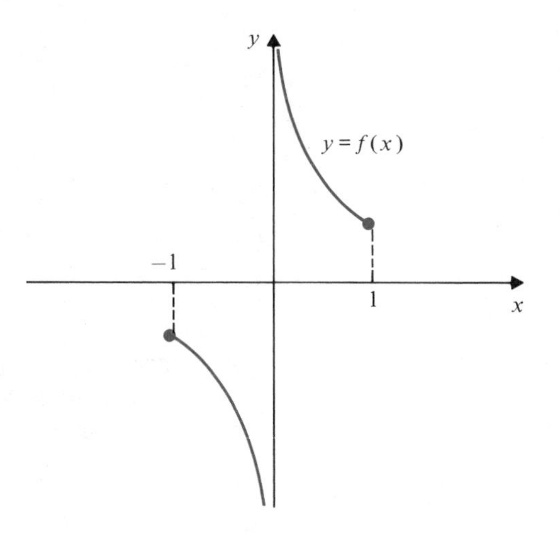

Figure 1.53

Example 5 Show that the cubic equation $x^3 + x^2 - 4 = 0$ has a solution x in the interval $(1, 2)$.

Solution The function $f(x) = x^3 + x^2 - 4$ is a polynomial, and so it is continuous everywhere. Since $f(1) = -2 < 0$ and $f(2) = 8 > 0$, the intermediate-value theorem assures us that there is at least one point x in the interval $(1, 2)$ such that $f(x) = 0$.

Example 6 Determine the intervals on which $f(x) = x^3 - 4x$ is positive and negative.

Solution Since $f(x) = x(x^2 - 4) = x(x - 2)(x + 2)$, $f(x) = 0$ only at $x = 0$, 2, and -2. Because f is continuous on the whole real line it must have constant sign on each of the intervals $(-\infty, -2)$, $(-2, 0)$, $(0, 2)$, and $(2, \infty)$. (This assertion follows from the intermediate-value theorem. If there were points a and b in, say $(0, 2)$ such that $f(a) < 0$ and $f(b) > 0$, then by that theorem there would exist c between a and b,

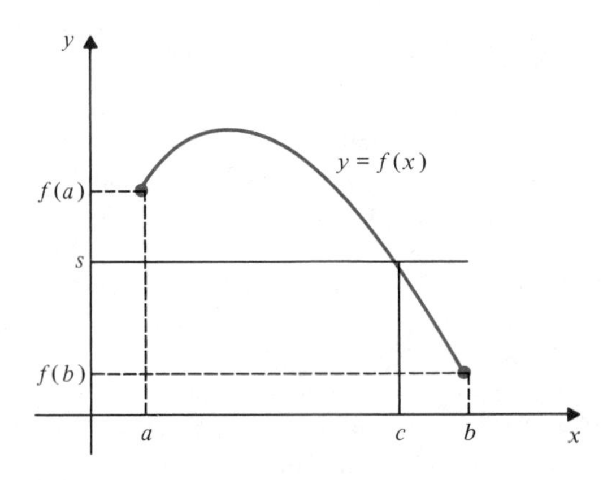

Figure 1.54

and therefore between 0 and 2, such that $f(c) = 0$, which is a contradiction.) Since $f(-3) = -15$, $f(-1) = 3$, $f(1) = -3$ and $f(3) = 15$, $f(x) < 0$ on $(-\infty, -2)$ and $(0, 2)$, and $f(x) > 0$ on $(-2, 0)$ and $(2, \infty)$.

EXERCISES

In Exercises 1–18 state where the given function is continuous and where it is discontinuous. At points of discontinuity state which of the conditions (i), (ii), and (iii) following Definition 12 fail.

1. $f(x) = 3x - 1$

2. $f(x) = \dfrac{1}{3x - 1}$

3. $f(x) = \sqrt{3x - 1}$

4. $f(x) = -\dfrac{1}{\sqrt{3x - 1}}$

5. $f(x) = x^2 + 1$

6. $f(x) = 7x^5 - 2x^3 + x - 250$

7. $f(x) = \dfrac{1}{x^2 - 1}$

8. $f(x) = \dfrac{x^5 + 3}{x^4 + x^3 - 6x^2}$

9. $f(x) = \dfrac{x + 2}{|x + 2|}$

10. $f(x) = \dfrac{x^2 + 1}{|x^2 + 1|}$

11. $f(x) = |x^2 - 1|$

12. $f(x) = \dfrac{x^2 - 1}{|x^2 - 1|}$

13. $f(x) = \begin{cases} -1, & \text{if } x > 1 \\ x, & \text{if } -1 \leq x \leq 1 \\ 1, & \text{if } x < -1 \end{cases}$

14. $f(x) = \begin{cases} x, & \text{if } x < 0 \\ x^2, & \text{if } x \geq 0 \end{cases}$

15. $f(x) = \begin{cases} 1 + x^2, & \text{if } x \neq 2 \\ 4.987, & \text{if } x = 2 \end{cases}$

16. $f(x) = \begin{cases} -3, & \text{if } x = -1 \\ \dfrac{x^2 - x - 2}{x + 1}, & \text{if } x \neq -1 \end{cases}$

17. $f(x) = \begin{cases} x, & \text{if } x \geq 0 \\ \dfrac{1}{x}, & \text{if } x < 0 \end{cases}$

18. $f(x) = \begin{cases} \dfrac{1}{x^4}, & \text{if } x \neq 0 \\ 0, & \text{if } x = 0 \end{cases}$

19. If the sum of two nonnegative numbers is 8, show that their product must be bounded. What is the largest possible value of their product?

20. Show that there is a point on the straight line with equation $y = (2 - 3x)/\sqrt{3}$ that is closest to the origin. Find that point. (*Hint:* Express the square of the distance from the point (x, y) to the origin as a function of x. Complete the square as in Example 3 in this section.)

Find the intervals in which the functions $f(x)$ in Exercises 21–24 are positive and negative.

21. $f(x) = x^2 + 4x + 3$

22. $f(x) = \dfrac{x^2 - 1}{x}$

23. $f(x) = \dfrac{x^2 + x - 2}{x^3}$

24. $f(x) = \dfrac{x^2 - 1}{x^2 - 4}$

Exercises 25–28 involve applications of the intermediate-value theorem. A number r is called a zero of the function $f(x)$ if $f(r) = 0$.

25. If $f(x) = x^3 + x - 1$, show that f has a zero between $x = 0$ and $x = 1$.

26. Show that $f(x) = x^3 - 15x + 1$ has at least three zeros in $[-4, 4]$.

27. Show that $F(x) = (x - a)(x - b) + x$ takes on the value $\dfrac{a + b}{2}$.

28. Suppose that $f(x)$ is continuous on $[0, 1]$ and satisfies $0 \leq f(x) \leq 1$ for every x in $[0, 1]$. Show that there exists c in $[0, 1]$ such that $f(c) = c$. (*Hint:* Consider $g(x) = f(x) - x$.)

29. If an even function is continuous on the right at $x = 0$, show that it is continuous at $x = 0$.

30. If an odd function is continuous on the right at $x = 0$, show that it is continuous at $x = 0$ and $f(0) = 0$.

2.1 Tangent Lines and Their Slopes
2.2 The Derivative
2.3 Differentiation Rules (Sums, Products, and Quotients)
2.4 The Chain Rule
2.5 Interpretations of the Derivative
2.6 Higher-Order Derivatives
2.7 Implicit Differentiation
2.8 The Mean-Value Theorem
2.9 Antiderivatives and Indefinite Integrals

2

Differentiation: Definition, Interpretation, and Techniques

There are two fundamental problems considered in calculus, the problem of slopes and the problem of areas. In the problem of slopes we are concerned with finding the straight line tangent to a given curve at a specified point on that curve. The solution of this problem is the subject of differential calculus. As we will see, it has many and varied applications in mathematics and other disciplines. The second problem, that of areas, is the subject of integral calculus, which we begin in Chapter 6.

2.1 TANGENT LINES AND THEIR SLOPES

In this section we will define precisely what we mean when we say that a straight line L is tangent to the curve C at the point P. The definition will in turn give us a method for finding L (that is, for finding its equation). It is often the case in mathematics that the most important step in the solution of a fundamental problem is making a suitable definition.

For simplicity, and to avoid certain problems best postponed until later, we will not deal now with the most general kinds of curves; we will deal with those that are the **graphs of continuous functions.** Let C be the graph of $y = f(x)$ and let P be the point (x_o, y_o) where $y_o = f(x_o)$, so that P lies on C. We assume that P is not an endpoint of C; hence C extends some distance to the left and right of P (see Fig. 2.1a).

How should we state precisely what we mean when we say that the line L is tangent to C at P? Past experience with tangent lines to circles does not seem to be of much help. A tangent line to a circle (Fig. 2.1b) has the following properties:

i) It meets the circle at only one point.
ii) The circle lies on one side of the line.
iii) The tangent is perpendicular to the line joining the center of the circle to the point of contact.

In general, curves do not have centers, so (iii) is useless for characterizing

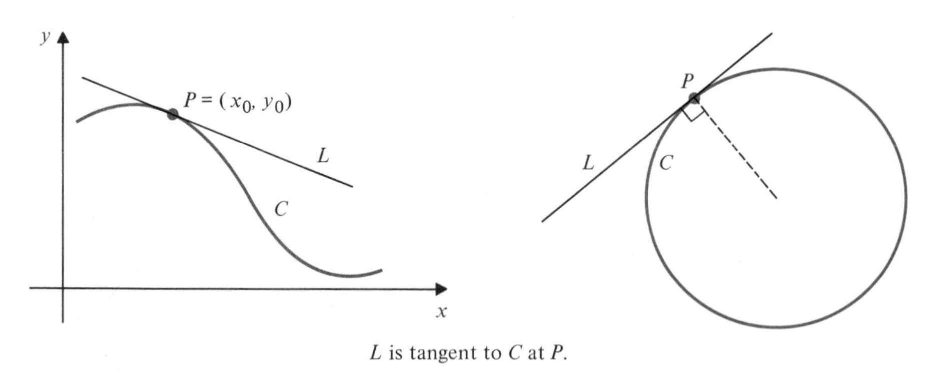

L is tangent to *C* at *P*.

(a) **Figure 2.1** (b)

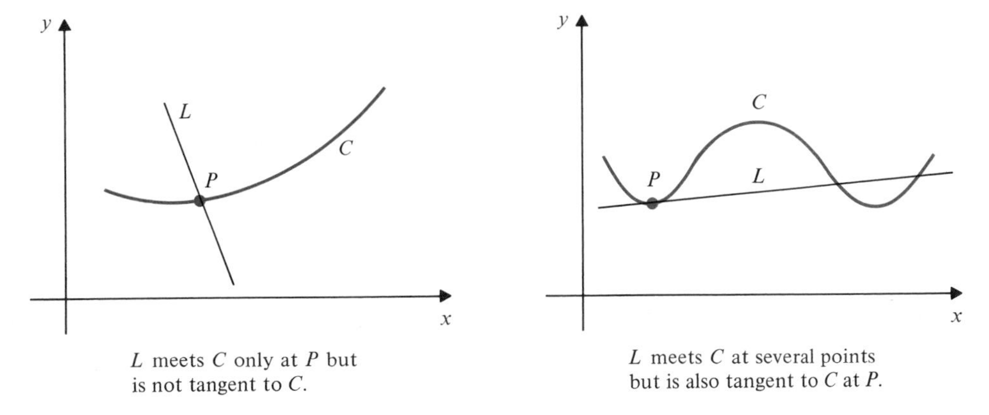

Figure 2.2

tangents to curves. The examples in Figs. 2.2 and 2.3 show that (i) and (ii) cannot be used to define tangency either.

A reasonable definition of tangency can be stated in terms of limits. If Q is a point on C different from P, then the *secant line PQ* rotates around P as Q moves along the curve. If the line L is tangent to C at P, then its slope is the limit of the slopes of these secant lines PQ as Q approaches P along C (see Fig. 2.4).

If C is the graph of $y = f(x)$ and $P = (x_o, f(x_o))$, then a different point Q on the graph will have a different x-coordinate, say $x_o + h$ where $h \neq 0$. Thus $Q = (x_o + h, f(x_o + h))$ and the slope of PQ is

$$\frac{f(x_o + h) - f(x_o)}{h}.$$

This expression is called the **Newton quotient** for f at x_o. Note that h may be positive or negative depending on whether Q is to the right or left of P.

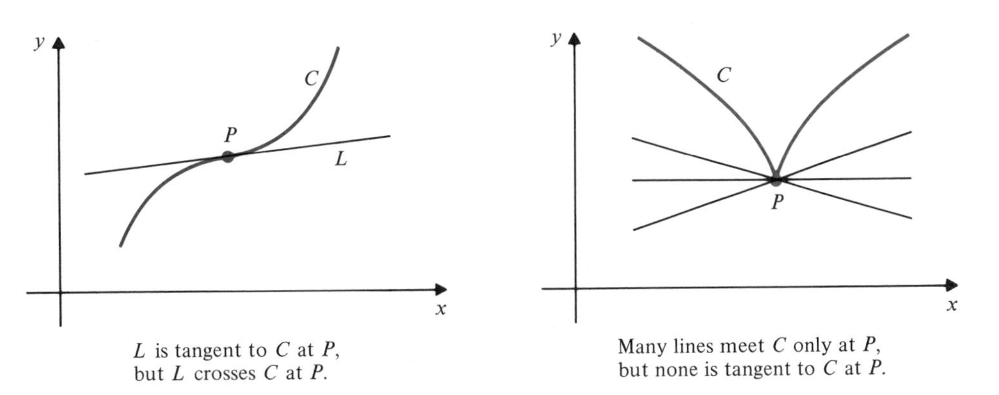

Figure 2.3

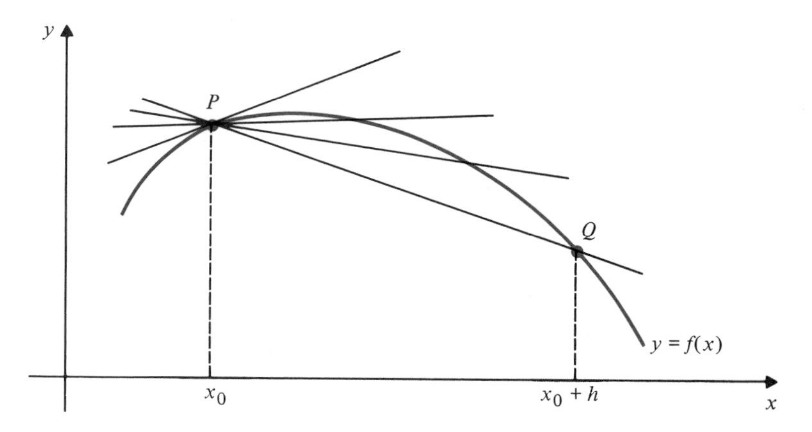

Figure 2.4

Definition 1

> Suppose that the function f is continuous at $x = x_o$. If $\lim_{h\to 0} [f(x_o + h) - f(x_o)]/h = m$ exists, then the straight line having slope m and passing through the point $P = (x_o, f(x_o))$, that is, the line with equation $y = f(x_o) + m(x - x_o)$, is tangent to the graph of $y = f(x)$ at P.
>
> If $\lim_{h\to 0} [f(x_o + h) - f(x_o)]/h = \infty$ (or $-\infty$), then the vertical straight line $x = x_o$ is tangent to $y = f(x)$ at P.
>
> If $\lim_{h\to 0} [f(x_o + h) - f(x_o)]/h$ fails to exist in any way other than by being either ∞ or $-\infty$, then the graph of $y = f(x)$ has no tangent line at P.

Example 1 Find the tangent line to $y = x^2$ at the point $(1, 1)$.

Solution Here $f(x) = x^2$ and $x_o = 1$, so the required tangent has slope

$$\lim_{h\to 0} \frac{f(1 + h) - f(1)}{h} = \lim_{h\to 0} \frac{(1 + h)^2 - 1}{h} = \lim_{h\to 0} \frac{2h + h^2}{h} = \lim_{h\to 0} 2 + h = 2.$$

Accordingly, the equation of the tangent line at $(1, 1)$ is $y = 1 + 2(x - 1)$ or $y = 2x - 1$. See Fig. 2.5.

Example 2 Find the tangent line to the curve $y = x^{1/3}$ at $x = 0$.

Solution Here $f(x) = x^{1/3}$ and $x_o = 0$, so the tangent has slope

$$\lim_{h\to 0} \frac{f(0 + h) - f(0)}{h} = \lim_{h\to 0} \frac{h^{1/3}}{h} = \lim_{h\to 0} h^{-2/3} = \infty.$$

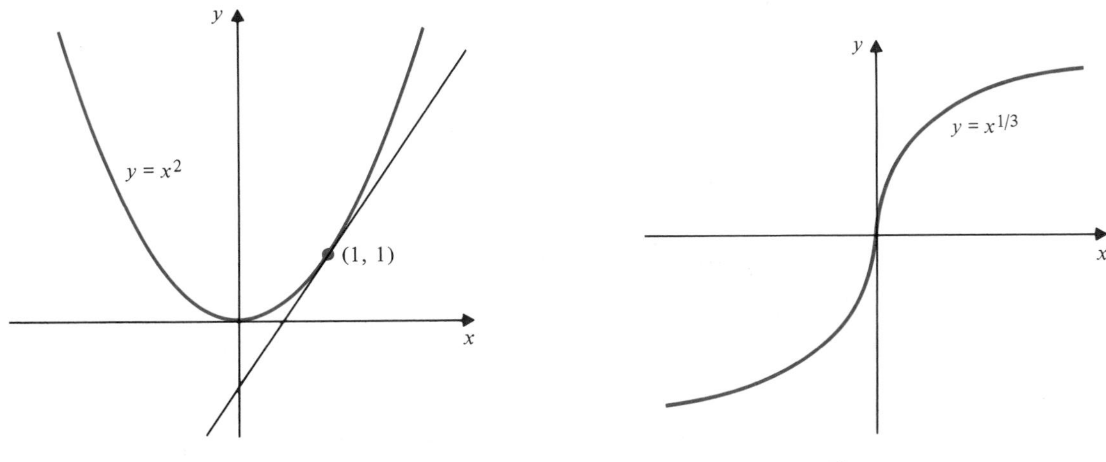

Figure 2.5

Figure 2.6

Thus the y-axis, $x = 0$ is the required tangent line, as shown in Fig. 2.6. Note that $h^{-2/3} = (h^{-1/3})^2 > 0$ for both $h > 0$ and $h < 0$, so the limit is ∞.

Example 3 Does the curve $y = x^{2/3}$ have a tangent line at (0, 0)?

Solution The Newton quotient here is

$$\frac{f(0 + h) - f(0)}{h} = \frac{h^{2/3}}{h} = h^{-1/3}.$$

Since $\lim_{h\to0+} h^{-1/3} = \infty$ and $\lim_{h\to0-} h^{-1/3} = -\infty$, the limit $\lim_{h\to0} h^{-1/3}$ does not exist.

The graph of $y = x^{2/3}$ has no tangent line at (0, 0) (see Fig. 2.7), so we say that $x = 0$ is a **singular point** of $x^{2/3}$. Because the curve $y = x^{2/3}$ suddenly reverses direction at (0, 0), and so has an infinitely sharp point there, we say that the curve has a **cusp** at the point (0, 0).

Example 4 Does the graph of $y = |x|$ have a tangent line at $x = 0$?

Solution The Newton quotient is

$$\frac{|0 + h| - |0|}{h} = \frac{|h|}{h} = \begin{cases} 1, \text{ if } h > 0 \\ -1, \text{ if } h < 0. \end{cases}$$

Since $\lim_{h\to0+} \dfrac{|h|}{h} = 1$ and $\lim_{h\to0-} \dfrac{|h|}{h} = -1$, once again the Newton quotient has no limit as $h \to 0$ and so $y = |x|$ has no tangent line at (0, 0). (See Fig. 2.8.) Again, $x = 0$ is a singular point of $|x|$, but the origin is not called a cusp here because it is not an infinitely sharp point.

We can see intuitively that curves have tangent lines only at points where they are smooth. The curves in Examples 3 and 4 above have tangent lines everywhere except at the origin, where they are not smooth.

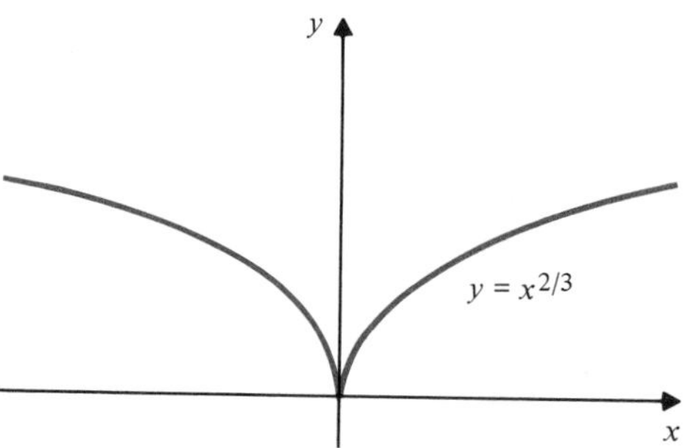

Figure 2.7

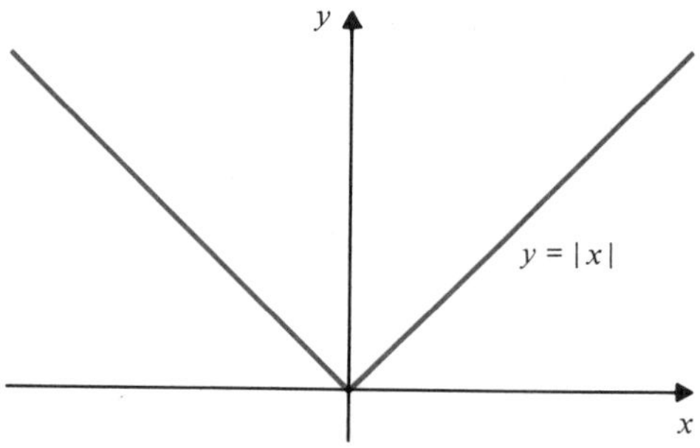

Figure 2.8

Definition 2

> The slope of a curve C at a point P is the slope of the tangent line to C at P if such a tangent line exists.

Example 5 Find the slope of the curve $y = x/(3x + 2)$ at the point $x = -2$.

Solution If $x = -2$, then $y = 1/2$, so the required slope is

$$\lim_{h \to 0} \frac{\dfrac{-2+h}{3(-2+h)+2} - \dfrac{1}{2}}{h} = \lim_{h \to 0} \frac{-4 + 2h - (-4 + 3h)}{2(-4+3h)h}$$

$$= \lim_{h \to 0} \frac{-1}{2(-4+3h)} = \frac{1}{8}.$$

Normals

If a curve C has a tangent line L at point P, then the straight line N through P perpendicular to L is called the **normal** to C at P. If L is horizontal, then N is vertical and so has infinite slope. If L is not horizontal, then the slope of N is the negative reciprocal of the slope of L:

$$\text{slope of the normal} = \frac{-1}{\text{slope of the tangent}}.$$

To see this suppose that L has slope m and that N has slope p. The right-angled triangle formed by L and N (in Fig. 2.9) and the vertical line lying h units to the right of P is divided by the horizontal line through P into two similar right-angled triangles. Comparing the sides of these triangles we obtain

$$\frac{mh}{h} = \frac{h}{-ph}.$$

Hence $p = -1/m$ as asserted.

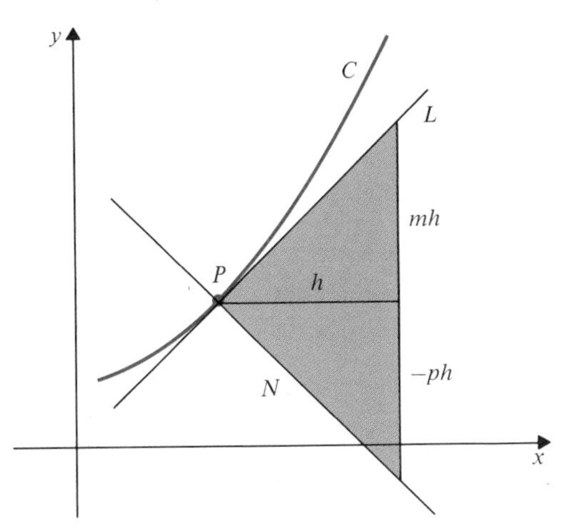

Figure 2.9

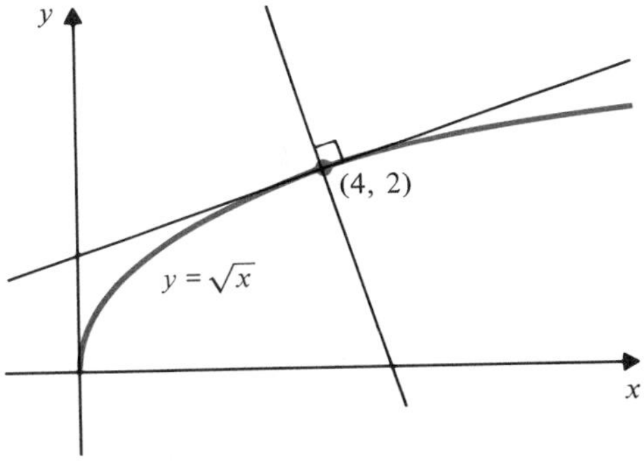

Figure 2.10

Example 6 Find an equation of the normal to $y = x^2$ at $(1, 1)$.

Solution By Example 1, the tangent line to $y = x^2$ at $(1, 1)$ has slope 2. Hence the normal has slope $-1/2$ and equation $y = 1 - (1/2)(x - 1)$ or $y = 3/2 - x/2$.

Example 7 Find the equation of the normal line to $y = x^{1/3}$ at the origin.

Solution By Example 2 the y-axis is tangent to $y = x^{1/3}$ at the origin. Hence the normal there is the x-axis, $y = 0$.

Example 8 Find equations of the straight lines tangent and normal to the curve $y = \sqrt{x}$ at the point $(4, 2)$.

Solution The slope of the tangent at $(4, 2)$ (see Fig. 2.10) is

$$\begin{aligned}\lim_{h\to 0}\frac{\sqrt{4+h}-2}{h} &= \lim_{h\to 0}\frac{(\sqrt{4+h}-2)(\sqrt{4+h}+2)}{h(\sqrt{4+h}+2)}\\ &= \lim_{h\to 0}\frac{4+h-4}{h(\sqrt{4+h}+2)}\\ &= \lim_{h\to 0}\frac{1}{\sqrt{4+h}+2} = \frac{1}{4}.\end{aligned}$$

Thus the tangent line has equation $y = 2 + \frac{1}{4}(x - 4)$ or $y = \frac{x}{4} + 1$, and the normal line has equation $y = 2 - 4(x - 4)$ or $y = -4x + 18$.

EXERCISES

In Exercises 1–11 find an equation of the straight line tangent to the given curve at the point indicated and also find an equation of the normal line there. Sketch the curve.

1. $y = 2x^2 - 5$ at $(2, 3)$

2. $y = x^2 - 2x + 2$ at $(1, 1)$

3. $y = 6 - x - x^2$ at $x = -2$

4. $y = x^3 + 8$ at $x = 0$

5. $y = x^3 + 8$ at $x = -2$

6. $y = \dfrac{x - 2}{x + 2}$ at $(0, -1)$

7. $y = \dfrac{1}{x^2}$ at $x = 3$

8. $y = \dfrac{1}{x^2 + 1}$ at $x = -1$

9. $y = \sqrt{x + 1}$ at $x = 3$

10. $y = \dfrac{1}{\sqrt{x}}$ at $x = 9$

11. $y = \dfrac{2x}{x + 2}$ at $x = 2$

12. $y = \sqrt{5 - x^2}$ at $x = 1$

13. $y = x^2$ at $x = x_o$

14. $y = \dfrac{1}{x}$ at $\left(a, \dfrac{1}{a}\right)$

15. $y = ax^2 + bx + c$ at $x = x_o$, $(a \neq 0)$

16. Find the slope of the curve $y = x^2 - 1$ at the point $x = x_o$. What is the equation of the tangent line to $y = x^2 - 1$ that has slope -3?

17. a) Find the slope of the curve $y = x^3$ at the point $x = a$.

b) Find the equations of the straight lines of slope 2 that are tangent to $y = x^3$.

18. Find all points on the curve $y = x^3 - 3x$ where the tangent line is parallel to the x-axis.

2.2 THE DERIVATIVE

A straight line has the property that its slope is the same at all points. For any other graph, however, the slope may vary from point to point. Thus the slope of the graph of $y = f(x)$ at the point x is itself a function of x. Wherever this slope exists (and is not infinite) we say that f is differentiable, and we call the slope the derivative of f.

Definition 3

If $\lim_{h \to 0} [f(x + h) - f(x)]/h = f'(x)$ exists as a finite real number, we say that the function f is **differentiable** at x. In this case the number $f'(x)$ is called the **derivative** of f at x, and the function f' is called the derivative of f.

The derivative of f is itself a function of x whose value at any point is the slope of the tangent line to the graph of f at that point. Thus the equation of the tangent line to $y = f(x)$ at $(x_o, f(x_o))$ is

$$y = f(x_o) + f'(x_o)(x - x_o).$$

The domain of f' may be smaller than the domain of f because it contains only those points in the domain of f at which f is differentiable. The graph of f' can often be sketched directly from that of f by visualizing slopes. The procedure of sketching the graph of f' by estimating slopes from the graph of f is called **graphical differentiation.** Indeed, the process of finding the derivative f' of a given function f

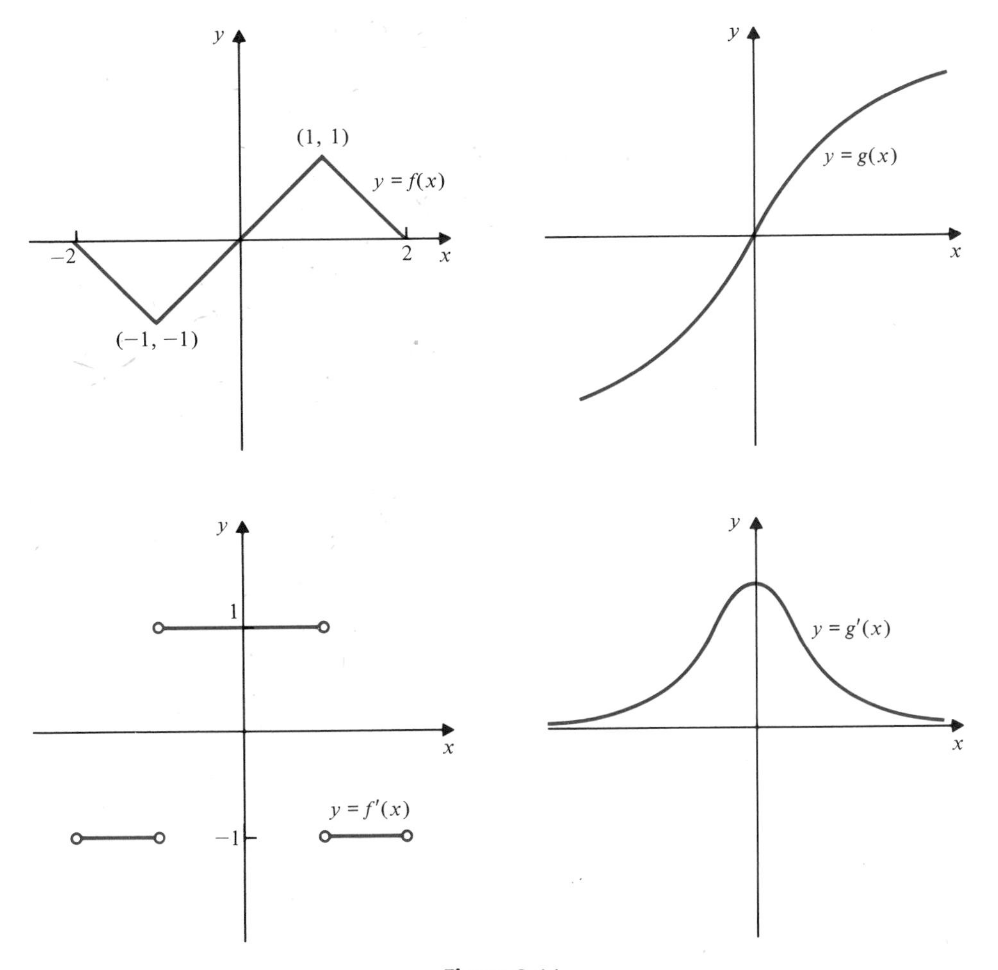

Figure 2.11

is called **differentiation.** In Fig. 2.11 the graphs of f' and g' were sketched by inspection of the graphs of f and g.

We now give several examples of the computation of derivatives algebraically from the definition of derivative. Some of these results are useful and are collected in a table at the end of the examples.

Example 1 $$f(x) = C \text{ (a constant function).}$$

It is evident graphically that $f'(x) = 0$ for all x. (See Figs. 2.12 and 2.13.) We check this algebraically:

$$\begin{aligned} f'(x) &= \lim_{h \to 0} \frac{f(x+h) - f(x)}{h} \\ &= \lim_{h \to 0} \frac{C - C}{h} = 0. \end{aligned}$$

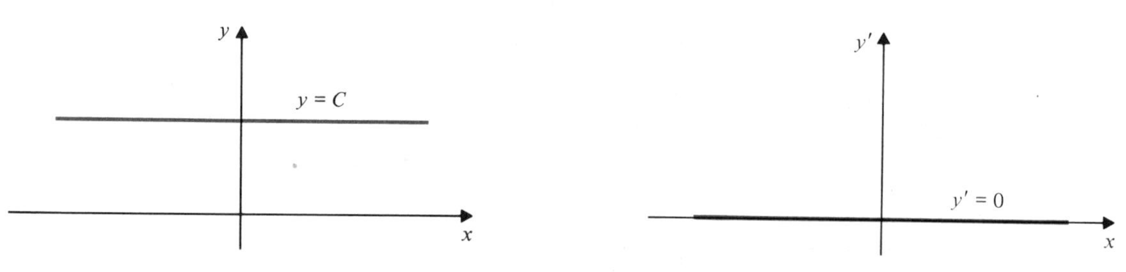

Figure 2.12

Figure 2.13

Example 2

$$f(x) = ax + b \ (a, b \text{ constants})$$

Again it is clear graphically that $f'(x) = a$ for all x. (See Figs. 2.14 and 2.15.) Algebraically,

$$\begin{aligned} f'(x) &= \lim_{h\to 0} \frac{f(x+h) - f(x)}{h} \\ &= \lim_{h\to 0} \frac{a(x+h) + b - (ax+b)}{h} \\ &= \lim_{h\to 0} \frac{ah}{h} = a. \end{aligned}$$

Example 3

$$f(x) = x^2$$

$$f'(x) = \lim_{h\to 0} \frac{(x+h)^2 - x^2}{h} = \lim_{h\to 0} \frac{2hx + h^2}{h} = \lim_{h\to 0} (2x + h) = 2x$$

Example 4

$$f(x) = \frac{1}{x} \quad (x \neq 0)$$

$$\begin{aligned} f'(x) &= \lim_{h\to 0} \frac{\dfrac{1}{x+h} - \dfrac{1}{x}}{h} \\ &= \lim_{h\to 0} \frac{\dfrac{x - (x+h)}{(x+h)x}}{h} \end{aligned}$$

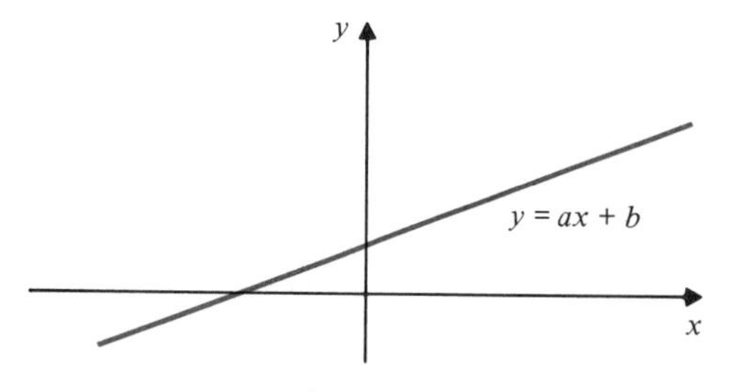

Figure 2.14

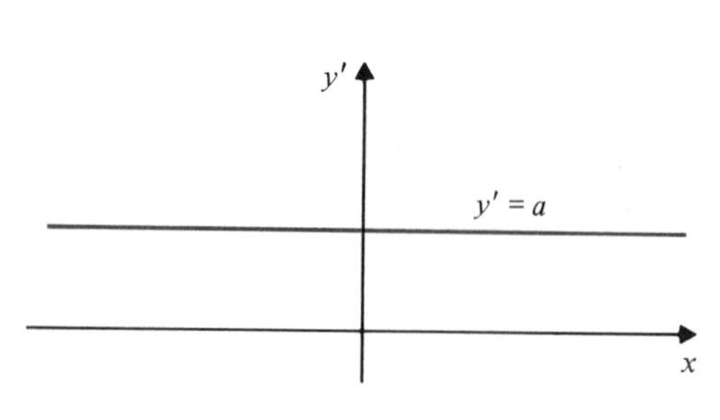

Figure 2.15

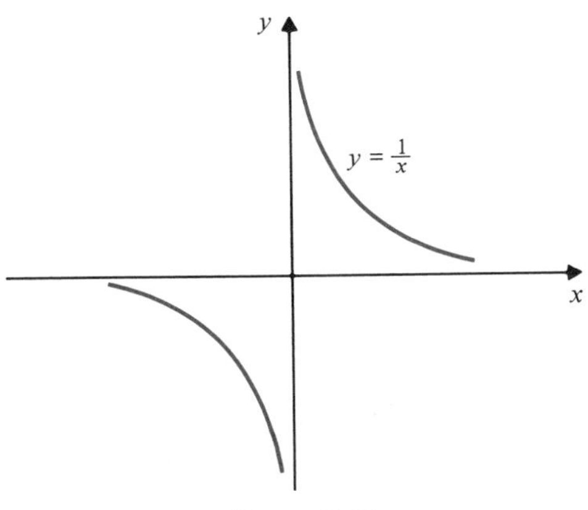

Figure 2.16

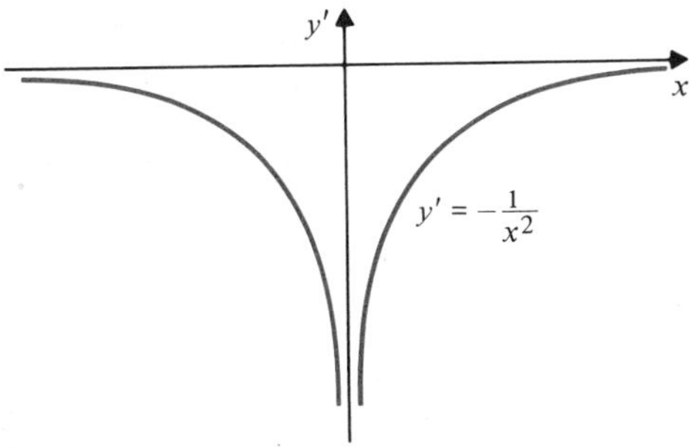

Figure 2.17

$$= \lim_{h\to 0} \frac{-h}{hx(x+h)} = -\frac{1}{x^2}$$

See Figs. 2.16 and 2.17.

Example 5

$$f(x) = \sqrt{x} \quad (x > 0)$$

$$f'(x) = \lim_{h\to 0} \frac{\sqrt{x+h} - \sqrt{x}}{h}$$

$$= \lim_{h\to 0} \frac{(\sqrt{x+h} - \sqrt{x})(\sqrt{x+h} + \sqrt{x})}{h(\sqrt{x+h} + \sqrt{x})}$$

$$= \lim_{h\to 0} \frac{x+h-x}{h(\sqrt{x+h} + \sqrt{x})}$$

$$= \lim_{h\to 0} \frac{1}{\sqrt{x+h} + \sqrt{x}} = \frac{1}{2\sqrt{x}}$$

See Figs. 2.18 and 2.19. Note that f is not differentiable at $x = 0$.

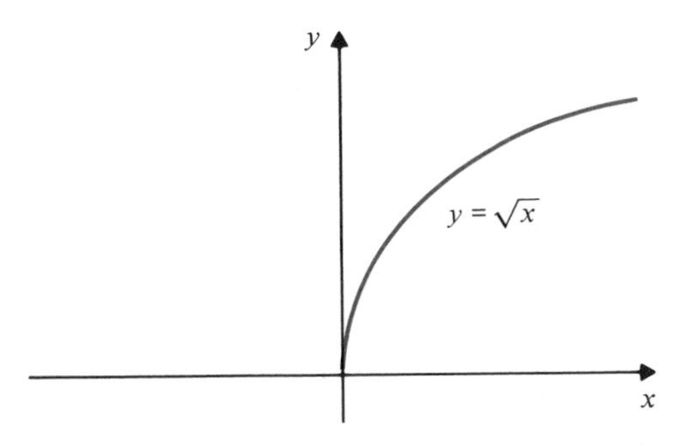

Figure 2.18

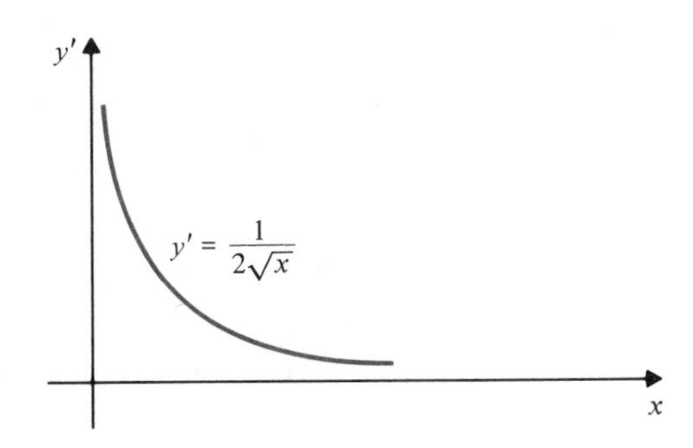

Figure 2.19

Example 6

$$f(x) = |x| = \begin{cases} x, & \text{if } x \geq 0 \\ -x, & \text{if } x < 0 \end{cases}$$

From Example 2, $f'(x) = 1$ if $x > 0$ and $f'(x) = -1$ if $x < 0$. Thus $f'(x) = \operatorname{sgn} x$ where sgn (x) is the signum function $x/|x|$:

$$\frac{d}{dx}|x| = \operatorname{sgn} x = \frac{x}{|x|} = \begin{cases} 1, & \text{if } x > 0 \\ -1, & \text{if } x < 0. \end{cases}$$

See Figs. 2.20 and 2.21. By Example 4 of Section 2.1, f is not differentiable at $x = 0$, so we leave sgn 0 undefined.

Table 1 Some Elementary Functions and Their Derivatives

$f(x)$	$f'(x)$
1	0
x	1
x^2	$2x$
$\frac{1}{x}$	$-\frac{1}{x^2}$ $(x \neq 0)$
$\sqrt{x}$	$\frac{1}{2\sqrt{x}}$ $(x > 0)$
$\|x\|$	$\operatorname{sgn} x$ $(x \neq 0)$

The first five functions $f(x)$ in the table are all of the form x^r for various values of r. The results in the table suggest that

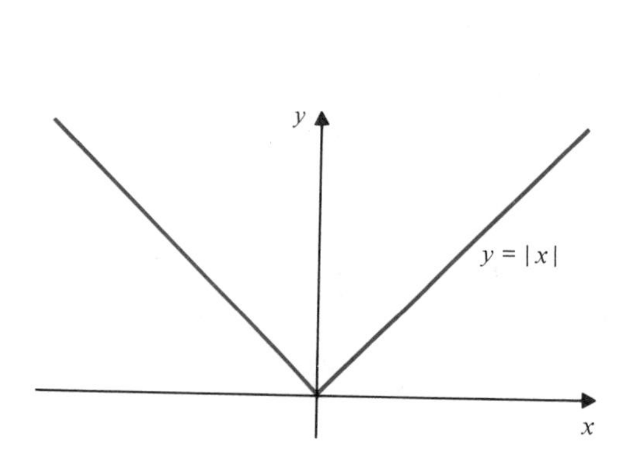

Figure 2.20

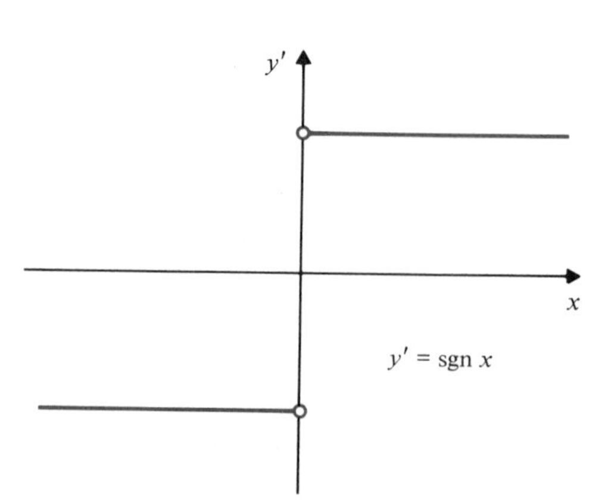

Figure 2.21

$$\text{if } f(x) = x^r, \text{ then } f'(x) = rx^{r-1} \qquad \text{(the general power rule).}$$

This formula is true whenever it makes sense, but we cannot give a complete proof yet. If $r = n$, a positive integer, we can prove it using the factorization of a difference of nth powers:

$$\begin{aligned} &a^n - b^n \\ &\quad = (a - b)(a^{n-1} + a^{n-2}b + a^{n-3}b^2 + \cdots + a^2b^{n-3} + ab^{n-2} + b^{n-1}). \end{aligned}$$

(Check this by multiplying out the right-hand side.) We have, for $f(x) = x^n$,

$$\begin{aligned} f'(x) &= \lim_{h\to 0} \frac{(x+h)^n - x^n}{h} \\ &= \lim_{h\to 0} \frac{h[(x+h)^{n-1} + (x+h)^{n-2}x + (x+h)^{n-3}x^2 + \cdots + x^{n-1}]}{h} = nx^{n-1}, \end{aligned}$$

as required. An alternative proof using the binomial theorem is suggested in Exercise 34 at the end of this section. Another is suggested as an example based on the product rule in Section 2.3. The factorization method used above can also be used to demonstrate the general power rule for the case $r = 1/n$. (See Exercise 33 at the end of this section.)

It is useful to have more than one notation for derivatives. If the function f is denoted by the dependent variable y, that is, if $y = f(x)$, then the derivative of f can be denoted in any of the following ways:

$$Dy = y' = \frac{dy}{dx} = \frac{d}{dx}f(x) = f'(x) = Df(x).$$

These should be read "the derivative of y" or "the derivative of $f(x)$." When the independent variable x appears explicitly as it does in the last four forms, we can say "the derivative of y with respect to x" or "the derivative of $f(x)$ with respect to x." Often the most convenient way of referring to the derivative of a function given explicitly as an expression in the variable x is to write d/dx in front of that expression. The symbol d/dx should be read "the derivative with respect to x of."

$$\frac{d}{dx}x^2 = 2x$$

$$\frac{d}{dx}\sqrt{x} = \frac{1}{2\sqrt{x}}$$

$$\frac{d}{dt}t^{100} = 100t^{99}$$

$$\text{if } y = u^3, \text{ then } \frac{dy}{du} = 3u^2$$

The value of the derivative of a function at a particular point x_o in its domain can also be expressed in several ways:

$$Dy\Big|_{x=x_o} = y'\Big|_{x=x_o} = \frac{dy}{dx}\Big|_{x=x_o} = \frac{d}{dx}f(x)\Big|_{x=x_o} = f'(x_o) = Df(x_o).$$

The symbol $\Big|_{x=x_o}$ is called an *evaluation symbol*. It signifies that the expression preceding it should be evaluated at $x = x_o$. Thus

$$\frac{d}{dx}x^4\Big|_{x=-1} = 4x^3\Big|_{x=-1} = 4(-1)^3 = -4.$$

Following are two more examples in which derivatives are computed from the definition.

Example 7 Use the definition of derivative to calculate $(d/dx)(x/(x^2+1))$.

Solution

$$\frac{d}{dx}\frac{x}{x^2+1} = \lim_{h\to 0}\frac{\dfrac{x+h}{(x+h)^2+1} - \dfrac{x}{x^2+1}}{h}$$

$$= \lim_{h\to 0}\frac{(x+h)(x^2+1) - x(x^2+2hx+h^2+1)}{h((x+h)^2+1)(x^2+1)}$$

$$= \lim_{h\to 0}\frac{1-x^2-hx}{((x+h)^2+1)(x^2+1)} = \frac{1-x^2}{(1+x^2)^2}$$

Example 8 Find an equation of the straight line tangent to the curve

$$y = \frac{\sqrt{x^2+3}}{x} \text{ at the point } (1, 2).$$

Solution The slope of the tangent line is

$$\frac{dy}{dx}\Big|_{x=1} = \lim_{h\to 0}\frac{\dfrac{\sqrt{(1+h)^2+3}}{1+h} - 2}{h}$$

$$= \lim_{h\to 0}\frac{(\sqrt{(1+h)^2+3} - 2(1+h))(\sqrt{(1+h)^2+3} + 2(1+h))}{h(1+h)(\sqrt{(1+h)^2+3} + 2(1+h))}$$

$$= \lim_{h\to 0}\frac{4+2h+h^2-4-8h-4h^2}{h(1+h)(\sqrt{(1+h)^2+3} + 2(1+h))}$$

$$= \lim_{h\to 0}\frac{-6-3h}{(1+h)(\sqrt{1+h)^2+3} + 2(1+h))} = -\frac{3}{2}.$$

Thus the tangent line has equation $y = 2 - (3/2)(x-1)$ or $y = 7/2 - 3x/2$.

Leibniz Notation and Differentials

The notations dy/dx and $(d/dx)f(x)$ are called Leibniz notations for the derivative, after Gottfried Wilhelm Leibniz (1646–1716), one of the creators of calculus who used such notations. The main ideas of calculus were discovered independently by Leibniz and Isaac Newton (1642–1727); the latter used notations similar to the prime (y') notations we use here. The Leibniz notation is suggested by the definition of derivative. The Newton quotient $[f(x + h) - f(x)]/h$ whose limit we take to find the derivative dy/dx can be written in the form $\Delta y/\Delta x$ where $\Delta y = f(x + h) - f(x)$ is the vertical displacement and $\Delta x = x + h - x$ is the horizontal displacement as we pass from the point $(x, f(x))$ to the point $(x + h, f(x + h))$ on the graph of f. (See Fig. 2.22.) Symbolically,

$$\frac{dy}{dx} = \lim_{\Delta x \to 0} \frac{\Delta y}{\Delta x}.$$

The Newton quotient $\Delta y/\Delta x$ is actually the quotient of two quantities, Δy and Δx. It is not at all clear, however, that the derivative dy/dx, the limit of $\Delta y/\Delta x$ as Δx approaches zero, can be regarded as a quotient. If y is a continuous function of x, then Δy approaches zero when Δx approaches zero, so dy/dx appears to be the quotient 0/0, which is meaningless. Nevertheless, it is sometimes useful to be able to refer to quantities dy and dx in such a way that their quotient is the derivative dy/dx. We can justify this by regarding dx as a new independent variable (called **the differential of** $\boldsymbol{x}$) and defining a new dependent variable dy (**the differential of** $\boldsymbol{y}$) as a function of x and dx by

$$dy = \frac{dy}{dx}\,dx = f'(x)dx.$$

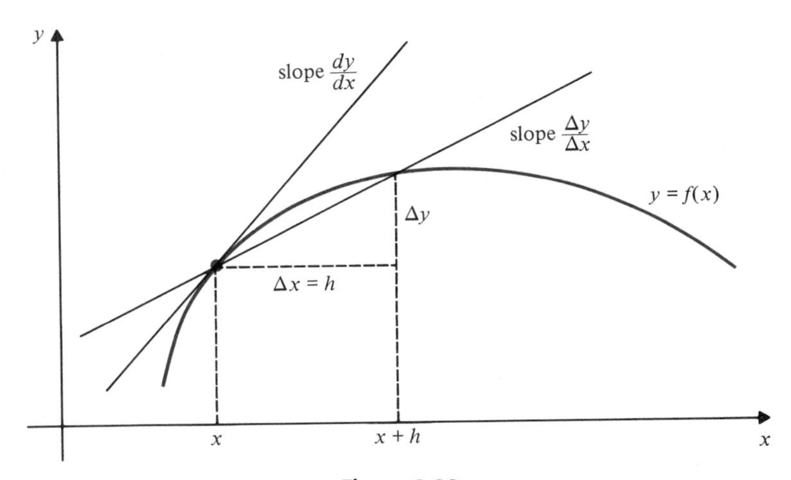

Figure 2.22

For example, if $y = x^2$, we can write $dy = 2xdx$ to mean the same thing as $dy/dx = 2x$. This *differential formalism* will be used for the interpretation and manipulation of integrals beginning in Chapter 6. We will make little use of it until then.

Note that, defined as above, differentials are merely variables that may or may not be small in absolute value. The differentials dy and dx were originally regarded (by Leibniz and his successors) as indefinitely small (but nonzero) quantities whose quotient dy/dx gave the slope of an infinitely short secant line to the graph of $y = f(x)$, that is, of the tangent line. It can be shown using the axioms for the real numbers that such "infinitesimal" quantities cannot exist (as real numbers). It is possible to extend the number system to contain infinitesimals and use these to develop calculus, but we will not take this approach here.

EXERCISES

In Exercises 1–18 calculate the derivative of the given function directly from the definition of derivative.

1. $y = x^2 - 3x$

2. $f(x) = 1 + 4x - 5x^2$

3. $f(x) = x^3$

4. $y = \frac{1}{2}x^3 - x$

5. $g(x) = \dfrac{2 - x}{2 + x}$

6. $s = \dfrac{1}{3 + 4t}$

7. $F(t) = \sqrt{2t + 1}$

8. $f(x) = \frac{3}{4}\sqrt{2 - x}$

9. $y = x + \dfrac{1}{x}$

10. $f(x) = x^2 - \dfrac{2}{x} + 7$

11. $z = \dfrac{s}{1 + s}$

12. $H(t) = \dfrac{1 + \sqrt{t}}{t}$

13. $y = \dfrac{1}{x^2}$

14. $y = \dfrac{1}{\sqrt{1 + x}}$

15. $f(t) = \dfrac{t^2 - 3}{t^2 + 3}$

16. $y = \dfrac{1 - 2x^2}{3 - 4x^2}$

17. $F(x) = \dfrac{1}{\sqrt{1 + x^2}}$

18. $f(x) = \sqrt{\dfrac{x - 1}{x + 1}}$

19. Using a calculator, find the slope of the secant line to $y = x^3 - 2x$ passing through the points corresponding to $x = 1$ and $x = 1 + \Delta x$ where

a) $\Delta x = 0.1$

b) $\Delta x = 0.01$

c) $\Delta x = 0.001$.

Also calculate $\left.\dfrac{d}{dx}(x^3 - 2x)\right|_{x=1}$.

20. Repeat Exercise 19 for the function $f(x) = \dfrac{1}{x}$ and the points $x = 2$ and $x = 2 + \Delta x$.

21.* Make use of the factoring of a difference of cubes

$$a^3 - b^3 = (a - b)(a^2 + ab + b^2)$$

to help you calculate the derivative of $f(x) = x^{1/3}$ directly from the definition of derivative.

Using the definition of derivative, find equations for the tangent lines to the curves in Exercises 22–25 at the points indicated.

22. $y = 5 + 4x - x^2$ at the point where $x = 2$

23. $y = \sqrt{x + 6}$ at the point $(3, 3)$

24. $y = \dfrac{t}{t^2 - 2}$ at the point where $t = -2$

25. $y = \dfrac{2}{t^2 + t}$ at the point $(1, 1)$

In Exercises 26–31 make use of the formulas for derivatives established in this section.

26. Find $\dfrac{d}{dx} x^{17}$.

27. Find $\dfrac{dy}{dt}$ if $y = t^{22}$.

28. Find an equation of the straight line tangent to the curve $y = \sqrt{x}$ at $x = x_o$.

29. Find an equation of the straight line normal to the curve $y = 1/x$ at the point where $x = x_o$.

30. There are two distinct straight lines that pass through the point $(1, -3)$ and are tangent to the curve $y = x^2$. Find their equations. (*Hint:* The points of tangency are not given; let them be denoted (a, a^2). Draw a sketch.)

31.* Show that there are two distinct tangent lines to the curve $y = x^2$ passing through the point (a, b) provided $b < a^2$. How many tangent lines to $y = x^2$ pass through (a, b) if $b = a^2$? If $b > a^2$?

32.* Show that the derivative of an odd differentiable function is even, and that the derivative of an even differentiable function is odd.

33.* Prove the case of the power rule for $(d/dx)x^r$ where $r = 1/n$, n being a positive integer. (*Hint:*

$$\frac{d}{dx} x^{1/n} = \lim_{h \to 0} \frac{(x + h)^{1/n} - x^{1/n}}{h} = \lim_{h \to 0} \frac{(x + h)^{1/n} - x^{1/n}}{((x + h)^{1/n})^n - (x^{1/n})^n}.$$

Apply the factorization of the difference of nth powers, $a^n - b^n$, to the denominator of the latter quotient.)

34.* Give a proof of the power rule $(d/dx)x^n = nx^{n-1}$ using the binomial expansion

$$(x + h)^n = x^n + \frac{n}{1} x^{n-1} h + \frac{n(n - 1)}{1 \times 2} x^{n-2} h^2 + \frac{n(n - 1)(n - 2)}{1 \times 2 \times 3} x^{n-3} h^3 + \cdots + h^n.$$

(See Appendix I.)

* Difficult and/or theoretical problems.

2.3 DIFFERENTIATION RULES (SUMS, PRODUCTS, QUOTIENTS)

If derivatives of functions could only be computed directly from the definition of the derivative, as in the examples of the previous section, calculus would indeed be a laborious subject. Fortunately there is an easier way. In the next two sections we present several general *differentiation rules* that enable us to calculate easily the derivatives of complicated combinations of functions from previous knowledge of the derivatives of the elementary functions from which they are constructed. For instance, we will be able to find the derivative of $x^2/\sqrt{x^2+1}$ if we know the derivatives of x^2 and $\sqrt{x}$.

Sums and Constant Multiples

If f and g are functions each differentiable at x and if C is a constant, then the functions $f + g$ and Cf are both differentiable at x and

$$(f+g)'(x) = f'(x) + g'(x) \quad \text{or} \quad \frac{d}{dx}(f(x)+g(x)) = \frac{d}{dx}f(x) + \frac{d}{dx}g(x),$$

$$(Cf)'(x) = Cf'(x), \quad \text{or} \quad \frac{d}{dx}(Cf(x)) = C\frac{d}{dx}f(x).$$

That is, the derivative of a sum of functions is the sum of their derivatives, and the derivative of a constant multiple of a function is the same constant multiple of the derivative of the function.

The proofs are straightforward, using corresponding properties of limits.

$$\begin{aligned}\frac{d}{dx}(f(x)+g(x)) &= \lim_{h\to 0}\frac{(f(x+h)+g(x+h))-(f(x)+g(x))}{h}\\ &= \lim_{h\to 0}\left(\frac{f(x+h)-f(x)}{h}+\frac{g(x+h)-g(x)}{h}\right)\\ &= \frac{d}{dx}f(x)+\frac{d}{dx}g(x)\\ \frac{d}{dx}(Cf(x)) &= \lim_{h\to 0}\frac{Cf(x+h)-Cf(x)}{h}\\ &= \lim_{h\to 0}C\frac{f(x+h)-f(x)}{h} = Cf'(x).\end{aligned}$$

Of course the rule for sums extends to any finite number of terms.

Example 1 a) $\frac{d}{dx}(x^3+x^2+x+1) = 3x^2+2x+1$

b) If $f(x) = 5\sqrt{x}+\frac{3}{x}-18$, then $f'(x) = \frac{5}{2\sqrt{x}}-\frac{3}{x^2}$.

c) If $y = \frac{1}{7}t^3 - \frac{1}{3}t^7$, then $\frac{dy}{dt} = \frac{3}{7}t^2 - \frac{7}{3}t^6$.

d) We find the equation of the tangent line to the curve $y = 3x^2 - 4/x$ at the point where $x = -2$. When $x = -2$, $y = 14$ and the slope of the curve is

$$\left.\frac{dy}{dx}\right|_{x=-2} = \left.\left(6x + \frac{4}{x^2}\right)\right|_{x=-2} = -11.$$

Thus the tangent line has equation $y = 14 - 11(x + 2)$ or $y = -11x - 8$.

The Product Rule

If f and g are functions each differentiable at x, then their product fg is also differentiable at x, and

$$(fg)' = f'g + fg', \text{ or } \frac{d}{dx}(f(x)g(x)) = \left(\frac{d}{dx}f(x)\right)g(x) + f(x)\left(\frac{d}{dx}g(x)\right).$$

In order to prove this rule we require the following rather obvious but important theorem.

Theorem 1 If f is differentiable at x, then f is continuous at x.

Proof We are given that f is differentiable at x, that is, that

$$\lim_{h\to 0}\frac{f(x+h) - f(x)}{h} = f'(x)$$

exists.

We wish to show that f is continuous at x, that is, that

$$\lim_{h\to 0} f(x+h) = f(x).$$

Using the laws of limits (Theorem 2 of Section 1.3), we have

$$\lim_{h\to 0} f(x+h) = \lim_{h\to 0}\left(f(x) + \frac{f(x+h) - f(x)}{h}h\right)$$
$$= f(x) + f'(x)(0) = f(x). \;\square$$

Proof of the Product Rule

$$\frac{d}{dx}(f(x)g(x)) = \lim_{h\to 0}\frac{f(x+h)g(x+h) - f(x)g(x)}{h}$$

To evaluate this we must rewrite it in the following way.

$$= \lim_{h\to 0}\frac{f(x+h)g(x+h) - f(x+h)g(x) + f(x+h)g(x) - f(x)g(x)}{h}$$

$$= \lim_{h \to 0} \left(f(x+h)\frac{g(x+h)-g(x)}{h} + g(x)\frac{f(x+h)-f(x)}{h} \right)$$

$$= f(x)g'(x) + g(x)f'(x)$$

where we have used the fact that g and f are differentiable, and f is continuous, as well as properties of limits from Theorem 2 of Section 1.3. □

Example 2 a) $\frac{d}{dx}(x^2+1)(x^3+4)) = (2x)(x^3+4) + (x^2+1)(3x^2) = 5x^4 + 3x^2 + 8x$. Of course, we could do this without the product rule by first multiplying the factors in the product:

$$\frac{d}{dx}((x^2+1)(x^3+4)) = \frac{d}{dx}(x^5 + x^3 + 4x^2 + 4) = 5x^4 + 3x^2 + 8x.$$

b) If $y = \left(2\sqrt{x} + \frac{3}{x}\right)\left(3\sqrt{x} - \frac{2}{x}\right)$, then

$$\frac{dy}{dx} = \left(\frac{1}{\sqrt{x}} - \frac{3}{x^2}\right)\left(3\sqrt{x} - \frac{2}{x}\right) + \left(2\sqrt{x} + \frac{3}{x}\right)\left(\frac{3}{2\sqrt{x}} + \frac{2}{x^2}\right)$$

$$= 6 - \frac{5}{2}x^{-3/2} + \frac{12}{x^3}.$$

c) The product rule can be used to show that $(d/dx)x^n = nx^{n-1}$ for $n = 2, 3, 4, \ldots$ Since $(d/dx)x = 1$ we have

$$\frac{d}{dx}x^2 = \frac{d}{dx}(xx) = (1)(x) + (x)(1) = 2x.$$

Therefore,

$$\frac{d}{dx}x^3 = \frac{d}{dx}(x^2x) = (2x)(x) + (x^2)(1) = 3x^2,$$

$$\frac{d}{dx}x^4 = \frac{d}{dx}(x^3x) = (3x^2)(x) + (x^3)(1) = 4x^3,$$

and so on in the obvious way. For an indication of how to base a formal proof on this observed pattern, see Appendix I.

d) The product rule can be extended to products of more than two factors:

$$\frac{d}{dx}(f(x)g(x)h(x)) = \frac{d}{dx}\Big((f(x)g(x))h(x)\Big)$$

$$= \left(\frac{d}{dx}(f(x)g(x))\right)h(x) + f(x)g(x)h'(x)$$

$$= f'(x)g(x)h(x) + f(x)g'(x)h(x) + f(x)g(x)h'(x).$$

In general, the derivative of a product of n functions will have n terms; each term will be the same product but with one of the factors differentiated.

$$\frac{d}{dx}(f_1(x)f_2(x)\ldots f_n(x))$$
$$= f_1'(x)f_2(x)\ldots f_n(x) + f_1(x)f_2'(x)\ldots f_n(x) + \ldots + f_1(x)f_2(x)\ldots f_n'(x)$$

The Reciprocal Rule

If f is differentiable at x and $f(x) \neq 0$ then $1/f$ is differentiable at x and

$$\left(\frac{1}{f}\right)' = -\frac{f'}{f^2}, \qquad \text{or} \qquad \frac{d}{dx}\frac{1}{f(x)} = -\frac{1}{(f(x))^2}\frac{d}{dx}f(x).$$

Proof

$$\frac{d}{dx}\frac{1}{f(x)} = \lim_{h\to 0}\frac{\dfrac{1}{f(x+h)} - \dfrac{1}{f(x)}}{h}$$
$$= \lim_{h\to 0}\frac{f(x) - f(x+h)}{hf(x+h)f(x)}$$
$$= \lim_{h\to 0}\left(-\frac{1}{f(x+h)f(x)}\right)\frac{f(x+h) - f(x)}{h}$$
$$= -\frac{1}{(f(x))^2}f'(x)$$

Again we have had to use the continuity of f (Theorem 1). □

Example 3 a) $\dfrac{d}{dx}\dfrac{1}{x^2+1} = -\dfrac{2x}{(x^2+1)^2}$

b) $\dfrac{d}{dx}x^{-1/2} = \dfrac{d}{dx}\dfrac{1}{\sqrt{x}} = -\dfrac{1/(2\sqrt{x})}{(\sqrt{x})^2} = -\dfrac{1}{2}x^{-3/2}$

Note that this result also fits the pattern of the general power rule, which we encountered in the previous section.

c) We an also extend the general power rule to arbitrary negative integer powers $r = -n$:

$$\frac{d}{dx}x^{-n} = \frac{d}{dx}\frac{1}{x^n} = -\frac{nx^{n-1}}{(x^n)^2} = -nx^{-n-1}.$$

For example,

$$\frac{d}{dx}\left(\frac{1}{x} + \frac{1}{x^2} + \frac{1}{x^3}\right) = \frac{d}{dx}(x^{-1} + x^{-2} + x^{-3})$$
$$= -x^{-2} - 2x^{-3} - 3x^{-4} = -\frac{1}{x^2} - \frac{2}{x^3} - \frac{3}{x^4}.$$

The Quotient Rule

If f and g are differentiable at x, and if $g(x) \neq 0$, then the quotient f/g is differentiable at x and

$$\left(\frac{f}{g}\right)' = \frac{gf' - fg'}{g^2}, \quad \text{or} \quad \frac{d}{dx}\frac{f(x)}{g(x)} = \frac{g(x)\left(\frac{d}{dx}f(x)\right) - f(x)\left(\frac{d}{dx}g(x)\right)}{(g(x))^2}.$$

Sometimes students have trouble remembering this rule because of the negative sign in the numerator. Try to remember and use it in the following form:

$$(\text{quotient})' = \frac{(\text{denominator}) \times (\text{numerator})' - (\text{numerator}) \times (\text{denominator})'}{(\text{denominator})^2}$$

Proof This is just an application of the product and reciprocal rules.

$$\frac{d}{dx}\left(\frac{f(x)}{g(x)}\right) = \frac{d}{dx}\left(f(x)\frac{1}{g(x)}\right) = f'(x)\frac{1}{g(x)} + f(x)\left(-\frac{g'(x)}{(g(x))^2}\right)$$
$$= \frac{g(x)f'(x) - f(x)g'(x)}{(g(x))^2}$$

Of course the reciprocal rule is really just a special case of the quotient rule. □

Example 4 a) $\dfrac{d}{dx}\dfrac{1 - x^2}{1 + x^2} = \dfrac{(1 + x^2)(-2x) - (1 - x^2)(2x)}{(1 + x^2)^2} = -\dfrac{4x}{(1 + x^2)^2}$

b) If $f(t) = \dfrac{\sqrt{t}}{3 - 5t}$, then

$$f'(t) = \frac{(3 - 5t)\frac{1}{2\sqrt{t}} - \sqrt{t}(-5)}{(3 - 5t)^2} = \frac{3 + 5t}{2\sqrt{t}(3 - 5t)^2}.$$

c) If $y = \dfrac{a + b\theta}{m + n\theta}$, then $\dfrac{dy}{d\theta} = \dfrac{(m + n\theta)(b) - (a + b\theta)(n)}{(m + n\theta)^2} = \dfrac{mb - na}{(m + n\theta)^2}$.

d) If $F(r) = \dfrac{(1 - 4r^3)\sqrt{r}}{2r^2 + 1}$, find $F'(1)$.

$$F'(1) = \left.\frac{(2r^2 + 1)\left(-12r^2\sqrt{r} + (1 - 4r^3)\frac{1}{2\sqrt{r}}\right) - (1 - 4r^3)\sqrt{r}(4r)}{(2r^2 + 1)^2}\right|_{r = 1}$$
$$= \frac{3(-12 - (3/2)) - (-3)(4)}{9} = -\frac{19}{6}$$

Note the use of the product rule for differentiating the numerator. Note also that the evaluation takes place immediately after the derivative is calculated, before any simplification takes place. It is easier to simplify an expression with numbers than with algebraic symbols.

In Examples 4(a) to 4(c) above, the differentiation was accomplished immediately (that is, after the first equal sign) and the result was then simplified by algebraic methods. The quotient rule often leads to fractions with numerators that look very complicated. These numerators can frequently be simplified by using algebra (expanding and factoring). You are well advised to attempt such simplification at all times; the usefulness of derivatives in applications of calculus usually depends on such simplifications.

EXERCISES

In Exercises 1–34 calculate the derivatives of the given functions.

1. $y = x^8 - x^4$

2. $y = 3 - 4x - 5x^2 + 6x^3$

3. $y = \dfrac{1}{3}x^3 - \dfrac{1}{2}x^2 + x - 1$

4. $y = 4x^{1/2} - \dfrac{5}{x}$

5. $f(x) = Ax^2 + Bx + C$

6. $f(x) = \dfrac{6}{x^3} + \dfrac{2}{x^2} - 2$

7. $g(t) = t^{1/3} + 2t^{1/4} + 3t^{1/5}$

8. $F(u) = \dfrac{Au + B}{C}$

9. $z = \dfrac{s^5 - s^3}{15}$

10. $y = x^{45} - x^{-45}$

11. $F(x) = (3x - 2)(1 - 5x)$

12. $y = \sqrt{x}\left(5 - x - \dfrac{1}{3}x^2\right)$

13. $g(t) = (t^2 + 1)(t^4 + 2)$

14. $f(r) = (r^{-2} + r^{-3})(r^2 + r^3 + 1)$

15. $f(x) = (1 + x)(1 + 2x)(1 + 3x)(1 + 4x)$

16. $y = (x^2 + 4)(\sqrt{x} + 1)(5x^{2/3} - 2)$

17. $y = \dfrac{1}{x^2 + 5x}$

18. $y = \dfrac{1}{3 - x}$

19. $f(t) = \dfrac{4}{2 - 5t}$

20. $g(y) = \dfrac{5}{1 - y^2}$

21. $y = \dfrac{t^3 + 4t^2}{7t}$

22. $y = \dfrac{3}{x + \sqrt{x}}$

23. $y = \dfrac{1}{(a^2 + b^2x^2)(c^2 - d^2x^2)}$

24. $x = \dfrac{y^2 - 3}{y^2}$

25. $y = \dfrac{x - 2}{x + 2}$

26. $f(x) = \dfrac{3 - 4x}{3 + 4x}$

27. $z = \dfrac{t^2 + 2}{t^2 - 2}$

28. $z = \dfrac{1 + \sqrt{t}}{1 - \sqrt{t}}$

29. $f(x) = \dfrac{x^3 - 4}{x + 1}$

30. $f(x) = \dfrac{ax + b}{cx + d}$

31. $F(t) = \dfrac{t^2 + 7t - 8}{t^2 - t + 1}$

32. $y = \dfrac{(x^2 + 1)(x^3 + 2)}{(x^2 + 2)(x^3 + 1)}$

33. $y = \dfrac{x}{x + \dfrac{1}{x+1}}$

34.* $f(x) = \dfrac{(\sqrt{x} - 1)(2 - x)(1 - x^2)}{\sqrt{x}(3 + 2x)}$

35. Where was Theorem 1 used in the proofs of the product rule and the reciprocal rule?

36. Find $\left.\dfrac{d}{dx}\dfrac{x^2 - 4}{x^2 + 4}\right|_{x=-2}$.

37. Find $\left.\dfrac{d}{dt}\dfrac{t(1 + \sqrt{t})}{5 - t}\right|_{t=4}$.

38. If $f(x) = \dfrac{\sqrt{x}}{x + 1}$, find $f'(2)$.

39. Find $F'(0)$ if $F(x) = (1 + x)(1 + 2x)(1 + 3x)$.

40. Find an equation of the tangent line to $y = 3 + \dfrac{4}{x}$ at $(-2, 1)$.

41. Find equations of the tangent and normal to $y = \dfrac{x + 1}{x - 1}$ at $x = 2$.

42. Find the points on the curve $y = x + 1/x$ where the tangent line is horizontal.

43. Find the equations of all horizontal lines that are tangent to the curve $y = x^2(4 - x^2)$.

44. Find the equation of the straight line that passes through the point $(0, b)$ and is tangent to the curve $y = 1/x$. Assume $b \neq 0$.

45.† Show that for any constant C, the function $y = Cx^k$ $(x > 0)$ satisfies the *differential equation* $xy' = ky$. Find a function y that is a solution of the *initial-value problem*

$$\begin{cases} xy' = 3y & (x > 0) \\ y|_{x=2} = -4. \end{cases}$$

2.4 THE CHAIN RULE

The last and most important of the differentiation rules is the chain rule, which enables us to differentiate compositions of functions whose derivatives we already know. The rule may be stated as follows: If the function g is differentiable at the point x and the function f is differentiable at the point $g(x)$, then the composition $f \circ g$ is differentiable at x and

$$(f \circ g)'(x) = f'(g(x))g'(x),$$

or, equivalently,

$$\frac{d}{dx}f(g(x)) = \left(\left.\frac{d}{du}f(u)\right|_{u=g(x)}\right)\left(\frac{d}{dx}g(x)\right).$$

† Problems relating to differential equations.

In terms of Leibniz notation, if $y = f(u)$ and $u = g(x)$, then $y = f(g(x))$ and, since $dy/du = f'(u) = f'(g(x))$ and $du/dx = g'(x)$,

$$\frac{dy}{dx} = \frac{dy}{du}\frac{du}{dx}.$$

This form of the chain rule can be interpreted as follows: as x changes, u changes du/dx times as fast as x, and y changes dy/du times as fast as u. Thus y changes $(dy/du)(du/dx)$ times as fast as x. It appears as though the symbol du cancels from the numerator and denominator of the two fractions, but this is not meaningful (at the moment) because dy/du was not defined as the quotient of the two things, but rather as a single quantity, the derivative of y with respect to u.

Of all the differentiation rules, the chain rule causes the most difficulty for students initially. We illustrate it with several examples before giving a proof. Think of composition as the nesting of one function inside another; the chain rule says to begin differentiating at the outside and work inward. In $f(g(x))$, f is the outside function and g is inside.

Example 1 a) If $y = \sqrt{3x - 2}$, then $\dfrac{dy}{dx} = \dfrac{1}{2\sqrt{3x-2}}\dfrac{d}{dx}(3x - 2) = \dfrac{3}{2\sqrt{3x-2}}$. Here we have $y = \sqrt{u}$ where $u = 3x - 2$, so that

$$\frac{dy}{dx} = \frac{dy}{du}\frac{du}{dx} = \frac{1}{2\sqrt{u}}(3) = \frac{3}{2\sqrt{3x-2}}.$$

b) In a similar vein we have

$$\frac{d}{dx}\sqrt{x^2+1} = \frac{1}{2\sqrt{x^2+1}}(2x) = \frac{x}{\sqrt{x^2+1}}.$$

As you read this, say the following to yourself: "The derivative of the square root of something is one over twice that thing multiplied by the derivative of that thing." In Examples 1(a) and 1(b) the outer function is square root.

c) If $f(x) = (ax + b)^n$, then $f'(x) = n(ax + b)^{n-1}(a) = an(ax + b)^{n-1}$. "The derivative of the nth power of something is n times the $(n - 1)$st power of that thing, multiplied by the derivative of that thing." In general, $(d/dx)(f(x))^n = n(f(x))^{n-1}f'(x)$.

d) If $h(t) = (t^3 + 3t + 4)^{1/3}$, then

$$h'(t) = \frac{1}{3}(t^3 + 3t + 4)^{-2/3}(3t^2 + 3) = \frac{t^2+1}{(t^3 + 3t + 4)^{2/3}}.$$

e) $$\frac{d}{dx}\left(1 + \frac{1}{\sqrt{7-5x}}\right)^{50} = \frac{d}{dx}(1 + (7 - 5x)^{-1/2})^{50}$$

$$= 50(1 + (7 - 5x)^{-1/2})^{49}\frac{d}{dx}(7 - 5x)^{-1/2}$$

$$= 50(1 + (7 - 5x)^{-1/2})^{49}\left(\frac{-1}{2}\right)(7 - 5x)^{-3/2}\frac{d}{dx}(7 - 5x)$$

$$= 50(1 + (7 - 5x)^{-1/2})^{49}\left(\frac{-1}{2}\right)(7 - 5x)^{-3/2}(-5)$$

$$= 125(7 - 5x)^{-3/2}\left(1 + \frac{1}{\sqrt{7 - 5x}}\right)^{49}$$

Several comments should be made on the final example. Observe that the chain rule was used two times: first to differentiate the 50th power, and then to differentiate the $-1/2$ power. Although we proceeded only one step at a time in the above calculation, it is usually better to try to accomplish the whole differentiation at one time as follows:

$$\frac{d}{dx}(1 + (7 - 5x)^{-1/2})^{50} = 50(1 + (7 - 5x)^{-1/2})^{49}\left(\frac{-1}{2}\right)(7 - 5x)^{-3/2}(-5)$$
$$= 125(7 - 5x)^{-3/2}(1 + (7 - 5x)^{-1/2})^{49}.$$

With a little practice one can learn to differentiate even very complicated expressions in one step. Note, however, that most applications of the chain rule lead to products that often need to be rewritten in a different order, as it is conventional to write factors in a product in order of increasing complexity.

Example 2 Note that the reciprocal rule of the previous section is just a special case of the chain rule with outer function $1/x$.

$$\frac{d}{dx}\frac{1}{f(x)} = \frac{-1}{(f(x))^2}f'(x) = \frac{-f'(x)}{(f(x))^2},$$

that is, since $(d/dx)(1/x) = -1/x^2$, the derivative of 1 over something is -1 over that something squared, multiplied by the derivative of that something.

Example 3 Recall that the derivative of the absolute value function is the signum function: $d/dx\,|x| = \operatorname{sgn} x = \dfrac{x}{|x|}$. It follows that

$$\frac{d}{dx}f(|x|) = f'(|x|)\operatorname{sgn} x, \text{ and}$$

$$\frac{d}{dx}|f(x)| = (\operatorname{sgn} f(x))f'(x).$$

For instance, if $y = |1 - x^2| + \sqrt{|3x + 2|}$, then

$$\frac{dy}{dx} = (\operatorname{sgn}(1 - x^2))(-2x) + \frac{1}{2\sqrt{|3x + 2|}}(\operatorname{sgn}(3x + 2))(3)$$

$$= -2x\operatorname{sgn}(1 - x^2) + \frac{3}{2}\frac{\operatorname{sgn}(3x + 2)}{\sqrt{|3x + 2|}}$$

With the exception of the absolute value function, we do not yet have much scope for applying the chain rule beyond powers and roots. In Chapter 3 we introduce many new and interesting functions, and we will calculate their derivatives. Among these functions will be $\sin x$, $\cos x$, and $\ln x$. We will show that these functions have the following derivatives:

$$\frac{d}{dx}\sin x = \cos x, \qquad \frac{d}{dx}\cos x = -\sin x, \qquad \text{and } \frac{d}{dx}\ln x = \frac{1}{x}.$$

For now let us take these derivatives as given and use them to provide some more examples of the use of the chain rule.

Example 4

a) $$\frac{d}{dx}\sin(ax^2 + b) = 2ax\cos(ax^2 + b)$$

b) $$\frac{d}{dx}\ln(1 + x^{1/2}) = \frac{1}{(1 + x^{1/2})}\frac{1}{2}x^{-1/2} = \frac{1}{2x^{1/2}(1 + x^{1/2})}$$

c) $$\begin{aligned}\frac{d}{dx}\sin(3\cos 5x) &= \cos(3\cos 5x)(-3\sin 5x)(5) \\ &= -15\sin 5x\cos(3\cos 5x)\end{aligned}$$

d) $$\begin{aligned}\frac{d}{dx}\cos\ln\sin\ln x &= (-\sin\ln\sin\ln x)\frac{1}{\sin\ln x}(\cos\ln x)\frac{1}{x} \\ &= -\frac{\cos\ln x}{x\sin\ln x}\sin\ln\sin\ln x\end{aligned}$$

e) $$\begin{aligned}\frac{d}{dx}(x\ln x)^2 &= 2x\ln x\left(\ln x + x\left(\frac{1}{x}\right)\right) \\ &= 2x(\ln x)^2 + 2x\ln x\end{aligned}$$

Derivatives of Inverse Functions

Recall that a one-to-one function f has an inverse function f^{-1} defined by

$$y = f^{-1}(x) \Leftrightarrow x = f(y).$$

Recall also that the graph of $y = f^{-1}(x)$ is the mirror image of the graph of $y = f(x)$ in the line $x = y$. (See Section 1.2 and Fig. 2.23.) If f is differentiable at y, and $f'(y) \neq 0$, then the graph of f has a nonhorizontal tangent line at the point $(y, f(y)) = (y, x)$. It follows that the graph of f^{-1} has a nonvertical tangent line at the point (x, y), and so f^{-1} must be differentiable at x. We can calculate $(d/dx)f^{-1}(x)$ using the chain rule as follows: Since $f(f^{-1}(x)) = x$, therefore $f'(f^{-1}(x))(f^{-1})'(x) = 1$. Hence,

$$\frac{d}{dx}f^{-1}(x) = (f^{-1})'(x) = \frac{1}{f'(f^{-1}(x))}.$$

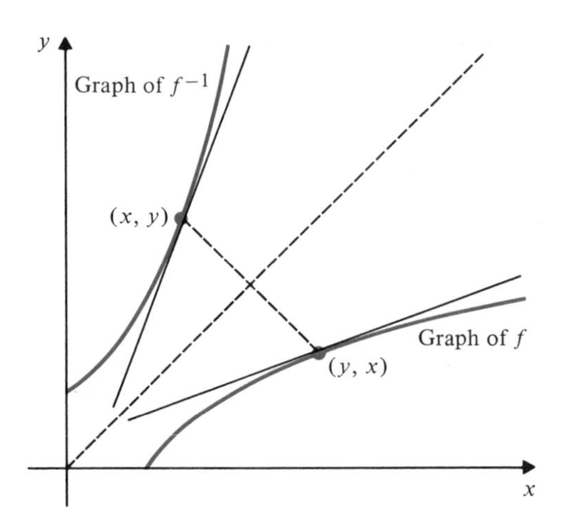

Figure 2.23

Example 5 If n is a positive integer, then $f(x) = x^{1/n}$ is the inverse of the function x^n. By the previous argument, f is differentiable for all $x > 0$. Since $(f(x))^n = x$ we have, differentiating this equation and using the chain rule on the left side,

$$n\,(f(x))^{n-1}f'(x) = 1,$$

so that

$$f'(x) = \frac{1}{n(f(x))^{n-1}} = \frac{1}{n}\frac{1}{x^{1-(1/n)}} = \frac{1}{n}x^{(1/n)-1},$$

which is what the general power rule would give with power $1/n$. In fact we can now extend the general power rule,

$$\frac{d}{dx}x^r = rx^{r-1},$$

to cover arbitrary rational exponents r. If $r = m/n$ where m and n are integers and $n > 0$, then

$$\frac{d}{dx}x^r = \frac{d}{dx}(x^m)^{1/n} = \frac{1}{n}(x^m)^{(1/n)-1}\,mx^{m-1} = \frac{m}{n}x^{(m/n)-1} = rx^{r-1}.$$

The general power rule is valid for every x for which both sides make sense. This is always the case if $x > 0$. If $r = m/n$ is a rational number with n an odd integer, then the rule is also valid for $x < 0$ and, if $r > 1$, it is valid for $x = 0$ as well.

Proof of the Chain Rule

Suppose that f is differentiable at the point $u = g(x)$ and that g is differentiable at x. Let the function $E(k)$ be defined by

$$E(0) = 0,$$
$$E(k) = \frac{f(u + k) - f(u)}{k} - f'(u), \qquad \text{if } k \neq 0.$$

By the definition of derivative, $\lim_{k\to 0} E(k) = f'(u) - f'(u) = 0 = E(0)$, so $E(k)$ is continuous at $k = 0$. Also,

$$f(u + k) - f(u) = (f'(u) + E(k))k.$$

In this last equation put $u = g(x)$ and $k = g(x + h) - g(x)$, so that $u + k = g(x + h)$, and obtain

$$f(g(x + h)) - f(g(x)) = \Big(f'(g(x)) + E(g(x + h) - g(x))\Big)(g(x + h) - g(x)).$$

Since g is differentiable at x, $\lim_{h\to 0}$, $[g(x + h) - g(x)]/h = g'(x)$. Also, since g is continuous at x by Theorem 1 in Section 2.3, we have

$$\lim_{h\to 0} E(g(x + h) - g(x)) = \lim_{k\to 0} E(k) = 0.$$

Hence,

$$\begin{aligned}\frac{d}{dx}f(g(x)) &= \lim_{h\to 0} \frac{f(g(x + h)) - f(g(x))}{h} \\ &= \lim_{h\to 0} \Big(f'(g(x)) + E(g(x + h) - g(x))\Big) \frac{g(x + h) - g(x)}{h} \\ &= (f'(g(x)) + 0)g'(x) = f'(g(x))g'(x),\end{aligned}$$

which was to be proved.

EXERCISES

Find the derivatives of the functions in Exercises 1–30.

1. $y = (2x + 3)^6$

2. $y = \left(1 - \dfrac{x}{3}\right)^{99}$

3. $f(x) = (4 - x^2)^{10}$

4. $F(t) = \dfrac{1}{3}(at^2 + bt + c)^{12}$

5. $y = (Ax + B)^r$

6. $y = \sqrt{x - 2}$

7. $y = \dfrac{3}{5 - 4x}$

8. $y = (2x^2 - 1)^{-3/2}$

9. $y = \sqrt{5x + 3}$

10. $y = (x^2 + 2x + 3)^{1/2}$

11. $y = (2 + x^3)^{-1/2}$

12. $y = \left(1 + \dfrac{1}{x}\right)^{10}$

13. $f(t) = (at^2 + bt + c)^{96}$

14. $z = (1 + x^{2/3})^{3/2}$

15. $y = \dfrac{\sqrt{x + 1}}{x + 2}$

16. $y = (x^2 + 9)\sqrt{x^2 + 3}$

17. $y = \dfrac{1}{2 + \sqrt{3x + 4}}$

18. $f(x) = \left(1 + \sqrt{\dfrac{x - 2}{3}}\right)^4$

19. $g(t) = \sqrt{(3 - t)^3(2 + t^2)^4}$

20. $y = \sqrt{4x + \dfrac{3}{x^2 + 1}}$

21. $z = \left(u + \dfrac{1}{u - 1}\right)^{-5/3}$

22. $f(y) = \left(\sqrt{y} + \dfrac{1}{1 + \sqrt{y^2 + 2}}\right)^2$

23. $y = \dfrac{1}{1 + \dfrac{1}{2 + \dfrac{3}{t}}}$

24. $y = \dfrac{x^5\sqrt{3 + x^6}}{(4 + x^2)^3}$

25. $f(t) = |2 + t^3|$

26. $y = 4x + |4x - 1|$

27. $y = (2 + |x|^3)^{1/3}$

28. $y = |1 - |x||$

29. $z = \sqrt{|x + 4|}$

30. $y = |\sqrt{x + 4}|$

31. Sketch the graphs of the functions in Exercises 25, 26, and 28.

In Exercises 32–35 express the derivative of the given function in terms of the derivative f' of the differentiable function f.

32. $f(2t + 3)$ **33.** $f(5 - x^2)$ **34.** $\left[f\left(\dfrac{x}{2}\right)\right]^3$ **35.** $f(3f(4t))$

36. Find $\left.\dfrac{d}{dx}\dfrac{\sqrt{x^2 - 1}}{x^2 + 1}\right|_{x=-2}$.

37. Find $\left.\dfrac{d}{dt}\sqrt{3t - 7}\right|_{t=3}$.

38. If $f(x) = \dfrac{1}{\sqrt{2x + 1}}$, find $f'(4)$

39. If $y = (x^3 + 9)^{17/2}$ find $\left.y'\right|_{x=-2}$.

40.* Find $F'(0)$ if $F(x) = (1 + x)(2 + x)^2(3 + x)^3(4 + x)^4$.

41.* Calculate y' if $y = (x + ((3x)^5 - 2)^{-1/2})^{-6}$. Try to do it all in one step.

Use the derivatives of sin, cos, and ln as provided in Example 4 to evaluate the derivatives in Exercises 42–47.

42. $\dfrac{d}{dx}\sin(2x)$

43. $\dfrac{d}{dx}(\cos(5 - 4x^3))^2$

44. $\dfrac{d}{dt}\ln(3t^2 + 6)$

45. $\dfrac{d}{dx}\ln(a\ln(bx + c))$

46. $\dfrac{d}{dx}\ln(2\ln(3\ln 4x))$

47. $\dfrac{d}{dx}\sin((\cos(\sin x))^2)$

In Exercises 48–51 find an equation of the tangent line to the given curve at the given point.

48. $y = \sqrt{1 + 2x^2}$ at $x = 2$

49. $y = (1 + x^{2/3})^{3/2}$ at $(1, 2\sqrt{2})$

50. $y = (ax + b)^8$ at $x = b/a$

51. $y = 1/(x^2 - x + 3)^{3/2}$ at $x = -2$

52.* Show that the derivative of the function $f(x) = (x - a)^m(x - b)^n$ vanishes at some point between a and b if m and n are positive integers.

53.* Let f^{-1} be the inverse of the one-to-one function $f(x) = x^3 + x$. Find $(f^{-1})'(10)$. (*Hint:* $f(2) = 10$.)

54.* Assume that the function $f(x)$ satisfies $f'(x) = \dfrac{1}{x}$, and that f is one-to-one. If $y = f^{-1}(x)$ show that $dy/dx = y$.

2.5 INTERPRETATIONS OF THE DERIVATIVE

When calculus is applied to other disciplines, mathematical functions are used to represent quantities appropriate to the particular problem under consideration. In such cases the derivatives of these functions may provide useful information about the problem. The mathematical interpretation of the derivative as the slope of the graph of the function suggests a more concrete interpretation of derivative as a rate of change of the function with respect to its independent variable: the slope of a straight line is the rate of change of vertical position with respect to horizontal position along the line.

Definition 4

> The **average rate of change** of $f(x)$ with respect to x over the interval $[a, b]$ is
> $$\frac{f(b) - f(a)}{b - a}.$$

The limit of the average rate of change as the interval over which the average is taken approaches zero in length is called the *instantaneous rate of change*, or simply the *rate of change*. The rate of change of $f(x)$ at $x = a$ is

$$\lim_{b \to a} \frac{f(b) - f(a)}{b - a} = \lim_{h \to 0} \frac{f(a + h) - f(a)}{h} = f'(a).$$

Definition 5

> The **rate of change** of $f(x)$ with respect to x at $x = x_o$ is $f'(x_o)$.

The sign of f' tells us whether $f(x)$ is increasing or decreasing as x increases. If $f'(x) > 0$ on an interval, then f is increasing as x increases on that interval; if $f'(x) < 0$, then $f(x)$ decreases as x increases. See Fig. 2.24.

At any point where $f'(x) = 0$ we say that f is **stationary** (that is, instantaneously it is neither increasing nor decreasing at that point, although it may be increasing or decreasing on intervals containing that point). Such points are called **critical points**

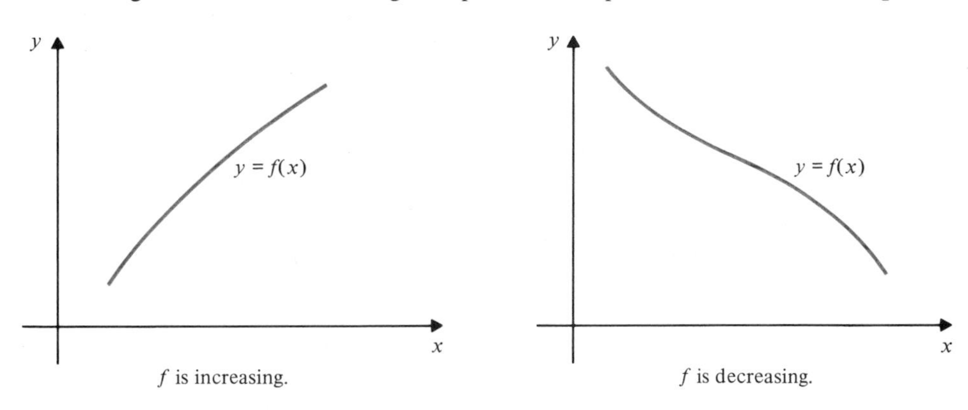

Figure 2.24

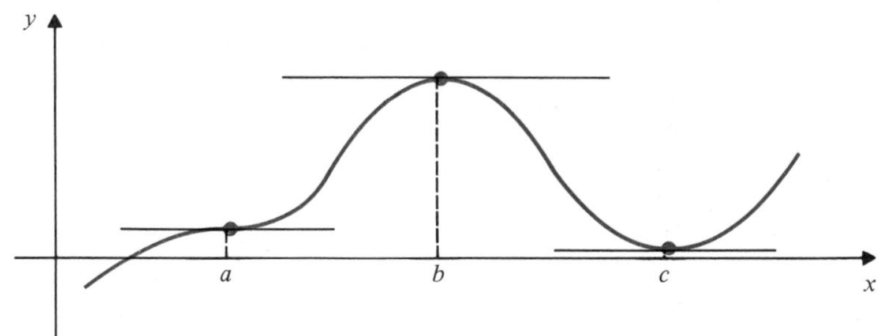

a, b, and c are critical points of f.

Figure 2.25

of f. At a critical point the tangent line to the graph $y = f(x)$ is horizontal, as shown in Fig. 2.25. Note that f may or may not have a maximum or minimum value at a critical point. These ideas will be examined more fully in Sections 2.8 and 4.1.

Velocity and Acceleration

Suppose that an object is moving along a straight line (say the x-axis) so that at time t its position is given by $x = f(t)$. The **velocity** of the object is the rate of change of its position with respect to time; the velocity at time t is

$$v(t) = \frac{dx}{dt} = f'(t).$$

If $v(t) > 0$ then x is increasing, so the object is moving to the right; if $v(t) < 0$ then x is decreasing, so the object is moving to the left. At a critical point of f, that is, a time t when $v(t) = 0$, the object is instantaneously at rest—at that instant it is not moving in either direction.

We distinguish between the term *velocity* (which involves direction of motion as well as rate) and **speed,** which does not involve direction. The speed is the absolute value of the velocity:

$$s(t) = |v(t)|.$$

The derivative of the velocity also has a useful interpretation. Being the rate of change of the velocity with respect to time, it is the **acceleration** of the moving object. The value of the acceleration at time t is

$$a(t) = v'(t) = \frac{dv}{dt} = \frac{d}{dt}\frac{dx}{dt}.$$

If $a(t) > 0$ the velocity is increasing, so the object is speeding up if it is moving to the right or slowing down if it is moving to the left. If $a(t) < 0$ the velocity is decreasing, so the object is slowing down if it is moving to the right and speeding up if it is moving to the left. If $a(t_o) = 0$ then the velocity is instantaneously unchanging at t_o. If $a(t) = 0$ on an interval of time, then the velocity is unchanging and therefore constant over that interval.

Example 1 A point P moves along the x-axis in such a way that its position at time t is given by

$$x = 2t^3 - 15t^2 + 24t.$$

Find the velocity and acceleration of P at time t. In what direction is P moving at time $t = 2$? What is its speed at that time, and is that speed increasing or decreasing? At what instant is P at rest? At what instant is the velocity stationary?

Solution The velocity and acceleration of P at time t are

$$v = \frac{dx}{dt} = 6t^2 - 30t + 24 \text{ and}$$

$$a = \frac{dv}{dt} = 12t - 30.$$

At $t = 2$ we have $v = -12$ and $a = -6$. Thus P is moving to the left with speed 12 and this speed is increasing. Since

$$v = 6(t^2 - 5t + 4) = 6(t - 1)(t - 4),$$

P is momentarily at rest at times $t = 1$ and $t = 4$. The velocity is stationary when $a = 0$, that is, when $t = 30/12 = 5/2$.

Example 2 An object is hurled upward from the roof of a 10-meter-high building at time $t = 0$. It rises and then falls back so that its height above the ground at time $t > 0$ is given by

$$y = -\frac{49}{10}t^2 + 8t + 10$$

until it strikes the ground. What is the greatest height above the ground that the object attains? With what speed does the object strike the ground?

Solution Refer to Fig. 2.26. The velocity at time t during flight is

$$v(t) = -\frac{98}{10}t + 8.$$

The object is rising for $t < 40/49$ since $v > 0$; the object is falling for $t > 40/49$ since $v < 0$. Thus the object is at its maximum height at time $t = 40/49$ and this maximum height is

$$y_{\max} = -\frac{49}{10}\left(\frac{40}{49}\right)^2 + 8\left(\frac{40}{49}\right) + 10 \simeq 13.265 \text{ meters (m)}.$$

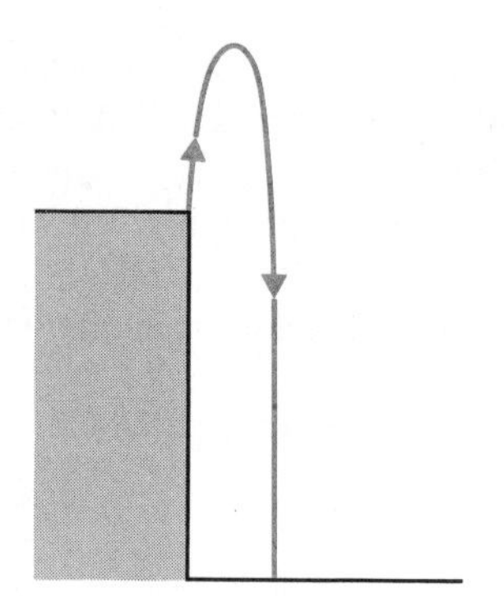

Figure 2.26

The time t at which the object strikes the ground is the positive root t of the quadratic equation ($y = 0$)

$$-49t^2 + 80t + 100 = 0,$$

namely,

$$t = \frac{-80 - \sqrt{6400 + 19600}}{-98} \approx 2.462 \text{ sec.}$$

The velocity at this time is $v = -(9.8)(2.462) + 8 \approx -16.12$ m/sec. Thus the object strikes the ground with a speed of about 16.12 m/sec.

Marginals

In economics the term *marginal* denotes the rate of change of a quantity with respect to its independent variable. The marginal cost of production in a manufacturing operation is the rate of change of the cost of production with respect to the number of units of product produced. It is sometimes loosely defined to be the extra cost of producing one more unit when a given number of units is being produced.

Example 3 In a manufacturing operation producing certain collectors' items, the cost of production has two components, a fixed cost of $\$A$ to set up the operation, and a cost of $\$B$ per item for the raw materials and labor. The price $\$P$ at which each item can be sold depends on the number n of items produced according to the formula $P = C - Dn$. (The more items produced, the less each is worth to a prospective buyer.) How many items should the manufacturer produce to maximize profit?

Solution The total cost of producing n items is $\$A + Bn$. These can then be sold for $\$Pn = Cn - Dn^2$ to generate a total profit of

$$\begin{aligned} T(n) &= Cn - Dn^2 - (A + Bn) \\ &= (C - B)n - Dn^2 - A. \end{aligned}$$

The marginal profit is

$$T'(n) = C - B - 2Dn.$$

If $n < (C - B)/2D$ the marginal profit is positive, so the total profit increases as n increases. If $n > (C - B)/2D$ the marginal profit is negative, so the total profit decreases as n increases. Accordingly, the profit will be greatest if $n = (C - B)/2D$ items are produced. (Note that we are using a ''real variable'' to represent an integer quantity in this question. If $(C - B)/2D$ is not an integer we would use the closest integer to it for the optimal number of items to be produced.)

EXERCISES

1. Find the rate of change of the area of a square with respect to the length of its side if the side length is 5 cm.
2. Find the rate of change of the area of a circle with respect to its radius when the radius is x meters.
3. Find the rate of change of the volume of a sphere $\left(V = \frac{4}{3}\pi r^3\right)$ with respect to its radius r when the radius is 20 cm.
4. What is the rate of change of the area of a square with respect to the diagonal of the square when the diagonal is 10 cm?
5. What is the rate of change of the circumference of a circle with respect to the area of the circle when the area is A sq cm?
6. Find the rate of change of the side of a cube with respect to the volume of the cube when the side of the cube is 30 cm.

In Exericises 7–11 a point moves along the x-axis so that its position x at time t is specified by the given function. In each case determine the following.

a) the time intervals on which the point is moving to the right
b) the time intervals on which the point is moving to the left
c) the times at which the velocity is zero
d) the acceleration when the velocity is zero
e) the average velocity over the time interval [0, 4]
f) the velocity at time $t = 2$

Make a sketch of the graph of x as a function of t in each case.

7. $x = t^2 - 4t + 3$

8. $x = 4 + 5t - t^2$

9. $x = t^3$

10. $x = t^3 - 4t + 1$

11. $x = \dfrac{t}{t^2 + 1}$

12. A ball is thrown upward from ground level with an initial speed of 100 ft/sec so that its height in feet after t sec is given by $y = 100t - 16t^2$. How high does the ball go? How fast does it strike the ground?

13. A ball is thrown downward from the top of a 100-meter-high tower with an initial speed of 2 m/sec. Its height in meters above the ground t sec later is $y = 100 - 2t - 4.9t^2$. How long does it take to reach the ground? At what instant is its velocity equal to its average velocity?

14.* The distance an aircraft travels along a runway before takeoff is given by $s = t^2$, where s is measured in meters from the starting point and t is measured in seconds from the time the brake is released. If the aircraft will become airborne when its velocity reaches

200 km/hr, how long will it take to become airborne, and what distance will it travel in that time?

15.* Show that if the position x of a moving point is given by a quadratic function of t, that is, if $x = At^2 + Bt + C$, then the average velocity over any time interval $[t_1, t_2]$ is equal to the instantaneous velocity at the midpoint of that time interval.

16.* The position of an object moving along the s-axis is given at time t by

$$s = \begin{cases} t^2 & \text{if } 0 \le t \le 2 \\ 4t - 4 & \text{if } 2 < t < 8 \\ -68 + 20t - t^2 & \text{if } 8 \le t \le 10. \end{cases}$$

Determine the velocity and acceleration at any time t. Is the velocity continuous? Is the acceleration continuous? What is the maximum velocity and when is it attained?

17. The cost $\$C$ of producing n widgets per month in a widget factory is known to be given by

$$C = \frac{80{,}000}{n} + 4n + \frac{n^2}{100}.$$

Find the marginal cost of production if the number of widgets manufactured each month is

a) 100 **b)** 300.

18.* In a mining operation the cost $\$C$ of extracting each ton of ore is given by

$$C = 10 + \frac{20}{x} + \frac{x}{1000},$$

where x is the number of tons extracted each day. (For small x, C decreases as x increases because of economies of scale, but for large x, C increases with x because of overloaded equipment and labor overtime.) If each ton of ore can be sold for \$13, how many tons should be extracted each day to maximize the daily profit of the mine?

19.* All 80 rooms in a motel will be rented each night if the manager charges \$20 or less per room. If he charges $\$(20 + x)$ per room, then $2x$ rooms will remain vacant ($x > 0$). If each rented room costs the manager \$5 per day and each unrented room \$1 per day in overhead, how much should the manager charge per room to maximize his daily profit?

20. Explain why describing the marginal cost of production as ''the extra cost of producing one more item'' is a ''loose'' interpretation of the definition of marginal cost of production.

2.6 HIGHER-ORDER DERIVATIVES

If the derivative $f'(x)$ of a function $f(x)$ is itself differentiable we can calculate its derivative, which we call the **second derivative** of f and denote by f'' or d^2f/dx^2:

$$f''(x) = \frac{d}{dx}f'(x) = \frac{d}{dx}\frac{d}{dx}f(x) = \frac{d^2}{dx^2}f(x).$$

If the function is denoted by a dependent variable, $y = f(x)$, then the second derivative can be denoted by y'' or d^2y/dx^2. Similarly one can consider third-, fourth-, and in general nth-order derivatives. The prime notation is inconvenient for higher-order derivatives, so we denote the order by a superscript in parentheses (to distinguish it from an exponent): the nth derivative of $y = f(x)$ is

$$f^{(n)}(x) = \frac{d^n}{dx^n}f(x) = y^{(n)} = \frac{d^ny}{dx^n}$$

and it is defined to be the derivative of the $(n - 1)$st derivative. For $n = 1, 2, 3$, primes are still normally used: $f^{(2)}(x) = f''(x)$, $f^{(3)}(x) = f'''(x)$. It is sometimes convenient to denote $f^{(0)}(x) = f(x)$, that is, to regard a function as its own zeroth-order derivative.

Example 1 Acceleration is the second derivative of position. If $x = f(t)$, then velocity $v = dx/dt = f'(t)$ and acceleration is

$$a = \frac{dv}{dt} = \frac{d^2x}{dt^2} = f''(t).$$

Example 2 Let $y = x^3$. Then $y' = 3x^2$, $y'' = 6x$, $y''' = 6$, $y^{(4)} = 0$, and all higher derivatives are zero.

In general, if $f(x) = x^n$ (n is a positive integer), then

$$\begin{aligned} f^{(k)}(x) &= n(n-1)(n-2)\ldots(n-(k-1))x^{n-k} \\ &= \begin{cases} \dfrac{n!}{(n-k)!}x^{n-k} & \text{if } 0 \le k \le n \\ 0 & \text{if } k > n, \end{cases} \end{aligned}$$

where $n!$ (called *n factorial*) is defined by

$$\begin{aligned} 0! &= 1 \\ 1! &= 1 \\ 2! &= 1 \times 2 = 2 \\ 3! &= 1 \times 2 \times 3 = 6 \\ 4! &= 1 \times 2 \times 3 \times 4 = 24 \\ &\vdots \\ n! &= 1 \times 2 \times 3 \times \cdots \times (n-1) \times n. \end{aligned}$$

It follows that if f is a polynomial of degree n, that is, if

$$f(x) = a_nx^n + a_{n-1}x^{n-1} + \cdots + a_1x + a_0,$$

where $a_n, a_{n-1}, \ldots, a_1, a_0$ are constants, then $f^{(k)}(x) = 0$ for $k > n$. For $k \le n$, $f^{(k)}$ is a polynomial of degree $n - k$.

Example 3 Calculate f', f'', and f''' for $f(x) = \sqrt{x^2 + 1}$.

Solution Since $f(x) = (x^2 + 1)^{1/2}$ we have

$$f'(x) = \frac{1}{2}(x^2 + 1)^{-1/2}(2x) = x(x^2 + 1)^{-1/2},$$

$$f''(x) = (x^2 + 1)^{-1/2} + x\left(\frac{-1}{2}\right)(x^2 + 1)^{-3/2}(2x)$$

$$= (x^2 + 1)^{-1/2} - x^2(x^2 + 1)^{-3/2},$$

$$f'''(x) = \frac{-1}{2}(x^2 + 1)^{-3/2}(2x) - 2x(x^2 + 1)^{-3/2} - x^2\left(\frac{-3}{2}\right)(x^2 + 1)^{-5/2}(2x)$$

$$= -3x(x^2 + 1)^{-3/2} + 3x^3(x^2 + 1)^{-5/2}.$$

Example 4 Find $y^{(n)}$ if $y = 1/(1 + x) = (1 + x)^{-1}$.

Solution

$$y' = -(1 + x)^{-2}$$

$$y'' = -(-2)(1 + x)^{-3} = 2(1 + x)^{-3}$$

$$y''' = 2(-3)(1 + x)^{-4} = -3!(1 + x)^{-4}$$

$$y^{(4)} = -3!(-4)(1 + x)^{-5} = 4!(1 + x)^{-5}$$

The pattern here is obvious. We conclude that

$$y^{(n)} = (-1)^n n!(1 + x)^{-n-1}.$$

Note the use of $(-1)^n$ to denote a positive sign if n is even and a negative sign if n is odd. Strictly speaking, we have not yet actually proved that the above formula is correct for every n, although it is for $n = 1, 2, 3$, and 4. To complete the proof, suppose that the formula is valid for some n. Consider $y^{(n+1)}$:

$$y^{(n+1)} = \frac{d}{dx}y^{(n)} = \frac{d}{dx}((-1)^n n!(1 + x)^{-n-1})$$

$$= (-1)^n n!(-n - 1)(1 + x)^{-n-2} = (-1)^{n+1}(n + 1)!(1 + x)^{-n-2}.$$

Thus if the formula for $y^{(n)}$ is correct, then the same formula gives $y^{(n+1)}$ if n is replaced with $n + 1$. Since the formula is known to be true for $n = 1$ it must therefore be true for every integer $n \geq 1$, for there can be no first value of n for which it is false. This technique of proof is called **mathematical induction.**

Mathematical induction is used to prove statements asserted to hold for all positive integers n from some integer n_o onward. (Usually n_o is 1.) The technique involves two steps:

i) Prove the statement for the lowest value of n ($n = n_o$).

ii) Assume the statement is true for some $n \geq n_o$ and deduce from that that it must also be true for the next integer, $n + 1$.

It follows that there can be no first value of n for which the statement is false; thus the statement is true for each $n \geq n_o$.

See Appendix I for more discussion of mathematical induction and further examples and exercises based on it.

Differential Equations and Initial-Value Problems

A differential equation is an equation involving one or more derivatives of an unknown function. Any function whose derivative satisfies the equation is called a solution of the differential equation. For instance, $y = x^3 - x$ is evidently a solution of the differential equation $y' = 3x^2 - 1$. This equation does not have only one solution; in fact, $y = x^3 - x + C$ is a solution for any value of the constant C. Sometimes we phrase exercises on differentiation in terms of showing that a given function satisfies a certain differential equation.

Example 5 Show that for any constants A and B, the function $y = Ax^3 + B/x$ satisfies the differential equation $x^2y'' - xy' - 3y = 0$ for any $x \neq 0$.

Solution If $y = Ax^3 + B/x$, then for $x \neq 0$ we have $y' = 3Ax^2 - B/x^2$ and $y'' = 6Ax + 2B/x^3$. Therefore,

$$x^2y'' - xy' - 3y = 6Ax^3 + \frac{2B}{x} - 3Ax^3 + \frac{B}{x} - 3Ax^3 - \frac{3B}{x} = 0,$$

as required.

As this example indicates, solutions to differential equations tend to involve arbitrary constants. (In general, there will be n arbitrary constants if the differential equation involves a derivative of order n.) Sometimes a problem provides, along with a differential equation, prescribed values for the solution function and its derivatives at a particular point (initial point). These values can be used to determine values for the arbitrary constants in the solution, and so yield a unique solution. Such problems are called *initial-value problems*.

Example 6 Use the result of Example 5 to solve the following initial-value problem.

$$\begin{cases} x^2y'' - xy' - 3y = 0 & (x > 0) \\ y(1) = 2 \\ y'(1) = -6 \end{cases}$$

Solution From Example 5 we have $y = Ax^3 + B/x$ as a solution for the differential equation. Setting $x = 1$ we get $2 = y(1) = A + B$. Since $y' = 3Ax^2 - B/x^2$, we also have $-6 = y'(1) = 3A - B$. Solving these two equations for A and B we get $A = -1$ and $B = 3$. Hence $y = -x^3 + 3/x$.

Differential equations and initial-value problems are of great importance in applications of calculus, and a large portion of the total mathematical endeavor of the last two hundred years has been devoted to their study. They are usually studied in separate courses on differential equations, but we mention them from time to time in this book, especially in exercises on differentiation.

EXERCISES

Find y', y'', and y''' for the functions in Exercises 1–6.

1. $y = x^2 - \dfrac{1}{x}$

2. $y = \sqrt{ax + b}$

3. $y = \dfrac{x - 1}{x + 1}$

4. $y = x^{1/3} - x^{-1/3}$

5. $y = x^{10} + 2x^8$

6. $y = (x^2 + 3)\sqrt{x}$

In Exercises 7–12 calculate enough derivatives of the given function to enable you to guess the general formula for $f^{(n)}(x)$. Then verify your guess using mathematical induction.

7. $f(x) = \dfrac{1}{x}$

8. $f(x) = \dfrac{1}{2 - x}$

9. $f(x) = \sqrt{x}$

10. $f(x) = \dfrac{1}{a + bx}$

11. $f(x) = x^{2/3}$

12.* $f(x) = \sqrt{1 - 3x}$

13.† Show that for any constants A and B the function $y = y(x) = Ax + B/x$ satisfies the *second-order differential equation* $x^2y'' + xy' - y = 0$ for $x \neq 0$. Find a function y satisfying the initial value problem:

$$\begin{cases} x^2y'' + xy' - y = 0 \qquad (x > 0) \\ y(1) = 2 \\ y'(1) = 4. \end{cases}$$

14.† Suppose that the quadratic equation $ar(r - 1) + br + c = 0$ has two distinct real roots, r_1 and r_2. Show that for any constants A and B the function $y = Ax^{r_1} + Bx^{r_2}$ satisfies, for $x > 0$, the differential equation $ax^2y'' + bxy' + cy = 0$.

Use the result of Exercise 14 to solve the initial-value problems in Exercises 15 and 16.

15.† $\begin{cases} x^2y'' - 6y = 0 \qquad (x > 0) \\ y(1) = 1 \\ y'(1) = 1 \end{cases}$

16.† $\begin{cases} 4x^2y'' + 4xy' - y = 0 \qquad (x > 0) \\ y(4) = 2 \\ y'(4) = -2 \end{cases}$

17.† Show that $y = x^{3/2}$ is a solution of the differential equation $y''y = \dfrac{3x}{4}$.

18. If f and g are twice-differentiable functions show that

$$(fg)'' = f''g + 2f'g' + fg''.$$

19. State and prove results analogous to that of Exercise 18 but for $(fg)^{(3)}$ and $(fg)^{(4)}$.

2.7 IMPLICIT DIFFERENTIATION

Suppose we wish to find the equation of the tangent line to the circle $x^2 + y^2 = 25$ at the point $(3, -4)$. As presented, this curve is not the graph of a function. (Vertical lines can meet the circle more than once.) It is, in fact, the union of the graphs of two functions: the upper semicircle, $y = \sqrt{25 - x^2}$ and the lower semicircle,

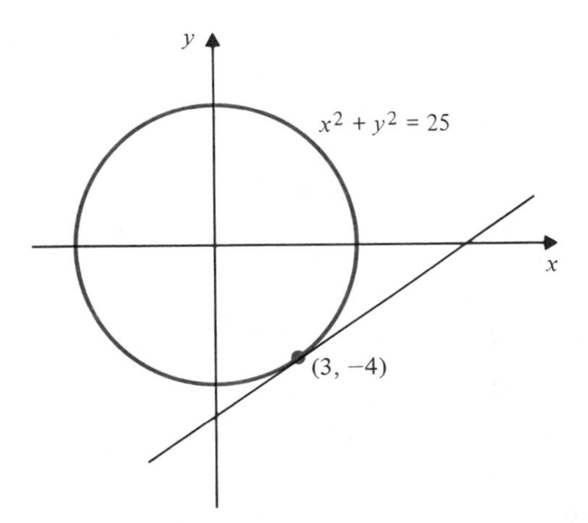

Figure 2.27

$y = -\sqrt{25 - x^2}$. As you can see in Fig. 2.27, $(3, -4)$ lies on the lower semicircle, so the slope of the required tangent line is

$$\frac{d}{dx}(-\sqrt{25 - x^2})\bigg|_{x=3} = \frac{-1}{2\sqrt{25 - x^2}}(-2x)\bigg|_{-x=3} = \frac{3}{4}.$$

Thus the tangent line has equation $y = -4 + (3/4)(x - 3)$.

There is another way to obtain the slope that does not require solving the given equation for y as a function of x. Part of the circle near $(3, -4)$ is certainly the graph of *some function* y. Let us differentiate the equation of the circle with respect to x, regarding y as a function of x:

$$\frac{d}{dx}(x^2 + y^2) = \frac{d}{dx}(25) = 0.$$

Since y is a function of x, the chain rule gives $(d/dx)y^2 = 2y(dy/dx)$, so

$$2x + 2y\frac{dy}{dx} = 0.$$

Hence, $dy/dx = -x/y$ and the required tangent line has slope

$$\frac{dy}{dx}\bigg|_{(3,\ -4)} = -\frac{3}{-4} = \frac{3}{4}.$$

This latter technique for finding the slope is called **implicit differentiation** because we are finding the derivative of a function y defined *implicitly* by the given equation rather than presented *explicitly* as a function of x. This method has the very great advantage that it can be used even if the given equation linking x and y cannot readily be solved for y as an explicit function of x. The price we have to pay for this advantage is that the derivative dy/dx so calculated is, in general, a function of both x and y, so both coordinates of a point must be known to find the slope there. This is to be expected because an equation in x and y can define more than one function of x. In the example of the circle in Fig. 2.27, $dy/dx = -x/y$ also gives the slope at points of the upper semicircle: at (3, 4) the slope is $-3/4$.

There is another, more subtle, and more important disadvantage to the use of implicit differentiation. When we differentiate an equation in x and y with respect to x, we are tacitly assuming that whatever function y of x we are considering is differentiable. Otherwise we would be attempting to calculate a quantity that does not exist.

We used the technique of implicit differentiation (as applied to the equation $x = f(y)$) to obtain the derivative of an inverse function $y = f^{-1}(x)$ in Section 2.4. At that time we were able to use a geometric argument to convince ourselves beforehand that the inverse function did indeed have a derivative. For more general equations in x and y, such geometric arguments will not suffice. It is possible to formulate and prove a theorem giving conditions under which an implicitly defined solution is differentiable. We cannot attempt to do this here; such a theorem is usually encountered in courses on advanced (multivariable) calculus. *When we use implicit differentiation here we will always be assuming, though we will not usually say so explicitly, that the required derivative exists.*

Example 1 Find dy/dx for the equation $x^3y + xy^3 = 2$.

Solution Differentiating the given equation with respect to x and remembering that y is a function of x, we get

$$3x^2y + x^3\frac{dy}{dx} + y^3 + 3xy^2\frac{dy}{dx} = 0.$$

Thus,

$$\frac{dy}{dx} = -\frac{3x^2y + y^3}{x^3 + 3xy^2}.$$

Example 2 Find the straight line tangent at the point $(-1, 2)$ to the curve

$$\frac{2y}{x^2} + \left(\frac{x}{y}\right)^2 = \frac{1}{4} - xy^2.$$

Solution Observe that the given point $(-1, 2)$ does indeed satisfy the given equation, so the point lies on the curve. Differentiate the equation of the curve with respect to x to get

$$\frac{x^2\left(2\dfrac{dy}{dx}\right) - 2y(2x)}{x^4} + 2\left(\frac{x}{y}\right)\frac{y - x\dfrac{dy}{dx}}{y^2} = -y^2 - 2xy\frac{dy}{dx}.$$

We now set $x = -1$ and $y = 2$ and solve for dy/dx:

$$2\frac{dy}{dx} + 8 + 2\left(\frac{-1}{2}\right)\frac{2 + \dfrac{dy}{dx}}{4} = -4 + 4\frac{dy}{dx},$$

$$12 - 2\frac{dy}{dx} - \frac{1}{2} - \frac{1}{4}\frac{dy}{dx} = 0,$$

$$\frac{23}{2} = \frac{9}{4}\frac{dy}{dx}.$$

Hence $dy/dx|_{(-1,\,2)} = 46/9$ and the tangent line has equation $y = 2 + (46/9)(x + 1)$.

Note that we substituted the coordinates of the given point as soon as we differentiated the given equation, and then we solved the resulting equation for the derivative. With numbers substituted for x and y it is usually much easier to solve for dy/dx than it would be with algebraic expressions. The appropriate order in which to perform the operations is

i) differentiate the given equation with respect to x,

ii) substitute any numerical coordinates,

iii) solve for the required slope.

Example 3 Calculate y' and y'' given that $xy + y^2 = 2x$.

Solution Differentiating with respect to x we obtain

$$y + xy' + 2yy' = 2.$$

Differentiating a second time with respect to x we obtain

$$y' + y' + xy'' + 2(y')^2 + 2yy'' = 0.$$

Solving the first equation for y' we get

$$y' = \frac{2 - y}{2y + x}.$$

Solving the second equation for y'' and substituting the above expression for y' we finally obtain

$$y'' = -\frac{2y' + 2(y')^2}{2y + x} = -2\left(\frac{2 - y}{2y + x}\right)\frac{1 + \dfrac{2 - y}{2y + x}}{2y + x}$$

$$= -2\frac{(2 - y)(y + x + 2)}{(2y + x)^3}.$$

Example 4 Show that for any constants A and B the curves $x^2 - y^2 = A$ and $xy = B$ intersect at right angles, that is, at any point where they intersect their tangents are perpendicular.

Solution The slope at any point on $x^2 - y^2 = A$ is given by $2x - 2yy' = 0$, or $y' = x/y$. The slope at any point on $xy = B$ is given by $y + xy' = 0$, or $y' = -y/x$. If the two curves (they are both hyperbolas if A and B are not 0) intersect at (x_o, y_o), then their slopes at that point are x_o/y_o and $-y_o/x_o$, respectively. Clearly these slopes are negative reciprocals, so the tangent line to one curve is the normal line to the other at that point. Hence the curves intersect at right angles.

EXERCISES

In Exercises 1–6 find dy/dx in terms of x and y.

1. $xy - x + 2y = 1$

2. $x^3y + xy^5 = 2$

3. $x^2y^3 = 2x - y$

4. $x^2 + 4(y - 1)^2 = 4$

5. $\dfrac{x - y}{x + y} = \dfrac{x^2}{y} + 1$

6. $x\sqrt{x + y} = 8 - xy$

In Exercises 7–10 find the equation of the tangent line at the given point.

7. $2x^2 + 3y^2 = 5$ at $(1, 1)$

8. $x^2y^3 - x^3y^2 = 12$ at $(-1, 2)$

9. $\dfrac{x}{y} + \left(\dfrac{y}{x}\right)^3 = 2$ at $(-1, -1)$

10. $x + 2y + 1 = \dfrac{y^2}{x - 1}$ at $(2, -1)$

In Exercises 11–14 find y'' in terms of x and y.

11. $xy = x + y$

12. $x^3 - y^2 + y^3 = x$

13. $x^3 - 3xy + y^3 = 1$

14. $x^2 + 4y^2 = 4$

15. For $x^2 + y^2 = a^2$ show that $y'' = -a^2/y^3$.

16. For $Ax^2 + By^2 = C$ show that $y'' = -AC/B^2y^3$.

17.* Show that the ellipse $x^2 + 2y^2 = 2$ and the hyperbola $2x^2 - 2y^2 = 1$ intersect at right angles.

18.* Show that the ellipse $x^2/a^2 + y^2/b^2 = 1$ and the hyperbola $x^2/A^2 - y^2/B^2 = 1$ intersect at right angles if $a^2 - b^2 = A^2 + B^2$. (This condition says that the ellipse and the hyperbola have the same foci.)

19.* Use implicit differentiation to find y' if $(x - y)/(x + y) = x/y + 1$. Now show that there are, in fact, no points on that curve, so the derivative you calculated is meaningless. This demonstrates the dangers of calculating something when you don't know it exists.

2.8 THE MEAN-VALUE THEOREM

Suppose that A and B are two points on a smooth curve. It seems obvious that there ought to be at least one point C on the curve between A and B such that the tangent line to the curve at C is parallel to the secant line AB. See Figure 2.28.

This intuitive principle is stated more precisely in the following theorem.

Theorem 2 (*Mean-Value Theorem*) Suppose that the function f is continuous on the closed, finite interval $[a, b]$ and that it is differentiable on the open interval (a, b). Then there exists a point c in the open interval (a, b) such that

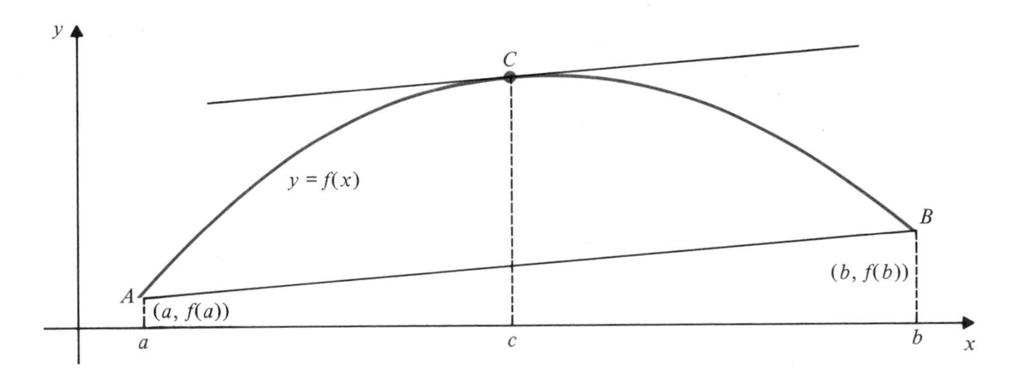

Figure 2.28

$$\frac{f(b) - f(a)}{b - a} = f'(c),$$

that is, the slope of the secant line joining the points $(a, f(a))$ and $(b, f(b))$ is equal to that of the tangent line to the curve $y = f(x)$ at the point $(c, f(c))$. □

The proof is somewhat involved and we defer it to the end of this section. For now we make several observations.

The hypotheses of the mean-value theorem are all necessary for the conclusion; if f fails to be continuous at even one point of $[a, b]$, or fails to be differentiable at even one point of (a, b), then there may be no point where the tangent line is parallel to the secant line AB. See Fig. 2.29.

The mean-value theorem gives no indication of how many points C there may be on the curve between A and B where the tangent is parallel to AB. If the curve is itself the straight line AB, then every point on the line between A and B has the required property. In general there may be more than one point (see Fig. 2.30); the mean-value theorem asserts only that there must be at least one.

The mean-value theorem gives us no information on how to find the point c, which it asserts exists. For some simple functions it is possible to calculate c, but doing so is usually of no practical value. As we shall see, the importance of the mean-value theorem lies in its use as a theoretical tool rather than as a practical tool. It belongs to a class of theorems called *existence theorems,* as does the intermediate-value theorem (Theorem 6 of Section 1.5). Such theorems assert the existence of something without telling you how to find it. Students sometimes complain that mathematicians worry too much about proving that a problem has a solution and not enough about how to find that solution. They argue: "If I can calculate a solution to a problem, then surely I do not need to worry about whether a solution exists." This is, however, false logic. Suppose we pose the problem "Find the largest positive integer." Of course this problem has no solution; there is no largest positive integer because we can add one to any integer and get a larger integer. Suppose, however, that we forget this and try to calculate a solution. We could proceed as follows:

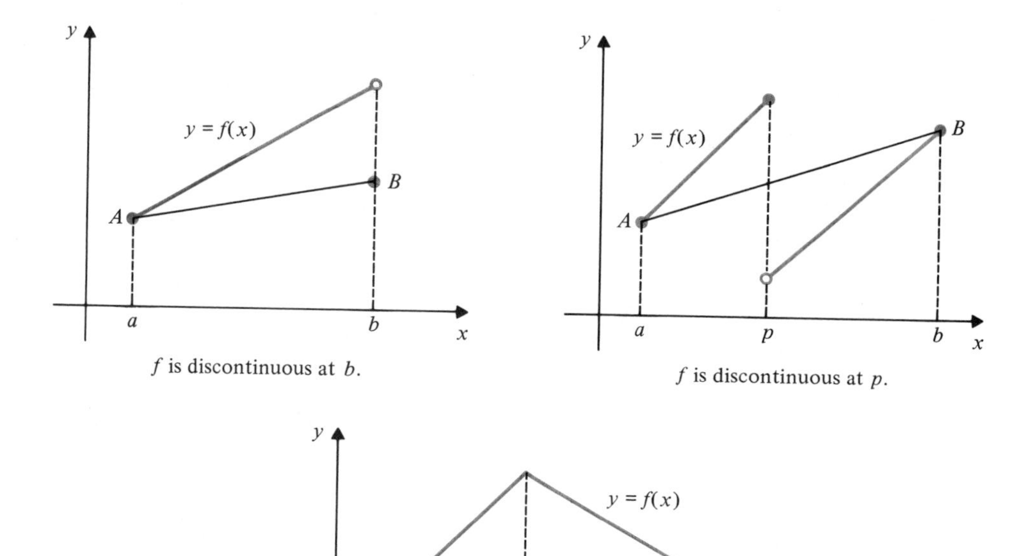

Figure 2.29

Let N be the largest positive integer.

Since 1 is a positive integer, $N \geq 1$.

Since N^2 is a positive integer, $N^2 \leq N$.

Thus $N(N - 1) \leq 0$ and we must have $N - 1 \leq 0$.

Hence $N \leq 1$. Therefore $N = 1$.

Therefore 1 is the largest positive integer.

The only error we have made here is in the assumption (in the first line) that the problem has a solution. It is partly to avoid logical pitfalls like this that mathematicians prove existence theorems. (See also Exercise 19 of Section 2.7.)

Example 1 Verify the conclusion of the mean-value theorem for $f(x) = \sqrt{x}$ on the interval $[a, b]$, where $0 \leq a < b$.

Solution The theorem asserts that there must be a number c satisfying $a < c < b$ and

$$\frac{1}{2\sqrt{c}} = f'(c) = \frac{f(b) - f(a)}{(b - a)} = \frac{\sqrt{b} - \sqrt{a}}{b - a} = \frac{1}{\sqrt{b} + \sqrt{a}}.$$

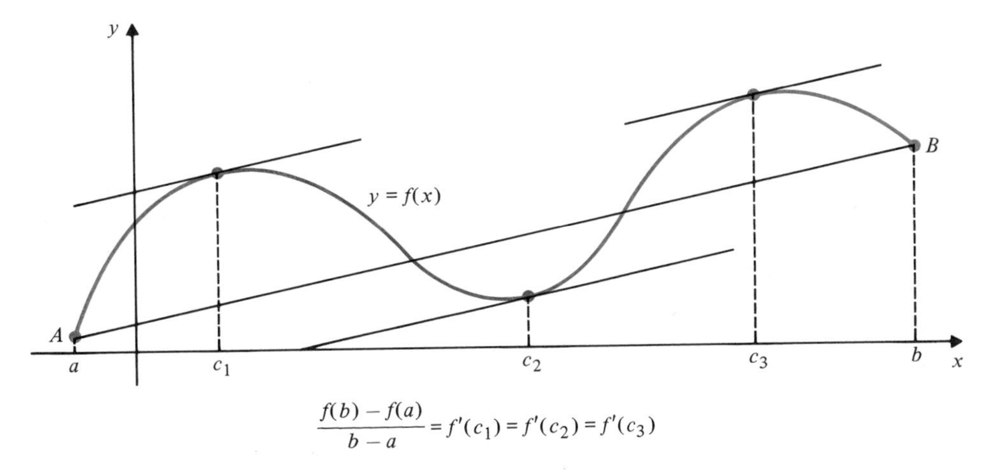

Figure 2.30

Clearly, $c = \left(\dfrac{\sqrt{b} + \sqrt{a}}{2}\right)^2$ satisfies the latter equation. Since $a < b$ we have

$$a = \left(\frac{\sqrt{a} + \sqrt{a}}{2}\right)^2 < c < \left(\frac{\sqrt{b} + \sqrt{b}}{2}\right)^2 = b,$$

so c lies in the required interval. (Note that in most significant applications of the mean-value theorem the point c cannot readily be calculated.)

Example 2 If $f'(x) = 0$ everywhere on an interval I, show that $f(x) = C$, a constant, on I.

Solution Pick any point x_o in I and let $C = f(x_o)$. If x is any other point of I, then by the mean-value theorem there exists a point c in I between x_o and x such that

$$\frac{f(x) - f(x_o)}{x - x_o} = f'(c).$$

But $f'(c) = 0$ because f' vanishes identically on I. Thus $f(x) = f(x_o) = C$ as claimed.

Example 3 Show that $\sqrt{1 + x} < 1 + x/2$ for all $x > 0$.

Solution Let $f(x) = \sqrt{1 + x}$. Then $f'(x) = 1/(2\sqrt{1 + x})$. Clearly $\sqrt{1 + x} > 1$ if $x > 0$, so $f'(x) < 1/2$ if $x > 0$. For any $x > 0$ we can apply the mean-value theorem to $f(x)$ on the interval $[0, x]$ and obtain a number c satisfying $0 < c < x$ for which

$$\frac{f(x) - f(0)}{x - 0} = f'(c),$$

that is,

$$\frac{\sqrt{1 + x} - 1}{x} = \frac{1}{2\sqrt{1 + c}} < \frac{1}{2}.$$

Hence $\sqrt{1 + x} - 1 < (1/2)x$, or $\sqrt{1 + x} < 1 + x/2$.

Increasing and Decreasing Functions

Definition 6

We say that a function f defined on an interval I is $\left\{\begin{array}{l}\text{increasing}\\ \text{nondecreasing}\\ \text{decreasing}\\ \text{nonincreasing}\end{array}\right\}$ on I provided that whenever x_1 and x_2 are points of I and $x_1 < x_2$, then
$$\left\{\begin{array}{l} f(x_1) < f(x_2)\\ f(x_1) \leq f(x_2)\\ f(x_1) > f(x_2)\\ f(x_1) \geq f(x_2)\end{array}\right\}$$

Figure 2.31 demonstrates such functions. Note the distinction between increasing and nondecreasing. A nondecreasing function (or a nonincreasing function) may be constant on all or part of an interval.

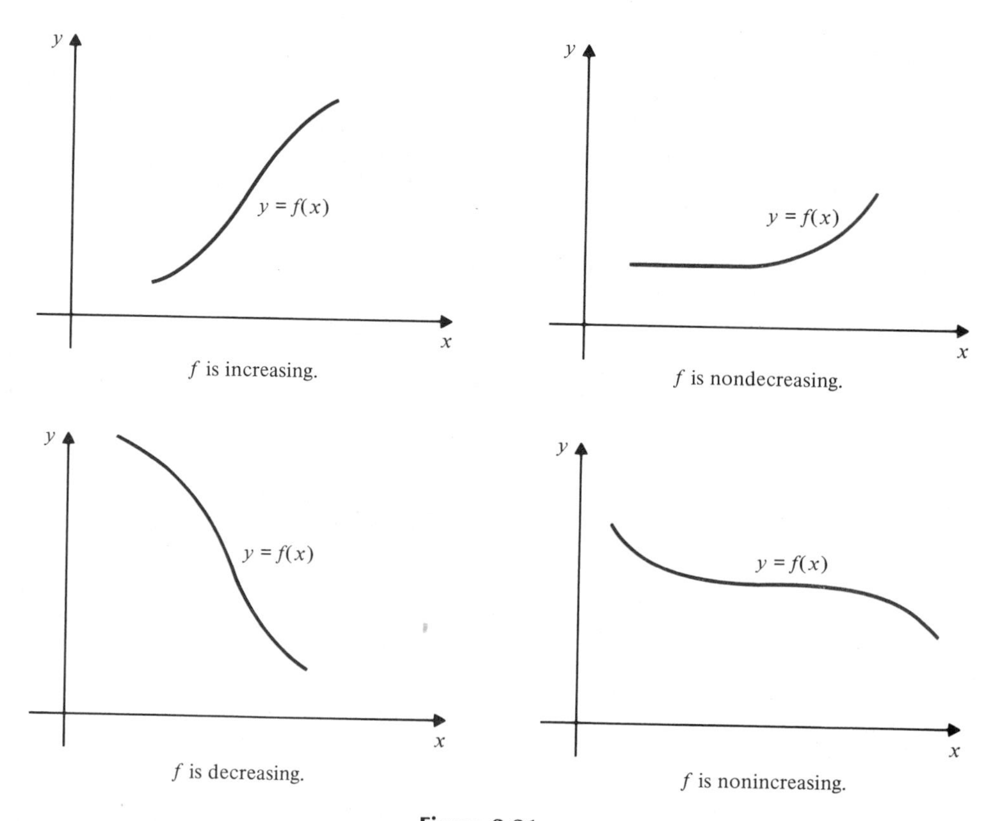

Figure 2.31

As we observed in Section 2.5, for a smooth function these properties can be determined by looking at the sign of the derivative of f on the interval in question.

Theorem 3 If f is differentiable at all points of the interval I, and if, for all x in I,

$$\begin{Bmatrix} f'(x) > 0 \\ f'(x) \geq 0 \\ f'(x) < 0 \\ f'(x) \leq 0 \end{Bmatrix}, \text{ then } f \text{ is } \begin{Bmatrix} \text{increasing} \\ \text{nondecreasing} \\ \text{decreasing} \\ \text{nonincreasing} \end{Bmatrix} \text{ on } I.$$

Proof Let x_1 and x_2 belong to I and suppose $x_1 < x_2$. By the mean-value theorem there is a point c between x_1 and x_2 such that

$$\frac{f(x_2) - f(x_1)}{x_2 - x_1} = f'(c).$$

Therefore, $f(x_2) = f(x_1) + f'(c)(x_2 - x_1)$. Since $x_2 - x_1 > 0$, the term $f'(c)(x_2 - x_1)$ will have the same sign as $f'(c)$. If $f'(x) > 0$ for all x in I, then $f'(c) > 0$ and $f(x_2) > f(x_1)$. Thus f is increasing on I. The other cases follow similarly. □

Example 4 Determine the intervals of increase and decrease of the function $f(x) = x^3 - 12x + 1$.

Solution $f'(x) = 3x^2 - 12 = 3(x - 2)(x + 2)$. Clearly, $f'(x) > 0$ if $x^2 > 4$, and $f'(x) < 0$ if $x^2 < 4$. Hence f is increasing on $(-\infty, -2)$ and on $(2, \infty)$, and is decreasing on $(-2, 2)$. At $x = -2$ and $x = 2$ we have $f'(x) = 0$, so these are both critical points of f. See Fig. 2.32. Since f is increasing to the left of -2 and decreasing to the right of -2, the point on the graph of f at $x = -2$ is higher than any other *nearby* points on the graph. Accordingly, we say that f has a *local maximum value* at $x = -2$. The local maximum value is $f(-2) = 17$. Similarly, because it is decreasing to the left of $x = 2$ and increasing to the right, f has a *local minimum value* of -15 at $x = 2$. All this information is most easily summarized in the chart given in Fig. 2.33, where symbols ↗ and ↘ are used to denote increasing and decreasing, and + and − to denote the signs of $f'(x)$.

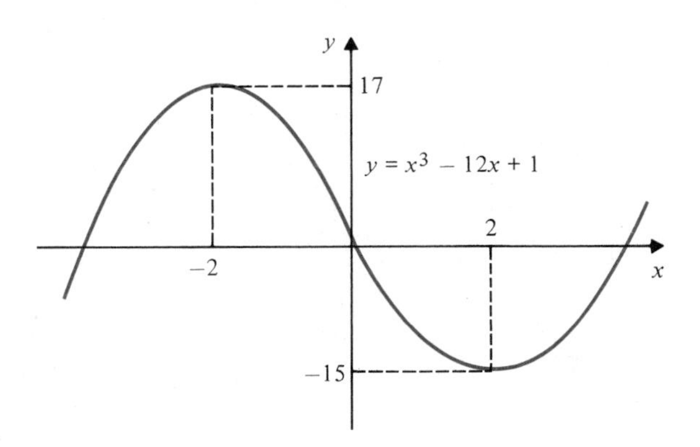

Figure 2.32

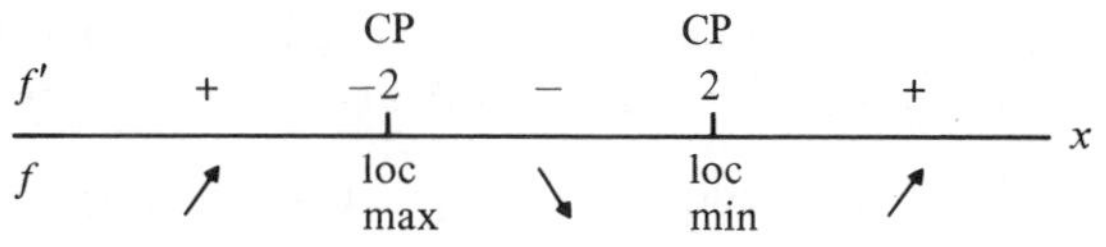

		CP		CP		
f'	+	-2	−	2	+	x
f	↗	loc max	↘	loc min	↗	

Figure 2.33

Example 5 Determine the critical points, local extreme values, and intervals of increase and decrease of the function

$$F(t) = \frac{t^3}{1 + t^4}.$$

Solution

$$F'(t) = \frac{(1 + t^4)3t^2 - t^3(4t^3)}{(1 + t^4)^2} = \frac{3t^2 - t^6}{(1 + t^4)^2} = \frac{t^2(\sqrt{3} + t^2)(\sqrt{3} - t^2)}{(1 + t^4)^2}$$

The critical points are $t = 0$, $t = 3^{1/4}$ and $t = -3^{1/4}$, and the sign of $F'(t)$ is solely determined by the factor $\sqrt{3} - t^2$, the other factors always being nonnegative. Figure 2.34 depicts the behavior of F in chart form.

Note that the critical point at $t = 0$ is neither a local maximum nor a local minimum. F is increasing both to the left and to the right of $t = 0$; in fact, F is increasing on the entire interval $(-3^{1/4}, 3^{1/4})$. The local minimum value of F at $t = -3^{1/4}$ is $F(-3^{1/4}) = (-1/4)(3^{3/4})$. The local maximum value occurs at $t = 3^{1/4}$ and is $F(3^{1/4}) = (1/4)(3^{3/4})$. The graph of F is sketched in Fig. 2.35. Because

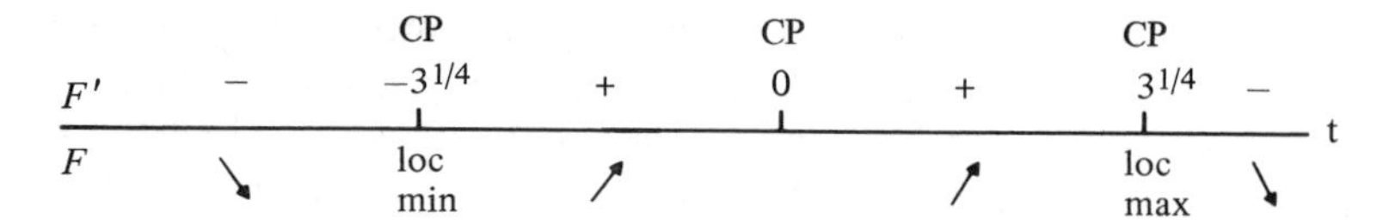

Figure 2.34

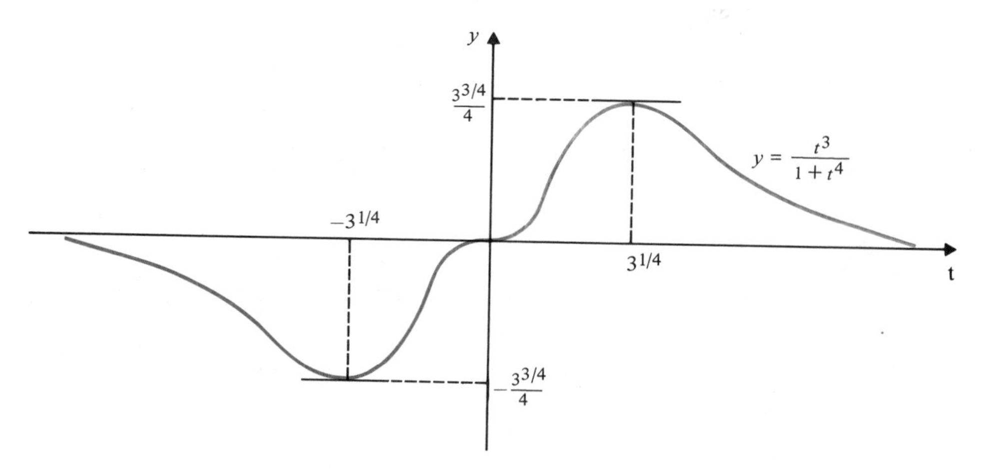

Figure 2.35

$$\lim_{t\to\pm\infty} \frac{t^3}{1 + t^4} = \lim_{t\to\pm\infty} \frac{1/t}{(1/t^4) + 1} = 0,$$

the graph of F draws closer and closer to the straight line $y = 0$ as t approaches $\pm\infty$. We therefore call that line a **horizontal asymptote** of the graph. (Asymptotes will be discussed in detail in Section 4.3.)

The organized use of calculus to determine the maximum and minimum values of functions and to sketch their graphs will be discussed in detail in Chapter 4. The two preceding examples suggest the methods to be used. We have presented them at this point to illustrate that the basic techniques are, in fact, derived from the mean-value theorem.

Proof of the Mean-Value Theorem

The mean-value theorem is one of those deeper results that is based on the completeness of the real number system via the fact that a continuous function on a closed, finite interval takes on a maximum and a minimum value (Theorem 5 of Section 1.5). Before giving the proof we establish the following preliminary result.

Theorem 4 If f is differentiable on an open interval (a, b), and if f achieves a local maximum (or minimum) value at the point c in (a, b), then $f'(c) = 0$, that is, c is a **critical point** of f.

Proof Suppose f has a local maximum value at c, where $a < c < b$. Then $f(x) \le f(c)$ for x nearby c, or, equivalently, $f(c + h) - f(c) \le 0$ for h near 0. If $h > 0$, then $[f(c + h) - f(x)]/h \le 0$ and so by Theorem 2(vi) of Section 1.3,

$$f'(c) = \lim_{h\to 0+} \frac{f(c + h) - f(c)}{h} \le 0.$$

Similarly, if $h < 0$, then $(f(c + h) - f(c))/h \ge 0$ so

$$f'(c) = \lim_{h\to 0-} \frac{f(c + h) - f(c)}{h} \ge 0.$$

Thus $f'(c) = 0$, as claimed. A similar proof works for a local minimum. □

Proof of the Mean-Value Theorem Suppose f satisfies the conditions of the mean-value theorem. Let

$$g(x) = f(x) - f(a) - \frac{f(b) - f(a)}{b - a}(x - a).$$

Clearly g is also continuous on $[a, b]$ and differentiable on (a, b). In addition, $g(a) = g(b) = 0$. We show that there is some point c in (a, b) where $g'(c) = 0$. Since

$$g'(x) = f'(x) - \frac{f(b) - f(a)}{b - a},$$

therefore

$$f'(c) = \frac{f(b) - f(a)}{b - a}$$

as required.

If $g(x) = 0$ for every x in $[a, b]$, then $g'(x)$ vanishes identically, so any c will do. Otherwise there is a point x_o in (a, b) where $g(x_o) \neq 0$. Without loss of generality we can assume that $g(x_o) > 0$. By Theorem 5(ii) of Section 1.5, there is some point c in $[a, b]$ such that $g(x) \leq g(c)$ for every x in $[a, b]$, that is, g has a maximum value at c. Since $g(c) \geq g(x_o) > 0$, c cannot be a or b, and c belongs to (a, b). Hence $g'(c) = 0$ by the preceding Theorem 4. This completes the proof. □

We remark that the special case of the mean-value theorem for a function g continuous on $[a, b]$, differentiable on (a, b), and satisfying $g(a) = g(b) = 0$ is called **Rolle's theorem.**

EXERCISES

In Exercises 1–6 illustrate the mean-value theorem by finding any points in the open interval (a, b) where the tangent line to $y = f(x)$ is parallel to the secant line joining $(a, f(a))$ and $(b, f(b))$.

1. $f(x) = Ax + B$ on $[a, b]$

2. $f(x) = x^2$ on $[a, b]$

3. $f(x) = x^3$ on $[a, b]$

4. $f(x) = \dfrac{1}{x}$ on $[1, 2]$

5. $f(x) = x^3 - 3x + 1$ on $[-2, 2]$

6. $f(x) = x^3 + 3x + 1$ on $[-2, 2]$

7. Use the method of Example 3 of this section to show that the inequality $\sqrt{1 + x} < 1 + x/2$ also holds for $-1 \leq x < 0$.

8. If r is a rational number satisfying $0 < r < 1$, show that $(1 + x)^r \leq 1 + rx$ for $-1 \leq x < \infty$.

9. If r is a rational number satisfying $r > 1$, show that $(1 + x)^r \geq 1 + rx$ for $-1 \leq x < \infty$.

10. If r is a rational number such that $r < 0$, show that $(1 + x)^r \geq 1 + rx$ for $-1 < x < \infty$.

Find the critical points, intervals of increase and decrease, and local extreme values for the functions in Exercises 11–16, and sketch their graphs.

11. $f(x) = x^2 + 2x + 2$

12. $f(x) = x^3 - 4x + 1$

13. $f(x) = x^3 + 4x + 1$

14. $f(x) = (x^2 - 4)^2$

15. $f(x) = \dfrac{1}{x^2 + 1}$

16. $f(x) = x^3(5 - x)^2$

17. If $f(x)$ is differentiable on an interval I and vanishes at $n \geq 2$ distinct points of I, prove that $f'(x)$ must vanish at at least $n - 1$ points in I.

18. If $f''(x)$ exists on an interval I and if f vanishes at at least three distinct points of I, prove that f'' must vanish at some point in I.

19.* Generalize Exercise 18 to a function for which $f^{(n)}$ exists in I and for which f vanishes at at least $n + 1$ distinct points in I.

20.* Suppose f is twice differentiable on an interval I (that is, f'' exists on I). Suppose that the points 0 and 2 belong to I and that $f(0) = f(1) = 0$ and $f(2) = 1$. Prove the following.

i) $f'(a) = \dfrac{1}{2}$ for some point a in I

ii) $f''(b) > \dfrac{1}{2}$ for some point b in I

iii) $f'(c) = \dfrac{1}{7}$ for some point c in I

2.9 ANTIDERIVATIVES AND INDEFINITE INTEGRALS

Definition 7 An **antiderivative** of a function f on an interval I is another function F whose derivative is equal to f on I:

$$F'(x) = f(x) \text{ for } x \text{ in } I.$$

For example, $F(x) = (1/2)x^2$ is an antiderivative of $f(x) = x$ on any interval, and $G(t) = 3t^{4/3} - 4t - 1/t$ is an antiderivative of $g(t) = 4t^{1/3} - 4 + t^{-2}$ on any interval.

Antiderivatives are not unique; indeed, if C is any constant, then $F(x) = (1/2)x^2 + C$ is an antiderivative of $f(x) = x$ on the real line. The following theorem shows that *all* antiderivatives are obtained in this way.

Theorem 5 If $F(x)$ and $G(x)$ are both antiderivatives of $f(x)$ on an interval I, then $F(x) = G(x) + C$ for all x in I, where C is some constant.

Proof Let $H(x) = F(x) - G(x)$ on I. Then $H'(x) = F'(x) - G'(x) = f(x) - f(x) = 0$ on I. It follows that $H(x) = C$, a constant, on I. (See Example 2 of Section 2.8.) Therefore, $F(x) = G(x) + C$ on I. □

It is essential for Theorem 5 that $F'(x) = G'(x)$ on an *interval*. If $F'(x) = G'(x)$ on a set S that is not an interval, it does not follow that $F(x) = G(x) + C$ on that set. For instance, $F(x) = 1$ and $G(x) = \operatorname{sgn} x$ both satisfy $F'(x) = G'(x) = 0$ for all $x \neq 0$, but there is no constant C such that $1 = \operatorname{sgn} x + C$ for all $x \neq 0$.

Theorem 5 shows that if $F(x)$ is any one particular antiderivative of $f(x)$ on an interval I then the most general antiderivative of $f(x)$ on I is $F(x) + C$ where C is an arbitrary constant.

Definition 8 The general antiderivative of a function f on an interval is called the **indefinite integral** of f on that interval and is denoted

$$\int f(x)\, dx.$$

The following are examples of indefinite integrals.

$$\int x dx = \frac{1}{2}x^2 + C$$

$$\int (x^3 - 5x^2 + 7)dx = \frac{1}{4}x^4 - \frac{5}{3}x^3 + 7x + C$$

$$\int \frac{dx}{x^2} = -\frac{1}{x} + C \qquad (x \neq 0)$$

All three formulas may be checked by differentiating the right-hand sides. The first two are valid on any interval, the third on any interval not containing $x = 0$.

Although we will not begin a detailed study of integration (antidifferentiation) until Chapter 6, at which point we will discover the real significance of the symbol $\int \ldots dx$, we can shed a little light on that symbol now by reconsidering the differential formalism introduced in Section 2.2. First observe that

$$\int dx = \int 1 dx = x + C,$$

so that, apart from the added constant C for the general antiderivative, the symbol $\int$ appears to be inverse to the differential symbol d; they cancel one another out. More generally, if $F'(x) = f(x)$ then $dF(x) = F'(x)dx = f(x)\,dx$, so

$$\int f(x)\,dx = \int dF(x) = F(x) + C$$

Again the integral $\int$ acts to "cancel" the differential.

The graphs of the different antiderivatives of the same function on the same interval are vertically displaced versions of the same curve, as shown in Fig. 2.36. In general only one of these curves will pass through any given point, so we can obtain a unique antiderivative of a given function on an interval by requiring the antiderivative to take on a prescribed value at a prescribed point. (Such problems are called *initial-value problems*.)

Example 1 Find the function $f(x)$ whose graph passes through the point $(2, 10)$ and whose derivative is $f'(x) = 6x^2 - 1$ for all real x.

Solution Since $f'(x) = 6x^2 - 1$ we have

$$f(x) = \int (6x^2 - 1)dx = 2x^3 - x + C$$

for some constant C. Since the graph of f passes through $(2, 10)$ we have

$$10 = f(2) = 16 - 2 + C.$$

Thus $C = -4$ and $f(x) = 2x^3 - x - 4$.

Example 2 Solve the initial-value problem

$$\begin{cases} y' = 3 - 4x \\ y(0) = -1. \end{cases}$$

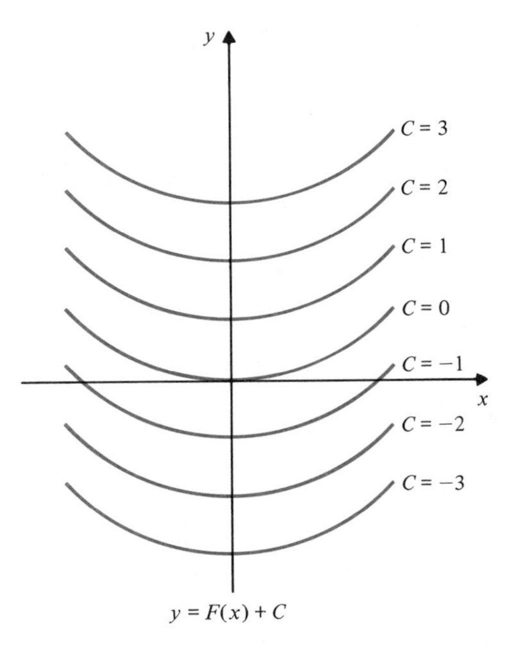

Figure 2.36

Solution

$$y = y(x) = \int (3 - 4x)dx = 3x - 2x^2 + C$$

$$-1 = y(0) = 3(0) - 2(0^2) + C$$

Therefore $C = -1$ and $y = 3x - 2x^2 - 1$.

Example 3 Find the function $f(t)$ whose derivative is $t^{-1/2} + 5t^{-3/2} + 2t^{2/3}$ and whose graph passes through the point (4, 1).

$$f(t) = \int (t^{-1/2} + 5t^{-3/2} + 2t^{2/3})dt$$

$$= 2t^{1/2} - 10t^{-1/2} + \frac{6}{5}t^{5/3} + C$$

$$1 = f(4) = 4 - 5 + \frac{6}{5}(4^{5/3}) + C$$

Hence, $C = 2 - \frac{6}{5}(4^{5/3})$ and

$$f(t) = 2t^{1/2} - 10t^{-1/2} + 2 + \frac{6}{5}(t^{5/3} - 4^{5/3}).$$

Example 4 An object falling freely near the surface of the earth is subject to a constant downward acceleration g, if the effect of air resistance is neglected. ($g \simeq 32$ ft/sec$^2 \simeq 9.8$ m/sec^2.) If we know the height y_o of the object at an initial time $t = 0$, and if we also know the vertical velocity v_o of the object at that time, the height $y(t)$ at other times while the object is still falling can be determined from the *second-order* initial-value problem:

$$\begin{cases} y''(t) = -g \\ y(0) = y_o \\ y'(0) = v_o. \end{cases}$$

We have

$$y'(t) = -\int g\,dt = -gt + C_1$$
$$v_o = y'(0) = 0 + C_1.$$

Thus $C_1 = v_o$.

$$y'(t) = -gt + v_o$$

$$y(t) = \int(-gt + v_o)dt = -\frac{1}{2}gt^2 + v_o t + C_2$$

$$y_0 = y(0) = 0 + 0 + C_2.$$

Thus $C_2 = y_o$. Finally,

$$y(t) = -\frac{1}{2}gt^2 + v_o t + y_o.$$

Example 5 A car is traveling at 72 km/hour. At a certain instant its brakes are applied to produce a constant deceleration of 0.8 m/sec². How far does the car travel before coming to a stop?

Solution Let $s(t)$ be the distance the car travels in the t seconds after the brakes are applied. Then $s''(t) = -0.8$ (m/sec²). The velocity at time t is given by

$$s'(t) = \int -0.8dt = -0.8t + C_1.$$

Since $s'(0) = 72$ km/hour $= 72 \times 1000/3600 = 20$ m/sec, we have $C_1 = 20$. Thus,

$$s'(t) = 20 - 0.8t$$

and

$$s(t) = \int(20 - 0.8t)dt = 20t - 0.4t^2 + C_2.$$

Since $s(0) = 0$ we have $C_2 = 0$ and $s(t) = 20t - 0.4t^2$. When the car has stopped, its velocity will be 0. Hence the stopping time is given by

$$0 = s'(t) = 20 - 0.8t,$$

that is, $t = 25$ (sec).

The distance traveled during deceleration is $s(25) = 250$ meters.

EXERCISES

In Exercises 1–16 find the given indefinite integrals.

1. $\int 5dx$

2. $\int x^2 dx$

3. $\int \sqrt{x}\, dx$

4. $\int x^{-1/2}\, dx$

5. $\int x^3 dx$

6. $\int x^r\, dx \qquad (r \neq -1, \text{ rational})$

7. $\int (a^2 - x^2)dx$

8. $\int (A + Bx + Cx^2)dx$

9. $\int (2x^{1/2} + 3x^{1/3})dx$

10. $\int 6(x^{-1/3} - x^{-4/3})dx$

11. $\int \left(\frac{x^3}{3} - \frac{x^2}{2} + x - 1\right)dx$

12. $30 \int (x^4 + x^5)dx$

13. $105 \int (1 + t^2 + t^4 + t^6)\, dt$

14.* $\int (ax + b)^r dx \qquad (r \neq -1)$

15.* $\int \frac{dx}{(1 + x)^2}$

16.* $\int \sqrt{2x + 3}\, dx$

17. Show that if a and b are constants, then

$$\int (af(x) + bg(x))dx = a\int f(x)dx + b\int g(x)dx.$$

In Exercises 18–27 find the solution $y = y(x)$ to the given initial-value problem. In what interval is the solution valid?

18.† $\begin{cases} y' = x - 2 \\ y(0) = 3 \end{cases}$

19.† $\begin{cases} y' = x^{-2} - x^{-3} \\ y(-1) = 0 \end{cases}$

20.† $\begin{cases} y' = 3\sqrt{x} \\ y(4) = 1 \end{cases}$

21.† $\begin{cases} y' = x^{1/3} \\ y(0) = 5 \end{cases}$

22.† $\begin{cases} y' = Ax^2 + Bx + C \\ y(1) = 1 \end{cases}$

23.† $\begin{cases} y' = x^{-9/7} \\ y(1) = -4 \end{cases}$

24.† $\begin{cases} y'' = 2 \\ y'(0) = 5 \\ y(0) = -3 \end{cases}$

25.† $\begin{cases} y'' = x^{-4} \\ y'(1) = 2 \\ y(1) = 1 \end{cases}$

26.† $\begin{cases} y'' = x^3 - 1 \\ y'(0) = 0 \\ y(0) = 8 \end{cases}$

27.† $\begin{cases} y'' = 5x^2 - 3x^{-1/2} \\ y'(1) = 2 \\ y(1) = 0 \end{cases}$

REVIEW EXERCISES FOR CHAPTER 2

1. Use the definition of derivative to find $f'(x)$ where $f(x) = (x - 2)/(3 + x^2)$.

2. Use the definition of derivative to find $f'(4)$ where $f(x) = (1 + \sqrt{x})/x$.

Find the derivative $\frac{dy}{dx}$ in Exercises 3–32.

3. $y = 2x + \frac{3}{x}$

4. $y = 16 - \sqrt{x}$

5. $y = \frac{x}{x - 1}$

6. $y = (x^3 + 1)(x^4 + 3)$

7. $y = (2x + 1)^{40}$

8. $y = \frac{1}{x^2 + 3x + 5}$

9. $xy + 2y^2 = 10$

10. $y = \frac{\sqrt{4 + x^2}}{x}$

11. $y = \frac{2 - x^2}{2 + x^2}$

12. $y = \left(\frac{x^3}{3} + \frac{x^2}{2} + x\right)^5$

13. $x^3 + y^3 = 6$

14. $\sqrt{x} + \sqrt{y} = 1$

15. $x^2 - y^2 = xy$

16. $y = x^2\sqrt{x^2 - 1}$

17. $y^5 = x^5 + 1$

18. $\frac{x - 1}{y + 1} = 1 + xy^2$

19. $y^2 = \frac{x}{2x + 1}$

20. $y = \frac{1}{x + \sqrt{x - 1}}$

21. $y = \frac{2}{(1 - \sqrt{2x + 1})^{3/2}}$

22. $y = \sqrt{5 + 4\sqrt{x + 3}}$

23. $\frac{x}{y} + \frac{2y}{3x} = 1$

24. $y = (x^2 + x + 1)\sqrt{x^2 + 2}$

25. $x + \sqrt{xy} = 1$

26. $y = \frac{x}{1 + \sqrt{x^2 + 1}}$

27. $y = \frac{x}{(2x + 3)^2}$

28. $y = |x^5 - x|$

29. $|x| + |y| = 2$

30. $y = \frac{(x^2 + 1)^3}{(3 - x^2)^2}$

31. $y = \frac{x\sqrt{x + 2}}{x + \sqrt{x + 2}}$

32. $y = \frac{x^2 + \sqrt{4 - x^2}}{x\sqrt{2 - x}}$

In Exercises 33–42 find an equation of the line tangent to the given curve at the given point.

33. $y = x^2 + x^{-2}$ at $(-1, 2)$

34. $y = \sqrt{9 - 2x}$ at $(0, 3)$

35. $y = \frac{5}{x^2 + 1}$ at $(2, 1)$

36. $y = \frac{2x - 3}{2x + 3}$ at $(0, -1)$

37. $x^3y + xy^2 = 10$ at $(2, 1)$

38. $y = \frac{3x}{\sqrt{x^2 + 5}}$ at $(2, 2)$

39. $2y^3 = xy + x^3$ at $(1, 1)$

40. $y = \frac{1}{(x - 2)^3}$ at $(3, 1)$

41. $y^2 = 2 + (x + 1)\sqrt{x + 4}$ at $(0, -2)$

42. $\frac{xy}{y - 1} = \frac{1}{2y} + x + 1$ at $(-1, -1)$

43. If $f(x) = \frac{1 - x}{1 + x}$, find $f''(x)$.

44. If $f(x) = \dfrac{x}{1 + 2x^2}$, find $f''(1)$.

45. If $f(x) = x^2 + x^{-4}$ find $f^{(n)}(x)$ for every positive integer n.

46. Find $f^{(n)}(x)$ if $f(x) = 1/(1 + 2x)^2$.

47. Find $\dfrac{d^2y}{dx^2}$ if $x^2 + y^3 = 1 + xy$.

48. Find $\dfrac{d^3y}{dx^3}$ if $2x^2 + 3y^2 = 5$.

49. What is the rate of change of the side of a cube with respect to the volume of the cube?

50. Find the velocity and acceleration at time t of an object moving along the x-axis if the position of the object at time t is $x = t^3(5 - t)^2$. When is the object moving to the left?

51. An object accelerates from rest with constant acceleration 4 cm/sec². How far has it traveled by the time its speed becomes 16 cm/sec?

52. Using the mean-value theorem, or otherwise, show that $\sqrt{4 + x} \le 2 + \dfrac{x}{4}$ if $x \ge 0$.

53. If f and g are differentiable functions satisfying $f(0) \le g(0)$ and $f'(x) \le g'(x)$ for $x > 0$, show that $f(x) \le g(x)$ for $x > 0$.

In Exercises 54–60 find the function $y = f(x)$ satisfying the given conditions.

54. $f'(x) = x^2 - x + 2, \quad f(1) = 3$

55. $\dfrac{dy}{dx} = \dfrac{1}{x^2}, \quad y = 1$ when $x = 2$

56. $f'(x) = \dfrac{1}{(x + 1)^2}, \quad f(0) = 0$

57. $f'(x) = |x|, \quad f(0) = 1$

58. $f''(x) = 4, \quad f'(0) = 3, \quad f(0) = 5$

59. $f''(x) = x, \quad f'(1) = 0, \quad f(1) = -2$

60. $f''(x) = x^{-1/2}, \quad f'(4) = 1, \quad f(4) = 0.$

3.1 The Circular (Trigonometric) Functions
3.2 Derivatives of the Trigonometric Functions
3.3 The Inverse Trigonometric Functions
3.4 The Natural Logarithm
3.5 The Exponential Function
3.6 General Exponentials and Logarithms
3.7 Growth and Decay Problems
3.8 The Hyperbolic Functions and Their Inverses

3 The Elementary Transcendental Functions

To this point we have dealt exclusively with algebraic functions, that is, functions involving only finitely many arithmetic operations (additions, subtractions, multiplications, and divisions) and the extraction of roots (fractional powers). Much of the importance of calculus and many of its most useful applications center on its ability to illuminate the behavior of certain classes of functions that are not algebraic in their construction but that arise naturally when we attempt to model concrete problems in mathematical terms. In this chapter we study the most important such classes, the circular (or trigonometric) functions and their inverses, and the logarithms and their inverses, the exponentials.

3.1 THE CIRCULAR (TRIGONOMETRIC) FUNCTIONS

Most students of calculus will be somewhat familiar with the trigonometric functions from their high school studies. Nevertheless we define them carefully and present their most important properties here. For purposes of calculus, all angles are measured in *radians* unless the contrary is explicitly stated.

Definition 1

Let C be the circle with equation $x^2 + y^2 = 1$ and A the point $(1, 0)$ on C. For any real number t let P_t be the point on C obtained by proceeding a distance $|t|$ from A measured along the circle, in a counterclockwise direction if $t > 0$ and in a clockwise direction if $t < 0$. Then the **radian measure** of angle AOP_t is t (see Fig. 3.1).

$$\text{angle } AOP_t = t \text{ radians}$$

If P_t has coordinates (x_t, y_t), then cosine t and sine t (abbreviated cos t and sin t) are defined by

$$\cos t = x_t$$
$$\sin t = y_t.$$

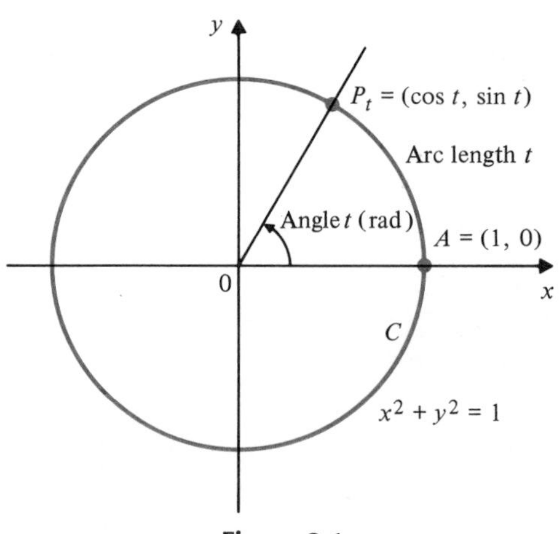

Figure 3.1

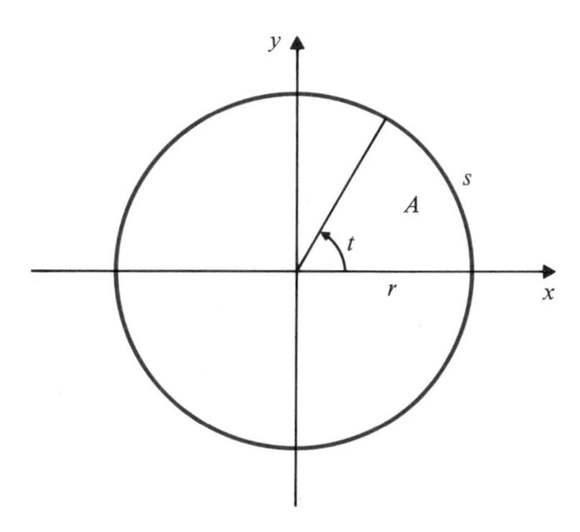

Figure 3.2

According to this definition any real number can be interpreted as a (signed) angle, and its cosine and sine can be determined. Since the circumference of the circle C is 2π, the relationship between radian and degree measure is 2π radians $=$ $360°$, or

$$\pi \text{ radians} = 180°.$$

Note that the unit radians is not usually written when an angle is expressed in radians: When we refer to the angle $\frac{\pi}{2}$ we mean $\frac{\pi}{2}$ radians or $90°$.

If an arc of length s on a circle of radius r subtends an angle t at the center of the circle, then, because the circumference of the whole circle is $2\pi r$, we must have

$$s = \frac{t}{2\pi}(2\pi r) = rt.$$

Similarly, the area of the circular sector with angle t is

$$A = \frac{t}{2\pi}(\pi r^2) = \frac{r^2 t}{2}.$$

See Fig. 3.2.

The functions cosine and sine are called **circular functions** because they are defined as coordinates of a point on a circle. Many properties of the trigonometric functions follow directly from the definitions we just presented. Here are the most important of these properties.

a) The Pythagorean Identity

$P_t = (\cos t, \sin t)$ lies on the circle $x^2 + y^2 = 1$; therefore, for any $t, x = \cos t$ and $y = \sin t$ satisfy $x^2 + y^2 = 1$. Hence,

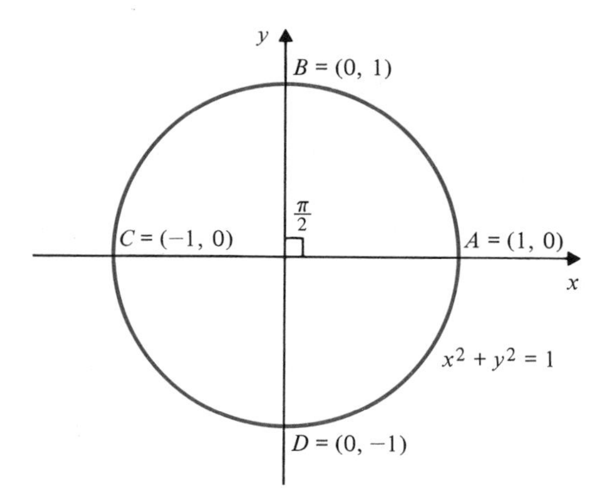

Figure 3.3

$$\cos^2 t + \sin^2 t = 1.$$

Note the use of the symbol $\cos^2 t$ to denote $(\cos t)^2$ and $\sin^2 t$ to denote $(\sin t)^2$.

b) Functions of Special Angles

Observe in Fig. 3.3 the coordinates of $A = P_0$, $B = P_{\pi/2}$, $C = P_\pi$, and $D = P_{-\pi/2}$.

$$\cos 0 = 1 \qquad \sin 0 = 0$$

$$\cos \frac{\pi}{2} = 0 \qquad \sin \frac{\pi}{2} = 1$$

$$\cos \pi = -1 \qquad \sin \pi = 0$$

$$\cos -\frac{\pi}{2} = 0 \qquad \sin -\frac{\pi}{2} = -1$$

Properties of isosceles right-angled triangles and equilateral triangles can also be used to obtain the coordinates of $P_{\pi/4}$, $P_{\pi/3}$ and $P_{\pi/6}$, as shown in Fig. 3.4.

c) Periodicity

The circle C has circumference 2π; therefore, $P_{t+2\pi} = P_t$ and

$$\cos(t + 2\pi) = \cos t$$
$$\sin(t + 2\pi) = \sin t,$$

that is, cosine and sine are **periodic functions** with *period* 2π.

d) Symmetry

For any t,

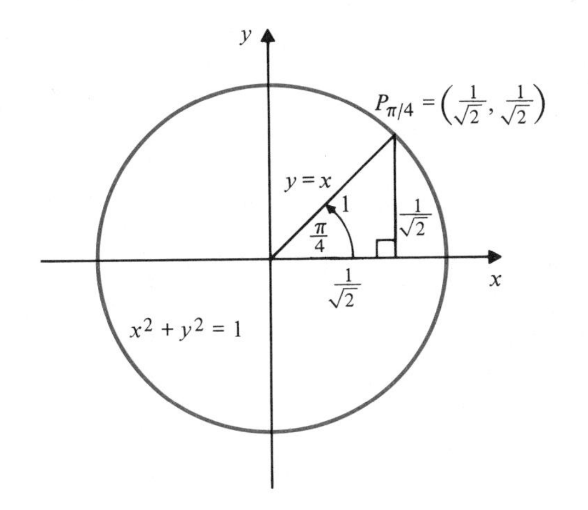

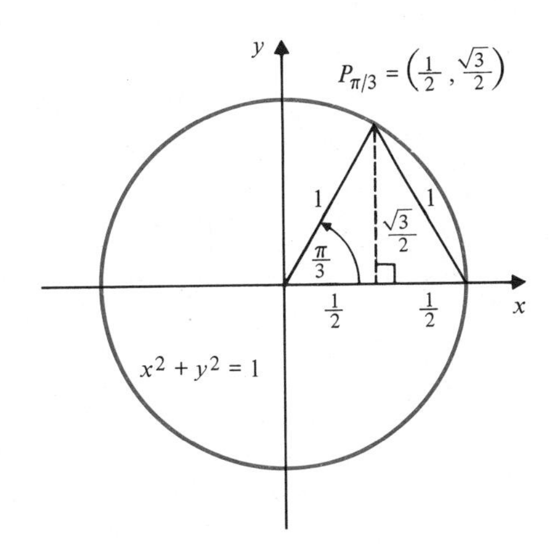

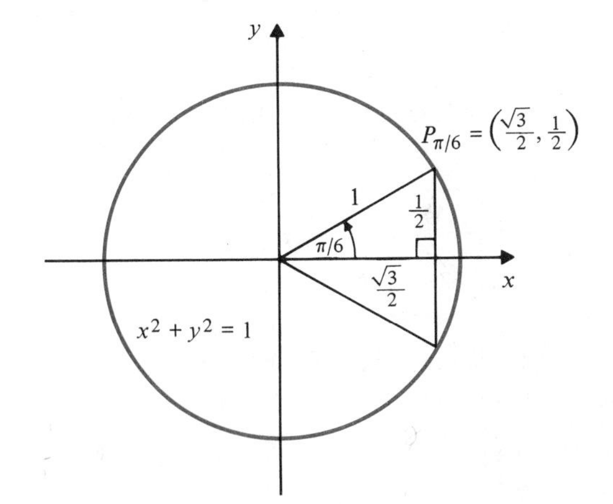

Degrees	0°	30°	45°	60°	90°
Radians	0	$\frac{\pi}{6}$	$\frac{\pi}{4}$	$\frac{\pi}{3}$	$\frac{\pi}{2}$
sine	0	$\frac{1}{2}$	$\frac{1}{\sqrt{2}}$	$\frac{\sqrt{3}}{2}$	1
cosine	1	$\frac{\sqrt{3}}{2}$	$\frac{1}{\sqrt{2}}$	$\frac{1}{2}$	0

Figure 3.4

$$\cos(-t) = \cos t$$
$$\sin(-t) = -\sin t$$

See Fig. 3.5. Thus cosine is an *even function* and sine is an *odd function*.

e) Complementary Angle Identities

The points P_t and $P_{(\pi/2)-t}$ are mirror images of each other in the line $y = x$ (see Fig. 3.6), so their x and y coordinates are reversed.

$$\cos\left(\frac{\pi}{2} - t\right) = \sin t$$
$$\sin\left(\frac{\pi}{2} - t\right) = \cos t$$

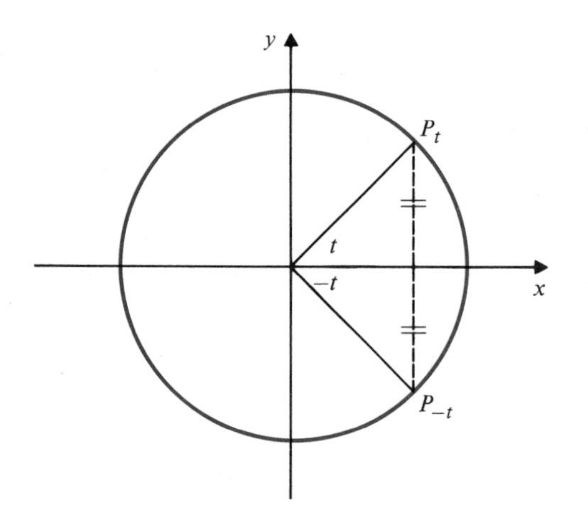

Figure 3.5

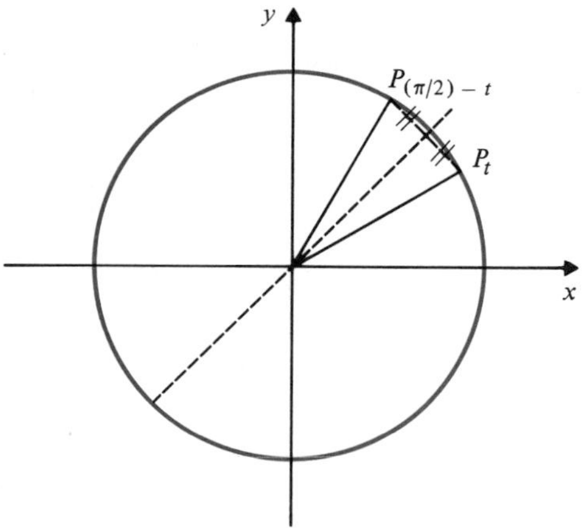

Figure 3.6

f) Supplementary Angle Identities

The points $P_{\pi-t}$ and P_t are mirror images of each other in the y-axis (see Fig. 3.7). Thus,

$$\cos(\pi - t) = -\cos t,$$
$$\sin(\pi - t) = \sin t.$$

The preceding properties of cosine and sine can be used to obtain the values of these functions at any multiple of $\pi/6$ or $\pi/4$. For example:

$$\cos\frac{5\pi}{6} = \cos\left(\pi - \frac{\pi}{6}\right) = -\cos\frac{\pi}{6} = -\frac{\sqrt{3}}{2} \quad \text{and}$$
$$\sin\frac{5\pi}{4} = \sin\left(\pi - \frac{5\pi}{4}\right) = \sin\left(-\frac{\pi}{4}\right) = -\sin\frac{\pi}{4} = -\frac{1}{\sqrt{2}}.$$

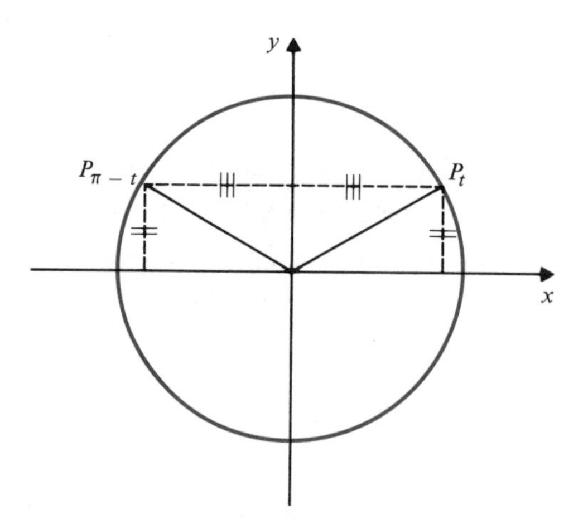

Figure 3.7

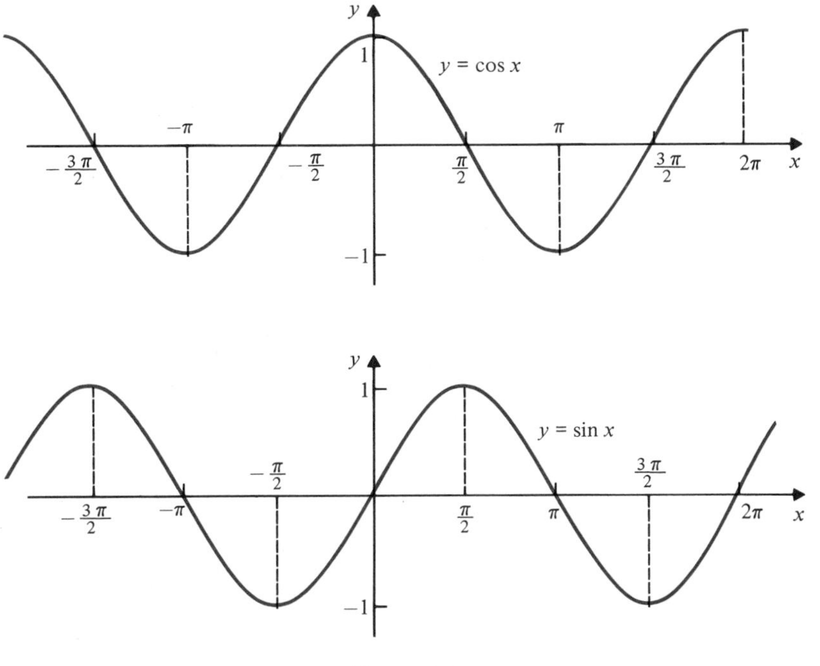

Figure 3.8

While decimal approximations to the values of sine and cosine can be found in mathematical tables or by using a scientific calculator, it is useful to remember the values for angles 0, $\pi/6$, $\pi/4$, $\pi/3$, and $\pi/2$ (as presented in the table accompanying Fig. 3.4). Values for any multiple of $\pi/6$ or $\pi/4$ can be readily determined by symmetry and consideration of which functions are positive in which quadrants.

When using a scientific calculator to calculate these functions, be sure you have selected the proper angular mode: degrees or radians. Many calculators also have a third mode, grads (100 grads = 90 degrees = $\pi/2$ radians). We will make no use of grads. The graphs of cosine and sine are shown in Fig. 3.8.

The Addition Formulas

Let s and t be real numbers and consider the points

$$\begin{aligned} P_t &= (\cos t, \sin t) \\ P_s &= (\cos s, \sin s) \\ P_{s-t} &= (\cos(s-t), \sin(s-t)) \\ A &= (1, 0) \end{aligned}$$

as represented in Fig. 3.9.

Angle $P_tOP_s = s - t$ radians = angle AOP_{s-t}; therefore,

$$(\text{distance } P_s \text{ to } P_t)^2 = (\text{distance } P_{s-t} \text{ to } A)^2.$$

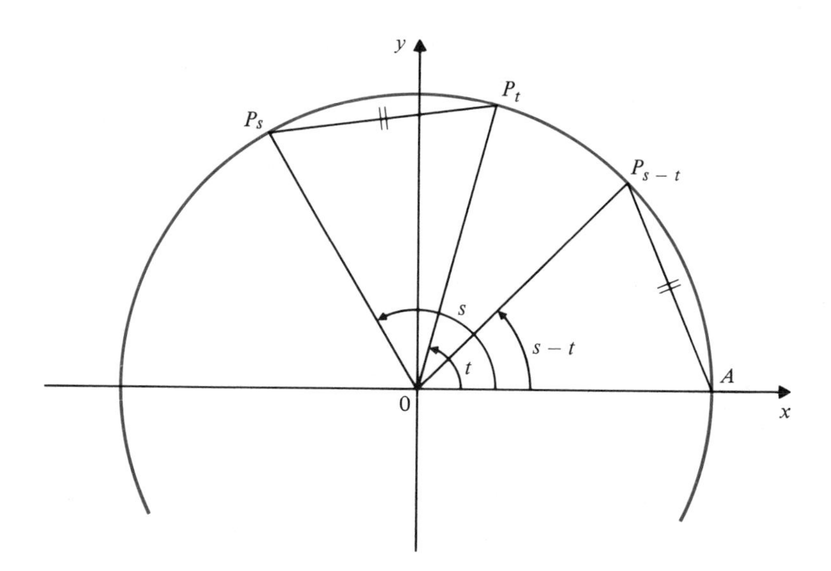

Figure 3.9

Expressing these squared distances in terms of coordinates, we obtain

$$(\cos s - \cos t)^2 + (\sin s - \sin t)^2 = (\cos(s - t) - 1)^2 + \sin^2(s - t),$$

$$\cos^2 s - 2\cos s \cos t + \cos^2 t + \sin^2 s - 2\sin s \sin t + \sin^2 t$$
$$= \cos^2(s - t) - 2\cos(s - t) + 1 + \sin^2(s - t).$$

Since $\cos^2 x + \sin^2 x = 1$ for every x, this reduces to

$$\cos(s - t) = \cos s \cos t + \sin s \sin t.$$

Replacing t with $-t$ in the formula above, and recalling that $\cos(-t) = \cos t$ and $\sin(-t) = -\sin t$, we have

$$\cos(s + t) = \cos s \cos t - \sin s \sin t.$$

Now we use the complementary angle formulas.

$$\begin{aligned}\sin(s + t) &= \cos\left(\frac{\pi}{2} - (s + t)\right)\\ &= \cos\left(\left(\frac{\pi}{2} - s\right) - t\right)\\ &= \cos\left(\frac{\pi}{2} - s\right)\cos t + \sin\left(\frac{\pi}{2} - s\right)\sin t\\ &= \sin s \cos t + \cos s \sin t\end{aligned}$$

That is, we have

$$\sin(s + t) = \sin s \cos t + \cos s \sin t$$

Finally, we replace t with $-t$ again to get

$$\sin(s - t) = \sin s \cos t - \cos s \sin t.$$

The four formulas obtained above are called the *addition formulas*.

Example 1 Compute $\cos \pi/12$.

Solution

$$\cos\frac{\pi}{12} = \cos\left(\frac{\pi}{3} - \frac{\pi}{4}\right) = \cos\frac{\pi}{3}\cos\frac{\pi}{4} + \sin\frac{\pi}{3}\sin\frac{\pi}{4}$$

$$= \left(\frac{1}{2}\right)\left(\frac{1}{\sqrt{2}}\right) + \left(\frac{\sqrt{3}}{2}\right)\left(\frac{1}{\sqrt{2}}\right) = \frac{1 + \sqrt{3}}{2\sqrt{2}}$$

From the addition formulas we obtain as special cases certain useful formulas called *half-angle formulas*. Put $s = t$ in the addition formulas for $\sin(s + t)$ and $\cos(s + t)$ to get

$$\begin{aligned} \sin 2t &= 2 \sin t \cos t \qquad \text{and} \\ \cos 2t &= \cos^2 t - \sin^2 t \\ &= 2\cos^2 t - 1 \qquad (\text{using } \sin^2 t + \cos^2 t = 1) \\ &= 1 - 2\sin^2 t \end{aligned}$$

From these latter,

$$\cos^2 t = \frac{1 + \cos 2t}{2}, \qquad \text{and } \sin^2 t = \frac{1 - \cos 2t}{2},$$

two formulas that often prove useful, especially when we later have to antidifferentiate certain combinations of trigonometric functions.

The Functions Tangent, Secant, Cotangent, and Cosecant

The remaining four trigonometric functions, tangent (tan), cotangent (cot), secant (sec), and cosecant (csc), are defined in terms of the two principal ones, cosine and sine. Their graphs are shown in Figs. 3.10 and 3.11.

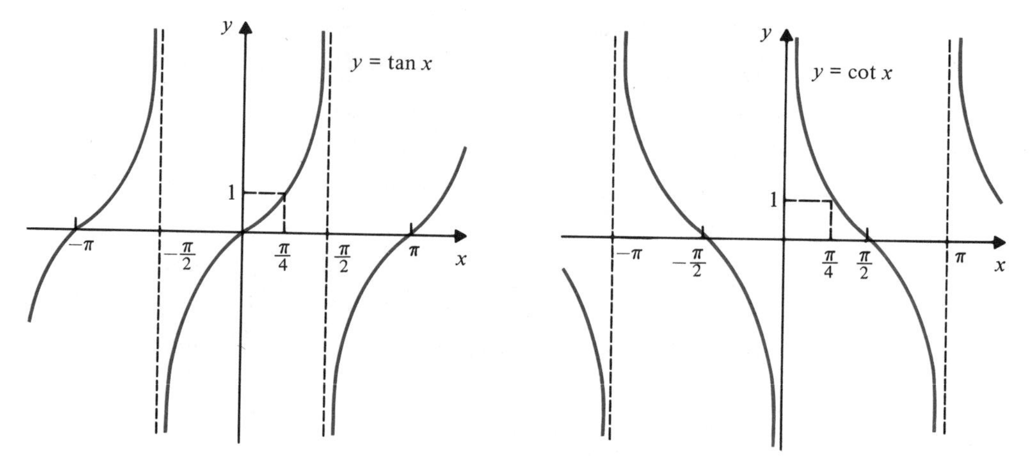

Figure 3.10

$$\tan t = \frac{\sin t}{\cos t}$$

$$\cot t = \frac{\cos t}{\sin t} = \frac{1}{\tan t}$$

$$\sec t = \frac{1}{\cos t}$$

$$\csc t = \frac{1}{\sin t}$$

Observe that each of these functions is not defined (and its graph has vertical asymptotes) at points where the function in the denominator of its defining fraction has value 0. Observe also that tangent, cotangent, and cosecant are odd functions and that secant is an even function. Since $|\sin x| \leq 1$ and $|\cos x| \leq 1$ for all x, $|\csc x| \geq 1$ and $|\sec x| \geq 1$ for all x where they are defined.

Like their reciprocals cosine and sine, the functions secant and cosecant are periodic with period 2π. However, tangent and cotangent are periodic with period π:

$$\tan (x + \pi) = \frac{\sin (x + \pi)}{\cos (x + \pi)} = \frac{\sin (\pi - (x + \pi))}{-\cos (\pi - (x + \pi))} = \frac{-\sin x}{-\cos x} = \tan x.$$

Division of the Pythagorean identity $\sin^2 x + \cos^2 x = 1$ by $\sin^2 x$ and $\cos^2 x$, respectively, leads to the two alternate versions of that identity:

$$1 + \cot^2 x = \csc^2 x, \qquad 1 + \tan^2 x = \sec^2 x.$$

Addition formulas for tangent and cotangent can be obtained from those for sine and cosine. For example,

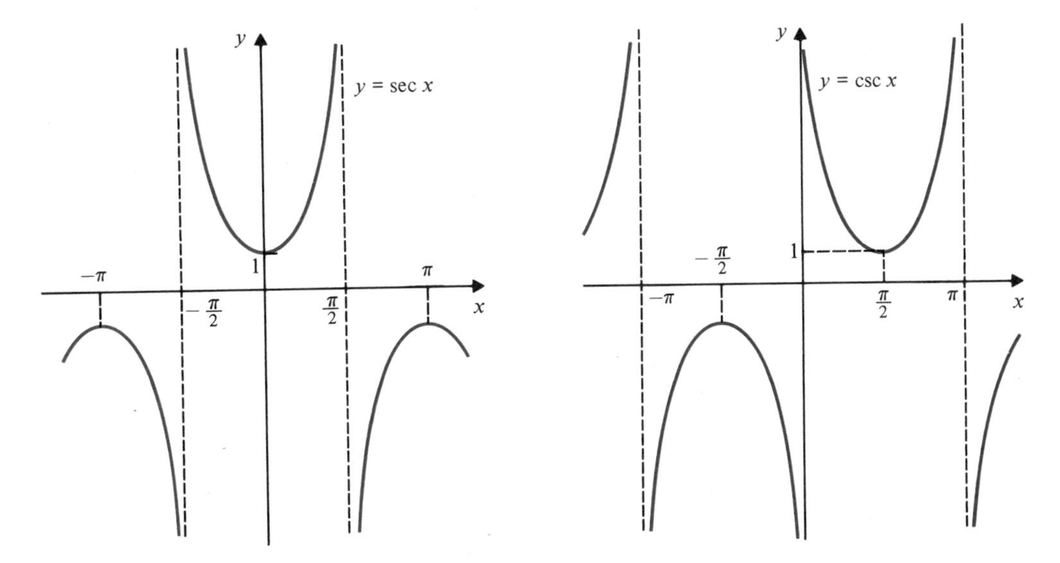

Figure 3.11

$$\tan(s+t) = \frac{\sin(s+t)}{\cos(s+t)} = \frac{\sin s \cos t + \cos s \sin t}{\cos s \cos t - \sin s \sin t}$$
$$= \frac{\tan s + \tan t}{1 - \tan s \tan t}.$$

Similarly,

$$\tan(s-t) = \frac{\tan s - \tan t}{1 + \tan s \tan t}.$$

Some Trigonometry

The trigonometric functions are so called because of their usefulness in expressing the relationships between the sides and angles of a triangle.

The sides of a right-angled triangle may be designated with respect to one of the acute angles of the triangle as hyp (hypotenuse), opp (side opposite the angle), and adj (side adjacent to the angle). If the angle is t, as shown in Fig. 3.12, then the triangle is similar to one with hypotenuse 1, side opposite $\sin t$, and side adjacent $\cos t$ (Fig. 3.13). Thus,

$$\sin t = \frac{\text{opp}}{\text{hyp}},$$
$$\cos t = \frac{\text{adj}}{\text{hyp}},$$
$$\tan t = \frac{\text{opp}}{\text{adj}}.$$

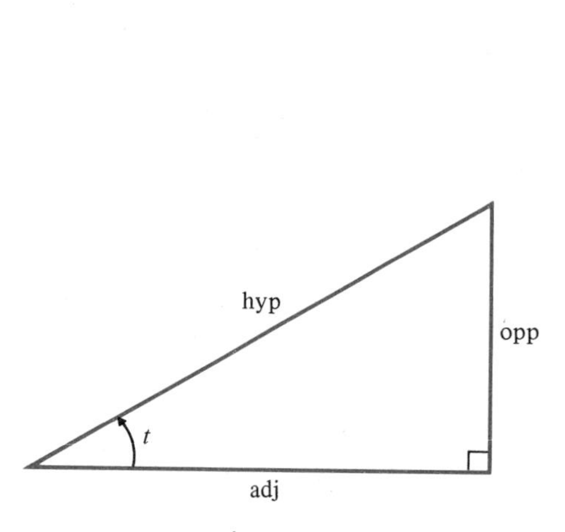

Figure 3.12

Figure 3.13

Example 2 Solve the given triangles for the unknown sides x and y.

Solution a)

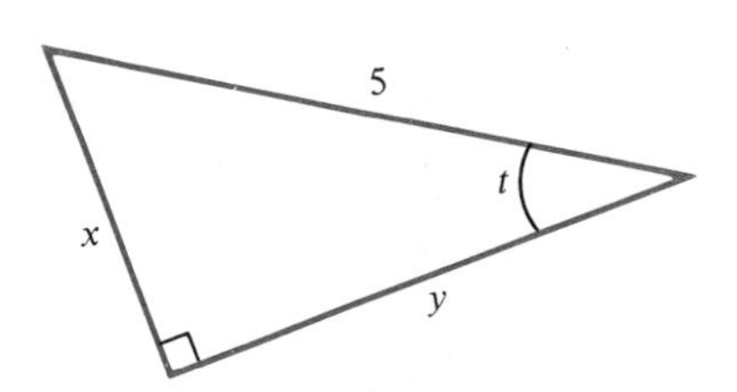

$$\frac{x}{5} = \sin t, \qquad \text{so } x = 5 \sin t$$

$$\frac{y}{5} = \cos t, \qquad \text{so } y = 5 \cos t$$

b)

$$\frac{x}{a} = \tan t, \qquad \text{so } x = a \tan t$$

$$\frac{y}{a} = \sec t, \qquad \text{so } y = a \sec t$$

For an arbitrary triangle ABC with sides a, b, and c and opposite angles A, B, and C, respectively, we note two laws that describe relationships in the triangle.

Sine Law: $$\frac{\sin A}{a} = \frac{\sin B}{b} = \frac{\sin C}{c}$$

Cosine Law: $$c^2 = a^2 + b^2 - 2ab \cos C$$

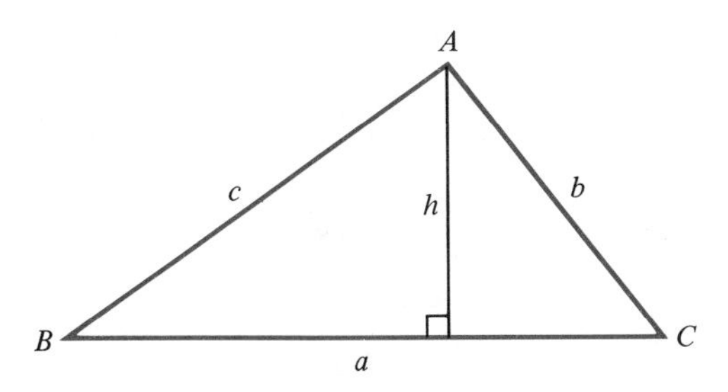

Figure 3.14

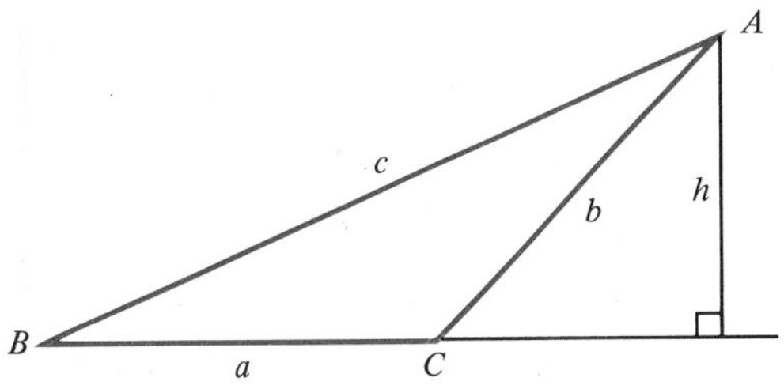

Figure 3.15

These may be proved as follows (see Figs. 3.14 and 3.15): Let h be the length of the perpendicular from A to the side BC. From right-angled triangles (and using $\sin(\pi - t) = \sin t$ if required) we get $c \sin B = h = b \sin C$. Thus $(\sin B)/b = (\sin C)/c$. By symmetry of the formulas (or by dropping a perpendicular to another side), both of these fractions must be equal to $(\sin A)/a$, so the sine law is proved. For the cosine law observe that

$$c^2 = \begin{cases} h^2 + (a - b\cos C)^2 & \text{if } C \le \dfrac{\pi}{2} \\[2ex] h^2 + (a + b\cos(\pi - C))^2 & \text{if } C > \dfrac{\pi}{2} \end{cases}$$

$$= h^2 + (a - b\cos C)^2 \qquad (\text{since } \cos(\pi - C) = -\cos C)$$

$$= b^2 \sin^2 C + a^2 - 2ab\cos C + b^2\cos^2 C$$

$$= a^2 + b^2 - 2ab\cos C. \ \square$$

EXERCISES

Find the values of the quantities in Exercises 1–15 by using various formulas presented in this section. Do not use tables or a calculator.

1. $\sin \dfrac{3\pi}{4}$ **2.** $\cos \dfrac{3\pi}{4}$ **3.** $\sin \dfrac{2\pi}{3}$ **4.** $\cos \dfrac{5\pi}{6}$

5. $\cos \dfrac{5\pi}{4}$ **6.** $\sin \dfrac{7\pi}{4}$ **7.** $\sin \dfrac{\pi}{12}$ **8.** $\cos \dfrac{5\pi}{12}$

9. $\sin \dfrac{5\pi}{12}$ **10.** $\sin \dfrac{7\pi}{12}$ **11.** $\cos \dfrac{7\pi}{12}$ **12.** $\sin \dfrac{4\pi}{3}$

13. $\cos \dfrac{7\pi}{6}$ **14.** $\sin \dfrac{7\pi}{6}$ **15.** $\sin \left(-\dfrac{\pi}{12}\right)$

In Exercises 16–27 express the given quantity in terms of $\sin x$ and $\cos x$.

16. $\sin(\pi + x)$ **17.** $\cos(\pi + x)$ **18.** $\sin(2\pi - x)$

19. $\cos(2\pi - x)$ **20.** $\sin\left(\dfrac{3\pi}{2} - x\right)$ **21.** $\cos\left(\dfrac{3\pi}{2} - x\right)$

22. $\cos\left(\dfrac{3\pi}{2} + x\right)$ **23.** $\sin\left(\dfrac{3\pi}{2} + x\right)$ **24.** $\tan x + \cot x$

25. $\sec x - \tan x$ **26.** $\sec^2 x + \csc^2 x$ **27.** $\dfrac{\tan x - \cot x}{\tan x + \cot x}$

In Exercises 28–31 prove the given identities.

28. $\dfrac{1 - \cos x}{\sin x} = \dfrac{\sin x}{1 + \cos x}$

29. $\dfrac{1 - \cos x}{\sin x} = \tan \dfrac{x}{2}$

30. $\dfrac{1 - \cos x}{1 + \cos x} = \tan^2 \dfrac{x}{2}$

31. $\dfrac{\cos x - \sin x}{\cos x + \sin x} = \sec 2x - \tan 2x$

32. Express $\sin 3x$ and $\cos 3x$ in terms of $\sin x$ and $\cos x$.

In Exercises 33–38 sketch the graph of the given function. What is the period of the function?

33. $\cos 2x$ **34.** $\sin \dfrac{x}{2}$ **35.** $\sin \pi x$

36. $\cos \dfrac{\pi x}{2}$ **37.** $\tan \pi x$ **38.** $\sec \dfrac{x}{3}$

In Exercises 39–44 $\sin x$, $\cos x$, or $\tan x$ is given. Find the other two functions if x lies in the specified interval.

39. $\sin x = \dfrac{3}{5}, \left[\dfrac{\pi}{2}, \pi\right]$

40. $\tan x = 2, \left[0, \dfrac{\pi}{2}\right]$

41. $\cos x = \dfrac{1}{3}, \left[-\dfrac{\pi}{2}, 0\right]$

42. $\cos x = -\dfrac{5}{13}, \left[\dfrac{\pi}{2}, \pi\right]$

43. $\sin x = \dfrac{-1}{2}, \left[\pi, \dfrac{3\pi}{2}\right]$

44. $\tan x = \dfrac{1}{2}, \left[\pi, \dfrac{3\pi}{2}\right]$

In Exercises 45–56 ABC is a triangle with right angle at C. The sides opposite angles A, B, and C are a, b, and c, respectively.

45. Find a and b if $c = 2$, $B = \dfrac{\pi}{3}$.

46. Find a and c if $b = 2$, $B = \dfrac{\pi}{3}$.

47. Find b and c if $a = 5$, $B = \dfrac{\pi}{6}$.

48. Express a in terms of A and c.

49. Express a in terms of A and b.

50. Express a in terms of B and c.

51. Express a in terms of B and b.

52. Express c in terms of A and a.

53. Express c in terms of A and b.

54. Express $\sin A$ in terms of a and c.

55. Express $\sin A$ in terms of b and c.

56. Express $\sin A$ in terms of a and b.

In Exercises 57–64 ABC is an arbitrary triangle with sides a, b, and c, opposite to angles A, B, and C, respectively. Find the indicated quantities. Use tables or a scientific calculator if necessary.

57. Find $\sin B$ if $a = 4$, $b = 3$, $A = \dfrac{\pi}{4}$.

58. Find $\cos A$ if $a = 2$, $b = 2$, $c = 3$.

59. Find $\sin B$ if $a = 2$, $b = 3$, $c = 4$.

60. Find c if $a = 2$, $b = 3$, $C = \dfrac{\pi}{4}$.

61. Find a if $c = 3$, $A = \frac{\pi}{4}$, $B = \frac{\pi}{3}$.

62. Find c if $a = 2$, $b = 3$, $C = 35°$.

63. Find b if $a = 4$, $B = 40°$, $C = 70°$.

64.* Find c if $a = 1$, $b = \sqrt{2}$, $A = 30°$. (There are two possible answers.)

65. Show that the area of triangle ABC is $(1/2)ab \sin C = (1/2)bc \sin A = (1/2)ca \sin B$.

66. Two guy wires stretch from the top T of a vertical pole to points B and C on the ground, where C is 10 m closer to the base of the pole than is B. If wire BT makes an angle of 35° with the horizontal, and wire CT makes an angle of 50° with the horizontal, how high is the pole?

67. Observers at positions A and B 2 km apart simultaneously measure the angle of elevation of a weather balloon to be 40° and 70°, respectively. If the balloon is directly above a point on the line segment from A to B, find the height of the balloon.

3.2 DERIVATIVES OF THE TRIGONOMETRIC FUNCTIONS

A careful inspection of the graphs of sine and cosine given in the previous section might lead one to guess the following formulas.

$$\frac{d}{dx}\sin x = \cos x$$
$$\frac{d}{dx}\cos x = -\sin x$$

In order to prove that these formulas are indeed correct we require the following result.

Lemma

$$\lim_{h\to 0}\frac{\sin h}{h} = 1$$

Proof Let h be a small positive angle, and represent h as angle AOP_h as in the definition at the beginning of Section 3.1 (see Fig. 3.16). Thus $A = (1, 0)$ and $P_h = (\cos h, \sin h)$. If C lies on the line through O and P_h and CA is perpendicular to OA, then

$$\frac{CA}{OA} = \frac{P_hB}{OB} = \frac{\sin h}{\cos h}.$$

Hence $C = (1, (\sin h/\cos h))$. Considering the diagram we can clearly assert that

$$\begin{aligned}\text{Area of triangle } P_hOB &< \text{Area of circular sector } P_hOA \\ &< \text{Area of triangle } COA.\end{aligned}$$

$$\frac{1}{2}\sin h\cos h < \frac{1}{2}h < \frac{1}{2}\frac{\sin h}{\cos h}.$$

* Difficult and/or theoretical problem.

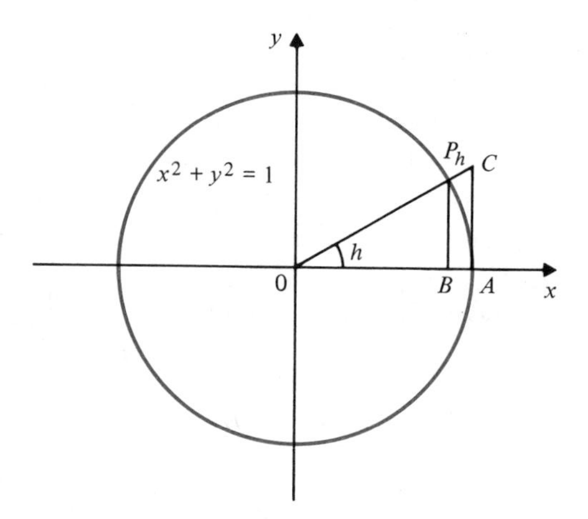

Figure 3.16

Since h is greater than 0, it follows that

$$\cos h < \frac{\sin h}{h} < \frac{1}{\cos h}.$$

Since $\lim_{h\to 0} \cos h = 1$ we have, by the squeeze theorem (Theorem 3 of Section 1.3),

$$\lim_{h\to 0+} \frac{\sin h}{h} = 1.$$

If $h < 0$ then $h = -k$ where $k > 0$, and

$$\lim_{h\to 0-} \frac{\sin h}{h} = \lim_{k\to 0+} \frac{\sin(-k)}{-k} = \lim_{k\to 0+} \frac{\sin k}{k} = 1,$$

because sine is an odd function. Hence

$$\lim_{h\to 0} \frac{\sin h}{h} = 1. \ \square$$

Another useful limit follows from this result:

$$\lim_{h\to 0} \frac{\cos h - 1}{h} = \lim_{h\to 0} -2\frac{\sin^2(h/2)}{h}$$

$$= \lim_{h\to 0} -\frac{\sin(h/2)}{h/2} \sin(h/2) = -1 \times 0 = 0.$$

We can now calculate the derivative of the sine function.

$$\frac{d}{dx} \sin x = \lim_{h\to 0} \frac{\sin(x+h) - \sin x}{h} \qquad \text{(definition of derivative)}$$

$$= \lim_{h\to 0} \frac{\sin x \cos h + \cos x \sin h - \sin x}{h} \qquad \text{(addition formula)}$$

$$= \lim_{h\to 0} \left(\sin x \frac{\cos h - 1}{h} + \cos x \frac{\sin h}{h} \right)$$

$$= (\sin x)(0) + (\cos x)(1) = \cos x \qquad \text{(the two limits above)}$$

We could calculate the derivative of cosine in the same manner, using the same two limits. An alternative method makes use of the complementary angle formulas:

$$\frac{d}{dx} \cos x = \frac{d}{dx} \sin \left(\frac{\pi}{2} - x \right) = -\cos \left(\frac{\pi}{2} - x \right) = -\sin x.$$

Example 1 Find the derivatives of $\cos(5x - 2)$, $\sin(x^2 + 1)$, $\sin^3 \pi x$, and $\cos(3\cos(5\cos 7x))$.

Solution

$$\frac{d}{dx} \cos(5x - 2) = (-\sin(5x - 2))(5) = -5\sin(5x - 2)$$

$$\frac{d}{dx} \sin(x^2 + 1) = (\cos(x^2 + 1))(2x) = 2x\cos(x^2 + 1)$$

$$\frac{d}{dx} \sin^3 \pi x = 3\sin^2 \pi x (\cos \pi x)(\pi) = 3\pi \sin^2 \pi x \cos \pi x$$

$$\frac{d}{dx} \cos(3\cos(5\cos 7x)) = [-\sin(3\cos(5\cos 7x))][-3\sin(5\cos 7x)][-5\sin 7x](7)$$

$$= -105 \sin 7x \sin(5\cos 7x) \sin(3\cos(5\cos 7x))$$

Example 2 Derive and then memorize the following formulas.

$$\frac{d}{dx} \tan x = \sec^2 x$$

$$\frac{d}{dx} \sec x = \sec x \tan x$$

$$\frac{d}{dx} \cot x = -\csc^2 x$$

$$\frac{d}{dx} \csc x = -\csc x \cot x$$

Solution

$$\frac{d}{dx} \tan x = \frac{d}{dx} \frac{\sin x}{\cos x} = \frac{\cos x \cos x - \sin x(-\sin x)}{\cos^2 x} = \frac{1}{\cos^2 x} = \sec^2 x,$$

using the quotient rule and the Pythagorean identity.

$$\frac{d}{dx} \cot x = \frac{d}{dx} \frac{1}{\tan x} = \frac{-1}{\tan^2 x} \sec^2 x = \frac{-1}{\sin^2 x} = -\csc^2 x,$$

$$\frac{d}{dx}\sec x = \frac{d}{dx}\frac{1}{\cos x} = \frac{-1}{\cos^2 x}(-\sin x) = \sec x \tan x,$$

$$\frac{d}{dx}\csc x = \frac{d}{dx}\frac{1}{\sin x} = \frac{-1}{\sin^2 x}\cos x = -\csc x \cot x,$$

by three applications of the reciprocal rule.

Observe that the functions sine, tangent, and secant do not have an explicit negative sign in their derivatives, while the "cofunctions" cosine, cosecant, and cotangent do. This is a useful mnemonic device.

Example 3 We tacitly assumed in the derivation of the derivatives of sine and cosine that x was measured in radians. (Where exactly was this used in the derivation?) Calculate the derivatives of $\sin(x^\circ)$, $\cos(x^\circ)$, and $\tan(x^\circ)$.

Solution Since $1^\circ = \dfrac{\pi}{180}$ radians, we have

$$\frac{d}{dx}\sin(x^\circ) = \frac{d}{dx}\sin\left(\frac{\pi x}{180}\right) = \frac{\pi}{180}\cos\left(\frac{\pi x}{180}\right) = \frac{\pi}{180}\cos(x^\circ).$$

Similarly,

$$\frac{d}{dx}\cos(x^\circ) = -\frac{\pi}{180}\sin(x^\circ) \qquad \text{and}$$

$$\frac{d}{dx}\tan(x^\circ) = \frac{\pi}{180}\sec^2(x^\circ).$$

Example 4 Find the equation of the straight line tangent to the curve

$$\sin\left(\frac{x}{y}\right) + \cos^2\left(\frac{x^2y}{9\pi}\right) = \frac{\sqrt{3}}{2} + \frac{1}{4}$$

at the point $(\pi, 3)$.

Solution First observe that the point $(\pi, 3)$ does indeed lie on the curve. Using implicit differentiation we get

$$\left(\cos\frac{x}{y}\right)\left(\frac{y - xy'}{y^2}\right) + 2\cos\left(\frac{x^2y}{9\pi}\right)\left(-\sin\left(\frac{x^2y}{9\pi}\right)\right)\frac{1}{9\pi}(2xy + x^2y') = 0.$$

Substituting $(x, y) = (\pi, 3)$, we have

$$\frac{1}{2}\left(\frac{1}{3} - \frac{\pi}{9}y'\right) + 2\left(\frac{1}{2}\right)\left(-\frac{\sqrt{3}}{2}\right)\frac{1}{9\pi}(6\pi + \pi^2y') = 0$$

$$\frac{1}{3} - \frac{2}{3}\sqrt{3} = y'\left(\frac{\pi}{9} + \frac{\pi}{9}\sqrt{3}\right).$$

Thus the slope of the required tangent line is

$$y'\Big|_{(\pi,\,3)} = -\frac{3}{\pi}\frac{2\sqrt{3}-1}{\sqrt{3}+1}$$

and its equation is

$$y - 3 = -\frac{3}{\pi}\left(\frac{2\sqrt{3}-1}{\sqrt{3}+1}\right)(x - \pi).$$

Finally, let us gather together the indefinite integrals that correspond to the differentiation formulas for the trigonometric functions.

$$\int \sin x\,dx = -\cos x + C$$

$$\int \cos x\,dx = \sin x + C$$

$$\int \sec^2 x\,dx = \tan x + C$$

$$\int \csc^2 x\,dx = -\cot x + C$$

$$\int \sec x \tan x\,dx = \sec x + C$$

$$\int \csc x \cot x\,dx = -\csc x + C$$

The Projectile Problem

Suppose that an object is thrown or fired at time $t = 0$ with an initial speed s_o in a direction making an angle α above the horizontal. The object will follow an arched path until it strikes the ground some distance away from its firing point (which we assume also to be at ground level). (See Fig. 3.17.)

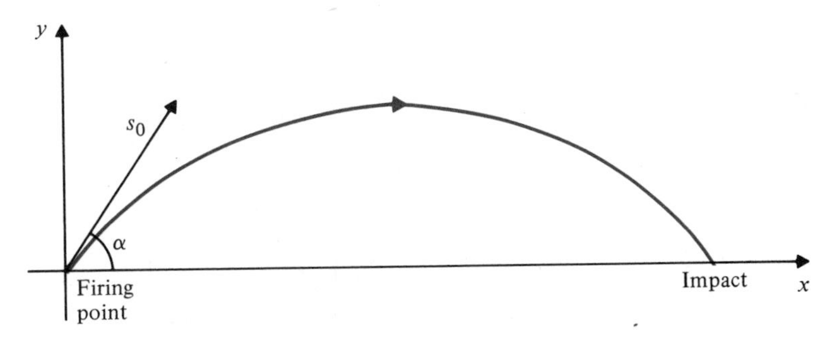

Figure 3.17

Assuming level ground, no air resistance, and s_o small enough that during its flight the object may be regarded as subject only to constant downward acceleration owing to gravity, we can find the equation of the trajectory (that is, path) of the object. Choosing the firing point as origin, the y-axis vertical, and the x-axis horizontal in the plane of motion, let (x, y) denote the position of the object at time $t > 0$.

The motion of the object can be regarded as resulting from two independent motions, one horizontal and the other vertical. Since the object experiences no horizontal forces during its flight, its horizontal acceleration is 0 (Newton's law of motion). Hence,

$$\frac{d^2x}{dt^2} = 0.$$

Thus the horizontal velocity is

$$\frac{dx}{dt} = C_1 \qquad \text{(a constant)}.$$

At $t = 0$ the horizontal velocity is $s_o \cos \alpha$, (See Fig. 3.18.) so $dx/dt = s_o \cos \alpha$. It follows that

$$x = \int s_o \cos \alpha \, dt = (s_o \cos \alpha)t + C_2.$$

$x = 0$ at time $t = 0$; therefore, $C_2 = 0$ and $x = (s_o \cos \alpha)t$.

During its flight the object experiences a constant downward acceleration owing to gravity, say

$$\frac{d^2y}{dt^2} = -g.$$

Thus

$$\begin{aligned}\frac{dy}{dt} &= -gt + C_3 \\ &= -gt + s_o \sin \alpha,\end{aligned}$$

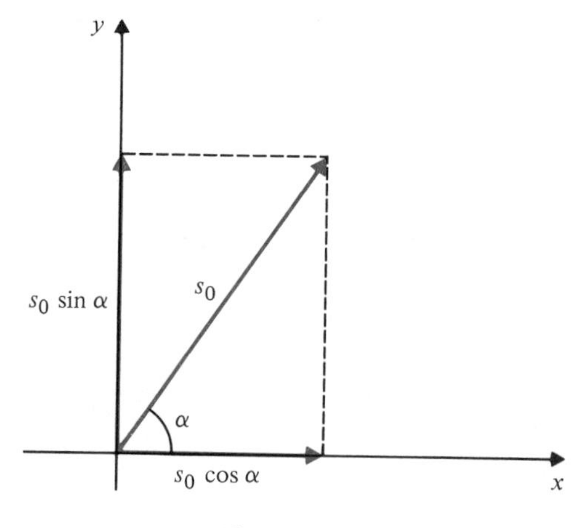

Figure 3.18

the vertical velocity being $s_o \sin \alpha$ at time $t = 0$. Finally,

$$y = -\frac{1}{2}gt^2 + (s_o \sin \alpha)t + C_4$$

$$= -\frac{1}{2}gt^2 + (s_o \sin \alpha)t$$

because $y = 0$ when $t = 0$.

The equations

$$x = (s_o \cos \alpha)t \qquad \text{and}$$

$$y = -\frac{1}{2}gt^2 + (s_o \sin \alpha)t$$

are called **parametric equations** of the trajectory of the object. The time t is called the **parameter.** The parametric equations give the position of the object at time t. The ordinary Cartesian equation of the trajectory can be found by eliminating the parameter between this pair of equations. Solving the first equation for $t = x/(s_o \cos \alpha)$ and substituting into the second equation we get

$$y = -\frac{gx^2}{2s_o^2 \cos^2 \alpha} + (\tan \alpha)x.$$

Because this equation is of the form $y = Ax^2 + Bx$ for certain constants A and B, the trajectory is a parabola.

We can use the parametric equations to calculate the maximum height, the time of flight, and the range (horizontal distance traveled by the projectile before impact).

Clearly the maximum height is attained when the vertical velocity is 0, that is, at time $t = (s_o \sin \alpha)/g$. Thus the maximum height is

$$y_{\max} = -\frac{g}{2}\left(\frac{s_o \sin \alpha}{g}\right)^2 + (s_o \sin \alpha)\frac{s_o \sin \alpha}{g}$$

$$= \frac{1}{2g} s_o^2 \sin^2 \alpha.$$

Since the projectile leaves ground level at time $t = 0$, the time of flight (that is, the elapsed time until impact with the ground) is the positive time t for which $y = 0$:

$$-\frac{1}{2}gt^2 + (s_o \sin \alpha)t = 0$$

$$t\left(s_o \sin \alpha - \frac{g}{2}t\right) = 0.$$

Thus the time of flight is $t_{\max} = (2s_o \sin \alpha)/g$. (Why do we ignore the root $t = 0$?) The range is the value of x at $t = t_{\max}$. Thus the range is

$$x_{\max} = (s_o \cos \alpha)\frac{2s_o \sin \alpha}{g} = \frac{s_o^2}{g} \sin 2\alpha.$$

For given initial speed s_o, the range x_{max} will be maximum if $\sin 2\alpha = 1$, that is, if $\alpha = \frac{\pi}{4} = 45°$. If the projectile is a shell being fired from a gun, this maximum range, s_o^2/g, increases with the square of the muzzle speed s_o.

Simple Harmonic Motion

If often happens in natural situations that a quantity displaced from an equilibrium value experiences a restoring force tending to move it back in the direction of its equilibrium. Besides the obvious examples of elastic motions in physics, one can imagine such a model applying, say, to a biological population in equilibrium with its food supply, or the price of a commodity in an elastic economy where increasing price causes increasing supply, which in turn causes decreasing demand and hence decreasing price.

In the simplest models, the restoring force is proportional to the amount of displacement from equilibrium. Such a force causes the quantity to oscillate sinusoidally, that is, to execute *simple harmonic motion*.

As a specific example, suppose a mass m is suspended by an elastic spring so that it hangs unmoving in its equilibrium position. If it is displaced an amount y from this position, a force $-ky$ $(k > 0)$ is exerted by the spring (Hooke's law), tending to restore the mass to its equilibrium position (see Fig. 3.19). Assuming the spring to be weightless, this force produces an acceleration

$$\frac{d^2y}{dt^2} = y''$$

for the mass m, given by $my'' = -ky$ (mass $\times$ acceleration $=$ force), or

$$y'' + \omega^2 y = 0, \qquad \omega^2 = \frac{k}{m}.$$

The second-order differential equation $y'' + \omega^2 y = 0$ is called the equation of simple harmonic motion. One can readily check by differentiation that if A, B, R, and t_o are any constants, then either of the functions

$$\begin{aligned} y &= A \cos \omega t + B \sin \omega t, \\ y &= R \sin (\omega(t - t_o)) \end{aligned}$$

is a general solution of the differential equation. These two are not really different functions; they are just different ways of writing the same function. If we expand the second version using the addition formula we get

$$\begin{aligned} y &= R \sin \omega t \cos \omega t_o - R \cos \omega t \sin \omega t_o \\ &= A \cos \omega t + B \sin \omega t, \end{aligned}$$

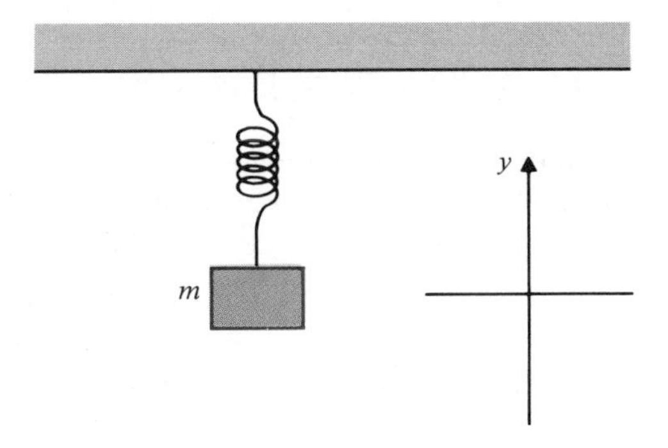

Figure 3.19

where $A = -R \sin \omega t_o$ and $B = R \cos \omega t_o$, or, equivalently, $R^2 = A^2 + B^2$ and $\tan \omega t_o = -A/B$.

The constants A and B are related to the position y_o and the velocity v_o of the mass m at time $t = 0$:

$$y_o = y(0) = A \cos 0 + B \sin 0 = A,$$
$$v_o = y'(0) = -A\omega \sin 0 + B\omega \cos 0 = B\omega,$$

$$y = y_o \cos \omega t + \frac{v_o}{\omega} \sin \omega t.$$

The constant $R = \sqrt{A^2 + B^2}$ is called the **amplitude** of the motion. Because $\sin x$ oscillates between -1 and 1, the displacement y varies between $-R$ and R. Note in Fig. 3.20 that the graph of the displacement as a function of time is that of a sine curve $y = \sin \omega t$ displaced t_o units to the right. The period of this curve is $T = 2\pi/\omega$, and we call $\omega = 2\pi/T$ the **circular frequency.** It is measured in radians

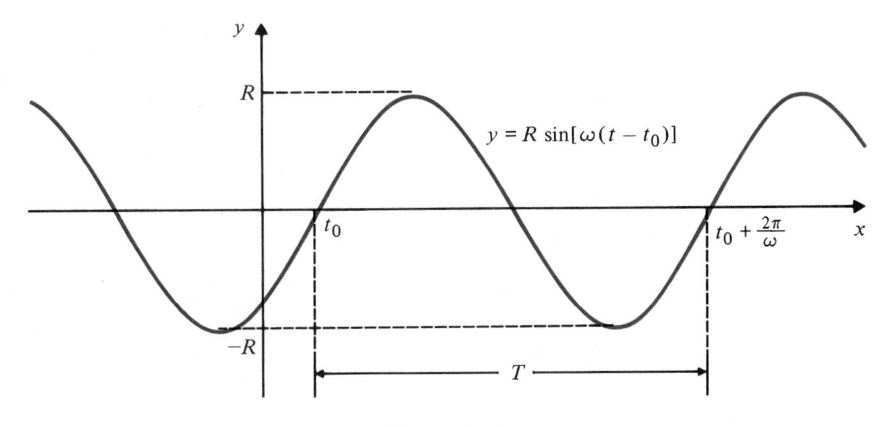

Figure 3.20

per second. The **frequency** of the motion, measured in Hertz (cycles/sec) is $\omega/2\pi = 1/T$, because 1 cycle = 1 revolution = 2π radians.

Example 5 Suppose that a 100-gm mass is suspended from a spring and that a force of 3×10^4 dynes (3×10^4 gm cm/sec^2) is required to produce a displacement from equilibrium of 1/3 cm. At time $t = 0$ the mass is pulled down 2 cm below equilibrium and released. Find its subsequent displacement at any time $t > 0$ and find the frequency of the motion.

Solution The spring constant k is determined from Hooke's law, $F = ky$:

$$3 \times 10^4 = \frac{1}{3}k,$$

so $k = 9 \times 10^4$ gm/sec^2. Hence the circular frequency is $\omega = \sqrt{k/m} = 30$ radians per second, and the frequency of motion is $\omega/2\pi = 15/\pi$ Hz.

Since the displacement at time $t = 0$ is $y_o = -2$ and the velocity at that time is $v_o = 0$ (the mass is released, not thrown), the subsequent displacement is

$$y = -2 \cos 30t \qquad (y \text{ in cm, } t \text{ in sec}).$$

EXERCISES

Differentiate the functions given in Exercises 1–30.

1. $y = \cos 3x$

2. $y = \sin 2x$

3. $y = \tan \pi x$

4. $y = \sin \dfrac{x}{5}$

5. $y = \cos \dfrac{2x}{\pi}$

6. $y = \sec ax$

7. $y = \cot (4 - 3x)$

8. $y = \sin (x^2 - 4x - 1)$

9. $y = \csc \dfrac{3}{x}$

10. $y = \sin (Ax + B)$

11. $f(x) = \cos (s - rx)$

12. $y = \tan (1 - x^2)$

13. $F(t) = \sin at \cos at$

14. $G(\theta) = \dfrac{\sin a\theta}{\cos b\theta}$

15. $u = \sec (x^2 - 4x)$

16. $f(t) = \sin^3\left(\cos \dfrac{t}{3}\right)$

17. $y = \dfrac{\tan^2 x}{x}$

18. $y = \cos (\sqrt{1 + \sin^2 x})$

19. $y = x \sin \dfrac{1}{x}$

20. $y = (1 - x)^2 \cos \dfrac{1}{x}$

21. $f(\theta) = \sqrt{\cos 2\theta}$

22. $y = \cos \sqrt{1 + x^4}$

23. $y = \sin t \cos 2t \tan 3t$

24. $y = \dfrac{\sin x + \cos x}{\cos x - \sin x}$

25. $f(t) = \sqrt{\dfrac{1 - \cos t}{1 + \cos t}}$

26. $u = \dfrac{\sin^2 x}{x^2}$

27. $y = (x^2 + 2)\cos(x^2 + 1)$

28. $G(t) = \sin\sqrt{|t|}$

29. $y = \cos^4\left(\frac{x-1}{2}\right) - \sin^4\left(\frac{x-1}{2}\right)$

30. $y = \sin^3(1 + \cos^4(2 + \tan 5x))$

31. Find an equation of the line tangent to the curve $y = \sin(x°)$ at the point where $x = 45°$.

32. Find an equation of the straight line normal to $y = \sec(x°)$ at the point where $x = 60°$.

33. Find the derivative of $f(x) = |\sin x|$ and sketch the graph of f.

34. If $z = \tan\frac{x}{2}$ show that $\frac{dx}{dz} = \frac{2}{1+z^2}$, $\sin x = \frac{2z}{1+z^2}$, and $\cos x = \frac{1-z^2}{1+z^2}$.

In Exercises 35–38 find an equation of the line tangent to the given curve at the given point.

35. $2x+y - \sqrt{2}\sin(xy) = \pi/2$ at $\left(\frac{\pi}{4}, 1\right)$

36. $\tan(xy^2) = \frac{2xy}{\pi}$ at $\left(-\pi, \frac{1}{2}\right)$

37. $x\sin(xy - y) = \frac{x-y}{y}$ at $(1, 1)$

38. $\cos\left(\frac{\pi y}{x}\right) = \frac{x^2}{y} - \frac{17}{2}$ at $(3, 1)$

39. If $y = \tan kx$, show that $y'' = 2k^2y(1 + y^2)$.

40. If $y = \sec kx$, show that $y'' = k^2y(2y^2 - 1)$.

41. Given that $\sin 2x = 2\sin x\cos x$, deduce that $\cos 2x = \cos^2 x - \sin^2 x$.

42. Given that $\cos 2x = \cos^2 x - \sin^2 x$, deduce that $\sin 2x = 2\sin x\cos x$.

43. Find a formula for the nth derivative of $\sin ax$.

44. Find a formula for the nth derivative of $\cos ax$.

45.* Find a formula for the nth derivative of $x\sin x$.

46.* Find a formula for the nth derivative of $x^2\cos x$.

Exercises 47–51 all refer to the differential equation

$$y'' + \omega^2 y = 0.$$

Together they show that $y = A\cos\omega t + B\sin\omega t$ is the *general solution* of this equation, that is, that every solution is of this form for some choice of the constants A and B.

47.† Show that if $y = f(x)$ and $y = g(x)$ are solutions of the differential equation, then so is $y = Af(x) + Bg(x)$ for any constants A and B.

48.† Show that $y = A\cos\omega x + B\sin\omega x$ is a solution.

49.† If $f(x)$ is any solution, show that $\omega^2(f(x))^2 + (f'(x))^2$ is constant.

50.† If $g(x)$ is a solution satisfying $g(0) = g'(0) = 0$, show that $g(x)$ is identically 0.

51.† Suppose that $f(x)$ is any solution of the differential equation. Show that $f(x) = A\cos\omega x + B\sin\omega x$, where $A = f(0)$ and $B\omega = f'(0)$. (*Hint:* Let $g(x) = f(x) - A\cos\omega x - B\sin\omega x$.)

† Problems relating to differential equations.

In Exercises 52–56 solve the given initial-value problems.

52.† $\begin{cases} y'' + 4y = 0 \\ y(0) = 2 \\ y'(0) = -5 \end{cases}$

53.† $\begin{cases} y'' + 100y = 0 \\ y(0) = 0 \\ y'(0) = 3 \end{cases}$

54.† $\begin{cases} y'' + \dfrac{1}{T^2}y = 0 \\ y(0) = A \\ y'(0) = 0 \end{cases}$

55.† $\begin{cases} y'' + y = 0 \\ y(2) = 3 \\ y'(2) = 0 \end{cases}$

56.† $\begin{cases} y'' + \omega^2 y = 0 \\ y(a) = A \\ y'(a) = B \end{cases}$

57. The acceleration owing to gravity is approximately 9.8 m/sec². Find the muzzle speed of a shell fired at an angle 30° above the horizontal if the shell strikes the ground at a point 2 km horizontally away from its firing point.

58. What mass should be suspended from the spring in Example 5 to provide a system whose natural frequency of oscillation is 10 Hz? Find the displacement of such a mass from its equilibrium position t sec after it is pulled down 1 cm from equilibrium and flicked upward with a speed of 2 cm/sec. What is the amplitude of this motion?

59. A mass of 400 gm suspended from a certain elastic spring will oscillate with a frequency of 24 Hz. What would be the frequency if the 400-gm mass were replaced with a 900-gm mass? A 100-gm mass?

3.3 THE INVERSE TRIGONOMETRIC FUNCTIONS

Defined on the whole real line, the trigonometric functions are not one-to-one and hence are not invertible. We can, however, suitably restrict their domains so that they become one-to-one. The inverse trigonometric functions are the inverses of these functions with restricted domains.

Definition 2

The **principal-value sine function,** denoted *Sine* (or *Sin*), is the restriction of sine to the interval $\left[-\dfrac{\pi}{2}, \dfrac{\pi}{2}\right]$.

$$\operatorname{Sin} x = \sin x, \qquad -\frac{\pi}{2} \le x \le \frac{\pi}{2}$$

Being increasing, Sine is one-to-one and has an inverse function called **Arcsine.** Arcsin x is sometimes also denoted $\sin^{-1} x$, although we will not use that notation in this book.

$$y = \operatorname{Arcsin} x \Leftrightarrow x = \operatorname{Sin} y$$

Note in Fig. 3.21 that

$$\text{domain of Arcsin} = [-1, 1] = \text{range of Sin},$$

$$\text{range of Arcsin} = \left[-\frac{\pi}{2}, \frac{\pi}{2}\right] = \text{domain of Sin}.$$

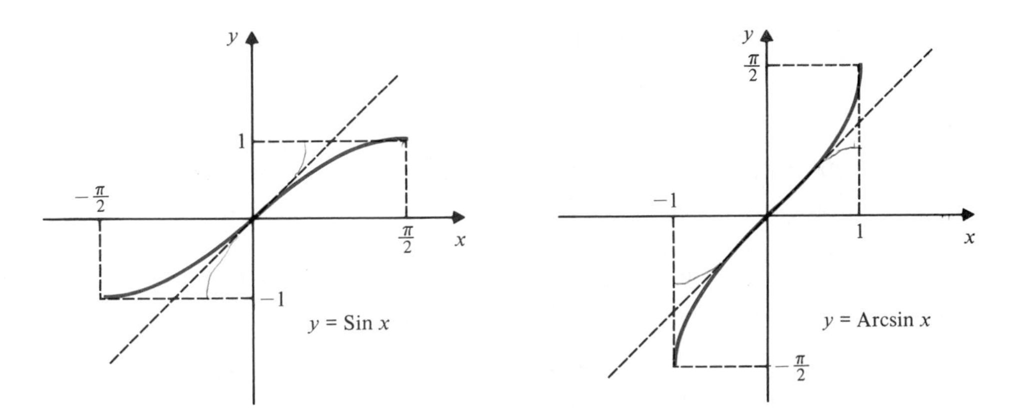

Figure 3.21

One should think of Arcsin x as "the angle whose Sine is x." Thus Arcsin $(1/2) = \pi/6$ and Arcsin $(-1) = -\pi/2$. The graph of $y =$ Arcsin x (or, equivalently, of $x =$ Sin y) is the mirror image of the graph of $y =$ Sin x (or, equivalently, $x =$ Arcsin y) in the line $y = x$.

We calculate the derivative of Arcsin by implicit differentiation:

$$y = \text{Arcsin}\, x \Leftrightarrow x = \text{Sin}\, y$$

$$\Leftrightarrow x = \sin y, \qquad -\frac{\pi}{2} \le y \le \frac{\pi}{2}$$

Thus

$$1 = \cos y \frac{dy}{dx}.$$

Since $\cos y$ is nonnegative in the interval $\left[-\frac{\pi}{2}, \frac{\pi}{2}\right]$, $\cos y = \sqrt{1 - \sin^2 y} = \sqrt{1 - x^2}$. Hence,

$$\frac{d}{dx} \text{Arcsin}\, x = \frac{1}{\sqrt{1 - x^2}}, \qquad -1 < x < 1.$$

Example 1 Find an equation of the straight line tangent to the curve $y = \text{Arcsin}\left(\frac{x}{2}\right)$ at the point $x = -1$.

Solution If $x = -1$, then $y = \text{Arcsin}\left(-\frac{1}{2}\right) = -\frac{\pi}{6}$. The slope of the tangent is

$$\frac{d}{dx} \text{Arcsin}\frac{x}{2}\bigg|_{x=-1} = \left(1 - \frac{x^2}{4}\right)^{-1/2}\left(\frac{1}{2}\right)\bigg|_{x=-1} = \frac{1}{\sqrt{3}}.$$

Thus the tangent has equation $y = -(\pi/6) + (1/\sqrt{3})(x + 1)$.

Example 2 Simplify the following expressions.

a) tan Arcsin x

b) Arcsin sin x

Solution a) If t = Arcsin x, then x = Sin t = sin t because t lies in the interval $\left[-\frac{\pi}{2}, \frac{\pi}{2}\right]$.
Since $\cos t \geq 0$ in this interval, we have, using the Pythagorean identity, $\cos t = \sqrt{1 - \sin^2 t} = \sqrt{1 - x^2}$. Hence

$$\tan \text{Arcsin } x = \tan t = \frac{\sin t}{\cos t} = \frac{x}{\sqrt{1 - x^2}} \qquad (-1 < x < 1).$$

This result can also be seen geometrically, as in Fig. 3.22. Label the sides of a right-angled triangle so that the sine of one of its acute angles, t, is x; for example, let the side opposite t be x and the hypotenuse 1. Then the side adjacent is $\sqrt{1 - x^2}$ and the value for tan Arcsin x = tan t follows by inspection.

b) We sketch the graph of $y = \sin x$ in Fig. 3.23 and color the part of the graph that is the graph of $y = \text{Sin } x$. Any real number x lies either in an interval of the form $[(2k - 1/2)\pi, (2k + 1/2)\pi]$ or one of the form $((2k + 1/2)\pi, (2k + 3/2)\pi)$ for some integer k. If $(2k - 1/2)\pi \leq x \leq (2k + 1/2)\pi$, then $x - 2k\pi$ lies in the interval $\left[-\frac{\pi}{2}, \frac{\pi}{2}\right]$. Since sine is periodic with period 2π, we have in this case

$$\text{Arcsin } \sin x = \text{Arcsin } \sin (x - 2k\pi) = \text{Arcsin Sin } (x - 2k\pi) = x - 2k\pi.$$

If $(2k + 1/2)\pi < x < (2k + 3/2)\pi$, then $(2k + 1)\pi - x$ lies in $(-\pi/2, \pi/2)$ and

$$\begin{aligned}
\text{Arcsin } \sin x &= \text{Arcsin } \sin (x - 2k\pi) && \text{(by periodicity)} \\
&= \text{Arcsin } \sin (\pi - (x - 2k\pi)) && \text{(supplementary angles)} \\
&= \text{Arcsin } \sin ((2k + 1)\pi - x) \\
&= \text{Arcsin Sin } ((2k + 1)\pi - x) = (2k + 1)\pi - x.
\end{aligned}$$

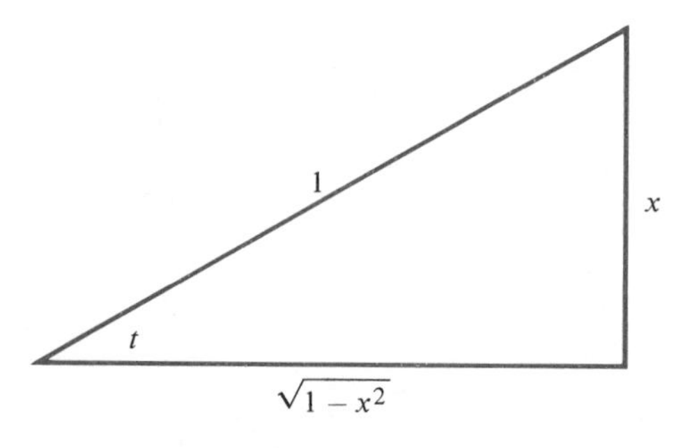

Figure 3.22

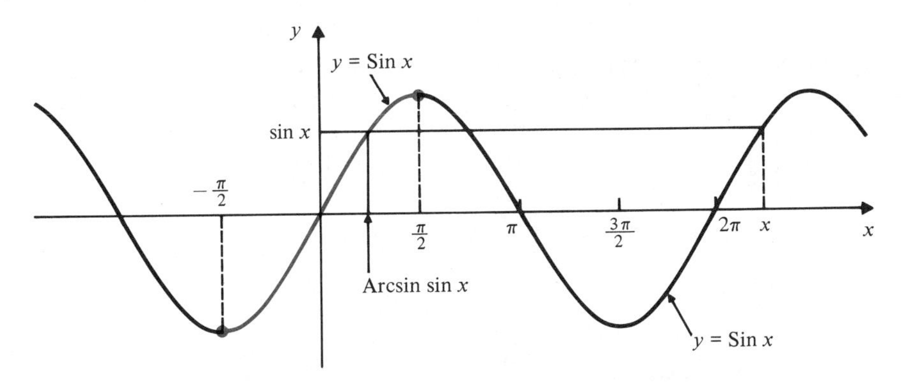

Figure 3.23

Thus for any integer k,

$$\text{Arcsin}\sin x = \begin{cases} x - 2k\pi & \text{if } \left(2k - \frac{1}{2}\right)\pi \le x \le \left(2k + \frac{1}{2}\right)\pi, \\ (2k + 1)\pi - x & \text{if } \left(2k + \frac{1}{2}\right)\pi < x < \left(2k + \frac{3}{2}\right)\pi. \end{cases}$$

For purposes of antidifferentiation, observe that

$$\frac{d}{dx}\text{Arcsin}\frac{x}{a} = \frac{1}{\sqrt{1 - \frac{x^2}{a^2}}}\frac{1}{a} = \frac{1}{\sqrt{a^2 - x^2}} \qquad (a > 0).$$

Hence,

$$\int \frac{dx}{\sqrt{a^2 - x^2}} = \text{Arcsin}\frac{x}{a} + C \qquad (a > 0).$$

Example 3 Find the solution y of the following initial-value problem.

$$\begin{cases} y' = \frac{4}{\sqrt{2 - x^2}} & (-\sqrt{2} < x < \sqrt{2}) \\ y(1) = 2\pi \end{cases}$$

Solution We have

$$y = 4\int \frac{dx}{\sqrt{2 - x^2}} = 4\,\text{Arcsin}\frac{x}{\sqrt{2}} + C$$

for some constant C. Also $2\pi = y(1) = 4\,\text{Arcsin}\,(1/\sqrt{2}) + C = 4\left(\frac{\pi}{4}\right) + C = \pi + C$. Thus $C = \pi$ and $y = 4\,\text{Arcsin}\,(x/\sqrt{2}) + \pi$.

Definition 3

The **principal-value tangent function,** denoted *Tangent* (or *Tan*), is the restriction of tangent to the interval $(-\pi/2, \pi/2)$, that is,

$$\text{Tan } x = \tan x, \qquad -\frac{\pi}{2} < x < \frac{\pi}{2}.$$

Tangent is increasing and has an inverse called **Arctangent** (or $\tan^{-1}$). See Fig. 3.24.

$$y = \text{Arctan } x \Leftrightarrow x = \text{Tan } y$$

$$\text{domain of Arctan} = (-\infty, \infty) = \text{range of Tan}$$
$$\text{range of Arctan} = (-\pi/2, \pi/2) = \text{domain of Tan}$$

$$y = \text{Arctan } x \Leftrightarrow x = \text{Tan } y$$
$$\Leftrightarrow x = \tan y, \qquad -\frac{\pi}{2} < y < \frac{\pi}{2}$$
$$1 = \sec^2 y \frac{dy}{dx} = (1 + \tan^2 y)\frac{dy}{dx} = (1 + x^2)\frac{dy}{dx}$$

Thus

$$\frac{d}{dx} \text{Arctan } x = \frac{1}{1 + x^2}$$

Note that

$$\frac{d}{dx} \text{Arctan } \frac{x}{a} = \frac{1}{1 + \dfrac{x^2}{a^2}} \frac{1}{a} = \frac{a}{a^2 + x^2}$$

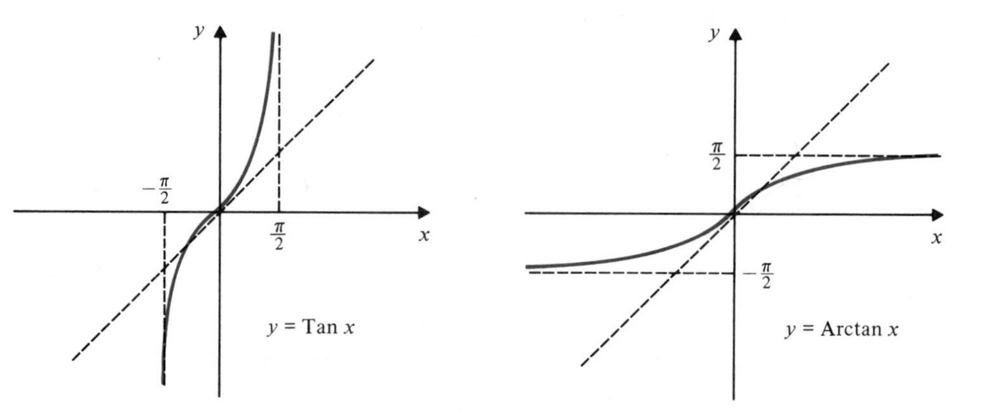

Figure 3.24

and hence

$$\int \frac{dx}{a^2 + x^2} = \frac{1}{a} \text{ Arctan } \frac{x}{a} + C.$$

Example 4 Evaluate

a) $\text{Arctan tan } \frac{\pi}{4}$

b) $\text{Arctan tan } \frac{3\pi}{4}$.

What is the value of Arctan tan x when x is not an odd multiple of $\pi/2$?

Solution

$$\text{Arctan tan } \frac{\pi}{4} = \text{Arctan } 1 = \frac{\pi}{4}$$

$$\text{Arctan tan } \frac{3\pi}{4} = \text{Arctan } (-1) = -\frac{\pi}{4}$$

If $y = \text{Arctan tan } x$, then $\text{Tan } y = \tan x = \tan(x + k\pi)$ for any integer k, since tangent is periodic with period π. But $\tan(x + k\pi) = \text{Tan}(x + k\pi)$ provided $-\pi/2 < x + k\pi < \pi/2$. Hence $\text{Arctan tan } x = x + k\pi$ where the integer k is chosen so that $-\pi/2 - k\pi < x < \pi/2 - k\pi$.

Example 5 Prove that

$$\text{Arctan } \frac{x-1}{x+1} = \text{Arctan } x - \frac{\pi}{4} \qquad \text{for } x > -1.$$

Solution Let $f(x) = \text{Arctan } \dfrac{x-1}{x+1} - \text{Arctan } x$. On the interval $(-1, \infty)$ we have

$$f'(x) = \frac{1}{1 + \left(\dfrac{x-1}{x+1}\right)^2} \frac{(x+1) - (x-1)}{(x+1)^2} - \frac{1}{1+x^2}$$

$$= \frac{2}{2 + 2x^2} - \frac{1}{1+x^2} = 0.$$

Hence $f(x) = C$ (constant) on that interval. We can find C by finding $f(0)$:

$$C = f(0) = \text{Arctan } (-1) - \text{Arctan } 0 = -\frac{\pi}{4}.$$

Hence the given identity holds on $(-1, \infty)$.

Definition 4

The **principal-value cosine function,** denoted *Cosine* (or *Cos*), is the restriction of cosine to the interval $[0, \pi]$.

$$\text{Cos}\, x = \cos x, \qquad 0 \leq x \leq \pi.$$

So defined, Cos is decreasing and has an inverse function, **Arccosine** (or $\cos^{-1}$). See Fig. 3.25.

$$y = \text{Arccos}\, x \Leftrightarrow x = \text{Cos}\, y$$

$$\begin{aligned} \text{domain of Arccos} &= [-1, 1] = \text{range of Cos} \\ \text{range of Arccos} &= [0, \pi] = \text{domain of Cos} \end{aligned}$$

By implicit differentiation, the derivative of Arccos is readily shown to be

$$\frac{d}{dx} \text{Arccos}\, x = \frac{-1}{\sqrt{1 - x^2}}, \qquad -1 < x < 1.$$

Note that for $-1 < x < 1$,

$$\frac{d}{dx}(\text{Arcsin}\, x + \text{Arccos}\, x) = \frac{1}{\sqrt{1 - x^2}} - \frac{1}{\sqrt{1 - x^2}} = 0.$$

Hence $\text{Arcsin}\, x + \text{Arccos}\, x = C$ (constant) on that interval. Since $\text{Arcsin}\, 0 = 0$ and $\text{Arccos}\, 0 = \pi/2$,

$$\text{Arcsin}\, x + \text{Arccos}\, x = \frac{\pi}{2}.$$

This identity also follows from the fact that $\text{Sin}\, x = \text{Cos}\,(\pi/2 - x)$ if $-\pi/2 \leq x \leq \pi/2$.

The remaining inverse trigonometric functions, Arcsec, Arccsc, and Arccot are of lesser importance as they can readily be expressed in terms of Arcsin, Arccos, and Arctan. For instance, defining $\text{Sec}\, x = 1/\text{Cos}\, x$ on $[0, \pi/2) \cup (\pi/2, \pi]$, and Arcsec as the inverse of Sec, we have

$$\begin{aligned} y = \text{Arcsec}\, x &\Leftrightarrow x = \text{Sec}\, y \\ &\Leftrightarrow \frac{1}{x} = \text{Cos}\, y \\ &\Leftrightarrow y = \text{Arccos}\, \frac{1}{x}. \end{aligned}$$

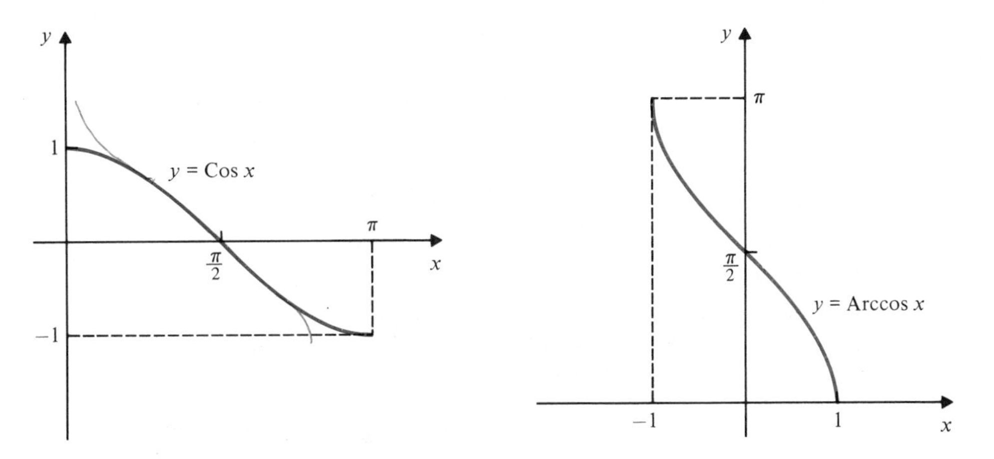

Figure 3.25

$$\text{Arcsec } x = \text{Arccos } \frac{1}{x}, \qquad x \leq -1 \text{ or } x \geq 1$$

See Fig. 3.26.

It follows that

$$\frac{d}{dx} \text{Arcsec } x = \frac{1}{|x|\sqrt{x^2 - 1}}, \qquad x < -1 \text{ or } x > 1.$$

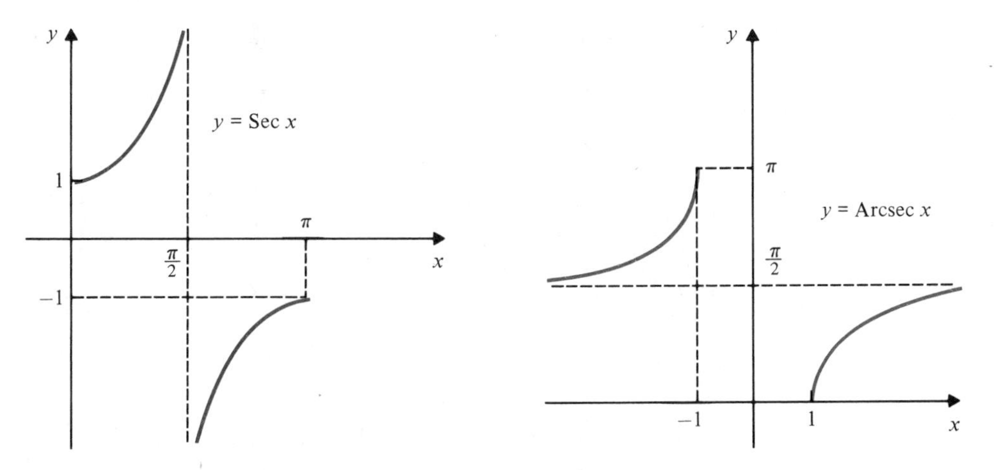

Figure 3.26

In a similar manner we can deal with

$$\text{Arccsc}\, x = \text{Arcsin}\, \frac{1}{x} \qquad \text{and}$$

$$\text{Arccot}\, x = \text{Arctan}\, \frac{1}{x}.$$

EXERCISES

In Exercises 1–9 evaluate the given expression.

1. $\text{Arcsin}\, \frac{\sqrt{3}}{2}$ **2.** $\text{Arccos}\left(\frac{-1}{2}\right)$ **3.** $\text{Arctan}\,(-1)$

4. $\text{Arcsec}\, \sqrt{2}$ **5.** $\text{Arctan}\, \tan \frac{2\pi}{3}$ **6.** $\sin \text{Arcsin}\, 0.71$

7. $\sin \text{Arccos}\, \frac{-1}{3}$ **8.** $\cos \text{Arctan}\, \frac{1}{2}$ **9.** $\tan \text{Arctan}\, 200$

In Exercises 10–20 simplify the given expression.

10. $\sin \text{Arcsin}\, x$ **11.** $\cos \text{Arccos}\, x$ **12.** $\tan \text{Arctan}\, x$

13. $\sin \text{Arccos}\, x$ **14.** $\cos \text{Arcsin}\, x$ **15.** $\cos \text{Arctan}\, x$

16. $\sin \text{Arctan}\, x$ **17.** $\tan \text{Arccos}\, x$ **18.** $\tan \text{Arcsec}\, x$

19.* $\text{Arccos} \cos x$ **20.*** $\text{Arcsin} \cos x$

21.* Sketch the graph of $y = \sin \text{Arcsin}\, x$ and the graph of $y = \text{Arcsin} \sin x$. What is the derivative of each of these functions?

22.* Sketch the graphs of $y = \tan \text{Arctan}\, x$ and $y = \text{Arctan} \tan x$ and find the derivative of each function.

In Exercises 23–38 differentiate the given function and simplify the answer whenever appropriate.

23. $y = \text{Arcsin}\, \frac{2x - 1}{3}$ **24.** $y = \text{Arctan}\,(ax + b)$

25. $y = \text{Arccos}\, \frac{x - b}{a}$ **26.** $f(x) = x\, \text{Arcsin}\, x$

27. $f(t) = t\, \text{Arctan}\, t$ **28.** $u = z^2\, \text{Arcsec}\,(1 + z^2)$

29. $F(x) = (1 + x^2)\, \text{Arctan}\, x$ **30.** $y = \text{Arcsin}\, \frac{a}{x}$

31. $G(x) = \frac{\text{Arcsin}\, x}{\text{Arcsin}\, 2x}$ **32.** $H(t) = \frac{\text{Arcsin}\, t}{\sin t}$

33. $f(x) = (\text{Arcsin}\, x^2)^{1/2}$ **34.** $y = \text{Arccos}\, \frac{a}{\sqrt{a^2 + x^2}}$

35. $y = \frac{\text{Arctan}\, \sqrt{x - 1}}{\sqrt{x - 1}}$ **36.** $y = \sqrt{a^2 - x^2} + a\, \text{Arcsin}\, \frac{x}{a} \quad (a > 0)$

37. $y = a\, \text{Arccos}\left(1 - \frac{x}{a}\right) - \sqrt{2ax - x^2} \quad (a > 0)$

38. $y = \sqrt{x} - \text{Arctan}\, \sqrt{x}$

39.* Show that the function $f(x)$ of Example 5 is also constant on the interval $(-\infty, -1)$. Find the value of the constant. (*Hint:* Find $\lim_{x\to-\infty} f(x)$.)

40.* Find the derivative of $f(x) = x - \text{Arctan}\tan x$. What does your answer imply about $f(x)$? Calculate $f(0)$ and $f(\pi)$. Is there a contradiction here?

41.* Define, appropriately, Csc x and Arccsc x and sketch their graphs. Find the derivative of Arccsc x.

42.* Define, appropriately, Cot x and Arccot x and sketch their graphs. Find the derivative of Arccot x.

43.* Show that Arcsin and Arctan are increasing functions and that Arccos is a decreasing function.

44.* The derivative of Arcsec x is positive for every x in the domain of Arcsec. Does this imply that Arcsec is increasing on its domain? Why?

In Exercises 45–48 solve the given initial-value problems.

45.† $\begin{cases} y' = \dfrac{1}{1+x^2} \\ y(0) = 1 \end{cases}$

46.† $\begin{cases} y' = \dfrac{1}{9+x^2} \\ y(3) = 2 \end{cases}$

47.† $\begin{cases} y' = \dfrac{1}{\sqrt{1-x^2}} \\ y(1/2) = 1 \end{cases}$

48.† $\begin{cases} y' = \dfrac{4}{\sqrt{25-x^2}} \\ y(0) = 0 \end{cases}$

3.4 THE NATURAL LOGARITHM

Consider the following table of derivatives of integer powers of x:

$f(x)$	$f'(x)$
$\vdots$	$\vdots$
x^3	$3x^2$
x^2	$2x$
$x^1 = x$	$1 = x^0$
$x^0 = 1$	0
$x^{-1} = \dfrac{1}{x}$	$-x^{-2} = \dfrac{-1}{x^2}$
$x^{-2} = \dfrac{1}{x^2}$	$-2x^{-3} = \dfrac{-2}{x^3}$
$\vdots$	$\vdots$

Every integer power of x appears somewhere in the left-hand column, but one power is conspicuously absent from the right-hand column, namely, $x^{-1} = 1/x$. It appears that we have not yet encountered any function whose derivative is $1/x$. We will remedy this situation by defining a new function, which we call ln x, in such a way that we will readily be able to show that it has derivative $1/x$.

You probably have encountered in your high school mathematics courses functions called logarithms. (Typically, we denote by $log_a\, x$—read "log of x to the base a"—

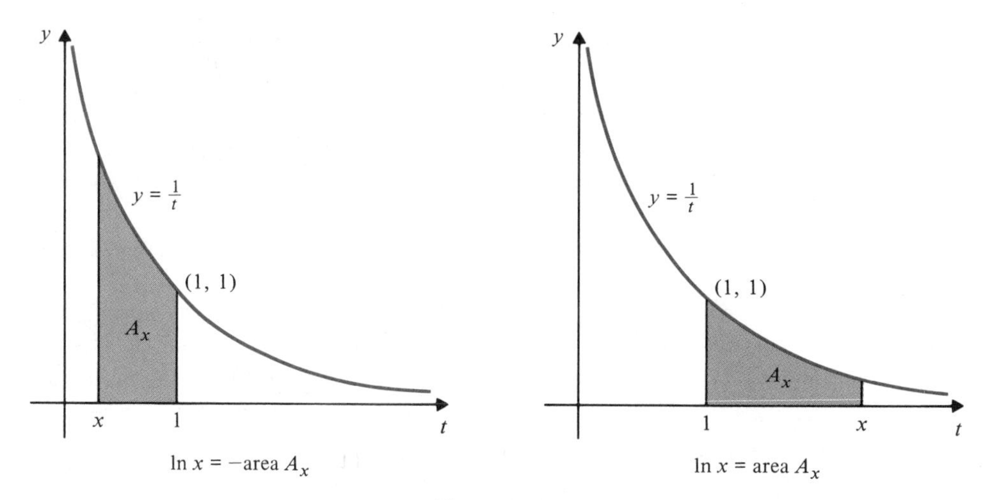

Figure 3.27

the number y that satisfies $a^y = x$.) On the surface, the following definition of $\ln x$ does not bear any resemblance to these logarithms you may have previously encountered. That $\ln x$ is, in fact, one of them will become obvious in the next section. Our definition is phrased in terms of area of a plane region. Area will be studied in detail in Chapter 6, but we assume you have an intuitive idea of what area means.

Definition 5

For $x > 0$, let A_x be the area of the plane region bounded by the curve $y = 1/t$, the t-axis, and the vertical lines $t = 1$ and $t = x$. The **natural logarithm of x,** $\ln x$ (pronounced ''lawn x''), is defined by

$$\ln x = \begin{cases} A_x & \text{if } x \geq 1, \\ -A_x & \text{if } 0 < x < 1, \end{cases}$$

as shown in Fig. 3.27.

Some books use the symbol $\log x$ to denote the natural logarithm. However, on most scientific calculators $\log x$ is used to denote $\log_{10} x$, a different function.

Observe that

$$\ln 1 = 0$$

and that

$$\begin{aligned} \ln x &> 0 \quad \text{if } x > 1, \\ \ln x &< 0 \quad \text{if } 0 < x < 1. \end{aligned}$$

Theorem 1

$$\frac{d}{dx} \ln x = \frac{1}{x} \qquad \text{if } x > 0$$

Proof For $x > 0$ and $h > 0$, $\ln (x + h) - \ln x$ is the area of the plane region bounded by $y = 1/t$, $y = 0$, and the vertical lines $t = x$ and $t = x + h$, that is, the shaded area in Fig. 3.28. Comparing this area with that of two rectangles, we see that

$$\frac{h}{x + h} < \text{shaded area} < \frac{h}{x}.$$

Hence the Newton quotient for $\ln x$ satisfies

$$\frac{1}{x + h} < \frac{\ln (x + h) - \ln x}{h} < \frac{1}{x}.$$

Letting h approach 0 from the right, we obtain (by the squeeze theorem, Theorem 3 of Section 1.3, but applied to one-sided limits)

$$\lim_{h \to 0+} \frac{\ln (x + h) - \ln x}{h} = \frac{1}{x}.$$

A similar argument shows that if $0 < x + h < x$, then

$$\frac{1}{x} < \frac{\ln (x + h) - \ln x}{h} < \frac{1}{x + h},$$

so that

$$\lim_{h \to 0-} \frac{\ln (x + h) - \ln x}{h} = \frac{1}{x}.$$

Combining these two one-sided limits we get the desired result:

$$\frac{d}{dx} \ln x = \lim_{h \to 0} \frac{\ln (x + h) - \ln x}{h} = \frac{1}{x}. \quad \square$$

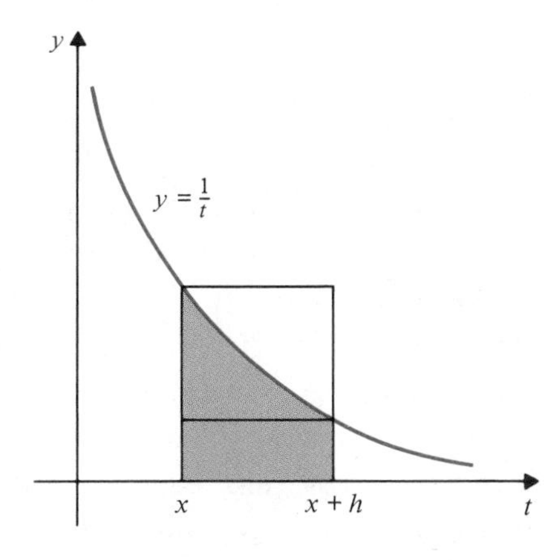

Figure 3.28

Note that although $\ln x$ is defined only for $x > 0$, it also provides an antiderivative for $1/x$ for $x < 0$, as follows:

$$\frac{d}{dx}\ln|x| = \frac{d}{dx}\ln(-x) = \frac{1}{-x}(-1) = \frac{1}{x} \quad (x < 0).$$

Hence

$$\frac{d}{dx}\ln|x| = \frac{1}{x} \qquad \text{for any } x \neq 0, \text{ and}$$

$$\int\frac{dx}{x} = \ln|x| + C \qquad (x \neq 0).$$

The two properties $d/dx \ln x = 1/x$ and $\ln 1 = 0$ are sufficient to characterize the function $\ln x$ completely. We shall deduce from these two properties that $\ln x$ has all the characteristics usually associated with logarithms.

Theorem 2 *Properties of the Natural Logarithm*

a) $\ln ax = \ln a + \ln x, \qquad (a > 0, x > 0)$

b) $\ln 1/x = -\ln x, \qquad (x > 0)$

c) $\ln a/x = \ln a - \ln x, \qquad (a > 0, x > 0)$

d) If $r = m/n$ is a rational number (m, n integers, $n \neq 0$), then

$$\ln x^r = r \ln x, \qquad (x > 0).$$

(This identity will also hold for arbitrary real r once we have defined what x^r means for such r. See Section 3.6.)

e) ln is increasing on $(0, \infty)$. Moreover,

$$\lim_{x\to 0+} \ln x = -\infty,$$

$$\lim_{x\to\infty} \ln x = \infty.$$

f) $\ln x \leq x - 1$ for $x > 0$

Proof a) $(d/dx)(\ln ax - \ln x) = (1/ax)(a) - 1/x = 0$ for $x > 0$. Therefore, $\ln ax - \ln x = C$ (constant). Substituting $x = 1$ gives $C = \ln a$.

b) $(d/dx)(\ln 1/x + \ln x) = 1/(1/x)(-1/x^2) + 1/x = 0$ for $x > 0$. Therefore, $\ln 1/x + \ln x = C$ (constant). Putting $x = 1$ we get $C = 0$.

c) $\ln a/x = \ln(a(1/x)) = \ln a + \ln 1/x = \ln a - \ln x$ by (a) and (b).

d) $(d/dx)(\ln x^r - r\ln x) = (1/x^r)(rx^{r-1}) - r/x = 0$ for $x > 0$. Hence $\ln x^r - r\ln x = C$ (constant). Putting $x = 1$ we get $C = 0$.

e) Since $d/dx \ln x = 1/x > 0$ on $(0, \infty)$, ln is increasing on that interval (Theorem 3 of Section 2.8). Since $\ln 2 > 0$, if $x > 2^n$ then $\ln x > \ln 2^n = n \ln 2$, which

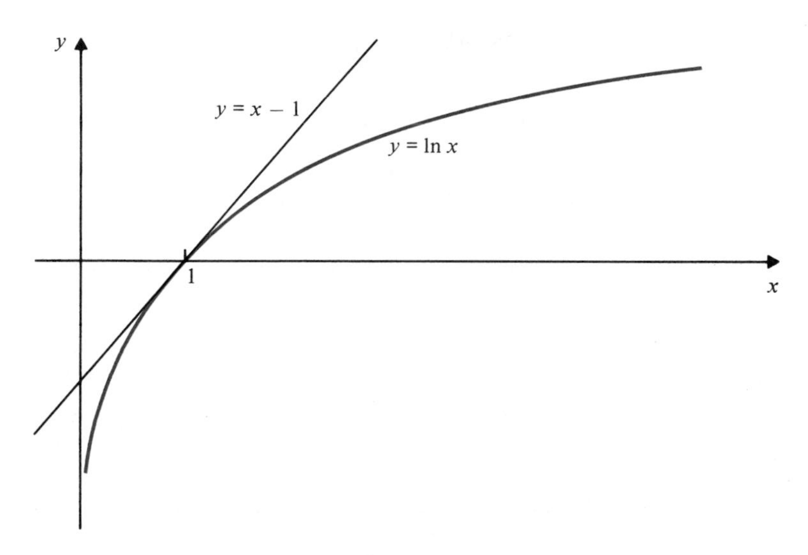

Figure 3.29

approaches infinity as n approaches infinity. Hence $\lim_{x\to\infty} \ln x = \infty$. Putting $x = 1/y$ we can now obtain $\lim_{x\to 0+} \ln x = \lim_{y\to\infty} \ln 1/y = -\lim_{y\to\infty} \ln y = -\infty$.

f) Let $g(x) = \ln x - (x - 1)$. Then $g'(x) = 1/x - 1 = (1 - x)/x$ for $x > 0$. Since $g'(x) > 0$ for $0 < x < 1$, g is increasing on $(0, 1)$. Since $g'(x) < 0$ for $1 < x < \infty$, g is decreasing there. Thus $g(x)$ has a maximum value when $x = 1$. Since $g(1) = 0$ we have $\ln x - (x - 1) \le 0$ for $x > 0$. Hence $\ln x \le x - 1$. Observe in Fig. 3.29 that $y = x - 1$ is tangent to the graph $y = \ln x$ at $(1, 0)$. □

Example 1 Simplify the following.

a) $4 \ln 2 + \ln 4 - 2 \ln 8$

b) $\ln (x^4 + 3x^2 + 2) + \ln (x^4 + 5x^2 + 6) - 4 \ln \sqrt{x^2 + 2}$.

Solution a)

$$\begin{aligned} 4 \ln 2 + \ln 4 - 2 \ln 8 &= \ln 2^4 + \ln 4 - \ln 8^2 \\ &= \ln 16 + \ln 4 - \ln 64 \\ &= \ln \frac{16 \times 4}{64} = \ln 1 = 0. \end{aligned}$$

b)

$$\begin{aligned} &\ln (x^4 + 3x^2 + 2) + \ln (x^4 + 5x^2 + 6) - 4 \ln \sqrt{x^2 + 2} \\ &= \ln \frac{(x^4 + 3x^2 + 2)(x^4 + 5x^2 + 6)}{(x^2 + 2)^2} \\ &= \ln \frac{(x^2 + 1)(x^2 + 2)(x^2 + 2)(x^2 + 3)}{(x^2 + 2)^2} \\ &= \ln (x^2 + 1)(x^2 + 3) = \ln (x^4 + 4x^2 + 3). \end{aligned}$$

Example 2 Find the following. Also restate your results in the form of indefinite integrals.

a) $\dfrac{d}{dx} \ln |\sec x|$

b) $\dfrac{d}{dx} \ln (x + \sqrt{x^2 + 1})$

Solution a) $\dfrac{d}{dx} \ln |\sec x| = 1/(\sec x) \sec x \tan x = \tan x$. Hence

$$\int \tan x \, dx = \ln |\sec x| + C.$$

b) $$\frac{d}{dx} \ln (x + \sqrt{x^2 + 1}) = \left(\frac{1}{x + \sqrt{x^2 + 1}}\right)\left(1 + \frac{2x}{2\sqrt{x^2 + 1}}\right)$$
$$= \left(\frac{1}{x + \sqrt{x^2 + 1}}\right)\left(\frac{\sqrt{x^2 + 1} + x}{\sqrt{x^2 + 1}}\right) = \frac{1}{\sqrt{x^2 + 1}}$$

Hence

$$\int \frac{dx}{\sqrt{x^2 + 1}} = \ln (x + \sqrt{x^2 + 1}) + C.$$

Example 3 Find the function $f(x)$ defined for $x < 0$ if $f(-2) = 5$ and $f'(x) = \dfrac{4}{x} - \dfrac{6}{x^2}$.

Solution We have

$$f(x) = 4\int \frac{dx}{x} - 6\int \frac{dx}{x^2} = 4 \ln |x| + \frac{6}{x} + C.$$

Substituting $x = -2$ we get

$$5 = f(-2) = 4 \ln |-2| + \frac{6}{-2} + C = 4 \ln 2 - 3 + C.$$

Hence $C = 8 - 4 \ln 2$ and

$$f(x) = 4 \ln |x| + \frac{6}{x} + 8 - 4 \ln 2$$
$$= 4 \ln \frac{|x|}{2} + \frac{6}{x} + 8 = 4 \ln \frac{-x}{2} + \frac{6}{x} + 8 \qquad (x < 0).$$

EXERCISES

In Exercises 1–6 simplify the given expressions.

1. $3 \ln 4 - 4 \ln 3$

2. $\ln (x + 1) + \ln (x - 1) - \ln (x^2 - 1)$

3. $2 \ln x + 5 \ln (x - 2)$

4. $4 \ln \sqrt{x} + 6 \ln (x^{1/3})$

5. $\ln (1 - \cos x) + \ln (1 + \cos x)$

6. $\ln (x^2 + 6x + 9)$

In Exercises 7–12 find the derivative of the given function. Whenever possible, simplify your answer.

7. $y = \ln(3x - 2)$

8. $y = \ln|3x - 2|$

9. $y = \ln(1 + x^5)$

10. $y = 2 \ln \sqrt{x^2 + 2}$

11. $y = \ln \ln x$

12. $y = \ln^2 \ln x$, that is, $(\ln \ln x)^2$

13. $y = \ln \ln^2 x$

14. $y = x \ln x - x$

15. $y = x^2 \ln x - \dfrac{x^2}{2}$

16. $y = x^n \ln x - \dfrac{x^n}{n}$

17. $y = \ln|\sin x|$

18. $y = \ln|\sec x + \tan x|$

19. $y = \ln|\csc x + \cot x|$

20. $y = \ln|x + \sqrt{x^2 - a^2}|$

21. $y = \ln(\sqrt{x^2 + a^2} - x)$

22. $y = (x + \sqrt{\ln x})^3$

23. $y = \ln\left(\dfrac{a + \sqrt{a^2 + x^2}}{x}\right)$

24. $y = \dfrac{\ln \ln \ln x}{\ln \ln x}$

25. State the results of Exercises 14–21 as indefinite integrals.

26.* Find the derivative of $f(x) = Ax \cos \ln x + Bx \sin \ln x$. Use the result to help you find the indefinite integrals

$$\int \cos \ln x \, dx \quad \text{and} \quad \int \sin \ln x \, dx.$$

27.† Show that $y = A \cos(\omega \ln x) + B \sin(\omega \ln x)$ is a solution of the differential equation $x^2y'' + xy' + \omega^2 y = 0$ for any constants A and B.

Exercises 28–30 are concerned with establishing the slowness of growth of $\ln x$ near infinity, and the slowness of growth of $|\ln x|$ near 0.

28.* For any rational number $r > 0$, let $s = r/2$. Show that

a) $\ln x < \dfrac{1}{s} x^s$ for $x > 0$. (*Hint:* Use parts (d) and (f) of Theorem 2.)

b) $\lim\limits_{x \to \infty} \dfrac{\ln x}{x^r} = 0.$

This result says that the natural logarithm grows more slowly than any positive power of x as x approaches infinity.

29.* Show that for any positive rational number r, $\lim_{x \to 0+} x^r \ln x = 0$. (Derive this from Exercise 28(b) by a change of variable.) We know that $\ln x$ approaches negative infinity as x approaches 0 from the right, but this exercise shows that that approach is slower than the growth of any negative power of x.

30.* Show that $\lim_{x \to \infty} \dfrac{(\ln x)^n}{x} = 0$ for any integer n.

3.5 THE EXPONENTIAL FUNCTION

Since ln is increasing on its domain $(0, \infty)$, it is one-to-one there and so has an inverse function. For the moment, let us call this inverse function exp.

$$y = \exp x \Leftrightarrow x = \ln y$$

Since $\ln 1 = 0$ we have

$$\exp 0 = 1.$$

Also

$$\text{domain of exp} = (-\infty, \infty) = \text{range of ln},$$
$$\text{range of exp} = (0, \infty) = \text{domain of ln}.$$

For every real number x we have

$$\ln \exp x = x,$$

and for every positive real number y we have

$$\exp \ln y = y.$$

The student who is familiar with both logarithms and the law of exponents from previous study of algebra will be aware that the inverse of a logarithm ought to be an "exponential" (that is, $x = \log_a y$ holds if and only if $a^x = y$). In order to show that $\exp x$ is such a function (corresponding to a certain value of the base a), we begin by showing that $\exp x$ satisfies one of the well-known laws of exponents, namely, $(a^x)^r = a^{rx}$. The other laws of exponents, such as $a^{x+y} = a^x a^y$ and $a^{-x} = 1/a^x$, will emerge later.

Theorem 3 For any rational number r,

$$(\exp x)^r = \exp rx.$$

This identity will remain valid for arbitrary real numbers r when we define what is meant by an arbitrary real (not necessarily rational) power in the next section.

Proof Let $y = (\exp x)^r$. By Theorem 2(d) of Section 3.4 we have $\ln y = r \ln \exp x = rx$. Hence, $y = \exp rx$, as required. □

Definition 6

$$\text{Let} \quad e = \exp 1.$$

The number e so defined satisfies $\ln e = 1$. It follows that the area between $y = 1/t$, $y = 0$, $x = 1$, and $x = e$ is 1 square unit, as shown in Fig. 3.30. Later we will have the means of computing e to any desired degree of accuracy. Like π, e is a nonrepeating, nonterminating decimal, that is, an irrational number:

$$e = 2.718281828459045. \ldots$$

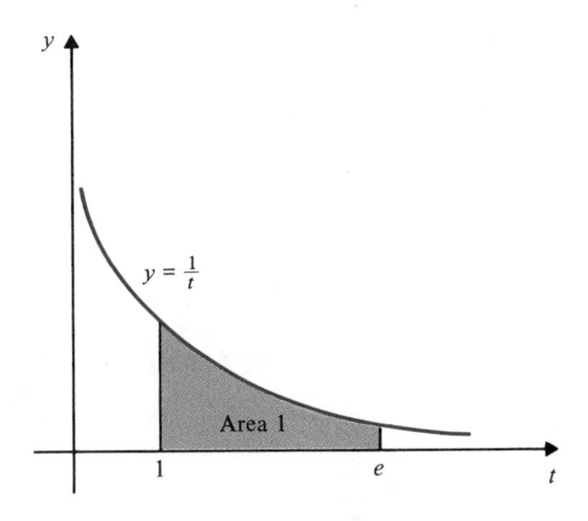

Figure 3.30

By Theorem 3 we have, for any rational number r,

$$\exp r = \exp 1r = (\exp 1)^r = e^r.$$

Consider the equation

$$\exp x = e^x.$$

Strictly speaking, the right-hand side makes sense *only when x is a rational number*: If $x = m/n$ where m and n are integers and $n > 0$, then $e^x = e^{m/n}$ just means the nth root of e^m. If x is not rational, it is not obvious what should be meant by e^x. However, being the inverse of $\ln x$, and therefore having the whole real line as domain, the function $\exp x$ is defined *for all real numbers x*. Since we have shown that $\exp x$ and e^x agree whenever the latter makes sense by previous standards, we can use this agreement to *define* e^x for all real x.

Definition 7

$$e^x = \exp x \qquad \text{for all real } x.$$

Hereafter we will normally use e^x to denote the exponential function that is inverse to $\ln x$:

$$y = e^x \Leftrightarrow x = \ln y.$$

The function e^x has all the properties we normally associate with powers of a fixed base.

Theorem 4

$$\left.\begin{aligned} &\text{a) } e^{u+v} = e^u e^v \\ &\text{b) } e^{-u} = \frac{1}{e^u} \\ &\text{c) } e^{u-v} = \frac{e^u}{e^v} \end{aligned}\right\} \text{Laws of Exponents}$$

Proof All these results follow from the corresponding properties of the natural logarithm. If $x = e^u$ and $y = e^v$, then $u = \ln x$ and $v = \ln y$, and so

$$e^{u+v} = e^{\ln x + \ln y} = e^{\ln xy} = xy = e^u e^v,$$

$$e^{-u} = e^{-\ln x} = e^{\ln (1/x)} = \frac{1}{x} = \frac{1}{e^u},$$

$$e^{u-v} = e^{\ln x - \ln y} = e^{\ln (x/y)} = \frac{x}{y} = \frac{e^u}{e^v}. \quad \square$$

Perhaps the most important property of the exponential function e^x is that it is equal to its own derivative. Let $y = e^x$ so that $x = \ln y$. By implicit differentiation we have

$$1 = \frac{1}{y}\frac{dy}{dx},$$

so that $dy/dx = y$, that is,

$$\frac{d}{dx} e^x = e^x.$$

The graph of $y = e^x$ is, of course, the mirror image of the graph of its inverse function, $y = \ln x$, in the line $y = x$. (See Fig. 3.31.) The graph has the property that the slope at any point is the height of that point above the x-axis.

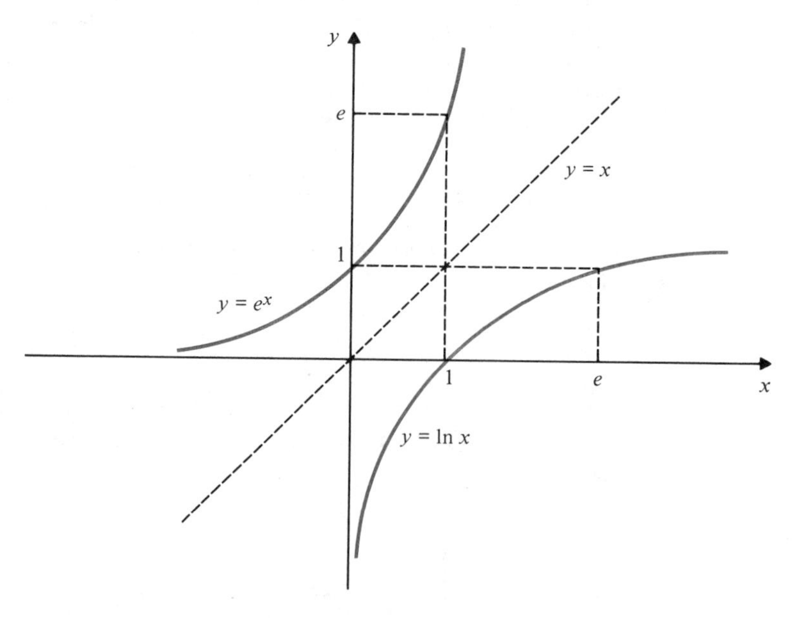

Figure 3.31

Since $\frac{d}{dx} e^{ax} = ae^{ax}$, it is also the case that

$$\int e^{ax}\, dx = \frac{1}{a} e^{ax} + C$$

or, in the special case $a = 1$,

$$\int e^{x}\, dx = e^{x} + C.$$

Example 1 Simplify the following.

a) $e^{3 \ln 2 - 2 \ln 3}$

b) $\ln \sqrt{(1/e^{4x})}$.

Solution a) $3 \ln 2 - 2 \ln 3 = \ln 2^3 - \ln 3^2 = \ln (8/9)$. Thus, $e^{3 \ln 2 - 2 \ln 3} = e^{\ln (8/9)} = \frac{8}{9}$.

b) $\sqrt{(1/e^{4x})} = (e^{-4x})^{1/2} = e^{-2x}$. Hence, $\ln \sqrt{(1/e^{4x})} = -2x$.

Example 2 Find $f^{(n)}(x)$ and $g^{(n)}(x)$ for $f(x) = e^x$ and $g(x) = e^{ax+b}$.

Solution We have

$$\begin{aligned} f'(x) &= e^x & \qquad g'(x) &= ae^{ax+b} \\ f''(x) &= e^x & g''(x) &= a^2e^{ax+b} \\ f^{(3)}(x) &= e^x & g^{(3)}(x) &= a^3e^{ax+b} \\ &\vdots & &\vdots \\ f^{(n)}(x) &= e^x & g^{(n)}(x) &= a^ne^{ax+b} \end{aligned}$$

Example 3 Find

$$\frac{d}{dt} e^{t^3+3t-5}.$$

Solution

$$\frac{d}{dt} e^{t^3+3t-5} = e^{t^3+3t-5}(3t^2 + 3) = 3(t^2 + 1)e^{t^3+3t-5}.$$

Example 4 Find all points at which the graph of $y = xe^{-x^2}$ has slope 0.

Solution The slope of $y = xe^{-x^2}$ at position x is given by

$$\frac{d}{dx} xe^{-x^2} = e^{-x^2} + xe^{-x^2}(-2x) = e^{-x^2}(1 - 2x^2).$$

Since $e^{-x^2} \neq 0$ for any x (the exponential function never takes the value 0), the slope is 0 only when $1 - 2x^2 = 0$, that is, if $x = \pm\frac{1}{\sqrt{2}}$. The corresponding values of y

are $y = \pm(1/\sqrt{2})e^{-1/2} = \pm 1/\sqrt{2e}$. Thus the slope is 0 at the points $(1/\sqrt{2}, 1/\sqrt{2e})$ and $(-1/\sqrt{2}, -1/\sqrt{2e})$.

Example 5 Show that $y = xe^{2x}$ is a solution of the differential equation $y'' - 4y' + 4y = 0$.

Solution

$$y' = e^{2x} + 2xe^{2x} = (1 + 2x)e^{2x}$$
$$y'' = 2e^{2x} + (1 + 2x)2e^{2x} = (4 + 4x)e^{2x}$$

Thus,

$$y'' - 4y' + 4y = (4 + 4x - 4(1 + 2x) + 4x)e^{2x} = 0e^{2x} = 0.$$

Example 6 Show that a differentiable function $y = f(x)$ satisfies the differential equation

$$\frac{dy}{dx} = ky \qquad (k \text{ a constant})$$

on an interval I *if and only if* $f(x) = Ce^{kx}$ on I, for some constant C.

Solution Suppose $y = Ce^{kx}$ on I. Then $\frac{dy}{dx} = Cke^{kx} = ky$ on I. Conversely, suppose that $y = f(x)$ satisfies $\frac{dy}{dx} = ky$ on I, that is, suppose $f'(x) = kf(x)$ on I. By the quotient rule,

$$\frac{d}{dx}\frac{f(x)}{e^{kx}} = \frac{e^{kx}f'(x) - f(x)ke^{kx}}{e^{2kx}} = \frac{f'(x) - kf(x)}{e^{kx}} = 0 \qquad \text{on } I.$$

Thus $(f(x))/(e^{kx}) = C$ (constant) on I and $f(x) = Ce^{kx}$.

Example 7 Solve the initial-value problem

$$\begin{cases} y' = -3y \\ y(0) = 7. \end{cases}$$

Solution By Example 6, $y = Ce^{-3x}$. But $7 = y(0) = Ce^0 = C$. Thus $y = 7e^{-3x}$.

The following theorem contains an interesting representation of e^x as a limit.

Theorem 5

$$e^x = \lim_{n\to\infty}\left(1 + \frac{x}{n}\right)^n \qquad \text{for every real } x$$

Proof If $x = 0$ there is nothing to prove; both sides of the identity are 1. If $x \neq 0$ let $h = x/n$. As n tends to infinity h approaches 0. Thus

$$\lim_{n\to\infty} \ln\left(1 + \frac{x}{n}\right)^n = \lim_{n\to\infty} n\ln\left(1 + \frac{x}{n}\right)$$
$$= \lim_{n\to\infty} x\,\frac{\ln\left(1 + \frac{x}{n}\right)}{\frac{x}{n}}$$

$$= x \lim_{h\to 0} \frac{\ln(1+h)}{h} \qquad \left(\text{where } h = \frac{x}{n}\right)$$

$$= x \lim_{h\to 0} \frac{\ln(1+h) - \ln 1}{h} \qquad (\text{since } \ln 1 = 0)$$

$$= x \frac{d}{dt} \ln t \bigg|_{t=1} \qquad (\text{by the definition of derivative})$$

$$= x \frac{1}{t}\bigg|_{t=1} = x.$$

Since ln is differentiable, it is continuous by Theorem 1 of Section 2.3. Hence

$$\ln\left(\lim_{n\to\infty}\left(1+\frac{x}{n}\right)^n\right) = x.$$

Since the exponential function is the inverse of the natural logarithm, we conclude

$$\lim_{n\to\infty}\left(1+\frac{x}{n}\right)^n = e^x. \ \square$$

In case $x = 1$ the formula given in Theorem 5 takes the following form

$$e = \lim_{n\to\infty}\left(1+\frac{1}{n}\right)^n.$$

We can use this formula to compute approximations to e, as follows.

n	$\left(1+\frac{1}{n}\right)^n$
1	2
10	2.59374 . . .
100	2.70481 . . .
1000	2.71692 . . .
10000	2.71815 . . .
100000	2.71827 . . .

In a sense we have cheated in deriving the numbers in this table. They were produced using the y^x function on a scientific calculator. However, this function is actually computed using the exponential function e^x, as we shall see in the next section. In any event, the formula in this table is not a very efficient way to calculate e to any great accuracy. A much better way is to use the series

$$e = 1 + \frac{1}{1!} + \frac{1}{2!} + \frac{1}{3!} + \frac{1}{4!} + \cdots$$
$$= 1 + 1 + \frac{1}{2} + \frac{1}{6} + \frac{1}{24} + \cdots,$$

which we establish in Chapter 9.

We conclude this section by stating a theorem to the effect that as x increases toward infinity, e^x grows faster than any power of x and also that as x decreases toward negative infinity, e^x decreases toward 0 faster than any negative power of x.

Theorem 6 For any integer n,

$$\lim_{x\to\infty} \frac{e^x}{x^n} = \infty.$$

Therefore also

$$\lim_{x\to-\infty} x^n e^x = 0. \ \square$$

We will not prove Theorem 6 at this time because a very simple proof can be given later as a consequence of l'Hôpital's rule, in Section 4.7. Note, however, that these properties of exponential are the appropriate "inverse" properties to the assertions made for the natural logarithm in Exercises 28 and 29 of Section 3.4. A proof of Theorem 6 can be based on these exercises. (See Exercises 57–59 in this section.)

EXERCISES

Simplify the expressions given in Exercises 1–8.

1. $e^3/\sqrt{e^5}$

2. $\ln (e^{1/2}e^{2/3})$

3. $e^{5 \ln x}$

4. $e^{(3 \ln 9)/2}$

5. $\ln \frac{1}{e^{3x}}$

6. $\ln (4e)^{1/2}$

7. $e^{3 \ln 5}e^{-\ln 25}$

8. $e^{2 \ln \cos x} + \ln^2 e^{\sin x}$

In Exercises 9–28 differentiate the given functions.

9. $y = e^{5x}$

10. $y = xe^x - x$

11. $y = \frac{x}{e^x}$

12. $y = x^2e^x$

13. $y = \frac{e^x + e^{-x}}{2}$

14. $y = e^{(x^2)}$

15. $y = e^{(e^x)}$

16. $y = \exp \exp \exp x$

17. $y = x^r e^{ax}$ (r rational)

18. $y = e^{ax} \ln x$

19. $y = \frac{e^x}{1 + e^x}$

20. $y = \frac{e^x - e^{-x}}{e^x + e^{-x}}$

21. $y = e^x \sin x$

22. $y = e^{-x} \cos x$

23. $y = xe^x \sin x$

24. $y = xe^{-x} \cos x \ln x$

25. $y = \sin e^x$

26. $y = \cos^2 e^x$

27. $y = \sqrt{1 + e^{2x}}$

28. $y = (x^3 + e^{3x})^{1/3}$

29. Find the nth derivative of $f(x) = xe^x$.

30.* Show that the nth derivative of $(ax^2 + bx + c)e^x$ is a function of the same form but with different constants.

31. Find the nth derivative of xe^{ax}.

32. Find the first four derivatives of e^{x^2}.

33. Let $f(x) = xe^{-x}$. Determine where f is increasing and where it is decreasing. Sketch the graph of f.

34. At what points does the graph $y = x^2e^{-x^2}$ have a horizontal tangent line?

35. Find an equation of the straight line tangent to the curve $y = e^{2x}$ at the point $(1, e^2)$.

36. Find the equation of a straight line of slope 4 that is tangent to the graph of $y = \ln x$.

37. Find an equation of the straight line tangent to the curve $y = e^x$ and passing through the origin. Sketch the curve and tangent.

38. Find an equation of the straight line tangent to the curve $y = \ln x$ and passing through the origin. Sketch the curve and tangent.

39. Find an equation of the straight line that is tangent to $y = e^x$ and that passes through the point $(a, 0)$. Sketch the curve and tangent.

40. On the same set of coordinate axes sketch the graphs of $y = e^x$, $y = e^{2x}$, and $y = e^{-x}$.

41. Find the slope of the curve

$$e^{xy} \ln \frac{x}{y} = x + \frac{1}{y}$$

at the point $\left(e, \dfrac{1}{e}\right)$.

42. Find an equation of the straight line tangent to the curve $xe^y + y - 2x = \ln 2$ at the point $(1, \ln 2)$.

43.* Let $F_{A,B}(x) = Ae^x \cos x + Be^x \sin x$. Show that

$$\frac{d}{dx} F_{A,B}(x) = F_{A+B,B-A}(x).$$

44.* Using the results of Exercise 43, find

a) $\dfrac{d^2}{dx^2} F_{A,B}(x)$

b) $\dfrac{d^3}{dx^3} e^x \cos x$

45.* Find $\dfrac{d}{dx}(Ae^{ax} \cos bx + Be^{ax} \sin bx)$ and use your result to find

$$\int e^{ax} \cos bx\, dx \qquad \text{and}$$

$$\int e^{ax} \sin bx\, dx$$

Exercises 46–48 constitute a proof of Theorem 6 of this section. They draw upon the corresponding results for the natural logarithm, Exercises 28–30 of Section 3.4, which should be done first.

46.* Use the result of Exercise 30 of Section 3.4 to show that for any integer n

$$\lim_{x\to\infty} \frac{x^n}{e^x} = 0.$$

47.* Show that $\lim_{x\to\infty} \dfrac{e^x}{x^n} = \infty$ for any integer n.

48.* Show that $\lim_{x\to-\infty} x^n e^x = 0$ for any integer n.

Solve the initial-value problems in Examples 49–51.

49.† $\begin{cases} y' + y = 0 \\ y(0) = 2 \end{cases}$

50.† $\begin{cases} y' - 4y = 0 \\ y(0) = -3 \end{cases}$

51.† $\begin{cases} 2y' + 5y = 0 \\ y(-1) = 3 \end{cases}$

52.† Show that y is a solution of the first-order *nonhomogenous* differential equation $y' + ky = g(x)$ if and only if $y = e^{-kx}u$ where $u'(x) = e^{kx}g(x)$.

Exercises 53–57 are concerned with solutions of the second-order differential equation with constant coefficients.

$$(*) \qquad ay'' + by' + cy = 0 \qquad (a, b, c \text{ constants}).$$

Associated with this differential equation is the corresponding quadratic equation

$$(**) \qquad ar^2 + br + c = 0,$$

which is called the indicial equation for (*) and which has roots

$$r = \frac{-b \pm \sqrt{b^2 - 4ac}}{2a}.$$

53.† Show that if $y = f(x)$ and $y = g(x)$ are solutions of (*), then $y = Af(x) + Bg(x)$ is also a solution for any choice of the constants A and B.

54.† Show that if r is a real root of the indicial equation (**), then $y = e^{rx}$ is a solution of the differential equation (*).

55.† Show that if $b^2 - 4ac > 0$, then (*) has solutions of the form $y = Ae^{r_1x} + Be^{r_2x}$ where r_1 and r_2 are distinct real solutions of (**).

56.† Show that if $b^2 - 4ac = 0$ so that (**) has a double root $r_1 = -b/2a$, then (*) has solutions of the form $y = Ae^{r_1x} + Bxe^{r_1x}$.

57.† Show that if $b^2 - 4ac < 0$ so that (**) does not have any real roots, then $y = e^{-(b/2a)x}u$ is a solution of (*) if and only if u satisfies $u'' + \omega^2 u = 0$ where $\omega^2 = (4ac - b^2)/4a^2$. Hence show that (*) has solutions of the form $y = e^{-(b/2a)x}(A \cos \omega x + B \sin \omega x)$.

The differential equation (*) is frequently encountered in mathematical models of electrical and mechanical systems. The coefficients a, b, and c are usually all positive numbers. If $b = 0$, the equation is just that of simple harmonic motion. If $b > 0$, the term by' represents a resistive force opposing any change in y. Such a term would be present in a spring mass system such as that considered in Section 3.2, provided there is resistance to the motion of the mass and that this resistance is proportional to the velocity. If b is small enough that $b^2 - 4ac < 0$, then oscillation still results, as is shown by Exercise 45, but the amplitude is exponentially damped. If $b^2 - 4ac > 0$, there are no oscillations and the system is called overdamped. The transition case $b^2 - 4ac = 0$ is called critically damped.

It can be shown that the solutions given in Exercises 55–57 are the general solutions of (*) for the cases considered. Use this result to solve the differential equations and initial-value problems in Exercises 58–65.

58.† $y'' + y' - 2y = 0$

59.† $y'' + 6y' + 9 = 0$

60.† $y'' + y' - y = 0$

61.† $y'' + 2y' + 2y = 0$

62.† $\begin{cases} y'' + y' = 0 \\ y(0) = 2 \\ y'(0) = 3 \end{cases}$

63.† $\begin{cases} y'' - 2y' + 5y = 0 \\ y(0) = 7 \\ y'(0) = 0 \end{cases}$

64.† $\begin{cases} y'' + 2y' + y = 0 \\ y(1) = 0 \\ y'(1) = 1 \end{cases}$

65.† $\begin{cases} y'' - 4y' + 8y = 0 \\ y(0) = 1 \\ y'(0) = 1 \end{cases}$

3.6 GENERAL EXPONENTIALS AND LOGARITHMS

As we mentioned in the previous section, the expression a^x is algebraically meaningful only if x is a rational number: If $x = m/n$ where m and n are integers and $n > 0$, then

$$a^x = a^{m/n} \text{ means } \sqrt[n]{a^m} \text{ or } (\sqrt[n]{a})^m.$$

If n is even, we require $a \geq 0$; if $m < 0$ we require $a \neq 0$. It is therefore not clear what, if any, meaning can be assigned to expressions such as $2^{\sqrt{2}}$, 5^π or 3^{-e} where the exponents are not rational.

The fact that for the particular base e we have been able to assign a meaning to e^x for *arbitrary real* x (as the inverse natural logarithm) enables us to define a^x for arbitrary x provided $a > 0$, and the new definition is consistent with the old one in terms of powers and roots if x is rational. Observe that if x is rational and $a > 0$, then

$$e^{x \ln a} = e^{\ln a^x} = a^x.$$

We use this identity to define a^x even if x is not rational.

Definition 8

For $a > 0$ and x real we define a^x by

$$a^x = e^{x \ln a}.$$

Evidently $a^0 = 1$ for every $a > 0$. It is easily checked that, defined as above, a^x has all the appropriate properties of an exponential function

$$a^{x+y} = a^x a^y$$

$$a^{-x} = \frac{1}{a^x}$$

$$a^{x-y} = \frac{a^x}{a^y}$$

$$(a^x)^y = a^{xy}$$

Note also that the rule

$$\ln a^x = x \ln a$$

now holds for any $a > 0$ and x real.

The derivative of a^x is readily calculated;

$$\frac{d}{dx} a^x = a^x \ln a.$$

We can also extend the general power rule to arbitrary exponents:

$$\frac{d}{dx} x^a = \frac{d}{dx} e^{a \ln x} = e^{a \ln x} \left(\frac{a}{x}\right) = \frac{ax^a}{x} = ax^{a-1}.$$

Be careful to distinguish between expressions like 3^x where the variable is in the exponent (exponential functions) and ones like x^3 where the variable is in the base (power functions):

$$\frac{d}{dx} 3^x = 3^x \ln 3, \qquad \frac{d}{dx} x^3 = 3x^2.$$

The function $y = a^x$ is constant, $y = 1$, if $a = 1$. If $a > 1$, then $dy/dx > 0$, so $y = a^x$ is increasing on $(-\infty, \infty)$. If $a < 1$, then $dy/dx < 0$ and $y = a^x$ is decreasing. In any case, the slope of $y = a^x$ at $x = 0$ is $\ln a$.

The graphs of some exponential functions are sketched in Fig. 3.32.

Since a^x is increasing or decreasing if $a \neq 1$, it is one-to-one and so has an inverse. The function inverse to a^x $(a \neq 1)$ is called $\log_a x$ (the logarithm of x to the base a).

$$y = \log_a x \Leftrightarrow x = a^y$$

It follows at once that if $y = \log_a x$ and hence $x = a^y$, then $\ln x = \ln a^y = y \ln a$, so that

$$\log_a x = \frac{\ln x}{\ln a}.$$

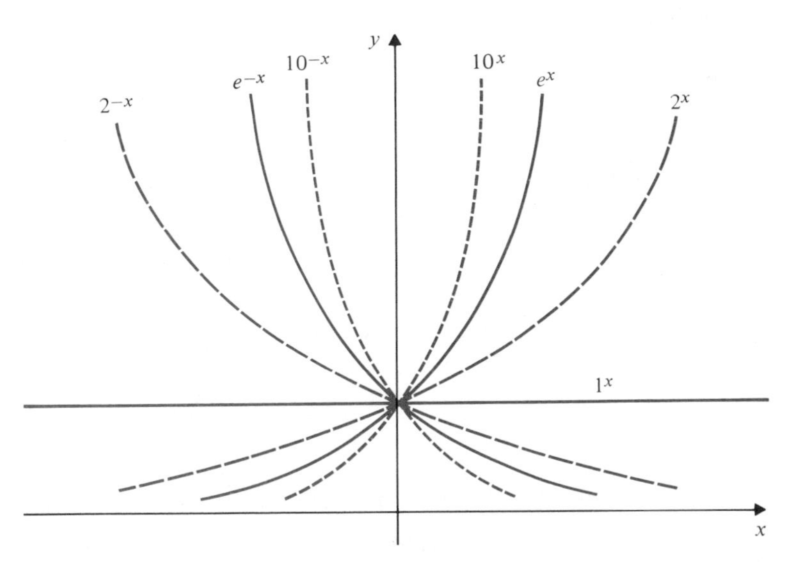

Some exponential functions

Figure 3.32

In particular,

$$\frac{d}{dx} \log_a x = \frac{1}{x \ln a}.$$

Little attention is usually paid to the function $\log_a$ because it is just a constant multiple of the natural logarithm $\ln = \log_e$. Scientific calculators often have two logarithm functions, the natural logarithm, $\ln x$, and the base 10 logarithm, $\log_{10} x$, called the *common logarithm,* which is usually denoted $\log x$. Before the advent of pocket calculators, tables of common logarithms and common antilogarithms (values of 10^x) were used to simplify numerical calculations involving multiplication, division, powers, and roots. Indeed, the slide rule, a simple mechanical forerunner of the pocket calculator, achieved multiplication and division by adding or subtracting distances on a logarithmic scale. Common logarithms still play a prominent role in certain scientific calculations, for example in the pH measure of acidity in chemistry.

Example 1 What is meant by 2^π? Find its value.

Solution $2^\pi = e^{\pi \ln 2}$. Using a scientific calculator we obtain

$$\begin{aligned} \ln 2 &= 0.6931471\ldots \\ 2^\pi = e^{\pi \ln 2} &= 8.8249778\ldots \end{aligned}$$

Most scientific calculators have a y^x function. This is actually calculated using the $\ln x$ and e^x functions in the calculator, as $e^{x \ln y}$.

Example 2 Solve the equation $3^{4x} = 2^{x+1}$.

Solution Taking natural logarithms of both sides we obtain

$$4x \ln 3 = (x + 1) \ln 2.$$

Hence

$$x = \frac{\ln 2}{4 \ln 3 - \ln 2} = \frac{\ln 2}{\ln (81/2)}.$$

Example 3 Show that $(\log_a b)(\log_b c) = \log_a c$.

Solution

$$(\log_a b)(\log_b c) = \frac{\ln b \ln c}{\ln a \ln b} = \frac{\ln c}{\ln a} = \log_a c.$$

Logarithmic Differentiation

Suppose we wish to differentiate a function of the form

$$y = (f(x))^{g(x)},$$

where the variable appears in both the base and the exponent so that neither the general power rule, $\frac{d}{dx} x^a = ax^{a-1}$, nor the exponential rule, $\frac{d}{dx} a^x = a^x \ln a$, can be directly applied. One method for finding the derivative of such a function is to express it in the form

$$y = e^{g(x) \ln f(x)}.$$

It follows that

$$\begin{aligned}\frac{dy}{dx} &= e^{g(x) \ln f(x)} \left(g'(x) \ln f(x) + g(x) \frac{1}{f(x)} f'(x) \right) \\ &= (f(x))^{g(x)} \left(g'(x) \ln f(x) + \frac{g(x) f'(x)}{f(x)} \right).\end{aligned}$$

The same results can be obtained by taking natural logarithms of both sides of the given equation and differentiating implicitly, as follows.

$$\begin{aligned}\ln y &= g(x) \ln f(x) \\ \frac{1}{y} \frac{dy}{dx} &= g'(x) \ln f(x) + \frac{g(x) f'(x)}{f(x)} \\ \frac{dy}{dx} &= (f(x))^{g(x)} \left(g'(x) \ln f(x) + \frac{g(x) f'(x)}{f(x)} \right)\end{aligned}$$

This latter technique is called *logarithmic differentiation*.

Example 4 Find y' if $y = x^x$ $(x > 0)$.

$$\ln y = x \ln x$$

Solution

$$\frac{1}{y}\frac{dy}{dx} = \ln x + \frac{x}{x} = 1 + \ln x$$

$$\frac{dy}{dx} = y(1 + \ln x) = x^x(1 + \ln x).$$

Example 5 Find $\frac{d}{dt}(\sin t)^{\ln t} \qquad (0 < t < \pi)$.

Solution Let $y = (\sin t)^{\ln t}$. Thus $\ln y = \ln t \ln \sin t$, and

$$\frac{1}{y}\frac{dy}{dt} = \frac{1}{t}\ln \sin t + \ln t \frac{\cos t}{\sin t}$$

$$\frac{dy}{dt} = y\left(\frac{\ln \sin t}{t} + \ln t \cot t\right)$$

$$= (\sin t)^{\ln t}\left(\frac{\ln \sin t}{t} + \ln t \cot t\right).$$

Logarithmic differentiation is also useful for finding the derivatives of functions expressed as products and quotients of many factors. Taking logarithms reduces these products and quotients to sums and differences.

Example 6 Differentiate $y = [(x + 1)(x + 2)(x + 3)]/(x + 4)$.

Solution $\ln y = \ln(x + 1) + \ln(x + 2) + \ln(x + 3) - \ln(x + 4)$. Thus

$$\frac{1}{y}y' = \frac{1}{x+1} + \frac{1}{x+2} + \frac{1}{x+3} - \frac{1}{x+4}$$

$$y' = \frac{(x+1)(x+2)(x+3)}{x+4}\left(\frac{1}{x+1} + \frac{1}{x+2} + \frac{1}{x+3} - \frac{1}{x+4}\right)$$

$$= \frac{(x+2)(x+3)}{x+4} + \frac{(x+1)(x+3)}{x+4} + \frac{(x+1)(x+2)}{x+4} - \frac{(x+1)(x+2)(x+3)}{(x+4)^2}.$$

Example 7 Find $\left.\frac{du}{dx}\right|_{x=1}$ if $u = \sqrt{(x + 1)(x^2 + 1)(x^3 + 1)}$

Solution

$$\ln u = (1/2)(\ln(x + 1) + \ln(x^2 + 1) + \ln(x^3 + 1))$$

$$\frac{1}{u}\frac{du}{dx} = \frac{1}{2}\left(\frac{1}{x+1} + \frac{2x}{x^2+1} + \frac{3x^2}{x^3+1}\right)$$

At $x = 1$ we have $u = \sqrt{8} = 2\sqrt{2}$. Hence

$$\left.\frac{du}{dx}\right|_{x=1} = \sqrt{2}\left(\frac{1}{2} + 1 + \frac{3}{2}\right) = 3\sqrt{2}.$$

EXERCISES

Simplify the expressions in Exercises 1–8.

1. $(\log_4 16)(\log_4 2)$

2. $\log_{15} 75 + \log_{15} 3$

3. $2^{\log_4 8}$

4. $\log_x x^3 - \log_{x+2} \dfrac{1}{(x+2)^3}$

5. $\log_6 9 + \log_6 4$

6. $\log_x y \log_y x$

7. $8^{-\log_4 x}$

8. $x^{1/(\log_a x)}$

Solve the equations in Exercises 9–16 for x.

9. $7^x = 8$

10. $2^{x+1} = 3^x$

11. $3^x = 9^{1-x}$

12. $\dfrac{1}{2^x} = \dfrac{5}{8^{x+3}}$

13. $5^{2x}3^{2x-1} = 15^{x-2}$

14. $2^{x^2-3} = 4^x$

15. $2x^x = 3x^{x-1}$

16.* $a^{x^3} = b^x$ (Consider all positive values of a and b.)

17. Evaluate the following using values for e^x and $\ln x$ obtained from a table or using a scientific calculator:

a) $3^{\sqrt{2}}$ **b)** $(\sqrt{2})^{\pi}$ **c)** π^e

Differentiate the functions in Exercises 18–31.

18. $y = 5^{2x+1}$

19. $f(x) = a^{bx+c}$

20. $y = 2^{(x^2-3x+8)}$

21. $g(x) = a^x x^a$

22. $f(x) = 10^{\ln \sin x}$

23. $h(t) = t^x - x^t$

24. $H(x) = t^x - x^t$

25. $y = \log_a (bx + c)$

26. $y = x^{\sqrt{x}}$

27. $y = x^{\ln x}$

28. $y = x^{1/(\ln x)}$

29. $y = (\sin x)^x$

30. $y = (\sec x)^{\tan x}$

31. $y = x^{(e^x)}$

32.* Differentiate

a) $y = (x^x)^x$ **b)** $y = x^{(x^x)}$

Which of these functions grows most rapidly near infinity?

Use logarithmic differentiation to find the required derivatives in Exercises 33–37.

33. $f(x) = (x-1)(x-2)(x-3)(x-4)$. Find $f'(x)$.

34. $f(x) = \dfrac{(x^2-1)(x^2-2)(x^2-3)}{(x^2+1)(x^2+2)(x^2+3)}$. Find $f'(2)$. Also find $f'(1)$.

35. $F(x) = \dfrac{\sqrt{1+x}\,(1-x)^{1/3}}{(1+5x)^{4/5}}$. Find $F'(0)$.

36. $y = \sqrt{\dfrac{x^2(1+x^2)}{(1+x^4)(1+x^6)}}$. Find $\dfrac{dy}{dx}$.

37. $g(x) = \dfrac{\sqrt{1+x}\,\sqrt{1+2x}\,\sqrt{1+3x}}{\sqrt{1+6x}}$. Find an equation of the line tangent to $y = g(x)$ at $x = 0$.

Exercises 38–43 deal with the second-order differential equation

$$(*) \qquad ax^2y'' + bxy' + cy = 0 \qquad (x \neq 0)$$

and its associated indicial equation (a quadratic equation in r)

$$(**) \qquad ar(r-1) + br + c = 0.$$

Differential equation (*) is called an Euler equation. Generally its solutions are valid in intervals not containing $x = 0$.

38.† Show that if $y = f(x)$ and $y = g(x)$ are solutions of (*), then so is $y = Af(x) + Bg(x)$ for any choice of constants A and B.

39.† Show that $y = |x|^r$ is a solution of (*) if r is a real solution of (**).

40.† If $(b-a)^2 > 4ac$, show that (*) has a solution of the form $y = A|x|^{r_1} + B|x|^{r_2}$ where r_1 and r_2 are distinct real solutions of (**).

41.† If $(b-a)^2 = 4ac$ so that (**) has a double root $r = (a-b)/2a$, show that (*) has a solution of the form $y = A|x|^r + B|x|^r \ln|x|$.

42.† If $(b-a)^2 < 4ac$ and $r = (a-b)/2a$ show that $y = |x|^r u$ satisfies (*) if and only if u satisfies

$$(***) \qquad x^2u'' + xu' + \omega^2 u = 0,$$

where $\omega^2 = (4ac - (a-b)^2)/4a^2$.

43.† Show that (***) has a solution of the form $y = A\cos\omega\ln|x| + B\sin\omega\ln|x|$ and hence that if $(b-a)^2 < 4ac$, then (*) has solution $y = A|x|^r\cos\omega\ln|x| + B|x|^r\sin\omega\ln|x|$ where $r = (a-b)/2a$ and $\omega^2 = (4ac - (a-b)^2)/4a^2$.

The solutions to (*) given in Exercises 40, 41, and 43 are general solutions, that is, every solution is of the specified form for some choice of the constants. Find general solutions to the differential equations in Exercises 44–49.

44.† $x^2y'' - xy' - 3y = 0$

45.† $x^2y'' + xy' - y = 0$

46.† $x^2y'' - xy' + y = 0$

47.† $x^2y'' + xy' + y = 0$

48.† $x^2y'' - xy' + 5y = 0$

49.† $x^2y'' + xy' = 0$

3.7 GROWTH AND DECAY PROBLEMS

Many natural processes involve quantities that increase or decrease at a rate proportional to their size. In a medium supplying adequate nourishment, for example, a culture of bacteria will grow at a rate proportional to its mass. The value of an investment bearing interest that is continually compounded increases at a rate proportional to that value. The temperature of a cup of coffee will decrease at a rate proportional to the amount by which that temperature exceeds the temperature of the surrounding air.

The mass of undecayed radioactive material in a sample decreases at a rate proportional to that mass.

All of these phenomena, and others exhibiting similar behavior, can be modeled mathematicaly in the same way. If $y = y(t)$ denotes the value of a quantity y at time t, and if y changes at a rate proportional to its size, then

$$\frac{dy}{dt} = ky,$$

where k is the constant of proportionality. We have already seen that the solution of this differential equation is

$$y(t) = Ce^{kt},$$

where C is a constant. If $k > 0$, then y is increasing with time; if $k < 0$, then y is decreasing as time increases. In either case we have

$$y(0) = Ce^0 = C,$$

so C is just the value of y at time $t = 0$. Thus

$$y(t) = y(0)e^{kt}.$$

Example 1 A certain cell culture grows at a rate proportional to the number of cells present. If the culture contains 500 cells initially, and 800 after 24 hours, how many cells will be present after a further 12 hours?

Solution Let $y(t)$ be the number of cells present t hours after there were 500 cells. Thus $y(0) = 500$ and $y(24) = 800$. Because $dy/dt = ky$, we have

$$y(t) = y(0)e^{kt} = 500e^{kt}.$$

Thus

$$800 = y(24) = 500e^{24k},$$

so

$$24k = \ln\frac{800}{500} = \ln\frac{8}{5}$$

and

$$y(t) = 500e^{(t/24)\ln(8/5)} = 500\left(\frac{8}{5}\right)^{t/24}.$$

We want to know y when $t = 36$:

$$y(36) = 500e^{(36/24)\ln(8/5)} \simeq 1012.$$

(We used a scientific calculator to evaluate the answer.) Thus 12 hours after 800 cells were in the culture, the number had grown to about 1012 cells.

Exponential growth is characterized by **fixed doubling time.** If T is the time at which y has doubled from its size at $t = 0$, then

$$2y(0) = y(T) = y(0)e^{kT},$$

so $e^{kT} = 2$. Hence

$$y(t + T) = y(0)e^{k(t+T)} = e^{kT}y(0)e^{kt} = 2y(t),$$

that is, T units of time are required for y to double from any value. Similarly, exponential decay involves fixed halving time (usually called **half-life**). If $y(T) = (1/2)y(0)$, then $e^{kT} = 1/2$ and

$$y(t + T) = y(0)e^{k(t+T)} = e^{kT}y(0)e^{kt} = \frac{1}{2}y(t).$$

Example 2 A radioactive material exhibits a half-life of 1200 years. What percentage of the original radioactivity of a sample is left after 10 years? How many years are required to reduce the radioactivity by 10 percent?

Solution Let $p(t)$ be the percentage of the original radioactivity left after t years. Thus $p(0) = 100$ and $p(1200) = 50$. Since the radioactivity decreases at a rate proportional to itself, $dp/dt = kp$ and

$$p(t) = 100e^{kt},$$
$$50 = p(1200) = 100e^{1200k},$$

so

$$k = -\frac{\ln 2}{1200}.$$

The percentage left after 10 years is

$$p(10) = 100e^{10k} = 100e^{-(1/120)\ln 2} \simeq 99.424.$$

If after t years 90 percent of the radioactivity is left, then

$$90 = 100e^{kt},$$
$$kt = \ln\frac{90}{100},$$
$$t = -\frac{1200}{\ln 2}\ln\frac{9}{10} \simeq 182.4,$$

so it will take a little over 182 years to reduce the radioactivity by 10 percent.

Sometimes an exponential growth or decay problem will involve a quantity that changes at a rate proportional to the difference between itself and a fixed value:

$$\frac{dy}{dt} = k(y - a).$$

In this case the difference $u = y - a$ should be used as the dependent variable, for it has the same rate of change as the original quantity y (that is, $du/dt = dy/dt$) and so satisfies

$$\frac{du}{dt} = ku.$$

Example 3 According to Newton's law of cooling, a hot object introduced into a cooler environment will cool at a rate proportional to the excess of its temperature above that of its environment. If a cup of coffee sitting in a room maintained at a temperature of 20°C cools from 80°C to 50°C in five minutes, how much longer will it take to cool to 40°C?

Solution Let $y(t)$ be the temperature of the coffee t minutes after it was 80°. Thus $y(0) = 80$ and $y(5) = 50$. Let $u(t) = y(t) - 20$, so $u(0) = 60$ and $u(5) = 30$. We have

$$\frac{du}{dt} = \frac{dy}{dt} = k(y - 20) = ku.$$

Thus

$$\begin{aligned} u(t) &= 60e^{kt}, \\ 30 = u(5) &= 60e^{5k}, \\ 5k &= \log\frac{1}{2} = -\ln 2. \end{aligned}$$

We wish to know t such that $y(t) = 40$, that is, $u(t) = 20$.

$$\begin{aligned} 20 = u(t) &= 60e^{-(t/5)\ln 2} \\ -\frac{t}{5}\ln 2 &= \ln\frac{20}{60} = -\ln 3, \\ t &= 5\frac{\ln 3}{\ln 2} \simeq 7.92 \end{aligned}$$

The coffee will take approximately 2.92 more minutes to cool from 50°C to 40°C.

Example 4 Suppose \$10,000 is invested for 1 year at simple interest with rate 8 percent. It will clearly grow to \$10,800 in that year. If instead the interest rate were 8 percent *compounded semiannually,* then the \$10,000 would grow in six months to \$10,000 × (1 + 4/100) = \$10,400, and then that amount would grow in the second six-month period to \$10,400 × (1 + 4/100) = \$10,816. Thus an interest rate of 8 percent compounded semiannually is equivalent to an effective annual rate of 8.16 percent (simple interest). Similarly, if the 8 percent is compounded quarterly, the \$10,000 would grow in one year to \$10,000 × $(1 + 2/100)^4$ = \$10,000 × 1.0824322 ≃ \$10824.32, for an effective yield of 8.24322 percent. Daily compounding would yield \$10,000 × $(1 + 8/36{,}500)^{365}$ = \$10,832.775, for an effective yield of 8.32775 percent.

In general, an interest rate of I percent per annum, compounded n times per year, will yield $\left(1 + \frac{I}{100n}\right)^n$ after one year, on an initial investment of \$1. Suppose that n approaches infinity. (This results in a situation in which the interest is being compounded *continuously* or *instantaneously*.) In this case the annual yield on \$1 will be

$$\lim_{n\to\infty}\left(1 + \frac{I}{100n}\right)^n = e^{I/100}$$

(by Theorem 5 of Section 3.5), and the effective annual interest rate is $100(e^{I/100} - 1)$. For $I = 8$ this rate is 8.3287066 . . . percent. For $I = 12$ it is 12.749685 . . . percent. For $I = 15$ it is 16.183424 . . . percent.

Logistic Growth

Sometimes a quantity that would tend to grow exponentially actually has its growth limited by external constraints. By virtue of its natural fertility, for example, an animal population may tend to exhibit exponential growth, but that growth may be limited by the level of the renewable food supply available. If the food supply were such that it would support only a constant population of size L, then it might be reasonable to assume that the size y of the population grows at a rate jointly proportional to y and to the amount by which y falls short of L:

$$\frac{dy}{dt} = ky(L - y).$$

This differential equation is the model for what is called **logistic growth,** that is, growth determined by supply of necessary resources. Observe that $dy/dt > 0$ if $0 < y < L$ and that this rate becomes small if y is small (there are few animals to reproduce) or if y is close to L (there are almost as many animals as the available resources can feed). Observe also that $dy/dt < 0$ if $y > L$; more animals die than are born, because of malnutrition or starvation, there being more animals than the resources can feed. Of course the steady-state populations $y = 0$ and $y = L$ are solutions of the differential equation. In order to find other solutions we write the equation in the form

$$\frac{L}{y(L - y)}\frac{dy}{dt} = kL,$$

or

$$\frac{1}{y}\frac{dy}{dt} + \frac{1}{L - y}\frac{dy}{dt} = kL$$

because $L/(y(L - y)) = (1/y) + (1/(L - y))$. The function $\ln y$ is evidently an antiderivative of $(1/y)(dy/dt)$, and $-\ln (L - y)$ is an antiderivative of $(1/(L - y))(dy/dt)$ provided $0 < y < L$. Thus

$$\ln y - \ln (L - y) = kLt + C$$

or

$$\frac{y}{L - y} = C_1 e^{kLt} \qquad (C_1 = e^C).$$

If $y(0) = y_0$ is the value of y at time $t = 0$, then $C_1 = y_0/(L - y_0)$ and so

$$\frac{y}{L - y} = \frac{y_0}{L - y_0} e^{kLt}.$$

Solving this equation for y leads to

$$y = \frac{Ly_0 e^{kLt}}{y_0 e^{kLt} + L - y_0} = \frac{Ly_0}{y_0 + (L - y_0)e^{-kLt}}.$$

Observe that, as expected,

$$\lim_{t \to \infty} y(t) = L, \qquad \lim_{t \to -\infty} y(t) = 0.$$

The solution given in the box above also holds for $y_0 > L$. However, the solution does not approach 0 as t approaches $-\infty$ in this case. It has a vertical asymptote at a certain negative value of t. (See Exercise 15 at the end of this section.) The graphs of solutions of the logistic equation for various positive values of y_0 are given in Fig. 3.33.

EXERCISES

1. Bacteria grow in a certain culture at a rate proportional to the amount present. If there are 1000 bacteria present initially and the amount doubles in 1 hour, how many will there be after a further $1\frac{1}{2}$ hours?
2. Sugar decomposes in water at a rate proportional to the amount still undecomposed. If there were 50 kg of sugar present initially, and at the end of 5 hours it is reduced to 20 kg, how long will it take until 90 percent of the sugar is decomposed?
3. A radioactive substance decays at a rate proportional to the amount present. If 30 percent of such a substance decays in 15 years, what is the half-life of the substance?
4. If the half-life of radium is 1690 years, what percentage of the amount present now will be remaining after (a) 100 years, (b) 1000 years?
5. An object is brought into a freezer maintained at a temperature of -5°C. If the object cools from 45° to 20° in 40 minutes, how many more minutes will it take to cool to 0°?
6. In a certain culture where the rate of growth of bacteria is proportional to the number present, the number triples in 3 days. If at the end of 7 days there are 10 million bacteria present in the culture, how many were present initially?
7. Use Newton's law of cooling to determine the reading on a thermometer 5 minutes after it is taken from a room at 72°F to the outdoors where the temperautre is 20°F, if the reading dropped to 48°F after one minute.

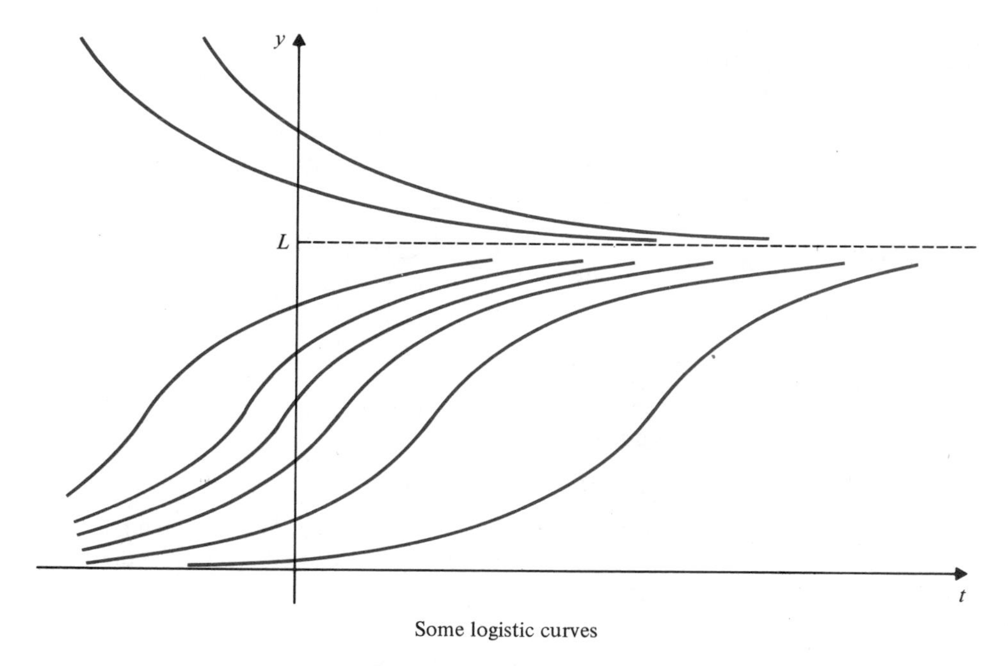

Some logistic curves

Figure 3.33

8. Find the half-life of a radioactive substance if after 1 year 99.57 percent of an initial amount still remains.

9. Money invested at compound interest (with instantaneous compounding) accumulates at a rate proportional to the amount present. If an initial investment of \$1000 grows to \$1500 in exactly 5 years, find (a) the doubling time for the investment, and (b) the effective annual rate of interest being paid.

10. When a simple electrical circuit containing inductance and resistance but no capacitance has the electromotive force removed, the rate of decrease of the current is proportional to the current. If the current is $I(t)$ amperes t seconds after cutoff, and if $I = 40$ when $t = 0$, and $I = 15$ when $t = 0.01$, find a formula for $I(t)$.

11. If the purchasing power of the dollar is decreasing at a rate of 9 percent annually, compounded continuously, how long will it take for the purchasing power to be reduced to 25 cents?

12. If a body in a room warms up from 5°C to 10°C in 4 minutes, and if the room is being maintained at 20°C, how much longer will the body take to warm up to 15°C?

13.* Suppose the quantity $y(t)$ exhibits logistic growth. If the values of $y(t)$ at times $t = 0$, $t = 1$, and $t = 2$ are y_0, y_1, and y_2, respectively, find a quadratic equation that must be satisfied by the limiting value L of $y(t)$. If $y_0 = 3$, $y_1 = 5$, and $y_2 = 6$, find L.

14.* Show that the solution of the logistic equation is increasing most rapidly when its value is $L/2$. (*Hint:* You do not need to use the formula for the solution to see this.)

15.* Show that the formula obtained for the solution to the logistic equation in the text is also valid if $y_0 > L$. In this case, on what interval is the solution valid? What happens to the solution as t approaches the left endpoint of this interval?

3.8 THE HYPERBOLIC FUNCTIONS AND THEIR INVERSES

The hyperbolic functions are certain combinations of exponential functions that have properties mimicking those of the circular functions.

Definition 9

For any real x the **hyperbolic cosine** ($\cosh x$) and the **hyperbolic sine** ($\sinh x$) are defined by

$$\cosh x = \frac{e^x + e^{-x}}{2}, \qquad \sinh x = \frac{e^x - e^{-x}}{2}.$$

All of the following properties of cosh and sinh follow directly from this definition.

i) $\cosh 0 = 1, \qquad \sinh 0 = 0$

ii) $\cosh(-x) = \cosh x$ (cosh is an even function)
$\sinh(-x) = -\sinh x$ (sinh is an odd function)

iii) $\dfrac{d}{dx}\cosh x = \sinh x, \qquad \dfrac{d}{dx}\sinh x = \cosh x$

iv) $\cosh^2 x - \sinh^2 x = 1$ for every x

We prove (iv).

$$\begin{aligned}\cosh^2 x - \sinh^2 x &= \left(\frac{e^x + e^{-x}}{2}\right)^2 - \left(\frac{e^x - e^{-x}}{2}\right)^2 \\ &= \frac{1}{4}(e^{2x} + 2 + e^{-2x} - (e^{2x} - 2 + e^{-2x})) \\ &= \frac{1}{4}(2 + 2) = 1.\end{aligned}$$

This identity is the reason these functions are called hyperbolic. For any t, the point $(\cosh t, \sinh t)$ lies on the rectangular hyperbola $x^2 - y^2 = 1$, just as the point $(\cos t, \sin t)$ lies on the circle $x^2 + y^2 = 1$. In the hyperbolic case there is no interpretation of t as an angle. However, the area of the **hyperbolic sector** bounded by $y = 0$, $x^2 - y^2 = 1$, and the ray from the origin to $(\cosh t, \sinh t)$ is equal to $t/2$, just as the area of the circular sector bounded by $y = 0$, $x^2 + y^2 = 1$, and the ray from the origin to the point $(\cos t, \sin t)$ is $t/2$. (This is illustrated in Fig. 3.34, although we cannot prove it yet.)

The following addition formulas for hyperbolic functions can also be verified directly from the definitions of cosh and sinh.

$$\begin{aligned}\sinh(x + y) &= \sinh x \cosh y + \cosh x \sinh y \\ \sinh(x - y) &= \sinh x \cosh y - \cosh x \sinh y \\ \cosh(x + y) &= \cosh x \cosh y + \sinh x \sinh y \\ \cosh(x - y) &= \cosh x \cosh y - \sinh x \sinh y\end{aligned}$$

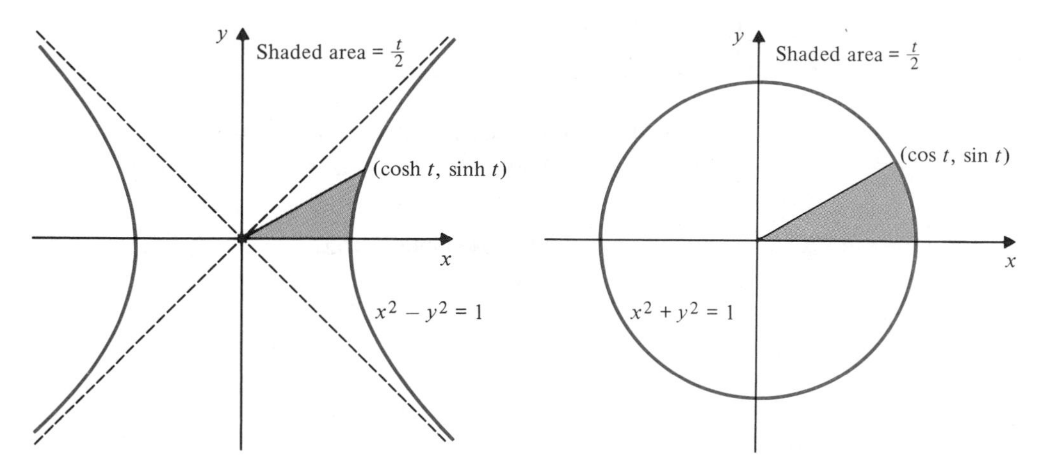

Figure 3.34

In particular,

$$\begin{aligned} \sinh 2x &= 2 \sinh x \cosh x \\ \cosh 2x &= \cosh^2 x + \sinh^2 x \\ &= 1 + 2 \sinh^2 x \\ &= 2 \cosh^2 x - 1. \end{aligned}$$

The graphs of cosh and sinh are illustrated in Fig. 3.35.

Definition 10

The remaining hyperbolic functions are defined as follows.

$$\tanh x = \frac{\sinh x}{\cosh x} = \frac{e^x - e^{-x}}{e^x + e^{-x}}$$

$$\coth x = \frac{\cosh x}{\sinh x} = \frac{e^x + e^{-x}}{e^x - e^{-x}}$$

$$\operatorname{sech} x = \frac{1}{\cosh x} = \frac{2}{e^x + e^{-x}}$$

$$\operatorname{csch} x = \frac{1}{\sinh x} = \frac{2}{e^x - e^{-x}}$$

Observe that

$$\lim_{x\to\infty} \tanh x = \lim_{x\to\infty} \frac{1 - e^{-2x}}{1 + e^{-2x}} = 1$$

$$\lim_{x\to-\infty} \tanh x = \lim_{x\to-\infty} \frac{e^{2x} - 1}{e^{2x} + 1} = -1,$$

so that the graph of $y = \tanh x$ has two horizontal asymptotes. The graph resembles that of Arctan in shape, as you can see in Fig. 3.36.

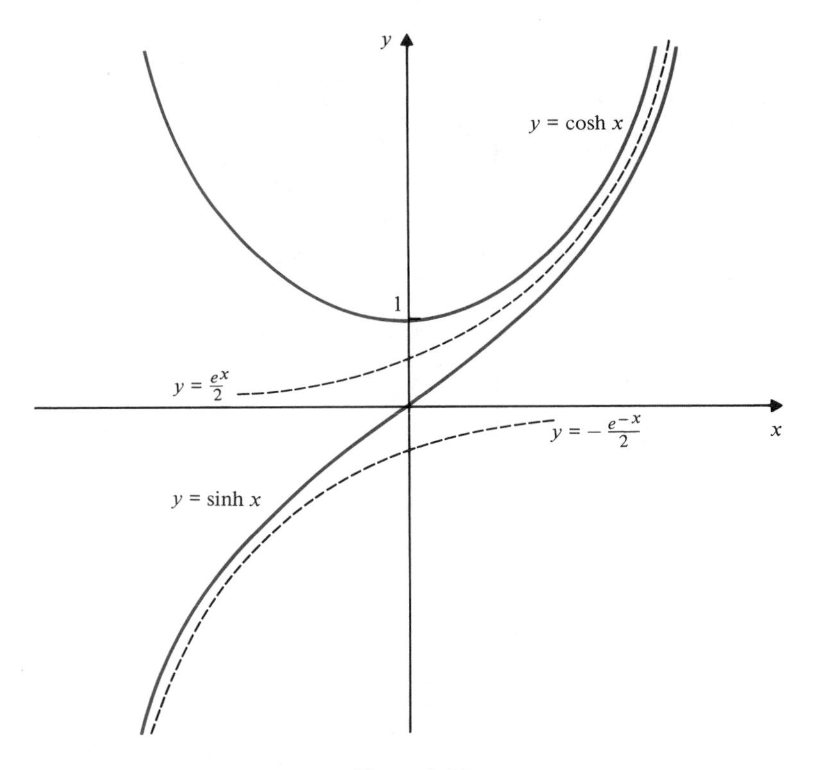

Figure 3.35

Since the hyperbolic functions are expressible in terms of exponentials, it is not surprising that their inverses are expressible in terms of logarithms. The functions sinh and tanh are increasing and therefore one-to-one and invertible on the whole real line. Their inverses are denoted $\sinh^{-1}$ and $\tanh^{-1}$, respectively.

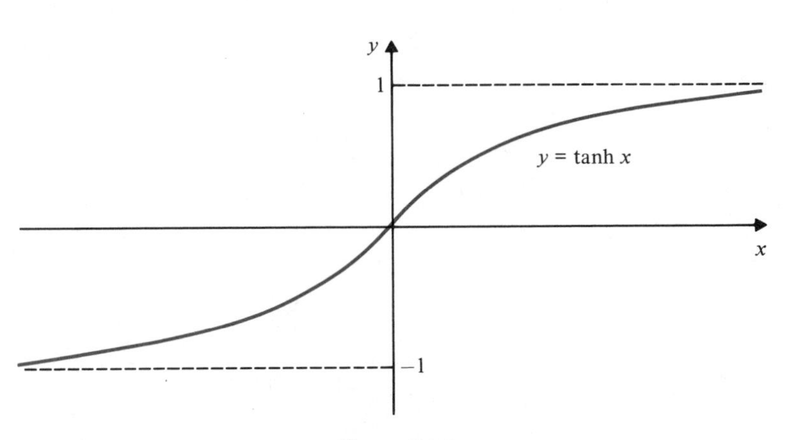

Figure 3.36

$$y = \sinh^{-1} x \Leftrightarrow x = \sinh y$$
$$y = \tanh^{-1} x \Leftrightarrow x = \tanh y$$

To express $y = \sinh^{-1} x$ in terms of logarithms, observe that

$$x = \sinh y = \frac{e^y - e^{-y}}{2} = \frac{e^{2y} - 1}{2e^y},$$

$$e^{2y} - 2xe^y - 1 = 0.$$

This is a quadratic equation in e^y and can be solved by the quadratic formula

$$e^y = \frac{2x \pm \sqrt{4x^2 + 4}}{2} = x + \sqrt{x^2 + 1}.$$

We need the positive square root since e^y cannot be negative. Hence

$$\sinh^{-1} x = \ln (x + \sqrt{x^2 + 1})$$

Similarly, we can express $y = \tanh^{-1}$ using log:

$$x = \tanh y = \frac{e^y - e^{-y}}{e^y + e^{-y}} = \frac{e^{2y} - 1}{e^{2y} + 1} \qquad (-1 < x < 1),$$

$$xe^{2y} + x = e^{2y} - 1,$$

$$e^{2y} = \frac{1 + x}{1 - x}, \qquad y = \frac{1}{2} \ln \left(\frac{1 + x}{1 - x}\right).$$

$$\tanh^{-1} x = \frac{1}{2} \ln \left(\frac{1 + x}{1 - x}\right) \quad (-1 < x < 1)$$

Since cosh is not one-to-one its domain must be restricted before an inverse can be defined. Let us define the principal value of cosh to be

$$\operatorname{Cosh} x = \cosh x \qquad (x \geq 0).$$

The inverse, Cosh^{-1}, is then defined by

$$y = \operatorname{Cosh}^{-1} x \Leftrightarrow x = \operatorname{Cosh} y \qquad (y \geq 0,\ x \geq 1).$$

As for $\sinh^{-1}$ we can obtain the formula

$$\operatorname{Cosh}^{-1} x = \ln (x + \sqrt{x^2 - 1}), \qquad (x \geq 1).$$

EXERCISES

1. Calculate the derivatives of all six hyperbolic functions using the definitions of those functions as given in this section. Express each derivative in terms of hyperbolic functions.
2. Verify the four addition formulas for cosh and sinh given in the text. Proceed by expanding the right-hand side of each in terms of exponentials.
3. Obtain addition formulas for $\tanh(x + y)$ and $\tanh(x - y)$ from those for sinh and cosh.
4. Sketch the graphs of $\coth x$, $\operatorname{sech} x$, and $\operatorname{csch} x$, showing any asymptotes.
5. Calculate the derivatives of $\sinh^{-1} x$, $\operatorname{Cosh}^{-1} x$ and $\tanh^{-1} x$. Hence express each of the indefinite integrals

$$\int \frac{dx}{\sqrt{x^2 + 1}} \qquad \int \frac{dx}{\sqrt{x^2 - 1}} \qquad \int \frac{dx}{x^2 - 1}$$

in terms of inverse hyperbolic functions.

6. Calculate the derivatives of $\sinh^{-1}(x/a)$, $\operatorname{Cosh}^{-1}(x/a)$ and $\tanh^{-1}(x/a)$ (where $a > 0$) and use your answers to provide appropriate indefinite integrals.
7. Simplify the following expressions.

a) $\sinh \ln x$ **b)** $\cosh \ln x$

c) $\tanh \ln x$ **d)** $\dfrac{\cosh \ln x + \sinh \ln x}{\cosh \ln x - \sinh \ln x}$

8.* Define $\operatorname{csch}^{-1} x$, find its domain, range, and derivative, and sketch its graph. Show that $\operatorname{csch}^{-1} x = \sinh^{-1}(1/x)$.

9.* Repeat Exercise 8 for the function $\coth^{-1} x$.

10.* Define $\operatorname{Sech} x$ at a suitably restricted version of $\operatorname{sech} x$ and repeat Exercise 8 for the function $\operatorname{Sech}^{-1} x$.

11.† Show that the functions

$f_{A,B}(x) = Ae^{kx} + Be^{-kx}$

$g_{C,D}(x) = C \cosh kx + D \sinh kx$

are both solutions of the differential equation $y'' - k^2y = 0$. (They are both general solutions.) Express $f_{A,B}$ in terms of g, and express $g_{C,D}$ in terms of f.

12.† Show that the function

$$h_{L,M}(x) = L \cosh k(x - a) + M \sinh k(x - a)$$

is also a solution of the differential equation in the previous exercise. Express $h_{L,M}$ in terms of the function f above.

13.† Solve the initial-value problem

$$\begin{cases} y'' - k^2y = 0 \\ y(a) = y_0 \\ y'(a) = v_0 \end{cases}$$

Express the solution in terms of the function $h_{L,M}$ of Exercise 12.

REVIEW EXERCISES FOR CHAPTER 3

Find dy/dx for Exercises 1–45.

1. $y = \sin 5x + \cos 3x$

2. $y = \sin^2 \frac{x}{2}$

3. $y = x \tan 2x$

4. $y = \sin^3 \cos^2 x$

5. $y = \sqrt{2 + \cos x}$

6. $y = \sec \frac{3}{x}$

7. $y = 7e^{-2x}$

8. $y = \ln (5x + 1)$

9. $y = \ln x + \ln \frac{1}{x}$

10. $y = e^{a - bx2}$

11. $y = \ln (a \cos x)$

12. $y = \sin (a \ln x)$

13. $y = \dfrac{\sin x}{\ln (1 + x)}$

14. $y = \dfrac{\ln x}{x}$

15. $y = x^3 e^{-3x}$

16. $y = e^{-x} \cos 2x$

17. $y = \text{Arcsin}\, \frac{x}{3}$

18. $y = x^2 e^{3x} \cos 4x$

19. $y = \dfrac{\text{Arctan}\, 2x}{1 + 4x^2}$

20. $y = x\, \text{Arcsin}\, x$

21. $y = \text{Arctan}\, (\sin x)$

22. $y = \sqrt{x}\, \text{Arcsin}\, \sqrt{x}$

23. $y = \cos x\, \text{Arcsin}\, x$

24. $y = e^{-2x}\, \text{Arctan}\, \frac{x}{5}$

25. $y = \dfrac{\text{Arcsin}\, x}{\text{Arctan}\, x}$

26. $y = \dfrac{e^x}{\sqrt{1 + e^{2x}}}$

27. $y = 2^{x2 - 1}$

28. $y = \dfrac{7^x}{x^7}$

29. $y = (2x)^{(3x)}$

30. $y = (\tan x)^{\text{Arctan}\, x}$

31. $y = (x^2 + 1)2^{\sqrt{x}}$

32. $y = (\ln x)^x$

33. $y = |\cos 2x|$

34. $x + \sin (xy) = 1$

35. $y^2 + \text{Arctan}\, \frac{x}{y} = x^2$

36. $e^{x(y-1)} = 1 + \ln y$

37. $y^2 \tan x = e^y$

38. $x^{xy} = y^x + 1$

39. $xy \tan (xy) = x + y$

40. $y = \sinh (\ln x)$

41. $y = \ln (\cosh x)$

42. $y = \dfrac{\sin x + \sinh x}{\cos x + \cosh x}$

43. $y = \ln \left(\dfrac{3 + \sqrt{9 + x^2}}{x} \right)$

44. $y = \ln |\sec x - \tan x|$

45. $y = \dfrac{xe^x \ln x}{(x + 1) \sin x}$

Without using tables or a calculator evaluate the expressions in Exercises 46–54.

46. $\sin\left(\frac{7\pi}{6}\right)$ **47.** $\cos\left(\frac{\pi}{12}\right)$ **48.** $\tan\left(-\frac{\pi}{3}\right)$

49. $\cos\left(\frac{\pi}{8}\right)$ **50.** $\sin\left(\frac{3\pi}{8}\right)$ **51.** $\text{Arcsin}\left(\cos\left(\frac{3\pi}{4}\right)\right)$

52. $e^{(3\ln 8 - 2\ln 4 - 4\ln 2)}$ **53.** $5^{-2/\ln 5}$ **54.** $e^{(1/e)(\ln 2^e)}$

Simplify the expressions in Exercises 55–60.

55. $(\cos x - \sin x)^2 + \sin 2x$ **56.** $\sqrt{1 + \cos x}$

57. $\sqrt{1 - \cos x}$ **58.** $\tan\left(\text{Arcsin}\frac{x}{3}\right)$

59. $\sin(\text{Arctan } 2x)$ **60.** $\sin(\text{Arcsin } x)$

Find the functions $y = f(x)$ satisfying the given conditions in Exercises 61–70.

61. $f'(x) = \sin(2x); f(0) = 1$

62. $\frac{dy}{dx} = \frac{1}{x}; y = 3$ when $x = e$

63. $f'(x) = \frac{3}{1 + x^2}; f(1) = 0$

64. $f'(x) = \frac{1}{\sqrt{1 - x^2}}; f(0) = 2$

65. $f''(x) = e^{-x}; f'(0) = 2; f(0) = 3$

66. $f''(x) = \frac{1}{x^2}; f(1) = f'(1) = 1$

67. $y' = 2y; y = 4$ when $x = 0$

68. $f'(x) = -f(x); f(1) = 1$

69. $y'' = -4y; y = 2$ and $y' = -1$ when $x = 0$

70. $y'' + y = 0; y'(0) = 0, y(0) = -3$

71. Find y'' if $y = e^x \sin x$.

72. Find $f^{(n)}(x)$ if $f(x) = e^{-kx}$.

73. Find $f^{(89)}(x)$ if $f(x) = \sin(3x)$

74. Find d^2y/dx^2 if $y = \text{Arcsin } x$.

75. Find d^3y/dt^3 if $y = \text{Arctan } t$.

Various Applications of Differentiation

4.1 Critical Points and Extreme Values

4.2 Concavity and Inflections

4.3 Sketching the Graph of a Function

4.4 Optimization Problems

4.5 Related Rates

4.6 Tangent-Line Approximations

4.7 Indeterminate Forms

We have now encountered all of the elementary functions of calculus and have developed all of the techniques required for differentiating them. It is time to turn our attention to some of the ways in which we can use derivatives. Some of these applications lie within mathematics itself. We will use derivatives to do the following:

i) locate extreme (maximum and minimum) values of a function,

ii) determine the shape of a curve,

iii) compute approximate values of functions, and locate approximately the zeros of functions, and

iv) calculate certain awkward limits.

Of course all of these mathematical applications are themselves applicable to problems arising outside of mathematics. We will consider in this chapter two kinds of problems to which the techniques of calculus can provide ready solutions:

i) optimization problems, in which some quantity is to be made largest or smallest, and

ii) related-rates problems, in which the rate of change of a quantity is expressed in terms of the rates of change of other quantities on which it depends.

In their most applied forms, these problems require translation into mathematical terms from some other context.

4.1 CRITICAL POINTS AND EXTREME VALUES

In Section 2.8 we considered some examples in which the first derivative was used to obtain information about the shape of the graph of a function; specifically we determined where the graph was rising and falling and consequently where the function had maximum and minimum values. This section reviews and formalizes these ideas.

We will consider functions defined on intervals, open or closed, or on finite unions of intervals. The domains of such functions do not contain isolated points. A point a in the domain of f is called an **endpoint of the domain** if for some $b \neq a$ either the interval (a, b) or the interval (b, a) lies outside the domain of f. For instance, $x = 0$ is an endpoint of the domain of $f(x) = \sqrt{x}$. Note that according to this definition $x = 0$ is not an endpoint of the domain of $g(x) = 1/\sqrt{x}$ since it does not belong to that domain.

Definition 1

f has the **absolute maximum value** (abs max) $f(x_0)$ at the point x_0 provided that

i) x_0 belongs to the domain of f, and

ii) $f(x) \leq f(x_0)$ for every x in the domain of f.

Similarly, f has the **absolute minimum value** (abs min) $f(x_1)$ at the point x_1 provided that

i) x_1 belongs to the domain of f, and

ii) $f(x) \geq f(x_1)$ for every x in the domain of f.

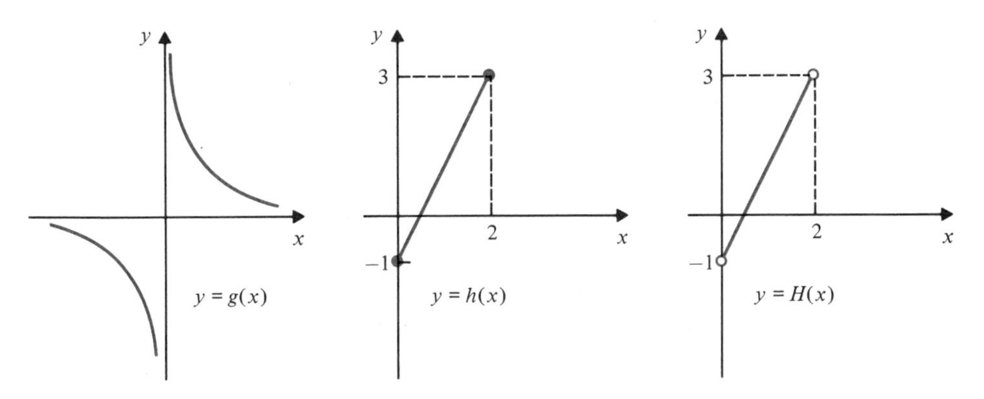

Figure 4.1

Geometrically, the graph of f is highest at x_0 and lowest at x_1.

We can also talk about the absolute maximum (or minimum) value of a function *f on an interval I* by restricting the domain of f to that interval.

Example 1 Consider the following functions.

a) $f(x) = \sin x$ on $(-\infty, \infty)$

b) $g(x) = \dfrac{1}{x}$ on $(-\infty, 0) \cup (0, \infty)$

c) $h(x) = 2x - 1$ on $[0, 2]$

d) $H(x) = 2x - 1$ on $(0, 2)$

Clearly $f(x)$ has absolute maximum value 1 at the points $x = (\pi/2) + 2k\pi$ ($k = 0, \pm 1, \pm 2, \ldots$) and absolute minimum value -1 at points $x = -(\pi/2) + 2k\pi$ ($k = 0, \pm 1, \pm 2, \ldots$). While a function can have at most one absolute maximum value and one absolute minimum value, both can occur at many points.

The graphs of g, h, and H are shown in Fig. 4.1. The function g has neither an abs max nor an abs min value. There are points as high or low as we please on its graph. The function h has an abs max value 3 at $x = 2$ and an abs min value -1 at $x = 0$. H has neither an abs max value nor an abs min value. For instance, the height of its graph gets as close as we like to 3 but is never actually 3 because 2 does not belong to the domain of H.

Example 2 Let $f(x) = x^2 - 4x$. Find the absolute maximum and absolute minimum values of $f(x)$ on the intervals $\mathbb{R}$, $(-\infty, 5]$, $[0, 2]$, $[1, 5]$, $[-1, 1]$ and $[3, 6]$.

Solution The graph of f is sketched in Fig. 4.2. It is apparent from the graph that on $\mathbb{R}$, $f(x)$ has no maximum value and has minimum value -4 at $x = 2$. This can also be seen by completing the square:

$$f(x) = (x - 2)^2 - 4 \qquad \geq -4 \text{ for all } x.$$

On $(-\infty, 5]$, max $f(x)$ does not exist; min $f(x) = -4$ at $x = 2$. On $[0, 2]$, max $f(x) = 0$ at $x = 0$; min $f(x) = -4$ at $x = 2$. On $[1, 5]$, max $f(x) = 5$ at $x = 5$; min $f(x) = -4$ at $x = 2$. On $[-1, 1]$, max $f(x) = 5$ at $x = -1$; min $f(x) = -3$ at $x = 1$. On $[4, 6]$, max $f(x) = 12$ at $x = 6$; min $f(x) = 0$ at $x = 4$.

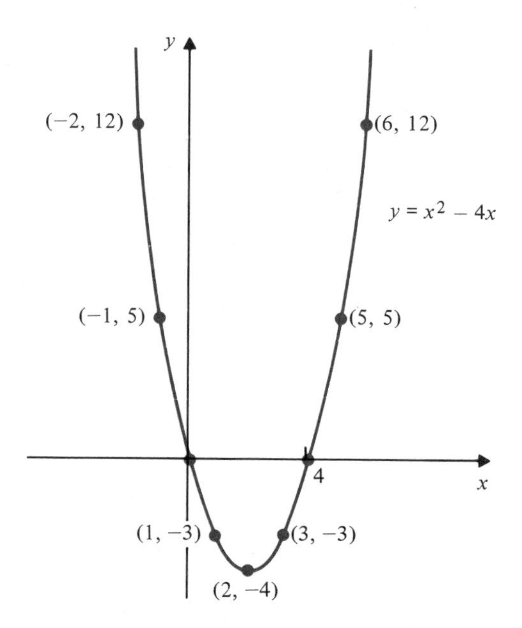

Figure 4.2

Since functions may or may not have absolute maximum or minimum values, it is useful to characterize situations in which we can be sure that a given function has a maximum or minimum value. One such situation is presented in the following theorem. Others are given in Exercises 54–56 at the end of this section.

Theorem 1 If f is a continuous function whose domain is a *closed,* finite interval, or a union of finitely many such intervals, then f has an absolute maximum value and an absolute minimum value. □

This is just a restatement of Theorem 5 of Section 1.5. The extension from one closed interval to a finite union is left as an exercise (Exercise 52 at the end of this section).

Definition 2

f has a **local maximum value** (loc max) $f(x_0)$ at the point x_0 provided

i) x_0 belongs to the domain of f and
ii) for some number $h > 0$ we have $f(x) \leq f(x_0)$ whenever x is in the domain of f and $|x - x_0| < h$.

Similarly, f has a **local minimum value** (loc min) $f(x_1)$ at x_1 provided

i) x_1 belongs to the domain of f and
ii) for some number $h > 0$ we have $f(x) \geq f(x_1)$ whenever x is in the domain of f and $|x - x_1| < h$.

Thus f has a loc max (or min) value at x if it has an abs max (or min) value at

x when restricted to points sufficiently near x. Geometrically, the graph of f is at least as high (or low) at x as it is at nearby points.

Example 3 Figure 4.3 shows the graph of a function f defined on the interval $[a, b]$. The function has local maximum values at $x = a$, $x = x_2$, and $x = x_4$ and local minimum values at $x = x_1$, $x = x_3$, and $x = b$. The abs max value of $f(x)$ is $f(x_2)$ and the abs min value is $f(b)$.

Evidently, if a function has an absolute maximum (or minimum) value, that value is the greatest (or least) of its local maximum (or minimum) values. Of course, a function may have local maximum or local minimum values without having any absolute maximum or absolute minimum. $f(x) = x^2 - 4x$ on $(-\infty, 5]$ (see Example 2) has a local maximum at $x = 5$ but no absolute maximum.

We now formalize two concepts introduced informally in Chapter 2.

Definition 3

> The point x_0 is called a **critical point** of f if $f'(x_0) = 0$.

Since functions do not possess derivatives at endpoints of their domains, a critical point cannot be an endpoint of the domain. In Example 3, $x = x_1$ and $x = x_4$ are critical points of f.

Definition 4

> The point x_0 is called a **singular point** of f provided
>
> i) x_0 is in the domain of f but is not an endpoint of the domain, and
>
> ii) $f'(x_0)$ does not exist.

For example, the functions $|x|$, $x^{1/3}$, and $x^{2/3}$ all have singular points at $x = 0$. (See Examples 2–4 of Section 2.1.) In Example 3 above $x = x_2$ and $x = x_3$ are singular points of f.

Theorem 2 If the function f is defined on an interval I and has a loc max (or loc min) value at point $x = x_0$ of I, then x_0 must be either an endpoint of I or a critical point or a singular point of f.

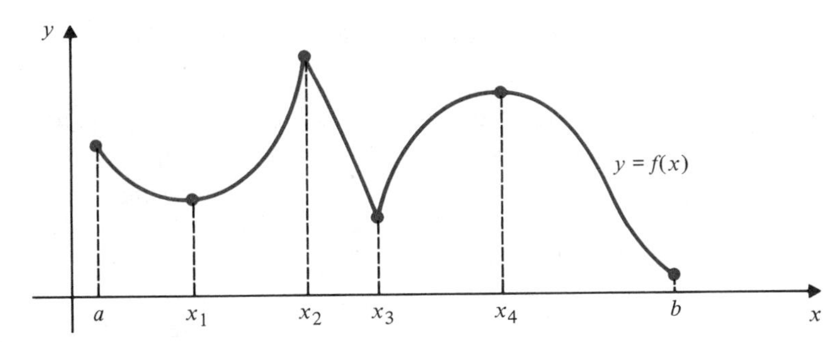

Figure 4.3

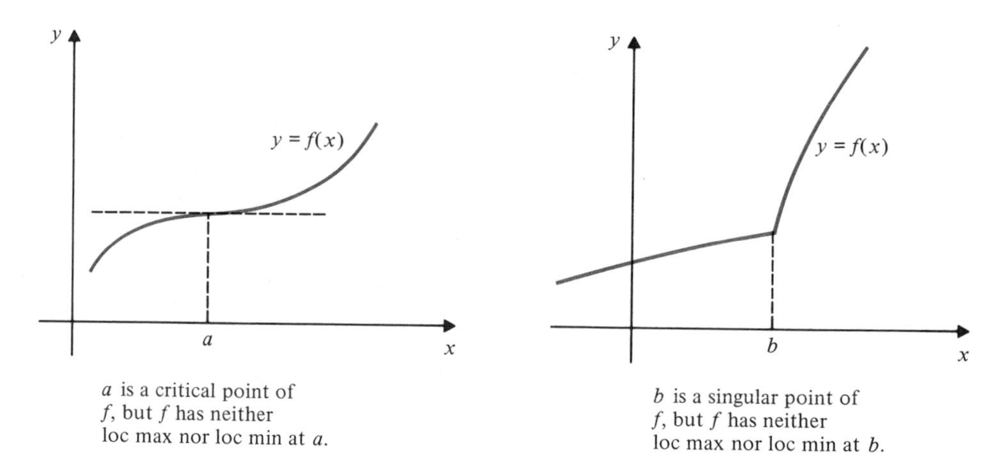

Figure 4.4

Proof Example 3 above shows that a function may have a loc max or loc min value at an endpoint, a critical point, or a singular point. If it has a loc max or loc min value at x_0, and x_0 is neither an endpoint nor a singular point, then $f'(x_0)$ exists. By Theorem 4 of Section 2.8, x_0 must be a critical point of f. □

Theorem 2 gives us a very powerful tool for finding local extreme values of continuous functions; we need only look at

i) endpoints of the domain,

iii) critical points, and

iii) singular points.

No local extreme values can exist anywhere else. However, not every one of these points will necessarily yield a local extreme value, as shown in Fig. 4.4. An endpoint of the domain need not be a loc max or loc min either. This is more difficult to illustrate. (See Exercise 53 at the end of this section.) Note that if a function is constant on an open interval, then, according to Definition 2, f has a loc max and a loc min value at every point of the interval.

The First-Derivative Test

Most functions you will encounter in elementary calculus have nonzero derivatives everywhere on their domains except possibly at a finite number of endpoints, critical points, and singular points. On intervals between these points the derivative exists and is not zero, so the function is either increasing or decreasing there. This information can be used, as in the examples of Section 2.8, to determine where maxinum and minimum values occur.

Theorem 3 (*The First-Derivative Test*) Suppose that the function f is continuous at x_0 and that x_0 is either a critical point or a singular point of f.

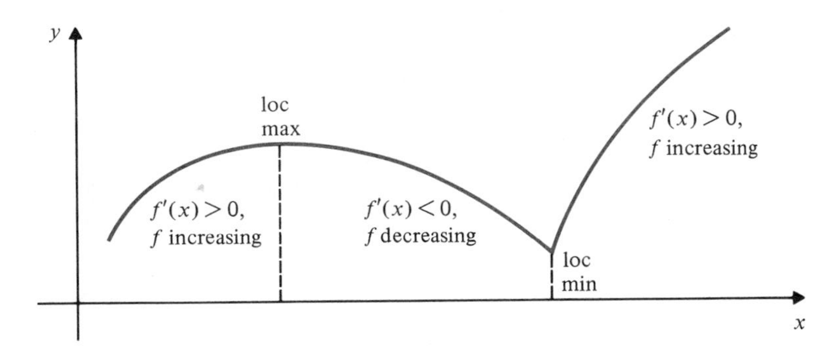

Figure 4.5

i) If $f'(x) > 0$ on some interval (x_1, x_0) and $f'(x) < 0$ on some interval (x_0, x_2), then f has a local maximum value at x_0.

ii) If $f'(x) < 0$ on some interval (x_1, x_0) and $f'(x) > 0$ on some interval (x_0, x_2), then f has a local minimum value at x_0.

Graphically, if the curve $y = f(x)$ rises to the left of x_0 and falls to the right, then f has a local maximum at x_0; if it falls to the left and rises to the right, then f has a local minimum of x_0. See Fig. 4.5. □

Proof Part (i). Let $x_1 < x < x_0$. Since f is continuous on $[x, x_0]$ and has (positive) derivative on (x, x_0), the mean-value theorem assures us that for some c in (x, x_0)

$$\frac{f(x) - f(x_0)}{x - x_0} = f'(c) > 0.$$

Since $x - x_0 < 0$, this implies that $f(x) - f(x_0) < 0$; that is, $f(x) < f(x_0)$. A similar argument shows that $f(x) < f(x_0)$ for x in (x_0, x_2). Therefore, $f(x) \le f(x_0)$ on an interval extending some distance to either side of x_0, and f must have a local maximum value at x_0.

The proof of part (ii) is similar. □

A result similar to Theorem 3 can be stated for endpoints. For instance, if x_0 is a right endpoint of the domain of f, and $f'(x) > 0$ on some interval (x_1, x_0), then f has a local maximum value at x_0.

Example 4 Locate and classify the extreme values of $f(x) = x^4 - 2x^2 - 3$ on the interval $[-2, 2]$. Sketch the graph of f.

Solution $f'(x) = 4x^3 - 4x = 4x(x - 1)(x + 1)$. The critical points are $x = 0, -1$, and 1. The corresponding values are $f(0) = -3$, $f(-1) = f(1) = -4$. There are no singular points. The endpoints are $x = -2$ and 2. The values here are $f(-2) = f(2) = 5$. The factorization of $f'(x)$,

$$f'(x) = 4x(x - 1)(x + 1),$$

is also convenient for determining the sign of $f'(x)$ on intervals between these endpoints

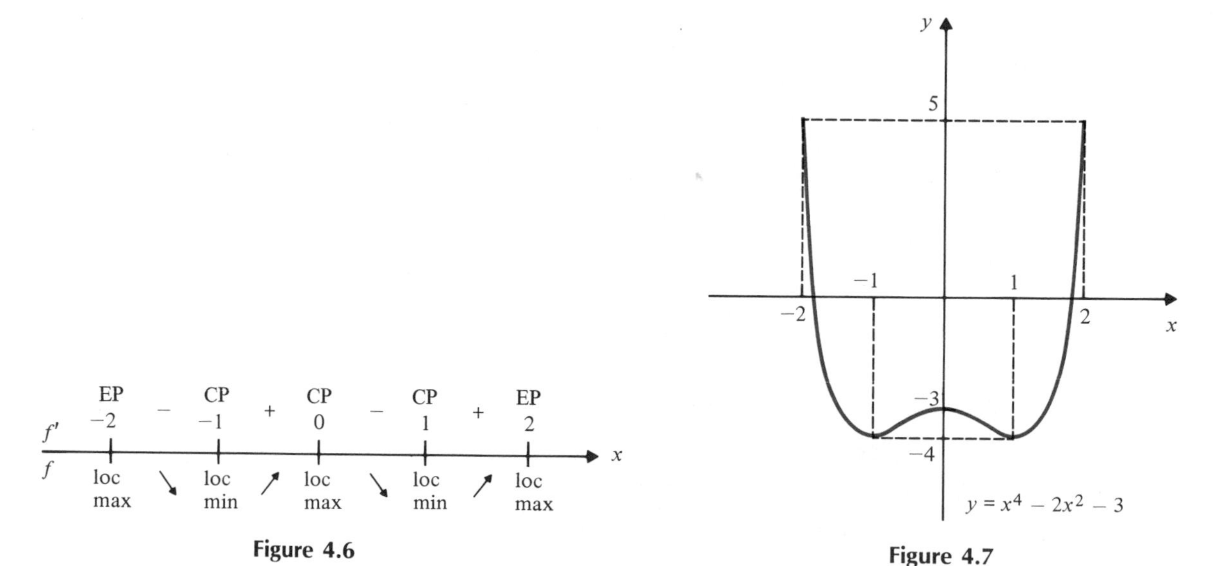

Figure 4.6

Figure 4.7

and critical points. On (1, 2) all three factors are positive, so $f'(x)$ is positive; on (0, 1) the factor $(x - 1)$ is negative while the other factors are still positive, and so $f'(x) < 0$ there. Similarly, $f'(x) > 0$ on $(-1, 0)$ because exactly two of its factors are negative there, and $f'(x) < 0$ on $(-2, -1)$ because three factors are negative on that interval. This information about the sign of $f'(x)$ on intervals between points where $f'(x)$ is 0 can also be obtained (provided $f'(x)$ is continuous as it is here) by using the intermediate-value theorem as in Example 6 of Section 1.5. We summarize the positive/negative properties of $f'(x)$ and the implied increasing/decreasing behavior of $f(x)$ in the chart in Fig. 4.6. The intervals of increase and decrease determine the local extreme values, as in the chart. The abs max of f is 5 (at the endpoints ± 2) and the abs min is -4 (at the critical points ± 1) (see Fig. 4.7).

Example 5 Locate and classify the extreme values of $F(x) = xe^x$. Sketch the graph of F.

Solution $F'(x) = e^x(x + 1)$. There are no endpoints or singular points, and the only critical point is $x = -1$, at which F has value $-1/e$. The chart in Fig. 4.8 shows that F has a loc min at $x = -1$ and no loc max.

Observe in Fig. 4.9 that

$$\lim_{x \to -\infty} xe^x = 0, \qquad \lim_{x \to \infty} xe^x = \infty.$$

Thus the x-axis is a horizontal asymptote of the graph of F at the left. Also F has abs min value $-1/e$ at $x = -1$.

Example 6 Find and classify the local extreme values of $f(x) = x - x^{2/3}$ on $[-1, 8]$. Sketch the graph of f.

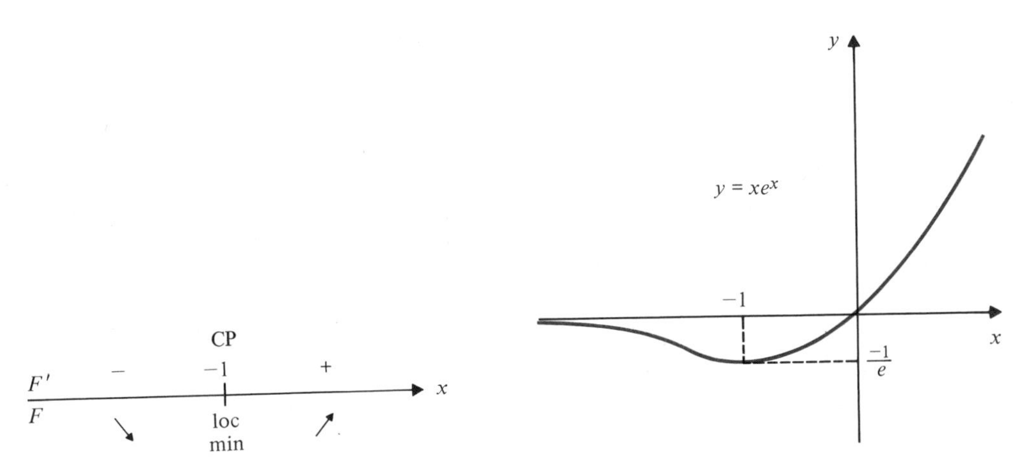

Figure 4.8

Figure 4.9

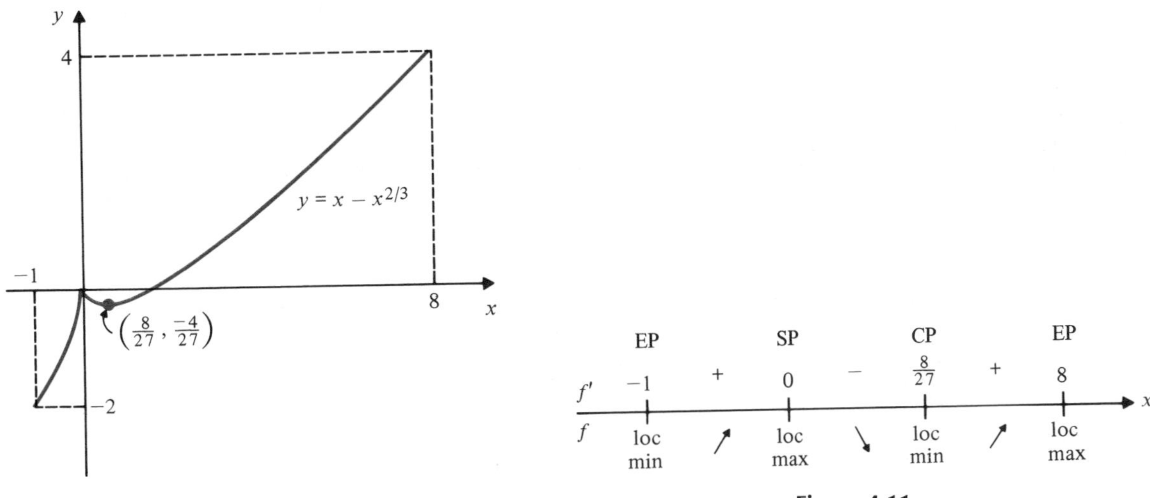

Figure 4.10

Figure 4.11

Solution $f'(x) = 1 - \dfrac{2}{3}x^{-1/3} = \left(x^{1/3} - \dfrac{2}{3}\right)\Big/ x^{1/3}$. There is a singular point, $x = 0$, and a critical pont, $x = 8/27$. The endpoints are $x = -1$ and $x = 8$. The values of f at these points are $f(-1) = -2$, $f(0) = 0$, $f(8/27) = -4/27$, and $f(8) = 4$ (see Fig. 4.10). There are two local minima and two local maxima, as indicated in Fig. 4.11. The abs max of f is 4 at $x = 8$; the abs min is -2 at $x = -1$.

EXERCISES

In Exercises 1–22 determine whether the given function has (a) any local extreme values, (b) an absolute maximum value, (c) an absolute minimum value.

1. $f(x) = x + 2$

2. $f(x) = x + 2$ on $[0, \infty)$

3. $f(x) = x + 2$ on $[-1, 1]$

4. $f(x) = x + 2$ on $(-\infty, 0)$

5. $f(x) = x + 2$ on $[-1, 1)$

6. $f(x) = x + 2$ on $(-1, 1]$

7. $f(x) = x^2 - 1$

8. $f(x) = x^2 - 1$ on $[-2, 3]$

9. $f(x) = x^2 - 1$ on $(-2, 3]$

10. $f(x) = x^2 - 1$ on $(2, 3)$

11. $f(x) = x^2 - 1$ on $(-2, 3)$

12. $f(x) = 1 - x^2$ on $(-1, 1)$

13. $f(x) = x^3 + x - 4$ on $[a, b]$

14. $f(x) = x^3 + x - 4$ on (a, b)

15. $f(x) = x^5 + x^3 + 2x$ on $[a, b)$

16. $f(x) = x^5 + x^3 + 2x$ on $(a, b]$

17. $f(x) = x^5 + x^3$ on $[-1, 1]$

18. $f(x) = \dfrac{1}{x - 1}$

19. $f(x) = \dfrac{1}{x^2 + 1}$

20. $f(x) = \dfrac{1}{x^2 + 1}$ on $[-1, 2]$

21. $f(x) = (x + 2)^{2/3}$

22. $f(x) = (x - 2)^{1/3}$

In Exercises 23–48 locate and classify all local extreme values of the given function. Determine whether any of these extreme values are absolute. Sketch the graph of the function.

23. $f(x) = 1 - x - x^2$

24. $f(x) = x^3 - 3x + 1$

25. $f(x) = (x^2 - 4)^2$

26. $f(x) = x(x - 1)^2$

27. $f(x) = x^4 + x$

28. $f(x) = x^3(x - 1)^2$

29. $f(x) = x^2(x - 1)^2$

30. $f(x) = \dfrac{\ln x}{x}$

31. $f(x) = \dfrac{x}{x^2 + 1}$

32. $f(x) = \dfrac{x^2}{x^2 + 1}$

33. $f(x) = \dfrac{x}{x^2 - 1}$

34. $f(x) = \dfrac{x^2}{x^2 - 1}$

35. $f(x) = xe^{-x}$

36. $f(x) = (x \ln x)^2$

37. $f(x) = e^{-x^2}$

38. $f(x) = xe^{-x^2}$

39. $f(x) = \cos^2 x$

40. $f(x) = x + \sin x$

41. $f(x) = |x + 1|$

42. $f(x) = |x^2 - 1|$

43. $f(x) = \sin |x|$

44. $f(x) = |\sin x|$

45.* $f(x) = (x - 1)^{2/3} - (x + 1)^{2/3}$

46.* $f(x) = (x - 1)^{1/3} + (x + 1)^{1/3}$

47.* $f(x) = x - x^{1/3}$

48.* $f(x) = (x - 1)^{1/3} + (x + 1)^{2/3}$

49.* Let $a < b < c$. If f is continuous on $[a, c]$ and $f'(x) > 0$ on (a, b) and on (b, c), prove that f is increasing on $[a, c]$.

50.* State a result analogous to that of Exercise 49 for a decreasing function.

51.* State a generalization of the results of Exercises 49 and 50 that applies to a function continuous on an interval with a derivative of constant sign at all but finitely many points of the interval.

52.* Prove Theorem 1 of this section. (*Hint:* Apply Theorem 5 of Section 1.5 to each separate closed interval of the finite union of such intervals that make up the domain of f. Use the maximum of the finitely many maximum values obtained this way.)

* Difficult and/or theoretical problems.

53.* Let

$$f(x) = \begin{cases} x \sin \dfrac{1}{x} & \text{if } x > 0, \\ 0 & \text{if } x = 0. \end{cases}$$

Show that f is continuous on $[0, \infty)$ and differentiable on $(0, \infty)$ but that it has neither a local maximum nor a local minimum value at the endpoint $x = 0$.

54.* Suppose that f is continuous on (a, b) where $-\infty \le a < b \le \infty$ and that

$$\lim_{x \to a+} f(x) = \lim_{x \to b-} f(x) = \infty.$$

Show that f has an absolute minimum value in (a, b).

55.* If the function f of Exercise 55 satisfies instead

$$\lim_{x \to a+} f(x) = \lim_{x \to b-} f(x) = -\infty,$$

show that f must have an absolute maximum value in (a, b).

56.* Suppose f is continuous on the interval (a, b) where $-\infty \le a < b \le \infty$, and suppose that

$$\lim_{x \to a+} f(x) = \lim_{x \to b-} f(x) = L$$

exists. Prove the following.

i) If $f(x) > L$ at any point of (a, b), then f has an absolute maximum value on (a, b).

ii) If $f(x) < L$ at any point of (a, b), then f has an absolute minimum value on (a, b).

iii) f must have at least one absolute extreme value on (a, b).

4.2 CONCAVITY AND INFLECTIONS

Like the first derivative, the second derivative of a function provides useful information about the shape of the graph; it determines whether the graph is bending upward (that is, has increasing slope) or bending downward (has decreasing slope) as we move along the graph toward the right.

Definition 5

> We say that the differentiable function f is **concave up** on an open interval I if the slope $f'(x)$ is an increasing function on I. Similarly, f is **concave down** on I if $f'(x)$ is a decreasing function on I.

Note that concavity is defined only for differentiable functions.

Example 1 The function f whose graph is shown in Fig. 4.12 is concave up on the interval (a, b) and concave down on the interval (b, c).

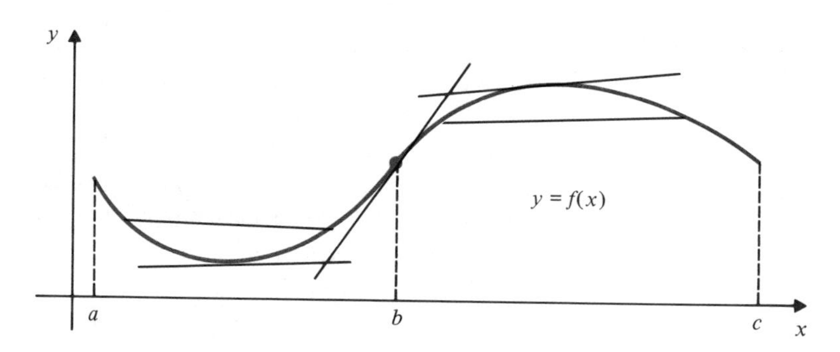

Figure 4.12

Some geometric observations can be made about concavity:

i) If f is concave up on an interval, then its graph lies above any tangent lines drawn at points in that interval, and any chord lines lie above the graph.

ii) If f is concave down on an interval, then its graph lies below any tangent lines drawn at points of that interval, and any chord lines lie below the graph.

iii) If the graph of f has a tangent line at a point, and if the concavity of f is opposite on opposite sides of that point, then the graph crosses that tangent line at that point. (This occurs at point b in Fig. 4.12. Such a point is called an *inflection point* of f.)

Definition 6

A point x_0 is called an **inflection point** of f if:

i) the graph of $y = f(x)$ has a tangent line at x_0, and

ii) the concavity of f is opposite on opposite sides of x_0.

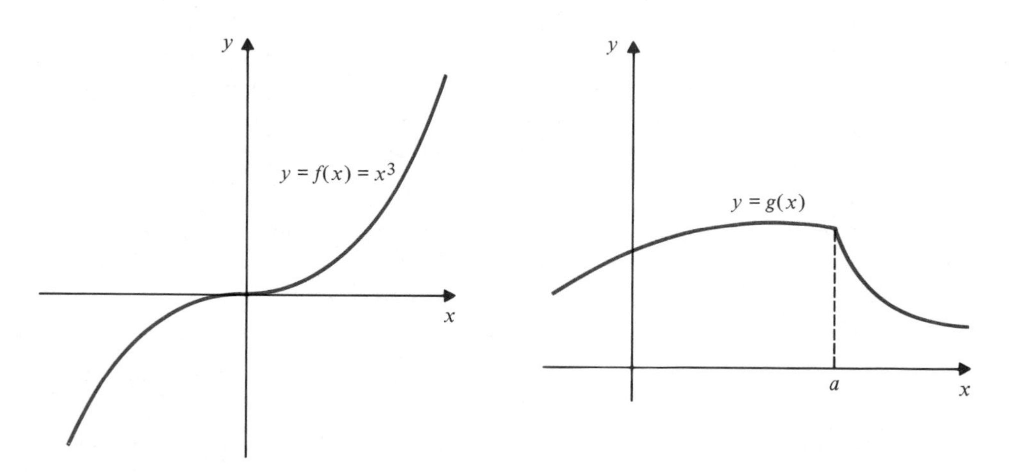

Figure 4.13

Note that (i) implies that either f is differentiable at x_0 or its graph has a vertical tangent line there and that (ii) implies that the graph crosses its tangent line at x_0.

Example 2 In Fig. 4.13, the function $f(x) = x^3$ has an inflection point at $x = 0$; the function is concave down on $x < 0$ and concave up on $x > 0$. The function g is concave down to the left of $x = a$ and concave up to the right of $x = a$, but $x = a$ is not an inflection point of g because the graph of g has no tangent line there. ($x = a$ is a singular point.)

An inflection point of a function f may or may not coincide with a critical point or a singular point. The point $x = 0$ is both a critical point and an inflection point for the function x^3 in Fig. 4.13. A function can have an inflection at a singular point only if its graph has a vertical tangent line there. For example, the graph of $y = x^{1/3}$, shown in Fig. 4.14, has a vertical tangent line at $x = 0$, and since it is concave up to the left and concave down to the right of $x = 0$, $x^{1/3}$ also has an inflection point at $x = 0$.

Theorem 4

i) If $f''(x) > 0$ on interval I, then f is concave up on I.

ii) If $f''(x) < 0$ on interval I, then f is concave down on I.

iii) If f has an inflection point at x_0 and $f''(x_0)$ exists, then $f''(x_0) = 0$.

Proof Parts (i) and (ii) follow from applying Theorem 3 of Section 2.8 to the derivative f' of f. If x_0 is an inflection point of f and $f''(x_0)$ exists, then f must be differentiable in an open interval containing x_0. Since f' goes from increasing to decreasing, or vice versa, as x passes through x_0, f' must have a loc max or loc min value at x_0. By Theorem 4 of Section 2.8, $f''(x_0) = 0$. □

Theorem 4 tells us that to find inflection points of a twice-differentiable function f we need only look at points where $f''(x) = 0$. Of course, not every such point has to be an inflection point. For example, $f(x) = x^4$, whose graph is shown in Fig. 4.15, does not have an inflection point at $x = 0$ even though $f''(0) = 12x^2|_{x=0} = 0$. In fact, x^4 is concave up on every interval.

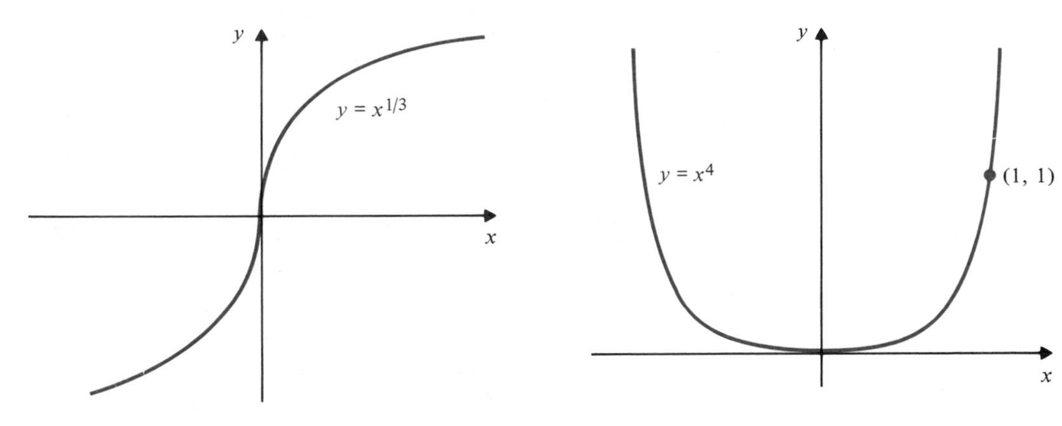

Figure 4.14 **Figure 4.15**

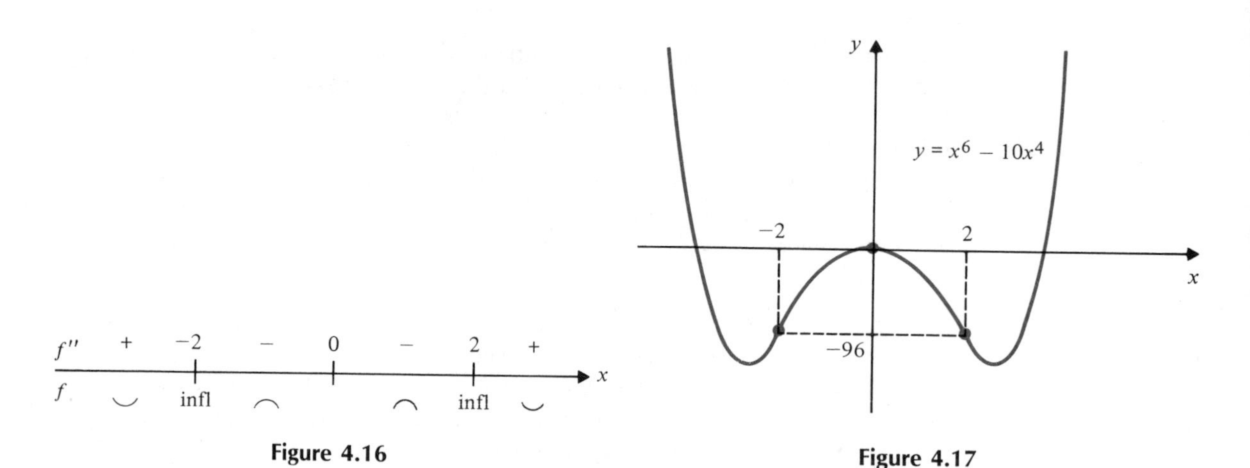

Figure 4.16

Figure 4.17

Example 3 Determine the intervals of concavity and the inflections of $f(x) = x^6 - 10x^4$.

Solution $f'(x) = 6x^5 - 40x^3$, $f''(x) = 30x^4 - 120x^2 = 30x^2(x - 2)(x + 2)$. Having factored $f''(x)$ in this manner, we can see that it vanishes only at $x = 0$, $x = -2$, and $x = 2$. On intervals of the real line separated by these points, the sign of $f''(x)$ can be determined from the signs of all its factors. No sign change occurs at $x = 0$ since $x^2 > 0$ for both positive and negative x, but the factor $x - 2$ changes sign at $x = 2$ and $x + 2$ changes sign at $x = -2$. As was the case for the first derivative, the sign information on $f''(x)$ and the consequent concavity of f can be conveniently conveyed in a chart. (See Fig. 4.16). Note that -2 and 2 are both inflection points but $x = 0$ is not because the concavity is downward on both sides of that point.

The graph of f is sketched in Fig. 4.17.

Example 4 Determine the intervals of increase and decrease, local extreme values, concavity, and inflections of $f(x) = x^4 - 2x^3 + 1$. Use the information to sketch the graph of f.

Solution

$$f'(x) = 4x^3 - 6x^2 = 2x^2(2x - 3) = 0 \text{ at } x = 0 \text{ and } x = 3/2.$$
$$f''(x) = 12x^2 - 12x = 12x(x - 1) = 0 \text{ at } x = 0 \text{ and } x = 1.$$

The behavior of f is summarized in Fig. 4.18.

$$f(0) = 1$$
$$f(1) = 0$$
$$f(3/2) = -11/16$$

f has a loc min value and abs min value of $-11/16$ at $x = 3/2$. There is a critical

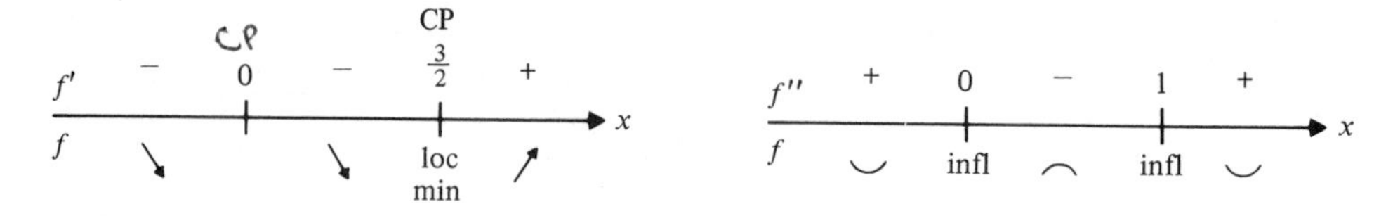

Figure 4.18

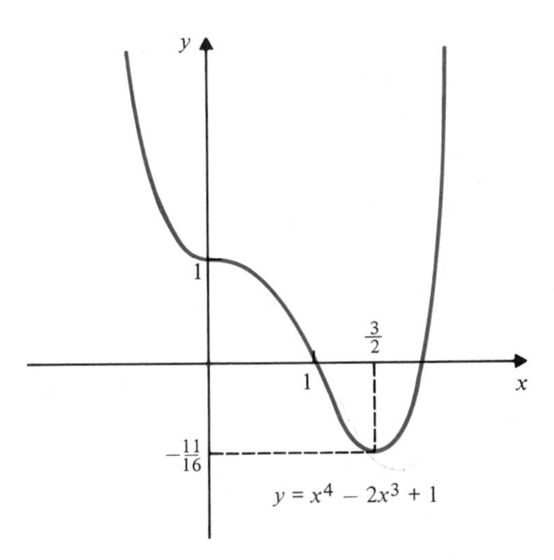

Figure 4.19

inflection point at (0, 1) and another inflection at (1, 0). The graph of f is sketched in Fig. 4.19.

The Second-Derivative Test

A function f will have a local maximum (or minimum) value at a critical point if its graph is concave downward (or upward) nearby. In fact, we can often use the value of the second derivative at the critical point to determine whether the function has a local maximum or a local minimum value there.

Theorem 5 (*The Second-Derivative Test*)

If $f'(x_0) = 0$ and $f''(x_0) < 0$, then f has a loc max value at x_0.
If $f'(x_0) = 0$ and $f''(x_0) > 0$, then f has a loc min value at x_0.

Proof Suppose that $f'(x_0) = 0$ and $f''(x_0) < 0$. Since

$$\lim_{h\to 0} \frac{f'(x_0 + h)}{h} = \lim_{h\to 0} \frac{f''(x_0 + h) - f'(x_0)}{h} = f''(x_0) < 0,$$

it follows that $f'(x_0 + h) < 0$ for all sufficiently small positive h, and $f'(x_0 + h) > 0$ for all sufficiently small negative h. By the first-derivative test (Theorem 3 of Section 4.1), f must have a local maximum value at x_0. The proof of the local minimum case is similar. □

Example 5 Find and classify the critical points of $f(x) = x^2e^{-x}$.

Solution

$$\begin{aligned} f'(x) &= (2x - x^2)e^{-x} = 0 \qquad \text{at } x = 0 \text{ and } x = 2 \\ f''(x) &= (2 - 4x + x^2)e^{-x} \\ f''(0) &= 2 > 0, \qquad f''(2) = -2e^{-2} < 0. \end{aligned}$$

Thus f has a loc min value at $x = 0$ and a loc max value at $x = 2$. See Fig. 4.20.

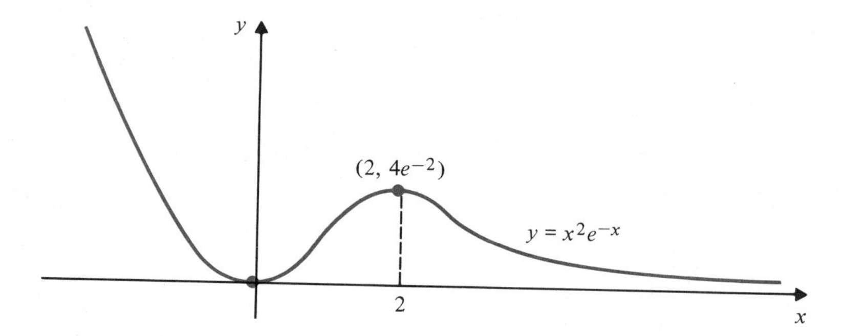

Figure 4.20

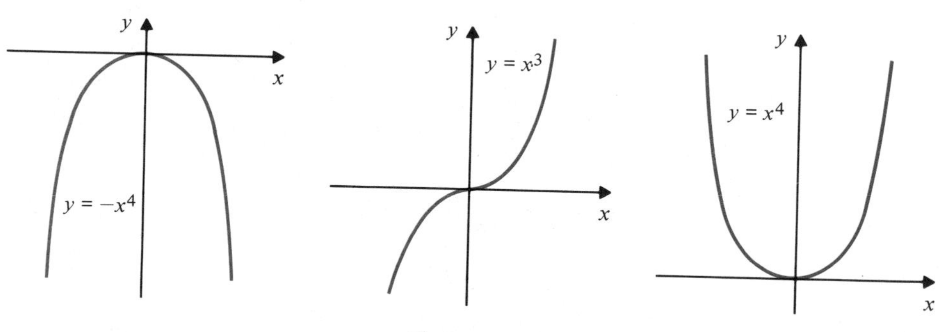

Figure 4.21

For many functions the second derivative may be more complicated to calculate than the first derivative, so the first-derivative test is likely to be of more use in classifying critical points than is the second-derivative test. Also note that the first-derivative test can locate local extreme values at singular points as well.

The second-derivative test makes no assertion about what can happen at x_0 if $f'(x_0) = 0$ and $f''(x_0) = 0$. In fact at such a point, f can have either a loc max or a loc min value, or it may have neither, but an inflection instead. The functions $-x^4$, x^4, and x^3 are examples of these three possibilities (see Fig. 4.21). (See also Exercise 48 at the end of this section.) It is possible to generalize the second-derivative test to obtain a higher derivative test to deal with some such situations. (See Exercise 49 at the end of this section.)

EXERCISES

In Exercises 1–30 determine the intervals of constant concavity of the given function and locate any inflection points.

1. $f(x) = \sqrt{x}$

2. $f(x) = (x + 2)^{1/3}$

3. $f(x) = 2x - x^2$

4. $f(x) = x^2 + 2x + 3$

5. $f(x) = x^2 + x^4$

6. $f(x) = 1 - x^2 + x^4$

7. $f(x) = x + x^3$

8. $f(x) = x - x^3$

9. $f(x) = 10x^3 - 3x^5$

10. $f(x) = 10x^3 + 3x^5$

11. $f(x) = 8x^3 - 6x + 1$

12. $f(x) = 8x^3 - 6x^2 + 1$

13. $f(x) = \sin x$

14. $f(x) = \cos 3x$

15. $f(x) = \tanh x$

16. $f(x) = x + \sin 2x$

17. $f(x) = (x^2 - 9)^2$

18. $f(x) = (x^2 - 4)^3$

19. $f(x) = \dfrac{x^3}{3} - 4x^2 + 12x - \dfrac{25}{3}$

20. $f(x) = e^{-x^2}$

21. $f(x) = xe^{-x^2}$

22. $f(x) = x^2e^{-x^2}$

23. $f(x) = (x + 1)(x - 1)^2$

24. $f(x) = (x - 1)^{1/3} + (x + 1)^{1/3}$

25. $f(x) = \dfrac{1}{x^2 + 1}$

26. $f(x) = \dfrac{x^2}{x^2 - 4}$

27. $f(x) = \dfrac{x}{x^2 - 1}$

28. $f(x) = \dfrac{x}{x^2 + 1}$

29. $f(x) = \ln(1 + x^2)$

30. $f(x) = (\ln x)^2$

31. Discuss the concavity of the linear function $f(x) = ax + b$. Are there any inflections?

In Exercises 32–45 find and classify all critical points. Use the second-derivative test.

32. $f(x) = 3x^3 - 36x - 3$

33. $f(x) = x(x - 2)^2 + 1$

34. $f(x) = 3x^5 - 65x^3 + 540x$

35. $f(x) = x(x - 2)^3 + 1$

36. $f(x) = x + \dfrac{4}{x}$

37. $f(x) = x^3 + \dfrac{1}{x}$

38. $f(x) = \dfrac{x}{2^x}$

39. $f(x) = \dfrac{x}{1 + x^2}$

40. $f(x) = xe^x$

41. $f(x) = x \ln x$

42. $f(x) = (x^2 - 4)^2$

43. $f(x) = (x^2 - 4)^3$

44. $f(x) = (x^2 - 3)e^x$

45. $f(x) = x^4e^{-2x^2}$

46.* Verify that if f is concave up on an interval then its graph lies above its tangent lines on that interval. (*Hint:* Suppose f is concave on an open interval containing x_0. Let $h(x) = f(x) - f(x_0) - f'(x_0)(x - x_0)$. Show that h has a local minimum value at x_0 and hence that $h(x) \geq 0$ on the interval. Show that $h(x) > 0$ if $x \neq x_0$.)

47.* Verify that the graph $y = f(x)$ crosses its tangent line at an inflection point. (*Hint:* Consider separately the cases where the tangent line is vertical and nonvertical.)

48. For $f_n(x) = x^n$ and $g_n(x) = -x^n (n = 2, 3, 4, \ldots)$ determine whether $x = 0$ is a loc max, a loc min, or an inflection point.

49.* Use your conclusions from Exercise 48 to suggest a generalization of the second-derivative test that applies when

$$f'(x_0) = f''(x_0) = \cdots = f^{(k-1)}(x_0) = 0, \qquad f^{(k)}(x_0) \neq 0,$$

for some $k \geq 2$.

50.* This problem shows that no test based solely on the signs of derivatives at x_0 can determine whether every function with critical point at x_0 has a loc max, a loc min, or an inflection point there. Let

$$f(x) = \begin{cases} e^{-1/x^2} & \text{if } x \neq 0, \\ 0 & \text{if } x = 0. \end{cases}$$

Prove the following.

i) $\lim_{x\to 0} x^{-n}f(x) = 0$ for $n = 0, 1, 2, 3, \ldots$

ii) $\lim_{x\to 0} P(1/x)f(x) = 0$ for every polynomial P.

iii) For $x \neq 0$, $f^{(k)}(x) = P_k(1/x)f(x)$ $(k = 1, 2, 3, \ldots)$ where P_k is a polynomial.

iv) $f^{(k)}(0)$ exists and equals 0 for $k = 1, 2, 3, \ldots$

v) f has a local minimum at $x = 0$; $-f$ has a local maximum at $x = 0$.

vi) If $g(x) = xf(x)$, then $g^{(k)}(0) = 0$ for every positive integer k and g has an inflection point at $x = 0$.

51.* A critical point of a function may be neither a local maximum nor a local minimum nor an inflection point. Show this by considering the following function.

$$f(x) = \begin{cases} x^2 \sin \dfrac{1}{x} & \text{if } x \neq 0 \\ 0 & \text{if } x = 0 \end{cases}$$

Show that $f'(0) = f(0) = 0$, so the x-axis is tangent to the graph of f at $x = 0$; but $f'(x)$ is not continuous at $x = 0$, so $f''(0)$ does not exist. Show that the concavity of f is not constant on any interval with endpoint 0.

4.3 SKETCHING THE GRAPH OF A FUNCTION

When sketching the graph of a function $y = f(x)$ we have three sources of useful information:

i) the function f itself, from which we determine the coordinates of some points on the graph, the symmetry of the graph, and any asymptotes,

ii) the first derivative, f', from which we determine the intervals of increase and decrease and any local extreme values, and

iii) the second derivative, f'', from which we determine concavity and inflections.

Items (ii) and (iii) have been investigated in the previous two sections. We consider here information coming from the function itself, and we illustrate the entire sketching procedure with several examples using all three sources of information.

A graph may be sketched by plotting the coordinates of many points on it and then joining them by a suitably smooth curve. This simplistic procedure is both tedious and uninformative; very often it can fail to reveal the most interesting aspects of the graph (singular points, extreme values, and so on). We could also compute the slope at each of the plotted points and, by drawing short line segments through these points with the appropriate slopes, ensure that the sketched graph passes through each plotted point with the correct slope. A better procedure is to obtain the coordinates of only a few points and use qualitative information from the first and second derivatives to determine the shape of the graph between these points.

Besides critical and singular points and inflections, a graph may have other "interesting" points. The **intercepts** (points at which the graph intersects the coordinate axes) are usually among these. When sketching any graph it is wise to try to find all such intercepts, that is, all points with coordinates $(x, 0)$ and $(0, y)$ that lie on the

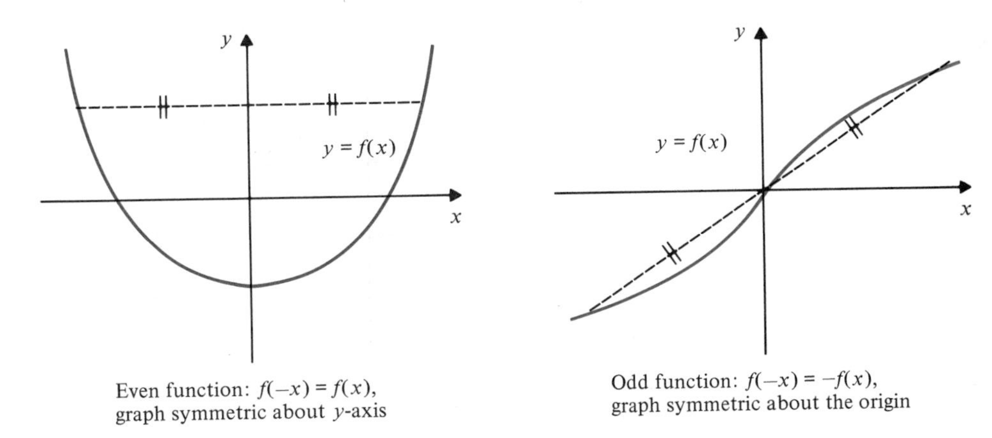

Even function: $f(-x) = f(x)$, graph symmetric about y-axis

Odd function: $f(-x) = -f(x)$, graph symmetric about the origin

Figure 4.22

graph. Of course, not every graph will have such points, and even when they do exist it may not always be possible to compute them exactly. Whenever a graph is made up of several disconnected pieces (as is often the case when there are vertical asymptotes), the coordinates of at least one point on each piece must be obtained. It can be very useful to determine the slopes at those points too.

In Section 1.2 we discussed odd and even functions and observed that odd functions have graphs that are symmetric about the origin, while even functions have graphs that are symmetric about the y-axis, as represented in Fig. 4.22. These symmetries can aid with the sketching of the graphs of such functions.

Some of the curves we have sketched in previous sections have had **asymptotes,** that is, straight lines to which the curve draws arbitrarily near as it recedes to infinite distance from the origin. Asymptotes are of three types: vertical, horizontal, and oblique.

> The graph of $y = f(x)$ has a vertical asymptote at $x = a$ if
>
> $$\lim_{x \to a-} f(x) = \pm\infty, \qquad \text{or } \lim_{x \to a+} f(x) = \pm\infty, \qquad \text{or both.}$$

This situation tends to arise when $f(x)$ is a fraction and $x = a$ is a zero of the denominator.

Example 1 $f(x) = 1/(x^2 - x)$ has vertical asymptotes at $x = 0$ and $x = 1$. (See Fig. 4.23.) Note that $\lim_{x\to 0-} 1/(x^2 - x) = \infty$, $\lim_{x\to 0+} 1/(x^2 - x) = -\infty$, $\lim_{x\to 1-} 1/(x^2 - x) = -\infty$, and $\lim_{x\to 1+} 1/(x^2 - x) = \infty$.

> The graph of $y = f(x)$ has a horizontal asymptote $y = L$ if
>
> $$\lim_{x \to \infty} f(x) = L, \qquad \text{or } \lim_{x \to -\infty} f(x) = L, \qquad \text{or both.}$$

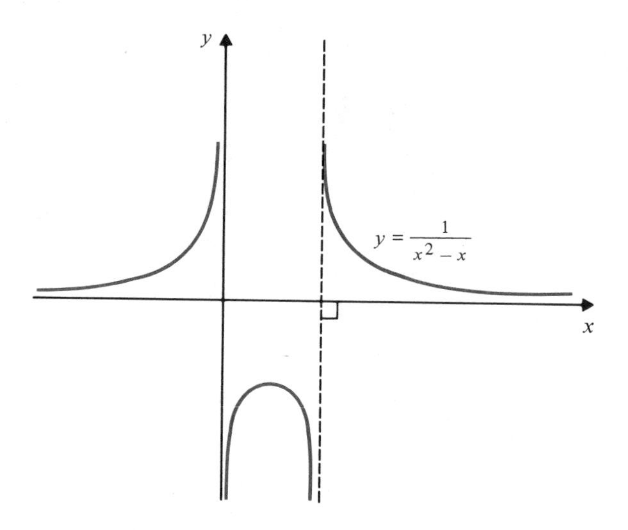

Figure 4.23

Example 2 a) $f(x) = 1/(x^2 - x)$ has horizontal asymptote $y = 0$; $\lim_{x \to \pm\infty} 1/(x^2 - x) = 0$.

b) $g(x) = 3x^2/(2x^2 + 1)$ has horizontal asymptote $y = 3/2$; $\lim_{x \to \pm\infty} 3x^2/(2x^2 + 1) = 3/2$. See Fig. 4.24.

In both (a) and (b) the asymptotes are "two-sided," that is, the graphs approach the asymptotes as x approaches both infinity and negative infinity. The function Arctan x has two one-sided horizontal asymptotes, $y = \pi/2$ (as $x \to \infty$) and $y = -(\pi/2)$ (as $x \to -\infty$) (see Fig. 4.25).

It can also happen that the graph of a function $f(x)$ approaches a nonhorizontal straight line as x approaches ∞ or $-\infty$ (or both). Such a line is called an *oblique asymptote* of the graph.

> The straight line $y = ax + b$ ($a \neq 0$) is an oblique asymptote of the graph of $y = f(x)$ if
>
> $$\lim_{x \to -\infty} (f(x) - (ax + b)) = 0, \quad \text{or} \quad \lim_{x \to \infty} (f(x) - (ax + b)) = 0, \quad \text{or both.}$$

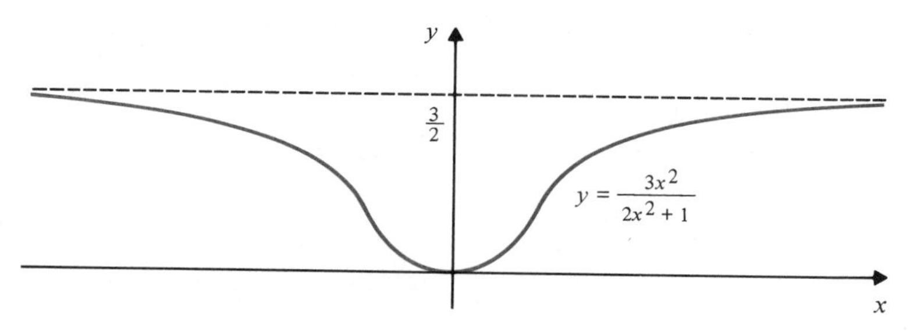

Figure 4.24

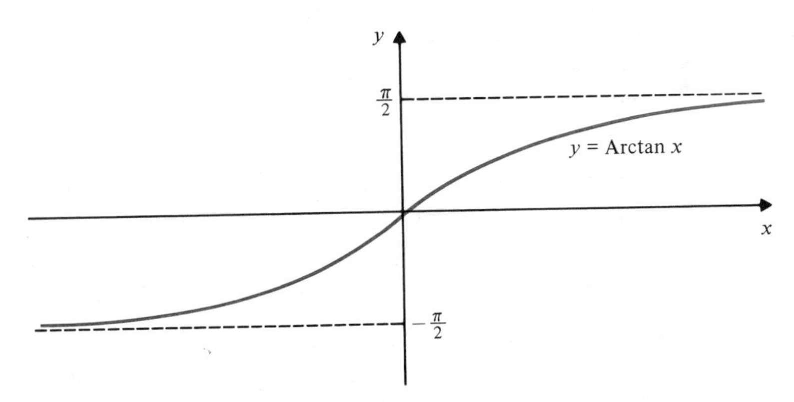

Figure 4.25

Example 3 Consider the function $f(x) = (x^2 + 1)/x = x + 1/x$, whose graph is given in Fig. 4.26. The straight line $y = x$ is an oblique asymptote of the graph of f because

$$\lim_{x \to \pm\infty} (f(x) - x) = \lim_{x \to \pm\infty} \frac{1}{x} = 0.$$

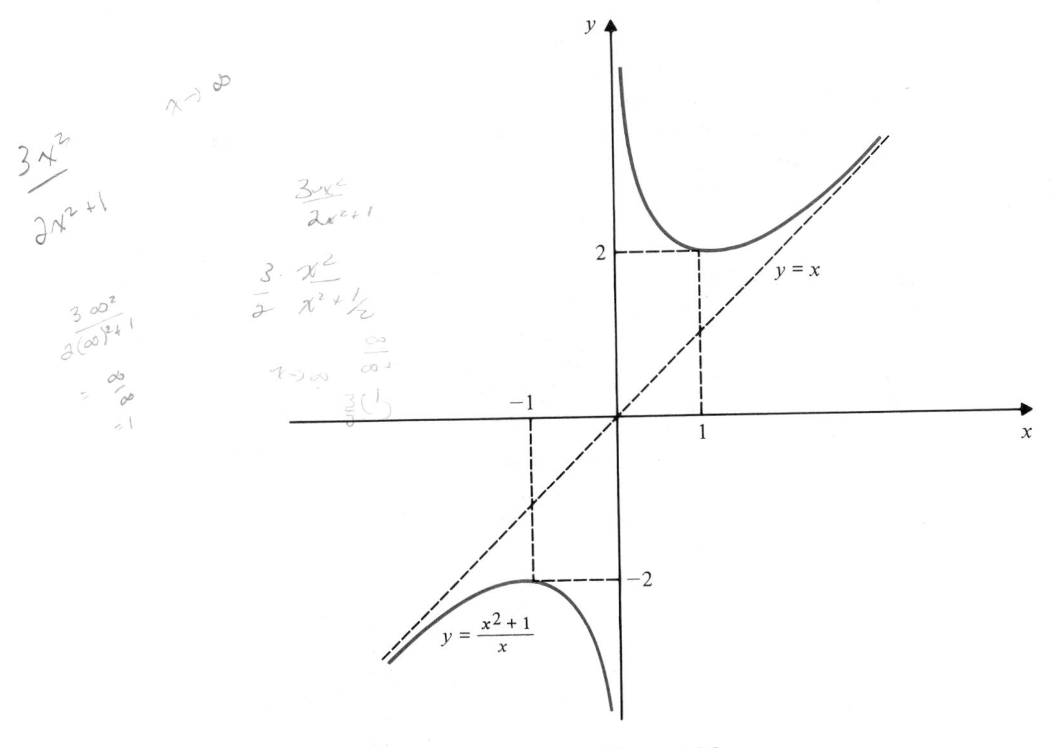

Figure 4.26

It is possible to be quite specific about the asymptotes of a rational function (that is, a quotient of polynomials). Suppose that $f(x) = P_n(x)/Q_m(x)$ where P_n and Q_m are polynomials of degree n and m, respectively. Suppose also that P_n and Q_m have no common linear factors. Then

i) The graph of f has vertical asymptotes at every position x such that $Q_m(x) = 0$.

ii) The graph of f has an oblique asymptote if $n = m + 1$. This asymptote can be found by dividing Q_m into P_n to obtain a linear quotient, $ax + b$ and remainder R, a polynomial of degree at most $m - 1$. That is,

$$f(x) = ax + b + \frac{R(x)}{Q_m(x)}.$$

The asymptote is $y = ax + b$.

iii) The graph of f has a horizontal asymptote $y = L$ $(L \neq 0)$ if $n = m$. Again, L can be found by long division or by simply evaluating the limit of $f(x)$ as x approaches ∞.

iv) The graph of f has a horizontal asymptote $y = 0$ if $n < m$.

We now illustrate the sketching procedure with several examples.

Example 4 Sketch the graph of

$$y = \frac{2x^3 + x^2 + 1}{x^2} = 2x + 1 + \frac{1}{x^2}.$$

Solution First calculate the derivatives y' and y'' and, where possible, factor them so that their zeros become obvious.

$$y' = 2 - \frac{2}{x^3} = \frac{2(x^3 - 1)}{x^3} = \frac{2(x - 1)(x^2 + x + 1)}{x^3}$$

$$y'' = \frac{6}{x^4}$$

From y: Symmetry: none (y is neither odd nor even).
Asymptotes: vertical $x = 0$,

oblique $y = 2x + 1$ $\left(\begin{array}{c} y - 2x - 1 = 1/x^2 \to 0 \\ \text{as } x \to \pm\infty \end{array}\right).$

Intercepts: evidently $y = 0$ if $x = -1$; point $(-1, 0)$.

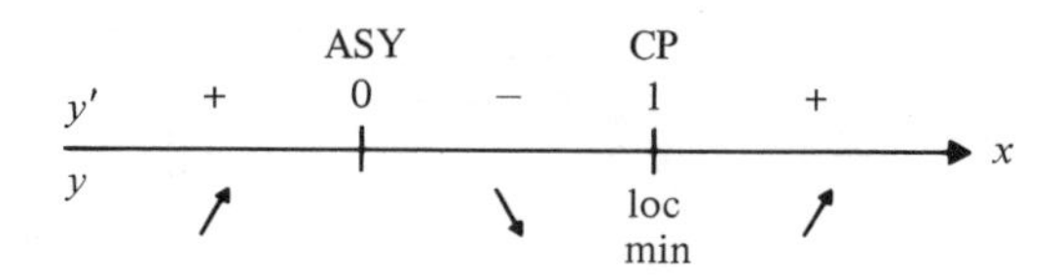

Figure 4.27

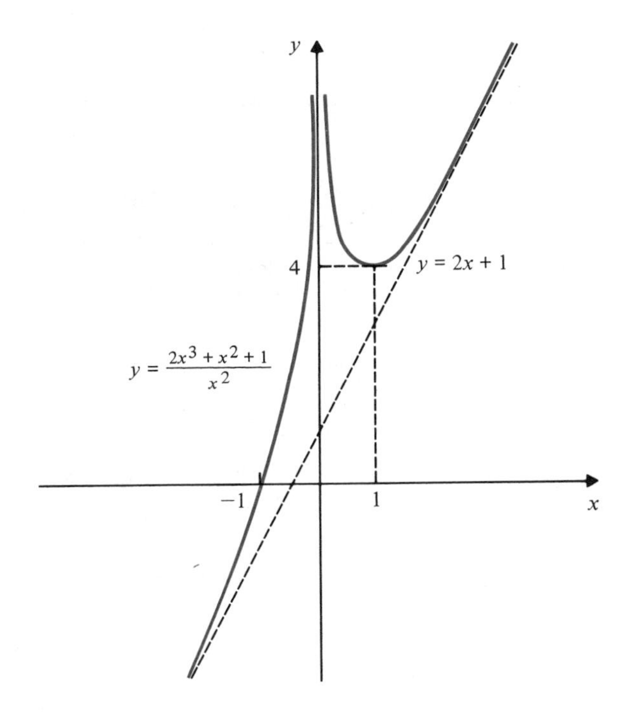

Figure 4.28

From y' (see Fig. 4.27): Critical point: $x = 1$; point $(1, 4)$; y' not defined if $x = 0$ but this is not a singular point (it is a vertical asymptote).

From y'': $y'' > 0$ everywhere except where not defined at $x = 0$. The graph is everywhere concave up. There are no inflections.

The graph is sketched in Fig. 4.28.

Example 5 Sketch

$$y = \frac{x^2 - 1}{x^2 - 4}.$$

Solution We have

$$y' = \frac{-6x}{(x^2 - 4)^2}, \qquad y'' = \frac{6(3x^2 + 4)}{(x^2 - 4)^3}.$$

From y: Symmetry: about the y-axis (y is even).
Asymptotes: horizontal $y = 1$ (at $\pm\infty$),
vertical $x = 2$ and $x = -2$.
Intercepts: $(0, 1/4)$, $(-1, 0)$, $(1, 0)$.
Other points: $(-3, 8/5)$, $(3, 8/5)$. (Since the two vertical asymptotes divide the graph into three parts we need points on each part. The outer parts require points with $|x| > 2$.)

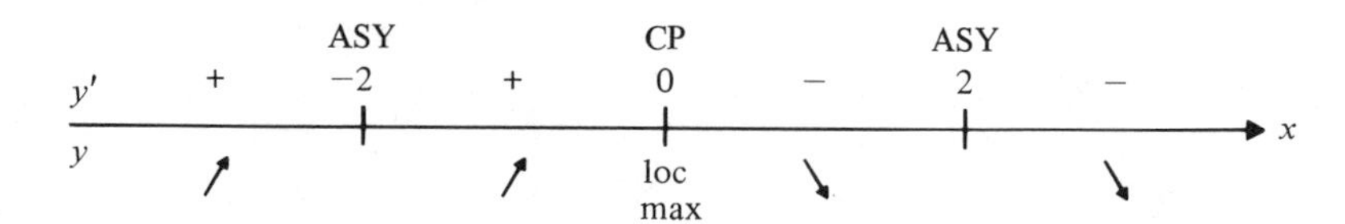

Figure 4.29

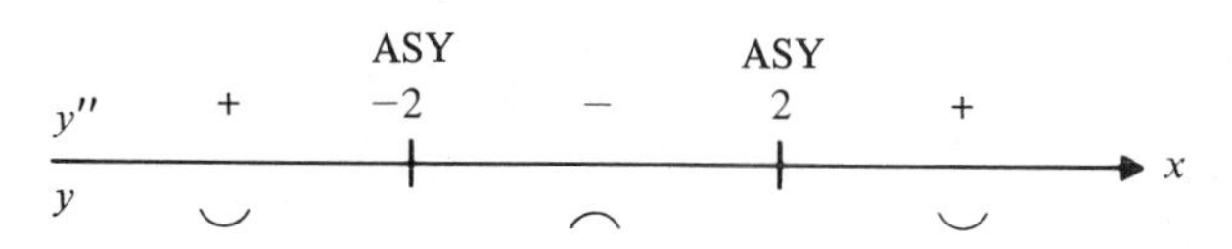

Figure 4.30

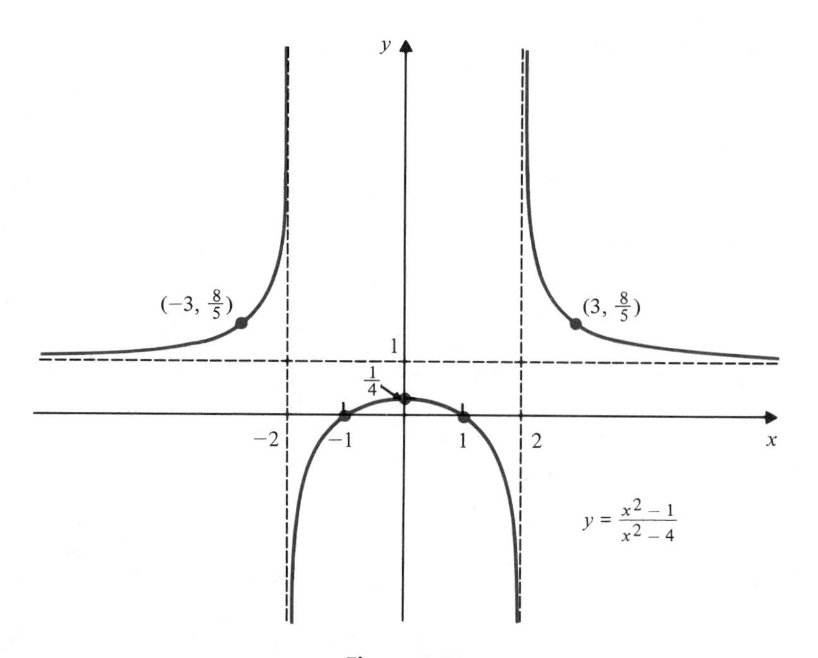

Figure 4.31

From y' (see Fig. 4.29): Critical point: $x = 0$; y' not defined at $x = 2$ or $x = -2$.

From y'' (see Fig. 4.30): $y'' = 0$ nowhere; y'' not defined at $x = 2$ or $x = -2$. There are no inflections

The graph is sketched in Fig. 4.31.

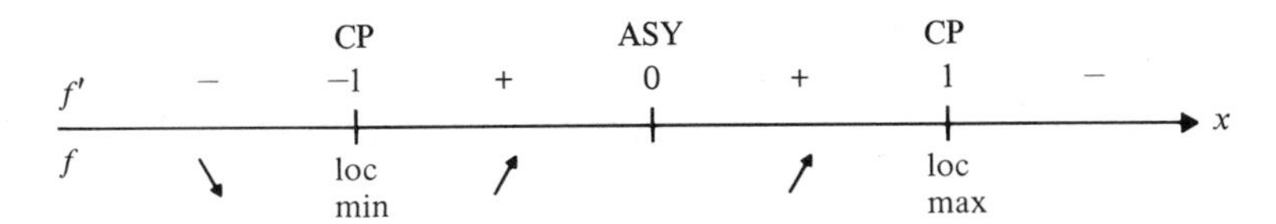

Figure 4.32

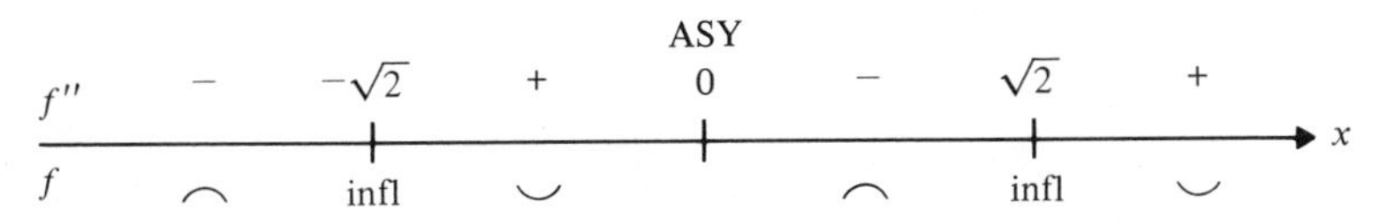

Figure 4.33

Example 6 Sketch the graph of the function

$$f(x) = \frac{3x^2 - 1}{x^3} = \frac{3}{x} - \frac{1}{x^3}.$$

Solution We have

$$f'(x) = -\frac{3}{x^2} + \frac{3}{x^4} = \frac{-3(x^2 - 1)}{x^4} = \frac{-3(x - 1)(x + 1)}{x^4},$$

$$f''(x) = \frac{6}{x^3} - \frac{12}{x^5} = \frac{6(x^2 - 2)}{x^5} = \frac{6(x - \sqrt{2})(x + \sqrt{2})}{x^5}.$$

From f: Symmetry: about the origin (f is odd).
Asymptotes: vertical $x = 0$,
horizontal $y = 0$ (at $\pm\infty$).
Intercepts: $(\pm 1/\sqrt{3}, 0) \simeq (\pm 0.58, 0)$.

From f' (see Fig. 4.32): Critical points: $x = -1$ and 1; points $(-1, -2)$, $(1, 2)$; f' not defined at $x = 0$ (vertical asymptote).

From f'' (see Fig. 4.33): $f''(x) = 0$ at $x = -\sqrt{2}$ and $x = \sqrt{2}$; not defined at $x = 0$.
Inflection points: $(\pm\sqrt{2}, \pm 5/2\sqrt{2}) \simeq (\pm 1.41, \pm 1.77)$.

The graph is sketched in Fig. 4.34.

Example 7 Sketch the graph of $y = xe^{-x^2/2}$.

Solution We have $y' = (1 - x^2)e^{-x^2/2}$, $y'' = x(x^2 - 3)e^{-x^2/2}$.

From y: Symmetry: about the origin (y is odd).
Asymptotes: horizontal $y = 0$. (Note that if $t = x^2/2$, then $|xe^{-x^2/2}| = \sqrt{2t}e^{-t} \to 0$ as $t \to \infty$ (hence as $x \to \pm\infty$), as can easily be deduced from Theorem 6 of Section 3.5.)
Intercepts: $(0, 0)$.

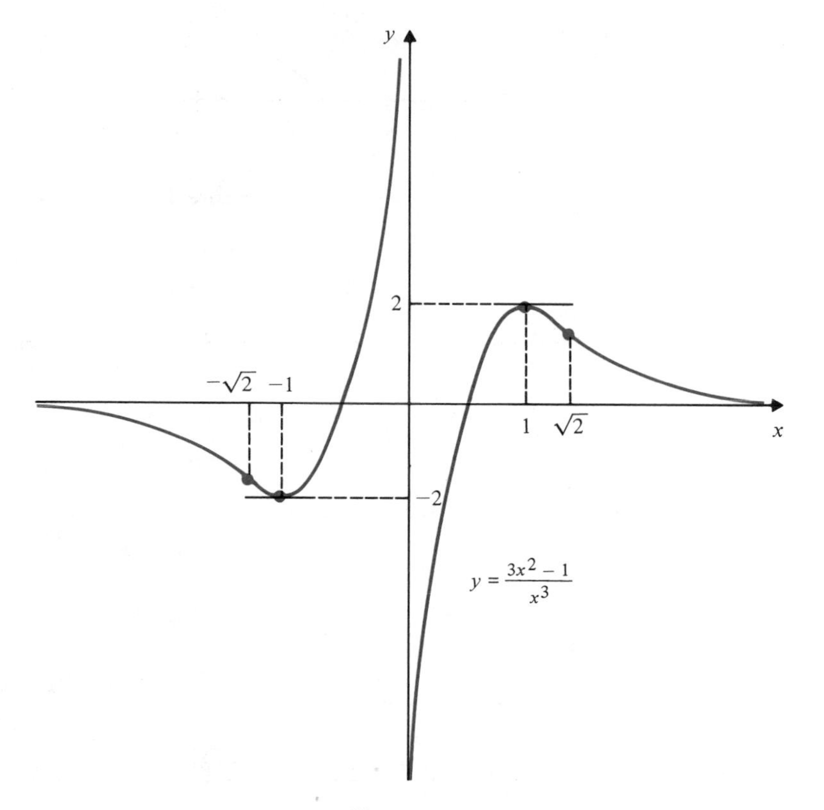

Figure 4.34

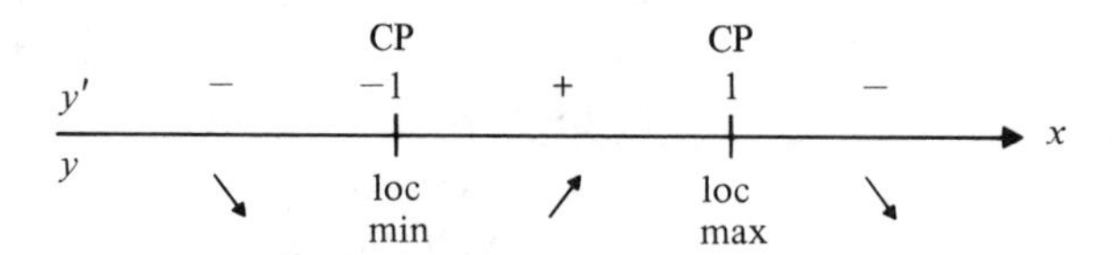

Figure 4.35

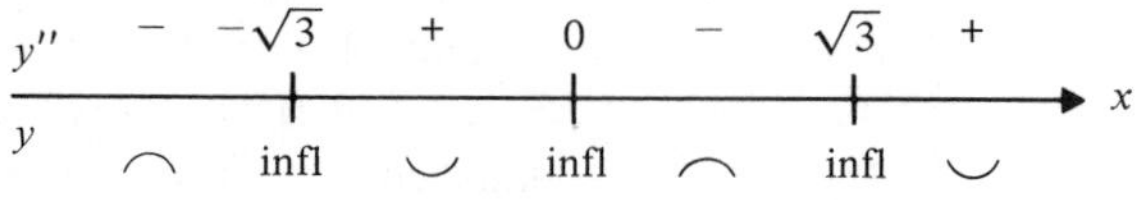

y″ − $-\sqrt{3}$ + 0 − $\sqrt{3}$ +

x

y ⌒ infl ⌣ infl ⌒ infl ⌣

Figure 4.36

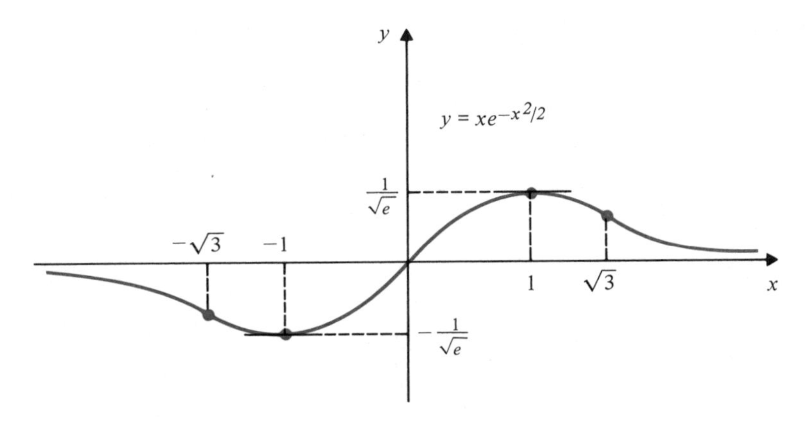

Figure 4.37

From y' (see Fig. 4.35): Critical points: $x = \pm 1$; points $(\pm 1, \pm 1/\sqrt{e}) \simeq (\pm 1, \pm 0.61)$.

From y'' (see Fig. 4.36): $y'' = 0$ at $x = 0$ and $x = \pm\sqrt{3}$; points $(0, 0)$, $(\pm\sqrt{3}, \pm\sqrt{3}e^{-3/2}) \simeq (\pm 1.73, \pm 0.39)$

The graph is sketched in Fig. 4.37.

Example 8 Sketch the graph of $f(x) = (x^2 - 1)^{2/3}$.

Solution

$$f'(x) = \frac{4}{3}\frac{x}{(x^2-1)^{1/3}}, \qquad f''(x) = \frac{4}{9}\frac{x^2-3}{(x^2-1)^{4/3}}.$$

From f: Symmetry: about the y-axis (f is an even function).
Asymptotes: none ($f(x)$ grows like $x^{4/3}$ as $x \to \pm\infty$).
Intercepts: $(\pm 1, 0)$, $(0, 1)$.

From f' (see Fig. 4.38): Critical points: $x = 0$.
Singular points: $x = \pm 1$.

From f'' (see Fig. 4.39): $f''(x) = 0$ at $x = \pm\sqrt{3}$; points $(\pm\sqrt{3}, 2^{2/3}) \simeq (\pm 1.73, 1.59)$; $f''(x)$ not defined at $x = 1$.

The graph is sketched in Fig. 4.40.

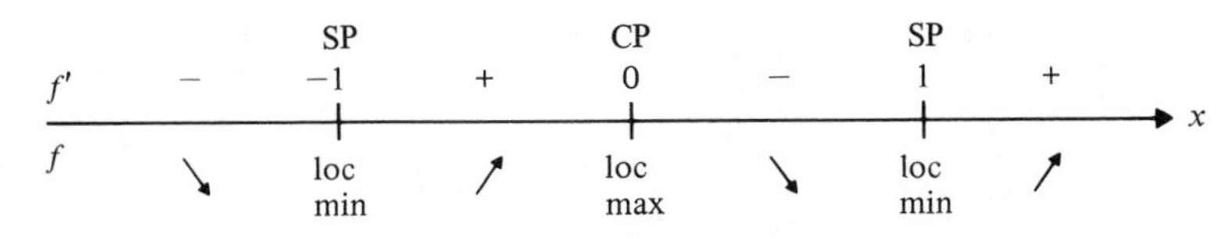

Figure 4.38

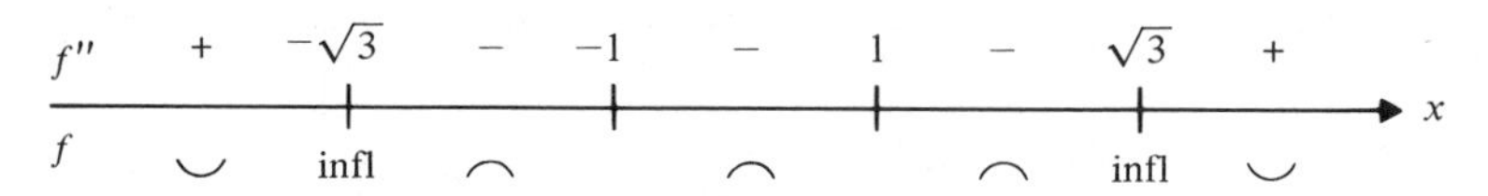

Figure 4.39

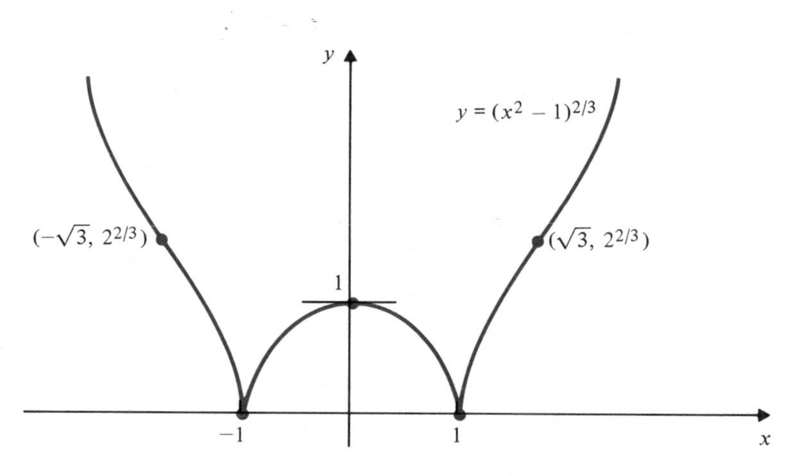

Figure 4.40

EXERCISES

In Exercises 1–35 sketch the graphs of the given functions, making use of any suitable information you can obtain from the function and its first and second derivatives.

1. $y = (x^2 - 1)^3$

2. $y = x(x^2 - 1)^2$

3. $y = \dfrac{2 - x}{x}$

4. $y = \dfrac{x - 1}{x + 1}$

5. $y = \dfrac{x^3}{1 + x}$

6. $y = \dfrac{1}{4 + x^2}$

7. $y = \dfrac{1}{2 - x^2}$

8. $y = \dfrac{x}{x^2 - 1}$

9. $y = \dfrac{x^2}{x^2 - 1}$

10. $y = \dfrac{x^3}{x^2 - 1}$

11. $y = \dfrac{x^3}{x^2 + 1}$

12. $y = \dfrac{x^2}{x^2 + 1}$

13. $y = \dfrac{x^2 - 4}{x + 1}$

14. $y = \dfrac{x^2 - 2}{x^2 - 1}$

15. $y = \dfrac{x^3 - 4x}{x^2 - 1}$

16. $y = \dfrac{x^2 - 1}{x^2}$

17. $y = \dfrac{x + 1}{x^2 - x}$

18. $y = \dfrac{(2 - x)^2}{x^3}$

19. $y = \dfrac{1}{x^3 - 4x}$

20. $y = \dfrac{x}{x^2 + x - 2}$

21. $y = \dfrac{x^4 - 4x^2 + 4}{x^3}$

22. $y = x + \sin x$

23. $y = x + 2 \sin x$

24. $y = e^{-x^2}$

25. $y = x^2 e^{-x^2}$

26. $y = e^{-x} \sin x \ (x \geq 0)$

27. $y = xe^x$

28. $y = x^2 e^x$

29. $y = \dfrac{\ln x}{x} \ (x > 0)$

30. $y = \dfrac{\ln x}{x^2} \ (x > 0)$

31. $y = \dfrac{1}{\sqrt{4 - x^2}}$

32. $y = \dfrac{x}{\sqrt{x^2 + 1}}$

33. $y = (x^2 - 1)^{1/3}$

34.* $y = x^{1/3} + (x - 1)^{2/3}$

35.* $y = (x - 1)^{2/3} - (x + 1)^{2/3}$

36.* What is $\lim_{x \to 0+} x \ln x$? $\lim_{x \to 0} x \ln |x|$? If $f(x) = x \ln |x|$ for $x \neq 0$ is it possible to define $f(0)$ in such a way that f is continuous on the whole real line? Sketch the graph of f.

4.4 OPTIMIZATION PROBLEMS

In this section we solve various word problems that, when translated into mathematical terms, require the finding of a maximum or minimum value of a function of one variable. Such problems can range from simple to very complex; they can be phrased in terminology appropriate to some other discipline or can be already partially translated into a more mathematical context. We consider a few examples before attempting to abstract any general principles for dealing with such problems.

Example 1 A rectangular enclosure is to be constructed having one side along an existing long wall and the other three sides fenced. If 100 meters (m) of fence are available, what is the largest possible area for the enclosure?

Solution This problem, like many others, is essentially a geometric one. A sketch should be made at the outset, as we have done in Fig. 4.41. Let the length and width of the enclosure be x and y meters, respectively, and let its area be A m². Thus $A = xy$.

Since the total length of the fence is 100 m, we have $x + 2y = 100$. A appears to be a function of two variables, x and y, but these variables are not independent; they are related by the *constraint* $x + 2y = 100$. This constraint equation can be solved for one variable in terms of the other, and A can therefore be written as a function of only one variable:

$$\begin{aligned} x &= 100 - 2y, \\ A &= A(y) = (100 - 2y)\, y = 100y - 2y^2. \end{aligned}$$

Evidently we require $y \geq 0$ and $y \leq 50$ (that is, $x \geq 0$) in order that the area make sense. (It would otherwise be negative.) Thus we must maximize the function $A(y)$ on the interval $[0, 50]$. Being continuous on this closed, finite interval, A must have a maximum value, by Theorem 1 of Section 4.1. Clearly $A(0) = A(50) = 0$ and $A(y) > 0$ for $0 < y < 50$. Hence the maximum cannot occur at an endpoint. Since A has no singular points, the maximum occurs at a critical point. For a critical point,

$$0 = A'(y) = 100 - 4y, \qquad y = 25.$$

Since A must have a maximum value and there is only one possible point where it can exist, the maximum must occur at $y = 25$. The greatest possible area for the enclosure is $A(25) = 1250\ \text{m}^2$.

Example 2 A lighthouse L is located on a small island 5 km north of a point A on a straight east-west shoreline. A cable is to be laid from L to point B on the shoreline 10 km east of A. The cable will be laid through the water in a straight line from L to a point C on the shoreline between A and B and from there to B along the shoreline. If the part of the cable lying in the water costs \$5000/km and the part along the shoreline costs \$3000/km, where should C be chosen to minimize the total cost of the cable? Where should C be chosen if B is only 3 km from A?

Solution Let C be x km from A toward B. Thus $0 \le x \le 10$. The length of LC is $\sqrt{25 + x^2}$ km and that of CB is $10 - x$ km, as illustrated in Fig. 4.42. Hence the total cost of the cable is \$$T$ where

$$T = T(x) = 5000\sqrt{25 + x^2} + 3000(10 - x), \qquad (0 \le x \le 10).$$

Being continuous on the closed, finite interval $[0, 10]$, T has a minimum value that may occur at one of the endpoints $x = 0$ or $x = 10$, or at a critical point of $T(x)$ in the interval $(0, 10)$. (There are no singular points.) To find any critical points, we set

$$0 = \frac{dT}{dx} = \frac{5000x}{\sqrt{25 + x^2}} - 3000.$$

Thus $(5x)/3 = \sqrt{25 + x^2}$ and $x^2 = 225/16$. Only one critical point, $x = 15/4$, lies in the interval $(0, 10)$. Since $T(0) = 55{,}000$, $T(15/4) = 50{,}000$, and $T(10) = 55{,}902$, the critical point evidently provides the minimum value for $T(x)$. Alternatively, the second derivative is $T''(x) = 125{,}000/(25 + x^2)^{3/2} > 0$, so the graph of T is concave

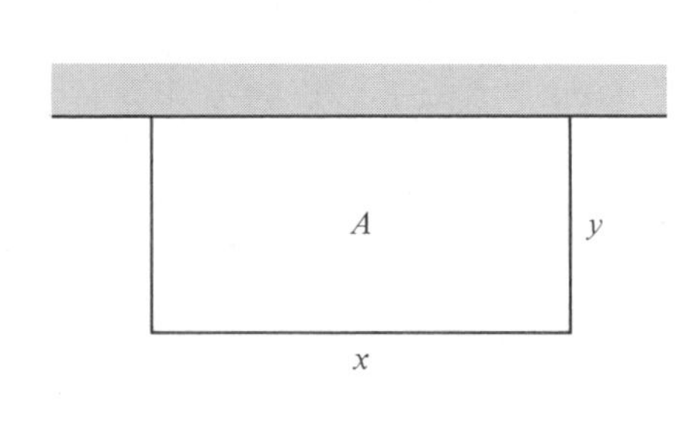

Figure 4.41

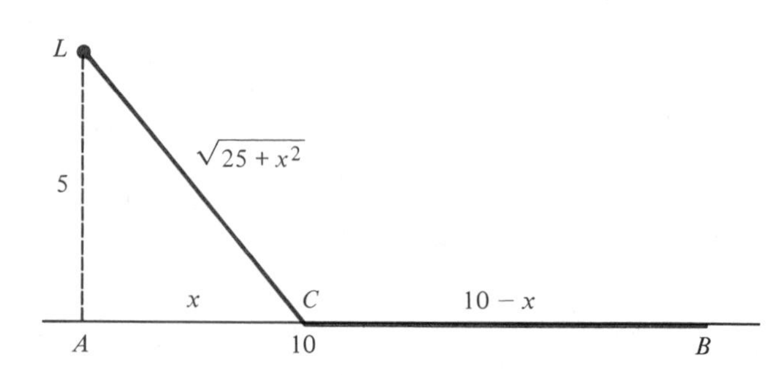

Figure 4.42

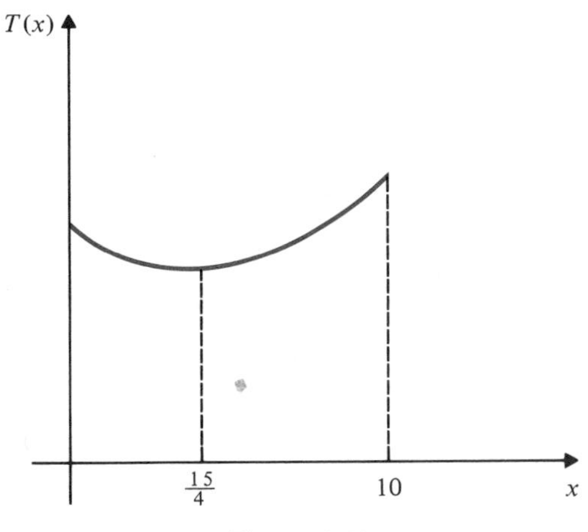

Figure 4.43

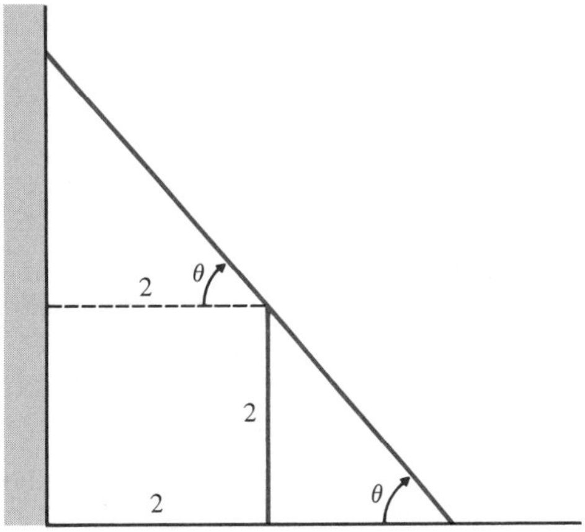

Figure 4.44

upward and the critical point yields a minimum (see Fig. 4.43). For minimal cost, C should be 3.75 km from A.

If B is 3 km from A, the corresponding function $T(x) = 5000\sqrt{25 + x^2} + 3000(3 - x)$ $(0 \le x \le 3)$ differs from the previous function only by an added constant (9000 rather than 30,000). It therefore has the same critical points, which do not lie in the interval $(0, 3)$. Since $T(0) > T(3)$, C should be chosen at B in this case; the cable should go straight from L to B.

Example 3 Find the length of the shortest ladder that can extend from a vertical wall, over a 2-meter-high fence located 2 meters away from the wall, to a point on the ground outside the fence.

Solution Let θ be the angle of inclination of the ladder, as shown in Fig. 4.44. Using the two right-angled triangles in the figure, we obtain the length L of the ladder as a function of θ:

$$L = L(\theta) = \frac{2}{\cos\theta} + \frac{2}{\sin\theta}$$

where $0 < \theta < \pi/2$. Since

$$\lim_{\theta\to(\pi/2)-} L(\theta) = \infty \qquad \text{and} \lim_{\theta\to 0+} L(\theta) = \infty,$$

any minimum value for $L(\theta)$ can occur only at a critical point. (There are no singular points in $(0, \pi/2)$. To find critical points, we set

$$0 = L'(\theta) = \frac{2\sin\theta}{\cos^2\theta} - \frac{2\cos\theta}{\sin^2\theta}.$$

It follows that $\tan^3\theta = 1$ and $\theta = \pi/4$.

This time we will use the first-derivative test to show that $\theta = \pi/4$ does indeed give the minimum value for $L(\theta)$. Observe that

$$L'(\theta) < 0 \text{ if } 0 < \theta < \frac{\pi}{4}, \text{ and}$$

$$L'(\theta) > 0 \text{ if } \frac{\pi}{4} < \theta < \frac{\pi}{2}.$$

Thus $\theta = \pi/4$ gives a local minimum and, being the only critical pont in $(0, \pi/2)$, the absolute minimum as well. The shortest ladder has length $L(\pi/4) = 4\sqrt{2}$ meters.

Suggested Procedures for Solving Optimization Problems

1. Define any symbols you wish to use that are not already specified in the statement of the problem.
2. Make a sketch if appropriate.
3. Express the quantity Q to be maximized or minimized as a function of one or more variables.
4. If Q depends on more than one variable (say n) find $n - 1$ equations (constraints) linking these variables. (If this cannot be done, the problem cannot be solved by single-variable techniques.)
5. Use the constraints to eliminate variables and hence express Q as a function of only one variable. Determine the interval(s) in which this variable must lie for the problem to make sense. Alternatively, regard the constraints as defining $n - 1$ of the variables, and hence Q, as functions of the remaining variable implicitly. (It is usually better to avoid this implicit method in an optimization problem if you can.)
6. Find all the local extreme values of Q, considering any critical points, singular points, or endpoints.
7. Give some justification that one particular value is the desired (absolute) extreme value. The first- or second-derivative test may be useful in this context, as may Theorem 1 of Section 4.1 or the results of Exercises 54–56 of Section 4.1. Probably the safest and most instructive way is to draw a very rough sketch of the graph of Q.
8. Make a concluding statement answering the question asked.

Example 4 Find the volume of the largest right circular cone that can be inscribed in a sphere of radius R.

Solution Let r, h, and V denote the radius, height, and volume of the cone, respectively. We have

$$V = \frac{1}{3}\pi r^2 h.$$

From the small right-angled triangle in Fig. 4.45,

$$(h - R)^2 + r^2 = R^2,$$

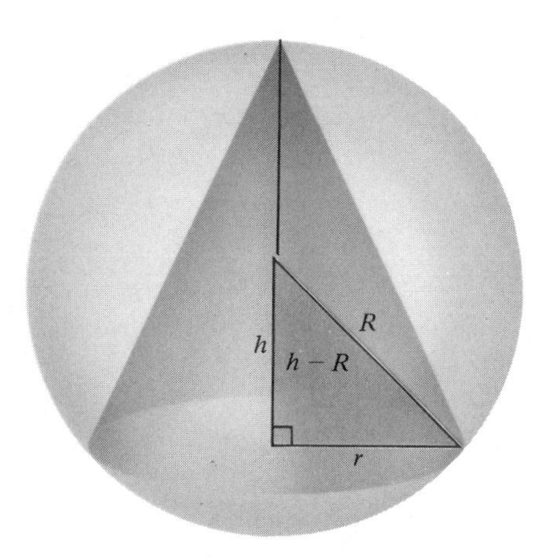

Figure 4.45

and so

$$V = \frac{\pi}{3}h(R^2 - (h - R)^2) = \frac{\pi}{3}(2Rh^2 - h^3).$$

Evidently $0 \leq h \leq 2R$ for any such cone. Since $V = 0$ when $h = 0$ or $h = 2R$, and $V > 0$ if $0 < h < 2R$, the maximum value of V must occur at a critical point. For a critical point,

$$0 = V'(h) = \frac{\pi}{3}(4Rh - 3h^2) = \frac{\pi}{3}h(4R - 3h),$$

$$h = 0 \text{ or } \frac{4R}{3}.$$

$V'(h) > 0$ if $0 < h < 4R/3$; $V'(h) < 0$ if $4R/3 < h < 2R$. Hence $h = 4R/3$ does indeed give the maximum value for V.

The volume of the largest cone that can be inscribed in a sphere of radius R is

$$V\left(\frac{4R}{3}\right) = \frac{\pi}{3}\left(2R\left(\frac{4R}{3}\right)^2 - \left(\frac{4R}{3}\right)^3\right) = \frac{32}{81}\pi R^3 \text{cubic units.}$$

Example 5 Find the most economical shape of a cylindrical tin can.

Solution This example is stated in a rather vague way. We must consider what is meant by "most economical" and even "shape." Wanting further information, we can take one of two points of view:

i) the volume of the tin can is to be regarded as given and we must choose the dimensions to minimize the total surface area, or

ii) the total surface area is given (we can use just so much metal) and we must choose the dimensions to maximize the volume.

We will discuss other possible interpretations later.

Since a cylinder is determined by its radius and height, its shape is determined

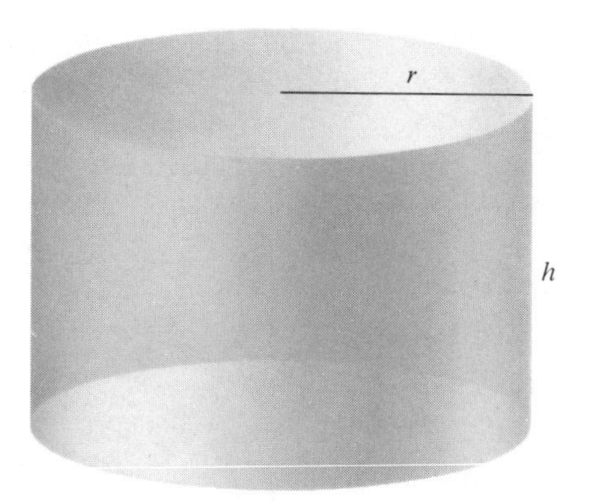

Figure 4.46

by the ratio radius:height (see Fig. 4.46). Let r, h, S, and V denote the radius, height, total surface area, and volume of the can, respectively. We have

$$\begin{aligned} V &= \pi r^2 h \\ S &= 2\pi r h + 2\pi r^2. \end{aligned}$$

The surface is made up of the cylindrical wall and circular discs for the top and bottom. Suppose we use interpretation (i); V is a given constant and S is to be minimized. Proceeding as in Example 1 to eliminate one variable, say h, we obtain

$$h = \frac{V}{\pi r^2},$$

$$S = S(r) = 2\pi r \frac{V}{\pi r^2} + 2\pi r^2 = \frac{2V}{r} + 2\pi r^2 \qquad (0 < r < \infty).$$

Evidently $\lim_{r \to 0+} S(r) = \lim_{r \to \infty} S(r) = \infty$. Thus any minimum value of $S(r)$ must occur at a critical point. To find any critical ponts,

$$0 = S'(r) = -\frac{2V}{r^2} + 4\pi r,$$

$$r^3 = \frac{V}{2\pi} = \frac{1}{2\pi} \pi r^2 h = \frac{1}{2} r^2 h.$$

Thus $h = 2r$ at the critical point of S. Since

$$S''(r) = \frac{4V}{r^3} + 4\pi > 0 \text{ if } r > 0,$$

the graph of $S(r)$ is concave up on $(0, \infty)$, and the critical point $r = h/2$ gives a local and absolute minimum value.

Thus, under interpretation (i), the most economical shape of a cylindrical tin can has diameter equal to height. We encourage you to solve the problem using interpretation (ii) and see that the same solution results.

There is another way to obtain the solution that shows directly that interpretations (i) and (ii) must give the same solution. Again we start from the two equations

$$V = \pi r^2 h, \qquad S = 2\pi rh + 2\pi r^2.$$

If we regard h as a function of r and differentiate implicitly, we obtain

$$\frac{dV}{dr} = 2\pi rh + \pi r^2 \frac{dh}{dr},$$

$$\frac{dS}{dr} = 2\pi h + 2\pi r \frac{dh}{dr} + 4\pi r.$$

Under interpretation (i) V is constant and we want a critical point of S; under interpretation (ii), S is constant and we want a critical point of V. In either case, $dV/dr = 0$ and $dS/dr = 0$. Hence both interpretations yield

$$2\pi rh + \pi r^2 \frac{dh}{dr} = 0,$$

$$2\pi h + 4\pi r + 2\pi r \frac{dh}{dr} = 0.$$

Elimination of dh/dr between these two equations again gives $h = 2r$.

Given the sparse information provided in the statement of the problem, interpretations (i) and (ii) are the best we can do. The problem could be made more meaningful economically (from the point of view, say, of a tin can manufacturer) if more elements were brought into it. For example:

a) Most cans use thicker material for the cylindrical wall than for the top and bottom discs. If the cylindrical wall material costs \$$A$ per unit area and that for the top and bottom costs \$$B$ per unit area, we might prefer to minimize the total cost for materials for a can of given volume.

b) The material for the cans is probably being cut out of sheets of metal. The cylindrical wall is made by bending up a rectangle, and rectangles can be cut from the sheet with little or no waste. There will, however, always be a proportion of material wasted when the discs are cut out. (We are assuming a large number of cans are to be made.) The exact proportion will depend on how the discs are arranged; two possible arrangements are shown in Fig. 4.47.

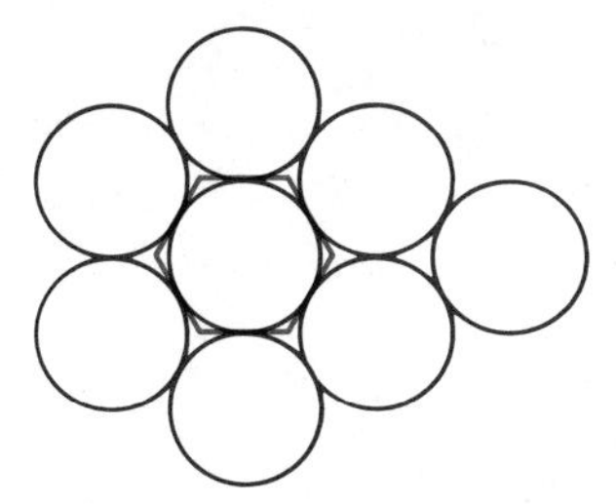

Hexagonal packing: Each disc uses up a hexagon.

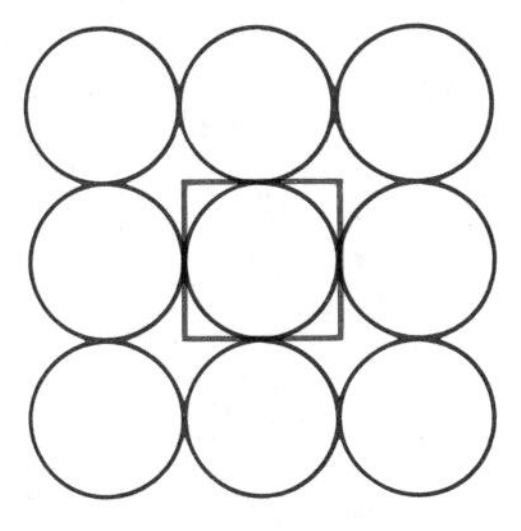

Square packing: Each disc uses up a square.

Figure 4.47

Any such modifications of the original problem will alter the optimal shape to some extent.

Example 6 A man can run twice as fast as he can swim. He is standing at point A on the edge of a circular swimming pool 40 m in diameter, and he wishes to get to the diametrically opposite point, B. He can run around the edge to point C, then swim directly from C to B. Where should C be chosen to minimize the total time taken to get from A to B?

Solution It is convenient to describe the position of C in terms of the angle AOC where O is the center of the pool (see Fig. 4.48). Let θ denote this angle. Clearly $0 \le \theta \le \pi$. (If $\theta = 0$ the man swims the whole way; if $\theta = \pi$ he runs the whole way.) Thus arc $AC = 20\theta$, and since angle $BOC = \pi - \theta$, chord $BC = 2LC = 40 \sin \dfrac{\pi - \theta}{2}$.

Suppose the man swims at a rate k m/sec and therefore runs at a rate $2k$ m/sec. If t is the total time he takes to get from A to B, then

$$t = t(\theta) = \text{time running} + \text{time swimming} = \frac{20\theta}{2k} + \frac{40}{k} \sin \frac{\pi - \theta}{2}.$$

(We are assuming that no time is wasted in jumping into the water at C.) The domain of t is $[0, \pi]$ and t has no singular points. Since t is continuous on a closed, finite interval it must have a minimum value, and that value must occur at a critical point or an endpoint. For critical points,

$$0 = t'(\theta) = \frac{10}{k} - \frac{20}{k} \cos \frac{\pi - \theta}{2}$$

$$\cos \frac{\pi - \theta}{2} = \frac{1}{2}, \quad \frac{\pi - \theta}{2} = \frac{\pi}{3}, \quad \theta = \frac{\pi}{3}.$$

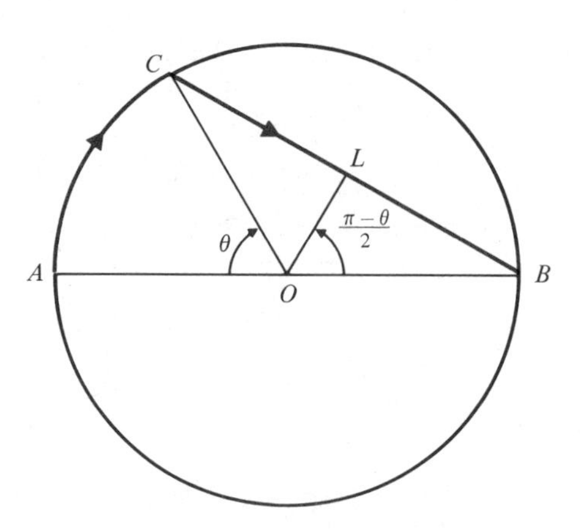

Figure 4.48

Observe, however, that $t''(\theta) = -\frac{10}{k}\sin\frac{\pi - \theta}{2} < 0$ on $(0, \pi)$, so $\theta = \frac{\pi}{3}$ gives, in fact, a local maximum rather than a local minimum to t. Thus the minimum time must occur at one of the endpoints $\theta = 0$ or $\theta = \pi$. Since $t(0) = 40/k$ and $t(\pi) = 10\pi/k < 40/k$, the man should run the entire distance.

This problem shows how important it is to check every candidate point to see whether it gives a loc max or loc min. Here the critical point yielded the worst possible strategy; running one-third of the way around and then swimming the remainder would take the greatest time, not the least. In the above solution we used the second-derivative test to test the critical point, but the first-derivative test could have been used as well. Both endpoints give local minima to t and only comparison of the actual values showed which was in fact the absolute minimum.

EXERCISES

1. Two nonnegative numbers have sum 7. What is the largest possible value for their product?

2. Two positive numbers have product 8. What is the smallest possible value of their sum?

3. Two nonnegative numbers have sum 60. What are the numbers if the product of one of them and the square of the other is maximal?

4. Two numbers have sum 16. What are the numbers if the product of the cube of one and the fifth power of the other is as large as possible?

5. Among all rectangles of given area show that the square has the least perimeter.

6. Among all isosceles triangles of given perimeter, show that the equilateral triangle has the greatest area.

7. Find the area of the largest rectangle that can be inscribed in a semicircle of radius R if one side of the rectangle lies along the diameter of the semicircle.

8. Find the largest possible perimeter of a rectangle inscribed in a semicircle of radius R if one side of the rectangle lies along the diameter of the semicircle. (It is interesting to note that the rectangle with the largest perimeter has a different shape than that with the largest area, obtained in Exercise 7.)

9. A rectangle with edges parallel to the coordinate axes is inscribed in the ellipse

$$\frac{x^2}{a^2} + \frac{y^2}{b^2} = 1.$$

Find the largest possible area for this rectangle.

10. Let ABC be a triangle right-angled at C and having area S. Find the maximum area of a rectangle inscribed in the triangle if

i) one corner of the rectangle lies at C, or

ii) one edge of the rectangle lies along the hypotenuse, AB.

11. A billboard is to be made with 100 m^2 of printed area, and with margins of 2 m at the top and bottom and 4 m on each side. Find the outside dimensions of the billboard if its total area is to be a minimum.

12. A rectangular box is to be made from a rectangular sheet of cardboard 70 cm by 150 cm by cutting equal squares out of the four corners and bending up the resulting four flaps to make the sides of the box. (The box has no top.) What is the largest possible volume of the box?

13. A 1-meter length of stiff wire is cut into two pieces. One piece is bent into a circle, the other piece into a square. Find the length of the part used for the square if the sum of the areas of the circle and the square is

a) maximum,

b) minimum.

14. A line segment has one end on the x-axis and the other end on the y-axis. Find the length of the shortest such segment that passes through the point $(9, \sqrt{3})$.

15.* Find the length of the longest beam that can be carried horizontally around the corner from a hallway of width a to a hallway of width b, as shown in Fig. 4.49. (Assume the beam has no width.)

16.* If the height of both hallways in Exercise 14 is c, and if the beam need not be carried horizontally, how long can it be and still get around the corner? (Assume the beam has no width or thickness.)

17. Find the shortest distance from the origin to the line $Ax + By = C$.

18. Find the shortest distance from the origin to the curve $x^2y^4 = 1$.

19. Find the shortest distance from the point (8, 1) to the curve $y = 1 + x^{3/2}$.

20. Find the dimensions of the largest right cylinder that can be inscribed in a sphere of radius R.

21.* Find the volume of the smallest right circular cone inside which a sphere of radius R can be put.

22. A box with square base and no top is to have a volume of 4 m³. Find the dimensions of the most economical box.

23. Refer to Example 5 in this section. Find the most economical shape of the tin can if the material used for the cylindrical walls is twice as expensive per unit area as that used for the top and bottom discs.

24. Suppose the discs to make the tops and bottoms of cans are cut from sheet metal in a square packing arrangement (see Fig. 4.47), and the waste is discarded. Find the most economical shape for the can in Exercise 23 under these circumstances.

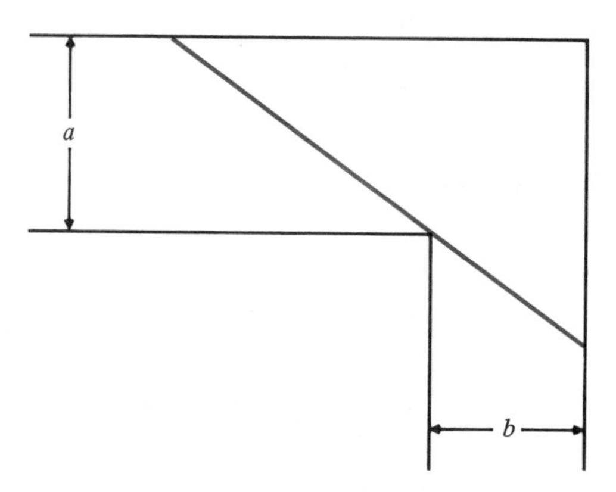

Figure 4.49

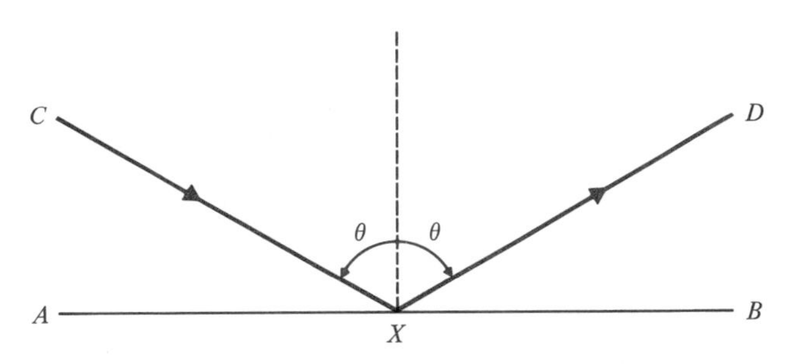

Figure 4.50

25.* Repeat Exercise 24, but use the hexagonal packing of the discs.

26. A fuel tank is made of a cylindrical part capped by hemispheres at each end. If the hemispheres are twice as expensive per unit area as the cylindrical wall, and if the volume of the tank is V, find the radius and height of the cylindrical part to minimize the total cost. The surface area of a sphere of radius r is $4\pi r^2$.

27. A window has perimeter 10 m and is in the shape of a rectangle with top edge replaced by a semicircle. Find the dimensions of the rectangle if the window admits the greatest amount of light.

28. Find the equation of the straight line of maximum slope tangent to the curve $y = 1 + 2x - x^3$.

29. A quantity Q grows according to the differential equation

$$\frac{dQ}{dt} = kQ^3(L - Q)^5 \qquad (k \text{ and } L \text{ positive constants}).$$

How large is Q when it is growing most rapidly?

30.* An enclosure is to be constructed having part of its boundary along an existing straight wall. The other part of the boundary is to be fenced in the shape of an arc of a circle. If 100 m of fencing are available, what is the area of the largest possible enclosure?

31. You are in a dune buggy in the desert 12 km due south of the nearest point A on a straight east-west road. You wish to get to point B on the road 10 km east of A. If your dune buggy can average 15 km/h traveling over the desert, and 39 km/h traveling on the road, toward what point on the road should you head in order to minimize your travel time to B?

32. Repeat Exercise 31, but assume that B is only 4 km from A.

33. A physical principle asserts that light travels in such a way that it requires the minimum possible time to get from one point to another. A ray of light from C reflects off a plane mirror AB at X and then passes through D (see Fig. 4.50). Show that the rays CX and XD make equal angles with the normal to AB at X. (*Remark:* You may wish to try and get a noncalculus solution in addition to the calculus solution by minimizing the travel time on CXD.)

34. If light travels with speed v_1 in one medium and speed v_2 in a second medium, and if the two media are separated by a plane interface, show that a ray of light passing from point A in one medium to point B in the other is bent at the interface in such a way that

$$\frac{\sin i}{\sin r} = \frac{v_1}{v_2},$$

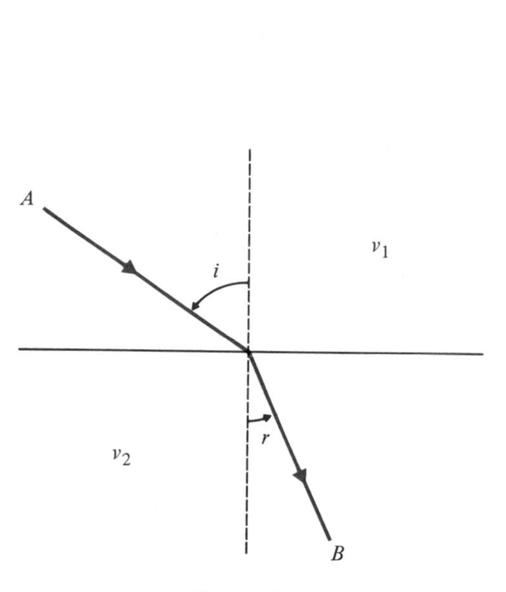

Figure 4.51

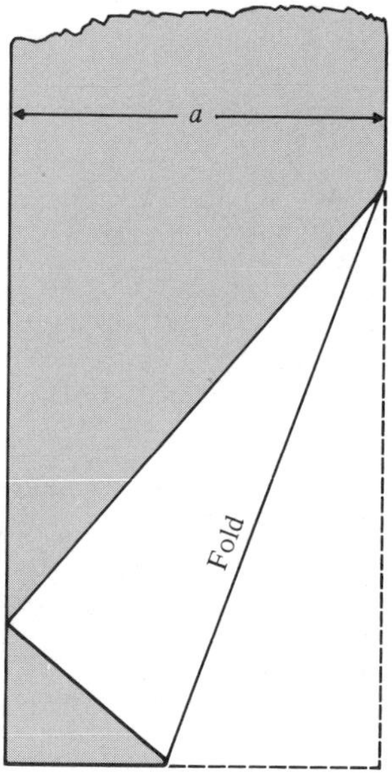

Figure 4.52

where i and r are the angles of incidence and refraction, as shown in Fig. 4.51. This is known as Snell's law. (Deduce it from the physical principle stated in Exercise 33.)

35. Find the dimensions of the circular cylinder of greatest volume that can be inscribed in a cone of base radius R and height H if the base of the cylinder lies in the base of the cone.

36. The stiffness of a wooden beam of rectangular cross section is proportional to the product of its width and the cube of its depth. Find the dimensions of the stiffest beam that can be cut out of a circular log of radius R.

37.* How far back from a mural should one stand to best view it if the mural is 10 ft high and the bottom of it is 2 ft above eye level?

38.* One corner of a strip of paper a cm wide is folded up so that it lies along the opposite edge, as shown in Fig. 4.52. Find the least possible length for the fold line.

4.5 RELATED RATES

When two or more quantities that change with time are linked by an equation, that equation can be differentiated with respect to time to yield an equation linking the rates of change of the quantities. Any one of these rates may then be determined when the others, and the values of the quantities themselves, are known. As in the previous section, we will consider a few examples before formulating a list of procedures for dealing with such problems.

Example 1 How fast is the area of a rectangle changing if one side is 10 cm long and is increasing at a rate of 2 cm/sec and the other side is 8 cm long and is decreasing at a rate of 3 cm/sec?

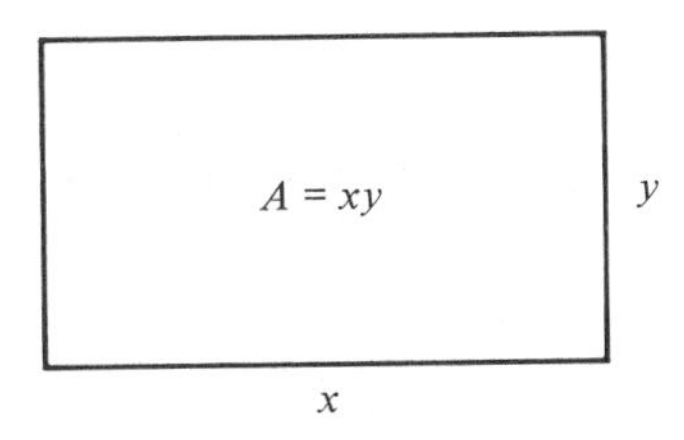

Figure 4.53

Solution Let the lengths of the sides of the rectangle at time t be x cm and y cm, respectively. Thus the area is $A = xy$ cm^2 at time t (see Fig. 4.53). We want to know the value of dA/dt when $x = 10$ and $y = 8$, given that $dx/dt = 2$ and $dy/dt = -3$. (Note the negative sign to indicate that y is decreasing. Since all the quantities in the equation $A = xy$ are functions of time, we can differentiate that equation implicitly with respect to time and obtain

$$\frac{dA}{dt} = \frac{dx}{dt}y + x\frac{dy}{dt}.$$

Thus

$$\left.\frac{dA}{dt}\right|_{\substack{x=10,\\ y=8}} = 2(8) + 10(-3) = -14.$$

At the time in question, the area is decreasing at a rate of 14 cm^2/sec.

Example 2 Air is being pumped into a spherical balloon. If the volume of the balloon is increasing at a rate of 20 cm^3/sec when the radius is 30 cm, how fast is the radius increasing at that time?

Solution Let r and V denote the radius and volume of the balloon, respectively, at time t. Then

$$V = \frac{4}{3}\pi r^3.$$

Differentiating with respect to time t (using the chain rule), we obtain

$$\frac{dV}{dt} = 4\pi r^2\frac{dr}{dt}.$$

We are given $dV/dt = 20$ when $r = 30$. At this time,

$$20 = 4\pi(900)\frac{dr}{dt},$$

so

$$\frac{dr}{dt} = \frac{1}{180\pi}.$$

The radius is increasing at $1/180\pi$ cm/sec when it is 30 cm.

Observe that while we wanted dr/dt ultimately, we did not solve the initial equation $V = (4/3)\pi r^3$ for r before differentiating. It is usually simpler to differentiate the relationship between the variables in its original form and then solve for the desired rate.

Example 3 A lighthouse L is located on a small island 2 km from the nearest point A on a long, straight shoreline. If the lighthouse lamp rotates at 3 revolutions per minute, how fast is the illuminated spot P on the shoreline moving along the shoreline when it is 4 km from A?

Solution Referring to Fig. 4.54, let

$$x = \text{distance } AP,$$
$$\theta = \text{angle } PLA.$$

Then

$$\frac{d\theta}{dt} = 3 \times 2\pi = 6\pi \text{ radians/min.}$$

Also $x = 2 \tan \theta$, so

$$\frac{dx}{dt} = 2 \sec^2 \theta \frac{d\theta}{dt} = 12\pi \sec^2 \theta.$$

When $x = 4$, $\tan \theta = 2$, so $\sec^2 \theta = 1 + \tan^2\theta = 5$. Hence P is traveling along the shoreline at a rate of 60π km/min when it is 4 km from A.

(Note that it was essential to convert the rate of change of θ from revolutions per minute to radians per minute. If θ were not measured in radians we could not assert that $(d/d\theta) \tan \theta = \sec^2 \theta$.)

Procedures for Related-Rates Problems

In the light of these examples we can formulate a few general procedures for dealing with related-rates problems similar to those listed for optimization problems in Section 4.4.

1. Define any symbols you intend to use that are not defined in the statement of the problem, and express given and required quantities and rates in terms of these symbols.

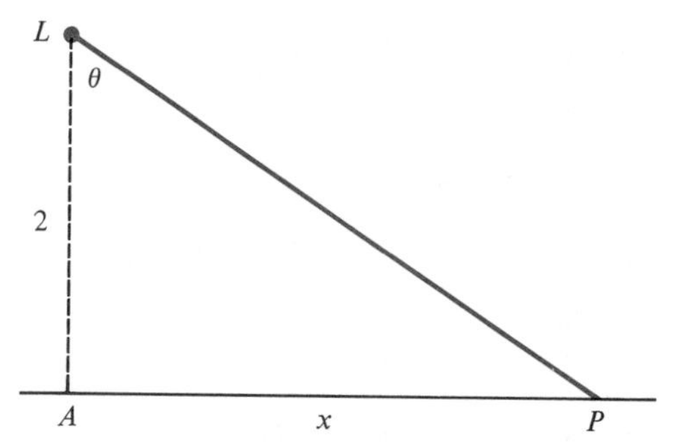

Figure 4.54

2. Make a sketch if appropriate.
3. Discover from a careful reading of the problem or consideration of the sketch an equation linking the quantity whose rate of change is to be found to other quantities whose rates are given or are easily determined. If more than one rate is to be found, more than one equation will be needed—one for each unknown rate.
4. Differentiate the equation or equations implicitly with respect to time, regarding all variable quantities as functions of time. The equations can be manipulated algebraically before the differentiation is performed (for instance, they could be solved for the quantities whose rates are to be found), but it is usually better to differentiate the equations as they are originally obtained and solve for the desired rates later.
5. Substitute any given values for the quantities and their rates and solve the resulting equation or equations for the unknown rates. It may be necessary to use the original (undifferentiated) equations to determine the values of some of the quantities from the given values of others.
6. Make a concluding statement answering the question asked.

Example 4 A water tank is in the shape of an inverted right circular cone with depth 5 m and top radius 2 m. Water leaks out of the tank at a rate proportional to the depth of water in the tank. When the water in the tank is 4 m deep it is leaking out at a rate $\frac{1}{12}$ m^3/min; how fast is the water level in the tank dropping at that time?

Solution Let r and h denote the surface radius and depth (both in meters) of water in the tank at time t. Thus the volume (V) (in m^3) of water in the tank at time t is

$$V = \frac{1}{3}\pi r^2 h.$$

Using similar triangles in Fig. 4.55 we can determine a relationship between r and h:

$$\frac{r}{h} = \frac{2}{5},$$

so

$$r = \frac{2h}{5}.$$

Thus

$$V = \frac{1}{3}\pi\left(\frac{2h}{5}\right)^2 h = \frac{4\pi}{75}h^3.$$

Differentiating this equation with respect to t we obtain

$$\frac{dV}{dt} = \frac{4\pi}{25}h^2\frac{dh}{dt}.$$

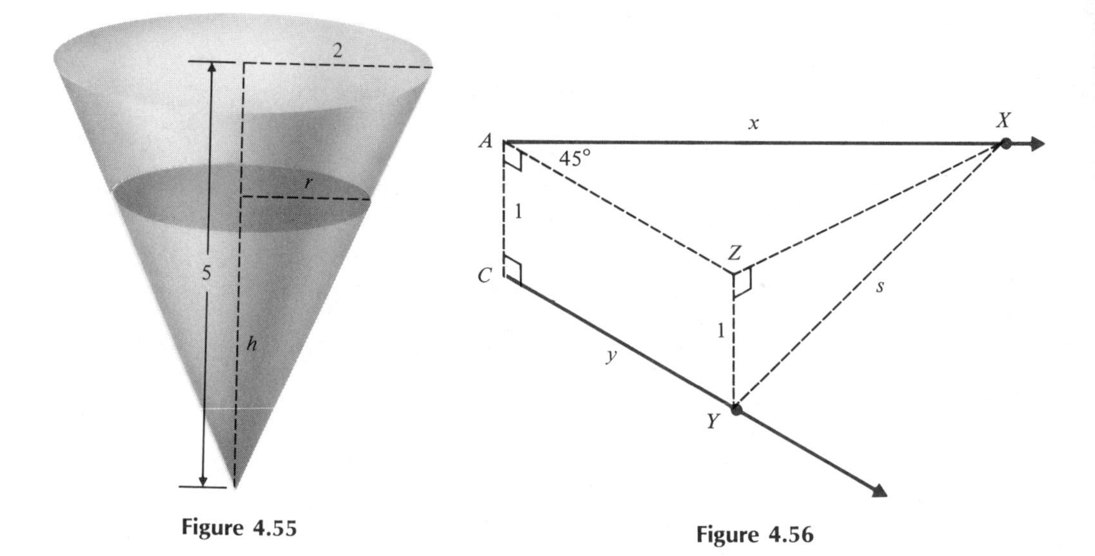

Figure 4.55

Figure 4.56

We are told that the volume of water in the tank is decreasing at a rate proportional to h, that is, $dV/dt = kh$, and that that rate is 1/12 m³/min when $h = 4$ m. Thus $-(1/12) = 4k$ and $k = -(1/48)$. Hence

$$-\frac{1}{48}h = \frac{dV}{dt} = \frac{4\pi}{25}h^2\frac{dh}{dt}.$$

It follows that

$$\frac{dh}{dt} = -\frac{25}{192\pi}\frac{1}{h}$$

and

$$\left.\frac{dh}{dt}\right|_{h=4} = -\frac{25}{768\pi}.$$

When the water in the tank is 4 m deep its level is dropping at a rate of $25/768\pi$ m/min $\simeq$ 1.036 cm/min.

Example 5 At a certain instant an aircraft flying due east at 400 km/h passes directly over a car traveling due southeast at 100 km/h on a straight, level road. If the aircraft is flying at an altitude of 1 km, how fast is the distance between the aircraft and the car increasing 36 sec after the aircraft passes directly over the car?

Solution A good diagram is essential here. Let time t be measured in hours from the time the aircraft was at position A directly above the car at position C. Let X and Y be the positions of the aircraft and the car, respectively, at time t. Let x = distance AX, y = distance CY, and s = distance XY (see Fig. 4.56). Let Z be the point 1 km above Y. Since angle $XAZ = 45°$, we have, using the Pythagorean theorem and cosine law,

$$s^2 = 1 + (ZX)^2 = 1 + x^2 + y^2 - 2xy\cos 45°$$
$$= 1 + x^2 + y^2 - \sqrt{2}xy;$$
$$2s\frac{ds}{dt} = 2x\frac{dx}{dt} + 2y\frac{dy}{dt} - \sqrt{2}y\frac{dx}{dt} - \sqrt{2}x\frac{dy}{dt}$$
$$= 400(2x - \sqrt{2}y) + 100(2y - \sqrt{2}x).$$

We are given $dx/dt = 400$ and $dy/dt = 100$. When $t = 1/100$ (that is, 36 sec after $t = 0$) we have $x = 4$ and $y = 1$. Hence

$$s^2 = 1 + 16 + 1 - 4\sqrt{2} = 18 - 4\sqrt{2}$$
$$s \simeq 3.5133.$$

Thus $\dfrac{ds}{dt} = \dfrac{1}{2s}(400(8 - \sqrt{2}) + 100(2 - 4\sqrt{2})) \simeq 322.86.$

The aircraft and the car are separating at a rate of approximately 323 km/h after 36 sec.

(Note that it was necessary to convert 36 seconds to hours in the solution. In general all measurements should be in compatible units.)

Example 6 A lump of modeling clay is being rolled out so that it maintains the shape of a circular cylinder. If the length is increasing at a rate proportional to itself, show that the radius is decreasing at a rate proportional to itself.

Solution If the radius, length, and volume of the cylinder are r, L, and V, respectively, then $V = \pi r^2 L$. The volume of clay remains constant, so

$$0 = \frac{dV}{dt} = \pi r^2\frac{dL}{dt} + 2\pi rL\frac{dr}{dt}.$$

If $dL/dt = kL$, then

$$kr^2L + 2rL\frac{dr}{dt} = 0,$$

so

$$\frac{dr}{dt} = -\frac{k}{2}r,$$

that is, r is decreasing at a rate proportional to itself.

EXERCISES

1. The radius of a circle is increasing at a rate of 4 cm/sec. How fast is the area of the circle increasing when the radius is 20 cm?
2. The area of a circle is decreasing at a rate of 2 cm²/min. How fast is the radius of the circle changing when the area is 100 cm²?
3. At a certain instant the length of a rectangle is 16 m and the width is 12 m. The width is

increasing at 3 m/sec. How fast is the length changing if the area of the rectangle is remaining constant?

4. The volume of a right circular cylinder is 60 cm^3 and is increasing at 2 cm^3/min at a time when the radius is 5 cm and is increasing at 1 cm/min. How fast is the height of the cylinder changing at that time?

5. How fast is the volume of a rectangular box changing when the length is 6 cm, the width is 5 cm, and the depth is 4 cm, if the length and depth are both increasing at a rate of 1 cm/sec and the width is decreasing at a rate of 2 cm/sec?

6. The area of a rectangle is increasing at a rate of 5 m^2/sec while the length is increasing at a rate of 10 m/sec. If the length is 20 m and the width is 16 m, how fast is the width changing?

7. A point moves on the curve $y = x^2$. How fast is y changing when $x = -2$ and x is decreasing at a rate 3?

8. A point is moving to the right along the first-quadrant portion of the curve $x^2y^3 = 72$. When the point has coordinates (3, 2) its horizontal velocity is 2 units/sec. What is its vertical velocity?

9. The point P moves so that at time t it is at the intersection of the curves $xy = t$ and $y = tx^2$. How fast is the distance of P from the origin changing at time $t = 2$?

10. A man 6 ft tall walks toward a lamppost on level ground at a rate of 2 ft/sec. If the lamp is 15 ft high on the post, how fast is the length of the man's shadow decreasing when he is 10 ft from the post? How fast is the shadow of his head moving at that time?

11. The top of a ladder 5 m long rests against a vertical wall. If the base of the ladder is being pulled away from the base of the wall at a rate of 1/3 m/sec, how fast is the top of the ladder slipping down the wall when it is 3 m above the base of the wall?

12. An aircraft is flying horizontally at a rate of 13 km/min at an altitude of 5 km. The aircraft passes directly over a radio beacon at 3:00 P.M. How fast is the distance between the aircraft and the beacon increasing 1 min later?

13. At 1:00 P.M. ship A is 25 mi due north of ship B. If ship A is sailing west at a rate of 16 mph and ship B is sailing south at 20 mph, find the rate at which the distance between the two ships is changing at 1:30 P.M.

14. What is the first time after 3 o'clock that the hands of the clock are together?

15. If the radius of a cylinder is a cm and is decreasing at a rate of k cm/sec, and if the volume of the cylinder is remaining constant at V cm^3, how fast is the length of the cylinder changing?

16. A balloon released at point A rises vertically with a constant speed of 5 m/sec. Point B is level with and 10 m distant from point A. How fast is the angle of elevation of the balloon at B changing when the balloon is 20 m above A?

17. Sawdust is falling onto a pile at a rate of 1/2 m^3/min. If the pile maintains the shape of a right circular cone with height equal to half the diameter of its base, how fast is the height of the pile increasing when it is 3 m?

18. A water tank is in the shape of an inverted right circular cone with top radius 10 m and depth 8 m. Water is flowing in at a rate of 1/10 m^3/min. How fast is the depth of water in the tank increasing when the water is 4 m deep?

19. Repeat Exercise 18 with the added assumption that water is leaking out of the bottom of the tank at a rate of $h^3/1000$ m^3/min when the depth of water in the tank is h m. How full can the tank get in this case?

20. How fast must you let out line if the kite you are flying is 30 m high, 40 m horizontally away from you, and moving horizontally away from you at a rate of 10 m/min?

21. You are riding on a Ferris wheel of diameter 20 m. The wheel is rotating at 1 revolution per min. How fast are you rising or falling when you are 6 m horizontally away from the vertical line passing through the center of the wheel?

22. An aircraft is 144 km east of an airport and is traveling west at 200 km/h. Simultaneously a second aircraft at the same altitude is 60 km north of the airport and traveling north at 150 km/h. How fast is the distance between the aircraft changing?

23. A boat is being pulled toward a pier by a rope attached to its bow. A person on the pier is pulling in the rope at a rate of 6 m/min. If the person's hands are 5 m higher than the bow of the boat, how fast is the boat moving toward the pier when there are still 13 m of rope out?

24. A straight highway and a straight canal intersect at right angles, the highway crossing over the canal on a bridge 20 m above the water. A boat traveling at 20 km/h passes under the bridge just as a car traveling at 80 km/h passes over it. How fast are the boat and car separating after one minute?

25. The cross section of a water trough is an equilateral triangle with top edge horizontal. If the trough is 10 m long and 30 cm deep, and if water is flowing in at a rate of 1/4 m^3/min, how fast is the water level rising when the water is 20 cm deep at the deepest?

26. A rectangular swimming pool is 8 m wide and 20 m long. Its bottom is a sloping plane, the depth increasing from 1 m at the shallow end to 3 m at the deep end. Water is draining out of the pool at a rate of 1 m^3/min. How fast is the surface of the water falling when the depth of water at the deep end is (a) 2.5 m? (b) 1 m?

27. Two crates, A and B, are on the floor of a warehouse. The crates are joined by a rope 33 ft long, each crate being hooked at floor level to an end of the rope. The rope is stretched tight and passes over a pulley P that is attached to a rafter 12 ft above a point Q on the floor directly between the two crates. If crate A is 5 ft from Q and is being pulled directly away from Q at a rate of 1/2 ft/sec, how fast is crate B moving toward Q?

28. Water is pouring into a leaky tank at a rate of 10 m^3/h. The tank is a cone with vertex down, 9 m in depth and 6 m in diameter at the top. The surface of water in the tank is rising at a rate of 20 cm/h when the depth is 6 m. How fast is the water leaking out at that time?

29.* Shortly after launch, a rocket is 100 km high and 50 km downrange. If it is traveling at 4 km/sec at an angle of 30° above the horizontal, how fast is its angle of elevation, as measured at the launch site, changing?

30. A lamp is 80 ft high on a pole. At time $t = 0$ a ball is dropped from a point level with the lamp and 20 ft away from it. The ball falls under gravity (acceleration 32 ft/sec^2) until it hits the ground. How fast is the shadow of the ball moving along the ground (a) two seconds after it is dropped? (b) just as the ball hits the ground?

31. A rocket blasts off at time $t = 0$ and climbs vertically with acceleration 20 ft/sec^2. The progress of the rocket is monitored by a tracking station located 4000 ft horizontally away from the launch pad. How fast is the tracking antenna rotating upward 10 sec after launch?

32. What is the maximum rate at which the antenna in Exercise 31 must be able to turn in order to track the rocket during its vertical ascent?

33.* You are in a tank (the military variety) moving down the y-axis toward the origin. At time $t = 0$ you are 4 km from the origin, and 10 min later you are 2 km from the origin. Your speed is decreasing; it is proportional to your distance from the origin. You know that an enemy tank is waiting somewhere on the positive x-axis, but there is a high wall along the curve $xy = 1$ (all distances in km) preventing you from seeing just where it is. How fast must your gun turret be capable of turning to maximize your chances of surviving the encounter?

4.6 TANGENT-LINE APPROXIMATIONS

The line tangent to the graph of $y = f(x)$ at $x = a$ (see Fig. 4.57) describes the behavior of that graph near the point $P = (a, f(a))$ better than does any other straight line through P. We exploit this fact by using the height to the tangent line to calculate approximate values for $f(x)$ for values of x near a. The tangent line has equation $y = f(a) + f'(a)(x - a)$, so the approximation is

$$f(x) \simeq f(a) + f'(a)(x - a).$$

Example 1 Find an approximate value for $\sqrt{26}$ using a suitable tangent line.

Solution If $f(x) = \sqrt{x}$, then $f'(x) = 1/(2\sqrt{x})$. Take $a = 25$ (the nearest point to 26 that is a perfect square), and obtain $f(a) = 5$, $f'(a) = 1/10$. Thus

$$\sqrt{x} = f(x) \simeq 5 + \frac{1}{10}(x - 25)$$

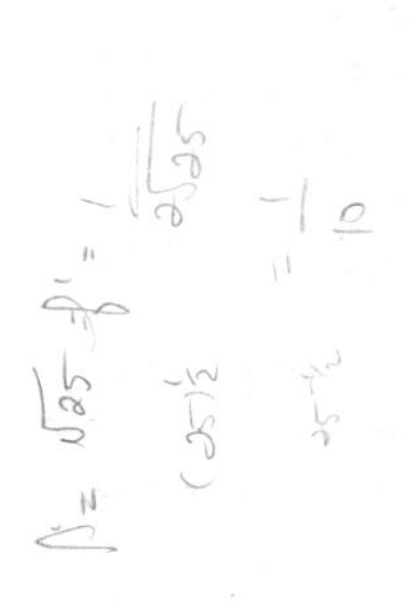

for x near 25, and $\sqrt{26} \simeq 5.1$. We can say a little more. Since f is increasing, $\sqrt{26} > 5$, and since $f''(x) = -1/4x^{3/2} < 0$ if $x > 0$, the graph of f is concave down and lies below its tangent lines. Hence $\sqrt{26} < 5.1$, so

$$5 < \sqrt{26} < 5.1.$$

(The actual value of $\sqrt{26}$ is 5.0990 . . ., but if we knew this we would have had no need of an approximation!)

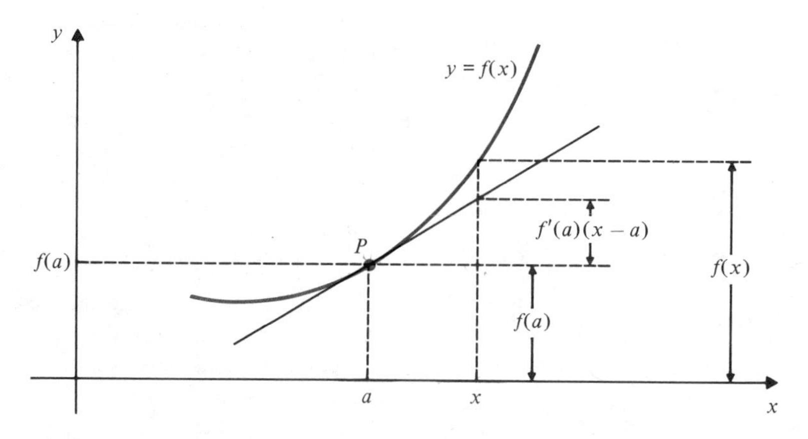

Figure 4.57

Example 2 Find an approximate value for $\cos 62°$.

Solution If $f(x) = \cos x$ and $a = 60° = \pi/3$, then $f(a) = \cos \pi/3 = 1/2$ and $f'(a) = -\sin(\pi/3) = -(\sqrt{3}/2)$. Hence

$$\cos x = f(x) \simeq \frac{1}{2} - \frac{\sqrt{3}}{2}\left(x - \frac{\pi}{3}\right)$$

for x near $\pi/3$. Since $62° = \pi/3 + 2\pi/180 = \pi/3 + \pi/90$, we have

$$\cos 62° = f\left(\frac{\pi}{3} + \frac{\pi}{90}\right) \simeq \frac{1}{2} - \frac{\sqrt{3}}{2}\frac{\pi}{90} \simeq 0.46977\ldots.$$

(We have used a calculator with square-root function to find the decimal value.) Again $f''(x) = -\cos x < 0$ for x near $\pi/3$, so the true value is less than the height to the tangent line: $\cos 62° < 0.46977\ldots$.

If we use the cosine function on a scientific calculator we can obtain the "true value" (actually, just another approximation, though presumably a better one): $\cos 62° = 0.46947\ldots$, but if we have such a calculator we don't need the approximation in the first place. Approximations are useful when there is no easy way to obtain the true value. If we don't know the true value, we would like to have some way of determining how good the approximation must be; that is, we want an *estimate for the error*.

The Error Estimate

If the tangent line at $x = a$ is used to approximate $f(x)$ near a, that is,

$$f(x) \simeq f(a) + f'(a)(x - a)$$

then the error $E(x)$ in this approximation is

$$\begin{aligned} E(x) &= \text{true value} - \text{approximate value} \\ &= f(x) - f(a) - f'(a)(x - a). \end{aligned}$$

The following theorem gives us a bound on the size of this error.

Theorem 6 If f is twice differentiable on an interval containing a and x, then there is some point X between a and x such that

$$E(x) = \frac{f''(X)}{2}(x - a)^2.$$

In particular, if $|f''(t)| \le K$ on an interval containing a and x, then

$$|E(x)| \le \frac{K}{2}(x - a)^2. \ \square$$

Before proving this theorem we illustrate its application to Examples 1 and 2 above.

Example 3 Obtain an estimate for the error in the approximation $\sqrt{26} \simeq 5.1$ obtained in Example 1.

Solution $f(x) = x^{1/2}$, $f'(x) = (1/2)x^{-1/2}$, $f''(x) = -(1/4)x^{-3/2}$. Clearly $|f''(t)| \leq (1/4)25^{-3/2} = 1/500$ if $t \geq 25$ (and in particular if $25 \leq t \leq 26$). Hence $|E(26)| \leq (1/1000) \times (26 - 25)^2 = 0.001$. Since concavity ($f''(x) < 0$) implies $\sqrt{26} \leq 5.1$, we can assert that $5.1 - 0.001 \leq \sqrt{26} \leq 5.1$, that is,

$$5.099 \leq \sqrt{26} \leq 5.1.$$

Example 4 Obtain an estimate for the error in the approximation to cos 62° obtained in Example 3.

Solution $f(x) = \cos x$, $f'(x) = -\sin x$, $f''(x) = -\cos x$. We have $|f''(t)| = |\cos t| \leq 1/2$ if $\pi/3 \leq t \leq \pi/2$ (and in particular if $60° \leq t \leq 62°$). Hence $|E(62°)| \leq (1/4)(\pi/90)^2 < 0.00031$. Again, concavity ($f''(x) < 0$) indicates that cos 62° is less than the approximate value calculated (0.46977), so

$$0.46977 - 0.00031 \leq \cos 62° \leq 0.46977,$$

that is,

$$0.46946 \leq \cos 62° \leq 0.46977.$$

Proof of Theorem 6 Let $R = 2E(x)/(x - a)^2$, and for any t let

$$g(t) = E(t) - \frac{R}{2}(t - a)^2 = f(t) - f(a) - f'(a)(t - a) - \frac{R}{2}(t - a)^2.$$

Clearly g is differentiable between a and x, and $g(a) = g(x) = 0$. By the mean-value theorem there is a number c between a and x such that

$$0 = g'(c) = f'(c) - f'(a) - R(c - a).$$

Since f' is also differentiable, there exists, again by the mean-value theorem, a number X between a and c such that

$$\frac{f'(c) - f'(a)}{c - a} = f''(X).$$

Hence $R = f''(X)$, and so $E(x) = (f''(X)/2)(x - a)^2$, as asserted. If $|f''(t)| \leq K$ on an interval containing a and x, then certainly $|f''(X)| \leq K$ and so

$$|E(x)| \leq \frac{K}{2}(x - a)^2. \square$$

Example 5 Find an approximate value for $(61)^{1/3}$ and a suitable bound for the error.

Solution Let $f(x) = x^{1/3}$; then $f'(x) = (1/3)x^{-2/3}$ and $f''(x) = -(2/9)x^{-5/3}$. We take $a = 64$ (the closest perfect cube) and so obtain

$$x^{1/3} = f(x) \simeq f(a) + f'(a)(x - a) = 4 + \frac{1}{48}(x - 64).$$

Thus $61^{1/3} \simeq 4 - 3/48 = 3.9375$.

Now $|f''(t)| = 2/(9t^{5/3}) \leq (2/9)27^{-5/3} \simeq 0.0009144$ if $t \geq 27$, and in particular if $61 \leq t \leq 64$. (Note that we could not use the value $t = 64$ to get an estimate this time because the values of $|f'(t)|$ for t to the left of 64 are greater than the value at 64. We went all the way down to 27 so as to hit another perfect cube. Clearly a better estimate could be found.) We have

$$|E(61)| \leq \frac{0.0009144}{2}(61 - 64)^2 < 0.0042.$$

Since $f''(x)$ is negative, f is concave down and the tangent line lies above the curve. Hence

$$3.9375 - 0.0042 < 61^{1/3} < 3.9375$$
$$3.9333 < 61^{1/3} < 3.9375.$$

(The actual value of $61^{1/3}$ is 3.9364 . . . .)

The error in the tangent-line approximation can be interpreted in terms of differentials (see Section 2.2) as follows. If $x - a = \Delta x = dx$, then the change in height to the graph of $y = f(x)$ as we pass from $x = a$ to $x = a + \Delta x$ is $f(a + \Delta x) - f(a) = \Delta y$, and the corresponding change in height to the tangent line is $f'(a)(x - a) = f'(a)\, dx$, which is just the value at $x = a$ of the differential $dy = f'(x)\, dx$. (See Fig. 4.58.) Thus

$$E(x) = \Delta y - dy.$$

The fact that the approximating line is tangent to the graph of f at $x = a$ implies that the error is small compared with Δx as Δx approaches 0. In fact,

$$\lim_{\Delta x \to 0} \frac{\Delta y - dy}{\Delta x} = \lim_{\Delta x \to 0}\left(\frac{\Delta y}{\Delta x} - \frac{dy}{dx}\right) = \frac{dy}{dx} - \frac{dy}{dx} = 0.$$

If $|f''(t)| \leq K$ near $t = a$, we can assert a stronger result than this, namely,

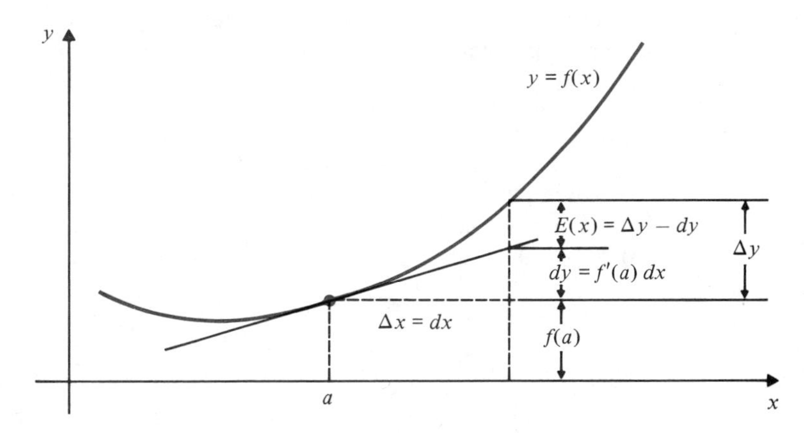

Figure 4.58

$$\left|\frac{\Delta y - dy}{(\Delta x)^2}\right| = \left|\frac{E(x)}{(\Delta x)^2}\right| \leq \frac{K}{2} = \text{constant},$$

so $|\Delta y - dy| \leq \text{const } (\Delta x)^2$.

Errors in Measurement

Suppose that a quantity x is obtained by measurement and that a second quantity y is determined as a function of x, that is, $y = f(x)$. Any error involved in the measurement of x will result in an error in the calculated value of y as well. It is convenient to use the differential $dx = \Delta x$ to denote the error involved in the measurement of x. The corresponding error in y is $\Delta y = f(x + dx) - f(x)$. The tangent-line approximation gives us a handy first-order approximation for Δy:

$$\Delta y \simeq dy = f'(x)\, dx.$$

The errors $dx = \Delta x$ and $dy \simeq \Delta y$ are absolute errors. Errors are frequently represented in relative terms, that is, as a fraction or percentage of the size of the quantity being expressed. The relative error in a measurement of x with absolute error dx is dx/x.

Example 6 The length x of the side of a square is measured with a 1 percent error. By approximately what percentage will the calculated area $A = x^2$ of the square be in error?

Solution We have $A = x^2$, so $\Delta A \simeq dA = 2x\, dx$. Thus

$$\frac{\Delta A}{A} \simeq \frac{2x\, dx}{x^2} = 2\frac{dx}{x}.$$

Since we are given that $dx/x = 1/100$ (that is, the error in x is 1 percent of x) we have $\Delta A/A \simeq 2/100$. Thus the area is in error by approximately 2 percent.

It must be stressed that $dA = 2x\, dx$ does not give the exact error in A. The exact error is $\Delta A = (x + dx)^2 - x^2 = 2x\,dx + (dx)^2$. Evidently if dx is small, then the term $(dx)^2$ is much smaller than the term $2x\, dx$, so the approximation of ΔA by dA is a good one. In Fig. 4.59, dA is the sum of the areas of two small rectangles of length x and width dx, while ΔA is this sum plus the area of a small square of side dx.

Example 7 The fraction of a radioactive sample remaining undecayed after one year is measured to be 0.998. If this measurement involves a possible error of up to 0.0001, find the approximate half-life of the sample and give an approximate maximum size for the error in this half-life.

Solution If $y(t)$ is the fraction of the original sample remaining undecayed after t years, then $y(t) = e^{kt}$. (See Section 3.7.) The constant k is determined from the fraction p remaining undecayed after one year: $p = e^k$, so $k = \ln p$.

For $p = 0.998$ and $dp = 0.0001$, we have $k = \ln 0.998$ and $dk = dp/p = 0.0001/0.998$. The half-life T of the sample is determined by $1/2 = e^{kT}$. Thus

$$T = \frac{1}{k}\ln\frac{1}{2} = -\frac{\ln 2}{k} \simeq 346.2 \text{ years}.$$

Now

$$\Delta T \simeq dT = \frac{dT}{dk}dk = \frac{\ln 2}{k^2}dk = -\frac{T}{k}\frac{dp}{p} \simeq 17.3.$$

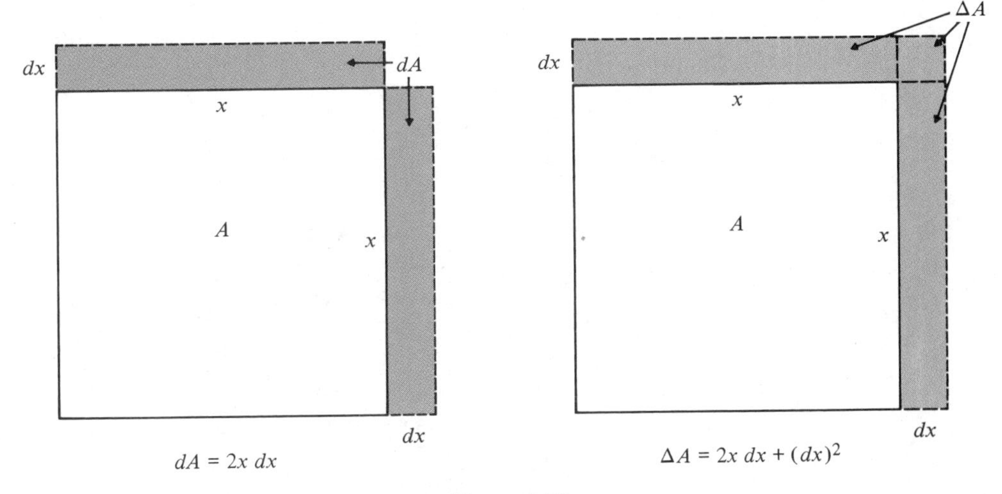

Figure 4.59

Thus the half-life of the sample is approximately 346.2 years with error no larger than about 17.3 years.

Newton's Method

Suppose that we wish to find a *root* or *zero* of the function $f(x)$, that is, a solution $x = r$ of the equation $f(x) = 0$. Unless f is a particularly simple function, for instance a linear or quadratic polynomial, we may not have any general procedure for finding its roots exactly.

Suppose that f is continuous on $[a, b]$ and that f has opposite signs at opposite ends of this interval; say $f(a) < 0$ and $f(b) > 0$. The intermediate-value theorem (Theorem 6 of Section 1.5) assures us that at least one root r must exist in $[a, b]$. (Refer to Fig. 4.60.) We can try to approximate this root as follows: Bisect $[a, b]$ at

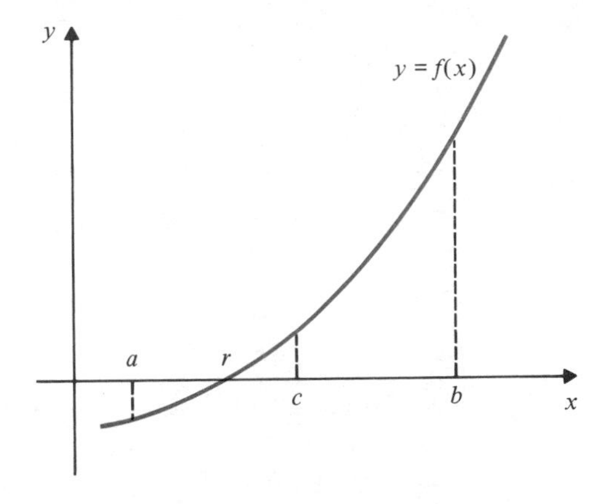

Figure 4.60

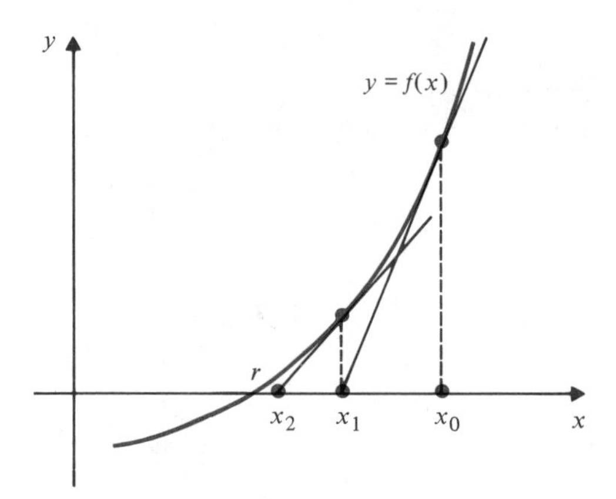

Figure 4.61

c and choose whichever interval, $[a, c]$ or $[c, b]$, is such that f has opposite signs at opposite ends as the new search interval. (If $f(c) = 0$ we have, of course, found our root.) This bisection process is repeated over and over, at each stage yielding an interval containing the root that is half as long as the interval at the previous stage. We stop when the interval is short enough that its midpoint must have the desired degree of closeness to the root. This method is called the bisection method. It always works (for continuous f), but it is slow and costly because many values of $f(x)$ must be computed before we achieve very much accuracy in locating the root.

If f is differentiable near the root r, then tangent lines can be used to produce a sequence of approximations to the root that approaches the root quite quickly. The idea is as follows. (See Fig. 4.61.) Make an initial guess at the root, say $x = x_0$. Let x_1 be the x-intercept of the tangent line to $y = f(x)$ at $x = x_0$. Under certain circumstances we can expect x_1 to be closer to the root than x_0 was. The process can be repeated over and over to get closer and closer; x_{n+1} is the x-intercept of the tangent line at $x = x_n$. The tangent line at $x = x_0$ has equation $y = f(x_0) + f'(x_0) \times (x - x_0)$. Since the point $(x_1, 0)$ lies on this line, we have

$$x_1 = x_0 - \frac{f(x_0)}{f'(x_0)},$$

and, in general,

$$x_{n+1} = x_n - \frac{f(x_n)}{f'(x_n)}.$$

One generally uses a calculator or computer to calculate the successive approximations $x_1, x_2, x_3, \ldots$ and observes whether they appear to converge to a limit. If $\lim_{n\to\infty} x_n = r$ exists, and if f/f' is continuous near r, then r must be a root of f

because

$$\lim_{n\to\infty} x_{n+1} = \lim_{n\to\infty} x_n - \lim_{n\to\infty} \frac{f(x_n)}{f'(x_n)}$$

$$r = r - \frac{f(r)}{f'(r)},$$

from which it follows that $f(r) = 0$. This method is called **Newton's method.**

Example 8 Use Newton's method to find the real root of the equation $x^3 + x - 1 = 0$ correct to six decimal places.

Solution We have $f(x) = x^3 + x - 1$ and $f'(x) = 3x^2 + 1$. Since $f'(x) > 0$ for all x, the graph of f is rising and so can cross the x-axis at only one root r. Since $f(0) = -1 < 0$ and $f(1) = 1 > 0$, r must lie in the interval $(0, 1)$. Let us make the initial guess $x_0 = 0.5$. Newton's formula here is

$$x_{n+1} = x_n - \frac{x_n^3 + x_n - 1}{3x_n^2 + 1} = \frac{2x_n^3 + 1}{3x_n^2 + 1}.$$

Thus

$$\begin{aligned} x_1 &= \frac{2(0.5)^3 + 1}{3(0.5)^2 + 1} = 0.7142857\ldots, \\ x_2 &= 0.6831797\ldots, \\ x_3 &= 0.6823284\ldots, \\ x_4 &= 0.6823278\ldots, \\ x_5 &= 0.6823278\ldots, \end{aligned}$$

Evidently $r = 0.682328$ correctly rounded to six decimal places.

Observe the behavior of the numbers x_n. By x_2 we have apparently picked up accuracy to two decimal places, and by x_3 to about five decimal places. It is characteristic of Newton's method that when one begins to get close to the root the convergence is very rapid.

Newton's method does not always work as well as it does in the preceding example. If the first derivative f' is very small near the root, or if the second derivative f'' is very large near the root, a single iteration of the formula can take us from quite close to the root to quite far away. Figure 4.62 illustrates this possibility.

The following theorem gives sufficient conditions for the Newton approximations to converge to a root r of f if the initial guess x_0 is sufficiently close to that root.

Theorem 7 Suppose that f, f', and f'' are continuous on an interval I containing x_n, x_{n+1}, and a root r of $f(x) = 0$. Suppose also that there exist constants K and $L > 0$ such that for all x in I

i) $|f''(x)| \leq K$, and

ii) $|f'(x)| \geq L$.

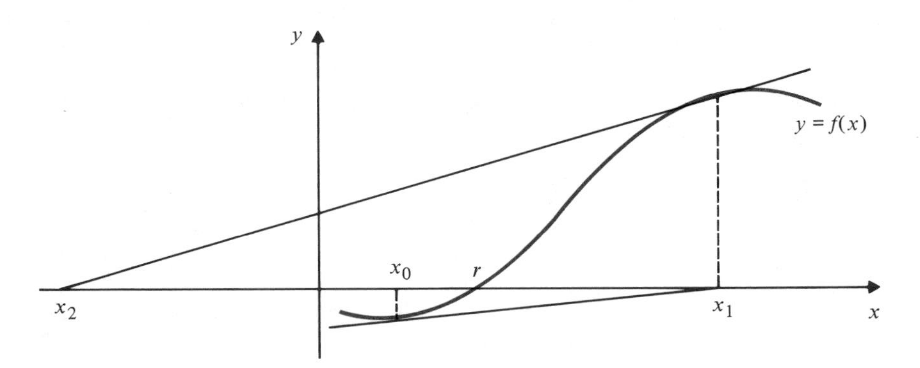

Figure 4.62

Then

a) $|x_{n+1} - r| \leq \dfrac{K}{2L}|x_{n+1} - x_n|^2$, and

b) $|x_{n+1} - r| \leq \dfrac{K}{2L}|x_n - r|^2$. □

We leave the proof of this theorem to you, in Exercises 28 and 29 at the end of this section. If $K/2L$ is not large (say $K/2L < 1$), this theorem states that x_n converges quickly to r once n becomes large enough that $|x_n - r| < 1$.

Theorem 7 is of considerable theoretical significance but little practical significance. In practice we compute successive approximations using Newton's formula and observe whether they seem to converge to a limit. If they do, and if the values of f at these approximations approach 0, we can be confident that we have located a root.

Example 9 Find the positive solution of the equation $e^x = 2 \cos x$.

Solution The graphs of e^x and $2 \cos x$ evidently cross at a point $x = r$ between $x = 0$ and $x = \pi/2 \simeq 1.4$ (see Fig. 4.63). Let us try $x_0 = 0.7$ as an initial guess. Let

$$f(x) = e^x - 2 \cos x.$$

Then

$$f'(x) = e^x + 2 \sin x.$$

Newton's formula becomes

$$x_{n+1} = x_n - \frac{e^{x_n} - 2 \cos x_n}{e^{x_n} + 2 \sin x_n}.$$

We calculate successive approximations using a scientific calculator (remembering that the angular mode must be set in radians).

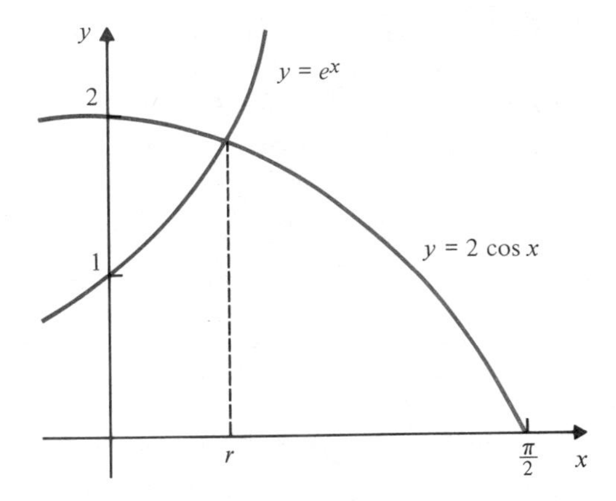

Figure 4.63

$$\begin{aligned} x_1 &= 0.5534098\ldots \\ x_2 &= 0.5398995\ldots \\ x_3 &= 0.5397851\ldots \\ x_4 &= 0.5397851\ldots \end{aligned}$$

The root is evidently 0.539785 rounded to six decimal places.

EXERCISES

In Exercises 1–12 use a suitable tangent-line approximation to determine approximations for the indicated values. In each case use concavity to determine the sign of the error, and obtain an estimate for the size of the error.

1. $\sqrt{50}$ **2.** $\sqrt{47}$ **3.** $\sqrt[4]{85}$

4. $\dfrac{1}{2.003}$ **5.** $e^{-1/10}$ **6.** $\sin 33°$

7. $\cos 46°$ **8.** $\sin \dfrac{\pi}{5}$ **9.** Arctan (1.05)

10. Arcsin (0.48) **11.** ln (0.94) **12.** cosh (0.02)

13. Approximately what percentage error can result in the calculation of the volume of a cube if the edge is measured to within 2 percent tolerance?

14. By approximately how much can the area of a circle be in error if the radius is 10 cm, with possible error of 0.005 cm? About how large is the relative error in this case?

15. If the radius and height of a circular cylinder are both measured with a relative error of less than 1 percent in absolute value, by about what percentage can the calculated volume of the cylinder be in error?

16. A spherical ball of ice melts so that its radius decreases from 20.00 cm to 19.80 cm in 1 hour. Approximately what volume of ice has melted in that hour?

In Exercises 17–24 find the indicated real roots by Newton's method. Make as good an initial guess x_0 as you can; sometimes a sketch is useful. Use a scientific calculator and express the root to six decimal places. If you have a programmable calculator you can program the Newton formula for the given function and obtain successive approximations very easily.

17. $x^3 - 2 = 0$ (Show there is only one real root.)

18. $x^3 + 3x - 8 = 0$ (Show there is only one real root.)

19. $x^3 + 3x^2 - 2 = 0$ (Find all three real roots.)

20. $x^5 + 5x - 3 = 0$ (Show that there is only one real root.)

21. $\sin x = 1 - x^2$ $(0 < x < \pi/2)$.

22. $\ln x = \dfrac{1}{x}$ **23.** $x = e^{-x}$ **24.** $x^2 = e^{-x^2}$ (Find both roots.)

25.* There is one solution x of the equation $\tan x = x$ in each interval $(-(\pi/2) + k\pi, \pi/2 + k\pi)$, $k = 0, 1, 2, \ldots$. For $k = 0$ the solution is clearly $x = 0$. Find the solutions for $k = 1$, 2, and 3. (*Hint:* Make a sketch of the graphs of $\tan x$ and x on the same set of axes to get an idea of the approximate positions of the roots.)

26.* A root x of an equation of the form $f(x) = x$ is called a *fixed point* of the function f. If f is continuous and increasing on an interval I, and has slope less than 1 wherever it is differentiable, and has a fixed point r in I, then r can be found by starting with an initial guess x_0 in I and calculating successive approximations

$$x_1 = f(x_0), \qquad x_2 = f(x_1), \qquad x_3 = f(x_2), \ldots$$

As Fig. 4.64 suggests, the x_n's converge to r. Use this method to find fixed points of the following functions.

a) $f(x) = e^{-x}$ on $x > 0$ (Compare with Exercise 23.)

b) $f(x) = k\pi + \text{Arctan } x$ $(k = 1, 2, 3)$ (Compare with Exercise 25.)

27.* Why does the method of Exercise 26 fail if the graph of the function f has slope greater than 1 near the fixed point? Make a sketch.

28.* Prove conclusion (a) of Theorem 7 as follows. Since x_{n+1} is the x-intercept of the tangent line to $y = f(x)$ at $x = x_n$, we have $f(x_n) + f'(x_n)(x_{n+1} - x_n) = 0$. Since $f(r) = 0$ also, we conclude that

$$f(x_{n+1}) - f(r) = f(x_{n+1}) - f(x_n) - f'(x_n)(x_{n+1} - x_n).$$

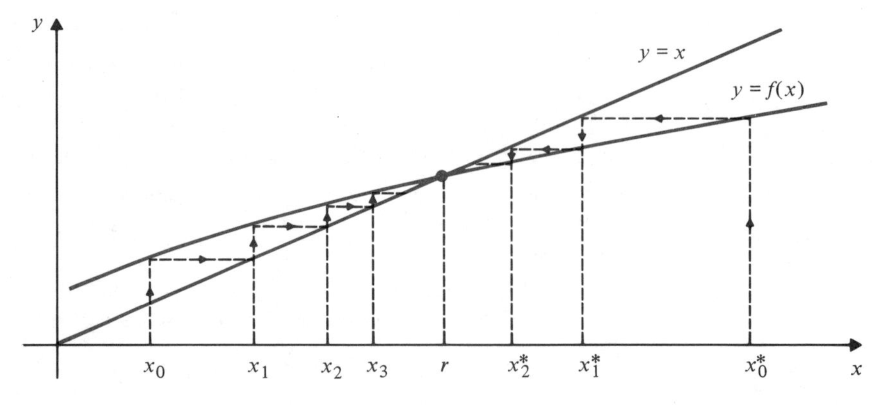

Figure 4.64

Apply the mean-value theorem to the left-hand side of this equation and Theorem 6 to the right-hand side, and thus complete the proof.

29.* Prove conclusion (b) of Theorem 7 as follows. Show that

$$x_{n+1} - r = \frac{f(r) - f(x_n) - f'(x_n)(x_n - r)}{f'(x_n)},$$

and then apply Theorem 6 to the right-hand side.

30.* Suppose f is three times differentiable on an interval containing a and x. Show that there exists X between a and x such that

$$f(x) = f(a) + f'(a)(x - a) + \frac{f''(a)}{2}(x - a)^2 + \frac{f'''(X)}{6}(x - a)^3.$$

(*Hint:* Mimic the proof of Theorem 6.) This shows that the error in the parabolic approximation

$$f(x) \simeq f(a) + f'(a)(x - a) + \frac{f''(a)}{2}(x - a)^2$$

is given by $E_2(x) = (f'''(X)/6)(x - a)^3$ for some X between a and x. If $|f'''(t)| \leq K$ on an interval containing a and x, then

$$|E_2(x)| \leq \frac{K}{6}|x - a|^3.$$

31. Use the parabolic approximation of Exercise 30 to find approximations for

a) $\sqrt{50}$

b) $e^{-1/10}$

In each case obtain an estimate for the error and compare your answers with those for Exercises 1 and 5.

32.* Generalize Exercise 30 to show for a function f that is $(n + 1)$ times differentiable on an interval containing a and x that

$$f(x) = f(a) + f'(a)(x - a) + \frac{f''(a)}{2!}(x - a)^2 + \cdots + \frac{f^{(n)}(a)}{n!} + E_n(x)$$

where

$$E_n(x) = \frac{f^{(n+1)}(X)}{(n + 1)!}(x - a)^{n+1}$$

for some X between a and x. This is known as Taylor's theorem with (Lagrange) remainder. We will study it in detail in Section 9.5.

4.7 INDETERMINATE FORMS

In section 3.2 we showed that

$$\lim_{x \to 0} \frac{\sin x}{x} = 1.$$

We could not readily see this by substituting $x = 0$ into the function $(\sin x)/x$ because, although both $\sin x$ and x are continuous at $x = 0$, nevertheless both vanish there. We call $(\sin x)/x$ an **indeterminate form** of type [0/0]. The limit of such an indeterminate form can be any number. For instance,

$$\lim_{x\to 0} \frac{kx}{x} = k, \qquad \lim_{x\to 0} \frac{|x|}{x^2} = \infty, \qquad \lim_{x\to 0} \frac{x^3}{x^2} = 0.$$

There are other types of indeterminate forms, specifically $[\infty/\infty]$, $[0 \cdot \infty]$, $[\infty - \infty]$, $[0^0]$, $[\infty^0]$, and $[1^\infty]$. Following are some examples.

$$\lim_{x\to 0} \frac{\ln x^2}{\dfrac{1}{x^2}} \qquad \text{type } [-\infty/\infty]$$

$$\lim_{x\to 0+} x \ln \frac{1}{x} \qquad \text{type } [0 \cdot \infty]$$

$$\lim_{x\to(\pi/2)-} \left(\tan x - \frac{1}{\pi - 2x}\right) \qquad \text{type } [\infty - \infty]$$

$$\lim_{x\to 0+} x^x \qquad \text{type } [0^0]$$

$$\lim_{x\to(\pi/2)-} (\tan x)^{\cos x} \qquad \text{type } [\infty^0]$$

$$\lim_{x\to\infty} \left(1 + \frac{1}{x}\right)^x \qquad \text{type } [1^\infty].$$

In this section we develop techniques called l'Hôpital's rules for evaluating some limits of indeterminate forms of the types [0/0] and $[\infty/\infty]$. The other types can usually be reduced to one of these two by algebraic manipulation and the taking of logarithms. (The Marquis de l'Hôpital, for whom these rules are named, lived before the French Revolution. The circumflex (ˆ) came into use in the French language after the Revolution. Thus the Marquis would have written his name l'Hospital.)

Before stating l'Hôpital's rules, we establish a generalized form of mean-value theorem.

The Generalized Mean-Value Theorem

Theorem 8 Suppose that functions f and g are continuous on the closed interval $[a, b]$ and differentiable on the open interval (a, b). Suppose also that $g'(x) \neq 0$ on (a, b), so $g(b) \neq g(a)$. Then there exists a point c in (a, b) such that

$$\frac{f(b) - f(a)}{g(b) - g(a)} = \frac{f'(c)}{g'(c)}.$$

Proof Let $h(x) = (f(b) - f(a))(g(x) - g(a)) - (g(b) - g(a))(f(x) - f(a))$. Clearly h is continuous on $[a, b]$ and differentiable on (a, b). Moreover, $h(a) = h(b) = 0$. By the mean-value theorem there exits a point c in (a, b) such that $h'(c) = 0$. Thus

$$(f(b) - f(a))g'(c) - (g(b) - g(a))f'(c) = 0,$$

and the conclusion follows. □

l'Hôpital's Rules

Theorem 9 (*The First l'Hôpital Rule*) Suppose the funtions f and g are differentiable in the interval (a, b), and $g'(x) \neq 0$ there. Suppose also

i) $\lim_{x\to a+} f(x) = \lim_{x\to a+} g(x) = 0$, and

ii) $\lim_{x\to a+} \dfrac{f'(x)}{g'(x)} = L$ (where L is finite or ∞ or $-\infty$).

Then

$$\lim_{x\to a+} \frac{f(x)}{g(x)} = L$$

also. Similar results hold if every $\lim_{x\to a+}$ is replaced by $\lim_{x\to b-}$ or even $\lim_{x\to c}$ where $a < c < b$. The cases $a = -\infty$ and $b = \infty$ are also allowed.

Proof We need prove only the case involving $\lim_{x\to a+}$. If $a < t < x < b$, we apply Theorem 8 on the interval $[t, x]$ to get

$$\frac{f(x) - f(t)}{g(x) - g(t)} = \frac{f'(c)}{g'(c)},$$

where c depends on t, and x and lies between t and x. If t and x both approach a from the right, then so does c. Hence

$$\lim_{x\to a+} \frac{f(x)}{g(x)} = \lim_{x\to a+} \lim_{t\to a+} \frac{f(x) - f(t)}{g(x) - g(t)}$$

$$= \lim_{x\to a+} \lim_{t\to a+} \frac{f'(c)}{g'(c)} = \lim_{c\to a+} \frac{f'(c)}{g'(c)} = L. \ \square$$

Note that in applying l'Hôpital's rules we calculate a quotient of derivatives, *not* the derivative of a quotient.

Example 1

$$\lim_{x\to 1} \frac{\ln x}{x^2 - 1} \qquad \left[\frac{0}{0}\right]$$

$$= \lim_{x\to 1} \frac{1/x}{2x} = \lim_{x\to 1} \frac{1}{2x^2} = \frac{1}{2}.$$

This example illustrates how calculations based on l'Hôpital's rule are carried out.

Having identified the limit as that of a [0/0] indeterminate form, we replace it by the limit of the quotient of derivatives; the existence of this latter limit justifies the equality. It is possible that the limit of the quotient of derivatives may still be indeterminate, in which case a second application of l'Hôpital's rule can be made. Such applications may be strung out until a limit can finally be extracted, which then justifies all the previous applications of the rule.

Example 2

$$\lim_{x\to 0} \frac{1 - e^{3x^2}}{x \sin 2x} \qquad \left[\frac{0}{0}\right]$$

$$= \lim_{x\to 0} \frac{-6xe^{3x^2}}{\sin 2x + 2x \cos 2x} \qquad \left[\frac{0}{0}\right]$$

$$= \lim_{x\to 0} \frac{-6e^{3x^2} - 36x^2e^{3x^2}}{4 \cos 2x - 4x \sin 2x} = \frac{-6 - 0}{4 - 0} = -\frac{3}{2}$$

Note that l'Hôpital's rule was used twice here.

Example 3

$$\lim_{x\to(\pi/2)-} \frac{2x - \pi}{\cos^2 x} \qquad \left[\frac{0}{0}\right]$$

$$= \lim_{x\to(\pi/2)-} \frac{2}{-2 \sin x \cos x} = -\infty$$

Example 4

$$\lim_{x\to 1+} \frac{x}{\ln x} = \infty$$

Note that l'Hôpital's rule was not used here at all; $x/(\ln x)$ is not an indeterminant form. (Had we tried to apply l'Hôpital's rule we would have been led to $\lim_{x\to 1+} (1/(1/x)) = 1$, an erroneous answer.)

Example 5

$$\lim_{x\to 0+} \left(\frac{1}{x} - \frac{1}{\sin x}\right) \qquad [\infty - \infty]$$

$$= \lim_{x\to 0+} \frac{\sin x - x}{x \sin x} \qquad \left[\frac{0}{0}\right]$$

$$= \lim_{x\to 0+} \frac{\cos x - 1}{\sin x + x \cos x} \qquad \left[\frac{0}{0}\right]$$

$$= \lim_{x\to 0+} \frac{-\sin x}{2 \cos x - x \sin x} = \frac{-0}{2} = 0.$$

Theorem 10 (*The Second l'Hôpital Rule*) Suppose that f and g are differentiable on the interval (a, b) and that $g'(x) \neq 0$ there. Suppose also

i) $\lim_{x\to a+} g(x) = \pm\infty$, and

ii) $\lim_{x\to a+} \dfrac{f'(x)}{g'(x)} = L$ (where L is finite, or ∞ or $-\infty$).

Then

$$\lim_{x \to a+} \frac{f(x)}{g(x)} = L \text{ also.}$$

Again, similar results hold for $\lim_{x \to b-}$ and for $\lim_{x \to c}$, and the cases $a = -\infty$ and $b = \infty$ are allowed.

Proof For $a < x < t < b$ we use Theorem 8 again to obtain

$$\frac{\dfrac{f(x)}{g(x)} - \dfrac{f(t)}{g(x)}}{\dfrac{g(x)}{g(x)} - \dfrac{g(t)}{g(x)}} = \frac{f(x) - f(t)}{g(x) - g(t)} = \frac{f'(c)}{g'(c)},$$

where c depends on x, and t and lies between x and t. Hold t fixed and let x approach a from the right. Since $\lim_{x \to a+} g(x) = \pm\infty$ we have

$$\lim_{x \to a+} \frac{f(t)}{g(x)} = 0 \qquad \text{and} \lim_{x \to a+} \frac{g(t)}{g(x)} = 0.$$

Hence

$$\lim_{x \to a+} \frac{f(x)}{g(x)} = \lim_{x \to a+} \frac{f'(c)}{g'(c)}.$$

Observe that the left side of this equation is independent of t. If we now let t approach a from the right (which forces c to do the same) we get

$$\lim_{x \to a+} \frac{f(x)}{g(x)} = \lim_{t \to a+} \lim_{x \to a+} \frac{f'(c)}{g'(c)} = \lim_{c \to a+} \frac{f'(c)}{g'(c)} = L,$$

as required. □

Example 6

$$\lim_{x \to \infty} \frac{x^2}{e^x} \qquad \left[\frac{\infty}{\infty}\right]$$

$$= \lim_{x \to \infty} \frac{2x}{e^x} \qquad \text{still} \left[\frac{\infty}{\infty}\right]$$

$$= \lim_{x \to \infty} \frac{2}{e^x} = 0$$

Similarly, one can prove that $\lim_{x \to \infty} x^n/e^x = 0$ for any positive integer n by repeated applications of l'Hôpital's rule and hence prove Theorem 6 of Section 3.5.

Example 7

$$\lim_{x \to 0+} x^a \ln x \qquad (a > 0) \qquad [-0 \cdot \infty]$$

$$= \lim_{x \to 0+} \frac{\ln x}{x^{-a}} \qquad \left[\frac{-\infty}{\infty}\right]$$

$$= \lim_{x \to 0+} \frac{1/x}{-ax^{-a-1}} = \lim_{x \to 0+} \frac{x^a}{-a} = 0$$

Example 8

$$\lim_{x\to 0+}\left(\frac{1}{x}-\frac{1}{e^x-1}\right) \qquad [\infty-\infty]$$

$$=\lim_{x\to 0+}\frac{e^x-1-x}{x(e^x-1)} \qquad \left[\frac{\infty}{\infty}\right]$$

$$=\lim_{x\to 0+}\frac{e^x-1}{e^x-1+xe^x} \qquad \left[\frac{\infty}{\infty}\right]$$

$$=\lim_{x\to 0+}\frac{e^x}{2e^x+xe^x}=\lim_{x\to 0+}\frac{1}{2+x}=\frac{1}{2}$$

Example 9

$$\lim_{x\to 0+} x^x \qquad [0^0]$$

Let $y = x^x$. $\lim_{x\to 0+} \ln y = \lim_{x\to 0+} x \ln x = 0$, by Example 7. Hence $\lim_{x\to 0+} x^x = \lim_{x\to 0+} y = e^0 = 1$.

Example 10

$$\lim_{x\to(\pi/2)-} (\tan x)^{\cos x} \qquad [\infty^0]$$

Let $y = (\tan x)^{\cos x}$.

$$\lim_{x\to(\pi/2)-} \ln y = \lim_{x\to(\pi/2)-} \cos x \ln \tan x \qquad [0\cdot\infty]$$

$$=\lim_{x\to(\pi/2)-}\frac{\ln \tan x}{\sec x} \qquad \left[\frac{\infty}{\infty}\right]$$

$$=\lim_{x\to(\pi/2)-}\frac{\dfrac{1}{\tan x}\sec^2 x}{\sec x\tan x}$$

$$=\lim_{x\to(\pi/2)-}\frac{\sec x}{\tan^2 x}=\lim_{x\to(\pi/2)-}\frac{\cos x}{\sin^2 x}=0$$

Thus $\lim_{x\to(\pi/2)-} (\tan x)^{\cos x} = e^0 = 1$.

Example 11 $\displaystyle\lim_{x\to\infty}\left(1+\sin\frac{3}{x}\right)^x \qquad [1^\infty]$

Let $y = \left(1+\sin\dfrac{3}{x}\right)^x$.

$$\lim_{x\to\infty}\ln y = \lim_{x\to\infty} x\ln\left(1+\sin\frac{3}{x}\right) \qquad [\infty\cdot 0]$$

$$=\lim_{x\to\infty}\frac{\ln\left(1+\sin\dfrac{3}{x}\right)}{\dfrac{1}{x}} \qquad \left[\frac{0}{0}\right]$$

$$= \lim_{x\to\infty} \frac{\dfrac{1}{1+\sin\dfrac{3}{x}}\left(\cos\dfrac{3}{x}\right)\left(-\dfrac{3}{x^2}\right)}{-\dfrac{1}{x^2}}$$

$$= \lim_{x\to\infty} \frac{3\cos\dfrac{3}{x}}{1+\sin\dfrac{3}{x}} = 3$$

Hence $\lim_{x\to\infty} (1 + \sin 3/x)^x = e^3$.

EXERCISES

Evaluate the limits in Exercises 1–34.

1. $\lim_{x\to 0} \dfrac{3x}{\tan 4x}$
2. $\lim_{x\to 2} \dfrac{\ln(2x-3)}{x^2-4}$
3. $\lim_{x\to 0} \dfrac{\sin ax}{\sin bx}$
4. $\lim_{x\to 0} \dfrac{1-\cos ax}{1-\cos bx}$
5. $\lim_{x\to 0} \dfrac{\text{Arcsin}\, x}{\text{Arctan}\, x}$
6. $\lim_{x\to 1} \dfrac{x^{1/3}-1}{x^{2/3}-1}$
7. $\lim_{x\to 0} x \cot x$
8. $\lim_{x\to 0} \dfrac{1-\cos x}{\ln(1+x^2)}$
9. $\lim_{t\to \pi} \dfrac{\sin^2 t}{t-\pi}$
10. $\lim_{x\to 0} \dfrac{10^x - e^x}{x}$
11. $\lim_{x\to \pi/2} \dfrac{\cos 3x}{\pi - 2x}$
12. $\lim_{x\to 1} \dfrac{\ln(ex)-1}{\sin \pi x}$
13. $\lim_{x\to\infty} x^2 \sin\dfrac{1}{x}$
14. $\lim_{x\to 0} \dfrac{x-\sin x}{x^3}$
15. $\lim_{x\to 0} \dfrac{x-\sin x}{x-\tan x}$
16. $\lim_{x\to 0} \dfrac{2-x^2-2\cos x}{x^4}$
17. $\lim_{x\to 0+} \dfrac{\sin^2 x}{\tan x - x}$
18. $\lim_{r\to\pi/2} \dfrac{\ln \sin r}{\cos r}$
19. $\lim_{t\to \pi/2} \dfrac{\sin t}{t}$
20. $\lim_{x\to 1-} \dfrac{\text{Arccos}\, x}{x-1}$
21. $\lim_{x\to\infty} x(2\,\text{Arctan}\, x - \pi)$
22. $\lim_{x\to\infty} x(1-\tanh x)$
23. $\lim_{x\to\infty} e^x(1-\tanh x)$
24. $\lim_{t\to(\pi/2)-} (\sec t - \tan t)$
25. $\lim_{t\to 0} \left(\dfrac{1}{t} - \dfrac{1}{te^{at}}\right)$
26. $\lim_{x\to 0+} x^{\sqrt{x}}$
27. $\lim_{x\to 0+} (\csc x)^{\sin^2 x}$
28. $\lim_{x\to 0} \left(\dfrac{1+x}{x} - \dfrac{1}{\ln(1+x)}\right)$

29. $\lim_{t\to 0} \frac{3 \sin t - \sin 3t}{3 \tan t - \tan 3t}$

30. $\lim_{x\to 0} \left(\frac{\sin x}{x}\right)^{1/x^2}$

31. $\lim_{t\to 0} (\cos 2t)^{1/t^2}$

32. $\lim_{x\to 0+} \frac{\csc x}{\ln x}$

33. $\lim_{x\to 1-} \frac{\ln \sin \pi x}{\csc \pi x}$

34. $\lim_{x\to 0} (1 + \tan x)^{1/x}$

35. Evaluate $\lim_{h\to 0} \frac{f(x + h) - 2f(x) + f(x - h)}{h^2}$ if f is twice differentiable.

36.* What is wrong with the following "proof" of Theorem 8? "By the mean-value theorem, $f(b) - f(a) = (b - a)f'(c)$ for some c between a and b, and similarly $g(b) - g(a) = (b - a)g'(c)$ for some such c. Hence $(f(b) - f(a))/(g(b) - g(a)) = f'(c)/g'(c)$, as required."

5.1 Parametric Curves

5.2 Smooth Curves

5.3 Velocity and Acceleration in the Plane

5.4 Polar Coordinates and Polar Curves

5 *Curves in the Plane*

Up to this point we have dealt almost exclusively with curves that are the graphs of functions in a Cartesian coordinate system. This class of curves is somewhat restricted; for instance, it excludes curves that intersect vertical lines more than once. Of course, in Section 2.7 we encountered curves of a more general nature that were graphs of equations in two variables, and we developed a technique there for finding their slopes.

In this chapter we consider curves that are specified parametrically, that is, for which the coordinates of a general point are given as functions of a separate independent variable called a parameter. We also consider curves that are graphs of equations in the polar coordinate system, in which points are located by distance and direction from the origin.

5.1 PARAMETRIC CURVES

Definition 1

> Let f and g be continuous functions defined on an interval I of the real line. The set of all points with coordinates (x, y), where $x = f(t)$ and $y = g(t)$ for some t in I, is called a **parametric curve** in the plane. The variable t is called the **parameter.**

The axis (real line) of the parameter t is considered to be distinct from the coordinate axes of the plane of the curve. In Section 3.2 the path of a fired projectile was presented as a parametric curve. As in that example, the parameter in a parametric curve frequently represents time; the parametric equations

$$x = f(t), \qquad y = g(t)$$

specify the coordinates of the position P_t at time t of a point moving in the plane. The curve is the path of the moving point (see Fig. 5.1). Because f and g are

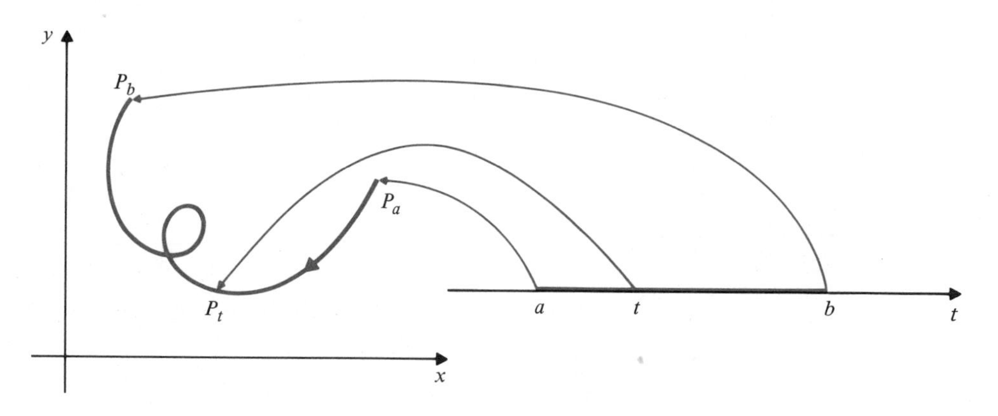

Figure 5.1

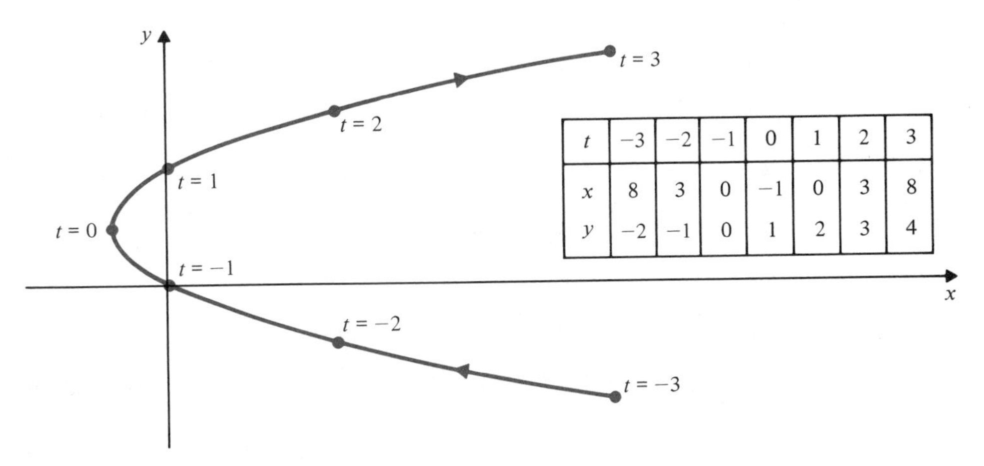

t	−3	−2	−1	0	1	2	3
x	8	3	0	−1	0	3	8
y	−2	−1	0	1	2	3	4

Figure 5.2

continuous, the curve has no breaks in it. (It can be drawn with a single continuous motion of the pen without removing the pen from the paper until the whole curve has been drawn.) Note that a parametric curve has a *direction* assigned to it (indicated, say, by an arrowhead), namely the direction corresponding to increasing values of the parameter.

Example 1 Sketch and identify the parametric curve

$$x = t^2 - 1, \qquad y = t + 1 \qquad (-\infty < t < \infty).$$

Solution We can construct a table of values of x and y for various values of t, thus getting the coordinates of a number of points on the curve. See Fig. 5.2. The curve seems to be part of a parabola with horizontal axis and vertex $(-1, 1)$. Alternatively, we can eliminate the parameter from the pair of parametric equations, thus producing a single equation in x and y whose graph is the desired curve:

$$t = y - 1, \qquad x = t^2 - 1 = (y - 1)^2 - 1 = y^2 - 2y.$$

The curve $x = y^2 - 2y$ is indeed a parabola.

Although the curve in this example is more easily identified when the parameter is eliminated, there is a loss of information in going to the nonparameteric form. Specifically, we lose the sense of the curve as the path of a moving point and hence also the direction of the curve. If the t in the parametric form represents time, the nonparametric equation $x = y^2 - 2y$ no longer tells us which point corresponds to any particular time t.

Example 2 The straight line passing through the two points $P_1 = (x_1, y_1)$ and $P_2 = (x_2, y_2)$ (see Fig. 5.3) has parametric equations

$$\left.\begin{aligned} x &= x_1 + t(x_2 - x_1) \\ y &= y_1 + t(y_2 - y_1) \end{aligned}\right\} \qquad (t \text{ real}).$$

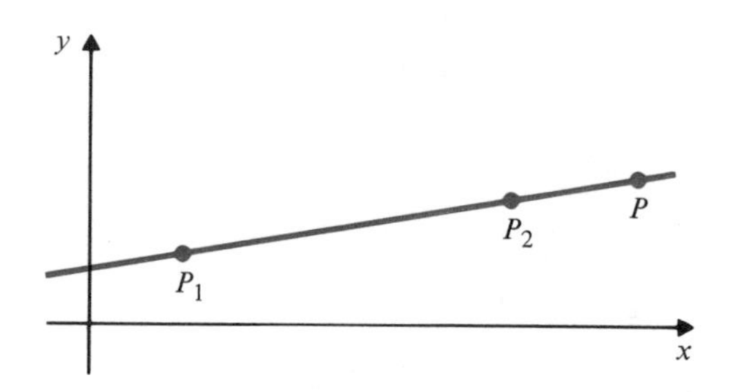

Figure 5.3

To see that these equations represent a straight line, note that

$$\frac{y - y_1}{x - x_1} = \frac{y_2 - y_1}{x_2 - x_1} = \text{constant} \qquad (\text{assuming } x_2 \neq x_1).$$

The point $P = (x, y)$ is at position P_1 when $t = 0$ and at P_2 when $t = 1$. If $t = 1/2$, then P is the midpoint between P_1 and P_2.

Example 3 Sketch the curve $x = 3 \cos t$, $y = 3 \sin t$ $(0 \leq t \leq 3\pi/2)$.

Solution Since $x^2 + y^2 = 9 \cos^2 t + 9 \sin^2 t = 9$, all points on the curve lie on the circle $x^2 + y^2 = 9$. As t increases from 0 through $\pi/2$ and π to $3\pi/2$, the point (x, y) moves from (3, 0) through (0, 3) and $(-3, 0)$ to $(0, -3)$. See Fig. 5.4.

Example 4 Sketch and identify the curve $x = a \cos s$, $y = b \sin s$ $(0 \leq s \leq 2\pi)$, where $a > b > 0$.

Solution Since

$$\frac{x^2}{a^2} + \frac{y^2}{b^2} = \cos^2 s + \sin^2 s = 1,$$

the curve is all or part of an ellipse with major axis from $(-a, 0)$ to $(a, 0)$ and minor axis from $(0, -b)$ to $(0, b)$. As s increases from 0 to 2π, the point (x, y) moves counterclockwise around the ellipse starting from $(a, 0)$ and returning to the same point. Thus the curve is the whole ellipse.

Figure 5.5 shows how the parameter s can be interpreted as an angle and how the points on the ellipse can be obtained using circles of radii a and b. Since the curve starts and ends at the same point, it is called a **closed curve.**

Example 5 Sketch the parametric curve

$$x = t^3 - 3t, \qquad y = t^2 \qquad (-2 \leq t \leq 2).$$

Solution We could eliminate the parameter and obtain

$$x^2 = t^2(t^2 - 3)^2 = y(y - 3)^2,$$

but this doesn't help much since we do not recognize this curve from its Cartesian equation. Instead let us obtain some points:

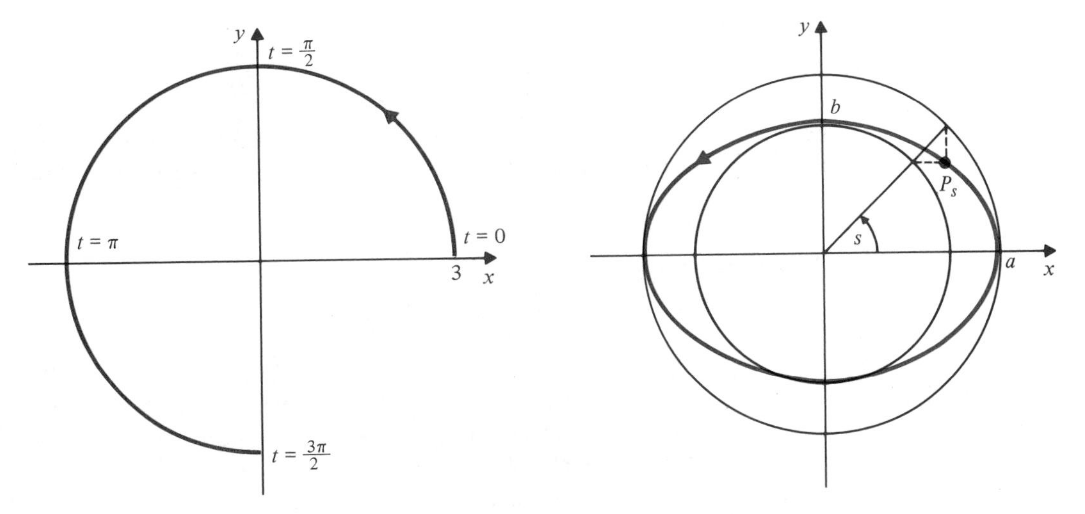

Figure 5.4

Figure 5.5

t	-2	$-\frac{3}{2}$	-1	$-\frac{1}{2}$	0	$\frac{1}{2}$	1	$\frac{3}{2}$	2
x	-2	$\frac{9}{8}$	2	$\frac{11}{8}$	0	$-\frac{11}{8}$	-2	$-\frac{9}{8}$	2
y	4	$\frac{9}{2}$	1	$\frac{1}{4}$	0	$\frac{1}{4}$	1	$\frac{9}{2}$	4

Note that the curve is symmetric about the y-axis because x is an odd function of t and y is an even function of t. (At t and $-t$, x has opposite values but y has the same value.)

The curve intersects itself on the y-axis (see Fig. 5.6). To find this self-intersection

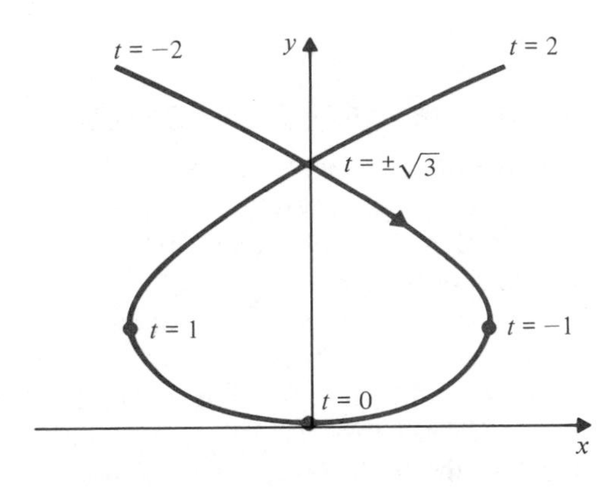

Figure 5.6

set $x = 0$:

$$0 = x = t^3 - 3t = t(t - \sqrt{3})(t + \sqrt{3}).$$

For $t = 0$ the curve is at $(0, 0)$, but for $t = \pm\sqrt{3}$ the curve is at $(0, 3)$. The self-intersection occurs because the curve passes through $(0, 3)$ for two different values of t.

General Plane Curves: Parametrizations

Definition 2

> A set S of points in the plane is called a **plane curve** if S is the parametric curve $x = f(t)$, $y = g(t)$ (t in I) for some choice of the continuous functions f and g and the interval I.

When we speak of a plane curve we are thinking only of the set of points, not any particular *parametrization* (that is, set of parametric equations) of the curve. Thus a plane curve has no distinguished direction.

Example 6 The circle $x^2 + y^2 = 1$ is a plane curve. The following are all possible parametrizations of C.

i) $x = \cos t,\ y = \sin t, \qquad (0 \le t \le 2\pi)$

ii) $x = \sin t^2,\ y = \cos t^2, \qquad (0 \le t \le \sqrt{2\pi})$

iii) $x = \cos(\pi t + 1),\ y = \sin(\pi t + 1), \qquad (-1 \le t \le 1)$

iv) $x = 1 - t^2,\ y = t\sqrt{2 - t^2}, \qquad (-\sqrt{2} \le t \le \sqrt{2})$

There are, of course, infinitely many other possible parametrizations of this curve.

Example 7 If f is a function continuous on an interval I then the graph of f is a plane curve. One obvious parametrization of this curve is

$$x = t, \qquad y = f(t), \qquad (t \text{ in } I).$$

Some Interesting Plane Curves

Example 8 (*The Involute of a Circle*) A string is wound around a fixed circle. One end is unwound in such a way that the part of the string not lying on the circle is extended in a straight line. The curve followed by this free end of the string is called an **involute** of the circle.

Suppose the circle has equation $x^2 + y^2 = a^2$ and suppose the end of the string being unwound starts at the point $A = (a, 0)$. At some subsequent time let P be the position of the end of the string and let T be the point where the string leaves the circle. Clearly PT is tangent to the circle at T. We shall parametrize the path of P in terms of the angle TOA, which we denote by t.

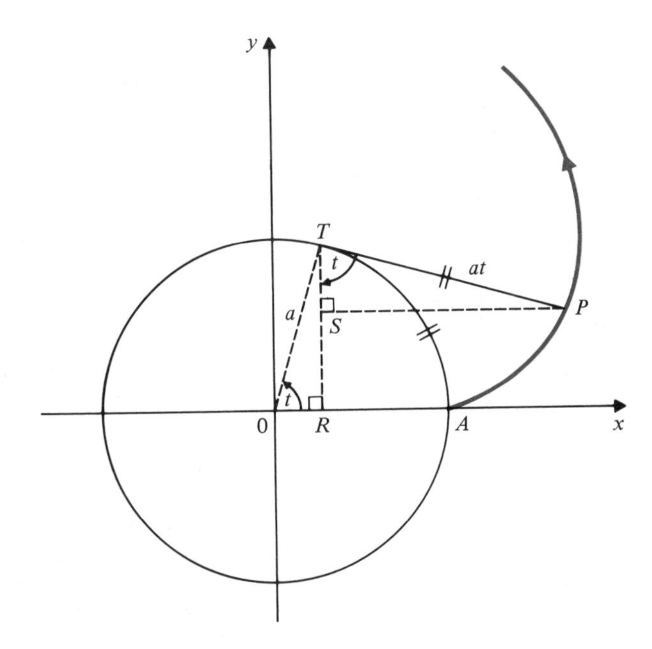

Figure 5.7

Let points R on OA and S on TR be as shown in Fig. 5.7. Clearly

$$OR = OT\cos t = a\cos t,$$

$$RT = a\sin t,$$

$$TP = \text{arc } TA = at \qquad \text{(the string does not stretch or slip on the circle)},$$

$$\text{angle } OTP = \frac{\pi}{2} \qquad \text{(a tangent to a circle is perpendicular to the radial line to the point of contact)},$$

$$\text{angle } STP = \text{angle } ROT = t \qquad \text{(similar triangles)},$$

$$SP = TP\sin t = at\sin t,$$

$$ST = at\cos t.$$

If P has coordinates (x, y), then $x = OR + SP$, $y = RT - ST$:

$$x = a\cos t + at\sin t, \qquad y = a\sin t - at\cos t \qquad (t \geq 0).$$

These are the parametic equations of the involute in terms of t.

Example 9 (*The Cycloid*) If a circle rolls without slipping along a straight line, the path followed by a point fixed on the circle is called a **cycloid.**

Suppose that the line is the x-axis, that the circle has radius a and lies above the

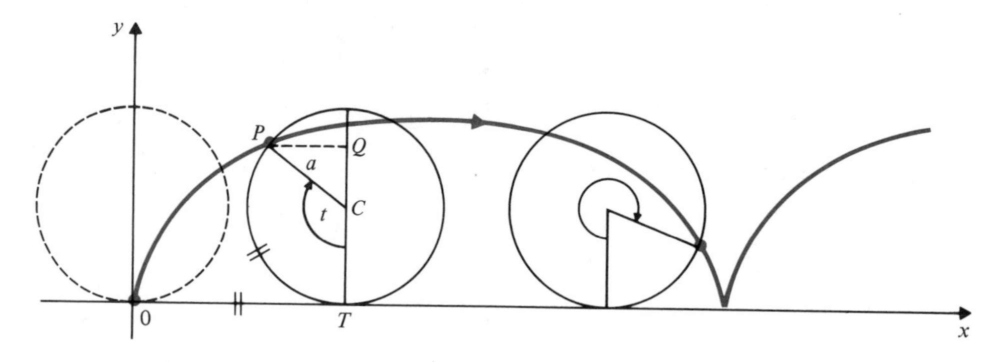

Figure 5.8

line, and that the point whose motion we follow is originally at the origin O. See Fig. 5.8.

The point has moved to position P when the circle has rolled through an angle t and is tangent to the line at T. Since no slipping occurs,

$$\text{segment } OT = \text{arc } PT = at.$$

Let PQ be perpendicular to TC, as shown in the figure. If P has coordinates (x, y), then

$$x = OT - PQ = at - a\sin(\pi - t) = at - a\sin t,$$
$$y = TC + CQ = a + a\cos(\pi - t) = a - a\cos t.$$

The parametric equations of the cycloid are therefore

$$x = a(t - \sin t), \qquad y = a(1 - \cos t).$$

EXERCISES

In Exercises 1–10 sketch the given parametric curve, showing its direction with an arrow. Eliminate the parameter to give an ordinary equation in x and y whose graph contains the parametric curve.

1. $x = t,\ y = 1 - t\ (0 \le t \le 1)$

2. $x = 2 - t,\ y = t + 1\ (0 \le t < \infty)$

3. $x = 1 + 2t,\ y = t^2\ (-\infty < t < \infty)$

4. $x = t^2,\ y = 3t^2 - 2\ (-a \le t \le a)$

5. $x = \dfrac{1}{t},\ y = t - 1\ (0 < t < 4)$

6. $x = \dfrac{1}{1 + t^2},\ y = \dfrac{t}{1 + t^2}\ (-\infty < t < \infty)$

7. $x = 3\sin 2t,\ y = 3\cos 2t\ \left(0 \le t \le \dfrac{\pi}{3}\right)$

8. $x = a\sec t,\ y = b\tan t\ \left(-\dfrac{\pi}{2} < t < \dfrac{\pi}{2}\right)$

9. $x = 3 \sin \pi t,\ y = 4 \cos \pi t\ (-1 \le t \le 1)$

10. $x = \cos \sin s,\ y = \sin \sin s\ (-\infty < s < \infty)$

11. $x = \cos^3 t,\ y = \sin^3 t\ (0 \le t \le 2\pi)$

12. Show that each of the following sets of parametric equations represents a different arc of the parabola $2(x + y) = 1 + (x - y)^2$.

i) $x = \cos^4 t,\ y = \sin^4 t$

ii) $x = \sec^4 t,\ y = \tan^4 t$

iii) $x = \tan^4 t,\ y = \sec^4 t$

13. Find a parametrization of the parabola $y = x^2$ using as parameter the slope of the tangent line at the general point.

14. Find a parametrization of the circle $x^2 + y^2 = R^2$ using as parameter the slope m of the line joining the general point to the point $(R, 0)$. Does this parametrization fail to give any point on the circle?

15.* Eliminate the parameter from the parametric equations

$$x = \frac{3t}{1 + t^3}, \qquad y = \frac{3t^2}{1 + t^3} \qquad (t \neq -1),$$

and hence find an ordinary equation in x and y for this curve (a *folium of Descartes*). The parameter t can be interpreted as the slope of the line joining the general point (x, y) to the origin. Sketch the curve and show that the line $x + y = -1$ is an asymptote.

16.* A railroad wheel has a flange extending below the level of the track on which the wheel rolls. If the radius of the wheel is a, and that of the flange is $b > a$, find parametric equations of the path of a point P at the circumference of the flange as the wheel rolls along the track. This curve is called a *prolate cycloid*. Note that for a portion of each revolution of the wheel, P is moving backward. Try to sketch the graph of a prolate cycloid.

17.* (*Hypocycloids*) If a circle of radius b rolls, without slipping, around the inside of a fixed circle of radius $a > b$, a point on the circumference of the rolling circle traces a curve called a *hypocycloid*. If the fixed circle is centered at the origin and the point tracing the curve starts at $(a, 0)$, show that the hypocycloid has parametric equations

$$x = (a - b)\cos t + b \cos\left(\frac{a - b}{b} t\right),$$

$$y = (a - b)\sin t - b \sin\left(\frac{a - b}{b} t\right),$$

where t is the angle between the positive x-axis and the line from the origin to the point at which the rolling circle touches the fixed circle.

If $a = 2$ and $b = 1$, show that the hypocycloid becomes a straight line segment.

If $a = 4$ and $b = 1$, show that the parametric equations of the hypoclycloid simplify to $x = 4\cos^3 t,\ y = 4 \sin^3 t$. This curve is called a hypocycloid of four cusps. (See Fig. 5.9.)

Hypocycloids resemble the curves produced by a popular children's toy called Spirograph, but Spirograph curves result from following a point inside the disc of the rolling circle rather than on its circumference, and they therefore do not have sharp cusps.

* Difficult and/or theoretical problems.

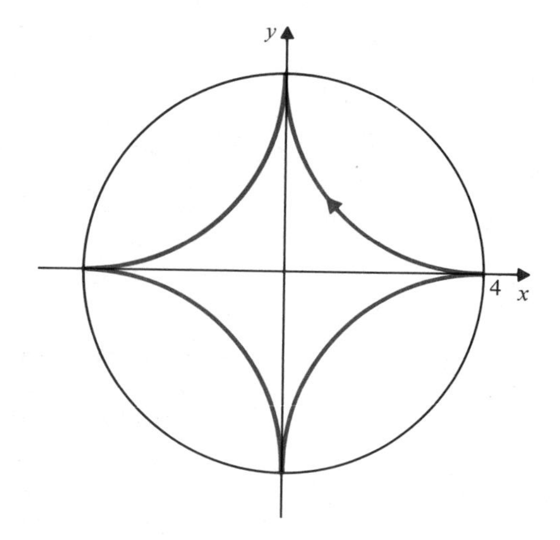

The hypocycloid $x = 4\cos^3 t$
$y = 4\sin^3 t$
or $x^{2/3} + y^{2/3} = 4^{2/3}$

Figure 5.9

5.2 SMOOTH CURVES

We say that a curve is *smooth* if it has a tangent line at each point P and this tangent turns in a continuous way as P moves along the curve. (That is, the angle between the tangent line at P and some fixed line, the x-axis say, is a continuous function of the position of P.)

If the curve C is the graph of a function f, then C is certainly smooth on any interval where the derivative $f'(x)$ exists and is a continuous function of x. It may also be smooth on intervals containing isolated singular points; for example, the curve $y = x^{1/3}$ is smooth everywhere even though dy/dx does not exist at $x = 0$. (See Example 2 of Section 2.1.)

For parametric curves $x = f(t)$, $y = g(t)$, the situation is more complicated. Even if f and g have continuous derivatives everywhere, such curves may fail to be smooth at certain points, specifically points where $f'(x) = g'(x) = 0$.

Example 1 Consider the parametric curve $x = t^2$, $y = t^3$. Even though $dx/dt = 2t$ and $dy/dt = 3t^2$ are continuous for all t, the curve is not smooth at $t = 0$. (See Fig. 5.10.) Observe that both dx/dt and dy/dt vanish at $t = 0$.

If we regard the parametric equations as specifying the position at time t of a moving point P, then the horizontal velocity dx/dt and the vertical velocity dy/dt are 0 at $t = 0$, so P is momentarily at rest at that instant.

The Slope of a Parametric Curve

Theorem 1 Suppose that f' and g' are continuous on an interval I. If $f'(t) > 0$ on I (or $f'(t) < 0$ on I), then the curve C with parametric equations $x = f(t)$, $y = g(t)$ is smooth for t in I and has, at the point with parameter value t, a tangent line with slope

$$\frac{dy}{dx} = \frac{g'(t)}{f'(t)}.$$

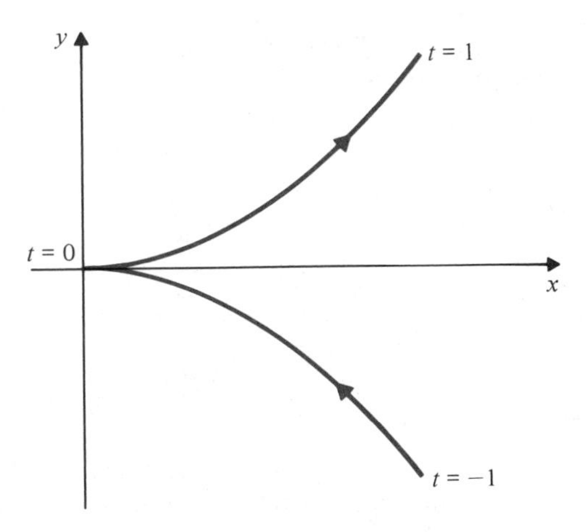

Figure 5.10

If $g'(t) > 0$ on I (or $g'(t) < 0$ on I), then C is smooth for t in I and has, at the point with parameter value t, a normal line with slope

$$-\frac{dx}{dy} = -\frac{f'(t)}{g'(t)}.$$

Thus C is smooth except possibly at points where $f'(t)$ and $g'(t)$ are both 0.

Proof If $f'(t) > 0$ on I (or $f'(t) < 0$ on I), then f is increasing (or decreasing) on I and so is one-to-one and invertible. The part of C corresponding to values of t in I has ordinary equation $y = g(f^{-1}(x))$ and hence slope

$$\frac{dy}{dx} = g'(f^{-1}(x))\frac{d}{dx}f^{-1}(x) = \frac{g'(t)}{f'(t)},$$

where we have used the result $(d/dx)f^{-1}(x) = \dfrac{1}{f'(f^{-1}(x))}$. (See Section 2.4.) This slope is a continuous function of t and so the tangent to C turns smoothly for t in I. The proof for $g'(t) > 0$ (or < 0) is similar. In this case the slope of the normal would be a continuous function of t, so it turns smoothly, and so does the tangent.

If f' and g' are continuous, and both vanish at some point t_o, then the curve $x = f(t)$, $y = g(t)$ *may or may not* be smooth around t_o. Example 1 was an example of a curve that was not smooth at such a point. □

Example 2 The curve $x = t^3$, $y = t^6$ is just the parabola $y = x^2$, so it is smooth everywhere though $dx/dt = 3t^2$ and $dy/dt = 6t^5$ both vanish at $t = 0$.

If f' and g' are continuous and not both 0 at t_o, then the parameteric equations

$$x = f(t_o) + sf'(t_o), \qquad y = g(t_o) + sg'(t_o)$$

represent (for $-\infty < s < \infty$) the tangent line to the parametric curve $x = f(t)$, $y = g(t)$ at the point $(f(t_o), g(t_o))$. The normal line there has parametric equations

$$x = f(t_o) + sg'(t_o), \qquad y = g(t_o) - sf'(t_o) \qquad (-\infty < s < \infty).$$

Example 3 Find the tangent and normal lines to the parametric curve $x = t^2 - t$, $y = t^2 + t$ at the point where $t = 2$.

Solution

$$\frac{dx}{dt} = 2t - 1 = 3 \qquad \text{at } t = 2$$

$$\frac{dy}{dt} = 2t + 1 = 5 \qquad \text{at } t = 2$$

Hence the tangent and normal lines have respective parameteric equations

$$\begin{cases} x = 2 + 3s \\ y = 6 + 5s \end{cases} \qquad \begin{cases} x = 2 + 5s \\ y = 6 - 3s \end{cases}$$

The concavity of a parametric curve can be determined using the second derivatives of the parametric equations. The procedure is just to calculate d^2y/dx^2 using the chain rule:

$$\frac{d^2y}{dx^2} = \frac{d}{dx}\frac{dy}{dx} = \frac{d}{dx}\frac{g'(t)}{f'(t)} = \frac{dt}{dx}\frac{d}{dt}\frac{g'(t)}{f'(t)}$$

$$= \frac{1}{\frac{dx}{dt}}\frac{f'(t)g''(t) - g'(t)f''(t)}{(f'(t))^2}$$

$$= \frac{f'(t)g''(t) - g'(t)f''(t)}{(f'(t))^3}.$$

Sketching Parametric Curves

As in the case of graphs of functions, derivatives provide useful information about the slope of a parametric curve. At points where $dy/dt = 0$ but $dx/dt \neq 0$, the tangent is horizontal; at points where $dx/dt = 0$ but $dy/dt \neq 0$, the tangent is vertical. For points where $dx/dt = dy/dt = 0$, anything can happen; it is wise to calculate left- and right-hand limits of the slope dy/dx as the parameter t approaches one of these points. Concavity can be determined using the technique or the formula given above. We illustrate these ideas by reconsidering two parametric curves that we encountered previously.

Example 4 Reconsider Example 5 of Section 5.1. The curve has parametric equations

$$x = t^3 - 3t, \qquad y = t^2 \qquad (-2 \leq t \leq 2).$$

We have

$$\frac{dx}{dt} = 3(t^2 - 1) = 3(t - 1)(t + 1), \qquad \frac{dy}{dt} = 2t.$$

The curve has a horizontal tangent at $t = 0$, that is, at $(0, 0)$, and vertical tangents at $t = \pm 1$, that is, at $(2, 1)$ and $(-2, 1)$. The rise/fall properties of the curve between these points can be summarized as in Fig. 5.11.

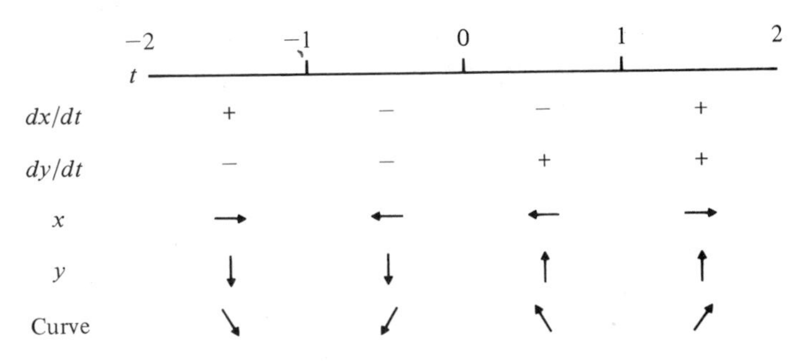

Figure 5.11

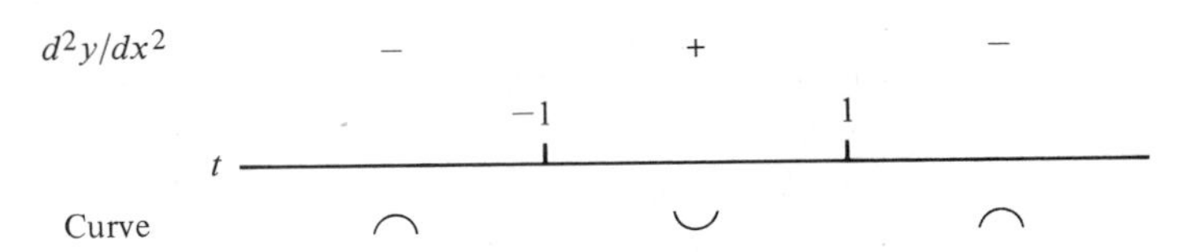

Figure 5.12

For concavity we deal directly with d^2y/dx^2. Since

$$\frac{d^2x}{dt^2} = 6t \text{ and } \qquad \frac{d^2y}{dt^2} = 2,$$

we have, using the formula obtained above,

$$\frac{d^2y}{dx^2} = \frac{3(t^2 - 1)2 - 2t(6t)}{[3(t^2 - 1)]^3} = -\frac{2}{9}\frac{t^2 + 1}{(t^2 - 1)^3},$$

which is never zero but which fails to be defined at $t = \pm 1$. The concavity is shown in Fig. 5.12.

The curve is sketched in Fig. 5.6.

Example 5 Consider the hypocycloid $x = 4 \cos^3 t$, $y = 4 \sin^3 t$ $(-\pi \le t \le \pi)$. We have

$$\frac{dx}{dt} = -12 \cos^2 t \sin t,$$

$$\frac{dy}{dt} = 12 \sin^2 t \cos t.$$

Both derivatives vanish at integer multiples of $\pi/2$. (See Figure 5.13.)

Since $dy/dx = (dy/dt)/(dx/dt) = -\tan t$, the slope approaches 0 as t approaches 0 or $\pm\pi$, and the absolute value of the slope approaches ∞ as t approaches $\pm(\pi/2)$. Also,

$$\frac{d^2y}{dx^2} = \frac{d}{dx}(-\tan t) = -\frac{1}{\dfrac{dx}{dt}} \sec^2 t = \frac{1}{12 \cos^4 t \sin t}.$$

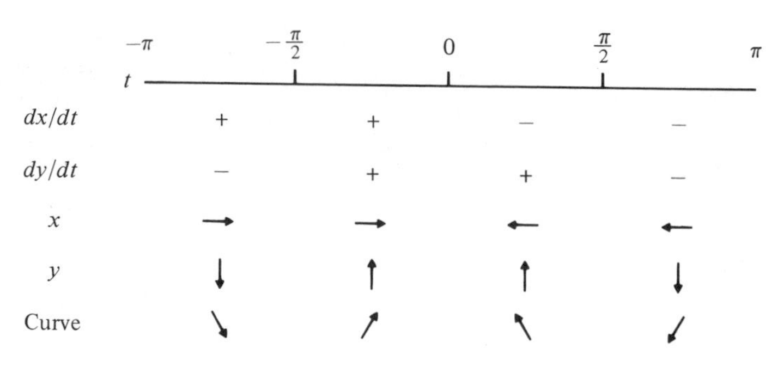

Figure 5.13

Clearly $d^2y/dx^2 > 0$ if $\sin t > 0$, that is, if $0 < t < \pi$, and $d^2y/dx^2 < 0$ if $\sin t < 0$, that is, if $-\pi < t < 0$.

The graph is drawn in Fig. 5.9, in Exercise 17 of Section 5.1.

EXERCISES

In Exercises 1–14 sketch the graphs of the given parameteric curves, making use of information from the first two derivatives. Unless otherwise stated, the parameter interval for each curve is $\mathbb{R}$.

1. $x = t^2 - 1,\ y = 2t + 1$

2. $x = t^2 - 2t,\ y = t^2 - 4t$

3. $x = 1 + \sin t,\ y = 1 + \cos^2 t$

4. $x = t^3,\ y = 3t^2 - 1$

5. $x = t^3 + 3t,\ y = t^2$

6. $x = t^3 - 3t - 2,\ y = t^2 - t - 2$

7. $x = \dfrac{t}{t^2 - 1},\ y = \dfrac{1}{t}$

8. $x = t^2 - 1,\ y = \dfrac{t}{1 + t^2}$

9. $x = t^3 - 3t,\ y = \dfrac{2}{1 + t^2}$

10. $x = t^2e^{-t},\ y = t^2e^t$

11. $x = 1 - \cos t - \sin t,\ y = 1 - \cos t + \sin t.$

12. $x = \sqrt{3}(1 - \cos t) - \sin t,\ y = \sqrt{3}(1 - \cos t) + \sin t.$

13. $x = \cos t + t \sin t,\ y = \sin t - t \cos t\ (t \geq 0)$. (See Example 8 of Section 5.1.)

14. $x = t - \sin t,\ y = 1 - \cos t$. (See Example 9 of Section 5.1.)

5.3 VELOCITY AND ACCELERATION IN THE PLANE

When we considered the projectile problem in Section 3.2 we decomposed the motion of the projectile into independent horizontal and vertical motions. In general, any motion in the plane can be decomposed into independent motions in perpendicular directions.

Suppose that $P = (x, y)$ is the position at time t of a point moving in the plane. The parametric equations of the path of the point,

$$x = f(t), \qquad y = g(t),$$

represent independent motions of the point in directions parallel to the x-axis and y-axis respectively; each equation determines a *component* of the overall motion. Similarly, the velocity and acceleration of the moving point have, at time t, two components. The x and y components of velocity are

$$v_x = \frac{dx}{dt} = f'(t), \qquad v_y = \frac{dy}{dt} = g'(t).$$

The x and y components of acceleration are

$$a_x = \frac{d^2x}{dt^2} = f''(t), \qquad a_y = \frac{d^2y}{dt^2} = g''(t).$$

In spite of the fact that we have used two quantities (v_x and v_y) to represent the velocity, we clearly think of this velocity as being a single quantity possessing two attributes, size and direction. (At a given time the point is moving in a given direction with a given speed.) In order to deal effectively with velocity and acceleration as single quantities, we need to develop the notion of vector.

Plane Vectors

A **vector** is a quantity that involves both *magnitude* (size or length) and *direction*. Such quantities are conveniently represented geometrically by arrows (directed line segments) and are often actually identified with these arrows. For instance, the vector $\overrightarrow{AB}$ is an arrow from the point A to the point B. We sometimes denote such a vector by a single symbol:

$$\mathbf{V} = \overrightarrow{AB}.$$

See Fig. 5.14. The magnitude of $\mathbf{V}$ is the length of the arrow and is denoted $|\mathbf{V}|$ or $|\overrightarrow{AB}|$.

While vectors have magnitude and direction, they do not generally have *position*, that is, they are not regarded as being in a particular place. We consider as equal two vectors $\mathbf{U}$ and $\mathbf{V}$ that have *the same length and the same direction*, even if their representative arrows do not coincide. The arrows must be parallel, have the same length, and point in the same direction. In Fig. 5.15, for example, if $ABYX$ is a parallelogram, then $\overrightarrow{AB} = \overrightarrow{XY}$.

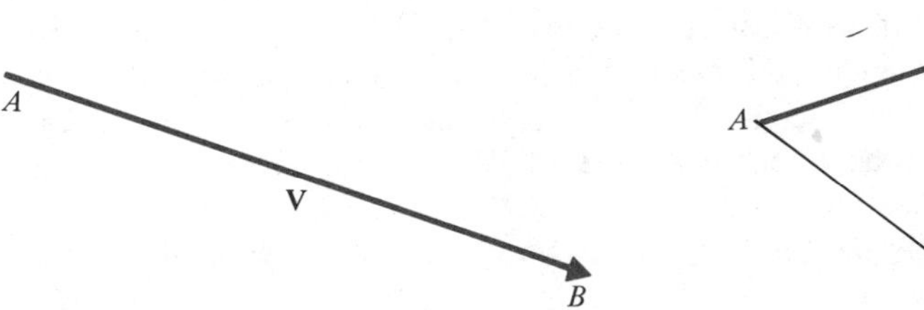

Figure 5.14

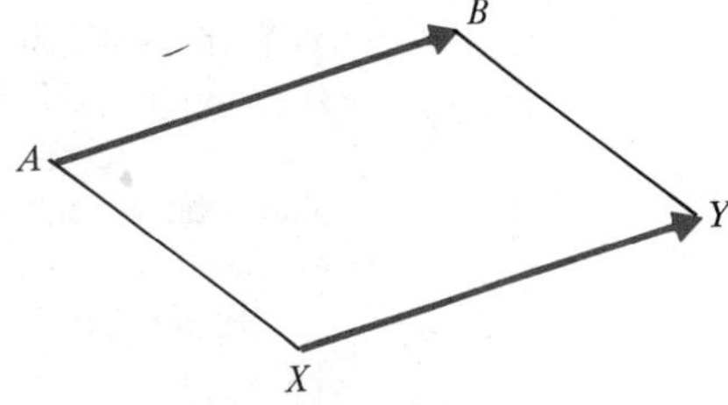

Figure 5.15

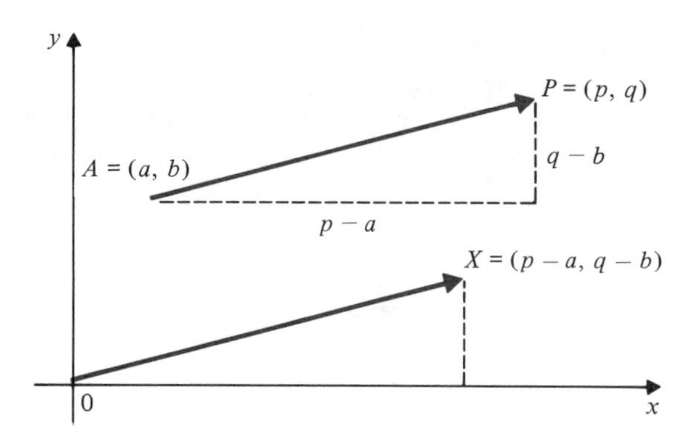

Figure 5.16

In this book we consider only plane vectors, that is, vectors whose representative arrows lie in a plane. If we introduce a Cartesian coordinate system into the plane we can talk about the x and y components of any vector. If $A = (a, b)$ and $P = (p, q)$, as shown in Fig. 5.16, then the x and y components of $\overrightarrow{AP}$ are, respectively, $p - a$ and $q - b$. Note that if O is the origin and X is the point $(p - a, q - b)$, then

$$|\overrightarrow{AP}| = \sqrt{(p - a)^2 + (q - b)^2} = |\overrightarrow{OX}|$$

$$\text{slope of } \overrightarrow{AP} = \frac{q - b}{p - a} = \text{slope of } \overrightarrow{OX}.$$

Hence $\overrightarrow{AP} = \overrightarrow{OX}$. In general, two vectors are equal if and only if they have the same x components and y components.

There are two important algebraic operations defined for vectors: addition and scalar multiplication.

Addition. Given two vectors **U** and **V**, their sum **U** + **V** is defined as follows. If an arrow representing **V** is placed with its tail at the head of an arrow representing **U**, then an arrow from the tail of **U** to the head of **V** represents **U** + **V**. Equivalently, if **U** and **V** have tails at the same point, then **U** + **V** is represented by an arrow with its tail at that point and its head at the opposite vertex of the parallelogram spanned by **U** and **V**. This is shown in Fig. 5.17.

Scalar Multiplication. If **V** is a vector and t is a real number (also called a *scalar*), then t**V** is a vector with magnitude $|t|$ times that of **V** and direction the same as **V** if $t > 0$, or opposite to that of **V** if $t < 0$. See Fig. 5.18. If $t = 0$, then t**V** has zero length and therefore no particular direction. It is the **zero vector,** denoted **O.**

Suppose that **U** has components x and y and that **V** has components a and b. Then the components of **U** + **V** are $x + a$ and $y + b$, and those of t**U** are tx and ty. See Fig. 5.19.

In the Cartesian plane we single out two particular vectors for special attention. They are

i) the vector **i** from the origin to the point (1, 0), and

ii) the vector **j** from the origin to the point (0, 1).

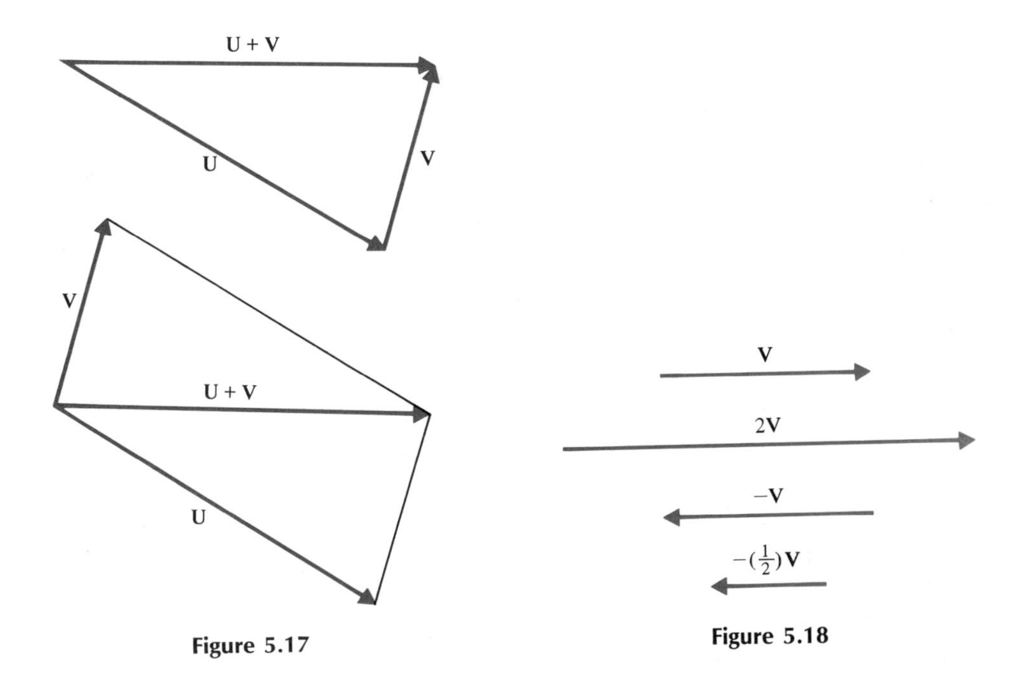

Figure 5.17

Figure 5.18

Thus **i** has components 1 and 0, and **j** has components 0 and 1. These vectors are called the **standard basis vectors** in the plane. If **U** is a vector with components x and y, then **U** can be expressed in the form

$$\mathbf{U} = x\mathbf{i} + y\mathbf{j}.$$

We say that we have written **U** as a **linear combination of the standard basis vectors.** (See Fig. 5.20.)

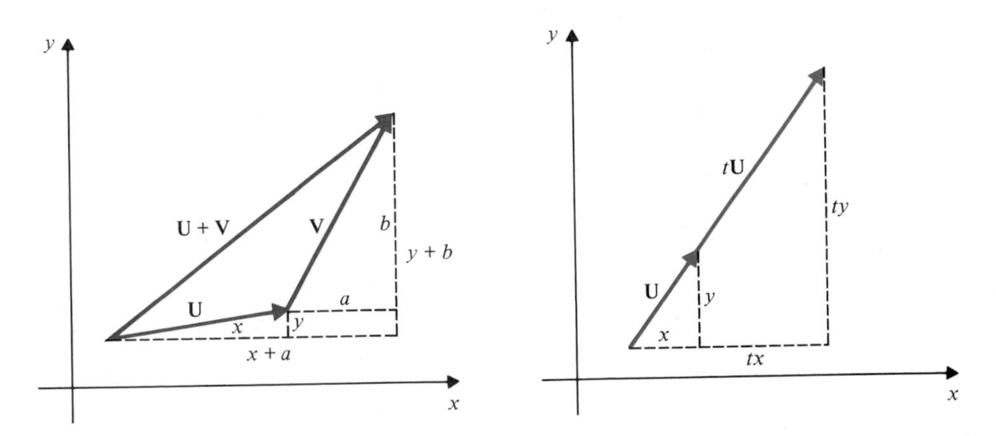

Figure 5.19

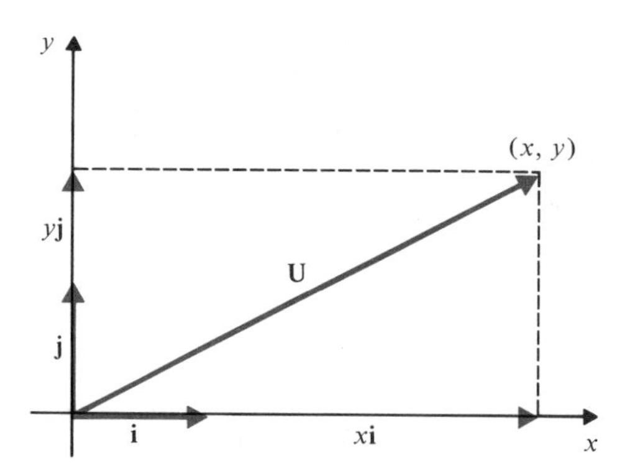

Figure 5.20

Example 1 If $A = (2, -1)$, $B = (-1, 4)$, and $C = (0, 2)$, express each of the following vectors as a linear combination of the standard basis vectors.

a) $\overrightarrow{AB}$ b) $\overrightarrow{BC}$ c) $\overrightarrow{AC}$ d) $\overrightarrow{AB} + \overrightarrow{BC}$ e) $2\overrightarrow{AC} - 3\overrightarrow{CB}$

f) $\dfrac{\overrightarrow{AB} + 2\overrightarrow{AC}}{3}$

Solution

a) $\overrightarrow{AB} = (-1 - 2)\mathbf{i} + (4 - (-1))\mathbf{j} = -3\mathbf{i} + 5\mathbf{j}$

b) $\overrightarrow{BC} = (0 - (-1))\mathbf{i} + (2 - 4)\mathbf{j} = \mathbf{i} - 2\mathbf{j}$

c) $\overrightarrow{AC} = (0 - 2)\mathbf{i} + (2 - (-1))\mathbf{j} = -2\mathbf{i} + 3\mathbf{j}$

d) $\overrightarrow{AB} + \overrightarrow{BC} = \overrightarrow{AC} = -2\mathbf{i} + 3\mathbf{j}$

e) $2\overrightarrow{AC} - 3\overrightarrow{CB} = 2(-2\mathbf{i} + 3\mathbf{j}) - 3(-\mathbf{i} + 2\mathbf{j}) = -\mathbf{i}$

f) $\dfrac{\overrightarrow{AB} + 2\overrightarrow{AC}}{3} = \dfrac{-3\mathbf{i} + 5\mathbf{j} + 2(-2\mathbf{i} + 3\mathbf{j})}{3} = \dfrac{-7}{3}\mathbf{i} + \dfrac{11}{3}\mathbf{j}.$

Implicit in the above example is the fact that the operations of addition and scalar multiplication obey appropriate algebraic rules, such as

$$\mathbf{U} + \mathbf{V} = \mathbf{V} + \mathbf{U}$$
$$(\mathbf{U} + \mathbf{V}) + \mathbf{W} = \mathbf{U} + (\mathbf{V} + \mathbf{W})$$
$$\mathbf{U} - \mathbf{V} = \mathbf{U} + (-1)\mathbf{V}$$
$$t(\mathbf{U} + \mathbf{V}) = t\mathbf{U} + t\mathbf{V}$$

We require one additional operation on vectors, the **dot product.** Given two vectors, $\mathbf{U} = x\mathbf{i} + y\mathbf{j}$ and $\mathbf{V} = a\mathbf{i} + b\mathbf{j}$, we define their dot product as the *number*

$$\mathbf{U} \cdot \mathbf{V} = xa + yb.$$

Note that the dot product of two vectors is not a vector but a number (scalar). Moreover, the arrows representing the vectors **U** and **V** are perpendicular if and only

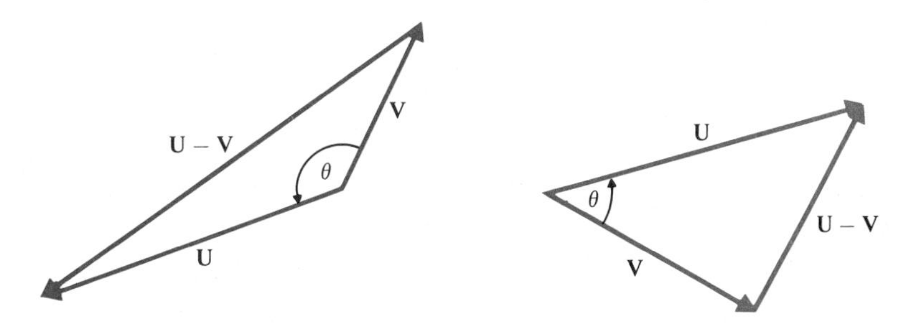

Figure 5.21

if $\mathbf{U} \cdot \mathbf{V} = 0$. To see this, observe that the slopes of the arrows **U** and **V** are y/x and b/a, respectively (assuming x and a are not 0), and the product of these slopes is -1 if and only if $xa + yb = 0$.

The dot product has the following algebraic properties

$$\begin{aligned} \mathbf{U} \cdot \mathbf{V} &= \mathbf{V} \cdot \mathbf{U} \\ \mathbf{U} \cdot (\mathbf{V} + \mathbf{W}) &= \mathbf{U} \cdot \mathbf{V} + \mathbf{U} \cdot \mathbf{W} \\ \mathbf{U} \cdot \mathbf{U} &= |\mathbf{U}|^2 \\ (t\mathbf{U}) \cdot \mathbf{V} &= \mathbf{U} \cdot (t\mathbf{V}) = t(\mathbf{U} \cdot \mathbf{V}), \end{aligned}$$

all of which are easily verified using the definition of dot product.

Finally we show that

$$\mathbf{U} \cdot \mathbf{V} = |\mathbf{U}||\mathbf{V}| \cos \theta,$$

where θ is the angle between the directions of **U** and **V** $(0 \le \theta \le \pi)$.
To see this, refer to Fig. 5.21 and apply the cosine law (see Section 3.1) to the triangle with the arrows **U, V,** and **U − V** as sides.

$$\begin{aligned} |\mathbf{U}|^2 + |\mathbf{V}|^2 - 2|\mathbf{U}||\mathbf{V}| \cos \theta &= |\mathbf{U} - \mathbf{V}|^2 = (\mathbf{U} - \mathbf{V}) \cdot (\mathbf{U} - \mathbf{V}) \\ &= \mathbf{U} \cdot (\mathbf{U} - \mathbf{V}) - \mathbf{V} \cdot (\mathbf{U} - \mathbf{V}) \\ &= \mathbf{U} \cdot \mathbf{U} - \mathbf{U} \cdot \mathbf{V} - \mathbf{V} \cdot \mathbf{U} + \mathbf{V} \cdot \mathbf{V} \\ &= |\mathbf{U}|^2 + |\mathbf{V}|^2 - 2\mathbf{U} \cdot \mathbf{V} \end{aligned}$$

Hence $|\mathbf{U}||\mathbf{V}| \cos \theta = \mathbf{U} \cdot \mathbf{V}$.

Example 2 The angle θ between the vectors $\mathbf{i} + \mathbf{j}$ and $\mathbf{i} - 3\mathbf{j}$ satisfies

$$\cos \theta = \frac{(\mathbf{i} + \mathbf{j}) \cdot (\mathbf{i} - 3\mathbf{j})}{|\mathbf{i} + \mathbf{j}|\ |\mathbf{i} - 3\mathbf{j}|} = \frac{1 - 3}{\sqrt{2}\sqrt{10}} = -\frac{2}{\sqrt{2}\sqrt{10}} = -\frac{1}{\sqrt{5}}.$$

Thus $\theta = \text{Arccos}\,(-1/\sqrt{5}) \simeq 116.565$ degrees.

Position, Velocity, and Acceleration as Vectors

A **position vector** is an arrow that has its tail at the origin; the head of the arrow indicates the position of some point in the plane, and the components of the vector are the coordinates of that point.

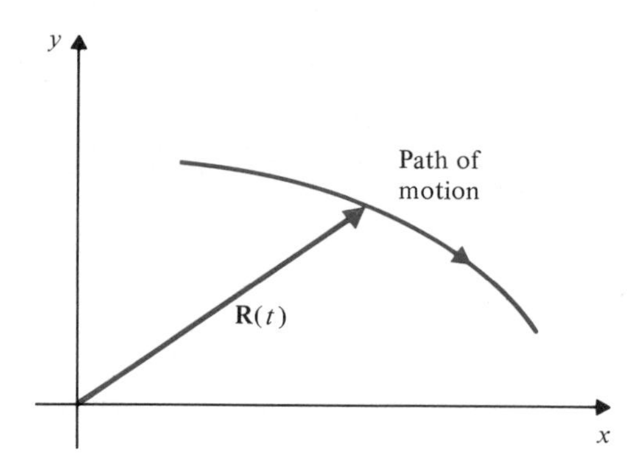

Figure 5.22

If a particle moves in the plane so that its position at time t is given by the parametric equations $x = f(t)$, $y = g(t)$, then that position can also be specified as a function of t using a position vector (see Fig. 5.22).

$$\mathbf{R} = x\mathbf{i} + y\mathbf{j}$$

or, showing dependence on t explicitly,

$$\mathbf{R}(t) = f(t)\mathbf{i} + g(t)\mathbf{j}.$$

The velocity of the particle at time t is also a vector since it embodies the direction as well as the speed of motion. The velocity is the rate of change of the position vector with respect to time:

$$\begin{aligned}
\mathbf{V}(t) = \frac{d\mathbf{R}}{dt} &= \lim_{h\to 0} \frac{\mathbf{R}(t+h) - \mathbf{R}(t)}{h} \\
&= \lim_{h\to 0} \left(\frac{f(t+h) - f(t)}{h}\mathbf{i} + \frac{g(t+h) - g(t)}{h}\mathbf{j} \right) \\
&= f'(t)\mathbf{i} + g'(t)\mathbf{j}.
\end{aligned}$$

As we might have anticipated, the components of the velocity vector are just the horizontal and vertical components of velocity, as considered earlier. The *speed* $s(t)$ of the particle at time t is the length of the velocity vector:

$$s(t) = |\mathbf{V}(t)| = \left| \frac{d\mathbf{R}}{dt} \right| = \sqrt{(f'(t))^2 + (g'(t))^2}.$$

The Newton quotient $(\mathbf{R}(t + h) - \mathbf{R}(t))/h$ is a vector along a secant line to the path of motion. Hence its limit, the velocity vector $\mathbf{V}(t)$ is tangent to the path of motion at the position $\mathbf{R}(t)$, as shown in Fig. 5.23. Evidently $\mathbf{V}(t)$ points in the direction in which the particle is moving at time t.

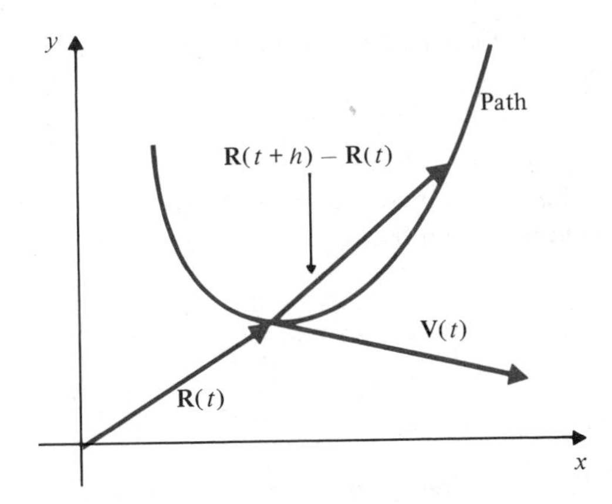

Figure 5.23

Similarly, the acceleration vector $\mathbf{A}(t)$ is the derivative of the velocity vector:

$$\mathbf{A}(t) = \frac{d\mathbf{V}}{dt} = \frac{d^2\mathbf{R}}{dt^2} = f''(t)\mathbf{i} + g''(t)\mathbf{j}.$$

The acceleration is also represented by an arrow with tail at $\mathbf{R}(t)$. The position of this arrow with respect to the velocity arrow indicates whether the speed is increasing or decreasing and which way the curve of motion is bending. If the angle between $\mathbf{V}$ and $\mathbf{A}$ at a point is less than $\pi/2$, then the speed is increasing; if it is greater than $\pi/2$, the speed is decreasing. (See Fig. 5.24 and Exercise 31 at the end of this section.)

The algebraic rules governing dot product ensure that the product rule for differentiation also holds for a dot product of vector functions: If $\mathbf{P}(t) = p(t)\mathbf{i} + q(t)\mathbf{j}$ and $\mathbf{U}(t) = u(t)\mathbf{i} + v(t)\mathbf{j}$, then

$$\begin{aligned}\frac{d}{dt}\mathbf{P}(t) \cdot \mathbf{U}(t) &= \frac{d}{dt}(p(t)u(t) + q(t)v(t)) \\ &= p'(t)u(t) + p(t)u'(t) + q'(t)v(t) + q(t)v'(t) \\ &= \left(\frac{d}{dt}\mathbf{P}(t)\right) \cdot \mathbf{U}(t) + \mathbf{P}(t) \cdot \left(\frac{d}{dt}\mathbf{U}(t)\right).\end{aligned}$$

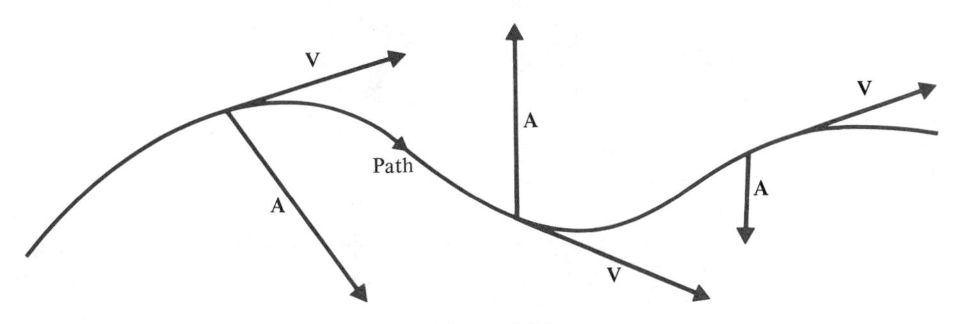

Figure 5.24

If the speed of a moving particle is constant over an interval of time, then

$$0 = \frac{d}{dt}|\mathbf{V}(t)|^2 = \frac{d}{dt}(\mathbf{V}(t) \cdot \mathbf{V}(t)) = \frac{d\mathbf{V}}{dt} \cdot \mathbf{V}(t) + \mathbf{V}(t) \cdot \frac{d\mathbf{V}}{dt} = 2\mathbf{A}(T) \cdot \mathbf{V}(t).$$

Thus $\mathbf{A}(t)$ is perpendicular to $\mathbf{V}(t)$. Conversely, if $\mathbf{A}$ is perpendicular to $\mathbf{V}$ over a time interval, then the speed is constant over that interval.

Example 3 An object moves so that its position at time t is given by

$$\mathbf{R} = \left(\frac{1}{12}t^3 - t + 1\right)\mathbf{i} + t\mathbf{j}.$$

Find its velocity and acceleration at time t and sketch the path followed by the object for $0 \leq t \leq 3$, showing position, velocity, and acceleration vectors at $t = 0, 1, 2,$ and 3.

Solution The following solution is illustrated in Fig. 5.25.

$$\mathbf{V} = \left(\frac{1}{4}t^2 - 1\right)\mathbf{i} + \mathbf{j}$$

$$\mathbf{A} = \frac{t}{2}\mathbf{i}$$

$$\mathbf{R}(0) = \mathbf{i}, \qquad \mathbf{V}(0) = -\mathbf{i} + \mathbf{j}, \qquad \mathbf{A}(0) = \mathbf{0}$$

$$\mathbf{R}(1) = \frac{\mathbf{i}}{12} + \mathbf{j}, \qquad \mathbf{V}(1) = \frac{-3}{4}\mathbf{i} + \mathbf{j}, \qquad \mathbf{A}(1) = \frac{\mathbf{i}}{2}$$

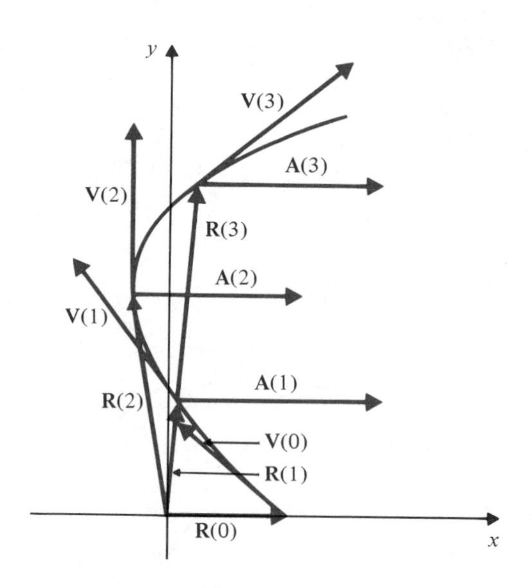

Figure 5.25

$$\mathbf{R}(2) = \frac{-1}{3}\mathbf{i} + 2\mathbf{j}, \qquad \mathbf{V}(2) = \mathbf{j}, \qquad \mathbf{A}(2) = \mathbf{i}$$

$$\mathbf{R}(3) = \frac{\mathbf{i}}{4} + 3\mathbf{j}, \qquad \mathbf{V}(3) = \frac{5}{4}\mathbf{i} + \mathbf{j}, \qquad \mathbf{A}(3) = \frac{3}{2}\mathbf{i}$$

Example 4 Find the motion of an object in the plane if its acceleration is proportional in magnitude to its position vector and is directed in the opposite direction. Assume that the position $\mathbf{R}_0$ and the velocity $\mathbf{V}_0$ at time $t = 0$ are given.

Solution The relation between acceleration and position can be expressed by the differential equation

$$\frac{d^2\mathbf{R}}{dt^2} = -k^2\mathbf{R}$$

where k^2 is a positive constant. This is simply the vector form of the equation of simple harmonic motion (see Section 3.2). Breaking it into separate equations for the x and y components of $\mathbf{R}$, we have

$$\frac{d^2x}{dt^2} = -k^2x, \qquad \frac{d^2y}{dt^2} = -k^2y.$$

Hence $x = a_1 \cos kt + b_1 \sin kt$, $y = a_2 \cos kt + b_2 \sin kt$, so

$$\mathbf{R} = \mathbf{A} \cos kt + \mathbf{B} \sin kt,$$

where $\mathbf{A} = a_1\mathbf{i} + a_2\mathbf{j}$ and $\mathbf{B} = b_1\mathbf{i} + b_2\mathbf{j}$. Since $\mathbf{R}_0 = \mathbf{R}(0) = \mathbf{A}$ and $\mathbf{V}_0 = \mathbf{V}(0) = (-k\mathbf{A} \sin kt + k\mathbf{B} \cos kt)|_{t=0} = k\mathbf{B}$, we have

$$\mathbf{R} = \mathbf{R}(t) = \mathbf{R}_0 \cos kt + \frac{1}{k}\mathbf{V}_0 \sin kt.$$

EXERCISES

1. Let $A = (-1, 2)$, $B = (2, 0)$, $C = (1, -3)$, $D = (0, -4)$. Express each of the following vectors as a linear combination of the standard basis vectors.

   a) $\overrightarrow{AB}$ **b)** $\overrightarrow{BA}$ **c)** $\overrightarrow{AC}$ **d)** $\overrightarrow{BD}$ **e)** $\overrightarrow{DA}$

   f) $\overrightarrow{AB} - \overrightarrow{BC}$ **g)** $\overrightarrow{AC} - 2\overrightarrow{AB} + 3\overrightarrow{CD}$ **h)** $\dfrac{\overrightarrow{AB} + \overrightarrow{AC} + \overrightarrow{AD}}{3}$

In Exercises 2–11, find $\mathbf{U} + \mathbf{V}$, $\mathbf{U} - \mathbf{V}$, $|\mathbf{U}|$, $|\mathbf{V}|$, $\mathbf{U} \cdot \mathbf{V}$, and the angle between $\mathbf{U}$ and $\mathbf{V}$. Sketch arrows for $\mathbf{U}$, $\mathbf{V}$, $\mathbf{U} + \mathbf{V}$ and $\mathbf{U} - \mathbf{V}$.

2. $\mathbf{U} = \mathbf{i}$, $\mathbf{V} = \mathbf{j}$
3. $\mathbf{U} = \mathbf{i} + \mathbf{j}$, $\mathbf{V} = \mathbf{i} - \mathbf{j}$
4. $\mathbf{U} = 3\mathbf{i}$, $\mathbf{V} = 4\mathbf{i} - 3\mathbf{j}$
5. $\mathbf{U} = \mathbf{i} - 2\mathbf{j}$, $\mathbf{V} = -\mathbf{i} + 3\mathbf{j}$
6. $\mathbf{U} = \mathbf{j}$, $\mathbf{V} = -\mathbf{i} - \mathbf{j}$
7. $\mathbf{U} = \mathbf{i} + 2\mathbf{j}$, $\mathbf{V} = 2\mathbf{i} - \mathbf{j}$
8. $\mathbf{U} = 4\mathbf{i} - 3\mathbf{j}$, $\mathbf{V} = 3\mathbf{i} + \mathbf{j}$
9. $\mathbf{U} = \mathbf{i} - \sqrt{3}\mathbf{j}$, $\mathbf{V} = \sqrt{3}\mathbf{i} - \mathbf{j}$
10. $\mathbf{U} = a\mathbf{i} + b\mathbf{j}$, $\mathbf{V} = 2b\mathbf{i} - 2a\mathbf{j}$
11. $\mathbf{U} = a\mathbf{i} + b\mathbf{j}$, $\mathbf{V} = c\mathbf{i} + d\mathbf{j}$
12. Use vectors to show that the triangle with vertices $(-1, 1)$, $(2, 5)$, and $(10, -1)$ is right-angled.

In Exercises 13–16 prove the stated geometric result using vectors.

13. The line segment joining the midpoints of two sides of a triangle is parallel to and half as long as the third side.

14. If P, Q, R, S are the midpoints of sides AB, BC, CD, and DA, respectively, of quadrilateral $ABCD$, prove that $PQRS$ is a parallelogram.

15. The diagonals of any parallelogram bisect each other.

16.* The medians of any triangle meet in a common point. (A median is a line joining one vertex to the midpoint of the opposite side. The common point is called the *centroid* of the triangle.)

In Exercises 17–24 **R** is the position of a point moving in the plane. Sketch the path of motion, and show on the sketch the velocity and acceleration vectors at time $t = 1$ and $t = 2$.

17. $\mathbf{R} = t\mathbf{i} - \frac{1}{2}t^2\mathbf{j}$

18. $\mathbf{R} = \sin\frac{\pi t}{2}\mathbf{i} + \cos\frac{\pi t}{2}\mathbf{j}$

19. $\mathbf{R} = (t^2 - 4t)\mathbf{i} + t^2\mathbf{j}$

20. $\mathbf{R} = t\mathbf{i} + (4t - t^3)\mathbf{j}$

21. $\mathbf{R} = (1/t)\mathbf{i} - t\mathbf{j}$

22. $\mathbf{R} = t\mathbf{i} + \sin t\mathbf{j}$

23. $\mathbf{R} = e^t\cos\frac{\pi t}{2}\mathbf{i} + e^t\sin\frac{\pi t}{2}\mathbf{j}$

24. $\mathbf{R} = \frac{1}{t}\cos\pi t\,\mathbf{i} + \frac{1}{t}\sin\pi t\,\mathbf{j}$

In Exercises 25–28 find the position vector at time t of a point that moves with given acceleration $\mathbf{A}(t)$ and has given position and velocity at the time indicated.

25. $\mathbf{A}(t) = 2\mathbf{i} - \mathbf{j},\ \mathbf{V}(0) = \mathbf{i} + \mathbf{j},\ \mathbf{R}(0) = -3\mathbf{j}$

26. $\mathbf{A}(t) = t\mathbf{i},\ \mathbf{V}(1) = \mathbf{0},\ \mathbf{R}(1) = -\mathbf{i} + \mathbf{j}$

27. $\mathbf{A}(t) = 6t\mathbf{i} - 6t^2\mathbf{j},\ \mathbf{V}(0) = 2\mathbf{i},\ \mathbf{R}(0) = \mathbf{0}$

28. $\mathbf{A}(t) = \sin t\,\mathbf{i} + e^{-t}\mathbf{j},\ \mathbf{V}(0) = \mathbf{i},\ \mathbf{R}(0) = \mathbf{0}$

29. Find the acceleration at time t of a point whose position is given by $\mathbf{R} = a(t - \sin t)\mathbf{i} + a(1 - \cos t)\mathbf{j}$. (The point is moving on a cycloid. See Example 9 of Section 5.1.) Show that the acceleration has constant magnitude. What is its direction at the cusps of the cycloid? At the peaks?

30. A point moves on the curve $y = e^x$ in the direction of increasing x. If the speed is constant (say k), find the velocity and acceleration vectors as functions of x.

31.* Let $\mathbf{V}(t)$ and $\mathbf{A}(t)$ be the velocity and acceleration, respectively, of an object moving in the plane. Show that the following conditions are equivalent:

i) the speed of the object is increasing at time t,

ii) $\mathbf{V}(t) \cdot \mathbf{A}(t) > 0$,

iii) the angle between $\mathbf{V}(t)$ and $\mathbf{A}(t)$ is less than $\pi/2$.

State an analogous result for decreasing speed.

5.4 POLAR COORDINATES AND POLAR CURVES

The **polar coordinate system** is an often useful alternative to the rectangular (Cartesian) coordinate system for describing the location of points in a plane. In the polar coordinate sytem there is an origin (or **pole**), O, and a **polar axis,** a ray (that is, a

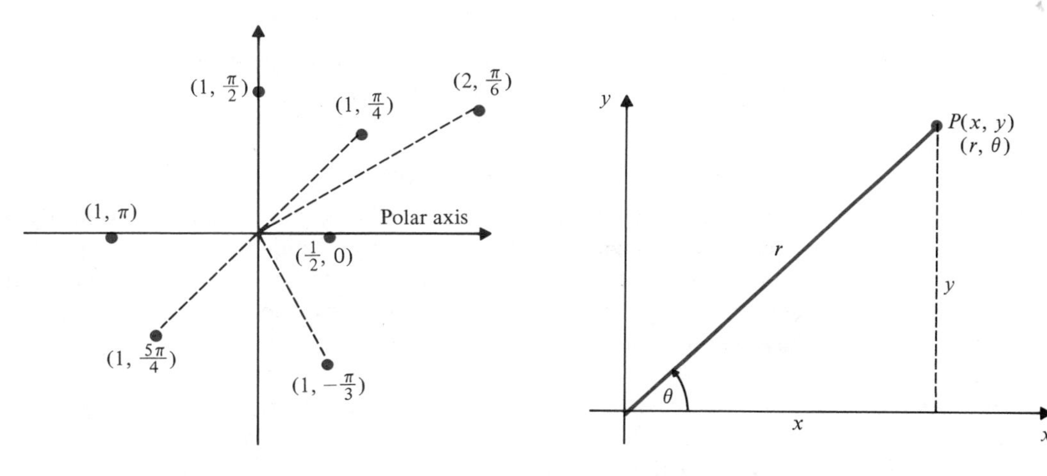

Figure 5.26

Figure 5.27

half-line) extending from O horizontally to the right. The position of any point P in the plane is then determined by its polar coordinates (r, θ), where

i) r is the distance from P to the origin, and
ii) θ is the angle that the ray OP makes with the polar axis (counterclockwise angles being considered positive).

Figure 5.26 shows some points with their polar coordinates. The rectangular axes x and y are often shown in the polar graph. The polar axis coincides with the positive x-axis.

Unlike rectangular coordinates, the polar coordinates of a point are not unique. Increasing (or decreasing) θ by an integer multiple of 2π does not change the distance from the origin, so $(r, \theta + 2\pi)$ denotes the same point as (r, θ). For instance,

$$\begin{aligned}(1, \pi/4) &= (1, 9\pi/4) = (1, -7\pi/4),\\ (2, \pi) &= (2, -\pi),\\ (2, 5\pi/4) &= (2, -3\pi/4),\\ (3, 0) &= (3, 2\pi).\end{aligned}$$

In addition, the origin O has polar coordinates (O, θ) for any value of θ. (If we go zero distance from O, the direction doesn't much matter.)

Sometimes we encounter points (r, θ) where $r < 0$. The appropriate interpretation for this "negative distance" r is that it represents a positive distance $-r$ measured in the *opposite direction* (that is, in the direction $\theta + \pi$):

$$(-r, \theta) = (r, \theta + \pi).$$

For example, $(-1, \pi/4) = (1, 5\pi/4)$.

If we wish to consider both rectangular and polar coordinate systems in the same plane, and choose the positive x-axis as the polar axis, then the relationship between the rectangular coordinates of a point and its polar coordinates are as shown in Fig. 5.27.

$$x = r\cos\theta, \qquad x^2 + y^2 = r^2$$
$$y = r\sin\theta, \qquad \tan\theta = \frac{y}{x}$$

A single equation in x and y generally represents some curve in the plane with respect to the rectangular coordinate system. Similarly, a single equation in r and θ generally represents a curve with respect to the polar coordinate system. The relationships given in the box above can be used to transform one representation of a curve into the other.

Example 1 The straight line $x - 2y = 5$ has polar equation $r(\cos\theta - 2\sin\theta) = 5$.

Example 2 The polar equation $r = 2a\cos\theta$ can be transformed to rectangular coordinates if we first multiply it by r:

$$r^2 = 2ar\cos\theta$$
$$x^2 + y^2 = 2ax$$
$$(x - a)^2 + y^2 = a^2$$

The given polar equation thus represents a circle with center $(a, 0)$ and radius a as shown in Fig. 5.28.

Some Polar Curves

We sketch the graphs of some fairly simple polar equations in Figs. 5.29–5.31. When the equation is of the form $r = f(\theta)$ we call its graph the polar graph of the function f. Some polar graphs can be recognized easily if the polar equation is transformed to rectangular form. For others, such a transformation does not help; the rectangular equation may be too complicated to be recognizable. In these cases one must resort to constructing a table of values and plotting points.

Example 3 The polar equations $r = a$ and $\theta = \beta$ (where a and β are constants) represent, respectively, a circle with radius $|a|$ centered at the origin and a line through the origin making angle β with the polar axis. Note in Fig. 5.29 that the line and the circle meet in two points, with polar coordinates (a, β) and $(-a, \beta)$. (If we disallow negative values of r, then $\theta = \beta$ represents a ray from the origin in the direction β.)

Example 4 Sketch and identify the curve $r = 2a\cos(\theta - \alpha)$.

Solution We proceed as in Example 2.

$$r^2 = 2ar\cos(\theta - \alpha) = 2ar\cos\alpha\cos\theta + 2ar\sin\alpha\sin\theta$$
$$x^2 + y^2 = 2a\cos\alpha\, x + 2a\sin\alpha\, y$$
$$x^2 - 2a\cos\alpha\, x + a^2\cos^2\alpha + y^2 - 2a\sin\alpha\, y + a^2\sin^2\alpha = a^2$$
$$(x - a\cos\alpha)^2 + (y - a\sin\alpha)^2 = a^2$$

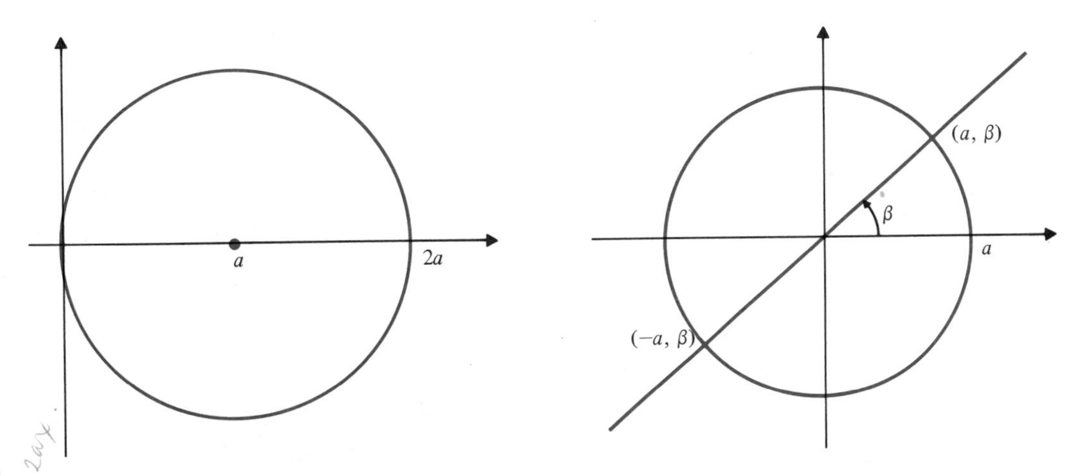

Figure 5.28 Figure 5.29

This is a circle of radius a passing through the origin (see Fig. 5.30). Its center has rectangular coordinates $(a \cos \alpha,\ a \sin \alpha)$ and hence polar coordinates (a, α). For $\alpha = \pi/2$ we have $r = 2a \sin \theta$ as the equation of a circle centered on the y-axis.

Example 5 Sketch the *cardioid* $r = a(1 - \cos \theta)$, $(a > 0)$.

Solution Transformation to rectangular coordinates is not much help here; the resulting equation is $(x^2 + y^2 + ax)^2 = a^2(x^2 + y^2)$. (Verify this.) Instead we make a table of values and plot some points.

θ	0	$\pm\frac{\pi}{6}$	$\pm\frac{\pi}{4}$	$\pm\frac{\pi}{3}$	$\pm\frac{\pi}{2}$	$\pm\frac{2\pi}{3}$	$\pm\frac{3\pi}{4}$	$\pm\frac{5\pi}{6}$	π
r	0	$0.13a$	$0.29a$	$0.5a$	a	$1.5a$	$1.71a$	$1.87a$	$2a$

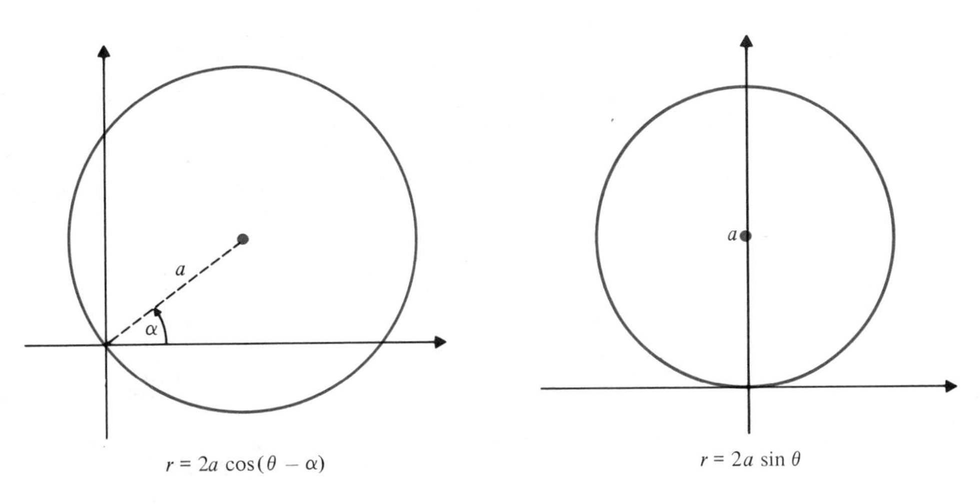

Figure 5.30

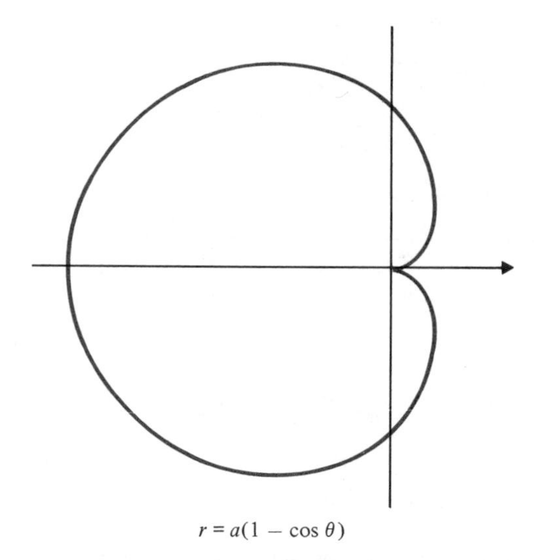

$r = a(1 - \cos\theta)$

Figure 5.31

Note the cusp at the origin in Fig. 5.31. In general a polar graph $r = f(\theta)$ approaches the origin from the direction θ for which $f(\theta) = 0$.

The equation $r = a(1 - \cos(\theta - \alpha))$ represents a cardioid of the same size and shape but rotated through an angle α about the origin. Its cusp is in the direction $\theta = \alpha$.

Illustrated in Fig. 5.32 are some other polar graphs best sketched by plotting some points. Observe that the two lemniscates have no points in the angular sectors where r^2 would have to be negative. While r can be negative, r^2 cannot.

Conics

Suppose a point moves in the plane so that the ratio of its distance from a fixed point F (the **focus**) to its distance from a fixed straight line L (the **directrix**) is a constant e (the **eccentricity**). The path of the point is a **conic**; it is an ellipse if $e < 1$, a parabola if $e = 1$, and a hyperbola if $e > 1$.

If we take the focus at the origin of a polar coordinate system and the directrix vertical and p units to the left of the focus, then an arbitrary point $P = (r, \theta)$ on the conic (see Fig. 5.33) satisfies

$$\frac{r}{p + r\cos\theta} = e$$

or, solving for r,

$$r = \frac{ep}{1 - e\cos\theta}.$$

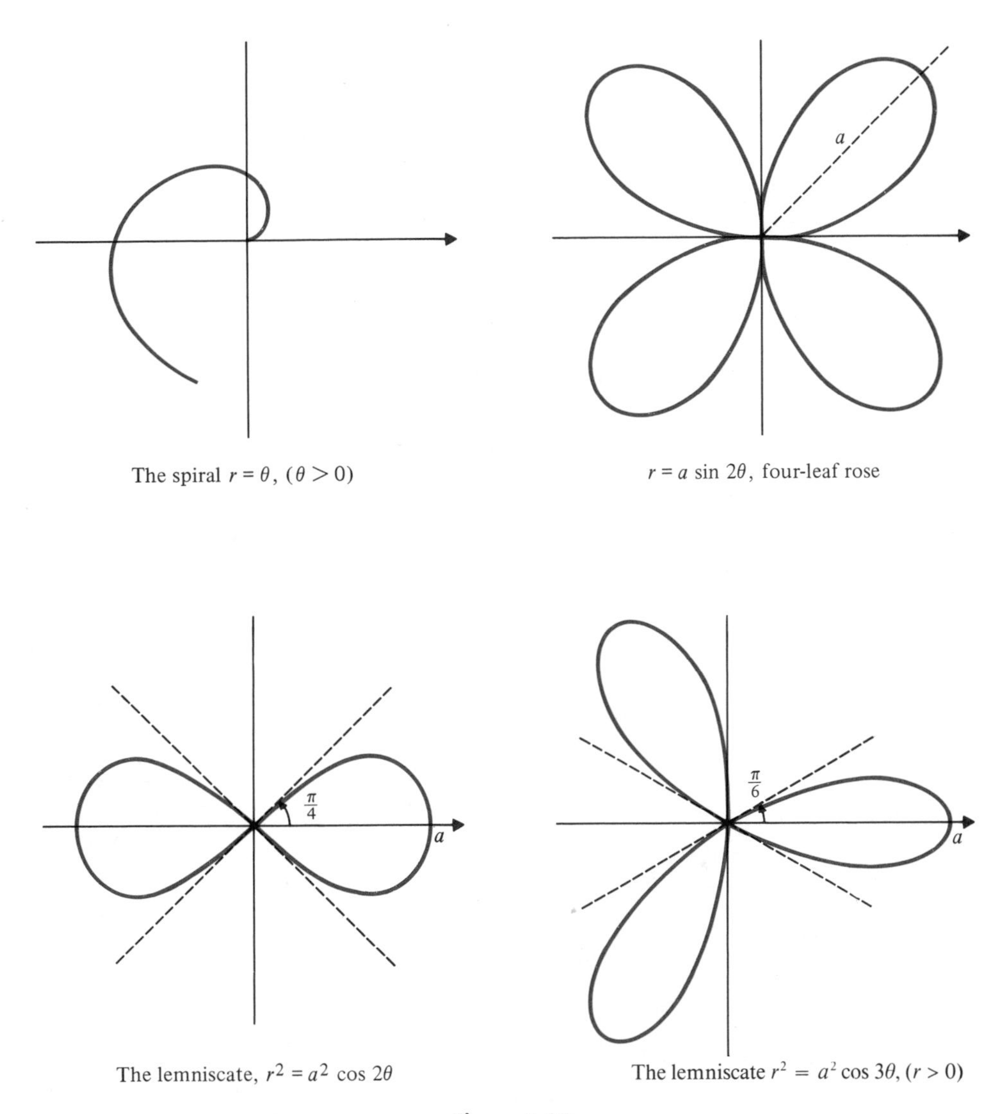

The spiral $r = \theta$, $(\theta > 0)$

$r = a \sin 2\theta$, four-leaf rose

The lemniscate, $r^2 = a^2 \cos 2\theta$

The lemniscate $r^2 = a^2 \cos 3\theta$, $(r > 0)$

Figure 5.32

For $e < 1$, this equation represents an ellipse with the major axis running from $(ep/(1 + e), \pi)$ to $(ep/(1 - e), 0)$, and with one focus at the origin and the second focus at $(2e^2p/(1 - e^2), 0)$. For $e = 1$ it is a parabola opening to the right with polar axis as axis, focus at the origin, and vertex at $(p/2, \pi)$. For $e > 1$ it is a hyperbola with vertices at $(-ep/(e - 1), 0) = (ep/(e - 1), \pi)$ and $(ep/(e + 1), \pi)$, one focus at the origin, and a second at the point $(2e^2p/(e^2 - 1), \pi)$. The center is at $(e^2p/(e^2 - 1), \pi)$ and the asymptotes pass through this point in the directions $\theta = \text{Arccos}\,(1/e)$, that is, the directions for which $1 - e \cos \theta = 0$. These three conics are shown in Fig. 5.34.

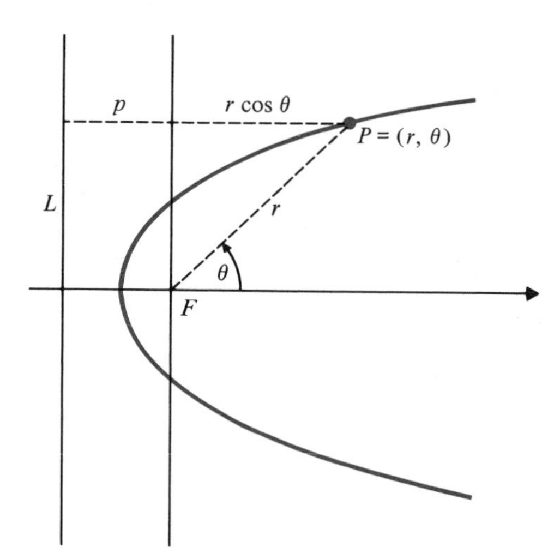

Figure 5.33

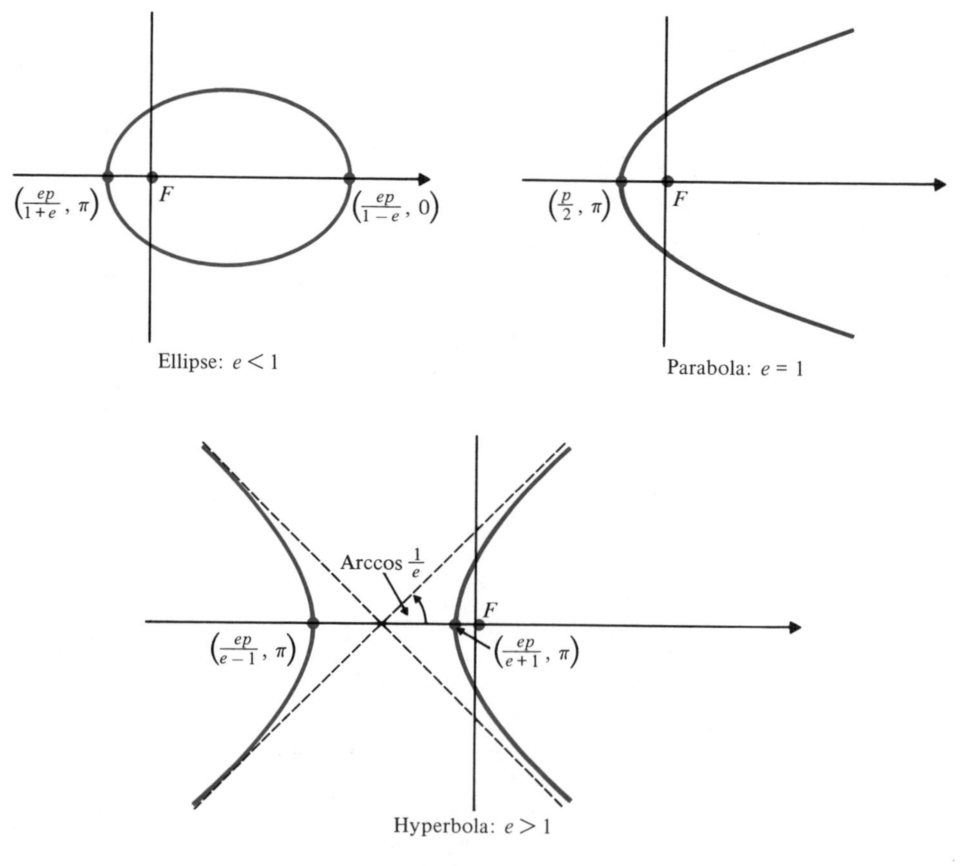

Figure 5.34

The Slope of a Polar Curve

The polar graph $r = f(\theta)$ can be regarded as a parametric curve in rectangular coordinates; the parameter is θ and the parametric equations are

$$x = r\cos\theta = f(\theta)\cos\theta$$
$$y = r\sin\theta = f(\theta)\sin\theta.$$

The slope of this curve at the point P with polar coordinates $(f(\theta), \theta)$ is

$$\frac{dy}{dx} = \frac{\dfrac{dy}{d\theta}}{\dfrac{dx}{d\theta}} = \frac{f'(\theta)\sin\theta + f(\theta)\cos\theta}{f'(\theta)\cos\theta - f(\theta)\sin\theta}.$$

If the curve $r = f(\theta)$ passes through the origin at $\theta = \theta_0$, then $f(\theta_0) = 0$. If $f'(\theta_0) \neq 0$, then

$$\left.\frac{dy}{dx}\right|_{\theta=\theta_0} = \frac{f'(\theta_0)\sin\theta_0}{f'(\theta_0)\cos\theta_0} = \tan\theta_0,$$

so the tangent line at the origin is $\theta = \theta_0$. If $f'(\theta_0) = f(\theta_0) = 0$, then, as described in Section 5.2, the curve may or may not be smooth at the origin.

Example 6 Find the points on the cardioid $r = 1 + \cos\theta$ where the tangent line is vertical or horizontal.

Solution

$$x = (1 + \cos\theta)\cos\theta$$

$$\frac{dx}{d\theta} = -\sin\theta\,(1 + 2\cos\theta)$$

$$y = (1 + \cos\theta)\sin\theta$$

$$\begin{aligned}\frac{dy}{d\theta} &= \cos\theta\,(1 + \cos\theta) - \sin^2\theta \\ &= 2\cos^2\theta + \cos\theta - 1 \\ &= (2\cos\theta - 1)(\cos\theta + 1)\end{aligned}$$

Note that

$$\frac{dx}{d\theta} = 0 \text{ at } \theta = 0,\ \pm\frac{2\pi}{3},\ \pi, \qquad \frac{dy}{d\theta} = 0 \text{ at } \theta = \pm\frac{\pi}{3},\ \pi.$$

The tangent line is vertical at $(2, 0)$ and $(1/2, \pm 2\pi/3)$. The tangent line is horizontal at $(3/2, \pm\pi/3)$. See Fig. 5.35. Since $dx/d\theta = dy/d\theta = 0$ at $\theta = \pi$, we must be more careful at this point. Indeed, the curve has a cusp there. However, the slope approaches 0 there as

$$\lim_{\theta\to\pi}\frac{dy}{dx} = \lim_{\theta\to\pi} -\frac{(2\cos\theta - 1)(\cos\theta + 1)}{(2\cos\theta + 1)\sin\theta} = -\frac{1}{3}\lim_{\theta\to\pi}\frac{-\sin\theta}{\cos\theta} = 0.$$

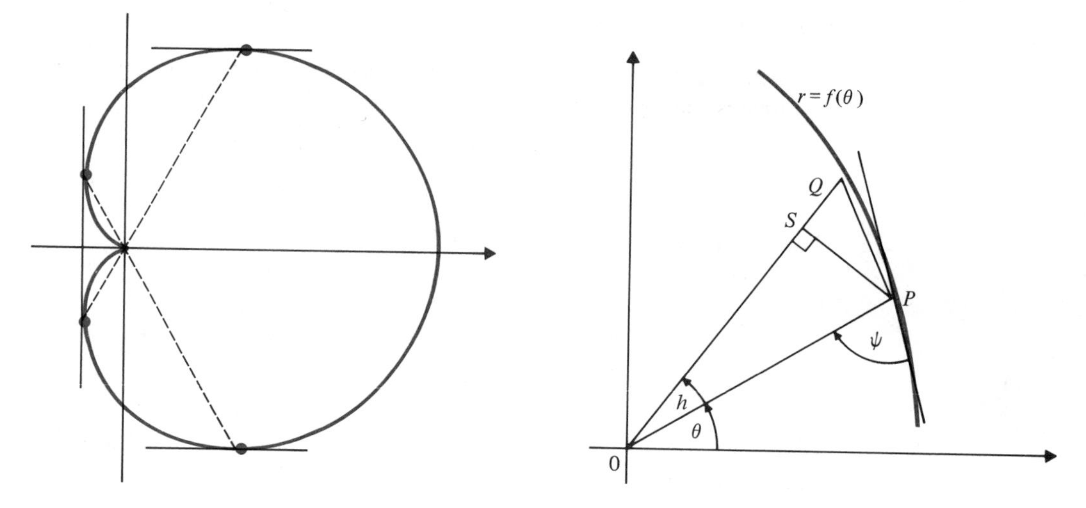

Figure 5.35

Figure 5.36

There is a simple formula that can be used to determine the direction of the tangent line to a polar curve $r = f(\theta)$ at a point $P = (r, \theta)$ other than the origin. Let Q be a point on the curve nearby P corresponding to polar angle $\theta + h$. Draw PS perpendicular to OQ, as shown in Fig. 5.36. Observe that $PS = f(\theta) \sin h$ and $SQ = OQ - OS = f(\theta + h) - f(\theta) \cos h$. If the tangent line to $r = f(\theta)$ at P makes angle ψ with the radial line OP, then evidently ψ is the limit of the angle SQP as $h \to 0$. Thus

$$\tan \psi = \lim_{h\to 0} \frac{PS}{SQ} = \lim_{h\to 0} \frac{f(\theta)\sin h}{f(\theta + h) - f(\theta)\cos h} \qquad \left[\frac{0}{0}\right]$$

$$= \lim_{h\to 0} \frac{f(\theta)\cos h}{f'(\theta + h) + f(\theta)\sin h} \qquad \text{(by l'Hôpital's rule)}$$

$$= \frac{f(\theta)}{f'(\theta)} = \frac{r}{\dfrac{dr}{d\theta}}.$$

The equation

$$\boxed{\tan \psi = \frac{r}{\dfrac{dr}{d\theta}}}$$

can be used to determine the angle between the tangent line at P and the radial line OP for any point P on the curve except the origin.

EXERCISES

In Exercises 1–12, transform the given polar equation to rectangular coordinates and hence identify the curve represented.

1. $r = 3 \sec \theta$

2. $r = -2 \csc \theta$

3. $r = \dfrac{5}{3 \sin \theta - 4 \cos \theta}$

4. $r = \sin \theta + \cos \theta$

5. $r^2 = \csc 2\theta$

6. $r = \sec \theta \tan \theta$

7. $r = \sec \theta\,(1 + \tan \theta)$

8. $r = 2(\cos^2 \theta + 4 \sin^2 \theta)^{-1/2}$

9. $r = \dfrac{1}{1 - \cos \theta}$

10. $r = \dfrac{2}{2 - \cos \theta}$

11. $r = \dfrac{2}{1 - 2 \sin \theta}$

12. $r = \dfrac{2}{1 + \sin \theta}$

In Exercises 13–22 sketch the polar graphs of the given equation.

13. $r = 1 - \sin \theta$

14. $r = 1 + 2 \cos \theta$

15. $r = 2 + \cos \theta$

16. $r = 1 - 2 \sin \theta$

17. $r = 2 \cos 2\theta$

18. $r = 2 \sin 2\theta$

19. $r = 2 \sin 3\theta$

20. $r = 2 \cos 4\theta$

21. $r^2 = 4 \sin 2\theta$

22. $r^2 = 4 \cos 3\theta$

23.* Sketch the graph of the equation $r = 1/\theta$. Show that this curve has a horizontal asymptote. Does $r = 1/(\theta - \alpha)$ have an asymptote?

24.* How many leaves does the curve $r = \cos n\theta$ have? The curve $r^2 = \cos n\theta$?

25. Sketch the polar graph $r = e^{\theta}$.

In Exercises 26–31 find all points on the given curve where the tangent line is horizontal, vertical, or does not exist.

26. $r = 2 \cos \theta$

27. $r = \cos \theta + \sin \theta$

28. $r = \sin 2\theta$

29. $r^2 = \cos 2\theta$

30. $r = 2(1 - \sin \theta)$

31. $r = e^{\theta}$

32. Determine the angles at which the straight line $\theta = \pi/4$ intersects the cardioid $r = 1 + \sin \theta$.

33.* At what points do the curves $r^2 = 2 \sin 2\theta$ and $r = 2 \cos \theta$ intersect? At what angle do the curves intersect at each of these points?

34.* At what points do the curves $r = 1 - \cos \theta$ and $r = 1 - \sin \theta$ intersect? At what angle do the curves interesect at each of these points?

6 Integration

6.1 Approximating Areas Using Rectangles
6.2 The Definite Integral
6.3 The Fundamental Theorem of Calculus
6.4 The Method of Substitution
6.5 Inverse Substitutions
6.6 The Method of Partial Fractions
6.7 Integration by Parts
6.8 Approximate Integration
6.9 Improper Integrals

The second fundamental problem addressed by calculus is the problem of areas, that is, the problem of determining the area of a region of the plane bounded by various curves. Like the problem of tangents considered in Chapter 2, the solution of the problem of areas necessarily involves the notion of limits. On the surface the problem of areas appears unrelated to the problem of tangents. However, we will see that the two problems are very closely related; one is the inverse of the other. Finding an area is equivalent to finding an antiderivative or, as we prefer to say, finding an integral.

The relationship between areas and antiderivatives is called the Fundamental Theorem of Calculus. When we have proved it we will be able to find areas at will, provided only that we can integrate (that is, antidifferentiate) the various functions we encounter.

We would clearly like to have at our disposal a "calculus" of integrals similar to the calculus of derivatives we developed in Chapter 2. We can find the derivative of any differentiable function using the differentiation rules established there. Unfortunately, integration is generally more difficult; indeed, some fairly simple functions are not themselves derivatives of simple functions. For example, $\exp(x^2)$ is not the derivative of any finite combination of elementary functions. Nevertheless, we expend considerable effort in this chapter to develop techniques for integrating as many functions as possible. Other functions such as $\exp(x^2)$ can be integrated using infinite series techniques, which we develop in Chapter 9.

6.1 APPROXIMATING AREAS USING RECTANGLES

We began the study of derivatives in Chapter 2 by defining what we mean when we say that a straight line is tangent to a curve at a particular point. It would be appropriate to begin the study of integrals by stating what we mean by the *area* of a plane region. However, a general definition of area is harder to give than one of tangency. Let us assume, therefore, that we know intuitively what area means and list some of its properties. Figure 6.1 contains illustrations of these properties.

i) The area of a rectangle of width w and height h is $A = wh$.
ii) The areas of congruent plane regions are equal.
iii) If region R is contained in region S, then the area of R is less than or equal to that of S.
iv) If R is a union of nonoverlapping regions, then the area of R is the sum of the areas of those regions.

Using these four properties we can calculate the area of any polygonal region, that is, a region bounded by straight line segments. However, if a region has a curved boundary its area can only be approximated by using rectangles (or triangles); calculating the exact area requires the evaluation of a limit.

You have no doubt encountered the formula for the area A of a circle of radius r: $A = \pi r^2$. You may be prepared to accept this formula on faith, but it is likely that you have not seen any proof of it. The number π is defined to be the ratio of the circumference of a circle to its diameter; it is not obvious that the ratio of the area of

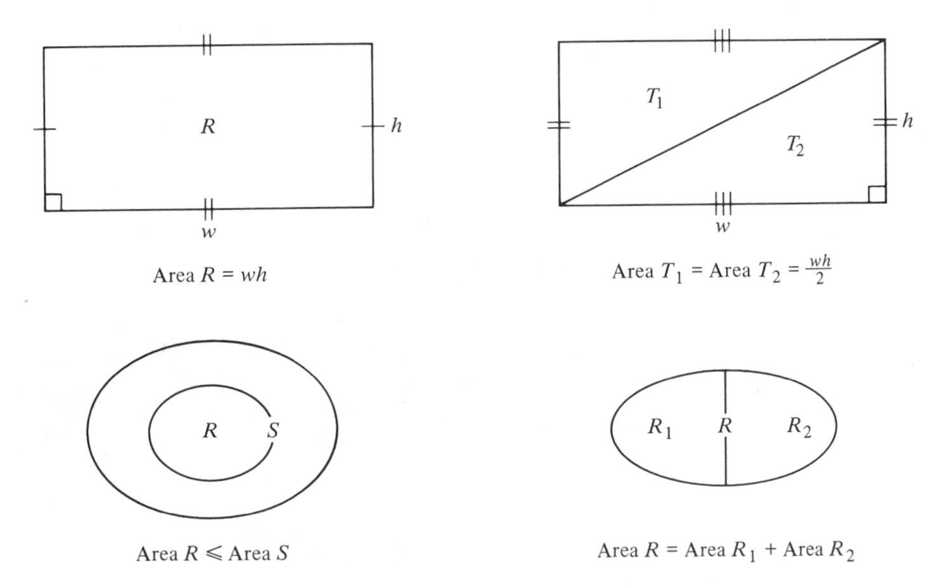

Figure 6.1

a circle to the square of its radius should be the same number. A method for proving this fact is suggested in Exercise 10 at the end of Section 6.1.

If f is a nonnegative-valued, continuous function defined on the interval $[a, b]$, the basic area problem is to find the area of the plane region R bounded by the graph of $y = f(x)$, the x-axis, and the vertical lines $x = a$ and $x = b$, as shown in Fig. 6.2.

We proceed as follows. Divide the interval $[a, b]$ into n equal subintervals, each of length $\Delta x = (b - a)/n$, using division points

$$a = x_0 < x_1 < x_2 < x_3 < \cdots < x_{n-1} < x_n = b.$$

Vertically above each subinterval $[x_{j-1}, x_j]$ $(1 \leq j \leq n)$ erect a rectangle whose base has length Δx and whose height is $f(x_j)$. These rectangles are shown shaded in Fig.

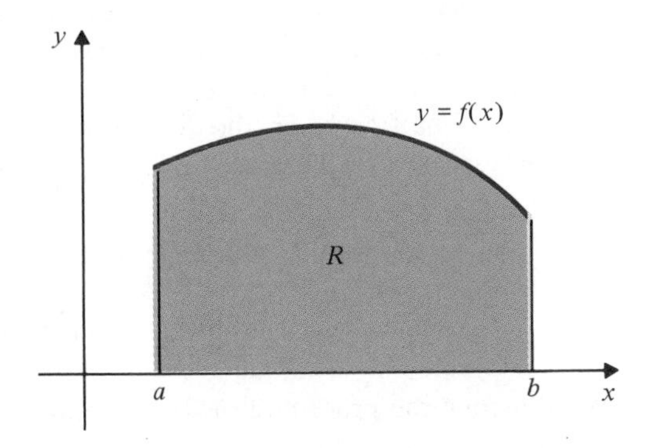

Figure 6.2

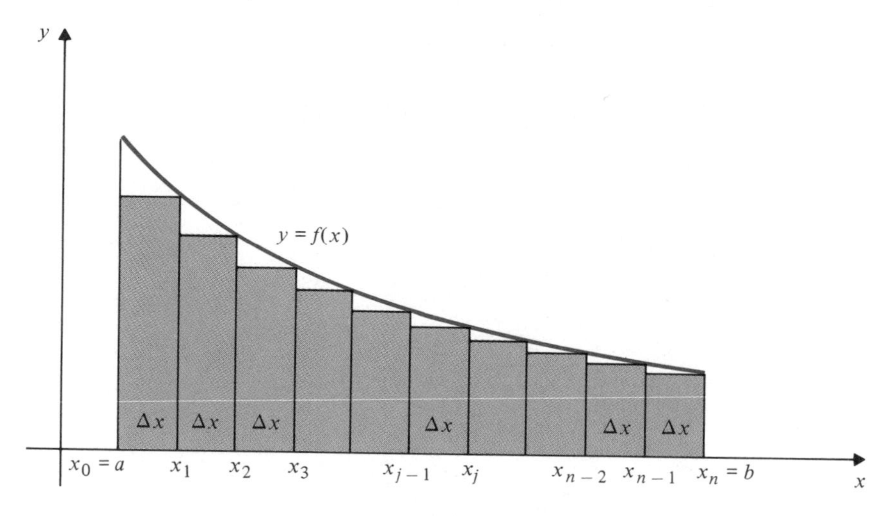

Figure 6.3

6.3 for a decreasing function f. For an increasing function the tops of these rectangles would lie above the graph of f rather than below it. The sum of the areas of these rectangles is

$$S_n = \sum_{j=1}^{n} f(x_j)\Delta x = f(x_1)\Delta x + f(x_2)\Delta x + f(x_3)\Delta x + \cdots + f(x_n)\Delta x.$$

Note the use of the *summation symbol* $\sum_{j=1}^{n} f(x_j)\Delta x$ as a shorthand notation for the sum of n similar quantities. It should be read "the sum of the numbers $f(x_j)\Delta x$ for $j = 1, 2, 3$, and so on up to and including n."

Evidently S_n is an approximation to the area of the region R under the graph of f; in the case of the decreasing function shown in Fig. 6.3, the approximation is smaller than the actual area of R, the total error being the sum of the areas of the n small, curved "triangles" above the rectangles and under the curve. It appears intuitively clear, at least for a decreasing function such as that in the figure (or else for an increasing function for which S_n will be larger than the area of R), that this total error decreases as n increases (even though there are more individual "triangles"), and in fact that the error approaches 0 as n approaches ∞. Observe in Fig. 6.4, for instance, that if we double the size of n, and thus replace each subinterval with two subintervals half as wide, the corresponding triangular errors decrease in total area. Thus it appears that the area A of R is given by

$$A = \lim_{n\to\infty} S_n = \lim_{n\to\infty} \sum_{j=1}^{n} f(x_j)\Delta x$$

$$= \lim_{n\to\infty} (f(x_1)\Delta x + f(x_2)\Delta x + \cdots + f(x_n)\Delta x).$$

We illustrate the procedure of determining such an area with Examples 1–3.

Example 1 Find the approximations S_2, S_4, and S_n for the area of the trapezoid bounded by $y = x + 1$, $y = 0$, $x = 0$, and $x = 2$. Use S_n to find the area of the trapezoid.

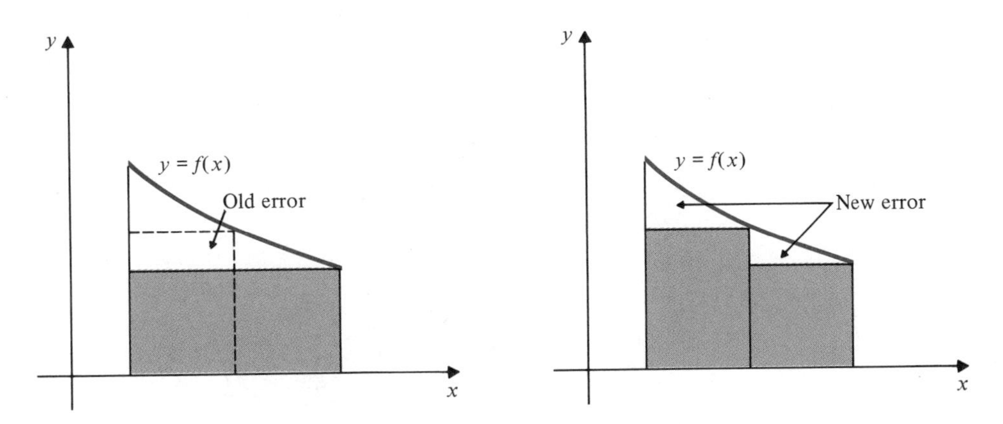

Figure 6.4

Solution The approximation S_2 is based on a division of the interval [0, 2] into the subintervals [0, 1] and [1, 2], each of width $\Delta x = 1$. S_2 is the sum of the areas of the two rectangles shown in Fig. 6.5:

$$S_2 = 1(1 + 1) + 1(2 + 1) = 5.$$

The approximation S_4 is the sum of the areas of four rectangles, each of width $\Delta x = 1/2$, based on the four subintervals [0, 1/2], [1/2, 1], [1, 3/2], and [3/2, 2] (see Fig. 6.6). Thus

$$\begin{aligned} S_4 &= \frac{1}{2}\left(\frac{1}{2} + 1\right) + \frac{1}{2}(1 + 1) + \frac{1}{2}\left(\frac{3}{2} + 1\right) + \frac{1}{2}(2 + 1) \\ &= \frac{3}{4} + 1 + \frac{5}{4} + \frac{3}{2} = \frac{9}{2}. \end{aligned}$$

Similarly, the approximation S_n is based on dividing the interval [0, 2] into n

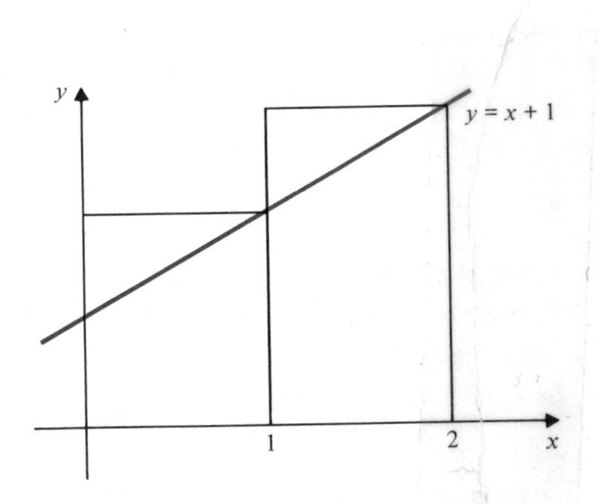

Figure 6.5

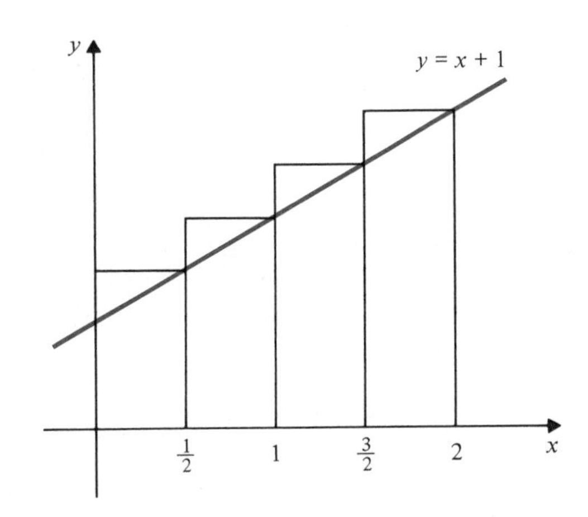

Figure 6.6

subintervals, each of length $\Delta x = 2/n$. These subintervals are therefore $[0, 2/n]$, $[2/n, 4/n]$, $[4/n, 6/n]$, . . ., $[(2(n-1))/n, 2n/n]$, as shown in Fig. 6.7.

The sum S_n of the areas of these rectangles is given by

$$\begin{aligned} S_n = \sum_{j=1}^{n} \frac{2}{n}\left(\frac{2j}{n}+1\right) &= \frac{2}{n}\left(\left(\frac{2}{n}+1\right)+\left(\frac{4}{n}+1\right)+\left(\frac{6}{n}+1\right)+\cdots+\left(\frac{2n}{n}+1\right)\right) \\ &= \frac{2}{n}\left(\frac{2}{n}(1+2+3+\cdots+n)+(1+1+1+\cdots+1)\right) \\ &= \frac{2}{n}\left(\frac{2}{n}\sum_{j=1}^{n} j + n\right) \\ &= \frac{4}{n^2}\sum_{j=1}^{n} j + 2. \end{aligned}$$

At this point we must evaluate the sum $\sum_{j=1}^{n} j = 1 + 2 + 3 + \cdots + n$. There are a variety of ways of proving the following formula:

$$\sum_{j=1}^{n} j = 1 + 2 + 3 + \cdots + n = \frac{n(n+1)}{2}.$$

One method is to use mathematical induction. (See Example 1 of Appendix 1.) Here we use a different method, one that does not require our knowing the formula in advance and that generalizes to summing higher powers of j. Start with the identity $(k+1)^2 - k^2 = 2k + 1$ and write it out for each integer k from 1 to n. Then add them up.

$$\begin{array}{lll} 2^2 - 1^2 & = 2(1) & +\ 1 \\ 3^2 - 2^2 & = 2(2) & +\ 1 \\ 4^2 - 3^2 & = 2(3) & +\ 1 \\ \vdots \quad \vdots & \quad \vdots & \quad \vdots \\ n^2 - (n-1)^2 & = 2(n-1) & +\ 1 \\ (n+1)^2 - n^2 & = 2(n) & +\ 1 \\ \hline (n+1)^2 - 1^2 & = 2\left(\sum_{j=1}^{n} j\right) & +\ n \end{array}$$

Note the cancellation that occurs when we add up the left-hand side: 2^2 cancels with -2^2, 3^2 with -3^2, and so on. The resulting equation can be solved for the required sum:

$$\sum_{j=1}^{n} j = \frac{(n+1)^2 - 1 - n}{2} = \frac{n(n+1)}{2}.$$

It follows that the area of the trapezoid is

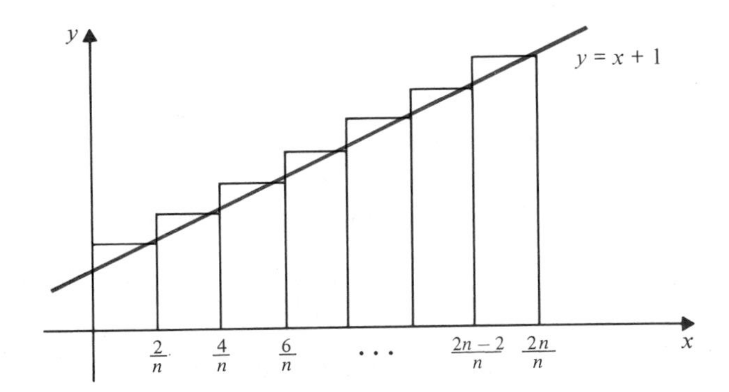

Figure 6.7

$$A = \lim_{n\to\infty} S_n = \lim_{n\to\infty}\left(\frac{4}{n^2}\frac{n(n+1)}{2} + 2\right) = \left(2\lim_{n\to\infty}\frac{n+1}{n}\right) + 2$$
$$= 2 + 2 = 4 \text{ square units}$$

We could have found this area, of course, by adding or subtracting the areas of triangles.

Example 2 Find the area of the region bounded by the curve $y = x^2$ and the straight lines $y = 0$, $x = 0$, and $x = b > 0$.

Solution The area A of the region is the limit of the sum S_n of areas of the rectangles shown in Fig. 6.8. Each rectangle has base of width $\Delta x = b/n$, and the height of the jth rectangle is $(jb/n)^2$. Thus

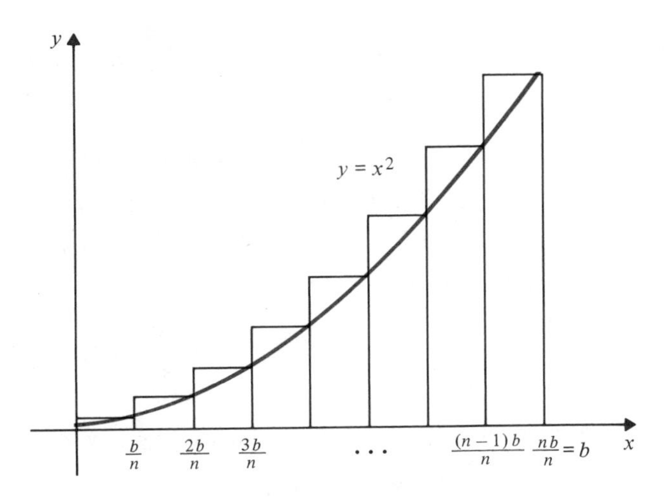

Figure 6.8

$$S_n = \frac{b}{n}\left(\left(\frac{b}{n}\right)^2 + \left(\frac{2b}{n}\right)^2 + \left(\frac{3b}{n}\right)^2 + \cdots + \left(\frac{nb}{n}\right)^2\right)$$
$$= \frac{b^3}{n^3}(1^2 + 2^2 + 3^2 + \cdots + n^2) = \frac{b^3}{n^3}\sum_{j=1}^{n} j^2.$$

Again we require a special sum, namely

$$\sum_{j=1}^{n} j^2 = 1^2 + 2^2 + 3^2 + \cdots + n^2.$$

The procedure we used in Example 1 will work here, too. This time we start with the identity $(k + 1)^3 - k^3 = 3k^2 + 3k + 1$. We write out this identity for $k = 1, 2, 3$, and so on up to n, and add them up.

$$\begin{aligned}
2^3 - 1^3 &= 3(1^2) + 3(1) + 1 \\
3^3 - 2^3 &= 3(2^2) + 3(2) + 1 \\
4^3 - 3^3 &= 3(3^2) + 3(3) + 1 \\
&\ \ \vdots \\
n^3 - (n-1)^3 &= 3(n-1)^2 + 3(n-1) + 1 \\
(n+1)^3 - n^3 &= 3(n^2) + 3(n) + 1 \\
\hline
(n+1)^3 - 1^3 &= 3\left(\sum_{j=1}^{n} j^2\right) + 3\left(\sum_{j=1}^{n} j\right) + n \\
&= 3\left(\sum_{j=1}^{n} j^2\right) + \frac{3n(n+1)}{2} + n
\end{aligned}$$

We have used $\sum_{j=1}^{n} j = (n(n + 1))/2$ as we obtained in the previous example. The final equation can be solved for the desired sum:

$$\sum_{j=1}^{n} j^2 = 1^2 + 2^2 + 3^2 + \cdots + n^2 = \frac{n(n+1)(2n+1)}{6}.$$

(This formula can also be established by mathematical induction.) Hence

$$S_n = \frac{b^3}{n^3}\,\frac{n(n+1)(2n+1)}{6} = \frac{b^3}{6}\,\frac{(n+1)(2n+1)}{n^2} = \frac{b^3}{6}\left(1 + \frac{1}{n}\right)\left(2 + \frac{1}{n}\right)$$

and the required area is

$$A = \lim_{n\to\infty} S_n = \frac{b^3}{6}(1)(2) = \frac{1}{3}b^3 \text{ square units.}$$

Example 3 Find the area of the region bounded by $y = e^{-x}$, $y = 0$, $x = a$, and $x = b > a$.

Solution Here the subdivision points are $x_0 = a$, $x_1 = a + \Delta x$, $x_2 = a + 2\Delta x$, . . . , $x_n = a + n\Delta x = b$, where $\Delta x = (b - a)/n$.

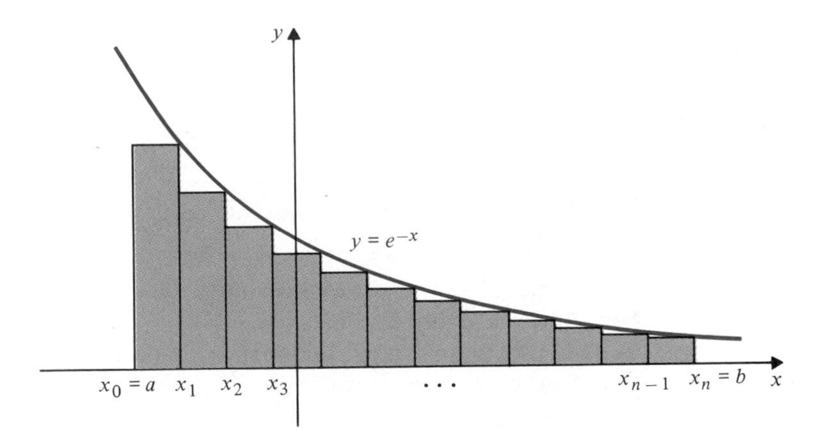

Figure 6.9

The sum S_n of the areas of the rectangles shaded in Fig. 6.9 is

$$
\begin{aligned}
S_n &= \Delta x\,(e^{-x_1} + e^{-x_2} + e^{-x_3} + \cdots + e^{-x_n}) \\
&= \frac{b-a}{n}(e^{-a-\Delta x} + e^{-a-2\Delta x} + e^{-a-3\Delta x} + \cdots + e^{-a-n\Delta x}) \\
&= \frac{b-a}{n} e^{-a-\Delta x}(1 + r + r^2 + \cdots + r^{n-1}),
\end{aligned}
$$

where $r = e^{-\Delta x}$. The sum $s = 1 + r + r^2 + r^3 + \cdots + r^{n-1}$ is called a geometric sum. It can be evaluated as follows:

$$
\begin{aligned}
s(1-r) &= s - rs \\
&= (1 + r + r^2 + \cdots + r^{n-1}) - (r + r^2 + r^3 + \cdots + r^n) \\
&= 1 - r^n.
\end{aligned}
$$

Thus

$$\sum_{j=0}^{n} r^j = 1 + r + r^2 + r^3 + \cdots + r^{n-1} = \frac{1-r^n}{1-r} \qquad (r \neq 1).$$

We now have

$$S_n = \frac{b-a}{n} e^{-a-\Delta x}\frac{1-e^{-n\Delta x}}{1-e^{-\Delta x}} = \frac{b-a}{n} e^{-a-((b-a)/n)}\frac{1-e^{-(b-a)}}{1-e^{-(b-a)/n}}.$$

The area of the specified region is the limit of S_n as n tends to infinity. It is convenient to substitute $y = (b-a)/n$ and take the limit as $y \to 0+$ instead:

$$A = \lim_{n\to\infty} S_n = \lim_{y\to 0+} y e^{-a-y}\frac{1-e^{-(b-a)}}{1-e^{-y}} \qquad \left(\text{multiply by } \frac{e^y}{e^y}\right)$$

$$= e^{-a}(1 - e^{a-b}) \lim_{y\to 0+} \frac{y}{e^y - 1} \qquad \left[\frac{0}{0}\right]$$

$$= (e^{-a} - e^{-b}) \lim_{y\to 0+} \frac{1}{e^y} = e^{-a} - e^{-b} \text{ square units.}$$

We have used l'Hôpital's rule to evaluate the limit.

The techniques used in Examples 1–3 can be extended to find areas under curves whose equations involve functions of the form x^n ($n = 1, 2, 3, \ldots$) or a^x ($a > 0$). For x^n the effort increases greatly with n because one must sum $\Sigma_{j=1}^{k} j^n$. If this were the only way to find areas, their computation would indeed be laborious. Fortunately, there is usually an easier way.

EXERCISES

Use the technqiues of Examples 1–3 to find the areas of the regions specified in Exercises 1–7.

1. Below $y = 3x$, above $y = 0$, from $x = 0$ to $x = 1$.
2. Below $y = ax + b$, above $y = 0$, between $x = c$ and $x = d$. Assume $ax + b \geq 0$ on $[c, d]$.
3. Below $y = x^2$, above $y = 0$, from $x = 1$ to $x = 3$.
4. Below $y = x^2 + 1$, above $y = 0$, from $x = 0$ to $x = a > 0$.
5. Below $y = 2^x$, above $y = 0$, from $x = -1$ to $x = 1$.
6. Below $y = a^x$ ($a > 0$, $a \neq 1$), above $y = 0$, from $x = 0$ to $x = b$.
7. Below $y = x^2 + 2x + 3$, above $y = 0$, from $x = -1$ to $x = 2$.
8. Below $y = \sqrt{x}$ above $y = 0$, between $x = 0$ and $x = 1$. (*Hint:* Approximate this area using horizontal rectangles.)

9.* **a)** Show that

$$\sum_{j=1}^{n} j^3 = 1^3 + 2^3 + 3^3 + \cdots + n^3 = \frac{n^2(n+1)^2}{4}.$$

b) Find the area of the region lying under $y = x^3$, above the x-axis, and between the vertical lines at $x = 0$ and $x = a > 0$.

10.* Here is a way to show that the area of a circle of radius r is $A = \pi r^2$.

a) In a circle of radius r inscribe a regular polygon of n sides. Show that the perimeter P_n and area A_n of this polygon are given by

$$P_n = 2rn \sin\frac{\pi}{n} \qquad \text{and} \qquad A_n = r^2 n \sin\frac{\pi}{n} \cos\frac{\pi}{n}.$$

b) Use the intuitive fact that the perimeter of the polygon approaches the circumference of the circle as $n \to \infty$ to show that

$$\lim_{n\to\infty} n \sin\frac{\pi}{n} = \pi.$$

* Difficult and/or theoretical problems.

c) Find the area A of the circle, observing that $A = \lim_{n\to\infty} A_n$.

11.* Identify the expression $S_n = \sum_{j=1}^{n} \frac{1}{n}\sqrt{1-(j/n)^2}$ as a sum of areas of rectangles approximating the area of a certain region in the plane. Hence evaluate $\lim_{n\to\infty} S_n$.

6.2 THE DEFINITE INTEGRAL

In this section we generalize the procedure used for finding areas in the previous section and use it to define the definite integral of a function f on an interval I. Let us assume, for the time being, that $f(x)$ is defined and continuous on the closed, finite interval $[a, b]$. We do not assume that $f(x)$ is nonnegative-valued.

Subdivide $[a, b]$ into n subintervals, each of length $\Delta x = (b - a)/n$, using the points

$$a = x_0 < x_1 < x_2 < x_3 < \cdots < x_{n-1} < x_n = b.$$

The jth subinterval is $[x_{j-1}, x_j]$ and it has length Δx. For $1 \le j \le n$ pick any point c_j in the jth subinterval; thus $x_{j-1} \le c_j \le x_j$. The sum

$$\begin{aligned} R_n(f) &= f(c_1)\Delta x + f(c_2)\Delta x + f(c_3)\Delta x + \cdots + f(c_n)\Delta x \\ &= \sum_{j=1}^{n} f(c_j)\Delta x \end{aligned}$$

is called a **Riemann sum** for f on $[a, b]$. Note in Fig. 6.10 that if $f(c_j) > 0$, then $f(c_j)\Delta x$ is the area of a rectangle with base Δx and height taken to a point on the graph $y = f(x)$ in the jth subinterval. If $f(c_j) < 0$, then $f(c_j)\Delta x$ is the negative of such an area.

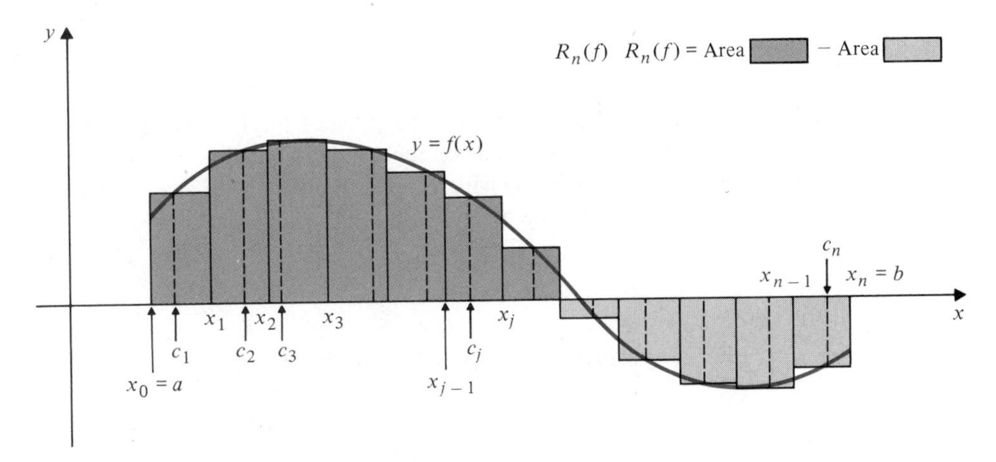

Figure 6.10

Since f is continuous on each subinterval $[x_{j-1}, x_j]$, it takes on maximum and minimum values at points of that interval. (See Theorem 5 of Section 1.5.) Thus there are numbers u_j and l_j in $[x_{j-1}, x_j]$ such that

$$f(l_j) \leq f(x) \leq f(u_j) \qquad \text{for } x_{j-1} \leq x \leq x_j.$$

Evidently,

$$f(l_j)\Delta x \leq f(c_j)\Delta x \leq f(u_j)\Delta x.$$

The particular Riemann sums

$$L_n(f) = f(l_1)\Delta x + f(l_2)\Delta x + \cdots + f(l_n)\Delta x = \sum_{j=1}^{n} f(l_j)\Delta x,$$

$$U_n(f) = f(u_1)\Delta x + f(u_2)\Delta x + \cdots + f(u_n)\Delta x = \sum_{j=1}^{n} f(u_j)\Delta x$$

are called *lower* and *upper* Riemann sums for f on $[a, b]$. See Fig. 6.11. Regardless of how the numbers c_j are chosen (with c_j in $[x_{j-1}, x_j]$), we have

$$L_n(f) \leq R_n(f) \leq U_n(f).$$

Definition 1

If $\lim_{n\to\infty} L_n(f) = \lim_{n\to\infty} U_n(f)$, then we say that the function f is **integrable** on $[a, b]$ and we denote the common value of these limits by the symbol $\int_a^b f(x)dx$, called the **definite integral** of f over the interval $[a, b]$.

$$\int_a^b f(x)dx = \lim_{n\to\infty} L_n(f) = \lim_{n\to\infty} U_n(f)$$

We stress that the definite integral of $f(x)$ over $[a, b]$ is a *number;* it is not a function of x. It depends on the numbers a and b and on the particular function f, but not on the variable x, which can be replaced with any other variable without changing the value of the integral:

$$\int_a^b f(x)dx = \int_a^b f(t)dt.$$

For all positive integers m and n we have

$$L_m(f) \leq \int_a^b f(x)dx \leq U_n(f).$$

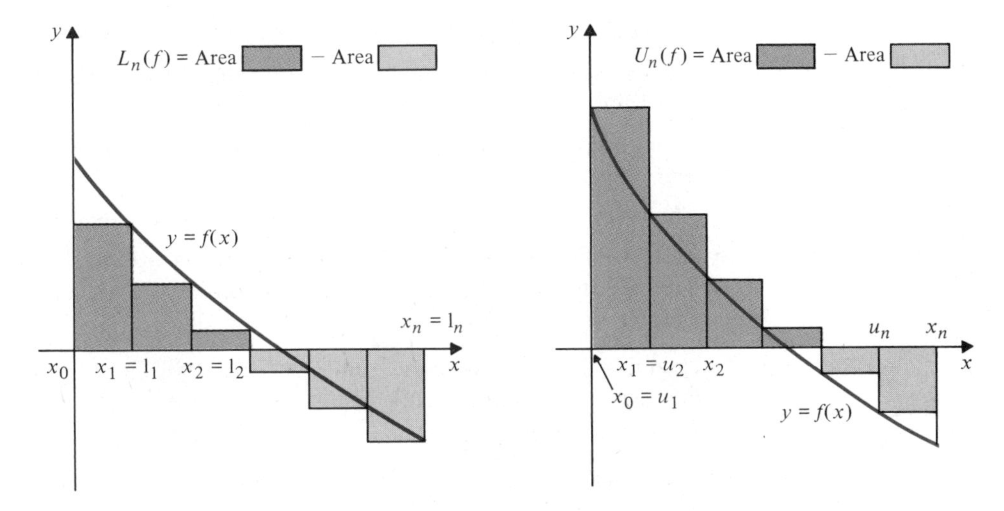

Figure 6.11

If $f(x) \geq 0$ on $[a, b]$, then we define the area of the region R bounded by the graph of $y = f(x)$, the x-axis and the lines $x = a$ and $x = b$ to be A square units, where $A = \int_a^b f(x)dx$. If $f(x) \leq 0$ on $[a, b]$, the area of R is $-\int_a^b f(x)dx$ square units. For general f, $\int_a^b f(x)dx$ is the area of that part of R lying above the x-axis less the area of that part lying below the x axis (see Fig. 6.12).

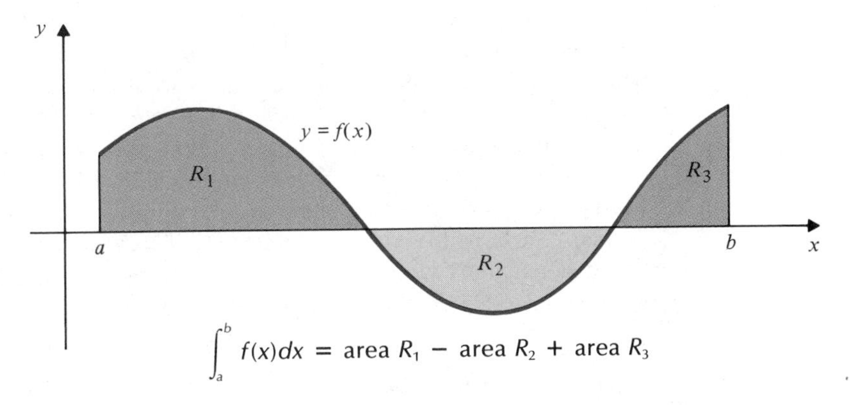

Figure 6.12

The various parts of the symbol $\int_a^b f(x)dx$ have their own names:

i) $\int$ is called the *integral sign;* it resembles the letter S since it represents the limit of a sum.

ii) a and b are called the *limits of integration;* a is the *lower limit,* b is the *upper limit.*

iii) The function f is the *integrand;* x is the *variable of integration.*

iv) dx is the *differential* of x. It replaces Δx in the Riemann sum. If an integrand depends on more than one variable, the differential tells you which one is the variable of integration.

You can think of $\int_a^b f(x)dx$ as a "sum" of "areas" of infinitely many rectangles with heights $f(x)$ and "infinitesimally small widths" dx.

Theorem 1 If f is continuous on $[a, b]$, then f is integrable on $[a, b]$. In this case,

$$\lim_{n\to\infty} R_n(f) = \int_a^b f(x)dx,$$

regardless of how the points c_j are chosen in constructing the Riemann sum $R_n(f)$. □

We cannot prove this theorem in its full generality. The proof involves subtle properties of the real numbers deriving from the completeness property. If f is increasing (or decreasing) on $[a, b]$ a proof can be given without too much difficulty. (It is given at the end of this section.) It follows that the theorem holds for any function that is the sum of an increasing function and a decreasing function. This class of functions includes any continuous functions we are likely to encounter in concrete applications of calculus but, unfortunately, does not include all continuous functions. A full proof of Theorem 1 can be found in Appendix 3.

The definition of integrability and the definite integral given above can be extended to a wider class than the continuous functions, and can also be phrased to allow for subdivisions having subintervals of unequal length. We could also allow $b < a$, in which case the Δx's in the Riemann sums would be negative. For functions that are not necessarily continuous everywhere, the definitions of upper and lower sums have to be modified because such f need not assume maximum and minimum values on closed intervals. However, if f is bounded on $[a, b]$, that is, if $|f(x)| \leq K$ for some constant K, then upper and lower sums still can be defined (see Appendix 3 for details). Such is the case for bounded, piecewise continuous functions, that is, bounded functions with finitely many points of discontinuity, as in Fig. 6.13. We examine integrals of such functions in Section 6.9.

Example 1 Show directly from the definition that $f(x) = x$ is integrable on $[a, b]$, and find $\int_a^b x\, dx$.

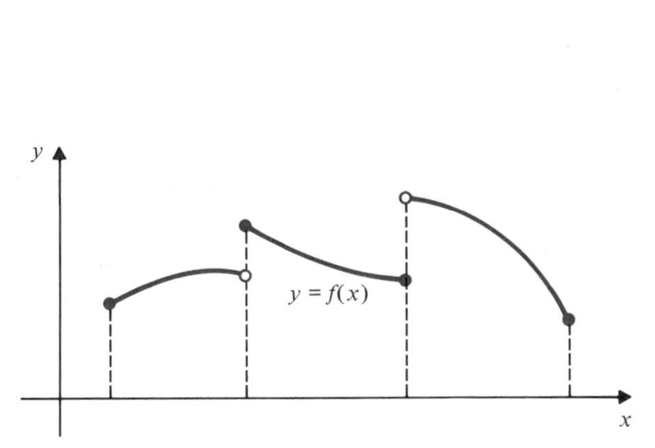

A piecewise continuous function
Figure 6.13

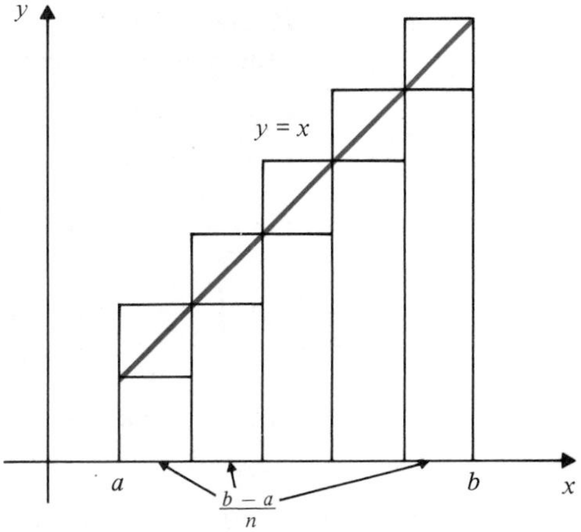

Figure 6.14

Solution We compute the upper and lower sums for f by making use of the identity $\Sigma_{j=1}^{n} j = (n(n+1))/2$ that we obtained earlier. See Fig. 6.14.

$$L_n(f) = \frac{b-a}{n}\left(a + \left(a + \frac{b-a}{n}\right) + \left(a + 2\frac{b-a}{n}\right) + \cdots + \left(a + (n-1)\frac{b-a}{n}\right)\right)$$

$$= \frac{b-a}{n}\left(na + \frac{b-a}{n}(0 + 1 + 2 + \cdots + (n-1))\right)$$

$$= (b-a)a + \frac{(b-a)^2}{n^2}\frac{(n-1)n}{2} = (b-a)a + \frac{(b-a)^2}{2}\left(1 - \frac{1}{n}\right)$$

$$U_n(f) = \frac{b-a}{n}\left(\left(a + \frac{b-a}{n}\right) + \left(a + 2\frac{b-a}{n}\right) + \cdots + \left(a + n\frac{b-a}{n}\right)\right)$$

$$= \frac{b-a}{n}\left(na + \frac{b-a}{n}(1 + 2 + 3 + \cdots + n)\right)$$

$$= (b-a)a + \frac{(b-a)^2}{n^2}\frac{n(n+1)}{2} = (b-a)a + \frac{(b-a)^2}{2}\left(1 + \frac{1}{n}\right).$$

Evidently,

$$\lim_{n\to\infty} L_n(f) = \lim_{n\to\infty} U_n(f) = a(b-a) + \frac{1}{2}(b-a)^2 = \frac{1}{2}(b^2 - a^2).$$

Thus $f(x) = x$ is integrable on $[a, b]$ and

$$\int_a^b x\,dx = \frac{1}{2}(b^2 - a^2).$$

Properties of the Definite Integral

Some of the most important properties of the definite integral are summarized in the following theorem.

Theorem 2 Let f and g be integrable on an interval containing the points a, b, and c. Then

a) $\int_a^a f(x)\, dx = 0.$

b) $\int_b^a f(x)\, dx = -\int_a^b f(x)\, dx.$

c) $\int_a^b (Af(x) + Bg(x))\, dx = A\int_a^b f(x)\, dx + B\int_a^b g(x)\, dx$ (A and B are constants).

d) $\int_a^b f(x)\, dx + \int_b^c f(x)\, dx = \int_a^c f(x)\, dx.$

e) If $f(x) \le g(x)$ on $a \le x \le b$, then

$$\int_a^b f(x)\, dx \le \int_a^b g(x)\, dx.$$

f) If $a \le b$, then

$$\left|\int_a^b f(x)\, dx\right| \le \int_a^b |f(x)|\, dx.$$

g) If f is an odd function (that is, $f(-x) = -f(x)$), then

$$\int_{-a}^a f(x)\, dx = 0.$$

h) If f is an even function (that is, $f(-x) = f(x)$), then

$$\int_{-a}^a f(x)\, dx = 2\int_0^a f(x)\, dx. \square$$

All these properties can be deduced from the definition of definite integral. Most of them should appear intuitively obvious if you regard the integrals as representing (signed) areas. For instance, properties (d) and (e) are properties (iv) and (iii), respectively, of areas mentioned at the beginning of Section 6.1 (see Fig. 6.15). Property (f) is a generalization of the triangle inequality for numbers: $|x + y| \le |x| + |y|$ or, more generally, $\left|\sum_{j=1}^n x_j\right| \le \sum_{j=1}^n |x_j|$. It follows from property (e), since $-|f(x)| \le f(x) \le |f(x)|$. The symmetry properties (g) and (h), which are illustrated in Fig. 6.16, will be particularly useful later.

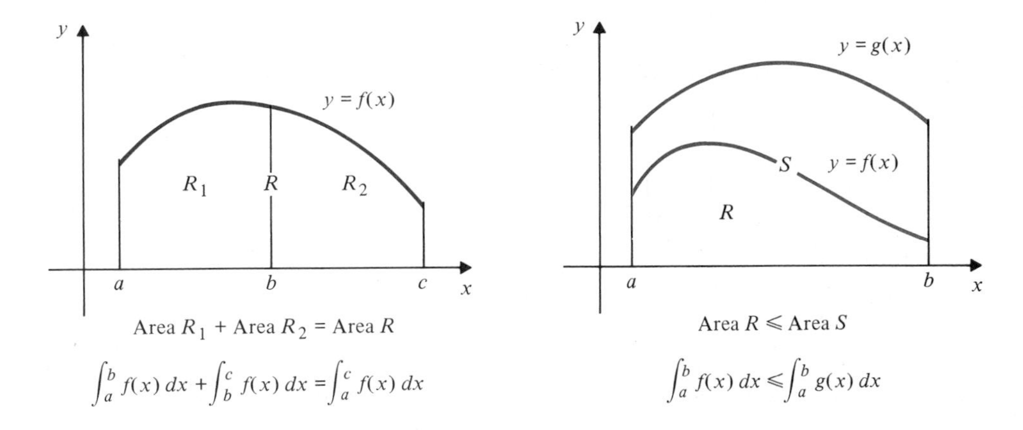

Figure 6.15

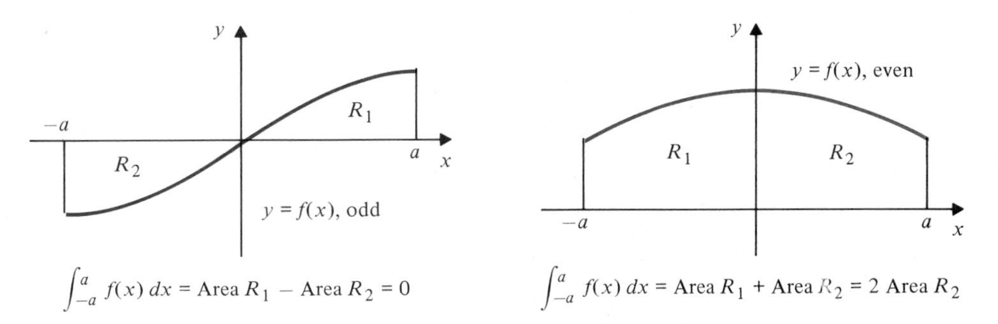

Figure 6.16

As yet we have no easy method for evaluating definite integrals. By using the above properties and interpreting the integrals as areas, however, we can sometimes find values of definite integrals by inspection.

Example 2 Figs. 6.17–6.19 illustrate parts (a)–(c), respectively.

a) $$\int_{-2}^{2} (1 + x)\,dx = \int_{-2}^{2} dx + \int_{-2}^{2} x\,dx$$
$$= 4 + 0 = 4$$

b) $$\int_{-3}^{3} \sqrt{9 - x^2}\,dx = 2\int_{0}^{3} \sqrt{9 - x^2}\,dx = \frac{9\pi}{2} \qquad \text{(area of a semicircle)}$$

c) $$\int_{0}^{3} (2 + x)\,dx = \int_{0}^{3} 2\,dx + \int_{0}^{3} x\,dx$$
$$= 6 + \frac{1}{2}(3)(3) = \frac{21}{2}$$

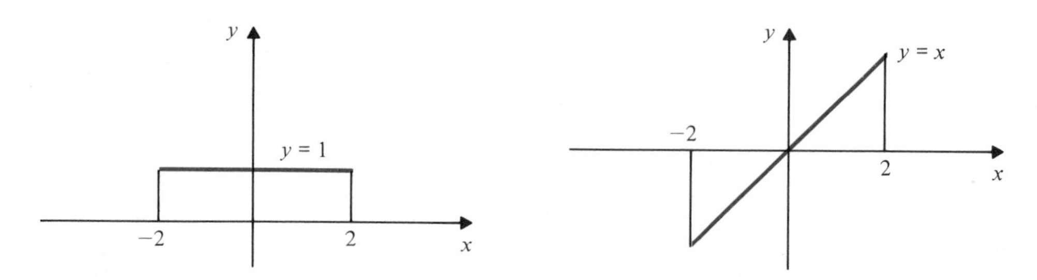

Figure 6.17

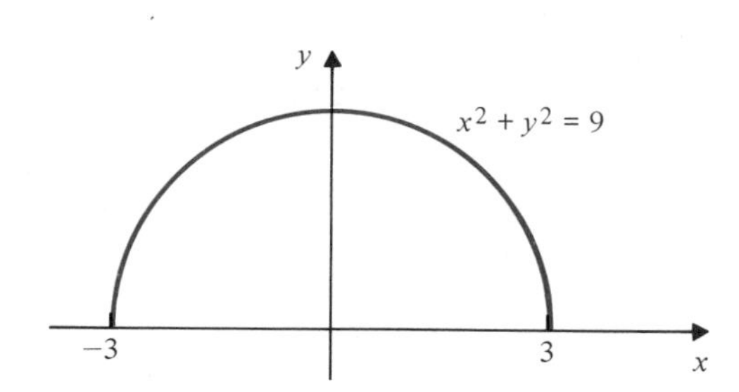

Figure 6.18

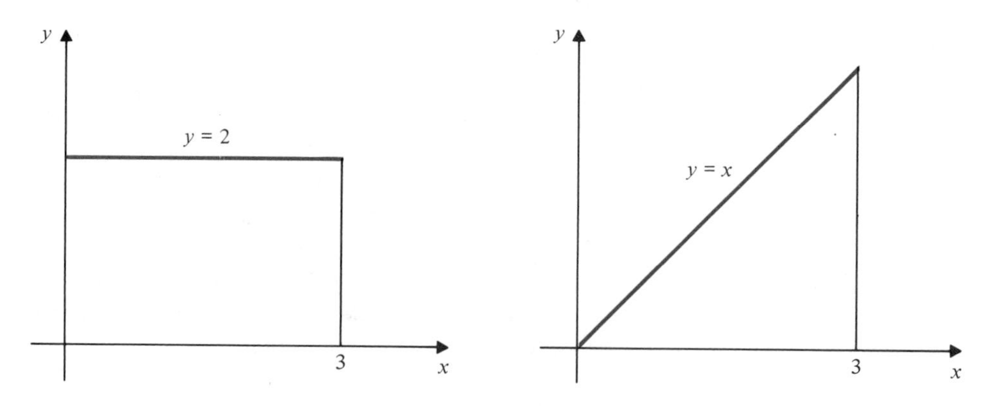

Figure 6.19

The Mean-Value Theorem for Integrals

Let f be a function continuous on the interval $[a, b]$. Then f assumes maximum and minimum values on that interval, say

$$m = f(x_0) \leq f(x) \leq f(x_1) = M \qquad \text{for all } x \text{ in } [a, b].$$

Thus

$$m(b - a) = L_1(f) \leq \int_a^b f(x)\, dx \leq U_1(f) = M(b - a)$$

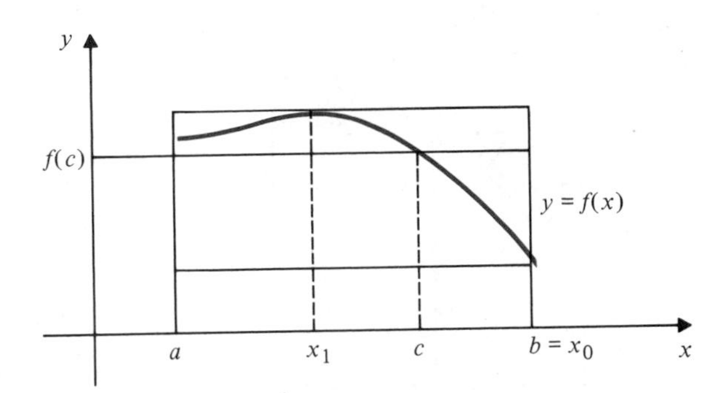

Figure 6.20

and

$$f(x_0) = m \leq \frac{1}{b - a} \int_a^b f(x)\, dx \leq M = f(x_1).$$

By the intermediate-value theorem (Theorem 6 of Section 1.5), $f(x)$ must take on every value between the two values $f(x_0)$ and $f(x_1)$ at some point between x_0 and x_1 (Fig. 6.20). Hence there is a number c between x_0 and x_1 such that

$$f(c) = \frac{1}{b - a} \int_a^b f(x)\, dx,$$

that is, $\int_a^b f(x)\, dx$ is equal to the area $(b - a)f(c)$ of a rectangle with base width $b - a$ and height $f(c)$ for some c between a and b. This is the mean-value theorem for integrals.

Theorem 3 (*The Mean-Value Theorem for Integrals*) If f is continuous on $[a, b]$, then there exists a point c in $[a, b]$ such that

$$\int_a^b f(x)\, dx = (b - a)f(c). \square$$

Observe in Fig. 6.21 that the area below the curve $y = f(x)$ and above the line $y = f(c)$ is equal to the area above $y = f(x)$ and below $y = f(c)$. In this sense $f(c)$ is the average value of the function $f(x)$ on the interval $[a, b]$. We adopt this point of view as a definition.

Definition 2 If f is integrable on $[a, b]$, then the **average value** of $f(x)$ on $[a, b]$ is

$$\frac{1}{b - a} \int_a^b f(x)\, dx.$$

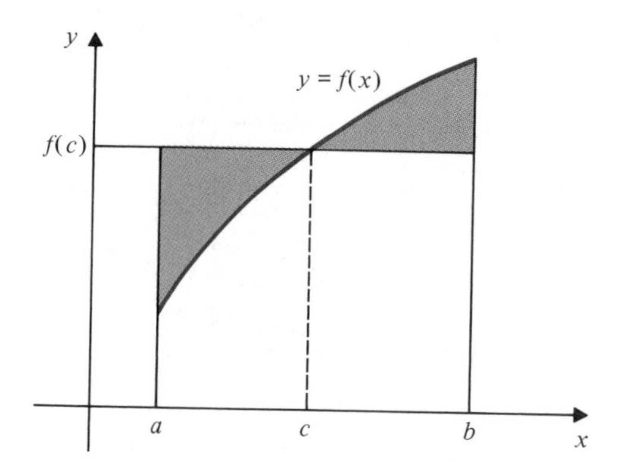

Figure 6.21

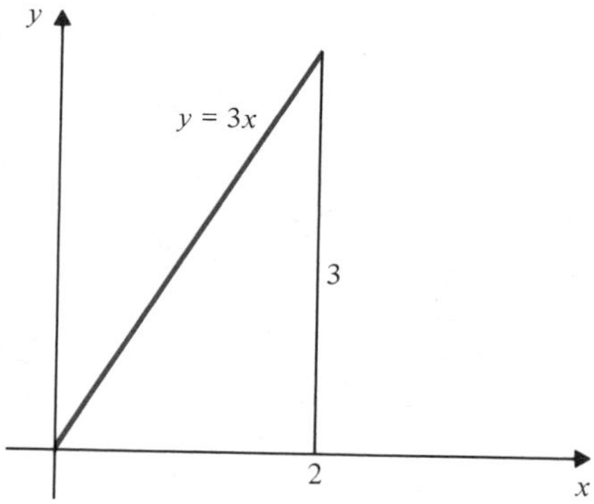

Figure 6.22

Example 3 The average value of $f(x) = 3x$ on $[0, 2]$ is

$$\frac{1}{2-0}\int_0^2 3x\,dx = \frac{1}{2}\frac{1}{2}(2)(6) = 3.$$

See Fig. 6.22.

Proof of Theorem 1 for an Increasing Function Suppose that $f(x)$ is continuous and increasing on $[a, b]$. Subdividing $[a, b]$ as in the definition of the definite integral (see Fig. 6.23), we observe that the upper and lower Riemann sums of f are given by

$$U_n = U_n(f) = \frac{b-a}{n}(f(x_1) + f(x_2) + f(x_3) + \cdots + f(x_n))$$

$$L_n = L_n(f) = \frac{b-a}{n}(f(x_0) + f(x_1) + f(x_2) + \cdots + f(x_{n-1})).$$

It follows that

$$0 \le U_n - L_n = \frac{b-a}{n}(f(x_n) - f(x_0)) = \frac{K}{n},$$

where $K = (b-a)(f(x_n) - f(x_0)) = (b-a)(f(b) - f(a))$.

Subdividing existing intervals into smaller ones causes lower sums to increase and upper sums to decrease, so for any positive integers m and n, $L_m \le L_{mn} \le U_{mn} \le U_n$.

Since every lower sum L_m satisfies $L_m \le U_1$, the completeness property for the real numbers (see Section 1.1) assures us that there exists a *smallest* real number I such that $L_m \le I$ for every m. Moreover, since every $L_m \le U_n$ for any n, we must have $U_n \ge I$ for every n. (If any U_n were less than I, then I would not be the smallest number greater than or equal to every L_m.) Thus for every n we have

$$I \le U_n = L_n + \frac{K}{n} \le I + \frac{K}{n}.$$

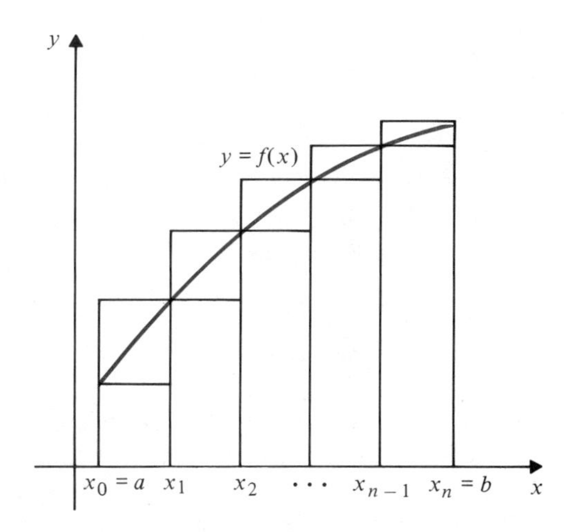

Figure 6.23

Hence $\lim_{n\to\infty} U_n = I$ by a version of the squeeze theorem. Similarly, for every m,

$$I \geq L_m = U_n - \frac{K}{n} \geq I - \frac{K}{n},$$

so $\lim_{m\to\infty} L_m = I$ by the squeeze theorem. Hence $\int_a^b f(x)\,dx$ exists and equals I. □

EXERCISES

Calculate the upper and lower Riemann sums $U_n(f)$ and $L_n(f)$ for the functions f, the intervals, and the values of n specified in Exercises 1–6.

1. $f(x) = x^2$ on $[0, 2]$, $n = 8$

2. $f(x) = x^3$ on $[0, 1]$, $n = 5$

3. $f(x) = e^x$ on $[-2, 2]$, $n = 4$

4. $f(x) = \ln x$ on $[1, 2]$, $n = 6$

5. $f(x) = \sin x$ on $[0, \pi]$, $n = 6$

6. $f(x) = \cos x$ on $[0, \pi]$, $n = 4$

7. Show that $f(x) = x^2$ is integrable on $[0, b]$ by evaluating the upper and lower sums $U_n(f)$ and $L_n(f)$ and taking their limits as n tends to infinity. What is the value of $\int_0^b x^2\,dx$?

8.* Repeat Exercise 7 for $f(x) = x^3$ on $[0, b]$.

9.* Repeat Exercise 7 for $f(x) = e^x$ on $[a, b]$.

Evaluate the integrals in Exercises 10–17 by interpreting them as areas.

10. $\int_{-2}^{2} (x + 2)\,dx$

11. $\int_0^2 (3x + 1)\,dx$

12. $\int_a^b x\,dx$

13. $\int_{-\sqrt{2}}^{0} \sqrt{2 - x^2}\,dx$

14. $\int_{-\pi}^{\pi} \sin(x^3)\,dx$

15. $\int_{-a}^{a} |x|\,dx$

16. $\int_{-1}^{1} (x^5 - 3x^3 + x)\, dx$

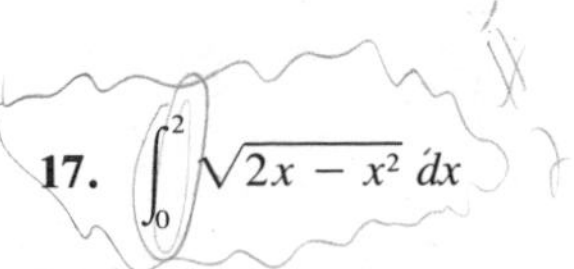

17. $\int_{0}^{2} \sqrt{2x - x^2}\, dx$

18. Find the average value of $f(x) = x^2$ over $[a, b]$.

19. Find the average value of $f(x) = (4 - x^2)^{1/2}$ over the interval $[0, 2]$.

20.* Modify the proof of Theorem 1 for the integrability of continuous increasing functions to cover the case of continuous decreasing functions.

6.3 THE FUNDAMENTAL THEOREM OF CALCULUS

In this section we demonstrate the relationship between the definite integral defined in Section 6.2 and the indefinite integral (or general antiderivative) introduced in Section 2.9. A consequence of this relationship is that we will be able to calculate the areas under graphs of functions whose antiderivatives we know.

Recall that in Section 3.4 we wanted to find a function whose derivative was $1/x$. We solved this problem by defining a function ($\ln x$) in terms of the area under the graph of $y = 1/x$. This idea motivates, and is a special case of, the following theorem.

Theorem 4 (*The Fundamental Theorem of Calculus*) Suppose that the function $f(x)$ is continuous on an interval I containing the point a. Let a function $F(x)$ be defined on I by

$$F(x) = \int_a^x f(t)\, dt.$$

Then F is differentiable on I and $F'(x) = f(x)$. That is,

$$\frac{d}{dx}\int_a^x f(t)\, dt = f(x).$$

If $G(x)$ is any antiderivative of $f(x)$ on I, that is, if $G'(x) = f(x)$ on I, then for any b in I we have

$$\int_a^b f(x)\, dx = G(b) - G(a).$$

Proof We calculate

$$F'(x) = \lim_{h\to 0}\frac{F(x+h) - F(x)}{h} = \lim_{h\to 0}\frac{1}{h}\left(\int_a^{x+h} f(t)\, dt - \int_a^x f(t)\, dt\right)$$

$$= \lim_{h\to 0}\frac{1}{h}\int_x^{x+h} f(t)\, dt \qquad \text{by Theorem 2(d)}$$

$$= \lim_{h\to 0} \frac{1}{h} hf(c) \qquad \text{for some } c \text{ between } x \text{ and } x + h \text{ (Theorem 3)}$$

$$= \lim_{c\to x} f(c) = f(x) \qquad \text{since } c \to x \text{ as } h \to 0 \text{ and } f \text{ is continuous.}$$

If also $G'(x) = f(x)$, then $F(x) = G(x) + C$ on I for some constant C (by Theorem 5 of Section 2.9). Hence

$$\int_a^x f(t)\, dt = F(x) = G(x) + C.$$

Let $x = a$ and obtain $0 = G(a) + C$, so $C = -G(a)$. Now let $x = b$ to get

$$\int_a^b f(t)\, dt = G(b) - G(a).$$

Of course we may now replace t with x (or any other variable) as the variable of integration on the left-hand side. □

Definition 3

> Hereafter we shall use the evaluation symbol
>
> $$F(x)\Big|_a^b$$
>
> to denote $F(b) - F(a)$.

Thus

$$\int_a^b f(x)\, dx = \left(\int f(x)\, dx\right)\Bigg|_a^b$$

where $\int f(x)\, dx$ denotes the indefinite integral or general antiderivative of f (see Section 2.9). When evaluating a definite integral this way we will omit $+C$ from the indefinite integral because it cancels out in the subtraction:

$$(F(x) + C)\Big|_a^b = F(b) + C - (F(a) + C) = F(b) - F(a) = F(x)\Big|_a^b.$$

Thus *any* antiderivative can be used to calculate the definite integral.

Example 1 $$\int_{-1}^2 (x^2 - 3x + 2)\, dx = \left(\frac{1}{3}x^3 - \frac{3}{2}x^2 + 2x\right)\Bigg|_{-1}^2$$

$$= \frac{1}{3}(8) - \frac{3}{2}(4) + 4 - \left(\frac{1}{3}(-1) - \frac{3}{2}(1) + (-2)\right) = \frac{9}{2}.$$

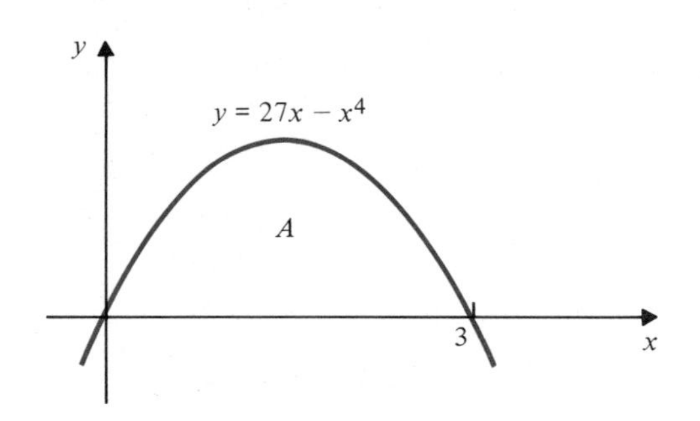

Figure 6.24

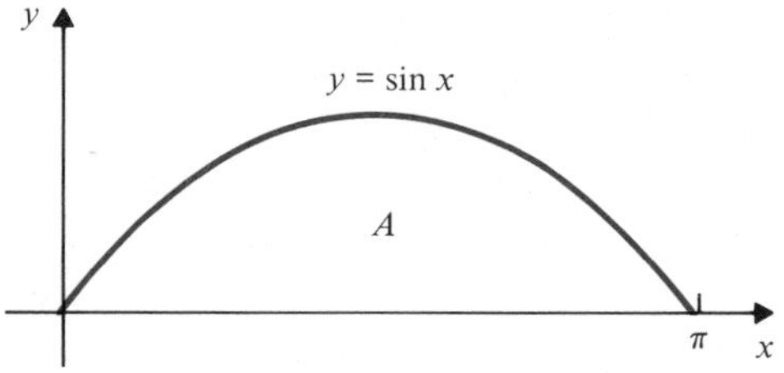

Figure 6.25

Example 2 Find the area of the plane region lying above the x-axis and under the curve $y = 27x - x^4$.

Solution We need to find the points where the curve $y = 27x - x^4$ meets the x-axis. These are solutions of the equation.

$$27x - x^4 = 0$$

$$x(27 - x^3) = 0$$

$$x(3 - x)(9 + 3x + x^2) = 0.$$

The only real roots are $x = 0$ and $x = 3$ (see Fig. 6.24). Hence the area of the region is given by

$$A = \int_0^3 (27x - x^4)dx = \frac{27}{2}x^2\bigg|_0^3 - \frac{1}{5}x^5\bigg|_0^3$$

$$= \frac{243}{2} - \frac{243}{5} = \frac{729}{10} \text{ square units.}$$

Example 3 Find the area under the curve $y = \sin x$, above $y = 0$ from $x = 0$ to $x = \pi$.

Solution The required area, illustrated in Fig. 6.25, is

$$A = \int_0^\pi \sin x\, dx = -\cos x\bigg|_0^\pi$$

$$= -(-1 - (1)) = 2 \text{ square units.}$$

Note that while the definite integral is a pure number, an area is a geometrical quantity implicitly involving units. Where the units of length along the x-axis and y-axis are not specified, areas should be quoted in square units.

Example 4 Find the area of the region R lying above the line $y = 1$ and below the curve $y = 5/(x^2 + 1)$.

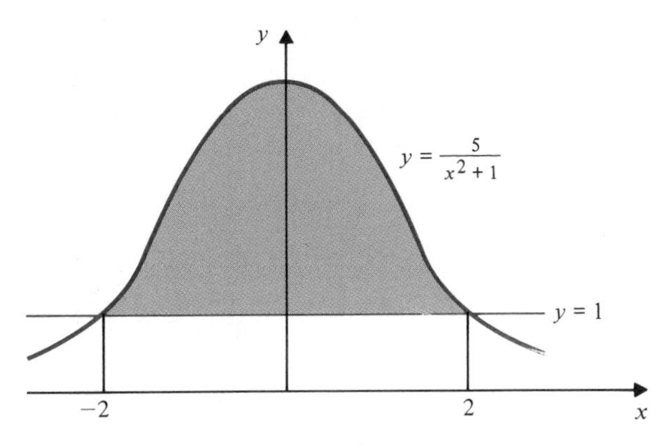

Figure 6.26

Solution The region is shaded in Fig. 6.26. To find the intersections of $y = 1$ and $y = 5/(x^2 + 1)$, we must solve these equations simultaneously:

$$1 = \frac{5}{x^2 + 1}$$

so $x^2 = 4$ and $x = \pm 2$.

The area of the shaded region is

$$A = \int_{-2}^{2} \frac{5}{x^2 + 1}\,dx - 4 = 2\int_{0}^{2} \frac{5}{x^2 + 1}\,dx - 4$$

$$= 10 \text{ Arctan } x \Big|_{0}^{2} - 4 = 10 \text{ Arctan } 2 - 4 \text{ square units.}$$

Example 5 Find the average value of $f(x) = x^4 - 4x^3 - x$ on the interval $[-2, 0]$.

Solution The average value is

$$\frac{1}{0 - (-2)} \int_{-2}^{0} (x^4 - 4x^3 - x)\,dx = \frac{1}{2}\left(\frac{x^5}{5} - x^4 - \frac{x^2}{2}\right)\Bigg|_{-2}^{0}$$

$$= -\frac{1}{2}\left(\frac{-32}{5} - 16 - 2\right) = \frac{61}{5}.$$

Beware of integrals of the form $\int_a^b f(x)\,dx$ where f is not continuous at all points of the interval $[a, b]$. The fundamental theorem does not apply in those cases.

Example 6 We know that $d/dx \ln |x| = 1/x$ if $x \neq 0$. It is *incorrect,* however, to state that

$$\int_{-1}^{1} \frac{dx}{x} = \ln |x| \Big|_{-1}^{1} = 0 - 0 = 0,$$

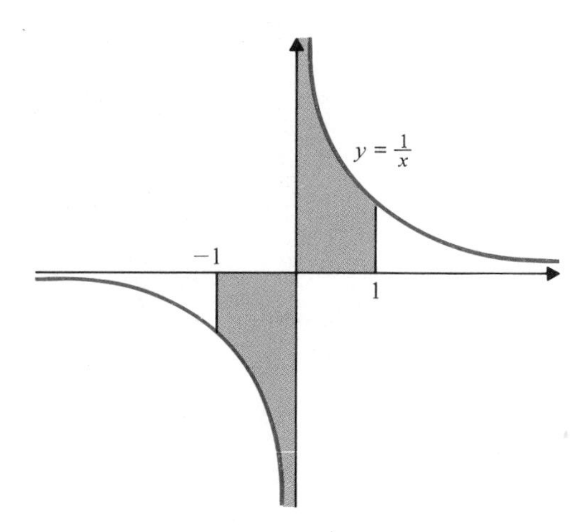

Figure 6.27

even though $1/x$ is an odd function. In fact, $1/x$ is discontinuous and unbounded on $[-1, 1]$, and it is not integrable on $[-1, 0)$ or $(0, 1]$ (Fig. 6.27). Observe that $\int_c^1 (1/x)\,dx = -\ln c$ approaches ∞ as c approaches 0 from the right. Integrals of this type are called *improper integrals*. We deal with them in Section 6.9.

Example 7 Find the derivatives of the following functions.

a) $F(x) = \displaystyle\int_x^3 e^{-t^2}\,dt$

b) $G(x) = \displaystyle\int_{-4}^{5x} e^{-t^2}\,dt$

c) $H(x) = \displaystyle\int_{x^2}^{x^3} e^{-t^2}\,dt$

Solution These are direct applications of the first conclusion of the fundamental theorem, together with the chain rule where appropriate

a) Observe that $F(x) = -\int_3^x e^{-t^2}\,dt$ (by Theorem 2(b) of Section 6.2). Therefore, by the fundamental theorem, $F'(x) = -e^{-x^2}$.

b) If $u = 5x$, then $G(x) = \displaystyle\int_{-4}^{u} e^{-t^2}\,dt$. Thus

$$G'(x) = \frac{dG}{du}\frac{du}{dx} = 5e^{-u^2}$$
$$= 5e^{-25x^2}.$$

c) $H(x) = \displaystyle\int_0^{x^3} e^{-t^2}\,dt - \int_0^{x^2} e^{-t^2}\,dt$. Hence (by two applications of the chain rule) $H'(x) = 3x^2e^{-x^6} - 2xe^{-x^4}$.

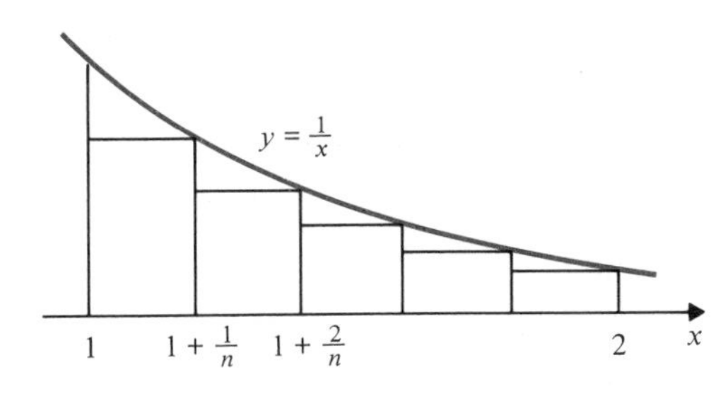

Figure 6.28

Example 8 Evaluate

$$\lim_{n\to\infty} \frac{1}{n}\left(\frac{n}{n+1} + \frac{n}{n+2} + \frac{n}{n+3} + \cdots + \frac{n}{n+n}\right).$$

Solution The sum looks as though it might be a Riemann sum for some function. Indeed, it is just

$$\frac{1}{n}\left(\frac{1}{1+\frac{1}{n}} + \frac{1}{1+\frac{2}{n}} + \frac{1}{1+\frac{3}{n}} + \cdots + \frac{1}{1+\frac{n}{n}}\right),$$

which is the lower sum $L_n(1/x)$ on the interval $[1, 2]$. See Fig. 6.28. Hence the limit of the sum is

$$\int_1^2 \frac{dx}{x} = \ln x \Big|_1^2 = \ln 2 - \ln 1$$
$$= \ln 2.$$

EXERCISES

Evaluate the definite integrals in Exercises 1–14.

1. $\int_0^1 (x^5 + x^3)\, dx$

2. $\int_0^4 \sqrt{x}\, dx$

3. $\int_{-\pi/4}^{-\pi/6} \cos x\, dx$

4. $\int_{-2}^{-1} \left(\frac{1}{x^2} - \frac{1}{x^3}\right) dx$

5. $\int_{-1}^{1} \frac{dx}{\sqrt{4 - x^2}}$

6. $\int_0^{\pi/3} \sec^2 \theta\, d\theta$

7. $\int_0^{\pi/4} \sec t \tan t\, dt$

8. $\int_{-2}^{0} \frac{dx}{4 + x^2}$

9. $\int_4^9 \left(\sqrt{x} - \frac{1}{\sqrt{x}}\right) dx$

10. $\int_{-2}^{2} (x^2 + 3)^2\, dx$

11. $\int_{-2}^{-1} \frac{dt}{t}$

12. $\int_0^3 \frac{2x}{x^2 + 1}\, dx$

13. $\int_{-1}^{1} 2^x \, dx$

14. $\int_{0}^{\pi/2} \sin t \cos t \, dt$

Find the area of the region R specified in Exercises 15–26. Sketch each region.

15. Bounded by $y = x^4$, $y = 0$, $x = 0$, and $x = 1$
16. Bounded by $y = 1/x$, $y = 0$, $x = e$, and $x = e^2$
17. Bounded by $y = x^{1/3} - x^{1/2}$, $y = 0$, $x = 0$, and $x = 1$
18. Bounded by $y = 5 - 2x - 3x^2$, $y = 0$, $x = -1$, and $x = 1$
19. Below $y = 3x - x^2$ and above $y = 0$
20. Below $y = \sqrt{x}$ and above $y = \dfrac{x}{2}$
21. Under $y = e^{-x}$ and above $y = 0$ from $x = -a$ to $x = 0$
22. Below $y = 1 - \cos x$ and above $y = 0$ between two consecutive intersections of these graphs
23. Above $y = x^2$ and to the right of $x = y^2$
24. Above $y = |x|$ and below $y = 12 - x^2$
25. Below $y = x^{-1/3}$ and above $y = 0$ from $x = 1$ to $x = 27$
26. Bounded by $y = x^{-3}$, $y = 0$, $x = -4$, and $x = -2$

In Exercises 27–31 find the average values of the given functions over the intervals specified.

27. $f(x) = e^{3x}$ over $[-2, 2]$
28. $f(x) = 1 + x + x^2 + x^3$ over $[0, 2]$
29. $f(x) = 1 + \sin x$ over $[0, \pi]$ and over $[-\pi, \pi]$
30. $f(x) = \sqrt{1 - x^2}$ over $[-1, 1]$
31. $g(t) = \begin{cases} 0 \text{ if } 0 \le t \le 1 \\ 1 \text{ if } 1 < t \le 3 \end{cases}$ over $[0, 3]$.

Find the indicated derivatives in Exercises 32–39.

32. $\dfrac{d}{dx} \displaystyle\int_{a}^{x} \frac{\sin t}{t} \, dt$

33. $\dfrac{d}{dx} \displaystyle\int_{x^2}^{0} \frac{\sin t}{t} \, dt$

34. $\dfrac{d}{dt} \displaystyle\int_{-\pi}^{t} \frac{\cos y}{1 + y^2} \, dy$

35. $\dfrac{d}{dt} \displaystyle\int_{2t}^{10} \sqrt{1 + s} \, ds$

36. $\dfrac{d}{dx} \displaystyle\int_{ax}^{bx} \frac{dt}{5 + t^4}$

37. $\dfrac{d}{d\theta} \displaystyle\int_{\sin\theta}^{\cos\theta} \frac{1}{1 - x^2} \, dx$

38. $F'(\sqrt{\pi})$ if $F(t) = \displaystyle\int_{0}^{t} \cos(x^2) \, dx$

39. $H'(1)$ if $H(x) = \displaystyle\int_{-x^2}^{x^2} e^{t^4 - 1} \, dt$

40.* Criticize the following erroneous calculation:

$$\int_{-1}^{1} \frac{dx}{x^2} = -\frac{1}{x}\bigg|_{-1}^{1} = -1 + \frac{1}{-1} = -2.$$

Exactly where does the error occur? Why is -2 an unreasonable value for the integral?

41.* Evaluate

$$\lim_{n\to\infty}\frac{1}{n}\left(\left(1+\frac{1}{n}\right)^5+\left(1+\frac{2}{n}\right)^5+\left(1+\frac{3}{n}\right)^5+\cdots+\left(1+\frac{n}{n}\right)^5\right).$$

42.* Evaluate

$$\lim_{n\to\infty}\frac{\pi}{n}\left(\sin\frac{\pi}{n}+\sin\frac{2\pi}{n}+\sin\frac{3\pi}{n}+\cdots+\sin\frac{n\pi}{n}\right).$$

43.* Evaluate

$$\lim_{n\to\infty}\left(\frac{n}{n^2+1}+\frac{n}{n^2+4}+\frac{n}{n^2+9}+\cdots+\frac{n}{2n^2}\right).$$

6.4 THE METHOD OF SUBSTITUTION

As we have seen, the evaluation of definite integrals is most easily carried out if we can antidifferentiate the integrand. In the next four sections we develop some methods for finding antiderivatives, that is, for integrating functions.

Let us begin by assembling some known indefinite integrals. These results have all emerged during our development of differentiation formulas. You should memorize them.

1. $\int (af(x)+bg(x))\,dx = a\int f(x)\,dx + b\int g(x)\,dx$
2. $\int dx = x + C$
3. $\int x\,dx = \frac{1}{2}x^2 + C$
4. $\int x^2\,dx = \frac{1}{3}x^3 + C$
5. $\int \frac{1}{x^2}\,dx = -\frac{1}{x} + C$
6. $\int \frac{1}{\sqrt{x}}\,dx = 2\sqrt{x} + C$
7. $\int x^r\,dx = \frac{1}{r+1}x^{r+1} + C \qquad (r \neq -1)$
8. $\int \frac{1}{x}\,dx = \ln|x| + C$
9. $\int \sin ax\,dx = -\frac{1}{a}\cos ax + C$

(continued)

10. $\displaystyle\int \cos ax\,dx = \frac{1}{a}\sin ax + C$

11. $\displaystyle\int \sec^2 ax\,dx = \frac{1}{a}\tan ax + C$

12. $\displaystyle\int \sec ax \tan ax\,dx = \frac{1}{a}\sec ax + C$

13. $\displaystyle\int \csc^2 ax\,dx = -\frac{1}{a}\cot ax + C$

14. $\displaystyle\int \csc ax \cot ax\,dx = -\frac{1}{a}\csc ax + C$

15. $\displaystyle\int \frac{1}{a^2 + x^2}\,dx = \frac{1}{a}\text{Arctan}\,\frac{x}{a} + C$

16. $\displaystyle\int \frac{1}{\sqrt{a^2 - x^2}}\,dx = \text{Arcsin}\,\frac{x}{a} + C \qquad (a > 0)$

17. $\displaystyle\int e^{ax}\,dx = \frac{1}{a}e^{ax} + C$

18. $\displaystyle\int b^{ax}\,dx = \frac{1}{a \ln b}b^{ax} + C$

19. $\displaystyle\int \cosh ax\,dx = \frac{1}{a}\sinh ax + C$

20. $\displaystyle\int \sinh ax\,dx = \frac{1}{a}\cosh ax + C$

Note that formulas 2–6 are special cases of formula 7, which holds on any interval where x^r makes sense. The linearity formula (formula 1) makes it possible to integrate sums and constant multiples of other functions.

Example 1

a) $\displaystyle\int (x^4 - 3x^3 + 8x^2 - 6x - 7)\,dx = \frac{x^5}{5} - \frac{3x^4}{4} + \frac{8x^3}{3} - 3x^2 - 7x + C$

b) $\displaystyle\int \left(5x^{3/5} - \frac{3}{2 + x^2}\right)dx = \frac{25}{8}x^{8/5} - \frac{3}{\sqrt{2}}\text{Arctan}\,\frac{x}{\sqrt{2}} + C$

c) $\displaystyle\int (4\cos 5x - 5\sin 3x)\,dx = \frac{4}{5}\sin 5x + \frac{5}{3}\cos 3x + C$

d) $\displaystyle\int \left(\frac{1}{\pi x} + a^{\pi x}\right)dx = \frac{1}{\pi}\ln|x| + \frac{1}{\pi \ln a}a^{\pi x} + C \qquad (x \neq 0)$

When an integral cannot be evaluated by inspection, as those in Example 1 can, we require one or more special techniques. The principal of these techniques is the *method of substitution,* the integral version of the chain rule. If we rewrite the chain rule,

$$\frac{d}{dx} f(g(x)) = f'(g(x))g'(x),$$

in integral form, we obtain

$$\int f'(g(x))g'(x)\, dx = f(g(x)) + C.$$

Observe that the following formalism would produce this latter formula even if we did not already know it was true:

Let $u = g(x)$. Then $du/dx = g'(x)$. Rewriting this in differential form, we have $du = g'(x)\, dx$. Thus

$$\int f'(g(x))g'(x)\, dx = \int f'(u)\, du = f(u) + C = f(g(x)) + C.$$

We recall that the equation $du = g'(x)\, dx$ is defined to be equivalent to $du/dx = g'(x)$.

Example 2 a) $$\int \frac{x}{x^2 + 1}\, dx$$ Let $u = x^2 + 1$.

Then $du = 2x\, dx$.

Thus $x\, dx = \frac{1}{2}\, du$.

$$= \frac{1}{2}\int \frac{du}{u} = \frac{1}{2} \ln |u| + C = \frac{1}{2} \ln (x^2 + 1) + C = \ln \sqrt{x^2 + 1} + C.$$

(Both versions of the final answer are equally acceptable.)

b) $$\int \frac{\sin (3 \ln x)}{x}\, dx$$ Let $u = 3 \ln x$.

Thus $du = (3/x)\, dx$.

$$= \frac{1}{3}\int \sin u\, du = -\frac{1}{3} \cos u + C = -\frac{1}{3} \cos (3 \ln x) + C.$$

c) $$\int e^x\sqrt{1 + e^x}\, dx$$ Let $v = 1 + e^x$.

Then $dv = e^x\, dx$.

$$= \int v^{1/2}\, dv = \frac{2}{3} v^{3/2} + C = \frac{2}{3} (1 + e^x)^{3/2} + C.$$

Sometimes the appropriate substitution is not as obvious as it was in (a)–(c), and it may be necessary to play with the integrand a bit to put it into a better form for substitution.

d) $$\int \frac{dx}{x^2 + 4x + 5} = \int \frac{dx}{(x + 2)^2 + 1}$$ Let $t = x + 2$, $dt = dx$.

$$= \int \frac{dt}{t^2 + 1}$$

$$= \text{Arctan } t + C = \text{Arctan } (x + 2) + C.$$

e) $$\int \frac{dx}{\sqrt{e^{2x} - 1}} = \int \frac{dx}{e^x\sqrt{1 - e^{-2x}}}$$

$$= \int \frac{e^{-x}\, dx}{\sqrt{1 - (e^{-x})^2}} \qquad \text{Let } u = e^{-x}, \quad du = -e^{-x}\, dx.$$

$$= -\int \frac{du}{\sqrt{1 - u^2}} = -\text{Arcsin } u + C = -\text{Arcsin } (e^{-x}) + C.$$

The method of substitution cannot be "forced" to work. There is no substitution that will do much good with the integral $\int x(2 + x^7)^{1/5}\, dx$, for instance. However, the integral $\int x^6(2 + x^7)^{1/5} dx$ is quite amenable to the substitution $u = 2 + x^7$. The substitution $u = g(x)$ is more likely to work if $g'(x)$ is a factor of the integrand.

The following theorem justifies the use of the method of substitution in definite integrals.

Theorem 5 Suppose that g is differentiable on $[a, b]$ and satisfies $g(a) = A$ and $g(b) = B$. Suppose that f is continuous on the range of g. Then

$$\int_a^b f(g(x))g'(x)\, dx = \int_A^B f(u)\, du.$$

Proof Let F be an antiderivative of f; $F'(u) = f(u)$. Then

$$\frac{d}{dx} F(g(x)) = F'(g(x))g'(x) = f(g(x))g'(x).$$

Thus

$$\int_a^b f(g(x))g'(x)\, dx = F(g(x))\Big|_a^b = F(g(b)) - F(g(a))$$

$$= F(B) - F(A) = F(u)\Big|_A^B = \int_A^B f(u)\, du. \quad \square$$

Example 3 Evaluate the integral

$$I = \int_0^8 \frac{\cos \sqrt{x + 1}}{\sqrt{x + 1}}\, dx.$$

First Solution Let $u = \sqrt{x + 1}$. Then $du = dx/(2\sqrt{x + 1})$. If $x = 0$, then $u = 1$; if $x = 8$, then $u = 3$. Thus

$$I = 2\int_1^3 \cos u\, du = 2\sin u\Big|_1^3 = 2\sin 3 - 2\sin 1.$$

Second Solution We use the same substitution as in the first solution, but we do not transform the limits of integration from x values to u values. Hence we must return to the variable x before substituting in the limits:

$$I = 2\int_{x=0}^{x=8} \cos u\, du = 2\sin u\Big|_{x=0}^{x=8} = 2\sin\sqrt{x+1}\Big|_0^8 = 2\sin 3 - 2\sin 1.$$

Note that the limits must be written $x = 0$ and $x = 8$ at any stage where the variable showing is not x. It would be wrong to write

$$I = 2\int_0^8 \cos u\, du$$

because this would imply that u goes from 0 to 8 rather than that x goes from 0 to 8. The first solution is shorter than, and therefore preferable to, the second solution. However, in cases where the transformed limits (u-limits) are very complicated to write, one might prefer to use the method of the second solution.

Example 4 Find the area of the region bounded by $y = (2 + \sin(x/2))^2\cos(x/2)$, $y = 0$, $x = 0$, and $x = \pi$.

Solution Because $\cos x/2 \geq 0$ when $0 \leq x \leq \pi$, the area is

$$A = \int_0^{\pi}\left(2 + \sin\frac{x}{2}\right)^2\cos\frac{x}{2}\,dx \qquad \text{Let } v = 2 + \sin\frac{x}{2},$$

$$dv = \frac{1}{2}\cos\frac{x}{2}\,dx.$$

$$= 2\int_2^3 v^2 dv = \frac{2}{3}v^3\Big|_2^3 = \frac{2}{3}(27 - 8) = \frac{38}{3} \text{ square units.}$$

Trigonometric Integrals

The method of substitution is often useful for evaluating trigonometric integrals. We begin by integrating the four trigonometric functions whose integrals we have not yet seen.

$$\int \tan x\, dx = \int \frac{\sin x}{\cos x}\,dx \qquad \begin{aligned} \text{Let } u &= \cos x \\ du &= -\sin x\, dx \end{aligned}$$

$$= -\int \frac{du}{u} = -\ln|u| + C$$

$$= -\ln|\cos x| + C = \ln\left|\frac{1}{\cos x}\right| + C = \ln|\sec x| + C.$$

$$\int \tan x\, dx = \ln |\sec x| + C.$$

Similarly, the substitution $u = \sin x$ leads to

$$\int \cot x\, dx = \int \frac{\cos x}{\sin x}\, dx = \ln |\sin x| + C.$$

Observe that

$$\frac{d}{dx} \ln |\sec x + \tan x| = \frac{\sec x \tan x + \sec^2 x}{\sec x + \tan x} = \sec x,$$

so

$$\int \sec x\, dx = \ln |\sec x + \tan x| + C.$$

This integral should be memorized, because nobody is very likely to guess that the integral can be done with the substitution $u = \sec x + \tan x$, provided that $\sec x$ is multiplied by $(\sec x + \tan x)/(\sec x + \tan x) = 1$ beforehand. Similarly

$$\int \csc x\, dx = -\ln |\csc x + \cot x| + C = \ln |\csc x - \cot x| + C.$$

(Show that the two answers given are equivalent!)

All four integrals given above are frequently encountered and should be committed to memory.

We now consider integrals of the form

$$\int \sin^m x \cos^n x\, dx.$$

If either m or n is an odd, positive integer, the integral can be done easily by substitution. If, say, $n = 2k + 1$ where k is an integer, then we can use the identity $\sin^2 x + \cos^2 x = 1$ to rewrite the integral in the form

$$\int \sin^m x\, (1 - \sin^2 x)^k \cos x\, dx,$$

which can be integrated using the substitution $u = \sin x$. Similarly, $u = \cos x$ can be used if m is an odd integer.

Example 5 a) $\int \sin^3 x \cos^8 x\, dx$

$$= \int (1 - \cos^2 x) \cos^8 x \sin x \, dx \qquad \begin{aligned} \text{Let } u &= \cos x, \\ du &= -\sin x \, dx. \end{aligned}$$

$$= -\int (1 - u^2) u^8 \, du = \int (u^{10} - u^8) \, du$$

$$= \frac{u^{11}}{11} - \frac{u^9}{9} + C = \frac{1}{11} \cos^{11} x - \frac{1}{9} \cos^9 x + C$$

b) $$\int \cos^5 ax \, dx = \int (1 - \sin^2 ax)^2 \cos ax \, dx \qquad \begin{aligned} \text{Let } u &= \sin ax, \\ du &= a \cos ax \, dx. \end{aligned}$$

$$= \frac{1}{a} \int (1 - u^2)^2 \, du = \frac{1}{a} \int (1 - 2u^2 + u^4) \, du$$

$$= \frac{1}{a} \left(u - \frac{2}{3} u^3 + \frac{1}{5} u^5 \right) + C$$

$$= \frac{1}{a} \left(\sin ax - \frac{2}{3} \sin^3 ax + \frac{1}{5} \sin^5 ax \right) + C$$

If the powers of $\sin x$ and $\cos x$ are both even, then we can make use of the trigonometric identities (see Section 3.1):

$$\cos^2 x = \frac{1}{2} (1 + \cos 2x), \qquad \sin^2 x = \frac{1}{2} (1 - \cos 2x).$$

Example 6 a) $$\int \cos^2 x \, dx = \frac{1}{2} \int (1 + \cos 2x) \, dx = \frac{x}{2} + \frac{1}{4} \sin 2x + C$$

$$= \frac{x}{2} + \frac{1}{2} \sin x \cos x + C$$

b) $$\int \sin^2 x \, dx = \frac{1}{2} \int (1 - \cos 2x) \, dx = \frac{x}{2} - \frac{1}{2} \sin x \cos x + C$$

c) $$\int \sin^4 x \, dx = \frac{1}{4} \int (1 - \cos 2x)^2 \, dx = \frac{1}{4} \int (1 - 2 \cos 2x + \cos^2 2x) \, dx$$

$$= \frac{x}{4} - \frac{1}{4} \sin 2x + \frac{1}{8} \int (1 + \cos 4x) \, dx$$

$$= \frac{x}{4} - \frac{1}{4} \sin 2x + \frac{x}{8} + \frac{1}{32} \sin 4x + C$$

$$= \frac{3}{8} x - \frac{1}{4} \sin 2x + \frac{1}{32} \sin 4x + C$$

(Note that there is no point in inserting the constant of integration C until the last integral has been evaluated.)

Using the identities $\sec^2 x = 1 + \tan^2 x$ and $\csc^2 x = 1 + \cot^2 x$ and one of the substitutions $u = \sec x$, $u = \tan x$, $u = \csc x$, or $u = \cot x$, we can evaluate integrals of the form

$$\int \sec^m x \tan^n x \, dx \qquad \text{or} \qquad \int \csc^m x \cot^n x \, dx,$$

unless m is odd and n is even. (If this is the case these integrals can be handled by integration by parts; see Section 6.7.)

Example 7 a) $\displaystyle\int \sec^2 x \tan^2 x \, dx$ Let $u = \tan x$, $du = \sec^2 x \, dx$.

$$= \int u^2 \, du = \frac{u^3}{3} + C = \frac{1}{3}\tan^3 x + C$$

b) $\displaystyle\int \sec^3 x \tan^3 x \, dx$

$$= \int \sec^2 x (\sec^2 x - 1) \sec x \tan x \, dx \qquad \text{Let } u = \sec x, \; du = \sec x \tan x \, dx.$$

$$= \int (u^4 - u^2) \, du = \frac{u^5}{5} - \frac{u^3}{3} + C = \frac{1}{5}\sec^5 x - \frac{1}{3}\sec^3 x + C.$$

EXERCISES

Evaluate the integrals in Exercises 1–52. Remember to include a constant of integration with the indefinite integrals. Your answers may appear different from those provided but still be correct. For example, evaluating $I = \int \sin x \cos x \, dx$ using the substitution $u = \sin x$ leads to the answer $I = 1/2 \sin^2 x + C$; using $u = \cos x$ leads to $I = -(1/2)\cos^2 x + C$; and rewriting $I = 1/2 \int \sin(2x)\, dx$ leads to $I = -(1/4)\cos(2x) + C$. These answers are all actually equivalent up to different choices for the constant of integration: $1/2 \sin^2 x = -(1/2)\cos^2 x + 1/2 = -(1/4)\cos(2x) + 1/4$.

You can always check your own answer to an indefinite integral by differentiating it to get back to the integrand. This is often easier than comparing your answer with the answer in the back of the book. You may find integrals that you can't do, but you should not make mistakes in those you can do, because the answer is so easily checked. (This is a good thing to remember in exams.)

1. $\displaystyle\int e^{5-2x}\, dx$

2. $\displaystyle\int \cos(ax + b)\, dx$

3. $\displaystyle\int \sqrt{3x + 4}\, dx$

4. $\displaystyle\int e^{2x} \sin(e^{2x})\, dx$

5. $\displaystyle\int \frac{x^2}{(x^3 + 2)^{5/2}}\, dx$

6. $\displaystyle\int (x + 2)(x^2 + 4x + 9)^{1/3}\, dx$

7. $\displaystyle\int \frac{x\, dx}{(4x^2 + 1)^5}$

8. $\displaystyle\int \frac{dx}{2x^2 + 3}$

9. $\displaystyle\int \sin x \sin(\cos x)\, dx$

10. $\displaystyle\int \frac{\sin\sqrt{x}}{\sqrt{x}}\, dx$

11. $\displaystyle\int x\, e^{x^2}\, dx$

12. $\displaystyle\int x^2 a^{x^3+1}\, dx \quad (a > 0,\ a \neq 1)$

13. $\displaystyle\int \frac{\cos x}{4 + \sin^2 x}\, dx$

14. $\displaystyle\int \frac{\sec^2 x}{\sqrt{1 - \tan^2 x}}\, dx$

15. $\displaystyle\int \frac{e^x + 1}{e^x - 1}\,dx$

16. $\displaystyle\int \frac{\ln t}{t}\,dt$

17. $\displaystyle\int \frac{ds}{\sqrt{4 - 5s}}$

18. $\displaystyle\int \frac{x + 1}{\sqrt{x^2 + 2x + 3}}\,dx$

19. $\displaystyle\int \frac{t\,dt}{\sqrt{4 - t^4}}$

20. $\displaystyle\int \frac{x^2\,dx}{2 + x^6}$

21. $\displaystyle\int \frac{dx}{e^x + 1}$

22. $\displaystyle\int \frac{dx}{e^x + e^{-x}}$

23. $\displaystyle\int \tan x \ln \cos x\,dx$

24. $\displaystyle\int \frac{x + 1}{\sqrt{1 - x^2}}\,dx$

25. $\displaystyle\int \frac{2t + 3}{t^2 + 9}\,dt$

26. $\displaystyle\int \frac{ax + b}{\sqrt{A^2 - B^2x^2}}\,dx$

27. $\displaystyle\int \frac{dx}{x^2 + 6x + 13}$

28. $\displaystyle\int \frac{dx}{\sqrt{4 + 2x - x^2}}$

29. $\displaystyle\int \sin^3 x \cos^5 x\,dx$

30. $\displaystyle\int \sin^4 t \cos^5 t\,dt$

31. $\displaystyle\int \sin ax \cos^2 ax\,dx$

32. $\displaystyle\int x \sin^2 (x^2) \cos^3 (x^2)\,dx$

33. $\displaystyle\int \sin^2 x \cos^2 x\,dx$

34. $\displaystyle\int \sin^4 x \cos^2 x\,dx$

35. $\displaystyle\int \sin^6 x\,dx$

36. $\displaystyle\int \cos^4 x\,dx$

37. $\displaystyle\int \sec^5 x \tan x\,dx$

38. $\displaystyle\int \sec^6 x \tan^2 x\,dx$

39. $\displaystyle\int \sqrt{\tan x}\,\sec^4 x\,dx$

40. $\displaystyle\int \sin^{-2/3} x \cos^3 x\,dx$

41. $\displaystyle\int \cos x \sin^4 (\sin x)\,dx$

42. $\displaystyle\int \frac{\sin^3 (\ln x) \cos^3 (\ln x)}{x}\,dx$

43. $\displaystyle\int \frac{\sin^2 x}{\cos^4 x}\,dx$

44. $\displaystyle\int \frac{\sin^3 x}{\cos^4 x}\,dx$

45. $\displaystyle\int \csc^5 x \cot^5 x\,dx$

46. $\displaystyle\int \frac{\cos^4 x}{\sin^8 x}\,dx$

47. $\displaystyle\int_0^4 x^3 (x^2 + 1)^{-1/2}\,dx$

48. $\displaystyle\int_1^{\sqrt{e}} \frac{\sin (\pi \ln x)}{x}\,dx$

49. $\displaystyle\int_0^{\pi/2} \sin^4 x\,dx$

50. $\displaystyle\int_{\pi/4}^{\pi} \sin^5 x\,dx$

51. $\displaystyle\int_e^{e^2} \frac{dt}{t \ln t}$

52. $\displaystyle\int_{\pi^2/16}^{\pi^2/9} \frac{2^{\sin\sqrt{x}} \cos \sqrt{x}}{\sqrt{x}}\,dx$

53. Use the identities $\cos 2\theta = 2\cos^2 \theta - 1 = 1 - 2\sin^2 \theta$ and $\sin \theta = \cos\left(\frac{\pi}{2} - \theta\right)$ to help you evaluate the following.

a) $\int_0^{\pi/2} \sqrt{1 + \cos x}\, dx$

b) $\int_0^{\pi/2} \sqrt{1 - \sin x}\, dx$

54. Find the area of the region bounded by $y = x/(x^2 + 16)$, $y = 0$, $x = 0$, and $x = 2$.

55. Find the area of the region bounded by $y = x/(x^4 + 16)$, $y = 0$, $x = 0$, and $x = 2$.

56. Express the area bounded by the ellipse $(x^2/a^2) + (y^2/b^2) = 1$ as a definite integral. Make a substitution that converts this integral into one representing the area of a circle, and hence evaluate it.

57.* **a)** Use the addition formulas for $\sin(x \pm y)$ and $\cos(x \pm y)$ to establish the following identities:

$$\cos x \cos y = \frac{1}{2}(\cos(x - y) + \cos(x + y)),$$

$$\sin x \sin y = \frac{1}{2}(\cos(x - y) - \cos(x + y)),$$

$$\sin x \cos y = \frac{1}{2}(\sin(x + y) + \sin(x - y)).$$

b) Use these results to calculate the following integrals:

$$\int \cos ax \cos bx\, dx, \qquad \int \sin ax \sin bx\, dx, \qquad \int \sin ax \cos bx\, dx$$

c) If m and n are integers, show that:

i) $\int_{-\pi}^{\pi} \cos mx \cos nx\, dx = 0$ if $m \neq n$,

ii) $\int_{-\pi}^{\pi} \sin mx \sin nx\, dx = 0$ if $m \neq n$,

iii) $\int_{-\pi}^{\pi} \sin mx \cos nx\, dx = 0$.

6.5 INVERSE SUBSTITUTIONS

The substitutions considered in the previous section were direct substitutions in the sense that one simplified an integrand by replacing an expression appearing in it with a single variable. In this section we consider the reverse approach; we replace the variable of integration with a function of a new variable. Such substitutions, called *inverse substitutions,* would appear on the surface to make the integral more complicated. That is, substituting $x = g(u)$ in the integral

$$\int_a^b f(x)\, dx$$

leads to the more "complicated" integral

$$\int_{x=a}^{x=b} f(g(u))g'(u)\,du = \int_{g^{-1}(a)}^{g^{-1}(b)} f(g(u))g'(u)\,du.$$

As we will see, however, sometimes such substitutions can actually simplify an integrand, transforming the integral into one that can be evaluated by inspection or to which other techniques can readily be applied.

The Inverse Trigonometric Substitutions

Three very useful inverse substitutions are

$$x = a \sin\theta, \qquad x = a\tan\theta, \qquad x = a\sec\theta.$$

These correspond to the direct substitutions

$$\theta = \text{Arcsin}\,\frac{x}{a}, \qquad \theta = \text{Arctan}\,\frac{x}{a}, \qquad \theta = \text{Arcsec}\,\frac{x}{a} = \text{Arccos}\,\frac{a}{x}.$$

> Integrals involving $\sqrt{a^2 - x^2}$ $(a > 0)$ can frequently be reduced to a simpler form by means of the substitution
>
> $$\theta = \text{Arcsin}\,x/a \text{ or, equivalently, } x = a\sin\theta.$$

Observe that $\sqrt{a^2 - x^2}$ makes sense only if $-a \le x \le a$, in which case we have $-\pi/2 \le \theta \le \pi/2$. Since $\cos\theta \ge 0$ for such θ, we have

$$\sqrt{a^2 - x^2} = \sqrt{a^2(1 - \sin^2\theta)} = \sqrt{a^2\cos^2\theta} = a\cos\theta.$$

The other trigonometric functions of θ can be recovered in terms of x by examining a right-angled triangle labeled to correspond to the substitution.

Example 1 Refer to Fig. 6.29.

$$\int \frac{dx}{(5 - x^2)^{3/2}} \qquad \text{Let } x = \sqrt{5}\sin\theta, \quad dx = \sqrt{5}\cos\theta\,d\theta.$$

$$= \int \frac{\sqrt{5}\cos\theta\,d\theta}{5^{3/2}\cos^3\theta}$$

$$= \frac{1}{5}\int \sec^2\theta\,d\theta = \frac{1}{5}\tan\theta + C = \frac{1}{5}\frac{x}{\sqrt{5 - x^2}} + C$$

Example 2 Find the area of the circular sector shaded in Fig. 6.30.

Solution The area is

$$A = 2\int_b^a \sqrt{a^2 - x^2}\,dx \qquad \text{Let } x = a\sin\theta, \quad dx = a\cos\theta\,d\theta.$$

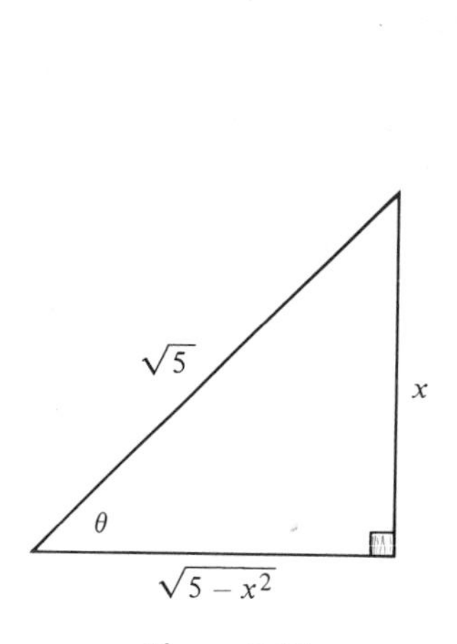

Figure 6.29

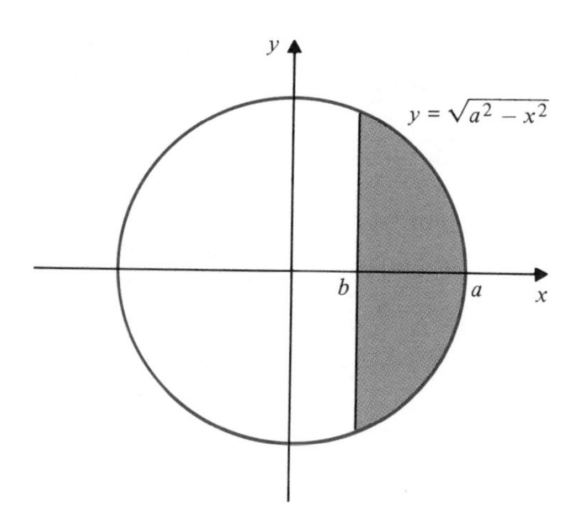

Figure 6.30

$$= 2\int_{x=b}^{x=a} a\cos\theta\, a\cos\theta\, d\theta$$

$$= a^2(\theta + \sin\theta\cos\theta)\Bigg|_{x=b}^{x=a} \qquad \text{(as in Example 6 of Section 6.4)}$$

$$= a^2\left(\text{Arcsin}\,\frac{x}{a} + \frac{x\sqrt{a^2 - x^2}}{a^2}\right)\Bigg|_b^a \qquad \text{(See Fig. 6.31)}$$

$$= \frac{\pi}{2}a^2 - a^2\,\text{Arcsin}\,\frac{b}{a} - b\sqrt{a^2 - b^2} \text{ square units.}$$

Integrals involving $\sqrt{a^2 + x^2}$ or $1/(x^2 + a^2)$ $(a > 0)$ are often simplified by the substitution

$$\theta = \text{Arctan}\,\frac{x}{a} \text{ or, equivalently, } x = a\tan\theta.$$

Again we have $-\pi/2 < \theta < \pi/2$, so $\sec\theta > 0$ and

$$\sqrt{a^2 + x^2} = a\sqrt{1 + \tan^2\theta} = a\sec\theta.$$

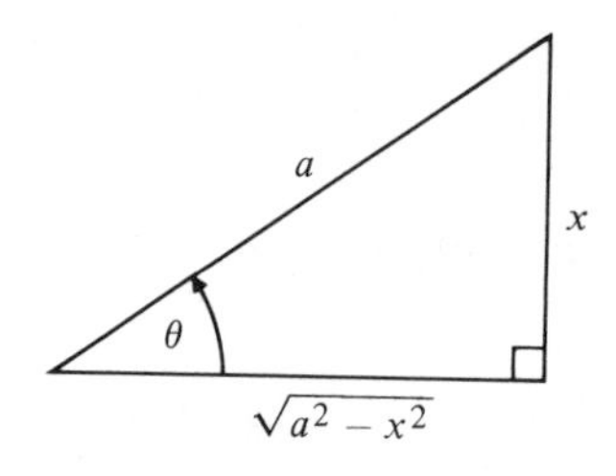

Figure 6.31

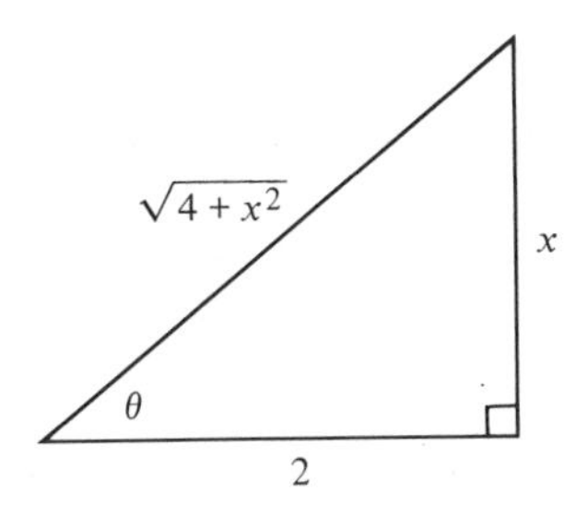

Figure 6.32

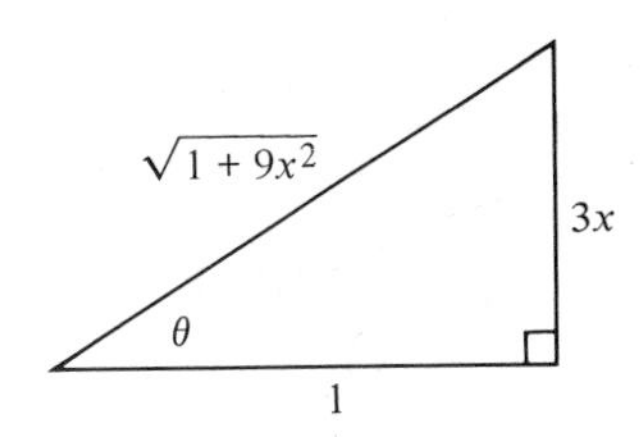

Figure 6.33

Example 3 Figures 6.32 and 6.33 illustrate parts (a) and (b), respectively.

a) $\displaystyle\int \frac{dx}{\sqrt{4+x^2}}$ Let $x = 2\tan\theta$, $dx = 2\sec^2\theta\, d\theta$.

$$= \int \frac{2\sec^2\theta\, d\theta}{2\sec\theta} = \int \sec\theta\, d\theta = \ln|\sec\theta + \tan\theta| + C$$

$$= \ln\left|\frac{\sqrt{4+x^2}}{2} + \frac{x}{2}\right| + C = \ln(\sqrt{4+x^2} + x) + C_1,$$

where $C_1 = C - \ln 2$. (Note that $\sqrt{4+x^2} + x > 0$ for all x, so we do not need an absolute value on it.)

b) $\displaystyle\int \frac{dx}{(1+9x^2)^2}$ Let $3x = \tan\theta$, $3\,dx = \sec^2\theta\, d\theta$, $1 + 9x^2 = \sec^2\theta$.

$$= \frac{1}{3}\int \frac{\sec^2\theta\, d\theta}{\sec^4\theta} = \frac{1}{3}\int \cos^2\theta\, d\theta = \frac{1}{6}\int (1 + \cos 2\theta)\, d\theta$$

$$= \frac{\theta}{6} + \frac{\sin 2\theta}{12} + C = \frac{\theta}{6} + \frac{1}{6}\sin\theta\cos\theta + C$$

$$= \frac{1}{6}\text{Arctan}(3x) + \frac{1}{6}\frac{3x}{\sqrt{1+9x^2}}\frac{1}{\sqrt{1+9x^2}} + C$$

$$= \frac{1}{6}\text{Arctan}(3x) + \frac{1}{2}\frac{x}{1+9x^2} + C$$

Integrals involving $\sqrt{x^2 - a^2}$ $(a > 0)$ can frequently be simplified by using the substitution

$$\theta = \text{Arcsec}\,\frac{x}{a} \quad \text{or, equivalently,} \quad x = a\sec\theta.$$

One must be more careful with this substitution. Although

$$\sqrt{x^2 - a^2} = a\sqrt{\sec^2\theta - 1} = a\sqrt{\tan^2\theta} = a\,|\tan\theta|,$$

it is not always true that $\tan\theta \geq 0$ on $[0, \pi/2) \cup (\pi/2, \pi]$, the range of the function Arcsec. Note that $\sqrt{x^2 - a^2}$ makes sense only if $x \geq a$ or $x \leq -a$. If $x \geq a$, then $0 \leq \theta < \pi/2$ and $\sqrt{x^2 - a^2} = a\tan\theta$. If $x \leq -a$, then $\pi/2 < \theta \leq \pi$ and $\sqrt{x^2 - a^2} = -a\tan\theta$.

Example 4 Find

$$I = \int \frac{dx}{\sqrt{x^2 - a^2}} \quad (a > 0).$$

Solution For the moment assume that $x \geq a$. If $x = a\sec\theta$, then $dx = a\sec\theta\tan\theta\, d\theta$ and $\sqrt{x^2 - a^2} = a\tan\theta$ (Fig. 6.34). Thus

$$I = \int \sec\theta\, d\theta = \ln|\sec\theta + \tan\theta| + C$$

$$= \ln\left|\frac{x}{a} + \frac{\sqrt{x^2 - a^2}}{a}\right| + C = \ln|x + \sqrt{x^2 - a^2}| + C_1,$$

where $C_1 = C - \ln a$. If $x \leq -a$, let $u = -x$ so that $u \geq a$. We have

$$I = -\int \frac{du}{\sqrt{u^2 - a^2}} = -\ln|u + \sqrt{u^2 - a^2}| + C_1$$

$$= \ln\left|\frac{1}{-x + \sqrt{x^2 - a^2}}\frac{x + \sqrt{x^2 - a^2}}{x + \sqrt{x^2 - a^2}}\right| + C_1$$

$$= \ln\left|\frac{x + \sqrt{x^2 - a^2}}{-a^2}\right| + C_1 = \ln|x + \sqrt{x^2 - a^2}| + C_2,$$

where $C_2 = C_1 - 2\ln a$. Thus in either case we have

$$I = \ln|x + \sqrt{x^2 - a^2}| + C.$$

Completing the Square

Quadratic expressions of the form $Ax^2 + Bx + C$ are often found in integrands. These can be written as sums or differences of squares using the procedure of completing the square. First factor out A so that the remaining expression begins with $x^2 + 2bx$, where $2b = B/A$. These are the first two terms of $(x + b)^2$.

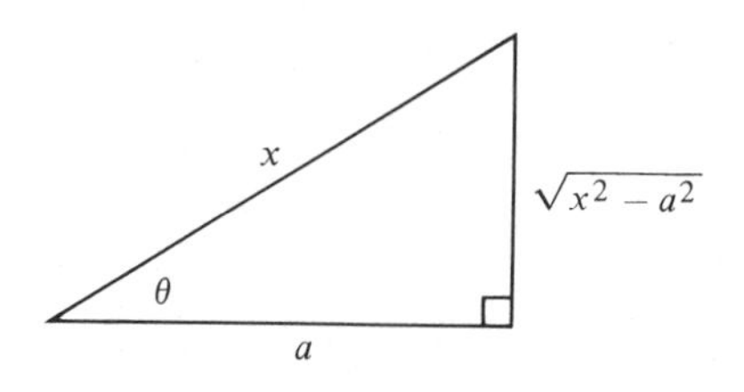

Figure 6.34

$$
\begin{aligned}
Ax^2 + Bx + C &= A\left(x^2 + \frac{B}{A}x + \frac{C}{A}\right) \\
&= A\left(x^2 + \frac{B}{A}x + \frac{B^2}{4A^2} + \frac{C}{A} - \frac{B^2}{4A^2}\right) \\
&= A\left(x + \frac{B}{2A}\right)^2 + \frac{4AC - B^2}{4A}
\end{aligned}
$$

The substitution $u = x + \dfrac{B}{2A}$ should then be made

Example 5 a)

$$
\begin{aligned}
\int \frac{dx}{\sqrt{2x - x^2}} &= \int \frac{dx}{\sqrt{1 - (1 - 2x + x^2)}} \\
&= \int \frac{dx}{\sqrt{1 - (x - 1)^2}} \qquad \text{Let } u = x - 1,\ du = dx. \\
&= \int \frac{du}{\sqrt{1 - u^2}} = \text{Arcsin}\, u + C = \text{Arcsin}\,(x - 1) + C
\end{aligned}
$$

b)

$$
\begin{aligned}
\int \frac{x\,dx}{4x^2 + 12x + 13} &= \int \frac{x\,dx}{4\left(x^2 + 3x + \frac{9}{4} + 1\right)} \\
&= \frac{1}{4}\int \frac{x\,dx}{\left(x + \frac{3}{2}\right)^2 + 1} \qquad \text{Let } u = x + (3/2),\ du = dx,\ x = (2u - 3)/2. \\
&= \frac{1}{4}\int \frac{u\,du}{u^2 + 1} - \frac{3}{8}\int \frac{du}{u^2 + 1} \qquad \text{In the first integral let } v = u^2 + 1,\ dv = 2u\,du. \\
&= \frac{1}{8}\int \frac{dv}{v} - \frac{3}{8}\text{Arctan}\, u \\
&= \frac{1}{8}\ln|v| - \frac{3}{8}\text{Arctan}\, u + C \\
&= \frac{1}{8}\ln(4x^2 + 12x + 13) - \frac{3}{8}\text{Arctan}\left(x + \frac{3}{2}\right) + C_1,
\end{aligned}
$$

where $C_1 = C - (\ln 4)/8$.

Other Substitutions

Integrals involving $\sqrt{ax + b}$ can sometimes be made simpler with the substitution $ax + b = u^2$.

Example 6

$$
\int \frac{dx}{1 + \sqrt{2x}} \qquad \text{Let } 2x = u^2,\ 2\,dx = 2u\,du.
$$

$$= \int \frac{u\,du}{1+u} = \int \frac{u+1-1}{1+u}\,du = \int \left(1 - \frac{1}{1+u}\right) du \qquad \begin{array}{l} \text{Let } v = 1+u, \\ dv = du. \end{array}$$

$$= u - \int \frac{dv}{v} = u - \ln|v| + C$$

$$= \sqrt{2x} - \ln(1+\sqrt{2x}) + C$$

If more than one fractional power is present it may be possible to eliminate all of them at once.

Example 7

$$\int \frac{dx}{x^{1/2}(1+x^{1/3})} \qquad \begin{array}{l} \text{Let } x = u^6, \\ dx = 6\,u^5\,du. \end{array}$$

$$= 6\int \frac{u^5\,du}{u^3(1+u^2)} = 6\int \frac{u^2}{1+u^2}\,du = 6\int \left(1 - \frac{1}{1+u^2}\right) du$$

$$= 6(u - \operatorname{Arctan} u) + C = 6(x^{1/6} - \operatorname{Arctan} x^{1/6}) + C$$

Finally, we mention a certain substitution that can transform an integral whose integrand is a rational function of $\sin\theta$ and $\cos\theta$ (that is, a quotient of polynomials in $\sin\theta$ and $\cos\theta$) into a rational function of x. The substitution is

$$x = \tan\frac{\theta}{2}.$$

Observe that

$$\cos^2\frac{\theta}{2} = \frac{1}{\sec^2\frac{\theta}{2}} = \frac{1}{1+\tan^2\frac{\theta}{2}} = \frac{1}{1+x^2},$$

so

$$\cos\theta = 2\cos^2\frac{\theta}{2} - 1 = \frac{2}{1+x^2} - 1 = \frac{1-x^2}{1+x^2}$$

$$\sin\theta = 2\sin\frac{\theta}{2}\cos\frac{\theta}{2} = 2\tan\frac{\theta}{2}\cos^2\frac{\theta}{2} = \frac{2x}{1+x^2}.$$

Also, $dx = \frac{1}{2}\sec^2\frac{\theta}{2}\,d\theta$, so

$$d\theta = 2\cos^2\frac{\theta}{2}\,dx = \frac{2\,dx}{1+x^2}.$$

Note that $\cos\theta$, $\sin\theta$, and $d\theta$ all involve only rational functions of x.

Example 8

$$\int \frac{d\theta}{2+\cos\theta} \qquad \text{Let } x = \tan(\theta/2) \text{, so } \cos\theta = (1-x^2)/(1+x^2),\ d\theta = (2\,dx)/(1+x^2).$$

$$= \int \frac{\dfrac{2\,dx}{1+x^2}}{2+\dfrac{1-x^2}{1+x^2}} = 2\int \frac{dx}{3+x^2} = \frac{2}{\sqrt{3}}\operatorname{Arctan}\frac{x}{\sqrt{3}} + C$$

$$= \frac{2}{\sqrt{3}}\operatorname{Arctan}\left(\frac{1}{\sqrt{3}}\tan\frac{\theta}{2}\right) + C.$$

EXERCISES

Evaluate the integrals in Exercises 1–38.

1. $\displaystyle\int \frac{dx}{\sqrt{1-4x^2}}$

2. $\displaystyle\int \frac{x^2\,dx}{\sqrt{1-4x^2}}$

3. $\displaystyle\int \frac{x\,dx}{\sqrt{1-4x^2}}$

4. $\displaystyle\int \frac{dx}{\sqrt{3+2x^2}}$

5. $\displaystyle\int \frac{x+1}{\sqrt{9-x^2}}\,dx$

6. $\displaystyle\int x^3\sqrt{9-x^2}\,dx$

7. $\displaystyle\int \frac{x^2}{\sqrt{9-x^2}}\,dx$

8. $\displaystyle\int \frac{dx}{x\sqrt{9-x^2}}$

9. $\displaystyle\int \frac{dx}{(a^2-x^2)^{3/2}}$

10. $\displaystyle\int \frac{x\,dx}{(a^2-x^2)^{3/2}}$

11. $\displaystyle\int \frac{dx}{(a^2+x^2)^{3/2}}$

12. $\displaystyle\int \frac{x^3\,dx}{(a^2+x^2)^{3/2}}$

13. $\displaystyle\int \frac{\sqrt{1-x^2}}{x^2}\,dx$

14. $\displaystyle\int \frac{dx}{(a^2+b^2x^2)^{3/2}}$

15. $\displaystyle\int \frac{dx}{4-x^2}$

16. $\displaystyle\int \frac{dx}{a^2-b^2x^2}$

17. $\displaystyle\int \frac{dx}{x\sqrt{x^2-a^2}}$

18. $\displaystyle\int \frac{dx}{x^2\sqrt{x^2-a^2}}$

19. $\displaystyle\int \frac{dx}{x^2+2x+10}$

20. $\displaystyle\int \frac{dx}{x^2+x+1}$

21. $\displaystyle\int \frac{dx}{(4x^2+4x+5)^2}$

22. $\displaystyle\int \frac{x\,dx}{x^2-2x+3}$

23. $\displaystyle\int \frac{x\,dx}{\sqrt{2ax-x^2}}$

24. $\displaystyle\int \frac{dx}{(4x-x^2)^{3/2}}$

25. $\displaystyle\int \frac{dx}{(3-2x-x^2)^{3/2}}$

26. $\displaystyle\int \frac{x\,dx}{(4x^2+12x+13)^{5/2}}$

27. $\displaystyle\int \frac{dx}{(x^2+1)^3}$

28. $\displaystyle\int \frac{dx}{(x^2+2x+2)^2}$

29. $\displaystyle\int \frac{dx}{2 + \sqrt{x}}$

30. $\displaystyle\int \frac{dx}{1 + x^{1/3}}$

31. $\displaystyle\int \frac{1 + x^{1/2}}{1 + x^{1/3}}\, dx$

32. $\displaystyle\int \frac{x\sqrt{2 - x^2}}{\sqrt{x^2 + 1}}\, dx$

33. $\displaystyle\int_{-\ln 2}^{0} e^x \sqrt{1 - e^{2x}}\, dx$

34. $\displaystyle\int_0^{\pi/2} \frac{\cos x}{1 + \sin^2 x}\, dx$

35. $\displaystyle\int_{-1}^{\sqrt{3}-1} \frac{dx}{x^2 + 2x + 2}$

36. $\displaystyle\int_1^3 \frac{dx}{\sqrt{4x - x^2}}$

37. $\displaystyle\int_1^2 \frac{dx}{x^2\sqrt{9 - x^2}}$

38. $\displaystyle\int_1^2 \frac{dx}{x^2\sqrt{9 + x^2}}$

In Exercises 39–41 evaluate the integral using the special substitution $x = \tan \dfrac{\theta}{2}$ as in Example 8.

39.* $\displaystyle\int \frac{d\theta}{2 + \sin\theta}$

40.* $\displaystyle\int_0^{\pi/2} \frac{d\theta}{1 + \cos\theta + \sin\theta}$

41.* $\displaystyle\int \frac{d\theta}{3 + 2\cos\theta}$

42. Find the area of the region bounded by $y = (2x - x^2)^{-1/2}$, $y = 0$, $x = 1/2$, and $x = 1$.

43. Find the area of the region lying below $y = 9/(x^4 + 4x^2 + 4)$ and above $y = 1$.

44. Find the average value of the function $f(x) = (x^2 - 4x + 8)^{-3/2}$ over the interval $[0, 4]$.

45.* Evaluate the integrals

$$\int \frac{dx}{\sqrt{x^2 - a^2}} \quad \text{and} \quad \int \frac{dx}{x^2\sqrt{x^2 - a^2}},$$

using the substitution $x = a \cosh u$. (*Hint:* Review the properties of the hyperbolic functions in Section 3.8.) This substitution is an alternative to $x = a \sec\theta$ when dealing with $\sqrt{x^2 - a^2}$.

6.6 THE METHOD OF PARTIAL FRACTIONS

In this section we are concerned with techniques for integrating rational functions, that is, for evaluating

$$\int \frac{P(x)}{Q(x)}\, dx,$$

where P and Q are polynomials. We need normally concern ourselves only with functions $P(x)/Q(x)$ where the *degree* of P is less than that of Q. (Recall that $P(x)$ is a polynomial of degree n if

$$P(x) = a_0 + a_1x + a_2x^2 + \cdots + a_nx^n$$

where $a_0, a_1, a_2, \ldots, a_n$ are constants and $a_n \neq 0$.) If the degree of P equals or exceeds the degree of Q then we can use long division or some equivalent procedure to express the fraction $P(x)/Q(x)$ as a polynomial quotient plus another fraction $R(x)/Q(x)$ where R, the remainder in the division, has degree less than that of Q.

Example 1 Evaluate $I = \int \frac{x^3 + 3x^2}{x^2 + 1}\,dx$.

Solution We use long division:

$$\begin{array}{r l} & x + 3 \\ x^2 + 1 \,\big) & x^3 + 3x^2 \\ & \underline{x^3 \qquad\quad + x} \\ & \qquad 3x^2 - x \\ & \qquad \underline{3x^2 \qquad + 3} \\ & \qquad\quad - x - 3 \end{array}$$

$$\frac{x^3 + 3x^2}{x^2 + 1} = x + 3 - \frac{x + 3}{x^2 + 1}.$$

Thus

$$\begin{aligned} I &= \int (x + 3)\,dx - \int \frac{x}{x^2 + 1}\,dx - 3\int \frac{dx}{x^2 + 1} \\ &= \frac{1}{2}x^2 + 3x - \frac{1}{2}\ln(x^2 + 1) - 3\,\text{Arctan}\,x + C. \end{aligned}$$

Example 2 Evaluate $I = \int \frac{x}{2x - 1}\,dx$.

Solution In this case the "long division" can be carried out by manipulation of the integrand:

$$\begin{aligned} I = \frac{1}{2}\int \frac{2x - 1 + 1}{2x - 1}\,dx &= \frac{1}{2}\int \left(1 + \frac{1}{2x - 1}\right) dx \\ &= \frac{1}{2}x + \frac{1}{4}\ln|2x - 1| + C. \end{aligned}$$

In the discussion that follows we always assume that the given rational function $R(x) = P(x)/Q(x)$ satisfies

$$\boxed{\text{degree of } P < \text{degree of } Q.}$$

The complexity of the problem of integrating $R(x)$ therefore depends on the degree of Q.

Suppose that $Q(x)$ has degree 1. Then $R(x) = c/(ax + b)$. The substitution $u = ax + b$ leads to

$$\int \frac{dx}{ax + b} = \frac{1}{a}\int \frac{du}{u} = \frac{1}{a}\ln|u| + C,$$

that is,

$$\int \frac{dx}{ax + b} = \frac{1}{a} \ln |ax + b| + C.$$

Now suppose that $Q(x)$ is quadratic, that is, has degree 2. We can assume for purposes of this discussion that $Q(x)$ is either of the form $x^2 + a^2$ or of the form $x^2 - a^2$, since completing the square and making the appropriate change of variable can always reduce a quadratic denominator to this form, as shown in the previous section. Since $P(x)$ can be at most a linear function, $P(x) = Ax + B$, we are led to consider the following four integrals:

$$\int \frac{x\,dx}{x^2 + a^2}, \quad \int \frac{x\,dx}{x^2 - a^2}, \quad \int \frac{dx}{x^2 + a^2}, \quad \text{and} \quad \int \frac{dx}{x^2 - a^2}.$$

(There are only two simpler forms if $a = 0$.) The first two integrals yield to the substitution $u = x^2 \pm a^2$; the third is a known integral:

$$\int \frac{x}{x^2 + a^2}\,dx = \frac{1}{2} \ln (x^2 + a^2) + C,$$

$$\int \frac{x}{x^2 - a^2}\,dx = \frac{1}{2} \ln |x^2 - a^2| + C,$$

$$\int \frac{1}{x^2 + a^2}\,dx = \frac{1}{a} \text{Arctan}\, \frac{x}{a} + C.$$

The fourth integral could be evaluated (for $|x| \le |a|$) using the substitution $x = a \sin \theta$. An alternative method is to write the integrand as a sum of two fractions with linear denominators:

$$\frac{1}{x^2 - a^2} = \frac{1}{(x - a)(x + a)} = \frac{A}{x - a} + \frac{B}{x + a} = \frac{Ax + Aa + Bx - Ba}{x^2 - a^2},$$

where we have added the two fractions together again in the last step. If this equation is to hold identically for all x (except $x = \pm a$), then the numerators on the left- and right-hand sides must be identical as polynomials in x. That is, $(A + B)x + (Aa - Ba) = 1 = 0x + 1$. Hence

$$A + B = 0 \quad \text{(the coefficient of } x\text{)}$$

$$Aa - Ba = 1 \quad \text{(the constant term)}$$

Solving this pair of linear equations for the unknowns A and B, we get $A = 1/(2a)$ and $B = -1/(2a)$. Therefore,

$$\int \frac{dx}{x^2 - a^2} = \frac{1}{2a} \int \frac{dx}{x - a} - \frac{1}{2a} \int \frac{dx}{x + a}$$

$$= \frac{1}{2a} \ln |x - a| - \frac{1}{2a} \ln |x + a| + C$$

$$= \frac{1}{2a} \ln \left| \frac{x - a}{x + a} \right| + C.$$

The technique used above, involving the writing of a complicated fraction as a sum of simpler fractions, is called the **method of partial fractions.** Suppose that a polynomial $Q(x)$ factors into a product of n *distinct* linear (degree 1) factors, say

$$Q(x) = a(x - x_1)(x - x_2) \cdots (x - x_n),$$

where $x_i \neq x_j$ if $i \neq j$, $1 \leq i, j \leq n$. If $P(x)$ is a polynomial of degree smaller than that of $Q(x)$, then $P(x)/Q(x)$ has a **partial fraction expansion** of the form

$$\frac{P(x)}{Q)x)} = \frac{A_1}{x - x_1} + \frac{A_2}{x - x_2} + \cdots + \frac{A_n}{x - x_n}$$

for certain values of the constants $A_1, A_2, \ldots, A_n$. We do not attempt to give any formal proof of this assertion here, it lies more properly in the domain of polynomial algebra. See Theorem 6 at the end of this section for a statement of a more general result.

Given that $P(x)/Q(x)$ has a partial fraction expansion as claimed above, there are two methods for determining the constants $A_1, A_2, \ldots, A_n$. The first of these methods, and the one that generalizes most easily to the more complicated expansions considered below, is to add up the fractions in the expansion, obtaining a new fraction $S(x)/Q(x)$ with numerator $S(x)$ a polynomial of degree one less than that of $Q(x)$. This new fraction will be identical to the original fraction $P(x)/Q(x)$ if S and P are identical polynomials. The constants $A_1, A_2, \ldots, A_n$ are determined by solving the n linear equations resulting from equating the coefficients of like powers of x in these two polynomials.

The second method involves the following observation: If we multiply the partial fraction expansion by $x - x_1$, we get

$$(x - x_1)\frac{P(x)}{Q(x)} = A_1 + A_2\frac{x - x_1}{x - x_2} + \cdots + A_n\frac{x - x_1}{x - x_n}.$$

All the terms on the right-hand side approach 0 as x approaches x_1 except the first term, A_1. Hence

$$A_1 = \lim_{x \to x_1} (x - x_1)\frac{P(x)}{Q(x)}.$$

Similarly,

$$A_j = \lim_{x \to x_j} (x - x_j) \frac{P(x)}{Q(x)} \qquad (1 \le j \le n).$$

Example 3 Evaluate

$$I = \int \frac{(x + 4)\, dx}{x^2 - 5x + 6}.$$

Solution

$$\frac{x + 4}{x^2 - 5x + 6} = \frac{x + 4}{(x - 2)(x - 3)} = \frac{A}{x - 2} + \frac{B}{x - 3}$$

We calculate A and B by both methods suggested above.

Method 1

Add the fractions to get

$$\frac{Ax - 3A + Bx - 2B}{(x - 2)(x - 3)}.$$

Thus

$$\begin{aligned} A + \ \ B &= 1, \\ -3A - 2B &= 4, \end{aligned}$$

and so $A = -6$, $B = 7$.

Method 2

$$A = \lim_{x \to 2} (x - 2) \frac{x + 4}{(x - 2)(x - 3)}$$

$$= \lim_{x \to 2} \frac{x + 4}{x - 3} = \frac{6}{-1} = -6$$

$$B = \lim_{x \to 3} (x - 3) \frac{x + 4}{(x - 2)(x - 3)}$$

$$= \lim_{x \to 3} \frac{x + 4}{x - 2} = 7$$

In either case we have

$$I = -6 \int \frac{dx}{x - 2} + 7 \int \frac{dx}{x - 3} = -6 \ln |x - 2| + 7 \ln |x - 3| + C.$$

Example 4 Evaluate

$$I = \int \frac{x^3 + 2}{x^3 - x}\, dx.$$

Solution Since the numerator does not have degree smaller than the denominator, we must write

$$I = \int \frac{x^3 - x + x + 2}{x^3 - x}\, dx = \int \left(1 + \frac{x + 2}{x^3 - x}\right) dx$$

$$= x + \int \frac{x + 2}{x^3 - x}\, dx.$$

Now we can use the method of partial fractions.

$$\frac{x + 2}{x^3 - x} = \frac{x + 2}{x(x - 1)(x + 1)} = \frac{A}{x} + \frac{B}{x - 1} + \frac{C}{x + 1}$$

$$= \frac{A(x^2 - 1) + B(x^2 + x) + C(x^2 - x)}{x(x - 1)(x + 1)}$$

We have

$$\begin{aligned} A + B + C &= 0 \qquad \text{(coefficient of } x^2\text{)} \\ B - C &= 1 \qquad \text{(coefficient of } x\text{)} \\ -A \qquad &= 2 \qquad \text{(constant term).} \end{aligned}$$

It follows that $A = -2$, $B = 3/2$, and $C = 1/2$. (We could also have found these by using the second method of Example 3: $A = \lim_{x\to 0} x(x + 2)/(x^3 - x)$ and so on.) Finally, we have

$$\begin{aligned} I &= x - 2\int \frac{dx}{x} + \frac{3}{2}\int \frac{dx}{x - 1} + \frac{1}{2}\int \frac{dx}{x + 1} \\ &= x - 2 \ln |x| + \frac{3}{2} \ln |x - 1| + \frac{1}{2} \ln |x + 1| + C \\ &= x - \ln \frac{\sqrt{|x - 1|^3 |x + 1|}}{x^2} + C. \end{aligned}$$

Either of the last two lines provides the correct answer. The second form is really no better than the first.

Next we consider a rational function whose denominator has a quadratic factor that is equivalent to a sum of squares and that cannot therefore be further factored into a product of real linear factors.

Example 5 Evaluate

$$I = \int \frac{2 + 3x + x^2}{x(x^2 + 1)}\, dx.$$

Solution Note that the numerator has degree 2 and the denominator degree 3, so no long division is necessary. If we expand the integrand as a sum of two simpler fractions, we want one with denominator x and one with denominator $x^2 + 1$. The appropriate form of the expansion turns out to be

$$\frac{2 + 3x + x^2}{x(x^2 + 1)} = \frac{A}{x} + \frac{Bx + C}{x^2 + 1} = \frac{A(x^2 + 1) + Bx^2 + Cx}{x(x^2 + 1)}.$$

(Note that corresponding to the quadratic denominator we use a linear numerator.) Equating coefficients in the two numerators, we obtain

$$\begin{aligned} A + B \qquad &= 1 \qquad \text{(coefficient of } x^2\text{)} \\ C &= 3 \qquad \text{(coefficient of } x\text{)} \\ A \qquad &= 2 \qquad \text{(constant term).} \end{aligned}$$

Hence $A = 2$, $B = -1$, and $C = 3$. We have, therefore,

$$\begin{aligned} I &= 2\int \frac{dx}{x} - \int \frac{x\,dx}{x^2 + 1} + 3\int \frac{dx}{x^2 + 1} \\ &= 2 \ln |x| - \frac{1}{2} \ln (x^2 + 1) + 3 \operatorname{Arctan} x + C. \end{aligned}$$

We remark that addition of the fractions is the only reasonable method for determining the constants A, B, and C here. We could determine A by a limit procedure such as in Example 3 above, but there is no simple equivalent way of finding B or C.

Example 6 Evaluate

$$I = \int \frac{dx}{x^3 + 1}.$$

Solution

$$\frac{1}{x^3+1} = \frac{1}{(x+1)(x^2-x+1)} = \frac{A}{x+1} + \frac{Bx+C}{x^2-x+1}$$
$$= \frac{A(x^2-x+1) + B(x^2+x) + C(x+1)}{(x+1)(x^2-x+1)}$$

$$\begin{aligned} A + B \quad &= 0 \qquad \text{(coefficient of } x^2) \\ -A + B + C &= 0 \qquad \text{(coefficient of } x) \\ A \quad + C &= 1 \qquad \text{(constant term)} \end{aligned}$$

Hence $A = 1/3$, $B = -(1/3)$, and $C = 2/3$. We have

$$I = \frac{1}{3}\int \frac{dx}{x+1} - \frac{1}{3}\int \frac{(x-2)}{x^2-x+1}\,dx$$

$$= \frac{1}{3}\ln|x+1| - \frac{1}{3}\int \frac{x - \frac{1}{2} - \frac{3}{2}}{\left(x-\frac{1}{2}\right)^2 + \frac{3}{4}}\,dx \qquad \text{Let } u = x - 1/2,\ du = dx.$$

$$= \frac{1}{3}\ln|x+1| - \frac{1}{3}\int \frac{u\,du}{u^2+\frac{3}{4}} + \frac{1}{2}\int \frac{du}{u^2+\frac{3}{4}}$$

$$= \frac{1}{3}\ln|x+1| - \frac{1}{6}\ln\left(u^2+\frac{3}{4}\right) + \frac{1}{2}\frac{2}{\sqrt{3}}\text{Arctan}\left(\frac{2u}{\sqrt{3}}\right) + C$$

$$= \frac{1}{3}\ln|x+1| - \frac{1}{6}\ln(x^2-x+1) + \frac{1}{\sqrt{3}}\text{Arctan}\left(\frac{2x-1}{\sqrt{3}}\right) + C.$$

We require one final refinement of the method of partial fractions. If any of the linear or quadratic factors of $Q(x)$ is *repeated* (say n times), then the partial fraction expansion of $P(x)/Q(x)$ requires n distinct fractions corresponding to that factor. The denominators of these fractions have exponents increasing from 1 to n. (See Theorem 6 at the end of this section.)

Example 7 Evaluate

$$I = \int \frac{dx}{x(x-1)^2}.$$

Solution The appropriate partial fraction expansion here is

$$\frac{1}{x(x-1)^2} = \frac{A}{x} + \frac{B}{x-1} + \frac{C}{(x-1)^2} = \frac{A(x^2-2x+1) + B(x^2-x) + Cx}{x(x-1)^2}.$$

Equating coefficients of x^2, x, and 1 in the numerators of both sides, we get

$$\left.\begin{array}{rl} A + B & = 0 \\ -2A - B + C & = 0 \\ A & = 1 \end{array}\right\} \Rightarrow A = 1, B = -1,\ C = 1.$$

Hence

$$I = \int \frac{dx}{x} - \int \frac{dx}{x-1} + \int \frac{dx}{(x-1)^2} = \ln|x| - \ln|x-1| - \frac{1}{x-1} + C$$

$$= \ln\left|\frac{x}{x-1}\right| - \frac{1}{x-1} + C.$$

Example 8 Evaluate

$$I = \int \frac{x^2+2}{4x^5+4x^3+x}\,dx\cdot$$

Solution

$$\frac{x^2+2}{4x^5+4x^3+x} = \frac{x^2+2}{x(2x^2+1)^2} = \frac{A}{x} + \frac{Bx+C}{2x^2+1} + \frac{Dx+E}{(2x^2+1)^2}$$

$$= \frac{A(4x^4+4x^2+1) + B(2x^4+x^2) + C(2x^3+x) + Dx^2 + Ex}{4x^5+4x^3+x}$$

Thus

$$\begin{array}{rll} 4A + 2B & = 0 & \text{(coefficient of } x^4) \\ 2C & = 0 & \text{(coefficient of } x^3) \\ 4A + B + D & = 1 & \text{(coefficient of } x^2) \\ C + E & = 0 & \text{(coefficient of } x) \\ A & = 2 & \text{(constant term).} \end{array}$$

Solving these equations, we get $A = 2$, $B = -4$, $C = 0$, $D = -3$, $E = 0$.

$$I = 2\int \frac{dx}{x} - 4\int \frac{x\,dx}{2x^2+1} - 3\int \frac{x\,dx}{(2x^2+1)^2} \qquad \begin{array}{l} \text{Let } u = 2x^2+1, \\ du = 4x\,dx. \end{array}$$

$$= 2\ln|x| - \int \frac{du}{u} - \frac{3}{4}\int \frac{du}{u^2}$$

$$= 2\ln|x| - \ln|u| + \frac{3}{4u} + C$$

$$= \ln\frac{x^2}{2x^2+1} + \frac{3}{4}\frac{1}{2x^2+1} + C$$

The following theorem summarizes the various aspects of the method of partial fractions.

Theorem 6 Let P and Q be real polynomials, that is, polynomials with real coefficients, and suppose that the degree of P is less than the degree of Q. Then

a) $Q(x)$ can be factored into the product of a constant, real linear factors, and real quadratic factors having no real roots:

$$Q(x) = a(x-x_1)^{m_1}(x-x_2)^{m_2} \cdots (x-x_j)^{m_j}(x^2+b_1x+c_1)^{n_1} \cdots (x^2+b_kx+c_k)^{n_k}.$$

The degree of Q is $m_1 + m_2 + \cdots + m_j + 2n_1 + \cdots + 2n_k$.

b) The rational function $P(x)/Q(x)$ can be expressed as a sum of partial fractions as follows:

i) corresponding to each factor $(x - a)^m$ of $Q(x)$ the expansion contains a sum of fractions of the form

$$\frac{A_1}{x-a} + \frac{A_2}{(x-a)^2} + \cdots + \frac{A_m}{(x-a)^m};$$

ii) corresponding to each factor $(x^2 + bx + c)^n$ of $Q(x)$ the expansion contains a sum of fractions of the form

$$\frac{B_1x + C_1}{x^2 + bx + c} + \frac{B_2x + C_2}{(x^2 + bx + c)^2} + \cdots + \frac{B_nx + C_n}{(x^2 + bx + c)^n}. \quad \square$$

The constants $A_1, A_2, \ldots, A_m, B_1, B_2, \ldots, B_n, C_1, C_2, \ldots, C_n$ can be determined by adding up the fractions in the decomposition and equating the coefficients of like powers of x in the numerator of the sum and in $P(x)$.

We will not attempt any proof of this theorem here; the proof belongs in a study of polynomial algebra. Proofs of part (a) can often be found in textbooks on complex analysis (calculus involving functions of a variable that is complex rather than real).

Also note that part (a) does not tell us how to find the factors of $Q(x)$; it tells us only what form they have. If we know the factorization of $Q(x)$, then the partial fraction expansion provided by part (b) enables us to integrate the rational function $P(x)/Q(x)$. As a final example, we illustrate the use of the method of partial functions in union with the method of substitution.

Example 9 Evaluate

$$I = \int \frac{dx}{x(x^2 + 4)^{3/2}}.$$

Solution Make the substitution $x^2 + 4 = u^2$ so that $x\,dx = u\,du$.

$$I = \int \frac{x\,dx}{x^2(x^2 + 4)^{3/2}} = \int \frac{u\,du}{(u^2 - 4)u^3} = \int \frac{du}{u^2(u^2 - 4)}$$

$$\frac{1}{u^2(u^2 - 4)} = \frac{A}{u} + \frac{B}{u^2} + \frac{C}{u-2} + \frac{D}{u+2}$$

$$= \frac{A(u^3 - 4u) + B(u^2 - 4) + C(u^3 + 2u^2) + D(u^3 - 2u^2)}{u^2(u^2 - 4)}$$

Thus

$$\left.\begin{array}{rl} A \qquad + C + D &= 0 \\ B + 2C - 2D &= 0 \\ -4A \qquad\qquad &= 0 \\ -4B \qquad\quad &= 1 \end{array}\right\} \Rightarrow A = 0, B = -\frac{1}{4}, C = \frac{1}{16}, D = -\frac{1}{16}.$$

$$\begin{aligned} I &= -\frac{1}{4}\int\frac{du}{u^2} + \frac{1}{16}\int\frac{du}{u-2} - \frac{1}{16}\int\frac{du}{u+2} \\ &= \frac{1}{4u} + \frac{1}{16}\ln\left|\frac{u-2}{u+2}\right| + C \\ &= \frac{1}{4}(x^2+4)^{-1/2} + \frac{1}{16}\ln\left|\frac{\sqrt{x^2+4}-2}{\sqrt{x^2+4}+2}\right| + C \end{aligned}$$

EXERCISES

Evaluate the integrals in Exercises 1–30.

1. $\displaystyle\int\frac{dx}{5-4x}$

2. $\displaystyle\int\frac{x^2}{x-4}\,dx$

3. $\displaystyle\int\frac{1}{x^2-9}\,dx$

4. $\displaystyle\int\frac{dx}{5-x^2}$

5. $\displaystyle\int\frac{dx}{a^2-x^2}$

6. $\displaystyle\int\frac{dx}{b^2-a^2x^2}$

7. $\displaystyle\int\frac{x^2\,dx}{x^2+x-2}$

8. $\displaystyle\int\frac{x\,dx}{3x^2+8x-3}$

9. $\displaystyle\int\frac{x-2}{x^2+x}\,dx$

10. $\displaystyle\int\frac{dx}{x^3+9x}$

11. $\displaystyle\int\frac{dx}{1-6x+9x^2}$

12. $\displaystyle\int\frac{dx}{2+6x+9x^2}$

13. $\displaystyle\int\frac{x^2\,dx}{6x-9x^2}$

14. $\displaystyle\int\frac{x^3+1}{12+7x+x^2}\,dx$

15. $\displaystyle\int\frac{dx}{x(x^2-a^2)}$

16. $\displaystyle\int\frac{dx}{x^4-a^4}$

17. $\displaystyle\int\frac{x^3\,dx}{x^3-a^3}$

18. $\displaystyle\int\frac{dx}{x^3+2x^2+2x}$

19. $\displaystyle\int\frac{dx}{x^3-4x^2+3x}$

20. $\displaystyle\int\frac{x^2+1}{x^3+8}\,dx$

21. $\displaystyle\int\frac{dx}{(x^2-1)^2}$

22. $\displaystyle\int\frac{x^2\,dx}{(x^2-1)(x^2-4)}$

23. $\displaystyle\int\frac{dx}{x^4-3x^3}$

24. $\displaystyle\int\frac{x\,dx}{(x^2-x+1)^2}$

25.* $\displaystyle\int\frac{t\,dt}{(t+1)(t^2+1)^2}$

26.* $\displaystyle\int\frac{dt}{(t-1)(t^2-1)^2}$

27.* $\displaystyle\int\frac{dx}{x(3+x^2)\sqrt{1-x^2}}$

28.* $\displaystyle\int\frac{dx}{e^{2x}-4e^x+4}$

29.* $\displaystyle\int \frac{d\theta}{\cos\theta\,(1 + \sin\theta)}$ **30.*** $\displaystyle\int \frac{d\theta}{\sin\theta\,(1 + \sin\theta)}$

31.* Suppose that P and Q are polynomials such that the degree of P is smaller than that of Q. If

$$\frac{P(x)}{Q(x)} = \frac{A_1}{x - x_1} + \frac{A_2}{x - x_2} + \cdots + \frac{A_n}{x - x_n}$$

where $x_i \neq x_j$ if $i \neq j$ $(1 \leq i, j \leq n)$, show that

$$A_j = \frac{P(x_j)}{Q'(x_j)} \qquad (1 \leq j \leq n).$$

This gives yet another method for computing the constants in a partial fraction expansion if the denominator factors completely into distinct linear factors.

6.7 INTEGRATION BY PARTS

Our final general method for antidifferentiation is called **integration by parts** or **partial integration.** Just as the method of substitution can be regarded as inverse to the chain rule for differentiation, so the method of integration by parts is inverse to the product rule for differentiation.

Suppose that $U(x)$ and $V(x)$ are two differentiable functions. According to the product rule,

$$\frac{d}{dx}(U(x)V(x)) = U(x)\frac{dV}{dx} + V(x)\frac{dU}{dx}.$$

Integrating both sides of this equation and transposing terms, we obtain

$$\int U(x)\frac{dV}{dx}\,dx = U(x)V(x) - \int V(x)\frac{dU}{dx}\,dx$$

or, more simply,

$$\boxed{\int U\,dV = UV - \int V\,dU.}$$

The above formula serves as a pattern for carrying out integration by parts, as we will see in the examples below. In each application of the method we break up the given integrand into a product of two pieces, U and V', where V' is readily integrated and where $\int VU'\,dx$ is usually (but not always) a "simpler" integral than $\int UV'\,dx$. The technique is called integration by parts because it replaces one integral with the sum of an integrated term and another integral that remains to be evaluated. That is, it accomplishes only "part" of the original integration.

Example 1

$$\int xe^x\,dx \qquad \begin{array}{ll} \text{Let } U = x & dV = e^x\,dx \\ \text{Then } dU = dx & V = e^x. \end{array}$$

$$= xe^x - \int e^x\,dx \qquad (\text{That is, } UV - \textstyle\int V\,dU.)$$

$$= xe^x - e^x + C.$$

Note the form in which the integration by parts is carried out. We indicate at the side what choices we are making for U and dV and then calculate dU and V from these. However, we do not actually substitute U and V into the integral; instead we use the formula $\int U\,dV = UV - \int V\,dU$ as a pattern or mnemonic device to replace the given integral by the equivalent partially integrated form on the second line.

Note also that had we included a constant of integration with V, (say, $V = e^x + K$), the constant would cancel out in the next step

$$\int xe^x dx = x(e^x + K) - \int (e^x + K)dx = xe^x + Kx - e^x - Kx + C = xe^x - e^x + C.$$

In general, do not include a constant of integration on the right-hand side until the last integral has been evaluated.

Example 2 a)

$$\int \ln x\,dx \qquad \begin{array}{ll} \text{Let } U = \ln x & dV = dx \\ dU = dx/x & V = x \end{array}$$

$$= x\ln x - \int x\frac{1}{x}\,dx$$

$$= x\ln x - x + C.$$

b)

$$\int x^2 \sin x\,dx \qquad \begin{array}{ll} \text{Let } U = x^2 & dV = \sin x\,dx \\ dU = 2x\,dx & V = -\cos x. \end{array}$$

$$= -x^2\cos x + 2\int x\cos x\,dx \qquad \begin{array}{ll} \text{Let } U = x & dV = \cos x\,dx \\ dU = dx & V = \sin x. \end{array}$$

$$= -x^2\cos x + 2(x\sin x - \int \sin x\,dx) = -x^2\cos x + 2x\sin x + 2\cos x + C$$

c)

$$\int x \operatorname{Arctan} x\,dx \qquad \begin{array}{ll} \text{Let } U = \operatorname{Arctan} x & dV = x\,dx \\ dU = dx/(1 + x^2) & V = (1/2)x^2. \end{array}$$

$$= \frac{1}{2}x^2 \operatorname{Arctan} x - \frac{1}{2}\int \frac{x^2}{1 + x^2}\,dx$$

$$= \frac{1}{2}x^2 \operatorname{Arctan} x - \frac{1}{2}\int \left(1 - \frac{1}{1 + x^2}\right) dx$$

$$= \frac{1}{2}x^2 \operatorname{Arctan} x - \frac{1}{2}x + \frac{1}{2}\operatorname{Arctan} x + C$$

d)

$$\int \operatorname{Arcsin} x\,dx \qquad \begin{array}{ll} \text{Let } U = \operatorname{Arcsin} x & dV = dx \\ dU = dx/(\sqrt{1 - x^2}) & V = x. \end{array}$$

$$= x \operatorname{Arcsin} x - \int \frac{x}{\sqrt{1 - x^2}}\,dx \qquad \begin{array}{l} \text{Let } u = 1 - x^2 \\ du = -2x\,dx. \end{array}$$

$$= x \operatorname{Arcsin} x + \frac{1}{2}\int u^{-1/2}\,du$$

$$= x \operatorname{Arcsin} x + u^{1/2} + C = x \operatorname{Arcsin} x + \sqrt{1 - x^2} + C.$$

e) $I = \displaystyle\int \sec^3 x\,dx$ Let $U = \sec x$, $dV = \sec^2 x\,dx$; $dU = \sec x \tan x\,dx$, $V = \tan x$

$$= \sec x \tan x - \int \sec x \tan^2 x\,dx$$

$$= \sec x \tan x - \int \sec x\,(\sec^2 x - 1)\,dx$$

$$= \sec x \tan x - \int \sec^3 x\,dx + \int \sec x\,dx$$

$$= \sec x \tan x - I + \ln|\sec x + \tan x|$$

This is an equation that can be solved for the desired integral I:

$$\int \sec^3 x\,dx = I = \frac{1}{2}\sec x \tan x + \frac{1}{2}\ln|\sec x + \tan x| + C.$$

Study the preceding examples carefully; they show the various ways in which integration by parts is used and give some insights into what choices should be made for U and dV in various situations. An improper choice can result in making an integral more rather than less complicated or difficult. Look for a factor of the integrand that is easily integrated, and include dx with that factor to make up dV. Then U is the remaining factor of the integrand. Sometimes it is necessary to take $dV = dx$ only. Where some choice is possible for dV, the choice should be made so that, if possible, $V\,dU$ is "simpler" (that is, easier to integrate) than $U\,dV$. The following are two useful rules of thumb.

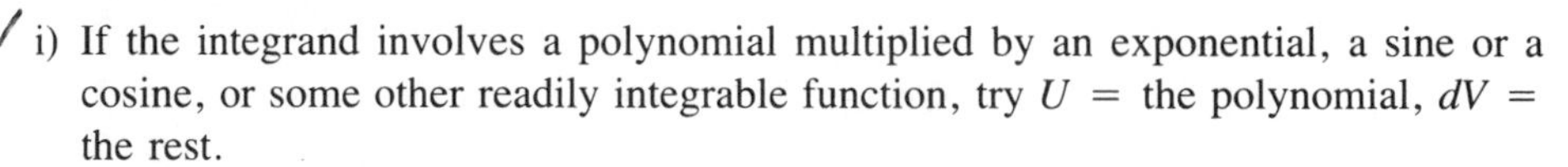

i) If the integrand involves a polynomial multiplied by an exponential, a sine or a cosine, or some other readily integrable function, try $U =$ the polynomial, $dV =$ the rest.

ii) If the integrand involves a logarithm, an inverse trigonometric function, or some other function that is not readily integrable but whose derivative is readily calculated, try that function for U and let dV equal the rest.

(Of course, these "rules" come with no guarantee. They may fail to be helpful if "the rest" is not of suitable form. There remain many integrals that cannot be evaluated by any of the standard techniques presented in this chapter.)

Example 2(e) illustrates a frequently occurring and very useful phenomenon. It may happen after one or two integrations by parts, with the possible application of some known identity, that the original integral reappears on the right-hand side. Unless its coefficient is 1, we then have an equation that can be solved for that integral.

Example 3 $I = \displaystyle\int e^{ax}\cos bx\,dx,\quad (b \neq 0)$ Let $U = e^{ax}$, $dV = \cos bx\,dx$; $dU = ae^{ax}\,dx$, $V = (1/b)\sin bx$.

$$I = \frac{1}{b} e^{ax} \sin bx - \frac{a}{b} \int e^{ax} \sin bx\, dx \qquad \text{Let } U = e^{ax} \quad dV = \sin bx\, dx$$
$$dU = ae^{ax}\, dx \quad V = -(1/b)\cos bx.$$

$$= \frac{1}{b} e^{ax} \sin bx - \frac{a}{b}\left(\frac{-1}{b} e^{ax} \cos bx + \frac{a}{b} \int e^{ax} \cos bx\, dx\right)$$

$$= \frac{1}{b} e^{ax} \sin bx + \frac{a}{b^2} e^{ax} \cos bx - \frac{a^2}{b^2} I$$

Thus

$$\left(1 + \frac{a^2}{b^2}\right) I = \frac{b}{b^2} e^{ax} \sin bx + \frac{a}{b^2} e^{ax} \cos bx + C_1$$

and

$$I = \frac{be^{ax} \sin bx + ae^{ax} \cos bx}{b^2 + a^2} + C.$$

Observe that after the first integration by parts we had an integral that was different from but no simpler than the original. At this point we might have become discouraged and given up on this method. However, perseverance proved worthwhile; a second integration by parts returned the original integral I in an equation that could be solved for I. Note also that, having chosen to let U be the exponential in the first integration by parts (we could have let it be the cosine), we made the same choice for U in the second integration by parts. Had we "switched horses in midstream" and decided to let U be the trigonometric function the second time, we would have obtained

$$I = \frac{1}{b} e^{ax} \sin bx - \frac{1}{b} e^{ax} \sin bx + I,$$

that is, we would have undone what we accomplished in the first step.

If we wish to evaluate a definite integral by the method of integration by parts we must remember to include the appropriate evaluation symbol with the integrated term.

Example 4

$$\int_1^e x^3 (\ln x)^2\, dx \qquad \text{Let } U = (\ln x)^2 \quad dV = x^3\, dx$$
$$dU = ((2\ln x)/x)\, dx \quad V = x^4/4.$$

$$= \frac{x^4}{4} (\ln x)^2 \Bigg|_1^e - \frac{1}{2} \int_1^e x^3 \ln x\, dx \qquad \text{Let } U = \ln x \quad dV = x^3\, dx$$
$$dU = dx/x \quad V = x^4/4.$$

$$= \frac{e^4}{4}(1) - 0 - \frac{1}{2}\left(\frac{x^4}{4} \ln x \Bigg|_1^e - \frac{1}{4} \int_1^e x^3\, dx\right)$$

$$= \frac{e^4}{4} - \frac{e^4}{8} + \frac{1}{8}\frac{x^4}{4} \Bigg|_1^e = \frac{e^4}{8} + \frac{e^4}{32} - \frac{1}{32} = \frac{5}{32} e^4 - \frac{1}{32}.$$

Reduction Formulas

Consider the problem of finding $\int x^{10} e^{-x}\,dx$. We can, of course, proceed by using integration by parts 10 times. Each time will reduce the exponent on x by 1. Since this is repetitive and laborious, we prefer the following approach. For $n \geq 1$ let

$$I_n = \int x^n e^{-x}\,dx \qquad \begin{aligned} \text{Let } U &= x^n & dV &= e^{-x}\,dx \\ dU &= nx^{n-1}\,dx & V &= -e^{-x} \end{aligned}$$

$$= -x^n e^{-x} + n\int x^{n-1} e^{-x}\,dx.$$

That is,

$$I_n = -x^n e^{-x} + nI_{n-1}.$$

This formula is called a **reduction formula** because it gives the value of the integral I_n in terms of I_{n-1}, an integral corresponding to a reduced value of the exponent n. We can apply the formula over and over (10 times) to get

$$\begin{aligned} &\int x^{10} e^{-x}\,dx \\ &= I_{10} = -x^{10}e^{-x} + 10I_9 \\ &= -x^{10}e^{-x} + 10(-x^9e^{-x} + 9I_8) \\ &= -e^{-x}(x^{10} + 10x^9) + 10\cdot 9(-x^8e^{-x} + 8I_7) \\ &= -e^{-x}(x^{10} + 10x^9 + 10\cdot 9x^8 + 10\cdot 9\cdot 8x^7) + 10\cdot 9\cdot 8\cdot 7I_6 \\ &= \cdots \\ &= -e^{-x}(x^{10} + 10x^9 + 10\cdot 9x^8 + 10\cdot 9\cdot 8x^7 + \cdots + 10\cdot 9\cdot 8\cdot \cdots \cdot 2x) + 10!I_0 \end{aligned}$$

Since $I_0 = \int e^{-x}\,dx = -e^{-x} + C$, we have

$$\int x^{10}e^{-x}\,dx = -e^{-x}\left(x^{10} + \frac{10!}{9!}x^9 + \frac{10!}{8!}x^8 + \cdots + \frac{10!}{1!}x + 10!\right) + C.$$

Example 5 Obtain and use a reduction formula to evaluate

$$I_n = \int_0^{\pi/2} \cos^n x\,dx \qquad (n \geq 0).$$

Solution Observe first that

$$I_0 = \int_0^{\pi/2} dx = \frac{\pi}{2}, \qquad I_1 = \int_0^{\pi/2} \cos x\,dx = \sin x\Big|_0^{\pi/2} = 1.$$

Now let $n \geq 2$:

$$I_n = \int_0^{\pi/2} \cos^{n-1} x \cos x\,dx \qquad \begin{aligned} \text{Let } U &= \cos^{n-1} x & dV &= \cos x\,dx \\ dU &= -(n-1)\cos^{n-2} x \sin x\,dx & V &= \sin x. \end{aligned}$$

$$= \sin x \cos^{n-1} x \Big|_0^{\pi/2} + (n-1)\int_0^{\pi/2} \cos^{n-2} x \sin^2 x \, dx$$

$$= 0 - 0 + (n-1)\int_0^{\pi/2} \cos^{n-2} x(1 - \cos^2 x)\, dx$$

$$= (n-1)I_{n-2} - (n-1)I_n.$$

Transposing the term $-(n-1)I_n$ we obtain $nI_n = (n-1)I_{n-2}$, or

$$I_n = \frac{n-1}{n} I_{n-2},$$

which is the required reduction formula. It is valid for $n \geq 2$. (Where have we used this assumption in deriving the formula?) If $n \geq 2$ is even, we have

$$I_n = \frac{n-1}{n} I_{n-2} = \frac{(n-1)(n-3)}{n(n-2)} I_{n-4} = \cdots = \frac{n-1}{n} \cdot \frac{n-3}{n-2} \cdot \frac{n-5}{n-4} \cdot \cdots \cdot \frac{5}{6} \cdot \frac{3}{4} \cdot \frac{1}{2} \cdot I_0$$

$$= \frac{n-1}{n} \cdot \frac{n-3}{n-2} \cdot \frac{n-5}{n-4} \cdot \cdots \cdot \frac{5}{6} \cdot \frac{3}{4} \cdot \frac{1}{2} \cdot \frac{\pi}{2}.$$

If $n \geq 2$ is odd, we have

$$I_n = \frac{n-1}{n} \cdot \frac{n-3}{n-2} \cdot \frac{n-5}{n-4} \cdot \cdots \cdot \frac{6}{7} \cdot \frac{4}{5} \cdot \frac{2}{3} \cdot I_1 = \frac{n-1}{n} \cdot \frac{n-3}{n-2} \cdot \frac{n-5}{n-4} \cdot \cdots \cdot \frac{6}{7} \cdot \frac{4}{5} \cdot \frac{2}{3}.$$

EXERCISES

Evaluate the integrals in Exercises 1–32.

1. $\int x \cos x \, dx$

2. $\int (x + a)e^{bx} \, dx$

3. $\int x^2 e^{kx} \, dx$

4. $\int (x^2 - x - 2) \sin 3x \, dx$

5. $\int x^5 e^{-x^2} \, dx$

6. $\int_0^1 \sqrt{x} \sin (\pi\sqrt{x}) \, dx$

7. $\int x^3 \ln x \, dx$

8. $\int x(\ln x)^3 \, dx$

9. $\int \text{Arctan}\, x \, dx$

10. $\int x^2 \,\text{Arctan}\, x \, dx$

11. $\int x \,\text{Arcsin}\, x \, dx$

12. $\int x^2 \,\text{Arcsin}\, x \, dx$

13. $\int \frac{\text{Arcsin}\, x}{\sqrt{1 - x^2}} \, dx$

14. $\int \frac{\text{Arctan}\, x}{1 + x^2} \, dx$

15. $\int_0^{\pi/4} \sec^5 x \, dx$

16. $\int \tan^2 x \sec x \, dx$

17. $\int e^{2x} \sin 3x \, dx$

18. $\int x e^{\sqrt{x}} \, dx$

19. $\displaystyle\int_{1/2}^{1} \frac{\text{Arcsin}\, x}{x^2}\, dx$

20. $\displaystyle\int \frac{\text{Arctan}\, x}{x^2}\, dx$

21. $\displaystyle\int x \sec^2 x\, dx$

22. $\displaystyle\int x \sin^2 x\, dx$

23. $\displaystyle\int \cos(\ln x)\, dx$

24. $\displaystyle\int_1^e \sin(\ln x)\, dx$

25. $\displaystyle\int \frac{\ln(\ln x)}{x}\, dx$

26. $\displaystyle\int_0^4 \sqrt{x} e^{\sqrt{x}}\, dx$

27. $\displaystyle\int \text{Arccos}\, x\, dx$

28. $\displaystyle\int x\, \text{Arcsec}\, x\, dx$

29. $\displaystyle\int_1^2 \text{Arcsec}\, x\, dx$

30. $\displaystyle\int x a^x\, dx \qquad (a > 0,\ a \neq 1)$

31. $\displaystyle\int \sqrt{9 + x^2}\, dx$

32. $\displaystyle\int \frac{x^2}{\sqrt{9 - x^2}}\, dx$

33.* Obtain a reduction formula for $I_n = \int (\ln x)^n\, dx$ and use it to evaluate I_4.

34.* Obtain a reduction formula for $I_n = \int \sin^n x\, dx$ $(n \geq 2)$ and use it to find I_7 and I_8.

35.* Obtain a reduction formula for $I_n = \int \sec^{2n+1} x\, dx$ $(n \geq 1)$ and use it to find I_3.

36.* By writing

$$I_n = \int \frac{dx}{(x^2 + a^2)^n} = \frac{1}{a^2} \int \frac{dx}{(x^2 + a^2)^{n-1}} - \frac{1}{a^2} \int x \frac{x}{(x^2 + a^2)^n}\, dx$$

and integrating the last integral by parts, using $U = x$, obtain a reduction formula for I_n. Use this formula to find I_3.

37.* If f is twice differentiable on $[a, b]$ and $f(a) = f(b) = 0$, show that

$$\int_a^b (x - a)(b - x) f''(x)\, dx = -2 \int_a^b f(x)\, dx.$$

(*Hint:* Use integration by parts on the left-hand side twice.) This formula will be used in the next section to construct an error estimate for the trapezoidal rule approximation formula.

Summary of Techniques of Integration

Students sometimes have difficulty deciding which method to use to evaluate a given integral. Often no one method will suffice to produce the whole solution, but one method may lead to a different, possibly simpler integral that can then be dealt with on its own merits. Here are a few guidelines:

1. First, and at all stages, be alert for simplifying substitutions. Even when these don't accomplish the whole integration, they can lead to integrals to which some other method can be applied.
2. If the integral involves a quadratic expression $Ax^2 + Bx + C$ with $A \neq 0$ and $B \neq 0$, complete the square. A simple substitution then reduces the quadratic expression to a sum or difference of squares.

3. Integrals of products of trigonometric functions can sometimes be accomplished or rendered simpler by the use of appropriate trigonometric identities such as $\sin^2 x + \cos^2 x = 1$, $\sec^2 x = 1 + \tan^2 x$, $\sin^2 x = (1/2)(1 - \cos 2x)$, and $\cos^2 x = (1/2)(1 + \cos 2x)$.
4. Integrals involving $(a^2 - x^2)^{1/2}$ can be transformed using $x = a \sin \theta$. Integrals involving $(a^2 + x^2)^{1/2}$ or $1/(a^2 + x^2)$ may yield to $x = a \tan \theta$. Integrals involving $(x^2 - a^2)^{1/2}$ can be transformed using $x = a \sec \theta$.
5. Use partial fractions to integrate rational functions whose denominators can be factored into real linear and quadratic factors. Remember to use long division first, if necessary, to reduce the fraction to one whose numerator has degree smaller than that of its denominator.
6. Use integration by parts for integrals of functions such as products of polynomials and transcendental functions, and for inverse trigonometric functions and logarithms. Be alert for ways of using integration by parts to get formulas representing complicated integrals in terms of simpler ones.

REVIEW EXERCISES ON TECHNIQUES OF INTEGRATION

Evaluate these integrals, which review the techniques of Sections 6.4–6.7.

1. $\displaystyle\int \frac{x\,dx}{2x^2 + 5x + 2}$

2. $\displaystyle\int \frac{x\,dx}{(x - 1)^3}$

3. $\displaystyle\int \sin^3 x \cos^3 x\,dx$

4. $\displaystyle\int \frac{(1 + \sqrt{x})^{1/3}}{\sqrt{x}}\,dx$

5. $\displaystyle\int \frac{3\,dx}{4x^2 - 1}$

6. $\displaystyle\int (x^2 + x - 2) \sin 3x\,dx$

7. $\displaystyle\int \frac{\sqrt{1 - x^2}}{x^4}\,dx$

8. $\displaystyle\int x^3 \cos (x^2)\,dx$

9. $\displaystyle\int \frac{x^2\,dx}{(5x^3 - 2)^{2/3}}$

10. $\displaystyle\int \frac{dx}{x^2 + 2x - 15}$

11. $\displaystyle\int \frac{dx}{(4 + x^2)^2}$

12. $\displaystyle\int (\sin x + \cos x)^2\,dx$

13. $\displaystyle\int 2^x\sqrt{1 + 4^x}\,dx$

14. $\displaystyle\int \frac{\cos x}{1 + \sin^2 x}\,dx$

15. $\displaystyle\int \frac{\sin^3 x}{\cos^7 x}\,dx$

16. $\displaystyle\int \frac{x^2\,dx}{(3 + 5x^2)^{3/2}}$

17. $\displaystyle\int e^{-x} \sin (2x)\,dx$

18. $\displaystyle\int \frac{2x^2 + 4x - 3}{x^2 + 5x}\,dx$

19. $\displaystyle\int \cos (3 \ln x)\,dx$

20. $\displaystyle\int \frac{dx}{4x^3 + x}$

21. $\displaystyle\int \frac{x \ln (1 + x^2)}{1 + x^2}\,dx$

22. $\displaystyle\int \sin^2 x \cos^4 x\,dx$

23. $\displaystyle\int \frac{x^2}{\sqrt{2 - x^2}}\,dx$

24. $\displaystyle\int \tan^4 x \sec x\,dx$

25. $\displaystyle\int \frac{x^2\,dx}{(4x + 1)^{10}}$

26. $\displaystyle\int x \operatorname{Arcsin} \frac{x}{2}\,dx$

27. $\displaystyle\int \sin^5 (4x)\,dx$

28. $\displaystyle\int \frac{dx}{x^5 - 2x^3 + x}$

29. $\displaystyle\int \frac{dx}{2 + e^x}$

30. $\displaystyle\int x^3 3^x\,dx$

31. $\displaystyle\int \frac{\sin^2 x \cos x}{2 - \sin x}\,dx$

32. $\displaystyle\int \frac{x^2 + 1}{x^2 + 2x + 2}\,dx$

33. $\displaystyle\int \frac{dx}{x^2\sqrt{1 - x^2}}$

34. $\displaystyle\int x^3 (\ln x)^2\,dx$

35. $\displaystyle\int \frac{x^3}{\sqrt{1 - 4x^2}}\,dx$

36. $\displaystyle\int \frac{e^{1/x}\,dx}{x^2}$

37. $\int \frac{x+1}{\sqrt{x^2+1}}\,dx$

38. $\int e^{(x^{1/3})}\,dx$

39. $\int \frac{x^3-3}{x^3-9x}\,dx$

40. $\int \frac{10^{\sqrt{x+2}}}{\sqrt{x+2}}\,dx$

41. $\int \sin^5 x \cos^9 x\,dx$

42. $\int \frac{x^2\,dx}{\sqrt{x^2-1}}$

43. $\int \frac{x\,dx}{x^2+2x-1}$

44. $\int \frac{2x-3}{\sqrt{4-3x+x^2}}\,dx$

45. $\int x^2 \text{ Arcsin } (2x)\,dx$

46. $\int \frac{\sqrt{3x^2-1}}{x}\,dx$

47. $\int \cos^4 x \sin^4 x\,dx$

48. $\int \sqrt{x-x^2}\,dx$

49. $\int \frac{dx}{(4+x)\sqrt{x}}$

50. $\int x \text{ Arctan } \frac{x}{3}\,dx$

51. $\int \frac{x^4-1}{x^3+2x^2}\,dx$

52. $\int \frac{dx}{x(x^2+4)^2}$

53. $\int \frac{\sin (2 \ln x)}{x}\,dx$

54. $\int \frac{\sin (\ln x)}{x^2}\,dx$

55. $\int \frac{e^{2 \text{ Arctan } x}}{1+x^2}\,dx$

56. $\int \frac{x^3+x-2}{x^2-7}\,dx$

57. $\int \frac{\ln (3+x^2)}{3+x^2}\,x\,dx$

58. $\int \cos^7 x\,dx$

59. $\int \frac{\text{Arcsin } (x/2)}{(4-x^2)^{1/2}}\,dx$

60. $\int \tan^4 (\pi x)\,dx$

61. $\int \frac{(x+1)\,dx}{\sqrt{x^2+6x+10}}$

62. $\int e^x(1-e^{2x})^{5/2}\,dx$

63. $\int \frac{x^3\,dx}{(x^2+2)^{7/2}}$

64. $\int \frac{x^2}{2x^2-3}\,dx$

65. $\int \frac{x^{1/2}}{1+x^{1/3}}\,dx$

66. $\int \frac{dx}{(2x+1)(x^2+x+1)^{1/2}}$

67. $\int \frac{1+x}{1+\sqrt{x}}\,dx$

68. $\int \frac{x\,dx}{4x^4+4x^2+5}$

69. $\int \frac{x\,dx}{(x^2-4)^2}$

70. $\int \frac{dx}{x^3+x^2+x}$

71. $\int x^2 \text{ Arctan } x\,dx$

72. $\int e^x \sec (e^x)\,dx$

73. $\int \frac{dx}{4 \sin x - 3 \cos x}$

74. $\int \frac{dx}{x^{1/3}-1}$

75. $\int \frac{dx}{\tan x + \sin x}$

76. $\int \frac{x\,dx}{\sqrt{3-4x-4x^2}}$

77. $\int \frac{\sqrt{x}}{1+x}\,dx$

78. $\int \sqrt{1+e^x}\,dx$

79. $\int \frac{x^4\,dx}{x^3-8}$

80. $\int xe^x \cos x\,dx$

6.8 APPROXIMATE INTEGRATION

Most of the applications of integration, both within and outside of mathematics, center on the definite integral

$$I = \int_a^b f(x)\,dx.$$

Thanks to the fundamental theorem of calculus we can evaluate such definite integrals by first finding the indefinite integral of f. This is why we have spent considerable time on developing techniques of integration. There are, however, two obstacles that can prevent our calculating I in this way:

i) Finding the indefinite integral of f may be impossible, or at least very difficult.

ii) We may not be given $f(x)$ explicitly as a function of x; for instance, $f(x)$ may be an unknown function whose values at certain points of the interval $[a, b]$ have been determined by experimental measurement.

In this section we shall investigate the problem of approximating the value of the definite integral I using only the values of $f(x)$ at finitely many points of $[a, b]$. Such an approximation is called **approximate** or **numerical integration,** or **numerical quadrature.**

There are many techniques for performing numerical integration. One simple method is to calculate a Riemann sum for the integral. We consider two somewhat better techniques in this section, the trapezoidal rule and Simpson's rule.

The Trapezoidal Rule

We assume that $f(x)$ is continuous on $[a, b]$ and subdivide $[a, b]$ into n subintervals of equal length $\Delta x = (b - a)/n$ using points

$$a = x_0 < x_1 < x_2 < \cdots < x_{n-1} < x_n = b.$$

Let us assume that we know the value of $f(x)$ at each of these points:

$$y_0 = f(x_0), \qquad y_1 = f(x_1), \qquad \ldots, \qquad y_n = f(x_n).$$

The trapezoidal rule approximates $\int_a^b f(x)\,dx$ by using straight line segments to approximate the graph of f, as shown in Fig. 6.35. In the first interval, $[x_0, x_1]$, we can approximate the graph of f by a straight line segment joining the points (x_0, y_0) and (x_1, y_1). The line has equation

$$y = A + B(x - x_0),$$

where $A = y_0$ and $B = (y_1 - y_0)/\Delta x$. Evidently $\int_{x_0}^{x_1} f(x)\,dx$ is approximated by

$$\int_{x_0}^{x_1} (A + B(x - x_0))\,dx = A(x_1 - x_0) + \frac{B}{2}(x_1 - x_0)^2 = \frac{\Delta x}{2}[2A + B\Delta x]$$

$$= \frac{\Delta x}{2}(y_0 + y_1).$$

If $f(x) \geq 0$ on $[x_0, x_1]$, this is just the area of the trapezoid T_1 with parallel sides of lengths y_0 and y_1 separated by a distance Δx (see Fig. 6.36). We can approximate the integral of f over any subinterval in the same way:

$$\int_{x_{k-1}}^{x_k} f(x)\,dx \simeq \frac{\Delta x}{2}(y_{k-1} + y_k) \qquad (1 \leq k \leq n).$$

It follows that the original integral I can be approximated by the sum of the "areas" of n trapezoids.

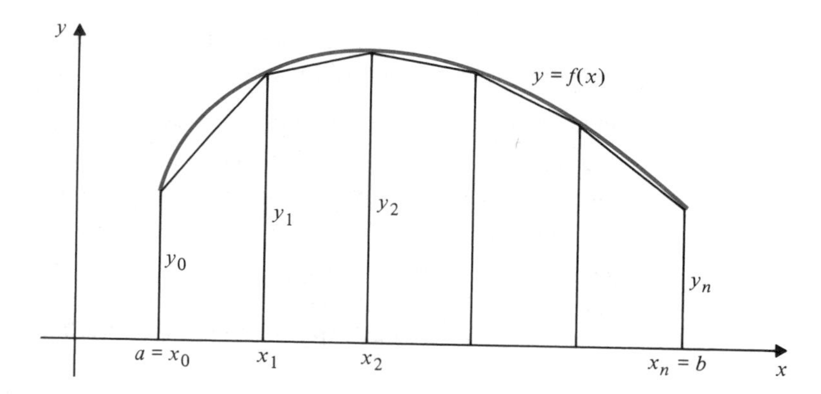

Figure 6.35

$$\int_a^b f(x)\,dx \simeq \frac{\Delta x}{2}[(y_0 + y_1) + (y_1 + y_2) + (y_2 + y_3) + \cdots + (y_{n-1} + y_n)]$$

$$= \Delta x\left(\frac{1}{2}y_0 + y_1 + y_2 + y_3 + \cdots + y_{n-1} + \frac{1}{2}y_n\right).$$

We call this the trapezoidal rule approximation to $\int_a^b f(x)\,dx$ based on n subintervals, and we denote it TRAP(n):

$$\int_a^b f(x)\,dx \simeq \text{TRAP}(n) = \Delta x\left(\frac{1}{2}y_0 + y_1 + y_2 + \cdots + y_{n-1} + \frac{1}{2}y_n\right).$$

Let us illustrate a trapezoidal rule approximation to a known integral:

$$I = \int_1^2 \frac{dx}{x} = \ln 2 = 0.693147\ldots.$$

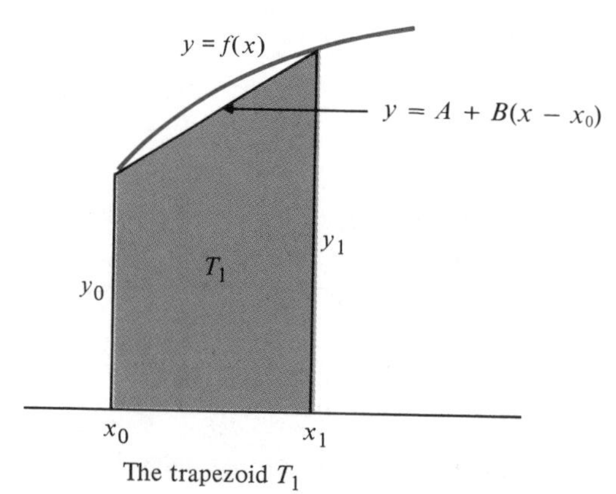

The trapezoid T_1

Figure 6.36

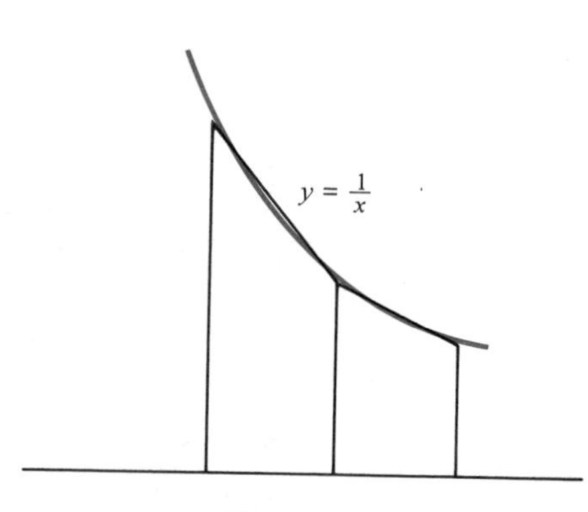

Figure 6.37

Example 1 Calculate TRAP(4) and TRAP(8) for the above integral I.

Solution The required values of the integrand are given in the following table.

x	1	9/8	5/4	11/8	3/2	13/8	7/4	15/8	2
$1/x$	1	8/9	4/5	8/11	2/3	8/13	4/7	8/15	1/2

For $n = 4$ we have $\Delta x = (2 - 1)/4 = 1/4$; for $n = 8$ we have $\Delta x = 1/8$.

$$\text{TRAP}(4) = \frac{1}{4}\left(\frac{1}{2}(1) + \frac{4}{5} + \frac{2}{3} + \frac{4}{7} + \frac{1}{2}\left(\frac{1}{2}\right)\right) = 0.6970\ldots$$

$$\text{TRAP}(8) = \frac{1}{8}\left(\frac{1}{2}(1) + \frac{8}{9} + \frac{4}{5} + \frac{8}{11} + \frac{2}{3} + \frac{8}{13} + \frac{4}{7} + \frac{8}{15} + \frac{1}{2}\left(\frac{1}{2}\right)\right) = 0.6941\ldots$$

Note that both approximations are larger than the actual value of I in this case. This fact could have been predicted because the graph of $y = 1/x$ is concave upward on $[1, 2]$ and so the tops of the approximating trapezoids lie above it (see Fig. 6.37).

The errors in the two approximations above are

$$I - \text{TRAP}(4) = \ln 2 - 0.6970\ldots \simeq -0.0038,$$
$$I - \text{TRAP}(8) = \ln 2 - 0.6941\ldots \simeq -0.001.$$

This example is somewhat artificial in the sense that we know the actual value of the integral so we really don't need an approximation. In practical applications of numerical integration we do not know the actual value. It is tempting to calculate several approximations for increasing values of n until the two most recent ones agree to within prescribed error tolerances. For example, we would have some inclination to claim that $\ln 2 \simeq 0.69$ or even 0.693 from examining TRAP(4) and TRAP(8). (TRAP(8) ought to be closer to the true value.) This approach cannot, however, be justified in general. The following theorem provides a bound for the error in the trapezoidal rule approximation.

Theorem 7 (*Error Estimate for the Trapezoidal Rule*) If f is twice differentiable on $[a, b]$ and satisfies $|f''(x)| \le K$ there, then

$$\left|\int_a^b f(x)\,dx - \text{TRAP}(n)\right| \le \frac{K(b-a)}{12}(\Delta x)^2,$$

where $\Delta x = (b - a)/n$. Note that this error bound decreases like the square of the subinterval length as n increases.

Proof Let $y = A + Bx$ be the straight line approximating $y = f(x)$ in the first subinterval, $[x_0, x_1]$, as considered above. Let

$$g(x) = f(x) - (A + Bx).$$

Then g is twice differentiable, $g''(x) = f''(x)$ and $g(x_0) = g(x_1) = 0$. Two integrations by parts (see Exercise 37 of Section 6.7) show that

$$\int_{x_0}^{x_1} f(x)\,dx - \frac{\Delta x}{2}(y_0 + y_1) = \int_{x_0}^{x_1} g(x)\,dx = -\frac{1}{2}\int_{x_0}^{x_1}(x - x_0)(x_1 - x)g''(x)\,dx$$

$$= -\frac{1}{2}\int_{x_0}^{x_1}(x - x_0)(x_1 - x)f''(x)\,dx.$$

Thus

$$\left|\int_{x_0}^{x_1} f(x)\,dx - \frac{\Delta x}{2}(y_0 + y_1)\right| \le \frac{1}{2}\int_{x_0}^{x_1}(x - x_0)(x_1 - x)|f''(x)|\,dx$$

$$\le \frac{K}{2}\int_{x_0}^{x_1}(-x^2 + (x_0 + x_1)x - x_0x_1)\,dx$$

$$= \frac{K}{12}(x_1 - x_0)^3 = \frac{K}{12}(\Delta x)^3.$$

(We have omitted the details in the evaluation of the last integral above.) A similar estimate holds on each subinterval $[x_{k-1}, x_k]$ $(1 \le k \le n)$. Therefore,

$$\left|\int_a^b f(x)\,dx - \mathrm{TRAP}(n)\right| = \left|\sum_{k=1}^{n}\left(\int_{x_{k-1}}^{x_k} f(x)\,dx - \frac{\Delta x}{2}(y_{k-1} + y_k)\right)\right|$$

$$\le \sum_{k=1}^{n}\left|\int_{x_{k-1}}^{x_k} f(x)\,dx - \frac{\Delta x}{2}(y_{k-1} + y_k)\right|$$

$$= \sum_{k=1}^{n}\frac{K}{12}(\Delta x)^3 = \frac{K}{12}n(\Delta x)^3 = \frac{K(b - a)}{12}(\Delta x)^2,$$

since $n\Delta x = b - a$. □

We illustrate this error estimate for the approximations of Example 1 above.

Example 2 Obtain bounds for $|I - \mathrm{TRAP}(4)|$ and $|I - \mathrm{TRAP}(8)|$ where $I = \int_1^2 dx/x$.

Solution If $f(x) = 1/x$, then $f'(x) = -1/x^2$ and $f''(x) = 2/x^3$. On $[1, 2]$ we have $|f''(x)| \le 2$ so we may take $K = 2$ in the estimate. Thus

$$|I - \mathrm{TRAP}(4)| \le \frac{K}{12}(b - a)(\Delta x)^2 = \frac{2}{12}(2 - 1)\left(\frac{1}{4}\right)^2 = 0.0104\ldots,$$

$$|I - \mathrm{TRAP}(8)| \le \frac{2}{12}(2 - 1)\left(\frac{1}{8}\right)^2 = 0.0026\ldots.$$

We have already calculated the exact errors in Example 1, and now we observe that they are well within these bounds.

Simpson's Rule

The trapezoidal rule approximation to $\int_a^b f(x)\,dx$ results from approximating the graph of f by straight line segments through adjacent pairs of data points on that graph. Intuitively we would expect to do better if we approximate the graph by a more general curve. Since straight lines are the graphs of linear functions, the next obvious generalization is to use the class of quadratic functions, that is, to approximate the graph of f by segments of parabolas. This is the basis of Simpson's rule.

Suppose that we are given three points in the plane, one on each of three equally spaced vertical lines, spaced, say, h units apart. If we choose the middle of these lines as the y-axis, then the coordinates of the three points will be, say, $(-h, y_L)$, $(0, y_M)$, and (h, y_R), as illustrated in Fig. 6.38.

Constants A, B, and C can be chosen so that the parabolic arc $y = A + Bx + Cx^2$ passes through these points; evidently

$$\left.\begin{aligned} y_L &= A - Bh + Ch^2 \\ y_M &= A \\ y_R &= A + Bh + Ch^2 \end{aligned}\right\} \Rightarrow A = y_M, \qquad 2Ch^2 = y_L - 2y_M + y_R.$$

Now we have

$$\begin{aligned} \int_{-h}^{h} (A + Bx + Cx^2)\,dx &= \left(Ax + \frac{B}{2}x^2 + \frac{C}{3}x^3\right)\Bigg|_{-h}^{h} = 2Ah + \frac{2}{3}Ch^3 \\ &= h\left(2y_M + \frac{1}{3}(y_L - 2y_M + y_R)\right) = \frac{h}{3}(y_L + 4y_M + y_R). \end{aligned}$$

Thus the area of the plane region bounded by the parabolic arc, the interval of length $2h$ on the x-axis, and the left and right vertical lines is equal to $(h/3)$ times the sum of the heights of the region at the left and right edges and four times the height at the middle.

Now suppose that we are given the same data for f as we were given for the trapezoidal rule, that is, we know the values $y_k = f(x_k)$ $(0 \le k \le n)$ at $n + 1$ equally

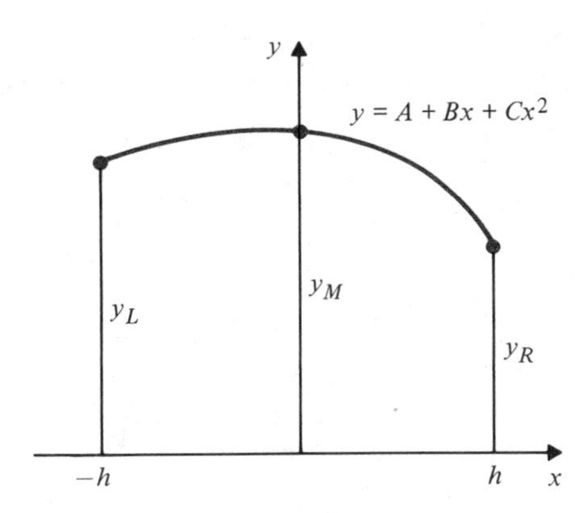

Figure 6.38

spaced points $a = x_0 < x_1 < x_2 < \cdots < x_{n-1} < x_n = b$, with spacing $\Delta x = (b - a)/n$. We can approximate the graph of f over *pairs* of the subintervals $[x_{k-1}, x_k]$ using parabolic segments and use the integrals of the corresponding quadratic functions to approximate the integrals of f over these subintervals. Since we need to use the subintervals two at a time we must assume that n is *even*. Using the integral computed for the parabolic segment above, we have

$$\int_{x_0}^{x_2} f(x)\,dx \simeq \frac{\Delta x}{3}(y_0 + 4y_1 + y_2)$$

$$\int_{x_2}^{x_4} f(x)\,dx \simeq \frac{\Delta x}{3}(y_2 + 4y_3 + y_4)$$

$$\vdots$$

$$\int_{x_{n-2}}^{x_n} f(x)\,dx \simeq \frac{\Delta x}{3}(y_{n-2} + 4y_{n-1} + y_n).$$

Adding these $(n/2)$ individual approximations we get the Simpson's rule approximation, SIMP(n), based on an even number n of subintervals.

$$\begin{aligned}\int_a^b f(x)\,dx &\simeq \frac{\Delta x}{3}(y_0 + 4y_1 + 2y_2 + 4y_3 + 2y_4 + \cdots + 2y_{n-2} + 4y_{n-1} + y_n)\\ &= \text{SIMP}(n)\\ &= \frac{\Delta x}{3}(y_{\text{"ends"}} + 4y_{\text{"odds"}} + 2y_{\text{"evens"}})\end{aligned}$$

Note that SIMP(n) requires no more data than TRAP(n); both require the values of $f(x)$ at $n + 1$ equally spaced points. However, SIMP(n) treats the data differently, weighting successive values either 1/3, 2/3, or 4/3. As we will see, this can produce a much better approximation to the integral of f.

Example 3 Calculate the approximations SIMP(4) and SIMP(8) for $I = \int_1^2 dx/x$ and compare them with the actual value $I = \ln 2 = 0.6931471\ldots$.

Solution Using the data provided in Example 1, we have

$$\text{SIMP}(4) = \frac{1}{12}\left(1 + 4\left(\frac{4}{5}\right) + 2\left(\frac{2}{3}\right) + 4\left(\frac{4}{7}\right) + \frac{1}{2}\right) = 0.6932539\ldots$$

$$\begin{aligned}\text{SIMP}(8) = \frac{1}{24}&\left(1 + 4\left(\frac{8}{9}\right) + 2\left(\frac{4}{5}\right) + 4\left(\frac{8}{11}\right) + 2\left(\frac{2}{3}\right)\right.\\ &\left. + 4\left(\frac{8}{13}\right) + 2\left(\frac{4}{7}\right) + 4\left(\frac{8}{15}\right) + \frac{1}{2}\right) = 0.6931545\ldots.\end{aligned}$$

The errors are

$$I - \text{SIMP}(4) \simeq -0.00011, \text{ and } I - \text{SIMP}(8) \simeq -0.000007,$$

evidently much smaller than the corresponding errors for TRAP(4) and TRAP(8).

Obtaining an error estimate for Simpson's rule is more difficult than it was for the trapezoidal rule. We state the appropriate estimate in the following theorem, but we do not attempt any proof. Proofs can be found in textbooks on numerical analysis.

Theorem 8 (*Error Estimate for Simpson's Rule*) If f is four times differentiable on $[a, b]$ and satisfies $|f^{(4)}(x)| \leq K$ there, then

$$\left| \int_a^b f(x)\,dx - \text{SIMP}(n) \right| \leq \frac{K(b-a)}{180}(\Delta x)^4,$$

where $\Delta x = (b - a)/n$. □

Observe that the error decreases as the fourth power of Δx as n increases. This accounts for the fact that SIMP(n) is a much better approximation than is TRAP(n), provided Δx is small and $|f^{(4)}(x)|$ is not unduly large compared to $|f''(x)|$. Note also that for any (even) n, SIMP(n) gives the exact value of the integral of any cubic function $f(x) = A + Bx + Cx^2 + Dx^3$, since $f^{(4)}(x) = 0$ identically for such f, and so we can take $K = 0$ in the error estimate.

Example 4 Obtain bounds for the absolute values of the errors in the approximations of Example 3.

Solution $f(x) = 1/x$, $f'(x) = -1/x^2$, $f''(x) = 2/x^3$, $f^{(3)}(x) = -6/x^4$, and so $f^{(4)}(x) = 24/x^5$. Clearly, $|f^{(4)}(x)| \leq 24$ on $[1, 2]$, so we can take $K = 24$ in the estimate of Theorem 8. We have

$$|I - \text{SIMP}(4)| \leq \frac{24(2-1)}{180}\left(\frac{1}{4}\right)^4 \simeq 0.00052,$$

$$|I - \text{SIMP}(8)| \leq \frac{24(2-1)}{180}\left(\frac{1}{8}\right)^4 \simeq 0.000032.$$

Again we observe that the actual errors are well within these bounds.

Example 5 Given the data in the following table for a function f, obtain the best trapezoidal rule approximation and the best Simpson's rule approximation that you can for $\int_0^{0.6} f(x)\,dx$. If it is known that $|f''(x)| \leq 22$ and $|f^{(4)}(x)| \leq 120$ on $[0, 0.6]$, obtain bounds for the errors in these approximations.

x	0	0.1	0.2	0.3	0.4	0.5	0.6
$f(x)$	1.0000	1.0005	1.0080	1.0405	1.1280	1.3125	1.6480

Solution The best approximations are obtained with $n = 6$, $\Delta x = 0.1$.

$$\text{TRAP}(6) = (0.1)\left(\frac{1}{2}(1.0000) + 1.0005 + 1.0080 + 1.0405 + 1.1280 + 1.3125 + \frac{1}{2}(1.6480)\right) = 0.68135$$

$$\text{SIMP}(6) = \frac{0.1}{3}(1.0000 + 1.6480 + 4(1.0005 + 1.0405 + 1.3125) + 2(1.0080 + 1.1280)) = 0.6778$$

The error bounds are

$$|I - \text{TRAP}(6)| \le \frac{22}{12}(0.6)(0.1)^2 = 0.011,$$

$$|I - \text{SIMP}(6)| \le \frac{120}{180}(0.6)(0.1)^4 = 0.00004.$$

EXERCISES

A scientific calculator can and should be used in some of these exercises. In Exercises 1–4, calculate the approximations TRAP(4), TRAP(8), SIMP(4), and SIMP(8) for the given integrals. Also calculate the exact value of each integral and so determine the exact error in each approximation. Compare these exact errors with the bounds for the size of the error supplied by Theorems 7 and 8.

1. $I = \int_0^2 (1 + x^2)\, dx$

2. $I = \int_0^1 e^{-x}\, dx$

3. $I = \int_0^{\pi/2} \sin x\, dx$

4. $I = \int_0^1 \frac{dx}{1 + x^2}$

5.*

a) Approximate $I = \int_0^1 e^{-x^2}$ using

i) the trapezoidal rule with $n = 10$,

ii) Simpson's rule with $n = 4$.

b) Find bounds for the absolute value of the error in each of the above approximations.

6.* Repeat Exercise 5 for $I = \int_1^2 \frac{\sin x}{x}\, dx$.

7. A function f has the following values.

x	0	0.2	0.4	0.6	0.8	1.0	1.2
$f(x)$	0.5000	0.4801	0.4211	0.3253	0.1967	0.0403	−0.1376

Find the best Simpson's rule approximation that you can for $I = \int_0^{1.2} f(x)\,dx$ using data from the table. If $|f^{(4)}(x)| \leq 2$ on $[0, 1.2]$, find a bound for the absolute value of the error.

8.* Compute the actual error in the approximation $\int_0^1 x^2\,dx \simeq \text{TRAP}(1)$ and use it to show that the constant 12 in the estimate of Theorem 7 cannot be improved. That is, show that the absolute value of the actual error is as large as allowed by that estimate.

9.* Compute the actual error in the approximation $\int_0^1 x^4\,dx \simeq \text{SIMP}(2)$ and use it to show that the constant 180 in the estimate of Theorem 8 cannot be improved.

10.* Since Simpson's rule is based on quadratic approximation, it is not surprising that it should give an exact value for an integral of $A + Bx + Cx^2$. It is more surprising that it is exact for a cubic function as well. Verify by direct calculation that $\int_0^1 x^3\,dx = \text{SIMP}(2)$.

6.9 IMPROPER INTEGRALS

Up to this point we have considered definite integrals of the form $I = \int_a^b f(x)\,dx$ where f is *continuous* on the *closed, finite* interval $[a, b]$. Since such a function is necessarily *bounded,* the integral I is necessarily a finite number; for positive f it corresponds to the area of a bounded region of the plane, that is, a region contained inside some disc of finite radius. Such integrals are also called **proper integrals.** We are now going to generalize the notion of definite integral to allow for two possibilities excluded in the situation described above:

i) We may have $a = -\infty$ or $b = \infty$ or both.

ii) f may become unbounded as x approaches a or b or both.

Definite integrals satisfying (i) are called **improper integrals of type I**; definite integrals satisfying (ii) are called **improper integrals of type II.** Either type of improper integral corresponds to the area of a region in the plane that "extends to infinity" in some direction and therefore is unbounded. As we will see, such integrals may or may not have finite values.

Example 1 Consider $I = \int_1^\infty (dx/x^2)$. This integral is improper of type I since the interval of integration is infinite. Evidently, I represents the area of the region S shaded in Fig. 6.39. It is not immediately obvious whether this region has finite area; the region has an infinitely long "spike" along the x-axis, but this spike becomes infinitely thin as x approaches ∞. In order to evaluate this improper integral, we interpret it as a limit of proper integrals over intervals $[1, R]$ as $R \to \infty$ (see Fig. 6.40).

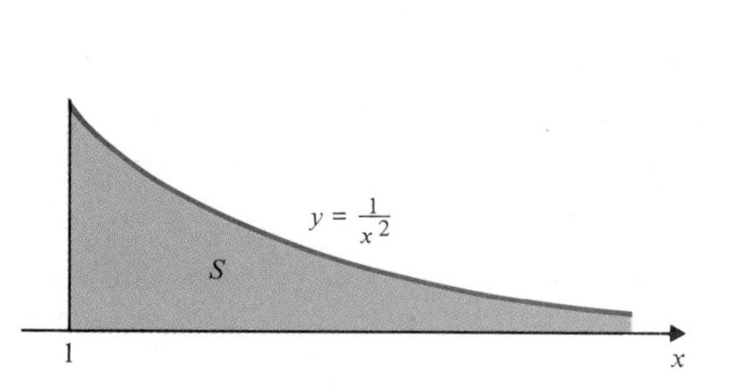

Figure 6.39

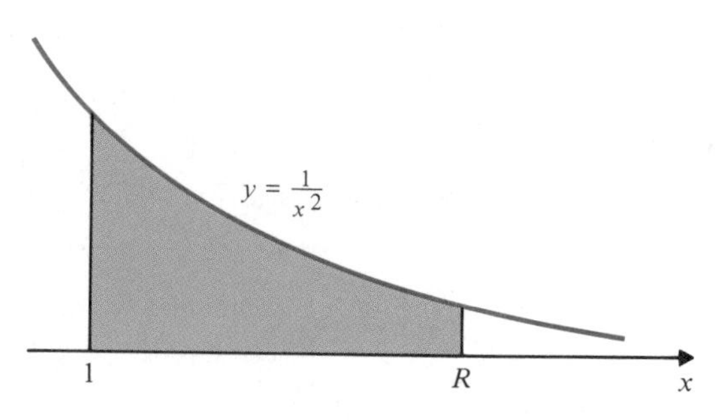

Figure 6.40

$$\int_1^\infty \frac{dx}{x^2} = \lim_{R\to\infty} \int_1^R \frac{dx}{x^2} = \lim_{R\to\infty} \left(-\frac{1}{x}\right)\Bigg|_1^R$$

$$= \lim_{R\to\infty} -\frac{1}{R} + 1 = 1$$

Since the limit exists (is finite), S has finite area, 1 square unit.

Definition 4

If f is continuous on $[a, \infty)$, we define the improper integral as the limit of proper integrals thus:

$$\int_a^\infty f(x)\, dx = \lim_{R\to\infty} \int_a^R f(x)\, dx.$$

If the limit is a finite number, we say that the improper integral **converges**; if the limit does not exist, or if it is ∞ or $-\infty$, we say that the improper integral **diverges** or **diverges to infinity** or **negative infinity.**

A similar definition holds for improper integrals of the form $\int_{-\infty}^b f(x)\, dx$. The integral $\int_{-\infty}^\infty f(x)\, dx$ is, for f continuous on the real line, improper of type I at both endpoints. We break it into two separate integrals:

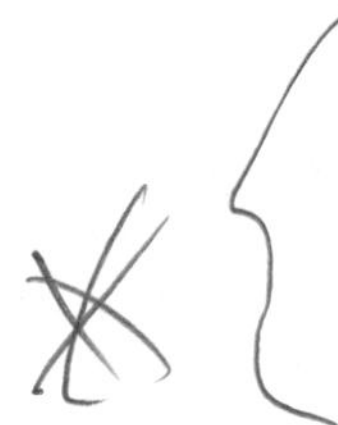

$$\int_{-\infty}^\infty f(x)\, dx = \int_{-\infty}^0 f(x)\, dx + \int_0^\infty f(x)\, dx.$$

The integral on the left converges if and only if each integral on the right converges.

Example 2 a)

$$\int_1^\infty \frac{dx}{x} = \lim_{R\to\infty} \int_1^R \frac{dx}{x} = \lim_{R\to\infty} \ln x \Bigg|_1^R = \lim_{R\to\infty} \ln R = \infty$$

This integral diverges to infinity. Note in Fig. 6.41 that, although the region whose area it represents looks similar to the region S of Example 1, this region has infinite area. Its spike is thicker than that of S for large x.

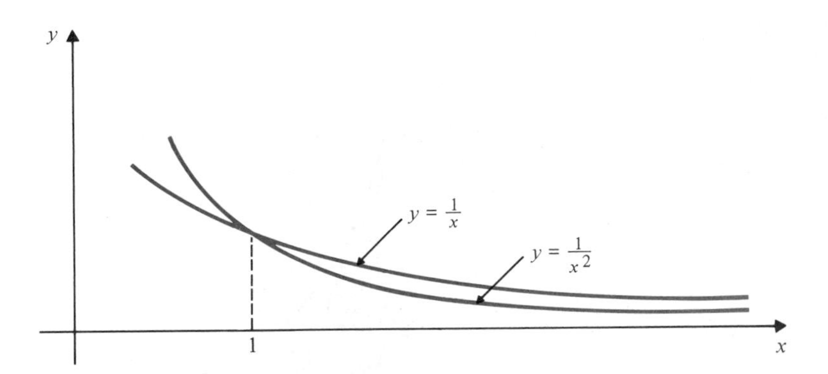

Figure 6.41

b) $$\int_{-\infty}^{\infty} \frac{dx}{1+x^2} = \int_{-\infty}^{0} \frac{dx}{1+x^2} + \int_{0}^{\infty} \frac{dx}{1+x^2} = 2\int_{0}^{\infty} \frac{dx}{1+x^2} \qquad \text{(by symmetry)}$$

$$= 2 \lim_{R\to\infty} \int_0^R \frac{dx}{1+x^2} = 2 \lim_{R\to\infty} \operatorname{Arctan} R = 2\left(\frac{\pi}{2}\right) = \pi$$

Perhaps the use of symmetry here requires some justification. At the time we used it we did not know whether each of the half-line integrals was finite or infinite. However, since both are positive, even if they are infinite their sum would still be twice one of them. If one had been positive and the other negative, we would not have been justified in canceling them to get 0 until we knew that they were finite. ($\infty + \infty = \infty$, but $\infty - \infty$ is not defined.) In any event, the given integral converges to π.

c) $$\int_0^{\infty} \cos x\, dx = \lim_{R\to\infty} \int_0^R \cos x\, dx = \lim_{R\to\infty} \sin R$$

This limit does not exist (and is not ∞ or $-\infty$), so all we can say is that the given integral diverges. See Fig. 6.42.

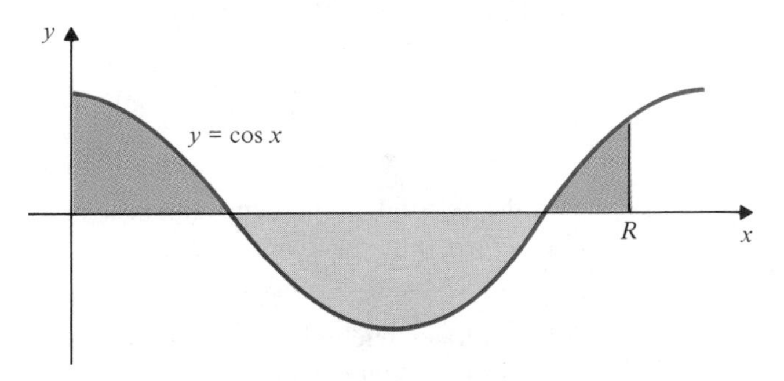

Figure 6.42

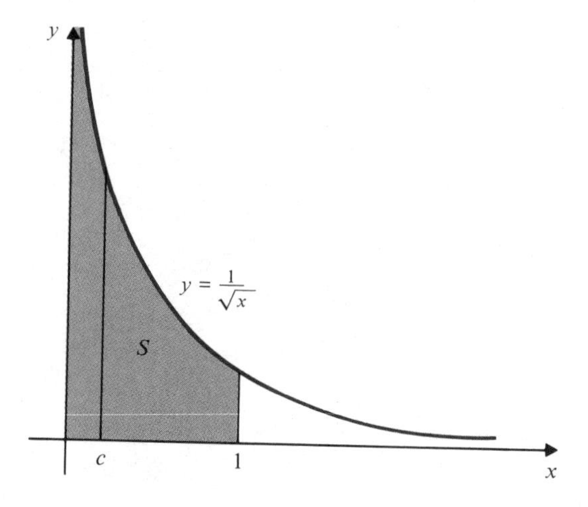

Figure 6.43

Example 3 Consider the integral $I = \int_0^1 dx/\sqrt{x}$. This integral is improper of type II since the integrand becomes unbounded near $x = 0$. Again the integral represents the area of a region S that is unbounded, having a spike extending to infinity along the y-axis, as shown in Fig. 6.43. As for improper integrals of type I, we express such integrals as limits of proper integrals.

$$I = \lim_{c \to 0+} \int_c^1 x^{-1/2}\, dx = \lim_{c \to 0+} 2x^{1/2} \Big|_c^1 = \lim_{c \to 0+} (2 - 2\sqrt{c}) = 2.$$

This integral converges, and S has finite area, 2 square units.

Definition 5

If f is continuous on the interval $(a, b]$ and possibly unbounded near a, we define the improper integral

$$\int_a^b f(x)\, dx = \lim_{c \to a+} \int_c^b f(x)\, dx.$$

Similarly, if f is continuous on $[a, b)$ and possibly unbounded near b, we define

$$\int_a^b f(x)\, dx = \lim_{c \to b-} \int_a^c f(x)\, dx.$$

Again the possibilities are convergence, divergence, divergence to infinity, and divergence to negative infinity.

While improper integrals of type I are always easily recognized because of the infinite limits of integration, improper integrals of type II can be somewhat harder to spot. You should be alert for singularities of integrands and especially points where they have vertical asymptotes. It may be necessary to break an improper integral into several improper integrals if it is improper at both endpoints or at points inside the interval of integration. For example,

$$\int_0^2 \frac{dx}{\sqrt{x|x-1|}} = \int_0^{1/2} \frac{dx}{\sqrt{x(1-x)}} + \int_{1/2}^1 \frac{dx}{\sqrt{x(1-x)}} + \int_1^2 \frac{dx}{\sqrt{x(x-1)}}.$$

Each integral on the right is improper because of a singularity of its integrand at one endpoint.

Example 4 a)

$$\int_0^1 \frac{dx}{x} = \lim_{c\to 0+} \int_c^1 \frac{dx}{x} = \lim_{c\to 0+} (-\ln c) = \infty$$

This integral diverges to infinity.

b)

$$\int_0^1 \frac{dx}{(1-x)^{1/3}} = \lim_{c\to 1-} \int_0^c \frac{dx}{(1-x)^{1/3}} \qquad \text{Let } u = 1 - x,\ du = -dx.$$

$$= \lim_{c\to 1-} -\int_1^{1-c} u^{-1/3}\,du = \lim_{c\to 1-} -\frac{3}{2}u^{2/3}\Big|_1^{1-c} = \frac{3}{2}$$

This integral converges to 3/2.

c)

$$\int_0^2 \frac{dx}{\sqrt{2x-x^2}} = \int_0^2 \frac{dx}{\sqrt{1-(x-1)^2}} \qquad \text{Let } u = x - 1,\ du = dx.$$

$$= \int_{-1}^1 \frac{du}{\sqrt{1-u^2}} = 2\int_0^1 \frac{du}{\sqrt{1-u^2}} \qquad \text{(by symmetry)}$$

$$= 2 \lim_{c\to 1-} \operatorname{Arcsin} u \Big|_0^c = 2 \lim_{c\to 1-} \operatorname{Arcsin} c = \pi$$

This integral converges to π.

Examples 4(b) and 4(c) show how a change of variable can be made before or after an improper integral is expressed as a limit of proper integrals.

d)

$$\int_0^1 \ln x\,dx = \lim_{c\to 0+} \int_c^1 \ln x\,dx$$

(See Example 2(a) of Section 6.7 for the evaluation of this integral.)

$$= \lim_{c\to 0+} (x\ln x - x)\Big|_c^1$$

$$= \lim_{c\to 0+} (-1 - c\ln c + c)$$

$$= -1 - \lim_{c\to 0+} \frac{\ln c}{1/c} \qquad \left[\frac{-\infty}{\infty}\right] \text{(Use l'Hôpital's rule.)}$$

$$= -1 - \lim_{c\to 0+} \frac{1/c}{-(1/c^2)} = -1 - \lim_{c\to 0+} (-c) = -1.$$

The integral converges to -1.

The following theorem summarizes the behavior of improper integrals of types I and II of powers of x.

Theorem 9 If $0 < a < \infty$, then

$$\int_a^\infty x^{-p}\,dx \begin{cases} \text{converges to } \dfrac{a^{1-p}}{p-1} & \text{if } p > 1 \\ \text{diverges to } \infty & \text{if } p \le 1 \end{cases}$$

$$\int_0^a x^{-p}\,dx \begin{cases} \text{converges to } \dfrac{a^{1-p}}{1-p} & \text{if } p < 1 \\ \text{diverges to } \infty & \text{if } p \ge 1. \ \square \end{cases}$$

The proof is left as an exercise (Exercise 17 at the end of this section). Note that $\int_0^\infty x^{-p}\,dx$ does not converge for any value of p.

Integrals of Bounded, Piecewise Continuous Functions

If $f(x)$ is bounded (i.e., $|f(x)| \le K$) and continuous on the half-open interval $[a, b)$ we may still treat the integral $\int_a^b f(x)\,dx$ as improper (even though it measures the "area" of a bounded region) and evaluate it as

$$\int_a^b f(x)\,dx = \lim_{c \to b-} \int_a^c f(x)\,dx.$$

See Fig. 6.44. Similarly, if $f(x)$ is bounded and continuous on $(a, b]$, then

$$\int_a^b f(x)\,dx = \lim_{c \to a+} \int_c^b f(x)\,dx.$$

See Fig. 6.45. Such integrals necessarily converge (we don't prove this here) and provide a means for integrating functions with finitely many points of discontinuity.

We say that $f(x)$ is **piecewise continuous** on (a, b) if it is continuous at all points of that interval except for a finite number of points. Suppose that $f(x)$ is both bounded and piecewise continuous on (a, b), having discontinuities only at the points $x_1 <$

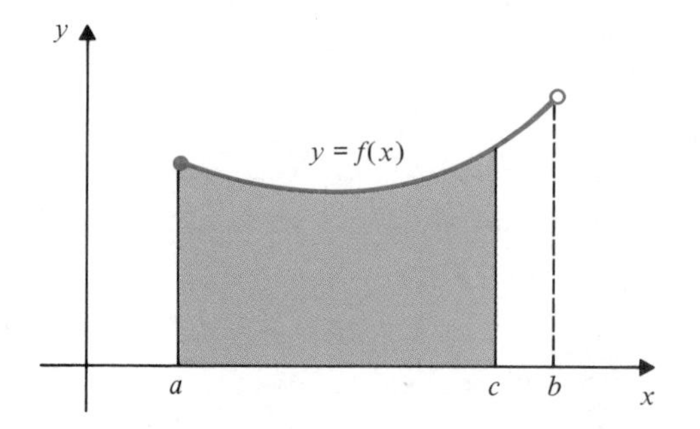

Figure 6.44

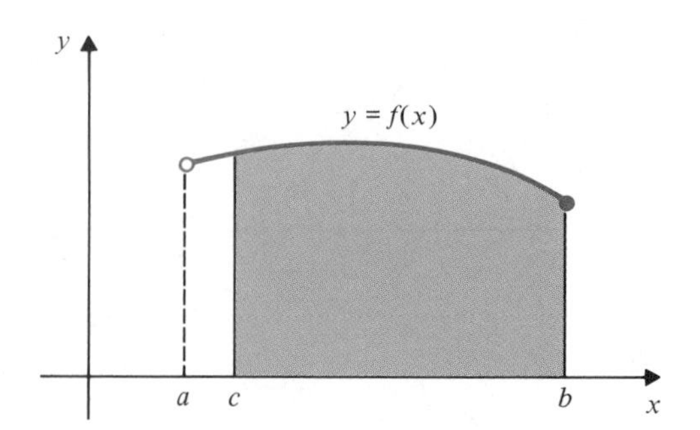

Figure 6.45

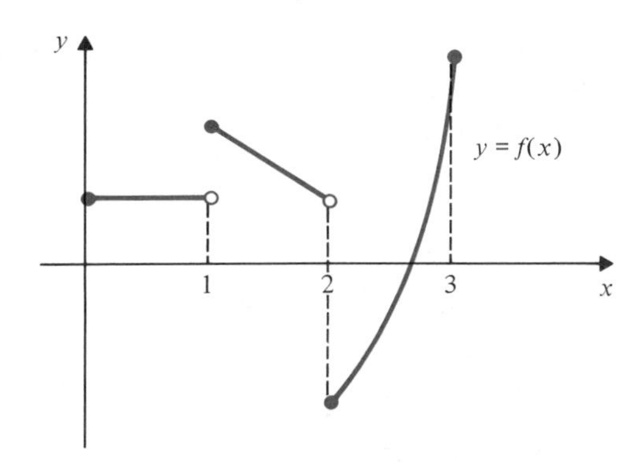

Figure 6.46

$x_2 < x_3 < \cdots < x_{n-1}$. We define $\int_a^b f(x)\,dx$ as the sum of integrals over open intervals where $f(x)$ is continuous:

$$\int_a^b f(x)\,dx = \sum_{j=1}^{n} \int_{x_{j-1}}^{x_j} f(x)\,dx \qquad (\text{where } x_0 = a, x_n = b).$$

Each integral in the sum may be improper at either endpoint.

Example 5 See Fig. 6.46. Let

$$f(x) = \begin{cases} 1 & \text{if } 0 \le x < 1 \\ 3 - x & \text{if } 1 \le x < 2 \\ x^2 - 6 & \text{if } 2 \le x \le 3. \end{cases}$$

Then

$$\int_0^3 f(x)\,dx = \int_0^1 dx + \int_1^2 (3 - x)\,dx + \int_2^3 (x^2 - 6)\,dx = 1 + \frac{3}{2} + \frac{1}{3} = \frac{17}{6}.$$

To integrate a function defined to be different continuous functions on different intervals, we merely add the integrals of the various component functions.

Estimating Convergence and Divergence

When an improper integral cannot be evaluated by the fundamental theorem of calculus because an antiderivative can't be found, we may still be able to determine whether the integral converges by comparing it with simpler integrals. The following theorem lies at the heart of this approach.

Theorem 10 (*The Comparison Theorem for Improper Integrals*) Let $-\infty \le a < b \le \infty$ and suppose that functions f and g are continuous on the interval (a, b) and satisfy $0 \le f(x) \le Kg(x)$ there for some positive constant K. If $\int_a^b g(x)\,dx$ converges, then so does $\int_a^b f(x)\,dx$. Equivalently, if $\int_a^b f(x)\,dx$ diverges to ∞, then so does $\int_a^b g(x)\,dx$.

Proof Since both integrands are nonnegative, there are only two possibilities for each integral: It can converge to a nonnegative number or diverge to ∞. Since $f(x) \leq Kg(x)$ on (a, b), it follows by Theorem 2(e) of Section 6.2 that if $a < r < s < b$, then

$$\int_r^s f(x)\,dx \leq K\int_r^s g(x)\,dx.$$

This theorem now follows by taking limits as $r \to a+$ and $s \to b-$. □

Example 6 Show that $\int_0^\infty e^{-x^2}\,dx$ converges and find an upper bound for its value.

Solution On $[0, 1]$ we have $0 < e^{-x^2} \leq 1$, so

$$0 < \int_0^1 e^{-x^2}\,dx \leq \int_0^1 dx = 1.$$

On $[1, \infty)$ we have $x^2 \geq x$ so that $0 < e^{-x^2} \leq e^{-x}$. Thus

$$0 < \int_1^\infty e^{-x^2}\,dx \leq \int_1^\infty e^{-x}\,dx = \lim_{R\to\infty} \left.\frac{e^{-x}}{-1}\right|_1^R$$

$$= \lim_{R\to\infty} \frac{1}{e} - \frac{1}{e^R} = \frac{1}{e}.$$

Hence $\int_0^\infty e^{-x^2}\,dx$ converges and its value is not greater than $1 + (1/e)$.

We remark that the above integral is in fact equal to $\sqrt{\pi}/2$ although we cannot prove this at this time. It is normally proved in advanced calculus courses using multivariable techniques.

Example 7 Determine whether

$$\int_0^\infty \frac{dx}{\sqrt{x + x^3}}$$

converges.

Solution The integral is improper of both types, so we write

$$\int_0^\infty \frac{dx}{\sqrt{x + x^3}} = \int_0^1 \frac{dx}{\sqrt{x + x^3}} + \int_1^\infty \frac{dx}{\sqrt{x + x^3}} = I_1 + I_2.$$

On $(0, 1]$ we have $\sqrt{x + x^3} > \sqrt{x}$, so

$$I_1 < \int_0^1 \frac{dx}{\sqrt{x}} = 2 \qquad \text{(by Example 3).}$$

On $[1, \infty)$ we have $\sqrt{x + x^3} > \sqrt{x^3}$, so

$$I_2 < \int_1^\infty x^{-3/2}\,dx = 2 \qquad \text{(by Theorem 9).}$$

Hence the given integral converges and its value is less than 4.

Example 8 Determine whether

$$\int_2^\infty \frac{dx}{\ln x}$$

converges.

Solution For $x > 2$ we have $\ln x < x$, so

$$\int_2^\infty \frac{dx}{\ln x} > \int_2^\infty \frac{dx}{x},$$

which diverges to infinity by Theorem 9. Hence the given integral diverges to infinity by Theorem 10.

EXERCISES

In Exercises 1–14 evaluate the given integral or show that it diverges.

1. $\displaystyle\int_0^\infty e^{-2x}\,dx$

2. $\displaystyle\int_{-\infty}^{-1} \frac{dx}{x^2+1}$

3. $\displaystyle\int_{-1}^{1} \frac{dx}{(x+1)^{2/3}}$

4. $\displaystyle\int_0^a \frac{dx}{a^2-x^2}$

5. $\displaystyle\int_0^{\pi/2} \frac{\cos x\,dx}{(1-\sin x)^{2/3}}$

6. $\displaystyle\int_1^\infty \frac{dx}{(x-1)^2}$

7. $\displaystyle\int_0^\infty x\,e^{-x}\,dx$

8. $\displaystyle\int_0^1 x\ln x\,dx$

9. $\displaystyle\int_0^1 \frac{dx}{\sqrt{x(1-x)}}$

10. $\displaystyle\int_0^\infty \frac{x}{1+2x^2}\,dx$

11. $\displaystyle\int_0^\infty \frac{x\,dx}{(1+2x^2)^{3/2}}$

12. $\displaystyle\int_0^{\pi/2} \sec x\,dx$

13. $\displaystyle\int_0^{\pi/2} \tan x\,dx$

14. $\displaystyle\int_e^\infty \frac{dx}{x(\ln x)^p}\quad (p>1)$

15. Find the area of the region that lies above $y = 0$, to the right of $x = 1$, and under the curve $y = (3/(x+2)) - (2/(x+1))$.

16. Find the area of the plane region that lies under the graph of $y = x^{-2}e^{-1/x}$, above the x-axis, and to the right of the y-axis.

17. Prove Theorem 9 in Section 6.9.

18. Evaluate $\int_{-1}^{1} (x \operatorname{sgn} x)/(x+2)\,dx$. Recall that $\operatorname{sgn} x = x/|x|$.

19. Evaluate $\int_0^2 x^2 \operatorname{sgn}(x-1)\,dx$.

20. Evaluate $\int_{-2}^{2} (x+1)\operatorname{sgn}(x^2-1)\,dx$.

21. Evaluate $\int_0^2 [x^2]\,dx$ where $[t]$ denotes the greatest integer $\leq t$.

In Exercises 22–31 state whether the given integral converges or diverges and justify your claim.

22. $\int_0^\infty \frac{x^2}{x^5 + 1}\,dx$

23. $\int_0^\infty \frac{dx}{1 + \sqrt{x}}$

24. $\int_0^\infty e^{-x^3}\,dx$

25. $\int_0^\pi \frac{\sin x}{x}\,dx$

26.* $\int_0^\infty \frac{|\sin x|}{x^2}\,dx$

27.* $\int_0^{\pi^2} \frac{dx}{1 - \cos\sqrt{x}}$

28.* $\int_0^\infty \frac{dx}{\sqrt{x} + x^2}$

29.* $\int_{-1}^1 \frac{e^x}{x + 1}\,dx$

30.* $\int_2^\infty \frac{dx}{\sqrt{x}\ln x}$

31.* $\int_0^\infty \frac{dx}{xe^x}$

32.* Given that $\int_0^\infty e^{-x^2}\,dx = \frac{1}{2}\sqrt{\pi}$, evaluate the following.

a) $\int_0^\infty x^2 e^{-x^2}\,dx$

b) $\int_0^\infty x^4 e^{-x^2}\,dx$

Applications of Integration

7.1 Areas of Plane Regions
7.2 Volumes
7.3 Arc Length and Surface Area
7.4 Geometric Applications for Polar and Parametric Curves
7.5 Mass, Moments, and Center of Mass
7.6 Centroids
7.7 Other Physical Applications
7.8 Probability
7.9 First-Order Separable and Linear Differential Equations

Numerous mathematical and physical quantities can be conveniently represented by definite integrals, both proper and improper. In addition to measuring plane areas, the problem that motivated the definition of the definite integral, we can use these integrals to express volumes of solids, lengths of curves, areas of surfaces, probabilities, forces, work, energy, pressure, and a variety of other quantities that are in one sense or another equivalent to areas under graphs.

In this chapter we develop a few of these applications.

7.1 AREAS OF PLANE REGIONS

In this section we review and extend the use of definite integrals to represent plane areas introduced in Chapter 6. Recall that the integral $\int_a^b f(x)\,dx$ measures the area between the graph of f and the x-axis from $x = a$ to $x = b$, but it counts as *negative* any part of this area that lies below the x-axis. (We are assuming that $a < b$.) In order to express the total area bounded by $y = f(x)$, $y = 0$, $x = a$, and $x = b$, counting all of the area positively, we can use the absolute value of f (see Fig. 7.1):

$$\int_a^b f(x)\,dx = A_1 - A_2$$

$$\int_a^b |f(x)| = A_1 + A_2.$$

Example 1 The area bounded by $y = \cos x$, $y = 0$, $x = 0$, and $x = 3\pi/2$ (see Fig. 7.2) is

$$A = \int_0^{3\pi/2} |\cos x|\,dx$$

$$= \int_0^{\pi/2} \cos x\,dx + \int_{\pi/2}^{3\pi/2} (-\cos x)\,dx$$

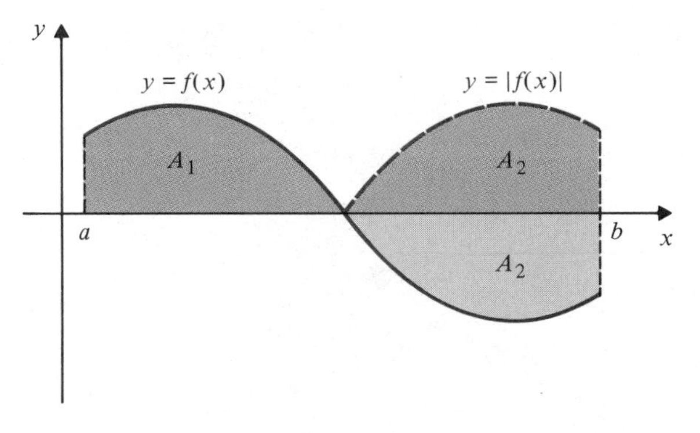

Figure 7.1

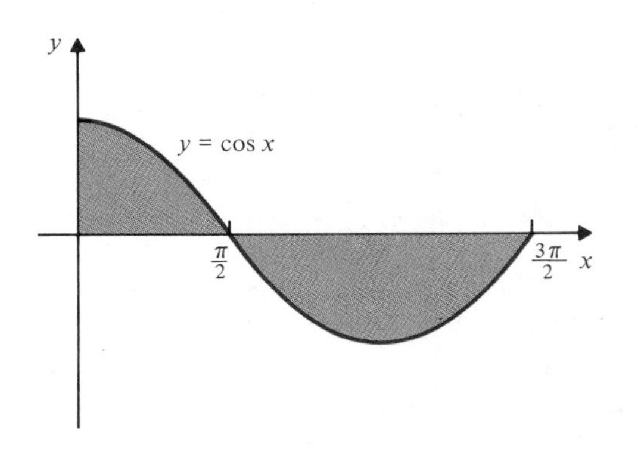

Figure 7.2

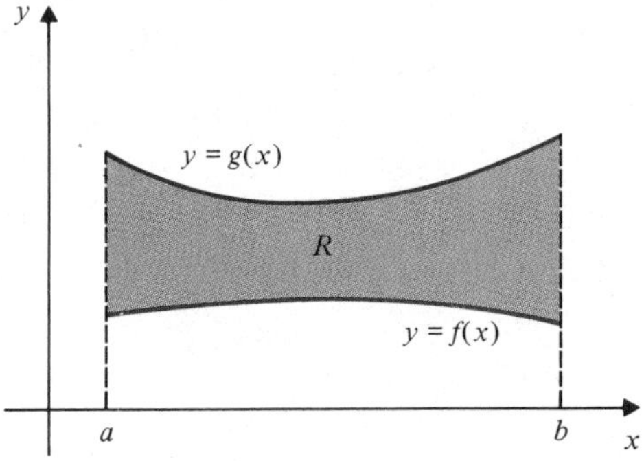

Figure 7.3

$$= \sin x \Big|_0^{\pi/2} - \sin x \Big|_{\pi/2}^{3\pi/2}$$

$$= (1 - 0) - (-1 - 1) = 3 \text{ sq units.}$$

Note that there is no "rule" for integrating $\int_a^b |f(x)|\, dx$; one must break the integral into a sum of integrals over intervals where $|f(x)| = f(x)$ and intervals where $|f(x)| = -f(x)$.

Areas Between Two Curves

Suppose that a plane region R is bounded by the graphs of two continuous functions, $y = f(x)$ and $y = g(x)$, and the vertical straight lines $x = a$ and $x = b$, as shown in Fig. 7.3. Assume that $a < b$ and that $f(x) \le g(x)$ on $[a, b]$, so the graph of f lies below that of g. If f is nonnegative-valued on $[a, b]$, the area A of R is evidently given by

$$A = \int_a^b g(x)\, dx - \int_a^b f(x)\, dx = \int_a^b (g(x) - f(x))\, dx.$$

It is useful to regard this formula as expressing A as the "sum" (that is, the integral) of *infinitely many differential elements of area*

$$dA = (g(x) - f(x))\, dx,$$

corresponding to values of x between a and b. Each such area element can be regarded as the area of an infinitely thin vertical rectangle of width dx and height $g(x) - f(x)$ (see Fig. 7.4). Even if f and g can take on negative values on $[a, b]$, but $f(x) \le g(x)$ there, this interpretation, and the resulting area formula

$$A = \int_a^b (g(x) - f(x))\, dx$$

remain valid.

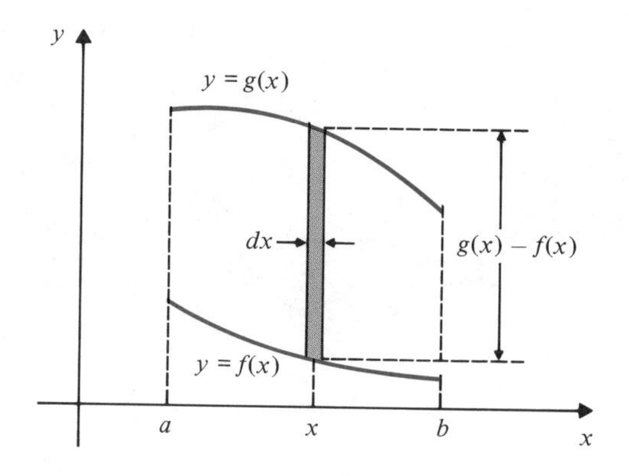

Figure 7.4

More generally, if the restriction $f(x) \le g(x)$ is removed, then the vertical rectangle of width dx at position x extending between the graphs of f and g has height $|f(x) - g(x)|$ and hence area $dA = |f(x) - g(x)|\,dx$ (see Fig. 7.5). Hence the total area lying between the graphs $y = f(x)$ and $y = g(x)$ and between the vertical lines $x = a$ and $x = b > a$ is given by

$$A = \int_a^b |f(x) - g(x)|\,dx.$$

Example 2 Find the area of the bounded, plane region lying between the curves $y = x^2 - 2x$ and $y = 4 - x^2$.

Solution For intersections of the curves we have $x^2 - 2x = y = 4 - x^2$, so $2x^2 - 2x - 4 = 0$. Thus $2(x - 2)(x + 1) = 0$ and $x = 2$ or -1. The region R is sketched in Fig. 7.6. (A sketch should always be made in problems of this sort.) Its area is

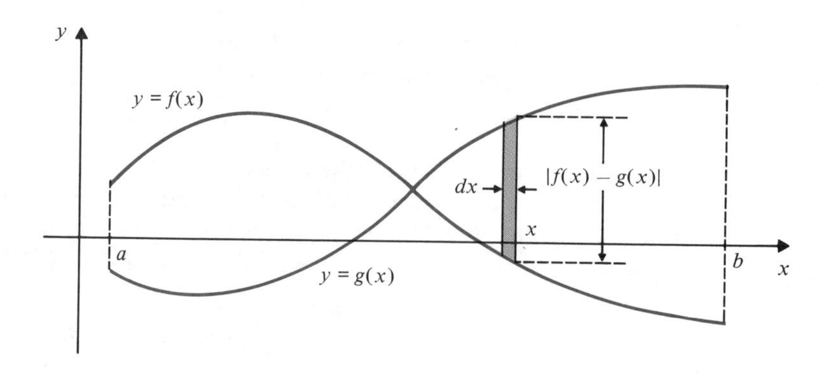

Figure 7.5

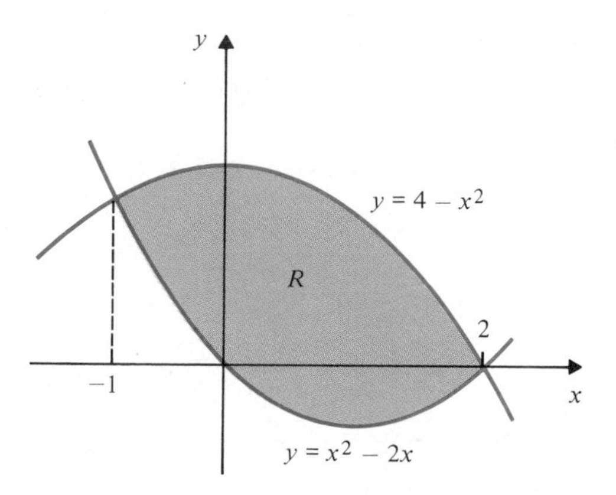

Figure 7.6

$$A = \int_{-1}^{2} ((4 - x^2) - (x^2 - 2x))\,dx$$

$$= \int_{-1}^{2} (4 - 2x^2 + 2x)\,dx = \left(4x - \frac{2}{3}x^3 + x^2\right)\Bigg|_{-1}^{2}$$

$$= 4(2) - \frac{2}{3}(8) + 4 - \left(-4 + \frac{2}{3} + 1\right)$$

$$= 9 \text{ sq units.}$$

Note that in representing the area as an integral it is essential that the height y to the lower curve be subtracted from the height y to the upper curve. Subtracting the wrong way would have produced a negative value for the area.

Example 3 Find the total area lying between the curves $y = \sin x$ and $y = \cos x$ from $x = 0$ to $x = 2\pi$.

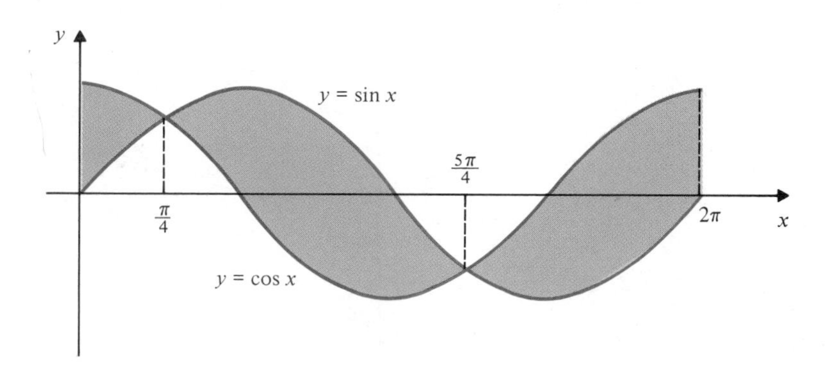

Figure 7.7

Solution The region is shaded in Fig. 7.7. Evidently the graphs of sine and cosine cross at $x = \pi/4$ and $x = 5\pi/4$. The required area is

$$A = \int_0^{\pi/4} (\cos x - \sin x)\, dx + \int_{\pi/4}^{5\pi/4} (\sin x - \cos x)\, dx + \int_{5\pi/4}^{2\pi} (\cos x - \sin x)\, dx$$

$$= (\sin x + \cos x)\Big|_0^{\pi/4} - (\cos x + \sin x)\Big|_{\pi/4}^{5\pi/4} + (\sin x + \cos x)\Big|_{5\pi/4}^{2\pi}$$

$$= (\sqrt{2} - 1) + (\sqrt{2} + \sqrt{2}) + (1 + \sqrt{2}) = 4\sqrt{2} \text{ sq units.}$$

The formula for area between curves also holds for unbounded regions; we can regard the area of such a region as the limit of areas of bounded regions and hence express the area as an improper integral.

Example 4 Find the area of the plane region R bounded by the curve $y = e^x/(1 + e^x)$ and the line $y = 1$ and lying to the right of the y-axis.

Solution The region R is shown in Fig. 7.8. Its area is

$$A = \int_0^{\infty} \left(1 - \frac{e^x}{1 + e^x}\right) dx = \int_0^{\infty} \frac{dx}{1 + e^x}$$

$$= \lim_{R\to\infty} \int_0^R \frac{e^{-x}}{e^{-x} + 1}\, dx.$$

Written in this last form, the integral can be evaluated via the substitution $u = e^{-x} + 1$, $du = -e^{-x}\, dx$. We have

$$A = -\lim_{R\to\infty} \int_2^{e^{-R}+1} \frac{du}{u} = -\lim_{R\to\infty} (\log(e^{-R} + 1) - \log 2) = \log 2 \text{ sq units.}$$

It is sometimes convenient to interchange the roles of x and y and integrate over an interval of the y-axis instead of the x-axis. The region R lying to the right of $x = f(y)$ and to the left of $x = g(y)$ and between the horizontal lines $y = c$ and $y =$

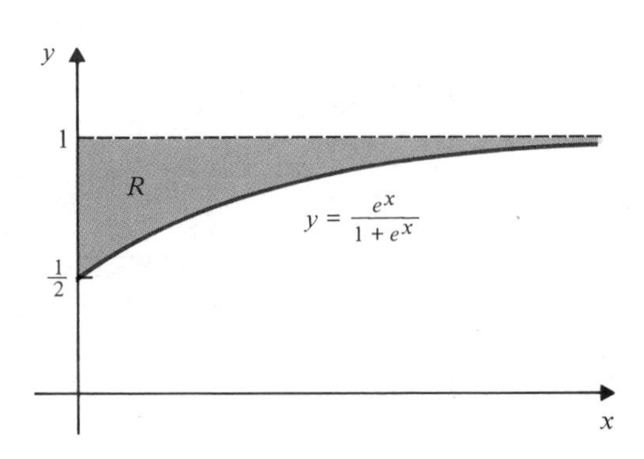

Figure 7.8

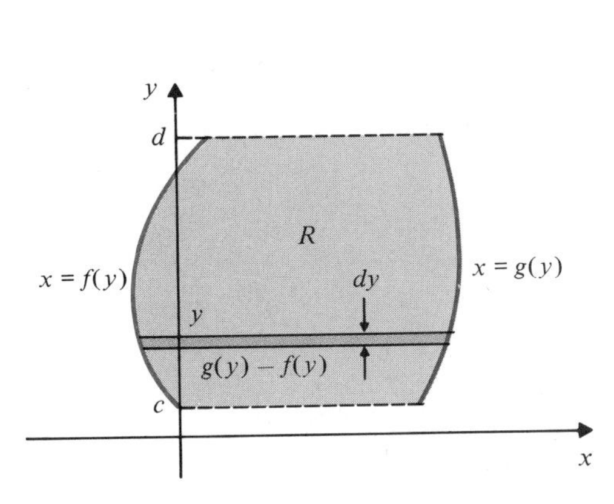

Figure 7.9

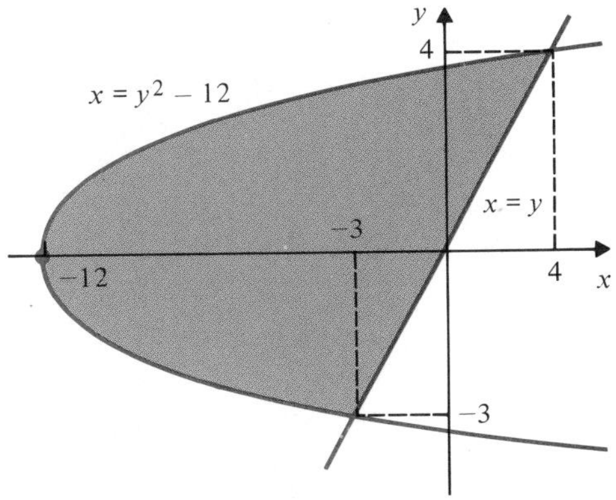

Figure 7.10

$d > c$ (see Fig. 7.9) has area element $dA = (g(y) - f(y))\, dy$ and its area is

$$A = \int_c^d (g(y) - f(y))\, dy.$$

Example 5 Find the area of the plane region lying to the right of the parabola $x = y^2 - 12$ and to the left of the straight line $y = x$, as illustrated in Fig. 7.10.

Solution For intersection of the curves:

$$y^2 - 12 = x = y$$

$$y^2 - y - 12 = 0$$

$$(y - 4)(y + 3) = 0 \Rightarrow y = 4 \text{ or } -3.$$

The area is

$$A = \int_{-3}^{4} (y - (y^2 - 12))\, dy = \left(\frac{y^2}{2} - \frac{y^3}{3} + 12y\right)\Bigg|_{-3}^{4} = \frac{343}{6} \text{ sq units.}$$

Of course, the same result could have been obtained by integrating in the x direction, but the integral would have been more complicated:

$$A = \int_{-12}^{-3} (\sqrt{12 + x} - (-\sqrt{12 + x}))\, dx + \int_{-3}^{4} (\sqrt{12 + x} - x)\, dx.$$

For regions bounded by curves that are the graphs of functions of y, it is usually easier to integrate in the y direction.

EXERCISES

In Exercises 1–24 find the area of the plane region bounded by the given curves. In each case you should make a sketch of the region.

1. $y = x$ and $y = x^2$

2. $y = \sqrt{x}$ and $y = x^2$

3. $y = x^2 - 5$ and $y = 3 - x^2$
4. $y = x^2 - 2x$ and $y = 6x - x^2$
5. $2y = 4x - x^2$ and $2y + 3x = 6$
6. $x - y = 7$ and $x = 2y^2 - y + 3$
7. $y = x^3$ and $y = x$
8. $y = x^3$ and $y = x^2$
9. $y = x^3$ and $x = y^2$
10. $x = y^2$ and $x = 2y^2 - y - 2$
11. $y = \dfrac{1}{x}$ and $2x + 2y = 5$
12. $y = (x^2 - 1)^2$ and $y = 1 - x^2$
13. $y = \dfrac{1}{2}x^2$ and $y = \dfrac{1}{x^2 + 1}$
14. $y = \dfrac{4x}{3 + x^2}$ and $y = 1$
15. $y = \dfrac{4}{x^2}$ and $y = 5 - x^2$
16. Above the curve $xy = 12$ $(x > 0)$ and inside the circle $x^2 + y^2 = 25$
17. $y = \sin x$ and $y = \cos x$ between two consecutive intersections of these curves
18.* $y = 4x/\pi$ and $y = \tan x$, between $x = 0$ and the first intersection of these curves to the right of $x = 0$
19. Inside the circle $x^2 + y^2 = a^2$ and above the line $y = b$ $(-a \le b \le a)$
20. To the left of $\dfrac{x^2}{a^2} + \dfrac{y^2}{b^2} = 1$ and to the right of the line $x = c$, where $-a \le c \le a$
21.* Below $y = e^{-x} \sin x$ and above $y = 0$ from $x = 0$ to $x = \pi$
22. Below $y = e^{-x}$ and above $y = e^{-2x}$ to the right of $x = 0$
23.* Below $y = 0$ and above $y = \ln x$, to the right of $x = 0$
24. Inside both of the circles $x^2 + y^2 = 1$ and $(x - 2)^2 + y^2 = 4$
25. Find the total area enclosed by the curve $y^2 = x^2 - x^4$.
26. Find the area of the closed loop of the curve $y^2 = x^4(2 + x)$ that lies to the left of the origin.
27. Find the area of the finite plane region that is bounded by the curve $y = e^x$, the line $x = 0$, and the tangent line to $y = e^x$ at $x = 1$.
28. Find the area of the finite plane region bounded by the curve $y = \ln x$, the line $y = 1$, and the tangent line to $y = \ln x$ at $x = 1$.
29.* Find the area of the finite plane region bounded by the curve $y = x^3$ and the tangent line to that curve at the point $(1, 1)$. (*Hint:* Find the other point at which that tangent line meets the curve.)

7.2 VOLUMES

In this section we show how the volumes of certain three-dimensional regions (or "solids") can be expressed as definite integrals and thereby evaluated.

Suppose that the cross-sectional area of a certain solid in every plane perpendicular to some fixed line is known. More specifically, suppose that the solid S lies between planes perpendicular to the x-axis at positions $x = a$ and $x = b$ and that the cross-

* Difficult and/or theoretical problems.

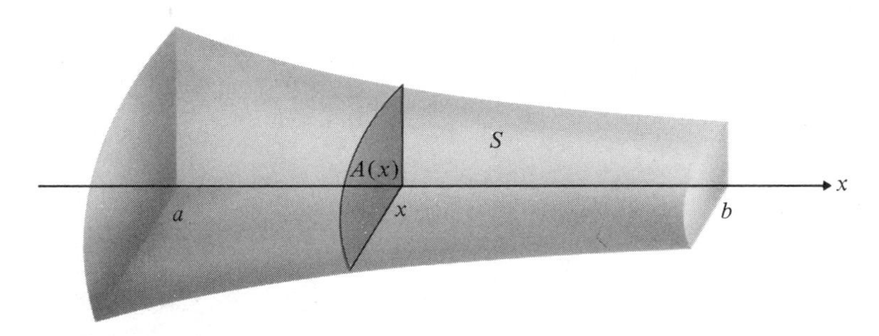

Figure 7.11

sectional area of S in the plane perpendicular to the x-axis at x is $A(x)$, where $a \le x \le b$. This is shown in Fig. 7.11. We assume that $A(x)$ is continuous on $[a, b]$.

We will show that the volume V of the solid S is given by

$$V = \int_a^b A(x)\,dx.$$

Subdivide $[a, b]$ into n equal subintervals of length $\Delta x = (b - a)/n$ using points $a = x_0 < x_1 < x_2 < \cdots < x_n = b$. For $1 \le k \le n$ let ΔV_k be the volume of S_k, that part of S lying between the planes at positions $x = x_{k-1}$ and $x = x_k$. S_k is a slice of the solid having thickness Δx. See Fig. 7.12. If $A(x) = A$ were constant through this slice, then the volume of the slice would be $\Delta V_k = A\Delta x$. Even if $A(x)$ is not constant but is continuous on $[x_{k-1}, x_k]$ there exists a point c_k in that interval such that

$$\Delta V_k = A(c_k)\Delta x.$$

This is an application of the intermediate-value theorem; $\Delta V_k/\Delta x$ lies between the least and greatest values of $A(x)$ in $[x_{k-1}, x_k]$. Hence the volume of S is

$$V = \sum_{k=1}^{n} \Delta V_k = \sum_{k=1}^{n} A(c_k)\Delta x.$$

Evidently the sum on the right is a Riemann sum for $\int_a^b A(x)\,dx$, and so taking the limit as $n \to \infty$ we get the desired result:

$$V = \int_a^b A(x)\,dx.$$

One can think of this formula as giving V as the sum of infinitely many differential elements of volume $dV = A(x)\,dx$ where each such element is the volume of a "slice" of S having area $A(x)$ and infinitesimal thickness dx. Hereafter we shall approach new volume formulas from this suggestive, if less rigorous, point of view.

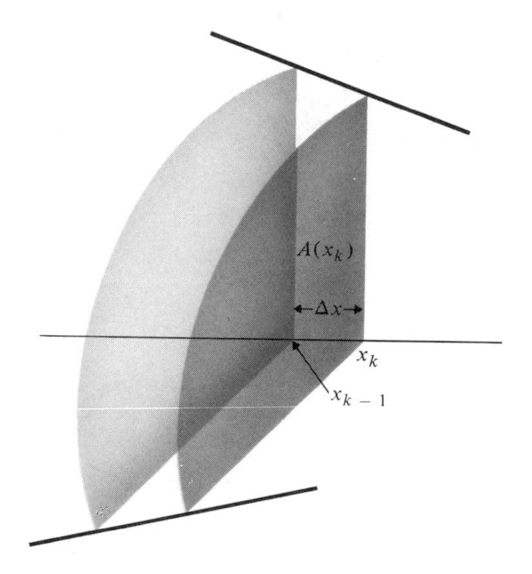

Figure 7.12

Example 1 If a pyramid has a polygonal base of area A and has height h measured perpendicular to the plane of the base, then cross sections of the pyramid in planes parallel to the base are similar polygons. If the origin is at the vertex of the pyramid and the x-axis is perpendicular to the base, then the cross section at position x is similar in shape to the base and has linear dimensions x/h times those of the base (see Fig. 7.13). Thus the area of the cross section at x is

$$A(x) = \left(\frac{x}{h}\right)^2 A.$$

The volume of the pyramid is therefore

$$V = \int_0^h \left(\frac{x}{h}\right)^2 A\,dx = \frac{A}{h^2}\frac{x^3}{3}\bigg|_0^h = \frac{1}{3}Ah \text{ cu units.}$$

A similar argument, using the same formula, holds for a cone, that is, a pyramid with a more general (curved) shape to its base, such as that in Fig. 7.14.

$$V = \frac{1}{3}Ah.$$

Example 2 The base of a certain solid is a circle of radius a. Each cross section of the solid perpendicular to a fixed diameter of the base is an isosceles triangle of height b. (Thus the shape of the solid is that of a tent with circular base and ridgepole at constant height above a diameter of the base.) Find the volume of the solid.

Solution Let the x-axis be the diameter perpendicular to which the cross sections are isosceles

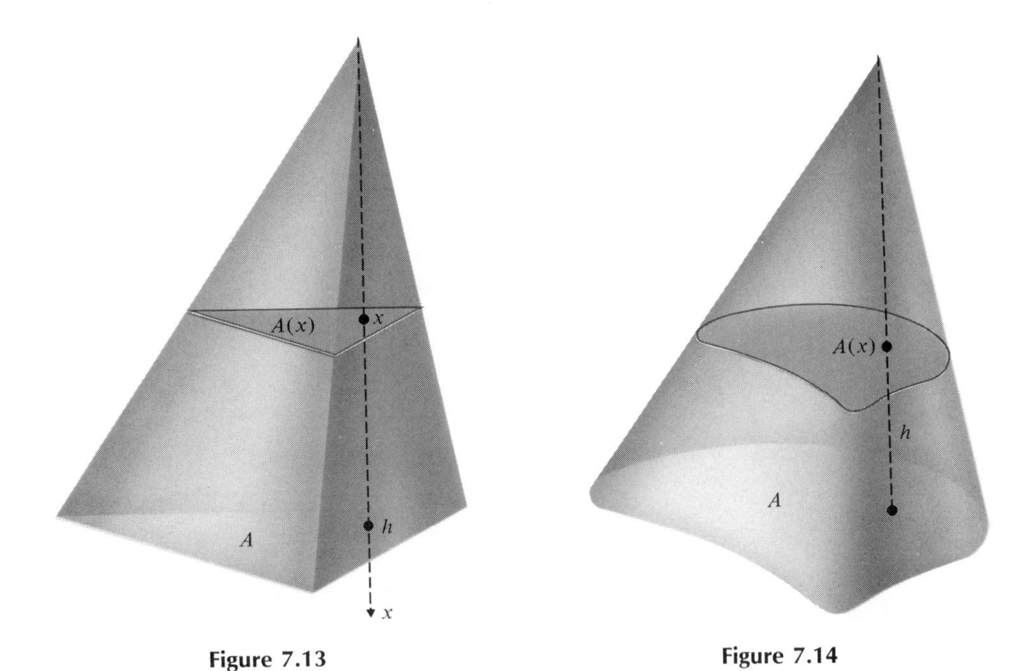

Figure 7.13 **Figure 7.14**

triangles. Refer to Fig. 7.15. The cross section at position x has base length $2\sqrt{a^2 - x^2}$, and so area

$$A(x) = \frac{1}{2}(2\sqrt{a^2 - x^2})b.$$

Thus the volume of the solid is

$$V = \int_{-a}^{a} b\sqrt{a^2 - x^2}\,dx = 2b\int_{0}^{a} \sqrt{a^2 - x^2}\,dx = 2b\,\frac{\pi a^2}{4} = \frac{\pi}{2}a^2b \text{ cu units.}$$

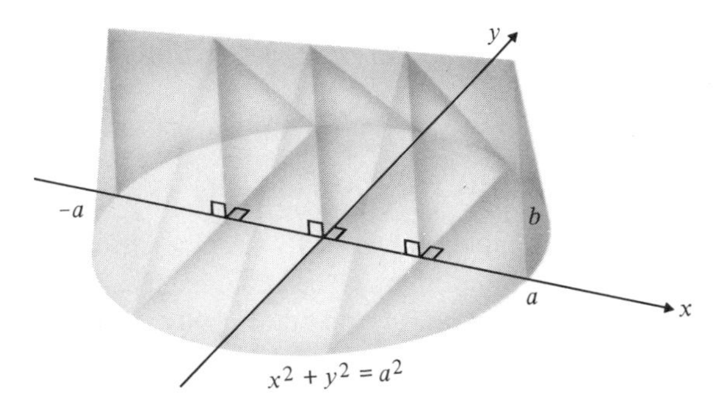

Figure 7.15

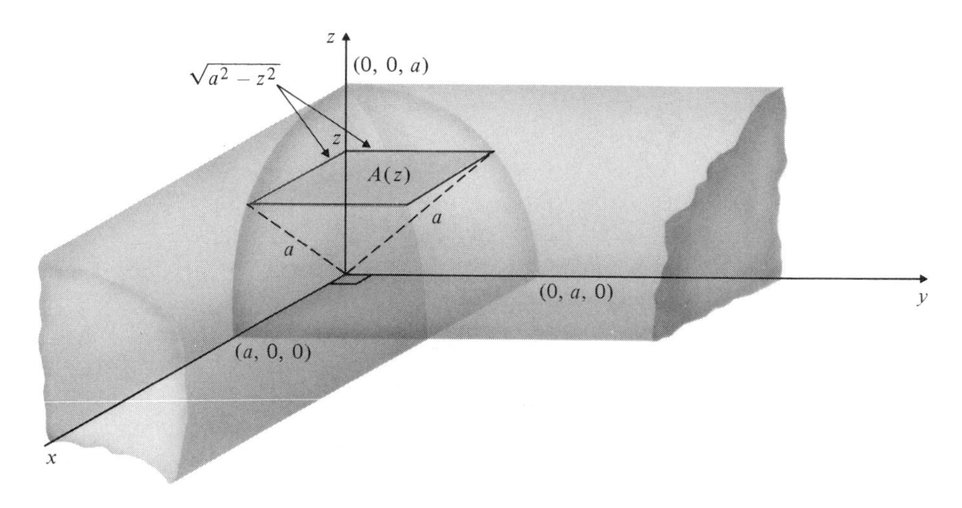

Figure 7.16

Example 3 Two circular cylinders, each having radius a, intersect so that their axes meet at right angles. Find the volume of the region lying interior to both cylinders.

Solution We represent the cylinders in a three-dimensional Cartesian coordinate system where the plane containing the x- and y-axes is horizontal and the z-axis is vertical. One-eighth of the solid is shown in Fig. 7.16, that part corresponding to all three coordinates of a spatial point being positive. The two cylinders have axes along the x-axis and y-axis, respectively. The cylinder with axis along the x-axis intersects the plane of the y- and z-axes in a circle of radius a. Similarly, the other cylinder meets the plane of the x- and z-axes in a circle of radius a. It follows that if the region lying inside both cylinders (and having $x \geq 0$, $y \geq 0$, and $z \geq 0$) is sliced by a horizontal plane at height z above the xy-plane, then the cross section in that plane is a square of side $\sqrt{a^2 - z^2}$ and has area $A(z) = a^2 - z^2$. The volume V of the whole region being eight times that of the part shown, we have

$$V = 8\int_0^a (a^2 - z^2)\,dz = 8\left(a^2 z - \frac{z^3}{3}\right)\Bigg|_0^a = \frac{16}{3}a^3 \text{ cu units.}$$

Solids of Revolution

Many common solids have circular cross sections in planes perpendicular to some axis. These solids may be thought of as being generated by a plane region rotating (in three dimensions) about an axis in that plane. Such solids are called **solids of revolution.** For example, a solid ball is generated by rotating a half-disc about the diameter of that half-disc. Similarly, a solid right-circular cone is generated by rotating a right-angled triangle about one of its legs.

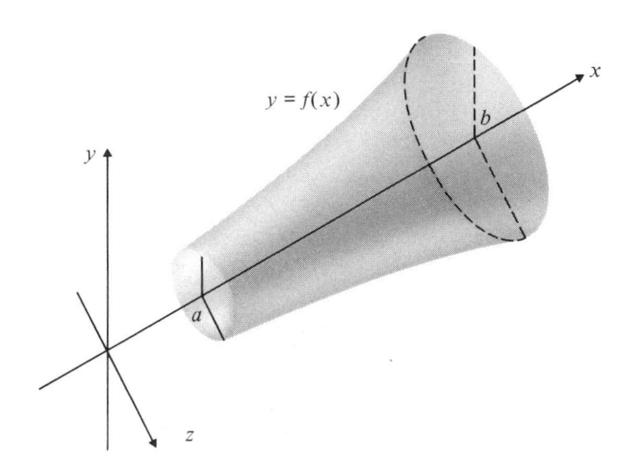

Figure 7.17

If the region R bounded by $y = f(x)$, $y = 0$, $x = a$, and $x = b$ is rotated about the x-axis, then the cross section of the solid generated in the plane at position x is a circle of radius $|f(x)|$. Thus the volume of such a solid of revolution is

$$V = \pi \int_a^b (f(x))^2 \, dx,$$

as shown in Fig. 7.17.

Example 4 Rotating the half-disc of radius a, $0 \le y \le \sqrt{a^2 - x^2}$, about the x-axis we generate a ball (Fig. 7.18) of radius a having volume

$$V = \pi \int_{-a}^{a} (\sqrt{a^2 - x^2})^2 \, dx = 2\pi \int_0^a (a^2 - x^2) \, dx$$

$$= 2\pi \left(a^2 x - \frac{x^3}{3} \right) \Bigg|_0^a = 2\pi \left(a^3 - \frac{1}{3} a^3 \right) = \frac{4}{3} \pi a^3 \text{ cu units.}$$

Example 5 Rotating the triangular region bounded by $y = 0$, $y = rx/h$, and $x = h$ about the x-axis we obtain a right-circular cone of base radius r and height h. (See Fig. 7.19.) This cone has volume

$$V = \pi \int_0^h \left(\frac{rx}{h} \right)^2 dx = \pi \left(\frac{r}{h} \right)^2 \frac{x^3}{3} \Bigg|_0^h = \frac{1}{3} \pi r^2 h \text{ cu units.}$$

(Note that this is just a special case of Example 1.)

Example 6 Find the volume of the infinitely long horn generated by rotating about the x-axis the region bounded by $y = 1/x$ and $y = 0$ and lying to the right of $x = 1$. The horn is illustrated in Fig. 7.20.

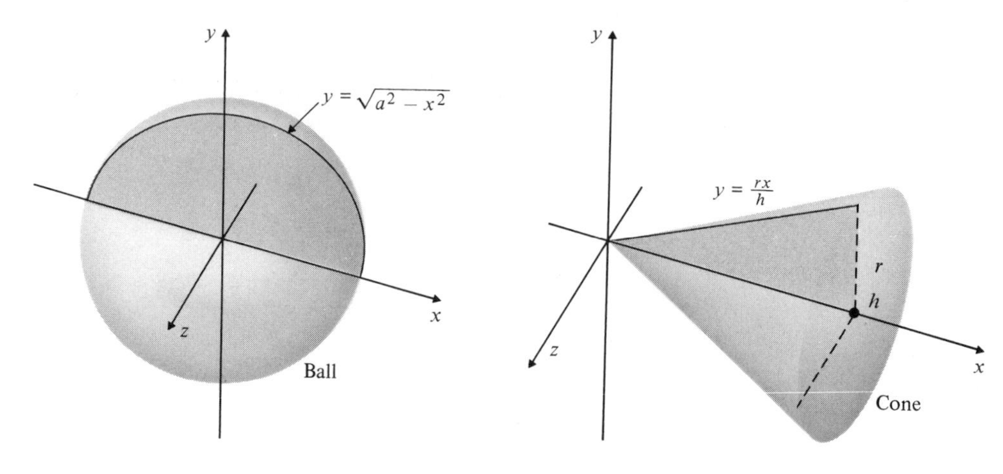

Figure 7.18

Figure 7.19

Solution The volume is

$$V = \pi \int_1^\infty \left(\frac{1}{x}\right)^2 dx = \pi \lim_{R\to\infty} \int_1^R \frac{1}{x^2}\, dx$$

$$= -\pi \lim_{R\to\infty} \frac{1}{x}\bigg|_1^R = \pi \text{ cu units.}$$

It is interesting to note that this finite volume arises from rotating a region that itself has infinite area: $\int_1^\infty dx/x = \infty$. We have a paradox: It takes infinitely much paint to paint the region, but only finitely much to fill the horn obtained by rotating the region. (How can you resolve this paradox?)

Example 7 A cylindrical hole of radius b is drilled through the center of a ball of radius $a > b$. Find the volume of the remaining part of the ball.

Solution For $-\sqrt{a^2 - b^2} \le x \le \sqrt{a^2 - b^2}$ the cross section in the plane perpendicular to the x-axis at position x is a disc with a hole through its center (see Fig. 7.21). The radius

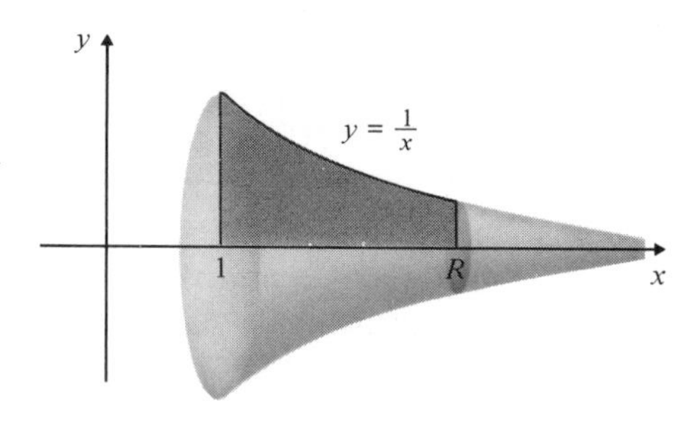

Figure 7.20

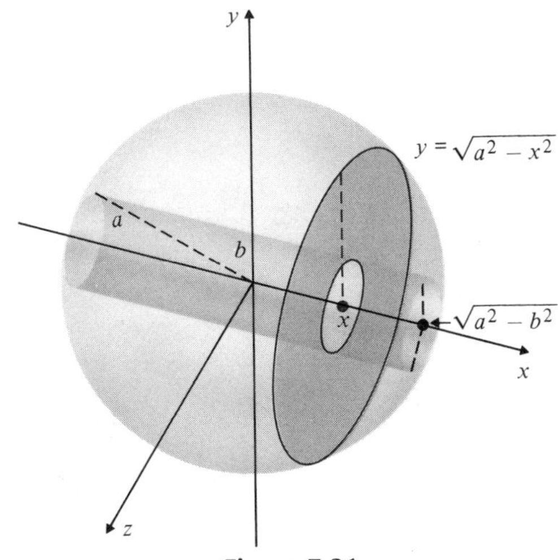

Figure 7.21

of the disc is $\sqrt{a^2 - x^2}$ and the radius of the hole is b. Thus the area of the cross section is

$$A(x) = \pi((a^2 - x^2) - b^2).$$

The volume of the remaining part of the ball is therefore

$$V = \pi \int_{-\sqrt{a^2-b^2}}^{\sqrt{a^2-b^2}} (a^2 - b^2 - x^2)\, dx = 2\pi \int_0^{\sqrt{a^2-b^2}} (a^2 - b^2 - x^2)\, dx$$

$$= 2\pi \left((a^2 - b^2)x - \frac{x^3}{3} \right) \Bigg|_0^{\sqrt{a^2-b^2}} = \frac{4\pi}{3} (a^2 - b^2)^{3/2} \text{ cubic units.}$$

Cylindrical Shells

Suppose that the region R bounded by $y = f(x)$, $y = 0$, $x = a \geq 0$, and $x = b > a$ is rotated about the y-axis to generate a solid of revolution S. In order to evaluate the volume of S using (plane) slices we would need to know the cross-sectional area $A(y)$ in each plane of height y, and this would entail solving the equation $y = f(x)$ for one or more solutions of the form $x = g(y)$. In practice this can be inconvenient or impossible.

An alternate approach is to slice S with a cylinder having axis along the y-axis, as shown in Fig. 7.22. If the radius of the cylinder is x, then the area of cross section of S in the cylinder is $A(x) = 2\pi x f(x)$. (We are assuming $f(x) \geq 0$ in $[a, b]$.) The corresponding differential element of volume of S is

$$dV = A(x)\, dx = 2\pi x f(x)\, dx,$$

which is the volume of a **cylindrical shell** of area $A(x)$ and infinitesimal thickness dx. The volume of S is the sum of these volume elements:

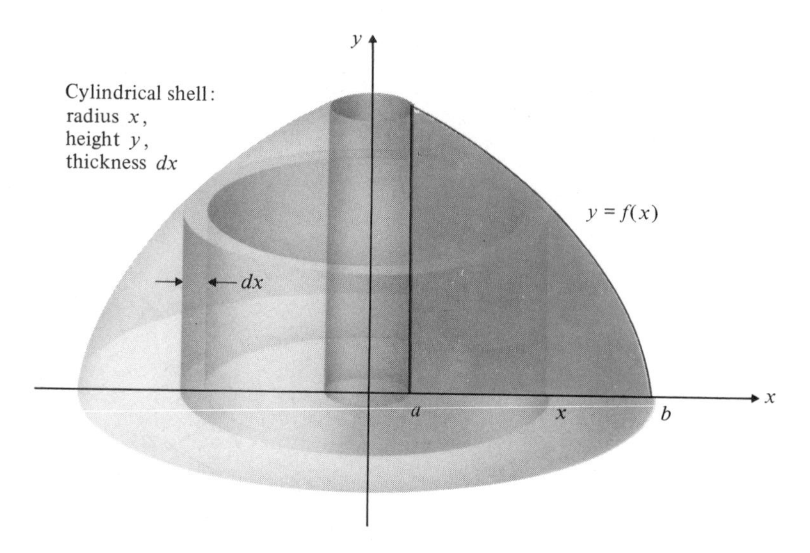

Figure 7.22

$$V = 2\pi \int_a^b xf(x)\,dx.$$

This formula can also be derived by other methods. While the "differential element" method used above seems less rigorous, it is both fast and geometrically suggestive. Integral formulas for various geometrical and physical quantities can often be obtained most quickly by considering the quantity as a suitable infinite sum of differential elements.

Example 8 A disc of radius a has center at the point $(b, 0)$ where $b > a$. The disc is rotated about the y-axis to generate a **torus** (a doughnut-shaped solid). This is illustrated in Fig. 7.23.

The volume of the torus is

$$V = 2 \times 2\pi \int_{b-a}^{b+a} x\sqrt{a^2 - (x - b)^2}\,dx \qquad \text{Let } u = x - b,\ du = dx.$$

$$= 4\pi \int_{-a}^{a} (u + b)\sqrt{a^2 - u^2}\,du$$

$$= 4\pi \int_{-a}^{a} u\sqrt{a^2 - u^2}\,du + 4\pi b \int_{-a}^{a} \sqrt{a^2 - u^2}\,du$$

$$= 0 + 4\pi b \frac{\pi a^2}{2} = 2\pi^2 a^2 b \text{ cu units.}$$

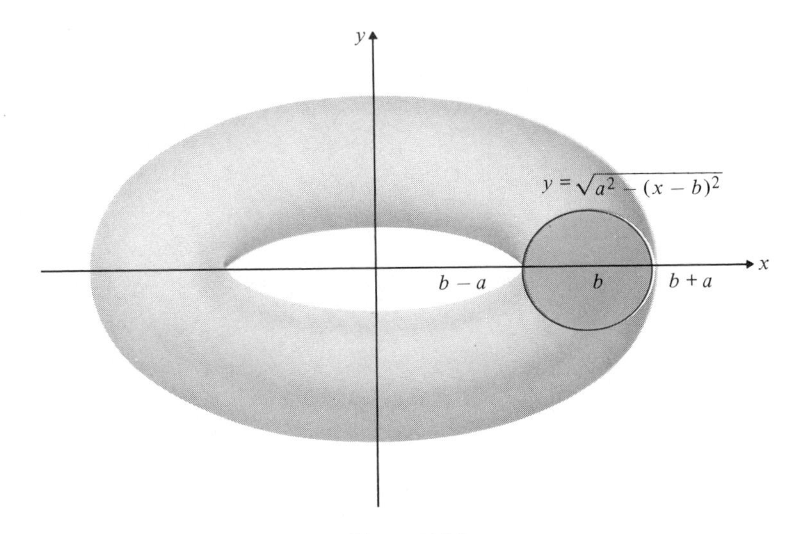

Figure 7.23

(The first of the final two integrals is 0 because the integrand is odd and the interval symmetric about 0; the second is the area of a semicircle.)

We have described two methods for determining the volume of a solid of revolution. The choice of method for a particular solid is usually dictated by the form of the equations defining the rotated region and by the axis of rotation. For very simple regions either method can be made to work easily, but it is usually best to use the method indicated in the following table.

If region R → is rotated about ↓	$y = f(x)$, region R over $a \le x \le b$	$x = g(y)$, region R over $c \le y \le d$
the x-axis	use plane slices $V = \pi \int_a^b (f(x))^2\,dx$	use cylindrical shells $V = 2\pi \int_c^d y g(y)\,dy$
the y-axis	use cylindrical shells $V = 2\pi \int_a^b x f(x)\,dx$	use plane slices $V = \pi \int_c^d (g(y))^2\,dy$

The methods of slicing and cylindrical shells can, of course, also be used when the axis of rotation does not coincide with one of the coordinate axes, as the following example illustrates.

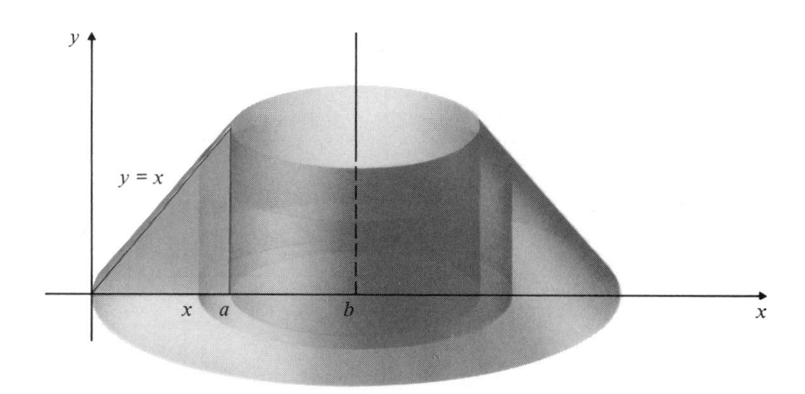

Figure 7.24

Example 9 The triangular region bounded by $y = x$, $y = 0$, and $x = a > 0$ is rotated about the line $x = b > a$. (See Fig. 7.24.) Find the volume of the solid so generated.

Solution Here the cylindrical cross section passing through x has radius $b - x$ and height x. Thus $dV = 2\pi(b - x)x\,dx$, and the volume of the solid is

$$V = 2\pi\int_0^a (b - x)x\,dx = 2\pi\left(\frac{bx^2}{2} - \frac{x^3}{3}\right)\Bigg|_0^a = \pi\left(a^2b - \frac{2a^3}{3}\right) \text{ cu units.}$$

EXERCISES

Find the volume of each solid S in Exercises 1–4 in two ways, using the method of slicing and also the method of cylindrical shells.

1. S is generated by rotating about the x-axis the region bounded by $y = x^2$, $y = 0$, and $x = 1$.
2. S is generated by rotating the region of Exercise 1 about the y-axis.
3. S is generated by rotating about the x-axis the region bounded by $y = x^2$ and $y = \sqrt{x}$ between $x = 0$ and $x = 1$.
4. S is generated by rotating the region of Exercise 3 about the y-axis.

Find the volumes of the solids obtained if the plane regions R described in Exercises 5–10 are rotated about (a) the x-axis, and (b) the y-axis.

5. R is bounded by $y = x(2 - x)$ and $y = 0$ between $x = 0$ and $x = 2$.
6. R is the finite region bounded by $y = x$ and $y = x^2$.
7. R is the finite region bounded by $y = x$ and $x = 4y - y^2$.

8. R is bounded by $y = 1 + \sin x$ and $y = 1$ from $x = 0$ to $x = \pi$.
9. R is bounded by $y = 1/(1 + x^2)$, $y = 2$, $x = 0$, and $x = 1$.
10. R is the finite region bounded by $y = 1/x$ and $x + y = 10/3$.
11. A solid has a circular base of radius r. All sections of the solid perpendicular to a particular diameter of the base are squares. Find the volume of the solid.
12. Repeat Exercise 11 but with sections that are equilateral triangles instead of squares.

13. A cylindrical hole is bored through the center of a ball of radius R. If the length of the hole is L, show that the volume of the remaining part of the ball depends only on L and not on R.

14. A cylindrical hole of radius a is bored through a solid right-circular cone of height h and base radius $b > a$. If the axis of the hole lies along that of the cone, find the volume of the remaining part of the cone.

15. Find the volume of the ellipsoid of revolution obtained by rotating the ellipse $(x^2/a^2) + (y^2/b^2) = 1$ about the x-axis.

16. A plane slices a ball of radius a into two pieces. If the plane passes b units away from the center of the ball (where $b < a$) find the volume of the smaller piece.

17. The elliptical disc bounded by the ellipse of Exercise 15 is rotated about the line $y = c > b$. Find the volume of the solid so generated.

18. A sphere of radius R and a right-circular cone of radius r and height $h > R$ intersect so that the vertex of the cone is at the center of the sphere. Find the volume of the region lying inside both the sphere and the cone.

19.* A solid has a circular base of radius r and a vertical cylindrical wall. Its top is a plane inclined at an angle to the horizontal. If the lowest and highest points on the top are at heights a and b, respectively, above the base, find the volume of the solid.

20.* Find the volume enclosed by the ellipsoid

$$\frac{x^2}{a^2} + \frac{y^2}{b^2} + \frac{z^2}{c^2} = 1.$$

(*Hint:* This is not a solid of revolution. As in Example 3, the z-axis is perpendicular to the plane of the x- and y-axes. Each plane $z = k$ $(-c \le k \le c)$ intersects the ellipsoid in an ellipse $(x/a)^2 + (y/b)^2 = 1 - (k/c)^2$. Thus $dV = dz \times$ the area of this ellipse.)

21.* The axes of two circular cylinders intersect at right angles. If the radii of the cylinders are a and b $(a > b > 0)$, show that the region lying inside both cylinders has volume

$$V = 8\int_0^b \sqrt{b^2 - z^2}\sqrt{a^2 - z^2}\, dz$$

(*Hint:* Review Example 3. Try to make a similar diagram, showing only one-eighth of the region. The integral is not easily evaluated.)

22. The region R bounded by $y = e^{-x}$ and $y = 0$ and lying to the right of $x = 0$ is rotated (a) about the x-axis, (b) about the y-axis. Find the volume of the solid of revolution generated in each case.

23. The region R bounded by $y = x^{-k}$ and $y = 0$ and lying to the right of $x = 1$ is rotated about the x-axis. Find all real values of k for which the solid so generated has finite volume.

24. Repeat Exercise 23 with rotation about the y-axis.

25.* Given that the surface area of a sphere of radius r is kr^2 for some constant k, express the volume of a ball of radius r as a "sum" of volume elements that are volumes of spherical shells of varying radii and infinitesimal thickness. Hence find k.

26.* The finite plane region bounded by the curve $xy = 1$ and the straight line $2x + 2y = 5$ is rotated about that line to generate a solid of revolution. Find the volume of that solid.

7.3 ARC LENGTH AND SURFACE AREA

Arc Length

If A and B are two points in the plane, let $|AB|$ denote the distance between A and B, that is, the length of the straight line segment AB.

Given a curve C joining A and B, we can form a polygonal line $P_0P_1P_2 \ldots P_{n-1}P_n$ by choosing points $A = P_0, P_1, P_2, \ldots, P_{n-1}$ and $P_n = B$ in order along the curve, as shown in Fig. 7.25. The lengths

$$L_n = \sum_{k=1}^{n} |P_{k-1}P_k|$$

of such polygonal lines generally increase as n increases and the lengths of all the individual segments $P_{k-1}P_k$ decrease toward 0. If L_n has a limit under these circumstances, we call that limit the arc length of the curve C.

It is possible to construct continuous curves that are bounded (that is, they do not go off to infinity anywhere) but that do not have finite arc lengths. To avoid such pathological examples, we assume that our curves are smooth, that is, they have continuously turning tangent lines. This condition suffices to guarantee the existence of a finite arc length between any two points on the curve. It is intuitively clear that if P and Q are points on a smooth curve, and arc (P, Q) denotes the arc length along the curve from P to Q (Fig. 7.26), then

$$\lim_{Q \to P} \frac{\text{arc}\,(P, Q)}{|PQ|} = 1.$$

Now suppose that C is the graph of a function f that has a continuous derivative f' on $[a, b]$. Then C certainly has a continuously turning tangent line there. Let P_x

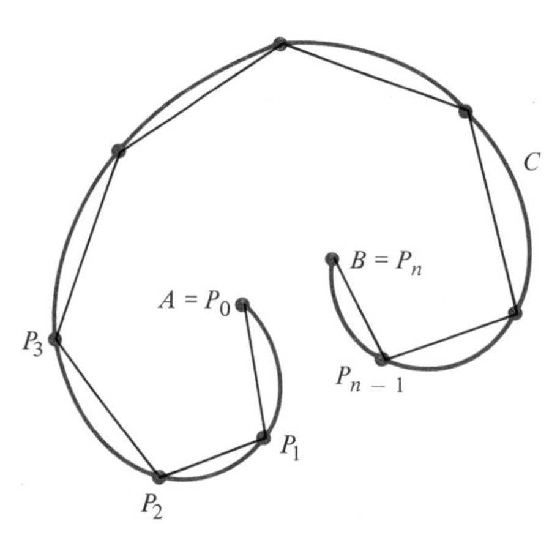

Figure 7.25

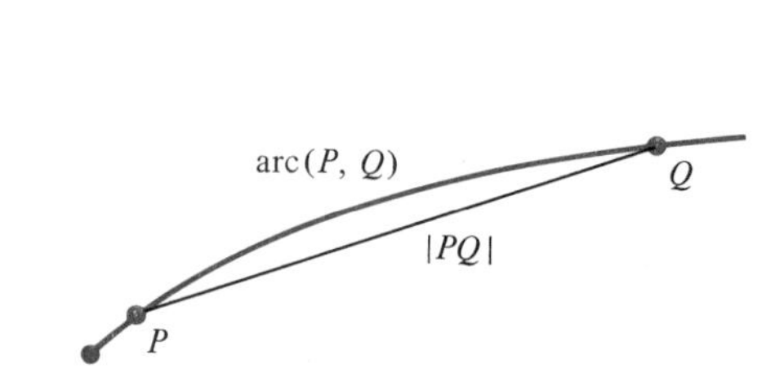

Figure 7.26

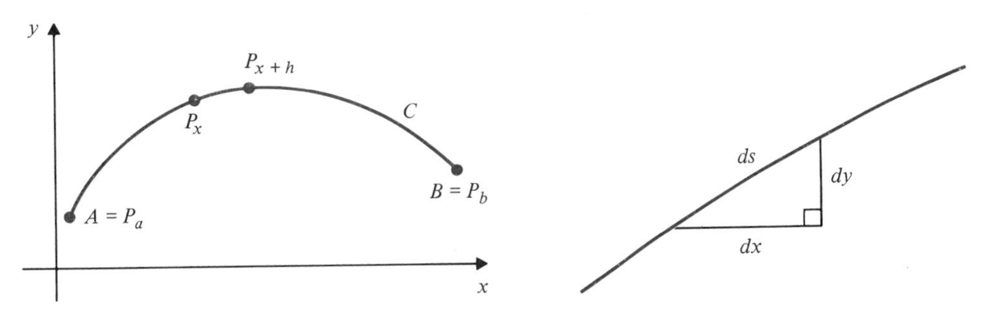

Figure 7.27

Figure 7.28

denote the point on C having coordinates $(x, f(x))$. See Fig. 7.27. If $s(x)$ denotes the arc length along C from $A = P_a$ to P_x, then

$$\begin{aligned}
\frac{s(x+h)-s(x)}{h} &= \frac{\operatorname{arc}(P_{x+h}, P_x)}{h} = \frac{\operatorname{arc}(P_{x+h}, P_x)}{|P_{x+h}P_x|}\frac{|P_{x+h}P_x|}{h} \\
&= \frac{\operatorname{arc}(P_{x+h}, P_x)}{|P_{x+h}P_x|}\frac{\sqrt{h^2+(f(x+h)-f(x))^2}}{h} \\
&= \frac{\operatorname{arc}(P_{x+h}, P_x)}{|P_{x+h}P_x|}\sqrt{1+\left(\frac{f(x+h)-f(x)}{h}\right)^2} \\
&\to (1)\sqrt{1+(f'(x))^2} \text{ as } h \to 0.
\end{aligned}$$

Thus $ds/dx = \sqrt{1+(f'(x))^2}$, and since $s(a) = 0$, the arc length s of C is $s(b)$, that is,

$$s = \int_a^b \sqrt{1+\left(\frac{dy}{dx}\right)^2}\,dx.$$

The formula for ds/dx can be written in the form $(ds/dx)^2 = 1 + (dy/dx)^2$ or, in terms of differentials,

$$(ds)^2 = (dx)^2 + (dy)^2.$$

This formula is easily remembered in terms of the "differential triangle" in Fig. 7.28, and it constitutes a useful mnemonic device for obtaining the formula for s:

$$s = \int_{x=a}^{x=b} ds = \int_a^b \frac{ds}{dx}\,dx = \int_a^b \sqrt{1+\left(\frac{dy}{dx}\right)^2}\,dx.$$

Example 1 Find the length of the curve $y = x^{2/3}$ from $x = 1$ to $x = 8$.

Solution Since $dy/dx = (2/3)x^{-1/3}$, the length of the curve is given by

$$s = \int_1^8 \sqrt{1 + \frac{4}{9}x^{-2/3}}\,dx = \int_1^8 \frac{\sqrt{9x^{2/3} + 4}}{3x^{1/3}}\,dx \qquad \begin{aligned} \text{Let } u &= 9x^{2/3} + 4, \\ du &= 6x^{-1/3}\,dx. \end{aligned}$$

$$= \frac{1}{18}\int_{13}^{40} u^{1/2}\,du = \frac{1}{27}u^{3/2}\Big|_{14}^{40} = \frac{40\sqrt{40} - 13\sqrt{13}}{27} \text{ units.}$$

Example 2 Find the length of the curve $y = x^4 + 1/(32x^2)$ from $x = 1$ to $x = 2$.

Solution Here $dy/dx = 4x^3 - 1/(16x^3)$ and

$$1 + \left(\frac{dy}{dx}\right)^2 = 1 + (4x^3)^2 - \frac{1}{2} + \left(\frac{1}{16x^3}\right)^2$$

$$= (4x^3)^2 + \frac{1}{2} + \left(\frac{1}{16x^3}\right)^2 = \left(4x^3 + \frac{1}{16x^3}\right)^2.$$

Hence the length of the curve is

$$s = \int_1^2 \left(4x^3 + \frac{1}{16x^3}\right) dx = \left(x^4 - \frac{1}{32x^2}\right)\Big|_1^2 = 16 - \frac{1}{128} - \left(1 - \frac{1}{32}\right)$$

$$= 15 + \frac{3}{128} \text{ units.}$$

The examples above are deceptively simple; the curves were chosen in such a way that the arc length integrals were easily evaluated. For instance, the number 32 in the curve in Example 2 was chosen just so the expression $1 + (dy/dx)^2$ would turn out to be a perfect square and its square root would cause no problems. Because of the square root in the formula, arc length problems for most curves lead to integrals that cannot be evaluated by elementary techniques.

Example 3 Find the circumference of the ellipse $(x^2/a^2) + (y^2/b^2) = 1$, where $a \geq b$. See Fig. 7.29.

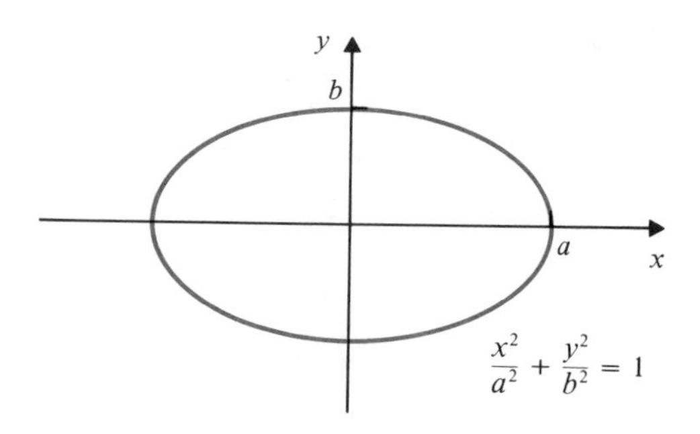

Figure 7.29

Solution Differentiating the equation of the ellipse with respect to x we obtain

$$\frac{2x}{a^2} + \frac{2y}{b^2}\,y' = 0.$$

Hence $y' = -b^2x/a^2y$ and

$$1 + (y')^2 = 1 + \frac{b^4}{a^4}\frac{x^2}{y^2} = 1 + \frac{b^4}{a^4}\frac{x^2}{b^2\left(1 - \dfrac{x^2}{a^2}\right)}$$

$$= 1 + \frac{b^2x^2}{a^2(a^2 - x^2)} = \frac{a^4 - a^2x^2 + b^2x^2}{a^2(a^2 - x^2)}$$

$$= \frac{a^2 - \dfrac{a^2 - b^2}{a^2}x^2}{a^2 - x^2}.$$

The circumference of the ellipse is four times the arc length of that part lying in the first quadrant, so

$$s = 4\int_0^a \frac{\sqrt{a^2 - \dfrac{a^2 - b^2}{a^2}x^2}}{\sqrt{a^2 - x^2}}\,dx \qquad \begin{aligned}\text{Let } x &= a \sin t,\\ dx &= a \cos t\, dt.\end{aligned}$$

$$= 4\int_0^{\pi/2} \sqrt{a^2 - (a^2 - b^2)\sin^2 t}\, dt$$

$$= 4a\int_0^{\pi/2} \sqrt{1 - \epsilon^2 \sin^2 t}\, dt = 4a\, E(\epsilon) \text{ units},$$

where $\epsilon = \sqrt{(a^2 - b^2)/a^2}$ is the eccentricity of the ellipse. The function $E(\epsilon)$ defined by the last integral is called a **complete elliptic integral.** It cannot be evaluated by elementry techniques for general ϵ, but tables of values (as a function of ϵ) can be found in collections of mathematical tables. Note that for $a = b$ we have $\epsilon = 0$, and the formula returns the circumference of a circle; $s = 4a\,(\pi/2) = 2\pi a$ units.

The arc length of (part of) a parabola can be found, with some effort, as we demonstrate in Example 4.

Example 4 Find the length of the parabola $y = x^2$ from the origin to the point (b, b^2) where $b > 0$.

Solution The length is

$$s = \int_0^b \sqrt{1 + 4x^2}\, dx \qquad \begin{aligned}\text{Let } 2x &= \tan\theta,\\ 2\,dx &= \sec^2\theta\, d\theta, \text{ see Fig. 7.30.}\end{aligned}$$

$$= \frac{1}{2}\int_{x=0}^{x=b} \sec^3\theta\, d\theta$$

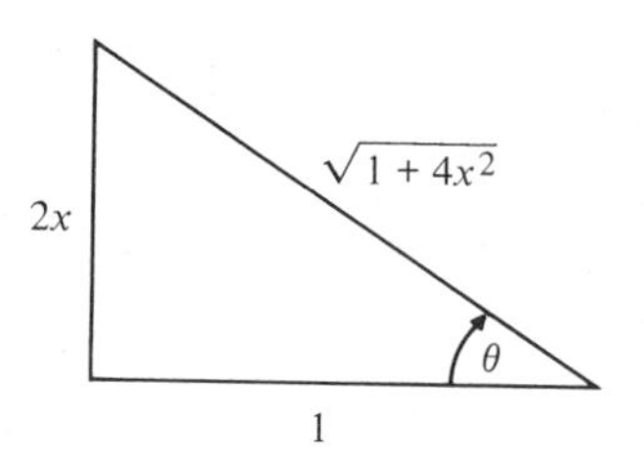

Figure 7.30

$$= \frac{1}{4}\left(\sec\theta\tan\theta + \ln|\sec\theta + \tan\theta|\right)\Bigg|_{x=0}^{x=b}$$

See Example 2e of Section 6.7.

$$= \frac{1}{4}\left(2x\sqrt{1+4x^2} + \ln|2x + \sqrt{1+4x^2}|\right)\Bigg|_{0}^{b}$$

$$= \frac{1}{2}b\sqrt{1+4b^2} + \frac{1}{4}\ln(2b + \sqrt{1+4b^2}) \text{ units.}$$

Areas of Surfaces of Revolution

When a plane curve is rotated (in three dimensions) about a line in the plane of the curve, it sweeps out a surface of revolution. For instance, a sphere of radius a is generated by rotating a semicircle of radius a about the diameter of that semicircle. The area of a surface of revolution can be found by integrating differential elements of surface area, dS, constructed by rotating about the given line differential elements ds of arc length along the curve. If the radius of rotation of an arc length element ds is r, then

$$dS = 2\pi r\, ds,$$

as illustrated in Fig. 7.31. If the smooth graph $y = f(x)$ $(a \le x \le b)$ is rotated about the x-axis, then $r = |y| = |f(x)|$ and $ds = \sqrt{1 + (f'(x))^2}$, so

$$S = 2\pi\int_{x=a}^{x=b} |y|\, ds = 2\pi\int_{a}^{b} |f(x)|\sqrt{1+(f'(x))^2}\, dx.$$

If the rotation is about the y-axis, then $r = |x|$ and

$$S = 2\pi\int_{x=a}^{x=b} |x|\, ds = 2\pi\int_{a}^{b} |x|\sqrt{1+(f'(x))^2}\, dx.$$

Example 5 Find the surface area of a sphere of radius a.

Solution Such a sphere can be generated by rotating about the x-axis the semicircle $y =$

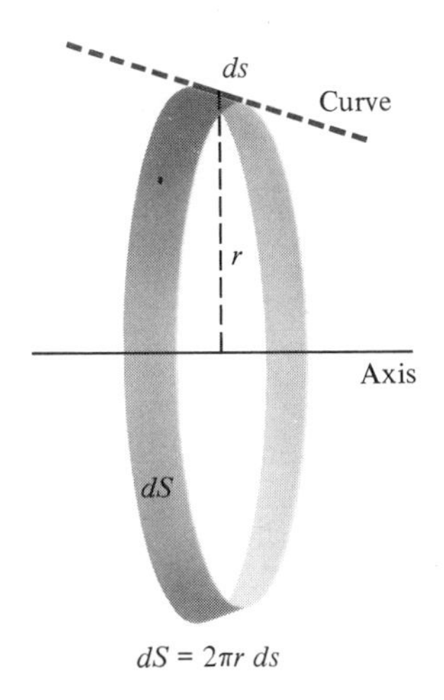

Figure 7.31

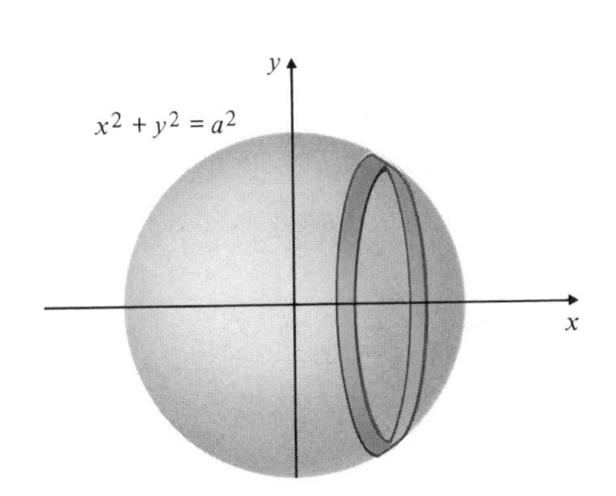

Figure 7.32

$\sqrt{a^2 - x^2}$ $(-a \le x \le a)$. See Fig. 7.32. Since we have

$$\frac{dy}{dx} = -\frac{x}{\sqrt{a^2 - x^2}} = -\frac{x}{y},$$

the area of the sphere is given by

$$S = 2\pi \int_{-a}^{a} y \sqrt{1 + \left(\frac{x}{y}\right)^2}\, dx = 4\pi \int_0^a \sqrt{y^2 + x^2}\, dx$$

$$= 4\pi a \int_0^a dx = 4\pi a x \Big|_0^a = 4\pi a^2 \text{ sq units.}$$

Example 6 Find the surface area of a parabolic reflector whose shape is obtained by rotating the parabolic arc $y = x^2$ $(0 \le x \le 1)$ about the y-axis, as illustrated in Fig. 7.33.

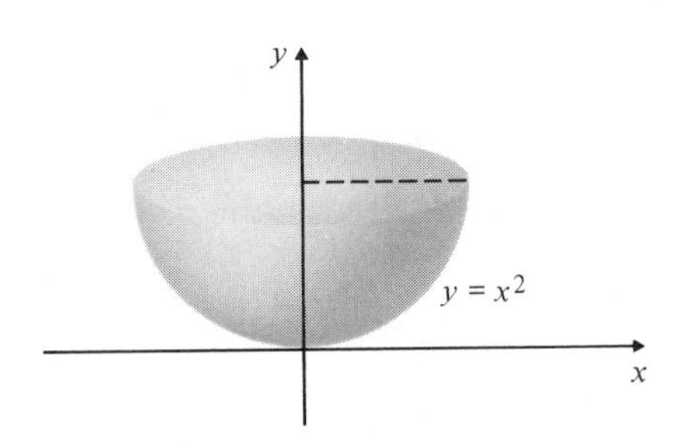

Figure 7.33

Solution The area is

$$S = 2\pi \int_0^1 x\sqrt{1 + 4x^2}\, dx \qquad \text{Let } u = 1 + 4x^2,\; du = 8x\, dx.$$

$$= \frac{\pi}{4}\int_1^5 u^{1/2}\, du$$

$$= \frac{\pi}{6}u^{3/2}\Big|_1^5 = \frac{\pi}{6}(5\sqrt{5} - 1) \text{ sq units.}$$

EXERCISES

In Exercises 1–10 find the lengths of the given curves.

1. $y = ax + b$ from $x = A$ to $x = B$

2. $y = x^{3/2}$ from $x = 1$ to $x = 9$

3. $y = (1/12)x^3 + (1/x)$ from $x = 1$ to $x = 4$

4. $y = (e^x + e^{-x})/2$ ($= \cosh x$) from $x = 0$ to $x = a$

5. $y = \ln(1 - x^2)$ from $x = -(1/2)$ to $x = 1/2$

6.* $y = \ln x$ from $x = 1$ to $x = e$

7.* $y = \sqrt{x}$ from $x = 0$ to $x = 4$

8. $y = x^2 - (\ln x)/8$ from $x = 1$ to $x = 2$

9. $y = \ln \cos x$ from $x = \pi/6$ to $x = \pi/4$

10. $y = \ln \dfrac{e^x - 1}{e^x + 1}$ from $x = 2$ to $x = 4$

11. Find the circumference of the closed curve $x^{2/3} + y^{2/3} = a^{2/3}$.

In Exercises 12–17 find the areas of the indicated surfaces of revolution.

12. The surface generated by rotating the curve of Exercise 2 about the y-axis

13. The surface generated by rotating the curve of Exercise 3 about the x-axis

14. The surface obtained by rotating the curve of Exercise 3 about the y-axis

15. The surface obtained by rotating the curve of Exercise 4 about the x-axis

16. The surface obtained by rotating the curve of Exercise 4 about the y-axis

17. The surface generated by rotating the curve of Exercise 6 about the y-axis

18. By rotating the line segment $y = hx/r$ $(0 \le x \le r)$ about the y-axis, find the area of the curved surface of a right-circular cone of base radius r and height h.

19. Find the surface area of the torus (doughnut) obtained by rotating the circle $x^2 + y^2 = a^2$ about the line $x = b > a$.

20.* The ellipse of Example 3 is rotated about the line $y = c > b$ to generate a doughnut with elliptical cross sections. Express the surface area of this doughnut in terms of the complete elliptic integral function $E(\epsilon)$ introduced in that example.

21.* A hollow container in the shape of an infinitely long horn is generated by rotating the curve $y = 1/x$ $(1 \le x < \infty)$ about the x-axis.

a) Find the volume of the container.

b) Show that the container has infinite surface area.

c) How do you explain the "paradox" that the container can be filled by a finite volume of paint but requires infinitely much paint to cover its surface?

22.* The curve $y = \ln x$ $(0 < x \leq 1)$ is rotated about the y-axis. Find the area of the horn-shaped surface so generated.

23.* For what real values of k does the surface generated by rotating the curve $y = x^k$ $(0 < x \leq 1)$ about the y-axis have finite surface area?

24.* If two parallel planes intersect a sphere, show that the area of that part of the sphere lying between the two planes depends only on the radius of the sphere and the distance between the planes and not on the position of the planes.

7.4 GEOMETRIC APPLICATIONS FOR POLAR AND PARAMETRIC CURVES

Like the material covered in Chapter 5 on which it depends, the material in this section is optional, and may be omitted from a first course in calculus.

Regions Bounded by Polar Curves

You may wish to review Section 5.4 before beginning this section. The basic area problem in polar coordinates is that of finding the area of the region R bounded by the polar graph $r=f(\theta)$ and the two rays $\theta=\alpha$ and $\theta=\beta$. We assume that $\beta > \alpha$ and that f is continuous on $[\alpha, \beta]$.

Let us subdivide the angle interval $[\alpha, \beta]$ into n equal subintervals of length $\Delta\theta = (\beta - \alpha)/n$, using points

$$\alpha = \theta_0 < \theta_1 < \theta_2 < \cdots < \theta_n = \beta.$$

The rays corresponding to these values of θ divide the region R into n sectors as shown in Fig. 7.34. The area ΔA_k of that part of R lying between the rays $\theta = \theta_{k-1}$ and $\theta = \theta_k$ lies between the areas of two circular sectors of angle $\Delta\theta$ and radii the least and greatest values, respectively, of $f(\theta)$ in $[\theta_{k-1}, \theta_k]$. See Fig. 7.35. By the

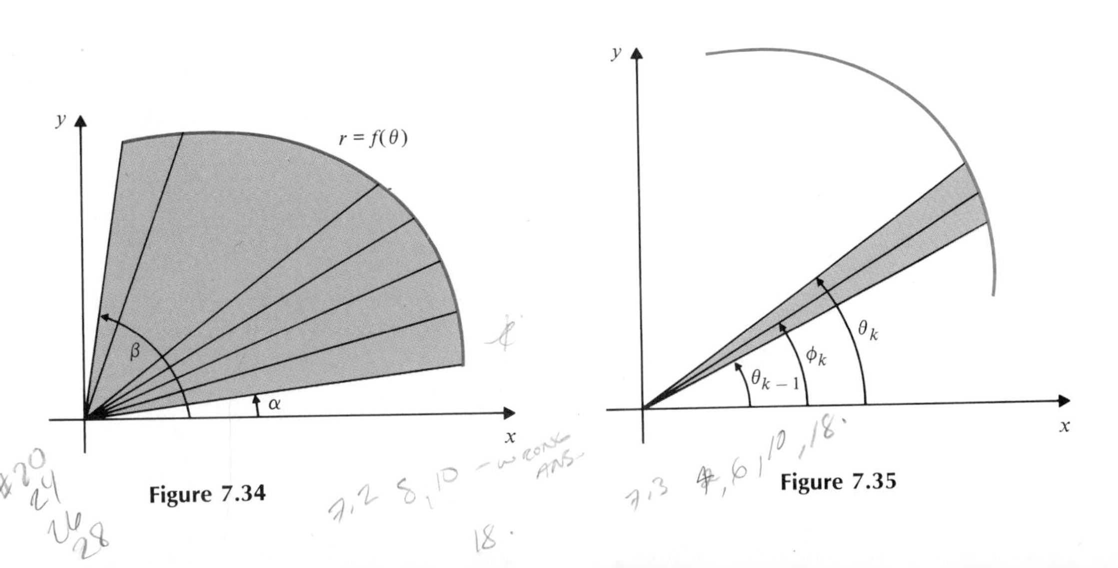

Figure 7.34

Figure 7.35

intermediate-value theorem,

$$\Delta A_k = \frac{1}{2}(f(\phi_k))^2\,\Delta\theta \text{ for some } \phi_k \text{ in } [\theta_{k-1}, \theta_k].$$

The area A of R is given by

$$A = \sum_{k=1}^{n} \frac{1}{2}(f(\phi_k))^2\,\Delta\theta.$$

Taking the limit as $n \to \infty$ and observing that the sum on the right is a Riemann sum for the integral, we obtain,

$$A = \frac{1}{2}\int_{\alpha}^{\beta}(f(\theta))^2\,d\theta = \frac{1}{2}\int_{\alpha}^{\beta} r^2\,d\theta.$$

We can think of this formula as giving A as an infinite sum of differential areas $dA = (1/2)r^2\,d\theta$, each such differential area being the area of a circular sector of radius r and angular width $d\theta$.

Example 1 Find the area bounded by the cardioid $r = a(1 + \cos\theta)$, as illustrated in Fig. 7.36.

Solution By symmetry, the area is twice that of the top half:

$$\begin{aligned}
A &= 2 \times \frac{1}{2}\int_0^{\pi} a^2(1 + \cos\theta)^2\,d\theta \\
&= a^2\int_0^{\pi}(1 + 2\cos\theta + \cos^2\theta)\,d\theta \\
&= a^2\int_0^{\pi}\left(1 + 2\cos\theta + \frac{1 + \cos 2\theta}{2}\right)d\theta \\
&= a^2\left(\frac{3}{2}\theta + 2\sin\theta + \frac{1}{4}\sin 2\theta\right)\Bigg|_0^{\pi} \\
&= \frac{3}{2}\pi a^2 \text{ sq units.}
\end{aligned}$$

Example 2 Find the total area of the region that lies outside the circle $r = 2$ and inside the lemniscate $r^2 = 8\sin 2\theta$.

Solution The region has two congruent parts, as shown in Fig. 7.37. We double the area of the part lying in the first quadrant. The intersections of the two curves are calculated (for the first quadrant only) as follows:

$$8\sin 2\theta = r^2 = 4$$

$$\sin 2\theta = \frac{1}{2} \Rightarrow 2\theta = \frac{\pi}{6} \text{ or } \frac{5\pi}{6}$$

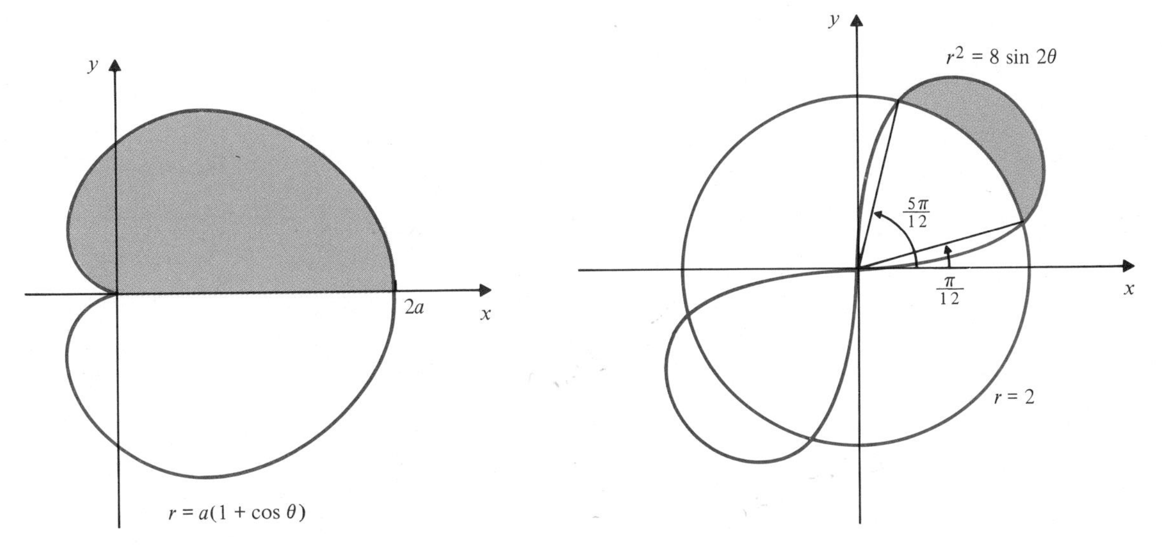

Figure 7.36

Figure 7.37

$$\Rightarrow \theta = \frac{\pi}{12} \text{ or } \frac{5\pi}{12}.$$

The required area is

$$\begin{aligned} A &= 2 \times \left(\frac{1}{2} \int_{\pi/12}^{5\pi/12} 8 \sin 2\theta \, d\theta - \frac{1}{2} \int_{\pi/12}^{5\pi/12} 4 \, d\theta \right) \\ &= -4 \cos 2\theta \Big|_{\pi/12}^{5\pi/12} - 4\theta \Big|_{\pi/12}^{5\pi/12} = -4 \left(-\frac{\sqrt{3}}{2} - \frac{\sqrt{3}}{2} \right) - 4 \left(\frac{\pi}{3} \right) \\ &= 4\sqrt{3} - \frac{4\pi}{3} \text{ sq units.} \end{aligned}$$

The methods of slicing and cylindrical shells can also be used to obtain volumes of solids of revolution when the plane region being rotated is bounded by a polar curve. The appropriate volume element dV can be obtained using a suitable diagram, but some care must be taken to ensure that dV is positive.

Example 3 Find the volume of the solid obtained by rotating the cardioid $r = 1 + \cos \theta$ about the x-axis. (See Fig. 7.38.)

Solution Evidently the entire solid is generated by rotating the upper half of the cardioid (corresponding to $0 \leq \theta \leq \pi$). If we rotate the whole cardioid we would sweep out the volume twice. The disc-shaped slice shown in Fig. 7.38 has radius $y = r \sin \theta = (1 + \cos \theta) \sin \theta$. Since

$$\begin{aligned} dx = \frac{d}{d\theta}(r \cos \theta) \, d\theta &= \frac{d}{d\theta}(\cos \theta + \cos^2 \theta) \, d\theta \\ &= -\sin \theta \, (1 + 2 \cos \theta) \, d\theta, \end{aligned}$$

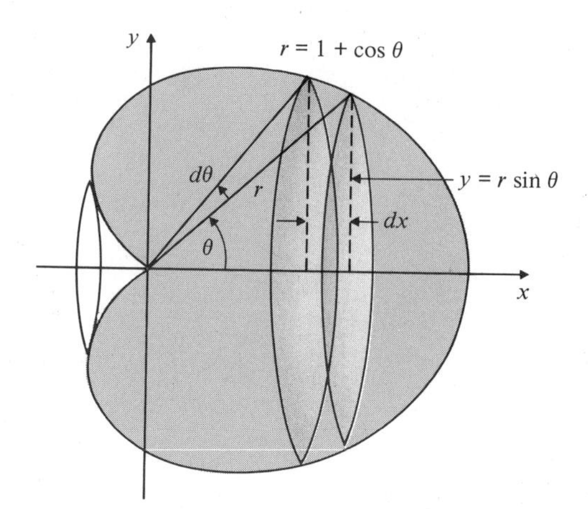

Figure 7.38

we have $dx \leq 0$ if $0 \leq \theta \leq 2\pi/3$, and we can use the volume element $dV = -\pi y^2\, dx$ to rotate the region lying between this part of the cardioid and the x-axis. We thus calculate the volume of a larger solid, that is, the cardioid of revolution with its cusp-shaped indentation on the left-hand side filled in. For $2\pi/3 \leq \theta \leq \pi$ we have $dx \geq 0$, so the same volume element is now negative. However, using this negative volume element for these values of θ produces the negative of the volume of the cusp-shaped indentation, so we can use dV over the whole interval $[0, \pi]$ to give the volume of revolution of the cardioid.

$$V = \pi\int_0^{\pi} \sin^3\theta\,(1+\cos\theta)^2(1+2\cos\theta)\,d\theta$$

$$= \pi\int_0^{\pi} (1-\cos^2\theta)(1+\cos\theta)^2(1+2\cos\theta)\sin\theta\,d\theta \qquad \begin{aligned}\text{Let } u &= \cos\theta\\ du &= -\sin\theta\,d\theta.\end{aligned}$$

$$= -\pi\int_1^{-1} (1-u^2)(1+u)^2(1+2u)\,du$$

$$= \pi\int_{-1}^{1} (1+4u+4u^2-2u^3-5u^4-2u^5)\,du$$

$$= 2\pi\int_0^{1} (1+4u^2-5u^4)\,du \qquad \text{The odd powers will give zeros.}$$

$$= 2\pi\left(1+\frac{4}{3}-1\right) = \frac{8\pi}{3} \text{ cu units}$$

Regions Bounded by Parametric Curves

You may wish to review Section 5.1 before proceeding with this material. Consider the parametric curve C with equations $x = f(t)$, $y = g(t)$ $(a \leq t \leq b)$, where f is differentiable and g is continuous on $[a, b]$. For the moment let us also assume that $f'(t) \geq 0$ and $g(t) \geq 0$ on $[a, b]$, so C has no points below the x-axis and is traversed from left to right as t increases from a to b.

For $a \leq T \leq b$, let $A(T)$ be the area of the region $R(T)$ bounded above by C, below by the x-axis, and at the left and right by $x = f(a)$ and $x = f(T)$. Imitating the proof of the fundamental theorem of calculus, we let

$$\Delta A = A(T + \Delta T) - A(T) \simeq g(T)(f(T + \Delta T) - f(T)),$$

the approximation resulting from replacing the shaded area in Fig. 7.39 with the area of the rectangle of height $g(T)$ and width $f(T + \Delta T) - f(T) = \Delta x$. Evidently this approximation becomes exact as $T \to 0$, and so

$$\frac{dA}{dT} = \lim_{\Delta T \to 0} \frac{\Delta A}{\Delta T} = \lim_{\Delta T \to 0} g(T)\frac{f(T + \Delta T) - f(T)}{\Delta T} = g(T)f'(T).$$

Since $A(a) = 0$, it follows that

$$A(T) = A(T) - A(a) = \int_a^T \frac{dA}{dt}\, dt = \int_a^T g(t)f'(t)\, dt.$$

Hence the area of the region bounded by C, the x-axis, and the vertical lines $x = f(a)$ and $x = f(b)$ is

$$A = \int_a^b g(t)f'(t)\, dt = \int_{t=a}^{t=b} y\, dx.$$

More generally, if we drop the restrictions $f'(t) \geq 0$ and $g(t) \geq 0$, the area of the region bounded by C, the x-axis, and the lines $x = f(a)$ and $x = f(b)$ is

$$A = \int_a^b |g(t)f'(t)|\, dt,$$

provided that either $f'(t) \geq 0$ on $[a, b]$ *or* $f'(t) \leq 0$ on $[a, b]$. If we omit the absolute

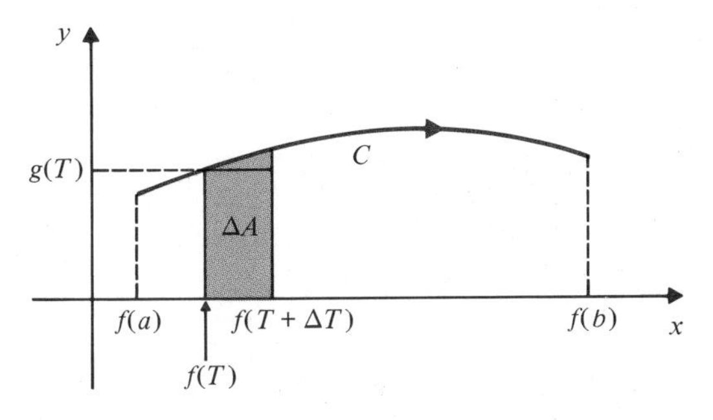

Figure 7.39

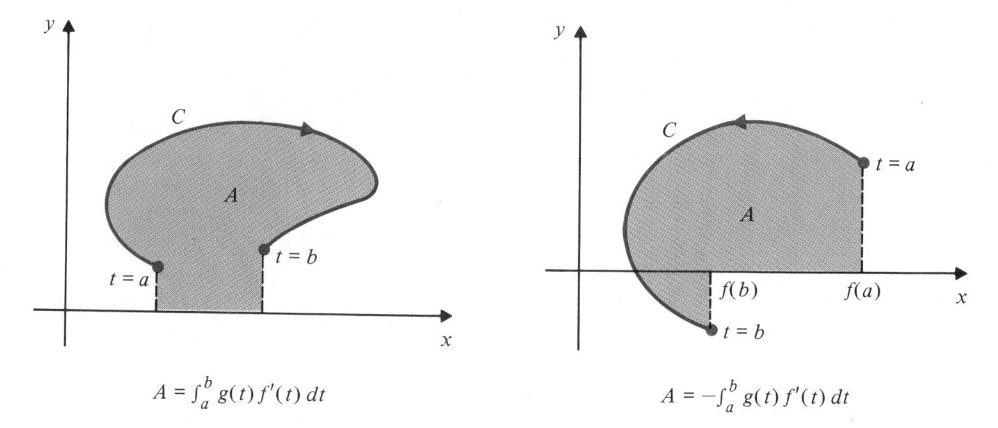

Figure 7.40

values on the integrand we obtain

$$\int_a^b g(t)f'(t)\,dt = A_1 - A_2,$$

where A_1 is the area lying vertically between C and that part of the x-axis consisting of points $x = f(t)$ such that $g(t)f'(t) \geq 0$ and A_2 is a similar area corresponding to points where $g(t)f'(t) < 0$. See Fig. 7.40. This formula is valid for arbitrary continuous g and differentiable f. In particular, if C is a non-self-intersecting closed curve, then the area of the region bounded by C is given by

$$A = \int_a^b g(t)f'(t)\,dt \qquad \text{if } C \text{ is traversed clockwise as } t \text{ increases,}$$

$$A = -\int_a^b g(t)f'(t)\,dt \qquad \text{if } C \text{ is traversed counterclockwise,}$$

both of which are illustrated in Fig. 7.41.

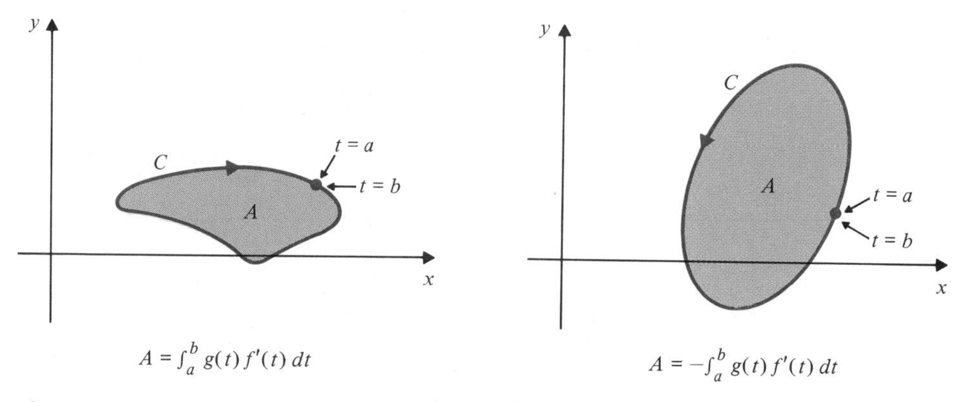

Figure 7.41

Example 4 Find the area bounded by the ellipse $x = a\cos t$, $y = b\sin t$ $(0 \le t \le 2\pi)$.

Solution This ellipse is traversed counterclockwise. (See Example 4 of Section 5.1.) The area enclosed is

$$A = -\int_0^{2\pi} b\sin t\,(-a\sin t)\,dt = \frac{ab}{2}\int_0^{2\pi}(1-\cos 2t)\,dt$$

$$= \frac{ab}{2}t\Big|_0^{2\pi} - \frac{ab}{4}\sin 2t\Big|_0^{2\pi} = \pi ab \text{ sq units.}$$

Example 5 Find the area above the x-axis and under one arch of the cycloid $x = at - a\sin t$, $y = a - a\cos t$.

Solution The cycloid is sketched in Example 9 of Section 5.1. One arch corresponds to $0 \le t \le 2\pi$. Since $y \ge 0$ and $dx/dt = a - a\cos t \ge 0$, the area under one arch is

$$A = \int_0^{2\pi} a^2(1-\cos t)^2 = a^2\int_0^{2\pi}\left(1 - 2\cos t + \frac{1+\cos 2t}{2}\right)dt$$

$$= a^2\left(t - 2\sin t + \frac{t}{2} + \frac{\sin 2t}{4}\right)\Bigg|_0^{2\pi} = 3\pi a^2 \text{ sq units.}$$

Similar arguments to those used above show that if f is continuous and g is differentiable, then we can also interpret

$$\int_a^b f(t)g'(t)\,dt = \int_{t=a}^{t=b} x\,dy = A_1 - A_2,$$

where A_1 is the area of the region lying *horizontally* between C and that part of the y-axis consisting of points $y = g(t)$ such that $f(t)g'(t) \ge 0$, and A_2 is the area of a similar region corresponding to $f(t)g'(t) < 0$. See Fig. 7.42.

Again we can find volumes of solids of revolution obtained by revolving regions bounded by parametric curves by using the methods of slicing and cylindrical shells.

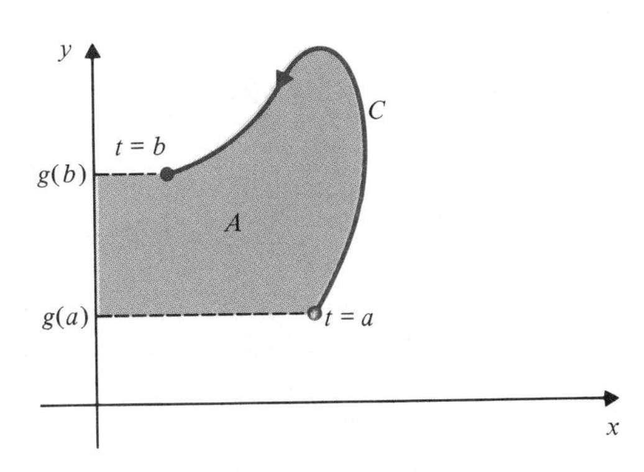

Figure 7.42

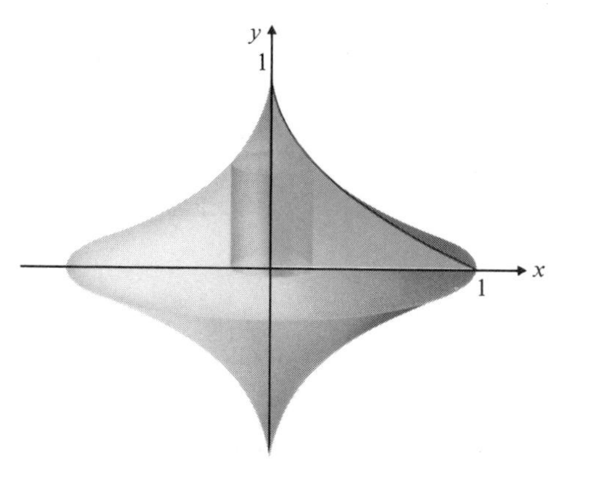

Figure 7.43

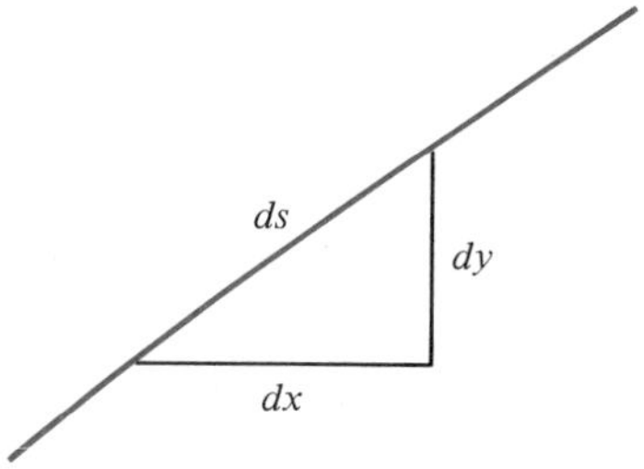

Figure 7.44

Example 6 Find the volume of the solid obtained by rotating about the y-axis the region lying to the right of that axis and to the left of the half-hypocycloid $x = \sin^3 t$, $y = \cos^3 t$ $(0 \le t \le \pi)$.

Solution We can double the volume obtained by rotating only the top half of the curve, corresponding to $0 \le t \le \pi/2$ (see Fig. 7.43). Using cylindrical shells we have

$$\begin{aligned}
V &= 2 \times 2\pi \int_{t=0}^{t=\pi/2} xy\,dx \\
&= 4\pi \int_0^{\pi/2} \sin^3 t \cos^3 t \, 3 \sin^2 t \cos t \, dt \\
&= 12\pi \int_0^{\pi/2} (1 - \cos^2 t)^2 \cos^4 t \sin t \, dt \qquad \text{Let } u = \cos t, \; du = -\sin t \, dt. \\
&= -12\pi \int_1^0 (1 - u^2)^2 u^4 \, du \\
&= 12\pi \int_0^1 (u^4 - 2u^6 + u^8)\, du = 12\pi\left(\frac{1}{5} - \frac{2}{7} + \frac{1}{9}\right) = \frac{32\pi}{105} \text{ cu units.}
\end{aligned}$$

Arc Lengths for Parametric and Polar Curves

Let C be a smooth parametric curve with equations

$$x = f(t), \qquad y = g(t), \qquad a \le t \le b,$$

where we assume that $f'(t)$ and $g'(t)$ are continuous on the interval $[a, b]$ and do not vanish simultaneously. Making use of the differential triangle with legs dx and dy and hypotenuse ds (see Fig. 7.44), we obtain

$$\left(\frac{ds}{dt}\right)^2 = \left(\frac{dx}{dt}\right)^2 + \left(\frac{dy}{dt}\right)^2$$

so that the length of the curve C is given by

$$s = \int_{t=a}^{t=b} ds = \int_a^b \frac{ds}{dt}\,dt = \int_a^b \sqrt{\left(\frac{dx}{dt}\right)^2 + \left(\frac{dy}{dt}\right)^2}\,dt.$$

Example 7 Find the length of the parametric curve

$$x = e^t \cos t, \qquad y = e^t \sin t, \qquad 0 \leq t \leq 2.$$

Solution We have

$$\frac{dx}{dt} = e^t(\cos t - \sin t), \qquad \frac{dy}{dt} = e^t(\sin t + \cos t).$$

Hence

$$\left(\frac{ds}{dt}\right)^2 = e^{2t}(\cos t - \sin t)^2 + e^{2t}(\sin t + \cos t)^2 = 2e^{2t}.$$

The length of the curve is therefore

$$s = \int_0^2 \sqrt{2e^{2t}}\,dt = \sqrt{2}\int_0^2 e^t\,dt = \sqrt{2}(e^2 - 1) \text{ units}.$$

The arc length formula for a polar curve can also be derived directly from that for parametric curves. The polar curve $r = f(\theta)$ $(\alpha \leq \theta \leq \beta)$ can be parametrized

$$x = r\cos\theta = f(\theta)\cos\theta, \qquad y = r\sin\theta = f(\theta)\sin\theta.$$

Thus

$$\frac{dx}{d\theta} = f'(\theta)\cos\theta - f(\theta)\sin\theta$$

$$\frac{dy}{d\theta} = f'(\theta)\sin\theta + f(\theta)\cos\theta.$$

Squaring and adding these, we obtain $(ds/d\theta)^2 = (f'(\theta))^2 + (f(\theta))^2$, so

$$s = \int_\alpha^\beta \sqrt{\left(\frac{dr}{d\theta}\right)^2 + r^2}\,d\theta.$$

This result also follows from consideration of a differential triangle, illustrated in Fig. 7.45:

$$(ds)^2 = (dr)^2 + (r\,d\theta)^2$$

$$\left(\frac{ds}{d\theta}\right)^2 = \left(\frac{dr}{d\theta}\right)^2 + r^2$$

$$ds = \sqrt{\left(\frac{dr}{d\theta}\right)^2 + r^2}\,d\theta$$

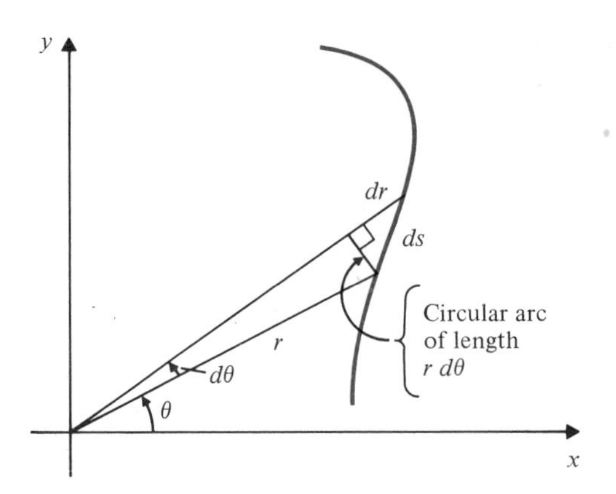

Figure 7.45

Example 8 Find the total length of the cardioid $r = a(1 + \cos\theta)$.

Solution The total length is twice the length from $\theta = 0$ to $\theta = \pi$:

$$\begin{aligned} s &= 2\int_0^{\pi} \sqrt{a^2 \sin^2\theta + a^2(1 + \cos\theta)^2}\,d\theta \\ &= 2\int_0^{\pi} \sqrt{2a^2 + 2a^2\cos\theta}\,d\theta \qquad (\text{but } 1 + \cos\theta = 2\cos^2(\theta/2)) \\ &= 2\sqrt{2}a\int_0^{\pi} \sqrt{2\cos^2\frac{\theta}{2}}\,d\theta = 4a\int_0^{\pi} \cos\frac{\theta}{2}\,d\theta = 8a\sin\frac{\theta}{2}\bigg|_0^{\pi} = 8a \text{ units.} \end{aligned}$$

Parametric and polar curves can be rotated around various axes to generate surfaces of revolution. The areas of these surfaces can be found by the same procedures used for graphs of functions, with the appropriate version of ds.

If the smooth parametric curve with equations

$$x = f(t), \qquad y = g(t), \qquad a \le t \le b,$$

is rotated about the x-axis, the area of S of the surface so generated is given by

$$S = 2\pi\int_{t=a}^{t=b} |y|\,ds = 2\pi\int_a^b |g(t)|\sqrt{(f'(t))^2 + (g'(t))^2}\,dt.$$

If the rotation is about the y-axis, then the area is

$$S = 2\pi\int_{t=a}^{t=b} |x|\,ds = 2\pi\int_a^b |f(t)|\sqrt{(f'(t))^2 + (g'(t))^2}\,dt.$$

If the polar curve $r = f(\theta)$ $(\alpha \le \theta \le \beta)$ is rotated about the x-axis, then the surface generated has area

$$S = 2\pi\int_{\theta=\alpha}^{\theta=\beta} |r\sin\theta|\,ds = 2\pi\int_{\alpha}^{\beta} |f(\theta)\sin\theta|\sqrt{(f'(\theta))^2 + (f(\theta))^2}\,d\theta.$$

If the rotation is about the y-axis, then

$$S = 2\pi \int_{\theta=\alpha}^{\theta=\beta} |r \cos\theta|\, ds = 2\pi \int_{\alpha}^{\beta} |f(\theta)\cos\theta| \sqrt{(f'(\theta))^2 + (f(\theta))^2}\, d\theta.$$

EXERCISES

In Exercises 1–11 sketch and find the areas of the given polar regions R.

1. R lies between the origin and the spiral $r = \sqrt{\theta}$, $0 \le \theta \le 2\pi$.

2. R lies between the origin and the spiral $r = \theta$, $0 \le \theta \le 2\pi$.

3. R is bounded by the curve $r^2 = a^2 \cos 2\theta$.

4. R is one leaf of the curve $r = \sin 3\theta$.

5. R is bounded by the curve $r = \cos 4\theta$.

6. R lies inside both of the circles $r = a$ and $r = 2a\cos\theta$.

7. R lies inside the cardioid $r = 1 - \cos\theta$ and outside the circle $r = 1$.

8. R lies inside the cardioid $r = a(1 - \sin\theta)$ and inside the circle $r = a$.

9. R lies inside the cardioid $r = 1 + \cos\theta$ and outside the circle $r = 3\cos\theta$.

10. R lies inside the lemniscate $r^2 = 2\cos 2\theta$ and outside the circle $r = 1$.

11. R is bounded by the smaller loop of the curve $r = 1 + 2\cos\theta$.

In Exercises 12–17 sketch and find the area of the region R described in terms of the given parametric curves.

12. R is the closed loop bounded by $x = t^3 - 4t$, $y = t^2$ $(-2 \le t \le 2)$.

13. R is bounded by the hypocycloid $x = a\cos^3\theta$, $y = a\sin^3\theta$ $(0 \le \theta \le 2\pi)$.

14. R is bounded by the coordinate axes and the parabolic arc $x = \sin^4 t$, $y = \cos^4 t$.

15. R is bounded by the x-axis, the hyperbola $x = \sec t$, $y = \tan t$, and the ray joining the origin to the point $(\sec t, \tan t)$.

16. R is bounded by $x = \cos s \sin s$, $y = \sin^2 s$ $(0 \le s \le \pi/2)$ and the y-axis.

17. R is bounded by the oval $x = (2 + \sin t)\cos t$, $y = (2 + \sin t)\sin t$.

18. Show that the region bounded by the x-axis and the hyperbola $x = \cosh t$, $y = \sinh t$ $(t > 0)$ and the ray from the origin to the point $(\cosh t, \sinh t)$ has area $t/2$ square units. This proves a claim made at the beginning of Section 3.8.

19. Find the volume of the solid obtained by rotating about the x-axis the region bounded by that axis and one arch of the cycloid (see Example 9 of Section 5.1) $x = at - a\sin t$, $y = a - a\cos t$.

20. Find the volume of the solid obtained by rotating about the y-axis the region bounded by one leaf of the polar curve $r = \sin 2\theta$.

Find the lengths of the curves in Exercises 21–23.

21. $x = 1 + t^3$, $y = 1 - t^2$, $-1 \le t \le 2$

22. $x = a\cos^3 t$, $y = a\sin^3 t$, $0 \le t \le 2\pi$

23. $x = \ln(1 + t^2)$, $y = 2\,\text{Arctan}\, t$, $0 \le t \le 1$

Find the lengths of the polar curves in Exercises 24–26.

24. $r = \theta^2$ from $\theta = 0$ to $\theta = \pi$

25. $r = e^{a\theta}$ from $\theta = -\pi$ to $\theta = \pi$

26. $r = a\theta$ from $\theta = 0$ to $\theta = 2\pi$

27. Show that the total arc length of the lemniscate $r^2 = \cos 2\theta$ is

$$4\int_0^{\pi/4} \sqrt{\sec 2\theta}\, d\theta.$$

28. One leaf of the lemniscate $r^2 = \cos 2\theta$ is rotated about a) the x-axis, b) the y-axis. Find the area of the surface generated in each case.

29. Find the area of the surface obtained by rotating one arch of the cycloid $x = at - a \sin t$, $y = a - a \cos t$ about the x-axis. (One arch corresponds to $0 \le t \le 2\pi$.)

30. Find the area of the surface obtained by rotating the curve of Exercise 25 about the y-axis.

7.5 MASS, MOMENTS, AND CENTER OF MASS

If a solid object is made out of a homogeneous material, we would expect different parts of the solid that have the same volume to have the same mass as well. We express this homogeneity by saying that the object has constant density, that density being the mass divided by the volume for the whole object or for any part of it. Thus a rectangular brick with dimensions 20 cm, 10 cm, and 8 cm would have volume $V = 20 \times 10 \times 8 = 1600$ cm^3 and, if it was made of material having constant density $\rho = 3$ gm/cm^3, it would have mass $m = \rho V = 3(1600) = 4800$ gm.

If the density of the material constituting a solid object is not constant but varies from point to point in the object, no such simple relationship exists between mass and volume. If the density $\rho = \rho(P)$ is a continuous function of position P we could subdivide the solid into many small volume elements and, by regarding ρ as approximately constant over any such element, determine the masses of all the elements and add them up to get the mass of the solid. The mass Δm of a volume element ΔV containing the point P would satisfy.

$$\Delta m \simeq \rho(P)\, \Delta V,$$

so the mass m of the solid can be approximated:

$$m = \Sigma \Delta m \simeq \Sigma \rho(P)\, \Delta V.$$

Such approximations become exact as we pass to the limit of differential elements $dm = \rho(P)\, dV$, so we expect to be able to calculate masses as integrals, that is, as the limits of such sums:

$$m = \int dm = \int \rho(P)\, dV.$$

Example 1 A solid vertical cylinder of height H cm and base area A cm^2 has density given by $\rho = \rho_o(1 + h)$ gm/cm^3 where h is the height in centimeters above the base and ρ_o is a constant. Find the mass of the cylinder.

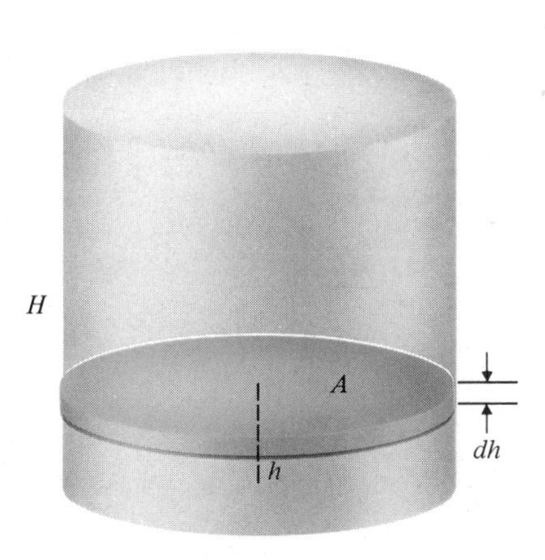

Figure 7.46

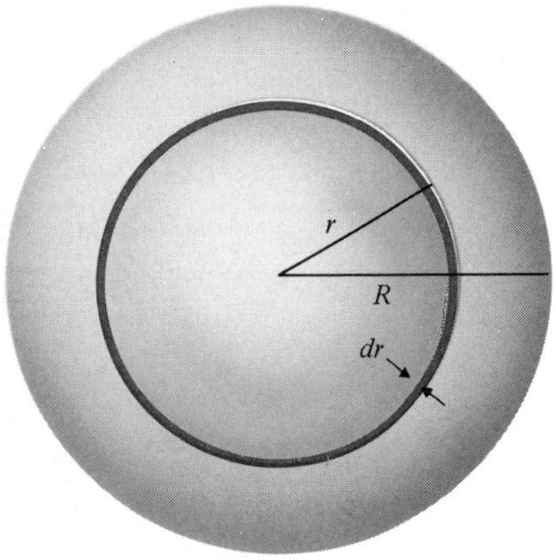

Figure 7.47

Solution See Fig. 7.46. A slice of the solid of thickness dh at height h above the base is a disc having volume

$$dV = A\ dh.$$

The mass of this volume element is

$$dm = \rho\ dV = \rho_o(1 + h)A\ dh.$$

Therefore, the mass of the cylinder is

$$m = \rho_o A \int_0^H (1 + h)\ dh = \rho_o A\left(H + \frac{H^2}{2}\right) \text{ gm}.$$

Example 2 A spherical planet of radius R km has density varying with distance r from the center according to the formula

$$\rho = \frac{\rho_o}{1 + r^2} \text{ kg/km}^3.$$

Find the mass of the planet.

Solution A spherical shell of thickness dr and radius r (see Fig. 7.47) has volume

$$dV = 4\pi r^2\ dr$$

and so has mass

$$dm = \rho\ dV = 4\pi\rho_o \frac{r^2}{1 + r^2}\ dr.$$

Hence the mass of the planet is

$$m = 4\pi\rho_o \int_0^R \frac{r^2}{1 + r^2}\ dr = 4\pi\rho_o \int_0^R \left(1 - \frac{1}{1 + r^2}\right) dr$$

$$= 4\pi\rho_o(r - \text{Arctan } r)\Big|_0^R$$

$$= 4\pi\rho_o(R - \text{Arctan } R) \text{ kg.}$$

Similar techniques can be applied to find masses of one- and two-dimensional objects such as wires and thin plates that have variable densities.

Example 3 A wire of variable composition is stretched along the x-axis from $x = 0$ to $x = L$ cm. Find the mass of the wire if the density at position x is $\rho(x) = x(L - x)$ gm/cm.

Solution An element of the wire of length dx at position x has mass $dm = \rho(x)\,dx = (Lx - x^2)\,dx$. Thus the mass of the wire is

$$m = \int_0^L (Lx - x^2)\,dx = \left(L\frac{x^2}{2} - \frac{x^3}{3}\right)\Big|_0^L = \frac{L^3}{6} \text{ gm.}$$

Example 4 Find the mass of a disc of radius a cm if the density at distance r cm from the center of the disc is $1/(r + 1)$ gm/cm^2.

Solution A thin ring of the disc having radius r and width dr (see Fig. 7.48) has area $dA = 2\pi r\,dr$ and hence has mass

$$dm = 2\pi r \frac{1}{r + 1}\,dr \text{ gm.}$$

The mass of the disc is therefore

$$m = \int_{r=0}^{r=a} dm = 2\pi\int_0^a \frac{r}{r + 1}\,dr = 2\pi\int_0^a \left(1 - \frac{1}{r + 1}\right) dr$$

$$= 2\pi(r - \ln|r + 1|)\Big|_0^a$$

$$= 2\pi(a - \ln(a + 1)) \text{ gm.}$$

Observe that in Example 3 the density ρ was specified in units of mass/length, that is, it was a line density, and the mass was determined from $dm = \rho\,dx$. In Example 4 the given density was an areal density (mass per unit area), so the mass was determined from $dm = \rho\,dA$. Distributions of mass along one-dimensional structures (lines or curves) necessarily lead to integrals of functions of one variable, but distributions of mass on a surface or in space can lead to integrals involving functions of more than one variable. Such integrals are studied in courses on advanced (multivariable) calculus. In Examples 1, 2, and 4 above the given densities were functions of only one variable, so these problems, though higher dimensional in nature, led to integrals of functions of only one variable and could be solved by the methods of this text.

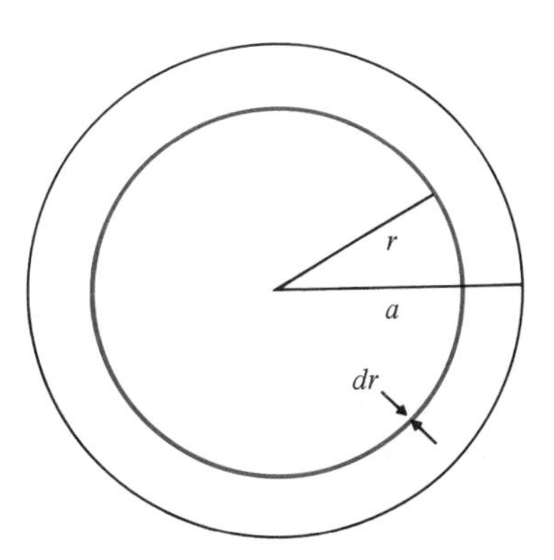

Figure 7.48

Moments and Centers of Mass

A mass m located at position x on the x-axis is said to have **moment** xm about the point 0 or, more generally, moment $(x - x_o)m$ about the point x_o. If several masses $m_1, m_2, m_3, \ldots, m_n$ are located at the points $x_1, x_2, x_3, \ldots, x_n$, respectively, then the total moment of the system of masses about the point x_o is the sum of the individual moments (see Fig. 7.49):

$$M = (x_1 - x_o)m_1 + (x_2 - x_o)m_2 + \cdots + (x_n - x_o)m_n = \sum_{j=1}^{n} (x_j - x_o)m_j.$$

The **center of mass** of the system of masses is the point $\bar{x}$ about which the system has **moment zero**. Thus

$$0 = \sum_{j=1}^{n} (x_j - \bar{x})m_j = \sum_{j=1}^{n} x_j m_j - \bar{x} \sum_{j=1}^{n} m_j,$$

and the center of mass is therefore given by

$$\bar{x} = \frac{\sum_{j=1}^{n} x_j m_j}{\sum_{j=1}^{n} m_j} = \frac{M}{m},$$

where m is the total mass of the system and M is the total moment about $x = 0$. If you think of the x-axis as being a weightless wire supporting the masses, $\bar{x}$ is the point at which the wire could be suspended and remain in perfect balance (equilibrium), not tipping either way. Even if the axis represents a nonweightless support, say a

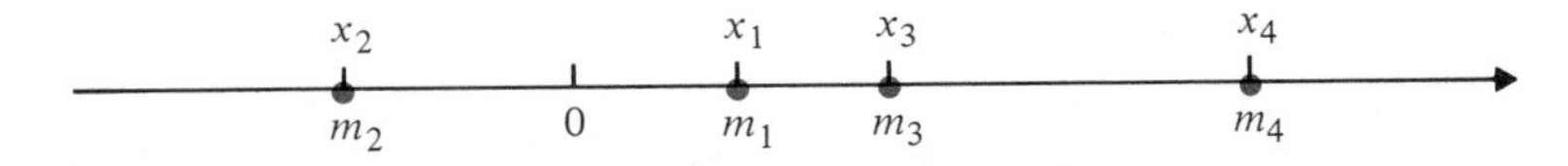

Figure 7.49

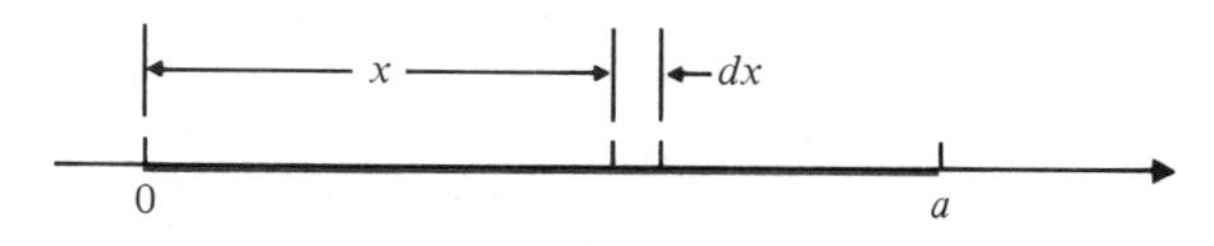

Figure 7.50

seesaw, supported at $x = \bar{x}$, it will remain balanced after the masses are added, provided it was balanced before they were placed on.

Now suppose that a one-dimensional distribution of mass with continuously variable line density $\rho(x)$ lies along the interval $[0, a]$ of the x-axis. An element of length dx at position x contains mass $dm = \rho(x)\,dx$ and so has moment $dM = x\,dm = x\rho(x)\,dx$ about the origin (see Fig. 7.50). The total moment about the origin is

$$M = \int_0^a x\rho(x)\,dx.$$

Since the total mass is

$$m = \int_0^a \rho(x)\,dx,$$

the center of mass is given by

$$\bar{x} = \frac{M}{m} = \frac{\int_0^a x\rho(x)\,dx}{\int_0^a \rho(x)\,dx}.$$

Example 5 At what point can the wire of Example 3 be suspended so that it will balance?

Solution In Example 3 we evaluated the mass of the wire to be $L^3/6$ gm. Its moment about $x = 0$ is

$$M = \int_0^L x\rho(x)\,dx = \int_0^L x^2(L - x)\,dx$$

$$= \left(\frac{x^3L}{3} - \frac{x^4}{4}\right)\Bigg|_0^L = \frac{L^4}{12}\text{ gm-cm.}$$

Thus the center of mass is $\bar{x} = (L^4/12)/(L^3/6) = L/2$. This wire will balance at its midpoint, at distance $L/2$ cm from the end $x = 0$, as shown in Fig. 7.51. The center of mass turned out to be the midpoint because the mass, though varying in density from point to point, is nevertheless distributed symmetrically about the midpoint $x = L/2$.

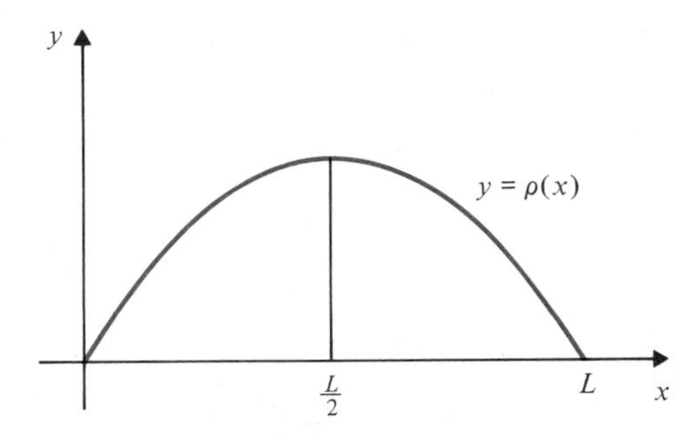

Figure 7.51

Such symmetry can frequently be useful in determining centers of mass without doing unnecessary integrals.

Example 6 Determine the center of mass of a wire of length L lying on the interval $[0, L]$ if the density at point x is $\rho(x) = x$.

Solution Here the mass is not distributed symmetrically; the density increases as we move to the right along the wire. Thus we expect the center of mass to be closer to the right end than to the left. The mass of the wire is

$$m = \int_0^L x\,dx = \frac{L^2}{2},$$

and the moment about the origin is

$$M = \int_0^L x^2\,dx = \frac{L^3}{3}.$$

Hence the center of mass is located at $\bar{x} = M/m = 2L/3$, two-thirds of the way along the wire from the left end.

The systems of mass considered above are all one-dimensional and lie along straight lines. If mass is distributed in a plane or in space, similar considerations prevail. For a system of masses m_1 at (x_1, y_1), m_2 at (x_2, y_2), . . . , m_n at (x_n, y_n), we define the **moment about the *y*-axis** to be

$$M_y = x_1m_1 + x_2m_2 + \cdots + x_nm_n = \sum_{j=1}^{n} x_jm_j,$$

and **the moment about the *x*-axis** to be

$$M_x = y_1m_1 + y_2m_2 + \cdots + y_nm_n = \sum_{j=1}^{n} y_jm_j.$$

The center of mass is the point $(\bar{x}, \bar{y})$ where

$$\bar{x} = \frac{M_y}{m} = \frac{\sum_{j=1}^{n} x_j m_j}{\sum_{j=1}^{n} m_j}, \qquad \bar{y} = \frac{M_x}{m} = \frac{\sum_{j=1}^{n} y_j m_j}{\sum_{j=1}^{n} m_j}.$$

Example 7 Find the center of mass of a rectangular plate occupying the region $0 \le x \le a$, $0 \le y \le b$ (all distances in centimeters) if the areal density of the material in the plate at position (x, y) is kx gm/cm^2.

Solution Since the density is independent of y and the rectangle is symmetric about the line $y = b/2$, the y-coordinate of the center of mass must be $\bar{y} = b/2$.

A thin vertical strip of width dx at position x (see Fig. 7.52) has mass $dm = bkx\,dx$ gm. The moment of this strip about the y-axis is $dM_y = x\,dm = kbx^2\,dx$. Hence the mass and moment about the y-axis of the whole plate are

$$m = kb\int_0^a x\,dx = \frac{kba^2}{2} \qquad M_y = kb\int_0^a x^2\,dx = \frac{kba^3}{3}.$$

Hence $\bar{x} = M_y/m = 2a/3$, and the center of mass of the plate is $(2a/3, b/2)$. The plate would be balanced if supported at this point.

For distributions of mass in three-dimensional space one defines, analogously, the moments M_{yz}, M_{zx}, and M_{xy} of the system of mass about the y-z-plane, the z-x-plane, and the x-y-plane, respectively. For a discrete system of masses m_1 at $(x_1, y_1, z_1), \ldots, m_n$ at (x_n, y_n, z_n), we have

$$M_{yz} = x_1m_1 + x_2m_2 + \cdots + x_nm_n,$$

with similar formulas for M_{zx} and M_{xy}. The center of mass is $(\bar{x}, \bar{y}, \bar{z}$ where $\bar{x} = M_{yz}/m$, $\bar{y} = M_{zx}/m$, and $\bar{z} = M_{xy}/m$, m being $m_1 + m_2 + \cdots + m_n$.

Example 8 Find the center of mass of a solid hemisphere of radius R if the density at height h above the base plane of the hemisphere is $\rho_o h$.

Solution The solid is symmetric about the vertical axis (say, the z-axis), and the density is constant in planes perpendicular to this axis. By this symmetry the center of mass must lie somewhere on this axis. A slice of the solid at height z above the base having thickness dz is a disc of radius $\sqrt{R^2 - z^2}$ (see Fig. 7.53). Its volume is $dV = \pi(R^2 - z^2)\,dz$, and its mass is $dm = \rho_o z\,dV = \rho_o\pi(R^2z - z^3)\,dz$. Its moment about the base plane (the x-y-plane) is $dM_{xy} = z\,dm = \rho_o\pi(R^2z^2 - z^4)\,dz$. The mass of the hemisphere is

$$m = \rho_o\pi\int_0^R (R^2z - z^3)\,dz = \rho_o\pi\left(\frac{R^2z^2}{2} - \frac{z^4}{4}\right)\Bigg|_0^R = \frac{\pi}{4}\rho_o R^4.$$

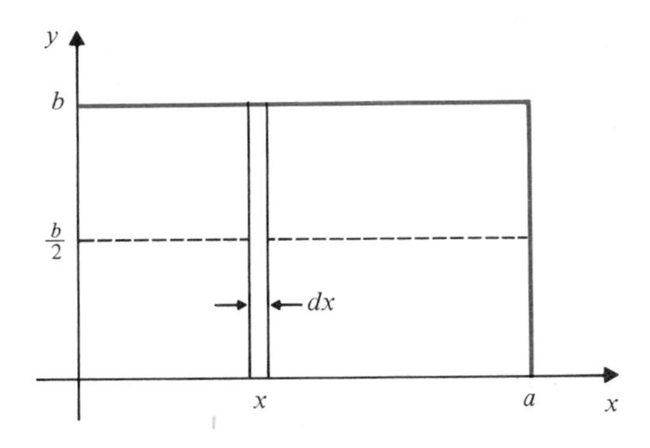

Figure 7.52

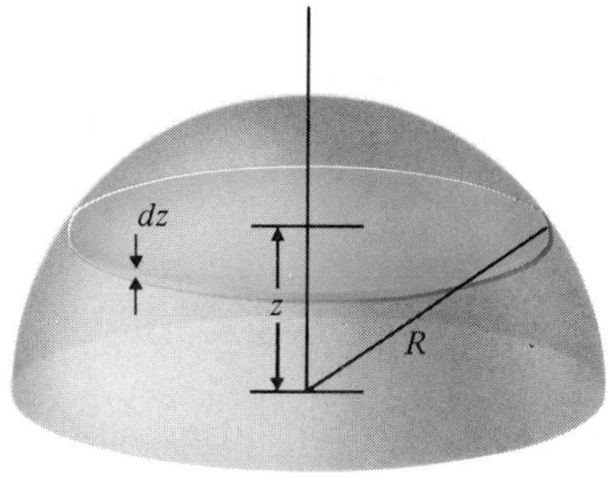

Figure 7.53

The moment of the hemisphere about the x-y-plane is

$$M_{xy} = \rho_o \pi \int_0^R (R^2 z^2 - z^4)\, dz = \rho_o \pi \left(\frac{R^2 z^3}{3} - \frac{z^5}{5} \right) \Bigg|_0^R = \frac{2\pi}{15} \rho_o R^5.$$

The center of mass therefore lies along the axis of symmetry of the hemisphere at height $\bar{z} = M_{xy}/m = 8R/15$ above the base of the hemisphere.

EXERCISES

Find the mass and center of mass for the systems in Exercises 1–12.

1. A straight wire of length L cm, where the density at distance x cm from one end is $\rho(x) = \sin \pi x/L$ gm/cm
2. The wire in Example 1 if it is bent in a quarter-circular arc centered at the origin and lying in the first quadrant of the x-y-plane
3. A plate occupying the region $0 \le y \le 4 - x^2$ if the areal density at (x, y) is ky
4. A right-triangular plate with legs 2 m and 3 m if the areal density at any point P is $5h$ kg/m^2, h being the distance of P from the shorter leg
5. A semicircular plate of radius a cm if the areal density at point P is y gm/cm^2 where y is the distance (in cm) of P from the base diameter of the semicircle
6. A square plate of edge a cm if the areal density at P is kx gm/cm^2 where x is the distance from P to one edge of the square
7. The plate in Exercise 6, but with areal density kr gm/cm^2 where r is the distance (in cm) from P to one of the diagonals of the square
8. A rectangular brick of dimensions 20 cm, 10 cm, and 5 cm if the density at P is kx gm/cm^3 where x is the distance from P to one of the 10×5 faces
9. A solid ball of radius R m if the density at P is z kg/m^3 where z is the distance from P to a plane at distance $2R$ m from the center of the ball
10. A right-circular cone of base radius a cm and height b cm if the density at point P is kz gm/cm^3 where z is the distance of P from the base of the cone
11. The cone of Exercise 10, but with density at P equal to ks gm/cm^3 where s is the distance

of P from the central axis of the cone. (*Hint:* First find the mass of a disc-shaped slice of the cone perpendicular to the axis of the cone. You will still need an integral to do this.)

12. A semicircular disc of radius R cm if the density at points at distance s from the midpoint of the base diameter (i.e., the center of the circle, half of which bounds the disc) is ks gm/cm^2

7.6 CENTROIDS

If matter is distributed uniformly in a system so that the density ρ is constant, then that density cancels out of the numerator and denominator in sum or integral expressions for coordinates of the center of mass. In such cases the center of mass depends only on the *shape* of the object, that is, on geometric properties of the region occupied by the object, and we call it the **center of gravity** or, more commonly, the **centroid** of the region.

Centroids are calculated using the same formulas as those used for centers of mass, except that the density (being constant) is taken to be unity, so the mass is just the length, area, or volume of the region.

Example 1 Find the centroid of the half-disc $-a \le x \le a,\ 0 \le y \le \sqrt{a^2 - x^2}$.

Solution By symmetry the x-coordinate of the centroid is $\bar{x} = 0$. To find $\bar{y}$ consider a horizontal strip at height y having width dy, as shown in Fig. 7.54. The moment of such a strip about the x-axis is

$$dM_x = 2y\sqrt{a^2 - y^2}\,dy$$

and the moment of the whole half-disc is

$$M_x = \int_0^a 2y\sqrt{a^2 - y^2}\,dy \qquad \begin{aligned} \text{Let } u &= a^2 - y^2, \\ du &= -2y\,dy. \end{aligned}$$

$$= -\int_{a^2}^{0} u^{1/2}\,du = -\frac{2}{3}u^{3/2}\Big|_{a^2}^{0} = \frac{2}{3}a^3.$$

Since the area of the half-disc is $(1/2)\pi a^2$, we have $\bar{y} = (2a^3/3)/(\pi a^2/2) = 4a/(3\pi)$. The centroid of the half-disc is $(0, 4a/(3\pi))$.

Example 2 Find the centroid of the semicircle $y = \sqrt{a^2 - x^2}$.

Solution Again $\bar{x} = 0$ by symmetry. A short arc of length ds at height y on the semicircle (see Fig. 7.55) has moment dM_x about the x-axis, where

$$dM_x = y\,ds = \sqrt{a^2 - x^2}\sqrt{1 + \left(\frac{dy}{dx}\right)^2}\,dx$$

$$= \sqrt{a^2 - x^2}\sqrt{1 + \frac{x^2}{a^2 - x^2}}\,dx$$

$$= a\,dx.$$

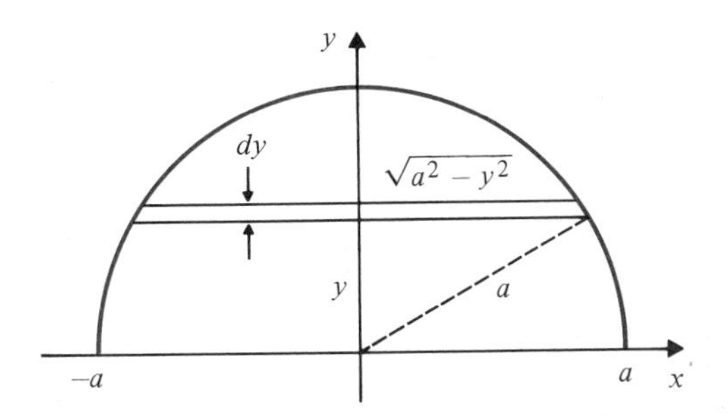

Figure 7.54

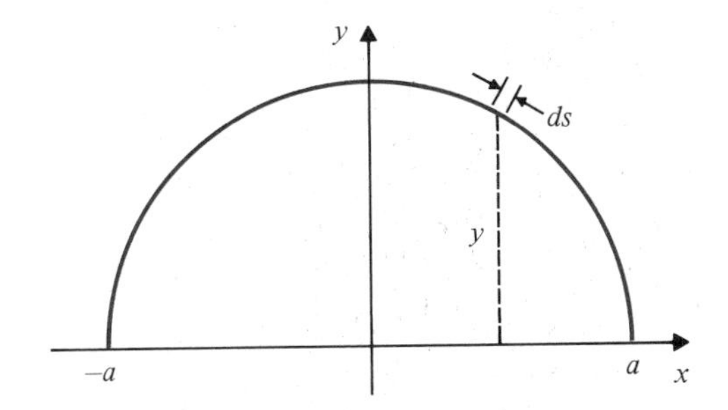

Figure 7.55

Thus

$$M_x = \int_{-a}^{a} a\,dx = ax\Big|_{-a}^{a} = 2a^2.$$

Since the length of the semicircle is πa we have $\bar{y} = M_x/\pi a = 2a/\pi$, and the centroid of the semicircle is $(0, 2a/\pi)$. Note that the centroid of a semicircle of radius a is not the same as that of a half-disc of radius a. Note also that the centroid of the semicircle does not lie on the semicircle itself.

Example 3 Show that the centroid of a triangle is at the point of intersection of the three medians of the triangle.

Solution We adopt a coordinate system such that the origin is at one vertex of the triangle and the x-axis lies along one side, as shown in Fig. 7.56. If this side has length b and lies at distance h from the third vertex of the triangle, then the area of the triangle is $A = bh/2$ sq units. A narrow strip of the triangle at distance y from the x-axis and having width dy has length $(b/h)(h - y)$ (by similar triangles) and so it has area $dA = (b/h)(h - y)\,dy$. The moment of this strip about the x-axis is $dM_x = y\,dA$, so the moment of the whole triangle about the x-axis is

$$M_x = \int_0^h \frac{b}{h}(hy - y^2)\,dy = \frac{b}{h}\left(\frac{h^3}{2} - \frac{h^3}{3}\right) = \frac{1}{6}bh^2.$$

Hence $\bar{y} = (bh^2/6)/(bh/2) = h/3$. The centroid lies on a line parallel to the base and one-third of the way from the base to the opposite vertex. The same argument works with any of the three sides of the triangle as base, so the centroid O lies at the intersection of three such lines, as shown in Fig. 7.57. We would like to show that O lies on each of the three medians of the triangle. (Recall that a median is a straight line joining one vertex to the midpoint of the opposite side.) Thus, if M is the point where AO meets BC, we want to show that $BM = MC$. Using various similar triangles in the diagram we can obtain

$$\frac{AF}{AB} = \frac{AX}{AY} = \frac{AO}{AM} = \frac{2}{3}; \qquad \frac{DM}{BM} = \frac{OM}{AM} = \frac{1}{3}; \qquad \frac{BD}{BC} = \frac{1}{3}.$$

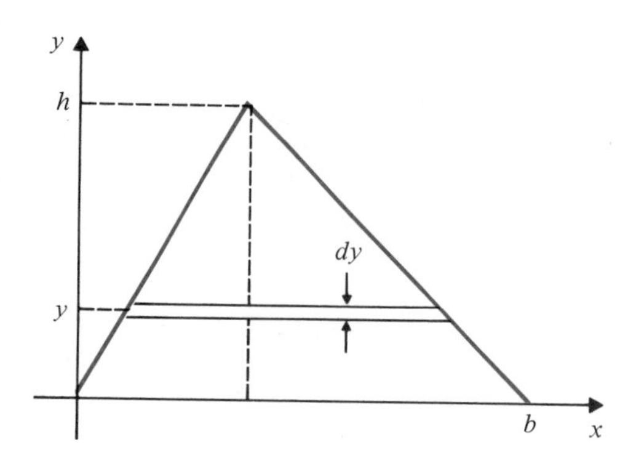

Figure 7.56

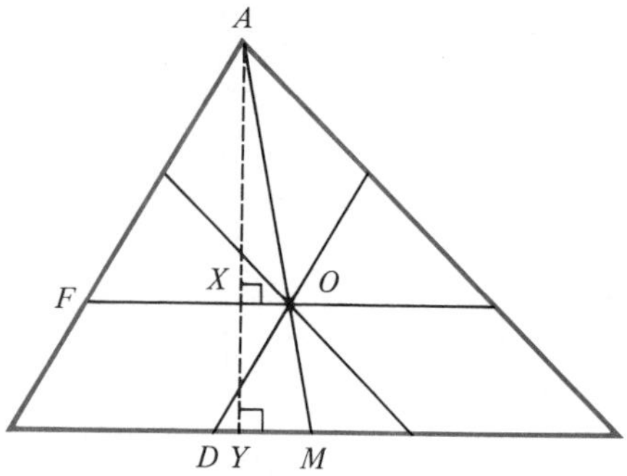

Figure 7.57

Hence $BM = BD + DM = (1/3)BC + (1/3)BM$ so $(2/3)BM = (1/3)BC$ and $BM = (1/2)BC = MC$. Therefore AOM is a median of the triangle. Similarly, O also lies on the other two medians of the triangle.

Example 4 Find the centroid of the solid region obtained by rotating about the y-axis the region lying between the x-axis and the parabola $y = 4 - x^2$.

Solution By symmetry the centroid of the parabolic solid will lie on the axis of symmetry, that is, on the y-axis. A thin disc-shaped slice of the solid at position y and having thickness dy (see Fig. 7.58) has volume

$$dV = \pi x^2\, dy = \pi(4 - y)\, dy$$

and has moment about the base plane

$$dM_b = y\, dV = \pi(4y - y^2)\, dy.$$

Hence the volume of the solid is

$$V = \pi\int_0^4 (4 - y)\, dy = \pi(16 - 8) = 8\pi \text{ cu units.}$$

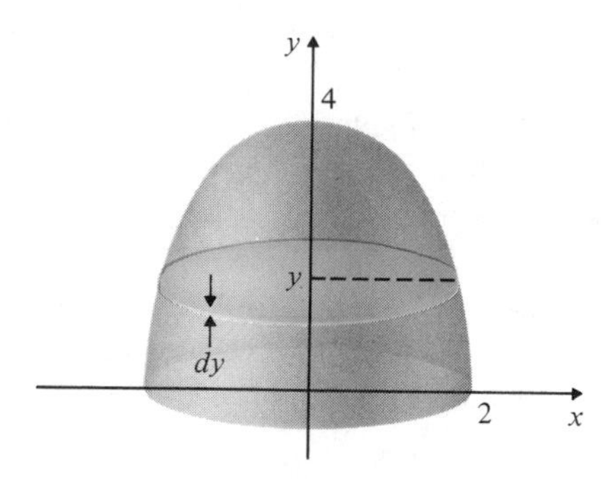

Figure 7.58

The moment of the solid about the base plane is

$$M_b = \pi \int_0^4 (4y - y^2)\,dy = \pi\left(32 - \frac{64}{3}\right) = \frac{32}{3}\pi.$$

Hence the centroid is located at $\bar{y} = 4/3$.

The Pappus Theorem

The following theorem relates volumes or surface areas of revolution to the centroid of the region or curve being rotated.

Theorem 1 (*Pappus*)

a) If a plane region R lies on one side of a line L in that plane, and if R is rotated about L to generate a solid of revolution, the volume V of that solid is the product of the area of R and the distance traveled by the centroid of R under the rotation; that is,

$$\boxed{V = 2\pi\bar{r}A,}$$

where A is the area of R and $\bar{r}$ is the perpendicular distance from the centroid of R to L.

b) If a plane curve C, lying on one side of a line L in the plane, is rotated about that line to generate a surface of revolution, then the area S of that surface is the length of C times the distance traveled by the centroid of C;

$$\boxed{S = 2\pi\bar{r}s,}$$

where s is the length of the curve C and $\bar{r}$ is the perpendicular distance from the centroid of C to the line L.

Proof a) Let us take L to be the x-axis and suppose that R lies above the x-axis so that $\bar{r} = \bar{y}$, the y-coordinate of the centroid of R. A thin strip of R at distance y from L generates, on rotation about L, a cylindrical shell of total length, say, $g(y)$, radius y, and volume $dV = 2\pi y\, g(y)\, dy$. See Fig. 7.59. If R lies between $y = a$ and $y = b$, the volume of the solid of revolution is

$$V = 2\pi \int_a^b y g(y)\,dy = 2\pi M_x,$$

where M_x is the moment of R about the x-axis. Since $\bar{y} = M_x/A$ we have $V = 2\pi A\bar{y}$ as asserted.

The proof of part (b) is similar and is left to the student as an exercise. □

As the following examples illustrate, Pappus's theorem can be used in two ways;

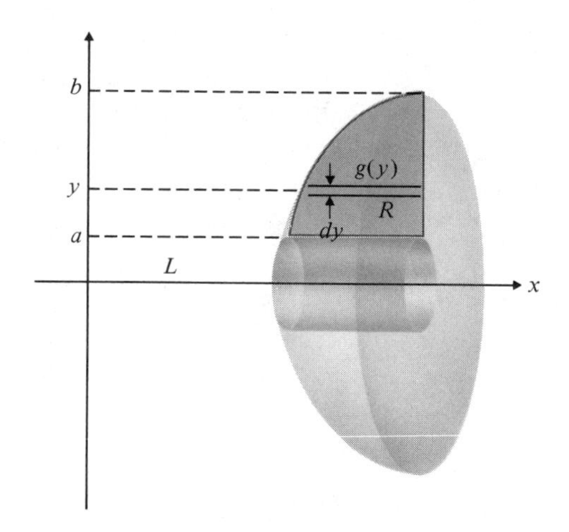

Figure 7.59

either the centroid can be determined when the appropriate volume or surface area is known, or the volume or surface area can be determined if the centroid of the rotating region or curve is known.

Example 5 The centroid of the semicircle $y = \sqrt{a^2 - x^2}$ lies on the axis of symmetry of that semicircle, so it is located at a point $(0, \bar{y})$. Since the semicircle has length πa units and generates, on rotation about the x-axis, a sphere having area $4\pi a^2$ sq units, we obtain, using part (b) of Pappus's theorem,

$$4\pi a^2 = 2\pi(\pi a)\bar{y},$$

so $\bar{y} = 2a/\pi$, as shown previously in Example 2.

Example 6 Rotating a disc of radius a about a line at distance $b > a$ from the center of the disc generates a solid torus (doughnut) of volume

$$V = 2\pi(\pi a^2)b = 2\pi^2 a^2 b \text{ cu units},$$

since the centroid of the disc is clearly at the center of the disc. The surface area of the torus (in case you want to have icing on the doughnut) is

$$S = 2\pi(2\pi a)b = 4\pi^2 ab \text{ sq units},$$

the surface being obtained by rotating a circle of radius a.

EXERCISES

Find the centroids of the geometric structures in Exercises 1–15.

1. The arc $y = (3/2)x^{2/3}$, $-1 \le x \le 1$

2. The quarter-circular arc $x^2 + y^2 = r^2$, $x \ge 0$, $y \ge 0$

3. The quarter-disc $x^2 + y^2 \le r^2$, $x \ge 0$, $y \ge 0$

4. The region $0 \le y \le 9 - x^2$

5. The trapezoid with vertices $(0, 0)$, $(0, 1)$, $(1, 1)$, and $(2, 0)$

6. The circular sector $x^2 + y^2 \leq r^2$, $0 \leq y \leq x$

7. The circular segment $0 \leq y \leq \sqrt{4 - x^2} - 1$

8. The semi-ellipse $0 \leq y \leq b\sqrt{1 - (x/a)^2}$

9. A hemispherical surface of radius r

10. A solid hemisphere of radius r

11. A conical surface of base radius r and height h

12. A solid cone of base radius r and height h

13. The plane region $0 \leq y \leq \sin x$, $0 \leq x \leq \pi$

14. The plane region $0 \leq y \leq \cos x$, $0 \leq x \leq \pi/2$

15. The solid obtained by rotating the plane region $0 \leq y \leq 1 - x^2$ about the x-axis

16. The line segment from $(1, 0)$ to $(0, 1)$ is rotated about the line $x = 2$ to generate part of a conical surface. Find the area of that surface.

17. The triangle with vertices $(0, 0)$, $(1, 0)$, and $(0, 1)$ is rotated about the line $x = 2$ to generate a certain solid. Find the volume of that solid.

18. An equilateral triangle of edge s cm is rotated about one of its edges to generate a solid. Find the volume and surface area of that solid.

19. Find the centroid of the infinitely long spike-shaped region lying between the x-axis and the curve $y = (x + 1)^{-3}$ and to the right of the y-axis.

20. Show that the curve $y = e^{-x^2}$ $(-\infty < x < \infty)$ generates a surface of finite area when rotated about the x-axis. What does this imply about the location of the centroid of this infinitely long curve?

7.7 OTHER PHYSICAL APPLICATIONS

Hydrostatic Pressure

The **pressure** (that is, force per unit area) that a liquid exerts on a surface immersed in it increases with depth and is proportional to depth. At a depth h the pressure is given by

$$p = \rho h,$$

where ρ is the density of the liquid. For water, $\rho \simeq 1$ g/cm^3 $= 1000$ kg/m^3 $= 1$ tonne/m^3 $\simeq 62.4$ lb/ft^3. Thus, for instance, the pressure at a depth of h m in water is

$$p = 1000h \text{ kg/m}^2 = 100h \text{ g/cm}^2.$$

The molecules in a liquid interact in such a way that the pressure at any depth acts equally in all directions; the pressure against a vertical surface is the same as that against a horizontal surface at the same depth. This is known as Pascal's principle.

The total force exerted by a liquid on a horizontal surface (say the bottom of a

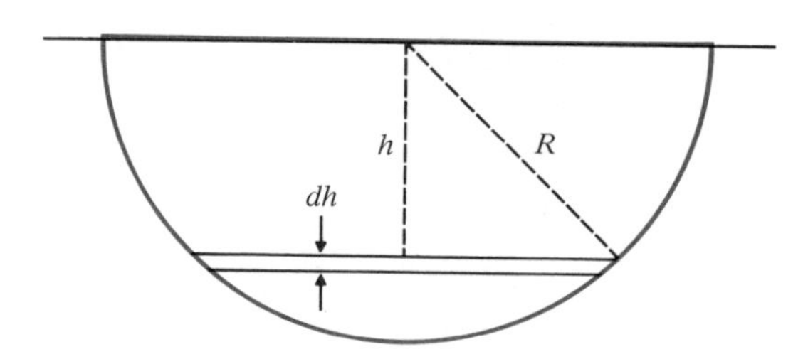

Figure 7.60

tank holding the liquid) is found by multiplying the area of that surface by the depth of the surface below the top of the liquid. For nonhorizontal surfaces, however, the pressure is not constant over the whole surface, and the total force cannot be determined so easily. In this case we divide the surface into area elements dA, each at some particular depth h, and we then sum (that is, integrate) the corresponding force elements $dF = \rho h\, dA$ to find the total force.

Example 1 A semicircular plate of radius R m is immersed vertically, curved edge down, in a tank of water, so that the top straight edge of the plate is at the surface of the water. Find the total force of the water against each face of the plate.

Solution A horizontal strip of the surface of the plate at depth h and having width dh (see Fig. 7.60) has length $2\sqrt{R^2 - h^2}$ m and hence area $dA = 2\sqrt{R^2 - h^2}\, dh$ m^2. The force of the water on this strip is

$$dF = \rho h\, dA = h\, dA \text{ tonnes,}$$

since $\rho = 1$ tonne/m^3. Thus the total force on one side of the plate is

$$F = 2\int_0^R h\sqrt{R^2 - h^2}\, dh \qquad \begin{aligned} \text{Let } u &= R^2 - h^2, \\ du &= -2h\, dh. \end{aligned}$$

$$= -\int_{R^2}^0 u^{1/2}\, du = -\frac{2}{3}u^{3/2}\bigg|_{R^2}^{0} = \frac{2}{3}R^3 \text{ tonnes.}$$

Example 2 Find the total force on a 100-m-long section of a dike of 10 m vertical height, if the surface holding back the water is inclined at an angle of 30° to the vertical and the water comes up to the top of the dike.

Solution The water in a horizontal layer of thickness dh at depth h actually makes contact with the section of dike along a slanted strip of width $dh \sec 30° = (2/\sqrt{3})\, dh$. See Fig. 7.61. The area of this strip is $dA = (200/\sqrt{3})\, dh$ m^2, and the force of water against the strip is $dF = (200/\sqrt{3})\, h\, dh$ tonnes. The total force on the dike section is therefore

$$F = \int_0^{10} \frac{200}{\sqrt{3}} h\, dh = \frac{200}{\sqrt{3}}\frac{h^2}{2}\bigg|_0^{10} = \frac{10{,}000}{\sqrt{3}} \text{ tonnes.}$$

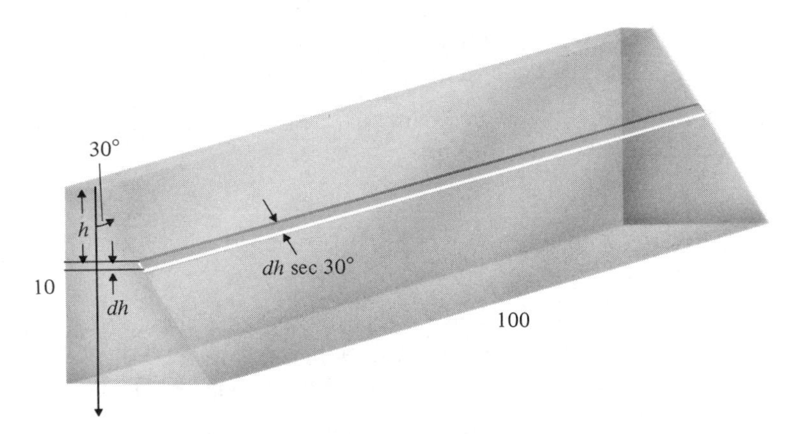

Figure 7.61

Work

When a force acts on an object to move that object, it is said to have done **work** on that object. The amount of work done is measured by the product of the force and the distance through which it moves the object, assuming that the force is constant as the object moves and that it is in the direction of the motion.

$$\text{Work} = \text{Force} \times \text{Distance}$$

Work is always related to a particular force. If other forces acting on an object cause it to move in a direction opposite to the force F, then work is said to have been done *against* the force F.

Suppose that a force moves an object from $x = a$ to $x = b$ on the x-axis and that the force varies continuously with the position x of the object; that is, $F = F(x)$ is a continuous function. The element of work done by the force in moving the object through a very short distance dx from x to $x + dx$ is $dW = F(x)\, dx$, so the total work done by the force is

$$W = \int_a^b F(x)\, dx.$$

Example 3 The force required to extend (or compress) an elastic spring to x units longer (or shorter) than its natural length is proportional to x (at least for sufficiently small values of x). This is known as Hooke's law. If we denote the force by $F(x)$, then

$$F(x) = kx,$$

where k is the *spring constant* for the particular spring. If a force of 10 kg is required to extend a certain spring to 4 cm longer than its natural length, how much work must be done to extend it that far?

Solution Here $F(x) = kx$ where, since $F(4) = 10$, we have $k = (5/2)$ kg/cm. The work done in extending the spring 4 cm is

$$W = \int_0^4 kx\, dx = k\frac{x^2}{2}\bigg|_0^4 = \frac{5}{2}\frac{16}{2} = 20 \text{ kg-cm}.$$

Example 4 Water fills a tank in the shape of a right-circular cone with top radius 3 m and depth 4 m. How much work must be done (against gravity) to pump all the water out of the tank over the top edge of the tank?

Solution A thin disc of water at height h above the vertex of the tank has radius r (see Fig. 7.62), where $r = (3/4)\, h$ (by similar triangles). The volume of this disc is

$$dV = \pi r^2\, dh = \frac{9}{16}\pi h^2\, dh \text{ m}^3$$

and its weight (mass) is

$$dM = \rho dV = \frac{9000}{16}\pi h^2\, dh \text{ kg}.$$

The water in this disc must be raised (against gravity) a distance $(4 - h)$ m by the pump. The work required to do this is

$$dW = \frac{9000}{16}\pi(4 - h)h^2\, dh \text{ kg-m}.$$

The total work that must be done to empty the tank is

$$W = \int_0^4 \frac{9000}{16}\pi(4h^2 - h^3)\, dh = \frac{9000}{16}\pi\left(\frac{4h^3}{3} - \frac{h^4}{4}\right)\bigg|_0^4$$

$$= 12{,}000\,\pi \text{ kg-m},$$

or 12π tonne-meters.

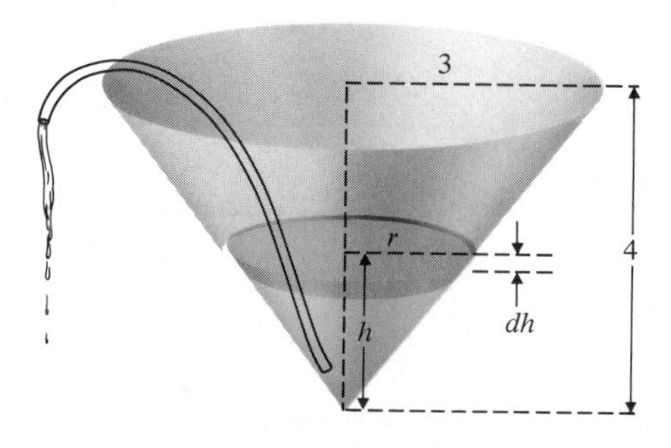

Figure 7.62

Example 5 The gravitational force attracting a mass m located at height h above the surface of the earth is given by

$$F(h) = \frac{Km}{(R+h)^2},$$

where R is the radius of the earth and K is a constant independent of m and h. Determine, in terms of K and R the work that must be done against gravity to raise an object from the surface of the earth to

a) a height H above the surface of the earth,

b) an infinite height above the surface of the earth.

Solution The work done to raise the mass m from height h to height $h + dh$ is

$$dW = \frac{Km}{(R+h)^2}\,dh.$$

The total work to raise it from height $h = 0$ to height $h = H$ is

$$W = \int_0^H \frac{Km}{(R+h)^2}\,dh = \left.\frac{-Km}{R+h}\right|_0^H = Km\left(\frac{1}{R} - \frac{1}{R+H}\right).$$

(If R and H are measured in kilometers and F is measured in kilograms, then W is measured in kilogram-kilometeres (kg-km).) The total work necessary to raise the mass m to infinite height is

$$W = \int_0^\infty \frac{Km}{(R+h)^2}\,dh = \lim_{H\to\infty} Km\left(\frac{1}{R} - \frac{1}{R+H}\right) = \frac{Km}{R}.$$

Potential and Kinetic Energy

Note that the units of work (force × distance) are the same as those of energy. Work done against a force may be regarded as storing up energy for future use or for conversion to other forms. Such stored energy is called **potential energy** (P.E.). For instance, in extending or compressing an elastic spring we are doing work against the tension in the spring and hence storing energy in the spring. When work is done against a (variable) force $F(x)$ to move an object from $x = a$ to $x = b$, the potential energy stored is

$$\text{P.E.} = -\int_a^b F(x)\,dx.$$

Since the work is being done against F, the signs of $F(x)$ and $b - a$ are opposite, so the integral is negative; the explicit negative sign is included so that the calculated potential energy will be positive.

One of the forms of energy into which potential energy can be converted is **kinetic energy** (K.E.), the energy of motion. If an object of mass m is moving with velocity v, it has kinetic energy

$$\text{K.E.} = \frac{1}{2} m v^2.$$

For example, if an object is raised and then dropped, it accelerates downward as more and more of the potential energy stored in it when it was raised is converted to kinetic energy.

Consider the change in potential energy stored in a mass m as it moves along the x-axis under the influence of a force $F(x)$:

$$\text{P.E.}(b) - \text{P.E.}(a) = -\int_a^b F(x)\, dx.$$

According to Newton's second law of motion, the force $F(x)$ causes the mass m to accelerate, with acceleration dv/dt given by

$$F(x) = m\frac{dv}{dt} \qquad (\text{force} = \text{mass} \times \text{acceleration}).$$

By the chain rule we can rewrite dv/dt in the form

$$\frac{dv}{dt} = \frac{dv}{dx}\frac{dx}{dt} = v\frac{dv}{dx},$$

so $F(x) = mv\dfrac{dv}{dx}\cdot$ Hence

$$\begin{aligned}\text{P.E.}(b) - \text{P.E.}(a) &= -\int_a^b mv\frac{dv}{dx}\, dx \\ &= -m\int_{x=a}^{x=b} v\, dv \\ &= -\frac{1}{2}mv^2\Big|_{x=a}^{x=b} \\ &= \text{K.E.}(a) - \text{K.E.}(b).\end{aligned}$$

Thus

$$\text{P.E.}(b) + \text{K.E.}(b) = \text{P.E.}(a) + \text{K.E.}(A).$$

This shows that the total energy (potential + kinetic) remains constant as the mass m moves under the influence of a force F, depending only on position. Such a force is said to be **conservative** and the above result is called the **law of conservation of energy.**

Example 6 We will use the result of Example 5 together with the following known values,

a) the radius of the earth is approximately $R = 6400$ km,

b) the acceleration of gravity at the surface of the earth is approximately $g = 9.8$ m/sec^2 $= 0.0098$ km/sec^2,

to determine the constant K in the gravitational force formula of Example 5, and hence to determine the escape velocity for a projectile fired from the surface of the earth. The escape velocity is the (minimum) speed that such a projectile must have at firing to ensure that it will continue to recede farther and farther from the surface of the earth and not fall back.

According to the formula of Example 5, the force of gravity on a mass m at the surface of the earth ($h = 0$) is

$$F = \frac{Km}{(R+0)^2} = \frac{Km}{R^2}.$$

According to Newton's second law of motion, this force is related to the acceleration of gravity there (g) by the equation $F = mg$. Thus

$$\frac{Km}{R^2} = mg$$

and $K = gR^2$.

According to the law of conservation of energy, the projectile must have sufficient kinetic energy at firing to do the work necessary to raise the mass m to infinite height (that is, to supply sufficient potential energy to raise it to that height). By the result of Example 5, this required energy is Km/R. If the initial velocity of the projectile is v, we want

$$\frac{1}{2}mv^2 \geq \frac{Km}{R}.$$

Thus v must satisfy

$$v \geq \sqrt{\frac{2K}{R}} = \sqrt{2gR} = \sqrt{2(0.0098)(6400)} \simeq 11.2 \text{ km/sec}.$$

Thus the escape velocity is approximately 11.2 km/sec (or 7 mi/sec). In this calculation we have neglected any air resistance near the surface of the earth. Such resistance depends on velocity rather than on position, so it is not a conservative force. The effect of such resistance would be to use up (convert to heat) some of the initial kinetic energy and so raise the escape velocity.

EXERCISES

1. A tank has a square base 2 m on each side and vertical sides 6 m high. If the tank is filled with water, find the total force exerted by the water on the bottom of the tank.

2. Find the total force exerted by the water in the tank in Exercise 1 on each of the four vertical walls of the tank.

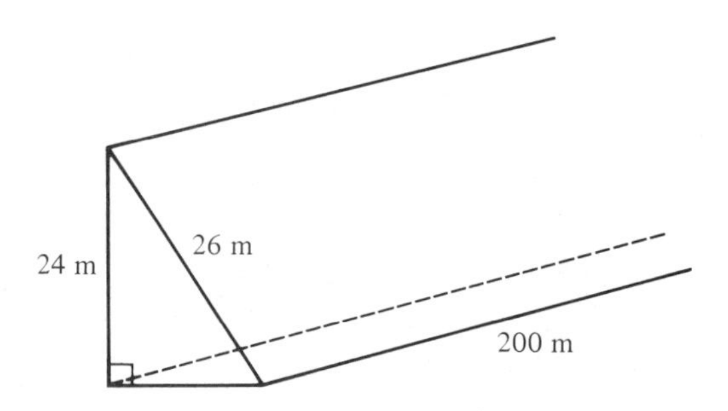

Figure 7.63

3. A swimming pool 20 m long and 8 m wide has a sloping plane bottom so that the depth of the pool at one end is 1 m and at the other end is 3 m. Find the total force exerted on the bottom of the pool if the pool is full of water.
4. A dam 200 m long and 24 m high presents a sloping face of 26 m slant height to the water in a reservoir behind the dam (see Fig. 7.63). If the surface of the water is level with the top of the dam, what is the total force of the water on the dam?
5. A pyramid with 4-m $\times$ 4-m square base and four equilateral triangular faces sits on the level bottom of a lake at a place where the lake is 10 m deep. Find the total force of the water on each of the triangular faces.
6. A lock on a canal has a gate in the shape of a vertical rectangle 5 m wide and 20 m high. If the water on one side of the gate comes up to the top of the gate and the water on the other side comes only 6 m up the gate, find the total force that must be exerted to hold the gate in place.
7. Find the total work that must be done to pump all the water in the tank of Exercise 1 out over the top of the tank.
8. Find the total work that must be done to pump all the water in the swimming pool of Exercise 3 out over the top edge of the pool.
9. Find the work that must be done to pump all the water in a full hemispherical bowl of radius a cm to a height h cm above the top of the bowl.
10. If 10 gm-cm of work must be done to compress an elastic spring to 3 cm shorter than its natural length, how much work must be done to compress it 1 cm further?
11. A bucket is raised vertically from ground level at a constant speed of 2 m/min by a winch. If the bucket weighs 1 kg and contains 15 kg of water when it starts up, but loses water by leakage at a rate of 1 kg/min thereafter, how much work must be done by the winch to raise the bucket to a height of 10 m?

7.8 PROBABILITY

Probability theory is a very important field of application of the definite integral. This subject cannot, of course, be developed thoroughly here—an adequate presentation requires one or more whole courses—but we can give a brief introduction that suggests some of the ways integrals occur in probability theory.

The **probability** of an event occurring is a real number between 0 and 1 that measures the proportion of times the event can be expected to occur in a large number of trials. If the occurrence of an event is certain, its probability is 1; if the event cannot possibly occur, its probability is 0. For example, the probability that a tossed coin will land heads is 1/2 because one would expect it to land heads about half the time if it were tossed a great many times. In such a tossing of a coin there are only two possible outcomes, heads or tails, each equally likely, that is, each having probability 1/2. For any toss, let $X = 0$ if the outcome is heads, and let $X = 1$ if the outcome is tails. X is called a **discrete random variable**. The probability that $X = 0$ is 1/2 and the probability that $X = 1$ is 1/2, so we write

$$\Pr(X = 0) = \frac{1}{2}, \qquad \Pr(X = 1) = \frac{1}{2}.$$

Note that $\Pr(X = 0) + \Pr(X = 1) = 1$ since it is certain that the coin will land either heads or tails.

Example 1 A single die is rolled so that it will show one of the numbers 1 to 6 on top when it stops. If X denotes the number showing on any roll, then X is a discrete random variable. Evidently no one value of X is any more likely than any other, so the probability that the number showing is x must be 1/6 for each possible value of x; that is,

$$\Pr(X = x) = \frac{1}{6} \qquad \text{for each } x \text{ in } \{1, 2, 3, 4, 5, 6\}.$$

The discrete random variable X is therefore said to be distributed **uniformly**. Again we note that

$$\sum_{n=1}^{6} \Pr(X = n) = 1,$$

reflecting the fact that the rolled die must certainly give one of the six possible outcomes. The probability that a roll will produce a value from 1 to 4 is

$$\Pr(1 \le X \le 4) = \sum_{n=1}^{4} \Pr(X = n) = \frac{1}{6} + \frac{1}{6} + \frac{1}{6} + \frac{1}{6} = \frac{2}{3}.$$

Now we consider an example with a continuous range of possible outcomes.

Example 2 Suppose that a needle is dropped at random on a flat table with a straight line drawn on it. For each drop, let X be the number of degrees in the (acute) angle that the needle makes with the line (see Fig. 7.64). Evidently X can take any real value in the interval $[0, 90]$, and X is called a **continuous random variable**. The probability that X takes on any particular real value is 0. (There are infinitely many real numbers in $[0, 90]$ and none is more likely than any other.) However, the probability that X lies in some interval, say $[10, 20]$, is the same as the probability that it lies in any other interval of the same length, in this case 10. Hence this probability is

$$\Pr(10 \le X \le 20) = \frac{10}{90} = \frac{1}{9}.$$

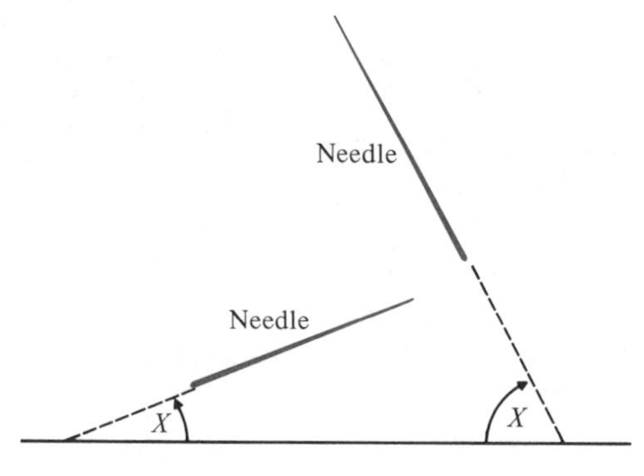

Figure 7.64

More generally, if $0 \leq x_1 \leq x_2 \leq 90$, then

$$\Pr(x_1 \leq X \leq x_2) = \frac{1}{90}(x_2 - x_1).$$

This situation can be conveniently represented as follows: Let $f(x)$ be defined on the interval $[0, 90]$, taking at each point the constant value $1/90$:

$$f(x) = \frac{1}{90}, \qquad 0 \leq x \leq 90.$$

The area under the graph of f is 1, and $\Pr(x_1 \leq X \leq x_2)$ is equal to the area under that part of the graph lying over the interval $[x_1, x_2]$ (see Fig. 7.65). The function $f(x)$ is called the **probability density function** for the random variable X. Since $f(x)$ is constant on its domain, X is said to be **uniformly distributed.**

In general, a function defined on an interval $[a, b]$ is a probability density function for a continuous random variable X distributed on $[a, b]$ if, whenever $a \leq x_1 \leq x_2 \leq b$ we have

$$\Pr(x_1 \leq X \leq x_2) = \int_{x_1}^{x_2} f(x)\, dx.$$

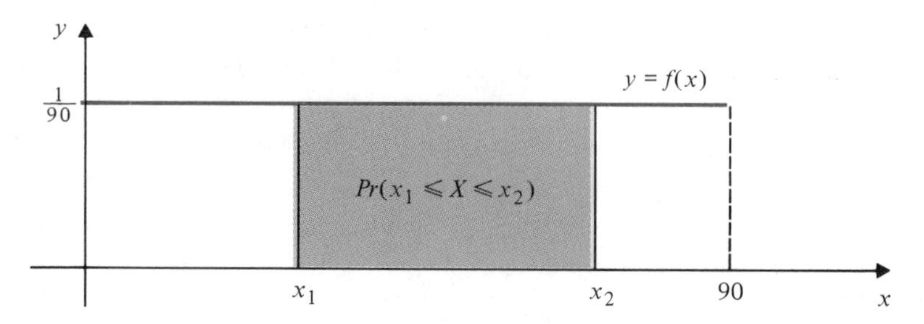

Figure 7.65

In order to be such a probability density function, f must satisfy two conditions:

a) $f(x) \geq 0 \quad$ on $[a, b]$, and

b) $\displaystyle\int_a^b f(x)\,dx = 1.$

That is, probability cannot be negative, and $\Pr(a \leq X \leq b) = 1$. These ideas extend to random variables distributed on semi-infinite or infinite intervals, but the integrals appearing will be improper in those cases.

In the example of the dropping needle, the probability density function was a horizontal straight line, and we termed such a probability distribution uniform. A uniform probability density function on the interval $[a, b]$ is

$$f(x) = \begin{cases} \dfrac{1}{b-a} & \text{if } a \leq x \leq b \\ 0 & \text{otherwise.} \end{cases}$$

Many other functions are commonly encountered as density functions for continuous random variables.

Example 3 The length of time T that any particular atom in a radioactive sample survives before decaying is a random variable taking values in $[0, \infty)$. It has been observed that the proportion of atoms that survive to time t grows small exponentially at t increases; thus

$$\Pr(T \geq t) = Ce^{-kt}.$$

Let f be the probability density function for the random variable T. Then

$$\int_t^\infty f(x)\,dx = \Pr(T \geq t) = Ce^{-kt}.$$

Differentiating this equation with respect to t (using the fundamental theorem of calculus), we obtain $-f(t) = -Cke^{-kt}$, so $f(t) = Cke^{-kt}$. C is determined by the requirement that $\int_0^\infty f(t)\,dt = 1$. We have

$$1 = Ck\int_0^\infty e^{-kt}\,dt = \lim_{R\to\infty} Ck\int_0^R e^{-kt}\,dt = -C\lim_{R\to\infty}(e^{-kR} - 1) = C.$$

Thus $C = 1$ and $f(t) = ke^{-kt}$. Note that $\Pr(T \geq (\ln 2)/k) = e^{-k(\ln 2)/k} = 1/2$, reflecting the fact that the half-life of such a radioactive sample is $(\ln 2)/k$.

Example 4 For what value of C is $f(x) = C(1 - x^2)$ a probability density function on $[-1, 1]$? If X is a random variable with this density what is the probability that $X \leq 1/2$?

Solution Evidently $f(x) \geq 0$ on $[-1, 1]$ if $C \geq 0$. Since

$$\int_{-1}^{1} f(x)\,dx = C\int_{-1}^{1} (1 - x^2)\,dx = 2C\left(x - \frac{x^3}{3}\right)\Bigg|_0^1 = \frac{4C}{3},$$

$f(x)$ will be a probability density function if $C = 3/4$. In this case

$$\begin{aligned}\Pr\left(X \leq \frac{1}{2}\right) &= \frac{3}{4}\int_{-1}^{1/2} (1 - x^2)\,dx = \frac{3}{4}\left(x - \frac{x^3}{3}\right)\Bigg|_{-1}^{1/2} \\ &= \frac{3}{4}\left(\frac{1}{2} - \frac{1}{24} - (-1) + \frac{-1}{3}\right) \\ &= \frac{27}{32}.\end{aligned}$$

Expectation, Mean, Variance, and Standard Deviation

Consider a simple gambling game in which the player pays the house C dollars for the privilege of rolling a single die and in which he wins X dollars where X is the number showing on top of the rolled die. In each game the possible winnings are 1, 2, 3, 4, 5 or 6 dollars, each with probability 1/6. In n games the player can expect to win about $n/6 + 2n/6 + 3n/6 + 4n/6 + 5n/6 + 6n/6 = 21n/6 = 7n/2$ dollars, so that his expected *average winnings per game* are 7/2 dollars, \$3.50. If $C > 3.5$, the player can expect, on average, to lose money. The amount 3.5 is called the **expectation** or **mean** of the discrete random variable X.

In general, if a random variable can take on values x_1 with probability p_1, x_2 with probability $p_2, \ldots,$ and x_n with probability p_n (where $p_1 + p_2 + \cdots + p_n = 1$), the mean μ or expectation $E(X)$ of that random variable X is given by

$$\mu = E(X) = \sum_{i=1}^{n} x_i p_i.$$

For a continuous random variable X with probability density function $f(x)$ on the interval $[a, b]$, the analogous definition of the mean, or expectation of X, is

$$\boxed{\mu = E(X) = \int_a^b xf(x)\,dx.}$$

(Note that in this usage $E(X)$ does not define a function of X but a constant (parameter) associated with the probability distribution for X. Note also that if $f(x)$ were a mass density such as that studied in Section 7.5, then μ would be the moment of the mass about 0 and, since the total mass would be $\int_a^b f(x)\,dx = 1$, μ would in fact be the center of mass.)

More generally, the **expectation** of any function $g(X)$ of the random variable X is

$$E(g(X)) = \int_a^b g(x)f(x)\,dx.$$

The **variance** σ^2 of a random variable X with density $f(x)$ on $[a, b]$ is the expectation of the square of the distance from X to the mean μ.

$$\sigma^2 = E((X - \mu)^2) = \int_a^b (x - \mu)^2 f(x)\,dx.$$

Since $\int_a^b f(x)\,dx = 1$, the expression above for σ^2 can be rewritten as follows:

$$\begin{aligned}\sigma^2 &= \int_a^b x^2 f(x)\,dx - 2\mu\int_a^b x f(x)\,dx + \mu^2\int_a^b f(x)\,dx \\ &= \int_a^b x^2 f(x)\,dx - 2\mu^2 + \mu^2 = E(X^2) - \mu^2,\end{aligned}$$

that is,

$$\sigma^2 = E(X^2) - \mu^2 = E(X^2) - (E(X))^2.$$

The **standard deviation** of the random variable X is the square root σ of the variance. Thus it is the square root of the mean square deviation of X from its mean:

$$\sigma = \left(\int_a^b (x - \mu)^2 f(x)\,dx\right)^{1/2} = \sqrt{E(X^2) - \mu^2}.$$

This standard deviation gives a measure of how spread out the probability distribution of X is. The smaller the standard deviation, the more concentrated is the area under the density curve around the mean, and so the smaller is the probability that a value of X will be far away from the mean. See Fig. 7.66.

Example 5 Find the mean and the standard deviation of a random variable X distributed uniformly on the interval $[a, b]$.

Solution The probability density function is $f(x) = 1/(b - a)$ on $[a, b]$, so the mean is given by

$$\mu = E(X) = \int_a^b \frac{x}{b - a}\,dx = \frac{1}{b - a}\frac{x^2}{2}\bigg|_a^b = \frac{1}{2}\frac{b^2 - a^2}{b - a} = \frac{b + a}{2}.$$

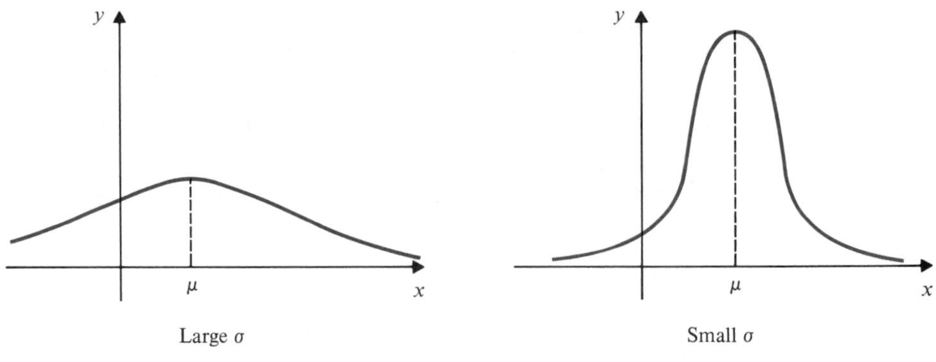

Densities with large and small standard deviations

Figure 7.66

Hence the mean is, as might have been anticipated, the midpoint of $[a, b]$ (see Fig. 7.67). The expectation of X^2 is given by

$$E(X^2) = \int_a^b \frac{x^2}{b-a}\,dx = \frac{1}{b-a}\frac{x^3}{3}\bigg|_a^b = \frac{1}{3}\frac{b^3-a^3}{b-a} = \frac{b^2+ab+a^2}{3}.$$

Hence the variance is

$$\sigma^2 = E(X^2) - \mu^2 = \frac{b^2+ab+a^2}{3} - \frac{b^2+2ab+a^2}{4} = \frac{(b-a)^2}{12},$$

and the standard deviation is

$$\sigma = \frac{b-a}{2\sqrt{3}} \simeq 0.29(b-a).$$

Observe that $\Pr(\mu - \sigma \le X \le \mu + \sigma) = \dfrac{1}{b-a}\dfrac{2(b-a)}{2\sqrt{3}} = \dfrac{1}{\sqrt{3}} \simeq 0.577.$

Example 6 Find the mean and the standard deviation of a random variable X distributed exponentially with density function $f(x) = ke^{-kx}$ on $[0, \infty)$.

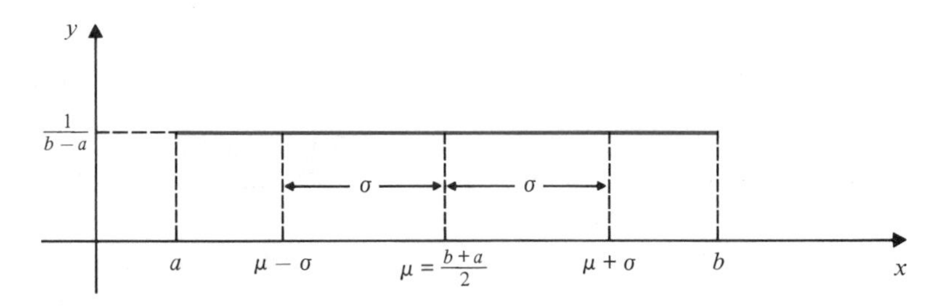

Figure 7.67

Solution We use integration by parts to find the mean:

$$\mu = E(X) = k\int_0^\infty xe^{-kx}\,dx$$

$$= \lim_{R\to\infty} k\int_0^R xe^{-kx}\,dx \qquad \text{Let } U = x, \quad dV = e^{-kx}\,dx, \quad dU = dx, \quad V = e^{-kx}/(-k).$$

$$= \lim_{R\to\infty}\left(-xe^{-kx}\Big|_0^R + \int_0^R e^{-kx}\,dx\right)$$

$$= \lim_{R\to\infty}\left(-Re^{-kR} - \frac{1}{k}(e^{-kR} - 1)\right) = \frac{1}{k} \qquad (\text{since } k > 0).$$

Thus the mean of the exponential distribution is $1/k$. This fact can be quite useful in determining the value of k for an exponentially distributed random variable. A similar integration by parts enables us to evaluate

$$E(X^2) = k\int_0^\infty x^2e^{-kx}\,dx = 2\int_0^\infty xe^{-kx}\,dx = \frac{2}{k^2},$$

so that the variance of the exponential distribution is

$$\sigma^2 = E(X^2) - \mu^2 = \frac{1}{k^2}$$

and the standard deviation is equal to the mean

$$\sigma = \mu = \frac{1}{k}.$$

Observe that

$$\begin{aligned}\Pr(\mu - \sigma \le X \le \mu + \sigma) &= \Pr(0 \le X \le 2/k)\\ &= k\int_0^{2/k} e^{-kx}\,dx\\ &= -e^{-kx}\Big|_0^{2/k}\\ &= 1 - e^{-2} \simeq 0.86\end{aligned}$$

is independent of the value of k. Exponential densities for small and large values of k are sketched in Fig. 7.68.

The Normal Distribution

The most important probability distributions are the so-called **normal** or **Gaussian** distributions. Such distributions govern the behavior of a great many random variables, in particular those associated with random errors in measurements. There is a family

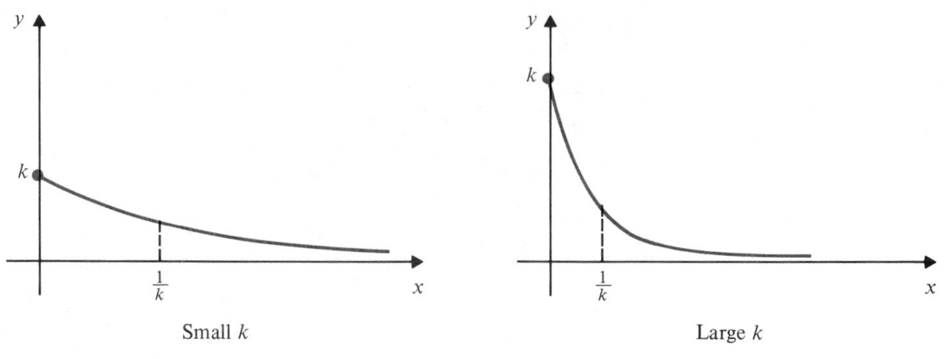

Small k Large k

Exponential densities, $y = ke^{-kx}$

Figure 7.68

of normal distributions, all related to the particular normal distribution called the **standard normal distribution,** which has probability density function

$$f(z) = \frac{1}{\sqrt{2\pi}} e^{-z^2/2}.$$

It is common to use z to denote the random variable in the standard normal distribution; the other normal distributions are obtained from this one by a change of variable. The graph of the standard normal density has a pleasant bell shape, as shown in Fig. 7.69.

As we have noted previously, the function e^{-z^2} has no elementary antiderivative, so the improper integral

$$I = \int_{-\infty}^{\infty} e^{-z^2/2}\, dz$$

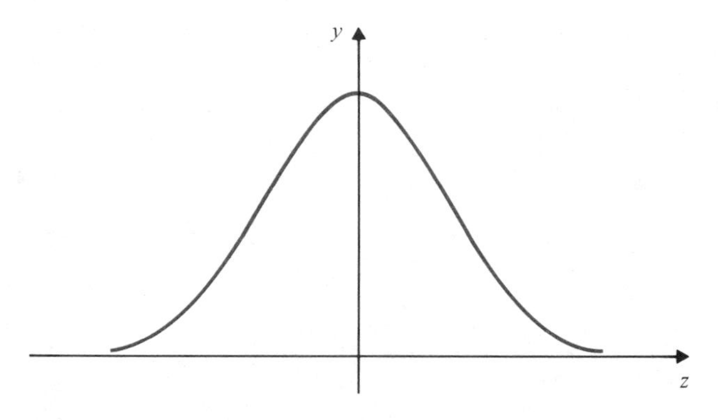

The standard normal density $y = \frac{1}{\sqrt{2\pi}} e^{-z^2/2}$

Figure 7.69

cannot be easily evaluated using the fundamental theorem of calculus, though it is easily shown to be a convergent improper integral. The integral can be evaluated by using techniques of advanced calculus involving integrals of functions of two variables, and the evaluation is usually given in multivariate calculus courses. The value is $I = \sqrt{2\pi}$, which ensures that the above-defined standard normal density $f(z)$ is indeed a probability density function:

$$\int_{-\infty}^{\infty} f(z)\,dz = \frac{1}{\sqrt{2\pi}}\int_{-\infty}^{\infty} e^{-z^2/2}\,dz = 1.$$

Since $ze^{-z^2/2}$ is an odd function of z, the mean of the standard normal distribution is 0:

$$\mu = E(Z) = \frac{1}{\sqrt{2\pi}}\int_{-\infty}^{\infty} ze^{-z^2/2}\,dz = \frac{1}{\sqrt{2\pi}}\lim_{R\to\infty}\int_{-R}^{R} ze^{-z^2/2}\,dz = 0.$$

We calculate the variance of the standard normal distribution using integration by parts as follows:

$$\sigma^2 = E(Z^2) = \frac{1}{\sqrt{2\pi}}\int_{-\infty}^{\infty} z^2\,e^{-z^2/2}$$

$$= \frac{1}{\sqrt{2\pi}}\lim_{R\to\infty}\int_{-R}^{R} z^2e^{-z^2/2}\,dz \qquad \begin{array}{ll} \text{Let } U = z, & dV = ze^{-z^2/2}\,dz, \\ dU = dz, & V = -e^{-z^2/2}. \end{array}$$

$$= \frac{1}{\sqrt{2\pi}}\lim_{R\to\infty}\left(-ze^{-z^2/2}\Big|_{-R}^{R} + \int_{-R}^{R} e^{-z^2/2}\,dz\right)$$

$$= \frac{1}{\sqrt{2\pi}}\lim_{R\to\infty}(-2R\,e^{-R^2/2}) + \frac{1}{\sqrt{2\pi}}\int_{-\infty}^{\infty} e^{-z^2/2}\,dz$$

$$= 0 + 1 = 1.$$

Hence the standard deviation of the standard normal distribution is 1.

Other normal distributions are obtained from the standard normal distribution by a change of variable. A random variable X on $(-\infty, \infty)$ is said to be *normally distributed with mean* μ and *standard deviation* σ (where μ is any real number and $\sigma > 0$) if its probability density function is

$$f_{\mu,\sigma}(x) = \frac{1}{\sigma}f\left(\frac{x-\mu}{\sigma}\right) = \frac{1}{\sigma\sqrt{2\pi}}e^{-(x-\mu)^2/2\sigma^2}.$$

See Fig. 7.70. Using the change of variable $z = (x - \mu)/\sigma$, $dz = (1/\sigma)\,dx$, we calculate

$$\int_{-\infty}^{\infty} f_{\mu,\sigma}(x)\,dx = \int_{-\infty}^{\infty}\frac{1}{\sigma}f\left(\frac{x-\mu}{\sigma}\right)dx = \int_{-\infty}^{\infty}\frac{1}{\sigma}f(z)\,\sigma\,dz = \int_{-\infty}^{\infty} f(z)\,dz = 1,$$

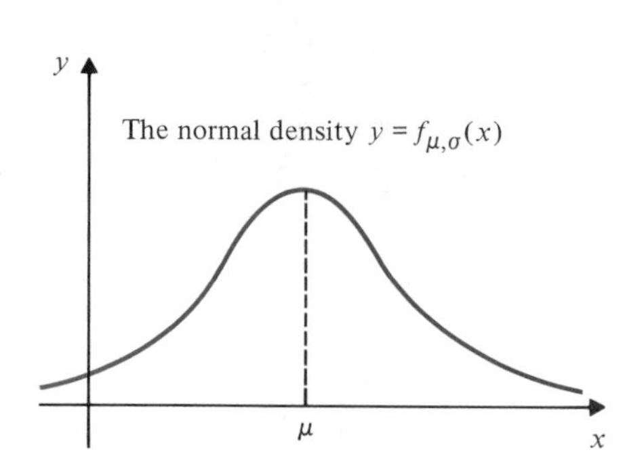

Figure 7.70

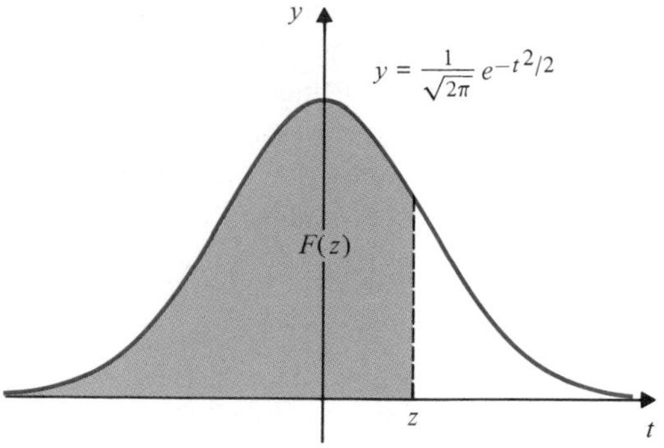

Figure 7.71

so $f_{\mu,\sigma}(x)$ is indeed a probability density function. Using the same change of variable we can readily show that

$$E(X) = \mu, \qquad \text{and} \qquad E((X - \mu)^2) = \sigma^2.$$

Hence $f_{\mu,\sigma}(x)$ does indeed have mean μ and standard deviation σ.

Because $e^{-z^2/2}$ cannot be readily antidifferentiated, we cannot carry out the determination of probabilities (that is, areas) for normal curves by using the fundamental theorem of calculus and a scientific calculator. Books on probability and statistics include tables of computed areas under the standard normal curve. Specifically, these tables usually provide values for what is called the **cumulative distribution function** of a random variable with standard normal distribution, that is, the function

$$F(z) = \frac{1}{\sqrt{2\pi}}\int_{-\infty}^{z} e^{-t^2/2}\,dt = \Pr(Z \le z),$$

the area under the standard normal density function from $-\infty$ up to z, as shown in Fig. 7.71.

For use in the following examples and exercises, we include an abbreviated version of such a table on the following page.

Example 7 If Z is a standard normal random variable find

a) $\Pr(-1.2 \le Z \le 2.0)$ and

b) $\Pr(Z \ge 1.5)$.

Solution a) $\Pr(-1.2 \le Z \le 2.0) = \dfrac{1}{\sqrt{2\pi}}\displaystyle\int_{-1.2}^{2.0} e^{-z^2/2}\,dz$

$$= F(2.0) - F(-1.2) \simeq 0.977 - 0.115 = 0.862$$

b) $\Pr(Z \ge 1.5) = 1 - \Pr(Z < 1.5) = 1 - F(1.5)$

$$\simeq 1 - 0.933 = 0.067$$

VALUES OF $F(z)$ (rounded to 3 decimal places)										
z	**0.0**	**0.1**	**0.2**	**0.3**	**0.4**	**0.5**	**0.6**	**0.7**	**0.8**	**0.9**
−3.0	0.001	0.001	0.001	0.001	0.000	0.000	0.000	0.000	0.000	0.000
−2.0	0.023	0.018	0.014	0.011	0.008	0.006	0.005	0.004	0.003	0.002
−1.0	0.159	0.136	0.115	0.097	0.081	0.067	0.055	0.045	0.036	0.029
−0.0	0.500	0.460	0.421	0.382	0.345	0.309	0.274	0.242	0.212	0.184
0.0	0.500	0.540	0.579	0.618	0.655	0.692	0.726	0.758	0.788	0.816
1.0	0.841	0.864	0.885	0.903	0.919	0.933	0.945	0.955	0.964	0.971
2.0	0.977	0.982	0.986	0.989	0.992	0.994	0.995	0.997	0.997	0.998
3.0	0.999	0.999	0.999	1.000	1.000	1.000	1.000	1.000	1.000	1.000

Example 8 A random variable X is distributed normally with mean 2 and standard deviation 0.4. Find

a) $\Pr(1.8 \le X \le 2.4)$ and

b) $\Pr(X > 2.4)$.

Solution Since X is distributed normally with mean 2 and standard deviation 0.4, $Z = (X - 2)/0.4$ is distributed according to the standard normal distribution (with mean 0 and standard deviation 1). Accordingly,

a) $$\Pr(1.8 \le X \le 2.4) = \Pr(-0.5 \le Z \le 1) = F(1) - F(-0.5) \simeq 0.841 - 0.309 = 0.532$$

b) $$\Pr(X > 2.4) = \Pr(Z > 1) = 1 - \Pr(Z \le 1) = 1 - F(1) \simeq 1 - 0.841 = 0.159.$$

EXERCISES

For each function $f(x)$ in Exercises 1–7 find the following:

a) the value of C for which f is a probability density on the given interval;

b) the mean, variance, and standard deviation of the probability density f;

c) $\Pr(\mu - \sigma \le X \le \mu + \sigma)$, that is, the probability that the random variable X is no further than one standard deviation away from its mean.

1. $f(x) = Cx$ on $[0, 3]$

2. $f(x) = Cx$ on $[1, 2]$

3. $f(x) = Cx^2$ on $[0, 1]$

4. $f(x) = C(x - x^2)$ on $[0, 1]$

5. $f(x) = C \sin x$ on $[0, \pi]$

6. $f(x) = C\,xe^{-kx}$ on $[0, \infty)$ $(k > 0)$

7. $f(x) = C\,xe^{-x^2}$ on $[0, \infty)$. (*Hint:* Use properties of the standard normal density to show that $\int_0^\infty e^{-x^2}\,dx = \sqrt{\pi}/2$.)

8. Is it possible for a random variable to be uniformly distributed on the whole real line? Explain why.

9. Carry out the calculations to show that the normal density $f_{\mu,\sigma}(x)$ defined in the text has mean μ and standard deviation σ.

10. Show that $f(x) = 2/(\pi(1 + x^2))$ is a probability density on $[0, \infty)$. Find the expectation of X for this density. If a machine generates values of a random variable X distributed with density $f(x)$, how much would you be willing to pay, per game, to play a game in which you actuate the machine to produce a value of X and win X dollars? Explain.

11. Calculate $\Pr(|X - \mu| \geq 2\sigma)$ for

a) the uniform distribution on $[a, b]$;

b) the exponential distribution with density $f(x) = ke^{-kx}$ on $[0, \infty)$;

c) the normal distribution with density $f_{\mu,\sigma}(x)$.

12. The length of time T (in hours) between malfunctions of a computer system is an exponentially distributed random variable. If the average length of time between successive malfunctions is 20 hours, find the probability that the system, having just had a malfunction corrected, will operate without malfunction for at least 12 hours.

13. The number X of meters of cable produced any day by a cable-making company is a normally distributed random variable with mean 5000 and standard deviation 200. On what fraction of the days the company operates will the number of meters of cable produced exceed 5500?

7.9 FIRST-ORDER SEPARABLE AND LINEAR DIFFERENTIAL EQUATIONS

This final section on applications of integration concentrates on applications of the indefinite integral rather than of the definite integral. We can use the techniques of integration developed in Chapter 6 to solve certain kinds of first-order differential equations that arise in a variety of modeling situations. We have already seen some examples of applications of differential equations to modeling growth and decay situations in Section 3.7. In particular, the method of partial fractions (though not by that name) was used in that section to solve the differential equation of logistic growth,

$$\frac{dy}{dt} = ky(L - y).$$

Separable Equations

The logistic equation is an example of a class of first-order differential equations called separable because when they are written in terms of differentials they can be separated with only the dependent variable on one side of the equation and only the independent variable on the other. Thus the logistic equation can be written in the form

$$\frac{dy}{y(L - y)} = k\,dt$$

and solved by integrating both sides.

Generally, separable equations are of the form

$$\frac{dy}{dx} = f(x)g(y),$$

and they are solved by being rewritten in the form

$$\frac{1}{g(y)}\,dy = f(x)\,dx$$

so that

$$\int \frac{1}{g(y)}\,dy = \int f(x)\,dx.$$

Example 1 Solve the equation $dy/dx = x/y$.

Solution We have $y\,dy = x\,dx$, so, on integrating both sides, we get $(1/2)y^2 = (1/2)x^2 + C$ or $y^2 - x^2 = C_1$ where $C_1 = 2C$ is an arbitrary constant. The solution curves are rectangular hyperbolas with asymptotes $y = x$ and $y = -x$.

Example 2 Find the function $y(x)$ that satisfies $dy/dx = x^2y^3$ and the initial condition $y(1) = 3$.

Solution We have $(1/y^3)dy = x^2\,dx$, so $\int dy/y^3 = \int x^2\,dx$ and

$$\frac{-1}{2y^2} = \frac{x^3}{3} + C.$$

Since $y = 3$ when $x = 1$, we have $-(1/18) = 1/3 + C$ and $C = -(7/18)$. It follows that

$$y(x) = \left(\frac{14}{18} - \frac{2}{3}x^3\right)^{-1/2}.$$

This solution is valid for $(2/3)x^3 < 14/18$, that is, for $x \leq (7/6)^{1/3}$.

Example 3 (*A Solution Concentration Problem*) Initially a tank contains 1000 l of brine with 50 kg of dissolved salt. If brine containing 10 gm of salt per liter is flowing into the tank at a constant rate of 10 l/min, if the contents of the tank are kept thoroughly mixed at all times, and if the solution also flows out at 10 l/min, how much salt remains in the tank at the end of 40 min?

Solution Let $x(t)$ be the number of kg of salt in solution in the tank after t minutes. Thus $x(0) = 50$. Salt is coming into the tank at a rate of $10 \times 10 = 100$ g/min $=$ 1/10 kg/min. At all times the tank contains 1000 l of liquid, so the concentration of salt in the tank at time t is $x/1000$ kg/l. Since the contents flow out at 10 l/min, salt is being removed at a rate of $10x/1000 = x/100$ kg/min. Therefore,

$$\frac{dx}{dt} = \text{rate in} - \text{rate out} = \frac{1}{10} - \frac{x}{100} = \frac{10 - x}{100}$$

or

$$\frac{dx}{10 - x} = \frac{dt}{100}.$$

Integrating both sides of this equation, we obtain

$$-\log|10 - x| = \frac{t}{100} + C.$$

Observe that $x(t) \neq 10$ for any finite time t (since log 0 is not defined). Since $x(0) = 50 > 10$, it follows that $x(t) > 10$ for all $t > 0$. ($x(t)$ is necessarily continuous so it cannot take any value less than 10 without somewhere taking the value 10 by the intermediate-value theorem.) Hence

$$\log(x - 10) = -\frac{t}{100} - C.$$

Since $x(0) = 50$ we have $-C = \log 40$ and

$$x = x(t) = 10 + 40e^{-t/100}.$$

After 40 min there will be $10 + 40e^{-0.4} \simeq 36.8$ kg of salt in the tank.

Example 4 (*A Rate of Reaction Problem*) In a chemical reaction that goes to completion in solution, one molecule of each of two reactants, A and B, combine to form each molecule of the product C. According to the law of mass action, the reaction proceeds at a rate proportional to the product of the concentrations of A and B in the solution. Thus if there were initially present a molecules/cm^3 of A and b molecules/cm^3 of B, then thereafter the number $x(t)$ of molecules/cm^3 of C present at time t is determined by the differential equation

$$\frac{dx}{dt} = k(a - x)(b - x).$$

We solve this equation by the technique of partial fraction decomposition under the assumption that $b \neq a$.

$$\int \frac{dx}{(a - x)(b - x)} = k\int dt = kt + C.$$

Since $1/(a - x)(b - x) = 1/(b - a)[1/(a - x) - 1/(b - x)]$ and since necessarily $x \leq a$ and $x \leq b$, we have

$$\frac{1}{b - a}[-\log(a - x) + \log(b - x)] = kt + C$$

or

$$\log\left(\frac{b - x}{a - x}\right) = (b - a)\,kt + C_1 \qquad (C_1 = (b - a)C).$$

By assumption, $x(0) = 0$, so $C_1 = \log b/a$ and

$$\log\frac{a}{b}\left(\frac{b - x}{a - x}\right) = (b - a)\,kt.$$

This equation can be solved for x to yield

$$x = x(t) = \frac{ab(e^{(b-a)kt} - 1)}{be^{(b-a)kt} - a}.$$

Example 5 Find a family of curves each of which intersects each parabola with equation of the form $y = Cx^2$ at right angles.

Solution The family of parabolas $y = Cx^2$ satisfies the differential equation

$$\frac{d}{dx}\frac{y}{x^2} = \frac{d}{dx}C = 0,$$

that is,

$$x^2\frac{dy}{dx} - 2xy = 0 \qquad \text{or} \qquad \frac{dy}{dx} = \frac{2y}{x}.$$

Any curve that meets the parabolas $y = Cx^2$ at right angles must, at any point (x, y) on it, have slope equal to the negative reciprocal of the slope of the particular parabola passing through that point. Thus such a curve must satisfy

$$\frac{dy}{dx} = -\frac{x}{2y}.$$

Separation of the variables leads to $2y\,dy = -x\,dx$, and integration of both sides then yields $y^2 = -(1/2)x^2 + C_1$ or $x^2 + 2y^2 = C_2 \quad (C_2 = 2C_1)$. This equation represents a family of ellipses centered at the origin. Each ellipse meets each parabola at right angles, as shown in the Fig. 7.72.

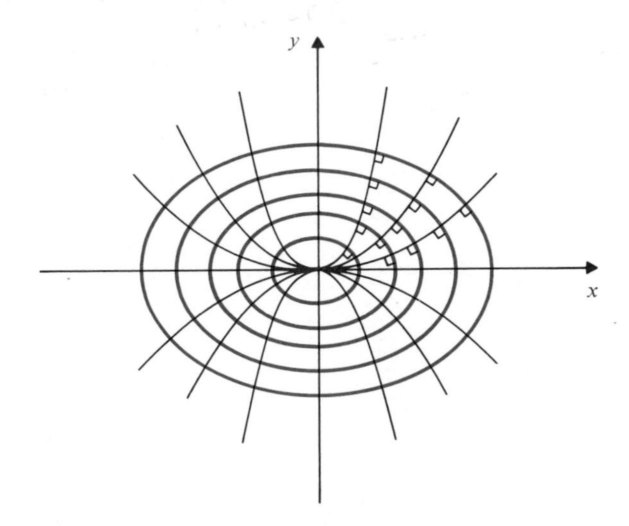

Figure 7.72

First-Order Linear Equations

A first-order linear differential equation is of the type

$$\frac{dy}{dx} + p(x)y = q(x),$$

where $p(x)$ and $q(x)$ are given functions, which we assume to be continuous. We can solve such equations (that is, find y as a function of x) by the following procedure.

Let $\mu(x)$ be any antiderivative of $p(x)$; thus $d\mu/dx = p(x)$. If $y = y(x)$ satisfies the given equation, then we calculate, using the product rule,

$$\begin{aligned}\frac{d}{dx}(e^{\mu(x)}y(x)) &= e^{\mu(x)}\frac{dy}{dx} + e^{\mu(x)}\frac{d\mu}{dx}y(x)\\ &= e^{\mu(x)}\left(\frac{dy}{dx} + p(x)y\right)\\ &= e^{\mu(x)}q(x).\end{aligned}$$

Therefore,

$$e^{\mu(x)}y(x) = \int e^{\mu(x)}q(x)\,dx$$

or

$$y(x) = e^{-\mu(x)}\int e^{\mu(x)}q(x)\,dx.$$

We reuse this method, rather than the final formula, in the examples below.

Example 6 Solve $dy/dx + y/x = 1$ for $x > 0$.

Solution Here we want $d\mu/dx = 1/x$, so $\mu(x) = \log x$ (for $x > 0$). Thus $e^{\mu(x)} = x$ and we calculate

$$\frac{d}{dx}(xy) = x\frac{dy}{dx} + y = x\left(\frac{dy}{dx} + \frac{y}{x}\right) = x,$$

and

$$xy = \int x\,dx = \frac{1}{2}x^2 + C.$$

Finally,

$$y = \frac{1}{x}\left(\frac{1}{2}x^2 + C\right) = \frac{x}{2} + Cx.$$

This function is a solution of the given equation for any value of constant C.

Example 7 Solve $dy/dx + xy = x^3$.

Solution Here $p(x) = x$ so $\mu(x) = x^2/2$. We calculate

$$\frac{d}{dx}(e^{x^2/2}y) = e^{x^2/2}\frac{dy}{dx} + e^{x^2/2}xy = x^3e^{x^2/2}.$$

Thus,

$$e^{x^2/2}y = \int x^3 e^{x^2/2}\,dx \qquad \begin{aligned} \text{Let } U &= x^2, & dV &= x\,e^{x^2/2}\,dx, \\ dU &= 2x\,dx, & V &= e^{x^2/2}. \end{aligned}$$

$$= x^2 e^{x^2/2} - 2\int x\,e^{x^2/2}\,dx$$

$$= x^2 e^{x^2/2} - 2\,e^{x^2/2} + C,$$

and, finally,

$$y = x^2 - 2 + Ce^{-x^2/2}.$$

Example 8 A savings account is opened with a deposit of A dollars. At any time t years thereafter, money is being continually deposited into the account at a rate of $(C + Dt)$ dollars per year. If interest is also being paid continually into the account at a rate of $100R\%$, find the balance $B(t)$ dollars in the account after t years. Illustrate the solution for the data $A = 5000$, $C = 1000$, $D = 200$, $R = 0.13$, and $t = 5$.

Solution Evidently B satisfies the differential equation

$$\frac{dB}{dt} = RB + (C + Dt)$$

or, equivalently, $dB/dt - RB = C + Dt$. This is a linear equation for B having $p(t) = -R$. Hence we may take $\mu(t) = -Rt$ so that $e^{\mu(t)} = e^{-Rt}$. We now calculate

$$\frac{d}{dt}(e^{-Rt}B(t)) = e^{-Rt}\frac{dB}{dt} - Re^{-Rt}B(t) = (C + Dt)\,e^{-Rt},$$

and

$$e^{-Rt}B(t) = \int (C + Dt)\,e^{-Rt}\,dt \qquad \begin{aligned} \text{Let } U &= C + Dt & dV &= e^{-Rt}\,dt \\ dU &= D\,dt & V &= -e^{-Rt}/R \end{aligned}$$

$$= -\frac{C + Dt}{R}e^{-Rt} + \frac{D}{R}\int e^{-Rt}\,dt$$

$$= -\frac{C + Dt}{R}e^{-Rt} - \frac{D}{R^2}e^{-Rt} + K \qquad (K = \text{constant}).$$

Hence

$$B(t) = -\frac{C + Dt}{R} - \frac{D}{R^2} + Ke^{Rt}.$$

Since $A = B(0) = -(C/R) - (D/R^2) + K$, $K = A + (C/R) + (D/R^2)$ and

$$B(t) = \left(A + \frac{C}{R} + \frac{D}{R^2}\right)e^{Rt} - \frac{C + Dt}{R} - \frac{D}{R^2}.$$

For the illustration $A = 5000$, $C = 1000$, $D = 200$, $R = 0.13$, and $t = 5$, we calculate, using a scientific calculator, $B(5) = 19{,}762.82$. Thus the account will contain \$19,762.82 after 5 years under these circumstances.

EXERCISES

Solve the differential equations in Exercises 1–14

1. $\dfrac{dy}{dx} = \dfrac{y}{2x}$

2. $\dfrac{dy}{dx} = \dfrac{3y - 1}{x}$

3. $\dfrac{dy}{dx} = \dfrac{x^2}{y^2}$

4. $\dfrac{dy}{dx} = x^2y^2$

5. $\dfrac{dY}{dt} = tY$

6. $\dfrac{dx}{dt} = e^x \sin t$

7. $\dfrac{dy}{dx} = 1 - y^2$

8. $\dfrac{dy}{dx} = 1 + y^2$

9. $\dfrac{dy}{dt} = 2 + e^y$

10. $\dfrac{dy}{dx} = y^2(1 - y)$

11. $\dfrac{dy}{dx} - \dfrac{2y}{x} = x^2$

12. $\dfrac{dy}{dx} + \dfrac{2y}{x} = \dfrac{1}{x^2}$

13. $\dfrac{dy}{dx} + y = e^x$

14. $\dfrac{dy}{dx} + y = x$

15. Why is the solution given for the chemical reaction rate problem in Example 4 not valid for $a = b$? Find the solution for the case $a = b$.

16. An object of mass m falling near the surface of the earth is retarded by air resistance proportional to its velocity so that, according to Newton's second law of motion,

$$m\frac{dv}{dt} = mg - kv,$$

where $v = v(t)$ is the velocity of the object at time t, and g is the acceleration of gravity near the surface of the earth. Assuming that the object falls from rest at time $t = 0$, that is, $v(0) = 0$, find the velocity $v(t)$ for any $t > 0$ (up until the object strikes the ground). Show $v(t)$ approaches a limit as $t \to \infty$. Do you need the explicit formula for $v(t)$ to determine this limiting velocity?

17. Repeat Exercise 16 except assuming that the air resistance is proportional to the square of the velocity so that the equation of motion is

$$m\frac{dv}{dt} = mg - kv^2.$$

18. Find the amount in a savings account after 1 year if the initial balance in the account was \$1000 and if interest is paid into the account continually at a variable rate $(1 + (t/40))\%$ per month, where t is the number of months after the account was opened.

19. Determine the balance $y(t)$ dollars in an investment account after t years if the initial balance is \$1000, if the interest is paid continuously into the account at a constant rate of 10% per annum, and if the account is being continuously depleted (by taxes, say) at a rate of $(1/1{,}000{,}000)y^2$ dollars per year. How large can the account grow? How long will it take the account to grow to half this balance?

20. Find the family of curves each of which intersects the curves of the family $xy = C$ at right angles.

21. Resolve the solution concentration problem in Example 3, changing the rate of inflow of brine into the tank to 12 l/min but leaving all the other data as they were in that example. Note that the volume of liquid in the tank is no longer constant as time increases.

8.1 Sequences and Convergence

8.2 Infinite Series

8.3 Convergence Tests for Positive Series

8.4 Absolute and Conditional Convergence

8.5 Estimating the Sum of a Series

8 Infinite Series

An infinite series is a sum involving infinitely many terms. Since addition is carried out on two numbers at a time, the evaluation of the sum of an infinite series necessarily involves the notion of limit. Complicated functions $f(x)$ can frequently be expressed as series of powers of x, that is, as polynomials of infinite degree. Since such series can be differentiated and integrated term by term, they play a very important role in the study of calculus. This role is developed in Chapter 9.

Much of the material in this chapter can be omitted without hindering the understanding of power series in Chapter 9. Section 8.2 should, however, not be omitted. Our presentation of power series will make some reference to the ratio test covered in Section 8.3 and the concepts of conditional and absolute convergence in Section 8.4, so these topics should at least be touched on. The discussion of infinite sequences and series of real numbers presented in this chapter involves limit and convergence notions similar to, but somewhat simpler than, those for functions and integrals. Accordingly, we proceed with slightly more rigor in developing them.

8.1 SEQUENCES AND CONVERGENCE

By a **sequence** (or **infinite sequence**) we mean an ordered list having a first element but no last element. For our purposes the elements (**terms**) of a sequence will always be real numbers, though much of our discussion could be applied to complex numbers as well. Examples of sequences are:

$\{1, 2, 3, 4, 5, \ldots\}$ the sequence of positive integers,

$\left\{-\frac{1}{2}, \frac{1}{4}, -\frac{1}{8}, \frac{1}{16}, \ldots\right\}$ the sequence of positive integer powers of $-\frac{1}{2}$.

It is convenient to list the terms of a sequence in braces { } as shown. The ellipsis (. . .) should be read "and so on."

The concept of infinite sequence is a special case of the concept of function. The sequence $\{a_1, a_2, a_3, a_4, \ldots\}$ can be regarded as a function a whose domain is the set of positive integers and that takes the value $a(n) = a_n$ at each integer n in the domain. A sequence can be specified in three ways:

i) We can list the first few terms followed by . . . *if the pattern is obvious.*

ii) We can provide a formula for the general term a_n as a function of n. Unless the contrary is stated, it is assumed that the first term is a_1, the second term a_2, and so on.

iii) We can provide a formula for the term a_n as a function of earlier terms a_1, $a_2, \ldots, a_{n-1}$ and specify a_1 so the process of computing higher terms can begin.

In each case it must be possible to determine any term of the sequence.

Example 1 a) $\{n\} = \{1, 2, 3, 4, 5, \ldots\}$

b) $\left\{\left(-\frac{1}{2}\right)^n\right\} = \left\{-\frac{1}{2}, \frac{1}{4}, -\frac{1}{8}, \frac{1}{16}, \ldots\right\}$

c) $\left\{\dfrac{n-1}{n}\right\} = \left\{0, \dfrac{1}{2}, \dfrac{2}{3}, \dfrac{3}{4}, \dfrac{4}{5}, \ldots\right\}$

d) $\{(-1)^{n-1}\} = \{1, -1, 1, -1, 1, \ldots\}$

e) $\{n^2/2^n\} = \left\{\dfrac{1}{2}, 1, \dfrac{9}{8}, 1, \dfrac{25}{32}, \dfrac{36}{64}, \dfrac{49}{128}, \ldots\right\}$

f) $\left\{\left(1 + \dfrac{1}{n}\right)^n\right\} = \left\{2, \left(\dfrac{3}{2}\right)^2, \left(\dfrac{4}{3}\right)^3, \left(\dfrac{5}{4}\right)^4, \ldots\right\}$

g) $\left\{\dfrac{\cos(n\pi/2)}{n}\right\} = \left\{0, -\dfrac{1}{2}, 0, \dfrac{1}{4}, 0, -\dfrac{1}{6}, 0, \dfrac{1}{8}, 0, \ldots\right\}$

h) $a_1 = 1, a_{n+1} = \sqrt{6 + a_n}, (n = 1, 2, 3, \ldots)$.
Thus $\{a_n\} = \{1, \sqrt{7}, \sqrt{6+\sqrt{7}}, \ldots\}$.

Note that there is no ''obvious'' formula for a_n as an explicit function of n in this case, but we can still calculate a_n for any desired value of n provided we first calculate all the earlier values $a_2, a_3, \ldots, a_{n-1}$.

In Examples 1(a)–1(g) the formulas on the left-hand sides define the general term of each sequence $\{a_n\}$ as an explicit function of n. In 1(h) we say the sequence $\{a_n\}$ is defined **recursively** or **inductively;** each term must be calculated from previous ones rather than directly as a function of n.

The following definitions introduce terminology used to describe various properties of sequences.

Definition 1

a) The sequence $\{a_n\}$ is **bounded below** by L, and L is a **lower bound** for $\{a_n\}$, if $a_n \geq L$ for every $n = 1, 2, 3, \ldots$. Similarly, the sequence is **bounded above** by M, and M is an **upper bound,** if $a_n \leq M$ for every n.

The sequence $\{a_n\}$ is **bounded** if it is both bounded above and bounded below. In this case there is a constant K such that $|a_n| \leq K$ for every $n = 1, 2, 3, \ldots$. (We may take K to be the maximum of $-L$ and M, that is, whichever of these numbers is larger.)

b) The sequence $\{a_n\}$ is **positive** if it is bounded below by zero, that is, if $a_n \geq 0$ for every $n = 1, 2, 3, \ldots$; it is **negative** if $a_n \leq 0$ for every n.

c) The sequence $\{a_n\}$ is **increasing** if $a_{n+1} \geq a_n$ for every $n = 1, 2, 3, \ldots$; it is **decreasing** if $a_{n+1} \leq a_n$ for every such n. The sequence is said to be **monotonic** if it is either increasing or decreasing.

d) The sequence $\{a_n\}$ is **alternating** if $a_n a_{n+1} < 0$ for every $n = 1, 2, \ldots$, that is, if the terms alternate in sign.

Example 2

a) The sequence $\{n\} = \{1, 2, 3, \ldots\}$ is increasing and bounded below; 1 is a lower bound, as is any smaller number. The sequence is not bounded above.

b) $\left\{\frac{n-1}{n}\right\} = \left\{0, \frac{1}{2}, \frac{2}{3}, \frac{3}{4}, \ldots\right\}$ is bounded. Evidently 0 is a lower bound and 1 is an upper bound. The sequence is positive and increasing.

c) $\left\{\left(-\frac{1}{2}\right)^n\right\} = \left\{-\frac{1}{2}, \frac{1}{4}, -\frac{1}{8}, \frac{1}{16}, \ldots\right\}$ is alternating and bounded. $\left(-\frac{1}{2}\right.$ is a lower bound; $\frac{1}{4}$ is an upper bound.$\left.\right)$

d) $\{(-1)^n n\} = \{-1, 2, -3, 4, -5, \ldots\}$ is alternating but not bounded above or below.

The sequence $\{n^2/2^n\} = \left\{\frac{1}{2}, 1, \frac{9}{8}, 1, \frac{25}{32}, \frac{36}{64}, \frac{49}{128}, \ldots\right\}$ is obviously positive therefore bounded below. It seems clear that from the fourth term on, all the terms are getting smaller. However, $a_2 > a_1$ and $a_3 > a_2$. Since $a_{n+1} \leq a_n$ only if $n \geq 3$, we say that this sequence is **ultimately decreasing.** The adverb *ultimately* is used to describe any termwise property of a sequence that the terms have from some point on, but not necessarily at the beginning of the sequence. Thus the sequence

$$\{n - 100\} = \{-99, -98, \ldots, -2, -1, 0, 1, 2, 3, \ldots\}$$

is *ultimately positive,* and the sequence

$$\left\{(-1)^n + \frac{4}{n}\right\} = \left\{3, 3, \frac{1}{3}, 2, -\frac{1}{5}, \frac{5}{3}, -\frac{3}{7}, \frac{3}{2}, \ldots\right\}$$

is *ultimately alternating* even though the first few terms do not alternate.

Central to the study of sequences is the notion of convergence. The concept of limit of a sequence is a special case of the concept of limit of a function $f(x)$ as $x \to \infty$. We say that the sequence $\{a_n\}$ **converges to the limit** L, and we write $\lim a_n = L$, provided the distance from a_n to L on the real line approaches 0 as n increases toward ∞. We state this definition somewhat more formally as follows:

Definition 2

> We say that the sequence $\{a_n\}$ converges to the limit L if for every positive real number ϵ there exists a number N depending on ϵ ($N = N(\epsilon)$) such that if $n > N$, then $|a_n - L| < \epsilon$.

This definition is illustrated in Fig. 8.1.

Every sequence $\{a_n\}$ must either **converge** to a finite limit L or **diverge.** That is, either $\lim a_n = L$ exists, or $\lim a_n$ does not exist. If $\lim a_n = \infty$, we can say that the sequence diverges to ∞; if $\lim a_n = -\infty$, we can say that it diverges to $-\infty$. If $\lim a_n$ simply does not exist (but is not ∞ or $-\infty$), we can only say that the sequence diverges.

Example 3 a) $\{(n - 1)/n\}$ converges to 1; $\lim \frac{n-1}{n} = \lim \left(1 - \frac{1}{n}\right) = 1.$

b) $\{n\} = \{1, 2, 3, 4, \ldots\}$ diverges to ∞.

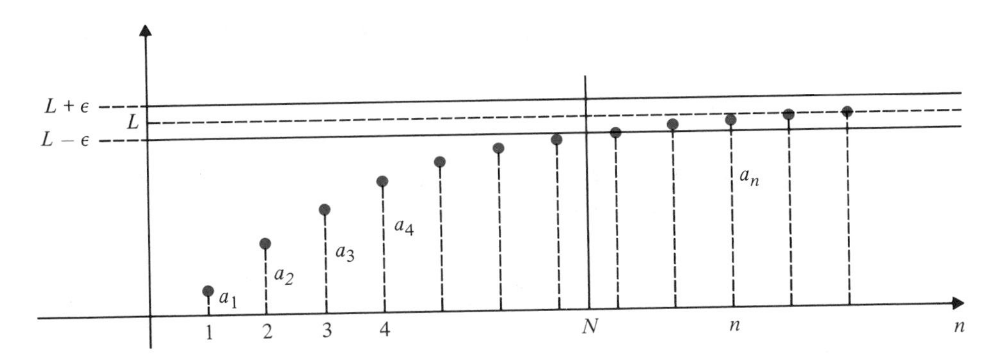

Figure 8.1

c) $\{-n\} = \{-1, -2, -3, -4, \ldots\}$ diverges to $-\infty$.

d) $\{(-1)^n\} = \{-1, 1, -1, 1, -1, \ldots\}$ simply diverges.

e) $\{(-1)^n n\} = \{-1, 2, -3, 4, -5, \ldots\}$ diverges (not to ∞ or $-\infty$ even though $\lim |a_n| = \infty$).

All the standard properties of limits (see Theorems 1–3 of Section 1.3) apply to the limits of sequences, with the appropriate changes of notation. Thus, if $\{a_n\}$ and $\{b_n\}$ converge, then

$$\lim (a_n \pm b_n) = \lim a_n \pm \lim b_n,$$

$$\lim ca_n = c \lim a_n,$$

$$\lim a_n b_n = (\lim a_n)(\lim b_n),$$

$$\lim (a_n / b_n) = (\lim a_n)/(\lim b_n) \qquad \text{assuming } \lim b_n \neq 0.$$

If $a_n \leq b_n$ ultimately, then $\lim a_n \leq \lim b_n$.

The limits of many explicitly defined sequences can be evaluated using these properties in a manner analogous to the methods used for limits of the form $\lim_{x\to\infty} f(x)$. See, for example, Examples 2–4 of Section 1.4

Example 4 a) $\{(2n^2 - n - 1)/(5n^2 + n - 3)\}$ converges to 2/5 since

$$\lim \frac{2n^2 - n - 1}{5n^2 + n - 3} = \lim \frac{2 - (1/n) - (1/n^2)}{5 + (1/n) - (3/n^2)} = \frac{2}{5}.$$

b) $\{\sqrt{n^2 + n} - n\}$ converges to 1/2 since

$$\lim (\sqrt{n^2 + n} - n) = \lim \frac{(\sqrt{n^2 + n} - n)(\sqrt{n^2 + n} + n)}{\sqrt{n^2 + n} + n}$$

$$= \lim \frac{n}{\sqrt{n^2 + n} + n} = \lim \frac{1}{\sqrt{1 + (1/n)} + 1} = \frac{1}{2}.$$

It often happens that $a_n = f(n)$ where f is a continuous function on $[1, \infty)$. In this case,

$$\lim a_n = \lim_{x\to\infty} f(x),$$

and the latter can be evaluated by standard techniques for functions, for instance, l'Hôpital's rules.

Example 5

$$\lim n \operatorname{Arctan}\frac{1}{n} = \lim_{x\to\infty} \frac{\operatorname{Arctan}\dfrac{1}{x}}{\dfrac{1}{x}} = \lim_{y\to 0+} \frac{\operatorname{Arctan} y}{y} \qquad \left[\frac{0}{0}\right]$$

$$= \lim_{y\to 0+} \frac{\dfrac{1}{y^2+1}}{1} = 1$$

Theorem 1 If $\{a_n\}$ converges, then $\{a_n\}$ is bounded.

Proof If $\lim a_n = L$, then, according to Definition 2, for $\epsilon = 1$ there exists a number N such that if $n > N$ then $|a_n - L| < 1$; therefore $|a_n| < 1 + |L|$ for such n. Let $M > N$ be an integer. If K denotes the largest of the numbers $|a_1|, |a_2|, \ldots, |a_{M-1}|, 1 + |L|$, then $|a_n| \le K$ for every $n = 1, 2, 3, \ldots$. Hence $\{a_n\}$ is bounded. □

The *completeness property* of the real number system (see Section 1.1) can be reformulated in terms of sequences to read as follows:

> If $\{a_n\}$ is bounded above and (ultimately) increasing, then it converges. The same conclusion holds if $\{a_n\}$ is bounded below and (ultimately) decreasing.

That is, a bounded, ultimately monotonic sequence is convergent. This is illustrated in Fig. 8.2.

Example 6 Let a_n be defined recursively by

$$a_1 = 1, \qquad a_{n+1} = \sqrt{6 + a_n} \qquad (n = 1, 2, 3, \ldots).$$

Show that $\lim a_n$ exists and find its value.

Solution Observe that $a_2 = \sqrt{6+1} = \sqrt{7} > a_1$. If $a_{k+1} > a_k$, then $a_{k+2} = \sqrt{6 + a_{k+1}} > \sqrt{6 + a_k} = a_{k+1}$, so $\{a_n\}$ is increasing, by induction. (See Appendix 1.) Now observe that $a_1 = 1 < 3$. If $a_k < 3$, then $a_{k+1} = \sqrt{6 + a_k} < \sqrt{6+3} = 3$, so $a_n < 3$ for every n, by induction. Since $\{a_n\}$ is increasing and bounded above, $\lim a_n = a$ exists, by completeness. Since $\sqrt{6 + x}$ is a continuous function of x, we have

$$a = \lim a_{n+1} = \lim \sqrt{6 + a_n} = \sqrt{6 + \lim a_n} = \sqrt{6 + a}.$$

Thus $a^2 = 6 + a$, or $a^2 - a - 6 = 0$, or $(a - 3)(a + 2) = 0$. Since $a_n \ge 1$ for every n, we must have $a \ge 1$. Therefore, $a = 3$, and $\lim a_n = 3$.

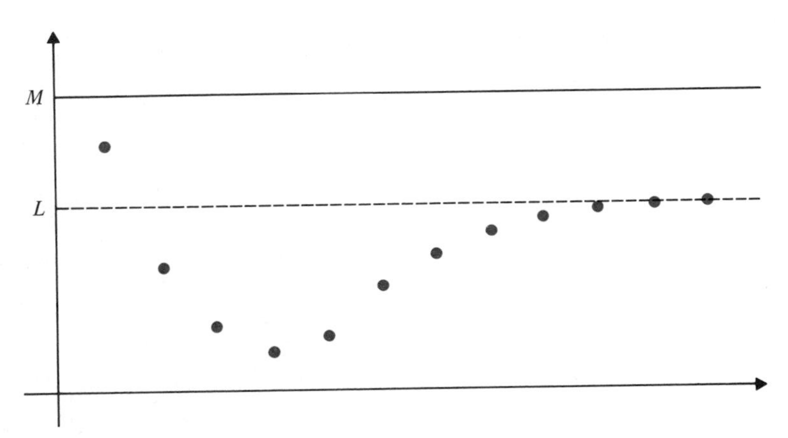

An ultimately increasing sequence that is bounded above.
M is an upper bound; L is the limit.

Figure 8.2

Example 7 Does $\left\{\left(1 + \frac{1}{n}\right)^n\right\}$ converge or not?

Solution We could expend considerable effort to show that the given sequence is, in fact, increasing and bounded above. (See Exercise 31 at the end of this section.) However, we already know the answer. By Theorem 5 of Section 3.5,

$$\lim \left(1 + \frac{1}{n}\right)^n = e^1 = e.$$

Theorem 2 If $\{a_n\}$ is (ultimately) increasing, then either it is bounded above and therefore convergent, or else it is not bounded above but diverges to infinity. □

The proof of this theorem is left as an exercise.

The following two examples find frequent application in the study of series.

Example 8

$$\lim x^n = 0 \qquad \text{if } |x| < 1.$$

To see this, observe that

$$\lim \ln |x|^n = \lim n \ln |x| = -\infty,$$

since $\ln |x| < 0$ when $|x| < 1$. Accordingly,

$$\lim |x|^n = \lim e^{\ln |x|^n} = e^{\lim \ln |x|^n} = e^{-\infty} = 0.$$

Example 9

$$\lim \frac{x^n}{n!} = 0 \qquad \text{for every real } x.$$

To see this, pick any x and let N be an integer such that $N > |x|$. If $n > N$ we have

$$\begin{aligned}\left|\frac{x^n}{n!}\right| &= \frac{|x|}{1}\frac{|x|}{2}\frac{|x|}{3}\cdots\frac{|x|}{N-1}\frac{|x|}{N}\frac{|x|}{N+1}\cdots\frac{|x|}{n}\\ &< \frac{|x|^{N-1}}{(N-1)!}\frac{|x|}{N}\frac{|x|}{N}\frac{|x|}{N}\cdots\frac{|x|}{N}\\ &= \frac{|x|^{N-1}}{(N-1)!}\left(\frac{|x|}{N}\right)^{n-N+1} = K\left(\frac{|x|}{N}\right)^n,\end{aligned}$$

where $K = \dfrac{|x|^{N-1}}{(N-1)!}\left(\dfrac{|x|}{N}\right)^{1-N}$ is a constant independent of n. Since $|x|/N < 1$ we have $\lim (|x|/N)^n = 0$ by Example 8 above. Thus $\lim |x^n/n!| = 0$, and so $\lim x^n/n! = 0$.

EXERCISES

In Exercises 1–12 determine whether the given sequence is

i) bounded (above or below),
ii) positive or negative (ultimately),
iii) increasing, decreasing, or alternating,
iv) convergent, divergent, divergent to ∞ or $-\infty$.

1. $\left\{\dfrac{2n^2}{n^2+1}\right\}$

2. $\left\{\dfrac{2n}{n^2+1}\right\}$

3. $\left\{4 - \dfrac{(-1)^n}{n}\right\}$

4. $\left\{\sin\dfrac{1}{n}\right\}$

5. $\left\{\dfrac{n^2-1}{n}\right\}$

6. $\{e^n/\pi^n\}$

7. $\{e^n/\pi^{n/2}\}$

8. $\left\{\dfrac{(-1)^n n}{e^n}\right\}$

9. $\left\{n\cos\left(\dfrac{n\pi}{2}\right)\right\}$

10. $\{1, 1, -2, 3, 3, -4, 5, 5, -6, \ldots\}$

11. $\{2^n/n^n\}$

12. $\{(n!)^2/(2n)!\}$

In Exercises 13–28 evaluate, wherever possible, the limit of the sequence $\{a_n\}$.

13. $a_n = \dfrac{5-2n}{3n-7}$

14. $a_n = \dfrac{n^2-4}{n+5}$

15. $a_n = \dfrac{n^2}{n^3+1}$

16. $a_n = (-1)^n\dfrac{n}{n^3+1}$

17. $a_n = \dfrac{n^2-2\sqrt{n}+1}{1-n-3n^2}$

18. $a_n = \dfrac{e^n-e^{-n}}{e^n+e^{-n}}$

19. $a_n = n\sin\dfrac{1}{n}$

20. $a_n = \left(\dfrac{n-3}{n}\right)^n$

21. $a_n = \dfrac{n}{\ln(n+1)}$

22. $a_n = \sqrt{n+1} - \sqrt{n}$

23. $a_n = n - \sqrt{n^2-4n}$

24. $a_n = \sqrt{n^2+n} - \sqrt{n^2-1}$

25. $a_n = \left(\frac{n-1}{n+1}\right)^n$

26. $a_n = \frac{(n!)^2}{(2n)!}$

27. $a_n = \frac{n^2 2^n}{n!}$

28. $a_n = \frac{\pi^n}{1 + 2^{2n}}$

29.* Let $a_1 = 1$ and $a_{n+1} = \sqrt{1 + 2a_n}$ $(n = 1, 2, 3, \ldots)$. Show that $\{a_n\}$ is increasing and bounded above. (*Hint:* Show that 3 is an upper bound.) Hence conclude that the sequence converges, and find its limit.

30.* Repeat Exercise 29 for the sequence defined by

$$a_1 = 3, \qquad a_{n+1} = \sqrt{15 + 2a_n}, \qquad n = 1, 2, 3, \ldots .$$

This time you will have to guess an upper bound.

31.* Let $a_n = \left(1 + \frac{1}{n}\right)^n$ so that $\log a_n = n \log\left(1 + \frac{1}{n}\right)$. Use properties of the logarithm function to show that

i) $\{a_n\}$ is increasing, and

ii) e is an upper bound for $\{a_n\}$.

32.* Prove Theorem 2 of Section 8.1. Also, state an analogous theorem pertaining to ultimately decreasing sequences.

33.* Which of the following statements are true and which are false? Justify your answers.

a) If $\lim a_n = \infty$ and $\lim b_n = L > 0$, then $\lim a_n b_n = \infty$.

b) If $\lim a_n = \infty$ and $\lim b_n = -\infty$, then $\lim (a_n + b_n) = 0$.

c) If $\lim a_n = \infty$ and $\lim b_n = -\infty$, then $\lim a_n b_n = -\infty$.

d) If neither $\{a_n\}$ nor $\{b_n\}$ converges, then $\{a_n b_n\}$ does not converge.

8.2 INFINITE SERIES

An **infinite series,** usually just called a **series,** is a formal sum of infinitely many terms; for instance

$$a_1 + a_2 + a_3 + a_4 + \cdots$$

is a series formed by adding the terms of the sequence $\{a_n\}$. This series is also denoted $\sum_{n=1}^{\infty} a_n$ or, when confusion is not likely to occur, $\sum a_n$.

$$\sum a_n = \sum_{n=1}^{\infty} a_n = a_1 + a_2 + a_3 + a_4 + \cdots$$

For example,

$$\sum \frac{1}{n} = 1 + \frac{1}{2} + \frac{1}{3} + \frac{1}{4} + \cdots$$

$$\sum \frac{(-1)^{n-1}}{2^{n-1}} = 1 - \frac{1}{2} + \frac{1}{4} - \frac{1}{8} + \frac{1}{16} - \cdots .$$

* Difficult and/or theoretical problems.

It is sometimes necessary or convenient to start the sum from some index other than 1; when we do we always show the index limits specifically.

$$\sum_{n=0}^{\infty} a^n = 1 + a + a^2 + a^3 + \cdots$$

$$\sum_{n=2}^{\infty} \frac{1}{\ln n} = \frac{1}{\ln 2} + \frac{1}{\ln 3} + \frac{1}{\ln 4} + \cdots$$

Note that the latter series would make no sense if we had started the sum from $n = 1$; the first term would have been undefined.

Addition is an operation that is carried out on two numbers at a time. If we want to calculate the finite sum

$$a_1 + a_2 + a_3,$$

we might proceed by adding $a_1 + a_2$ and then adding this sum to a_3, or else we might first add $a_2 + a_3$ and then add a_1 to this sum. Of course the associative law for addition assures us we will get the same answer both ways. This is the reason the symbol $a_1 + a_2 + a_3$ makes sense; we would otherwise have to write $(a_1 + a_2) + a_3$ or $a_1 + (a_2 + a_3)$. This reasoning extends to any *finite sum* $a_1 + a_2 + \cdots + a_n$, but it is not obvious what should be meant by an *infinite sum:*

$$a_1 + a_2 + a_3 + a_4 + \cdots.$$

We no longer have any assurance that the terms can be added up in any order to yield the same sum. The interpretation we place on the infinite sum is that of adding from left to right as suggested by

$$\cdots((((a_1 + a_2) + a_3) + a_4) + a_5) + \cdots.$$

We formalize this in the following definition.

Definition 3

Corresponding to the infinite series $\sum_{n=1}^{\infty} a_n$ we form the **sequence** $\{s_n\}$ of **partial sums** of the series:

$$s_1 = a_1$$
$$s_2 = a_1 + a_2$$
$$s_3 = a_1 + a_2 + a_3$$
$$\vdots$$
$$s_n = a_1 + a_2 + a_3 + \cdots + a_n = \sum_{j=1}^{n} a_j.$$
$$\vdots$$

We say that the series $\sum_{n=1}^{\infty} a_n$ **converges to the sum** s if $\lim s_n = s$, that is, the series converges if its sequence of partial sums converges.

In this case we write

$$\sum_{n=1}^{\infty} a_n = \lim s_n = s$$

and call s the **sum** of the series.

Similarly, a series is said to diverge to infinity, diverge to negative infinity, or simply diverge if its sequence of partial sums does so. It must be stressed that the convergence of the series $\Sigma_{n=1}^{\infty} a_n$ depends on the convergence of the sequence $\{s_n\} = \{\Sigma_{j=1}^{n} a_j\}$, not that of the sequence $\{a_j\}$.

Example 1 (*Geometric Series*) A series of the form

$$\sum_{n=1}^{\infty} ar^{n-1} = a + ar + ar^2 + ar^3 + \cdots$$

is called a **geometric series.** The number a is the first term. The number r is called the common ratio of the series since it is the value of the ratio of the $(n + 1)$st term to the nth term:

$$\frac{ar^n}{ar^{n-1}} = r \qquad n = 1, 2, 3, \cdots$$

The nth partial sum of a geometric series is calculated as follows.

$$\begin{aligned} s_n &= a + ar + ar^2 + ar^3 + \cdots + ar^{n-1} \\ rs_n &= \phantom{a + {}} ar + ar^2 + ar^3 + \cdots + ar^{n-1} + ar^n \end{aligned}$$

The second equation is obtained by multiplying the first by r. Subtracting these two equations (note the cancellations), we get $(1 - r)s_n = a - ar^n$. Hence, if $r \neq 1$,

$$s_n = a\frac{1 - r^n}{1 - r}.$$

If $|r| < 1$, then $\lim r^n = 0$, so $\lim s_n = a/(1 - r)$. If $r > 1$, then $\lim r^n = \infty$, and $\lim s_n = \infty$ if $a > 0$ or $\lim s_n = -\infty$ if $a < 0$. The same conclusion holds if $r = 1$ since $s_n = na$ in this case. If $r \leq -1$, $\lim r^n$ does not exist and neither does $\lim s_n$. Hence we conclude that

$$\sum_{n=1}^{\infty} ar^{n-1} \begin{cases} \text{converges to } \dfrac{a}{1-r} & \text{if } -1 < r < 1 \\ \text{diverges to } \infty & \text{if } r \geq 1 \text{ and } a > 0 \\ \text{diverges to } -\infty & \text{if } r \geq 1 \text{ and } a < 0 \\ \text{diverges} & \text{if } r \leq -1. \end{cases}$$

Example 2 a) $1 + \frac{1}{2} + \frac{1}{4} + \frac{1}{8} + \cdots = \sum_{n=1}^{\infty} \left(\frac{1}{2}\right)^{n-1} = \frac{1}{1 - \frac{1}{2}} = 2$

b) $\pi - e + \frac{e^2}{\pi} - \frac{e^3}{\pi^2} + \cdots = \sum_{n=1}^{\infty} \pi\left(-\frac{e}{\pi}\right)^{n-1} = \frac{\pi}{1 - \left(-\frac{e}{\pi}\right)} = \frac{\pi^2}{\pi + e}$ since $e/\pi < 1$.

c) $1 + 2^{1/2} + 2 + 2^{3/2} + \cdots = \sum_{n=1}^{\infty} (\sqrt{2})^{n-1}$ diverges to ∞ since $\sqrt{2} > 1$.

d) $1 - 1 + 1 - 1 + 1 - \cdots = \sum_{n=1}^{\infty} (-1)^{n-1}$ diverges

e) Let $x = 0.323232\ldots = 0.\overline{32}$; then

$$x = \frac{32}{100} + \frac{32}{100^2} + \frac{32}{100^3} + \cdots = \frac{32}{100}\frac{1}{1 - \frac{1}{100}} = \frac{32}{99}.$$

This is an alternative to the method of Section 1.1 for representing repeating decimals as quotients of integers.

Example 3 Consider the series

$$\sum_{n=1}^{\infty} \frac{1}{n(n+1)} = \frac{1}{1 \times 2} + \frac{1}{2 \times 3} + \frac{1}{3 \times 4} + \frac{1}{4 \times 5} + \cdots.$$

Since $\frac{1}{n(n+1)} = \frac{1}{n} - \frac{1}{n+1}$ we can write the partial sum s_n in the form

$$\begin{aligned} s_n &= \frac{1}{1 \times 2} + \frac{1}{2 \times 3} + \frac{1}{3 \times 4} + \cdots + \frac{1}{(n-1)n} + \frac{1}{n(n+1)} \\ &= \left(1 - \frac{1}{2}\right) + \left(\frac{1}{2} - \frac{1}{3}\right) + \left(\frac{1}{3} - \frac{1}{4}\right) + \cdots + \left(\frac{1}{n-1} - \frac{1}{n}\right) + \left(\frac{1}{n} - \frac{1}{n+1}\right) \\ &= 1 - \left(\frac{1}{2} - \frac{1}{2}\right) - \left(\frac{1}{3} - \frac{1}{3}\right) - \cdots - \left(\frac{1}{n} - \frac{1}{n}\right) - \frac{1}{n+1} \\ &= 1 - \frac{1}{n+1}. \end{aligned}$$

Therefore $\lim s_n = 1$ and the series converges to 1:

$$\sum_{n=1}^{\infty} \frac{1}{n(n+1)} = 1.$$

This is an example of a **telescoping series.** Other examples can be found in the exercises at the end of this section. As indicated in this example, the method of partial fractions can be a useful tool for series as well as for integrals.

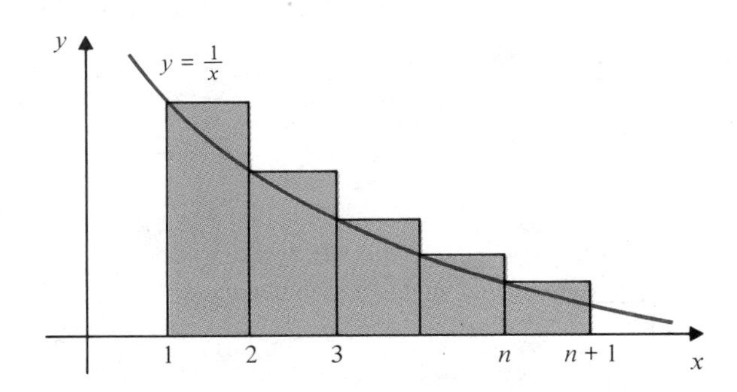

Figure 8.3

Example 4 The series

$$\sum_{n=1}^{\infty} \frac{1}{n} = 1 + \frac{1}{2} + \frac{1}{3} + \frac{1}{4} + \cdots$$

is called a **harmonic series.** Its partial sums s_n are given by

$$s_n = 1 + \frac{1}{2} + \frac{1}{3} + \cdots + \frac{1}{n}$$

$$= \text{sum of areas of rectangles shaded in Fig. 8.3.}$$

$$> \text{area under } y = \frac{1}{x} \text{ from 1 to } n + 1.$$

Thus

$$s_n > \int_1^{n+1} \frac{dx}{x} = \ln (n + 1).$$

Therefore $\lim s_n = \infty$.

$$\sum_{n=1}^{\infty} \frac{1}{n} = 1 + \frac{1}{2} + \frac{1}{3} + \cdots$$

diverges to infinity.

We now collect some useful theoretical results concerning series.

Theorem 3 If Σa_n converges, then $\lim a_n = 0$.

Proof If $s_n = a_1 + a_2 + \cdots + a_n$, then $s_n - s_{n-1} = a_n$. If Σa_n converges, then $\lim s_n = s$ exists, and $\lim s_{n-1} = s$ also. Hence $\lim a_n = s - s = 0$. □

Theorem 3 is very important to the understanding of infinite series. Students often err either in forgetting that a series cannot converge if its terms do not approach zero or in confusing this result with its *converse,* which is false. The converse would say that if $\lim a_n = 0$, then Σa_n must converge. The harmonic series is a counterexample showing the falsehood of this assertion:

$$\lim \frac{1}{n} = 0, \quad \text{but} \quad \sum \frac{1}{n} \text{ diverges to infinity.}$$

When considering whether a given series converges, the first question you should ask yourself is "Does the nth term approach 0 as n approaches ∞?" If the answer is no, then the series cannot converge. If the sequence of terms $\{a_n\}$ tends to a nonzero limit L, then Σa_n diverges to infinity if $L > 0$ and diverges to negative infinity if $L < 0$.

Example 5 a) $\sum \frac{n}{2n+1}$ diverges to infinity since $\lim \frac{n}{2n+1} = 1/2 > 0$.

b) $\Sigma(-1)^n n \sin(1/n)$ diverges since

$$\lim \left| (-1)^n n \sin \frac{1}{n} \right| = \lim \frac{\sin(1/n)}{1/n} = 1 \neq 0.$$

It is only the ultimate behavior of $\{a_n\}$ that determines whether Σa_n converges. Any finite number of terms can be dropped from the beginning of a series without affecting the convergence; the convergence depends only on the "tail" of the series.

Theorem 4 $\Sigma_{n=1}^{\infty} a_n$ converges if and only if $\Sigma_{n=N}^{\infty} a_n$ converges (N is any positive integer). □

Theorem 5 If $\{a_n\}$ is ultimately positive, then Σa_n must either converge (if its partial sums are bounded above) or diverge to infinity (if its partial sums are not bounded above). □

The proofs of these two theorems are left as exercises at the end of this section. Theorem 6 is just a reformulation of standard laws of limits.

Theorem 6 If $\Sigma_{n=1}^{\infty} a_n$ and $\Sigma_{n=1}^{\infty} b_n$ converge to A and B, respectively, then

a) $\Sigma_{n=1}^{\infty} ca_n$ converges to cA (c is a constant);

b) $\Sigma_{n=1}^{\infty} (a_n \pm b_n)$ converges to $A \pm B$;

c) if $a_n \leq b_n$ for all $n = 1, 2, 3, \ldots$ then $A \leq B$. □

EXERCISES

In Exercises 1–14 find the sum of the given series, or show that the series diverges (possibly to infinity or negative infinity). Exercises 7–10 are telescoping series and should be done by partial fractions as suggested in Example 3 in the text.

1. $\frac{1}{3} + \frac{1}{9} + \frac{1}{27} + \cdots = \sum_{n=1}^{\infty} \frac{1}{3^n}$

2. $3 - \frac{3}{4} + \frac{3}{16} - \frac{3}{64} + \cdots = \sum_{n=1}^{\infty} 3\left(-\frac{1}{4}\right)^{n-1}$

3. $\displaystyle\sum_{n=5}^{\infty} \frac{1}{(2+\pi)^{2n}}$

4. $\displaystyle\sum_{n=0}^{\infty} \frac{5}{10^{3n}}$

5. $\displaystyle\sum_{n=2}^{\infty} \frac{(-5)^n}{8^{2n}}$

6. $\displaystyle\sum_{n=0}^{\infty} \frac{1}{e^n}$

7. $\displaystyle\sum_{n=1}^{\infty} \frac{1}{n(n+2)} = \frac{1}{1 \times 3} + \frac{1}{2 \times 4} + \frac{1}{3 \times 5} + \cdots$

8. $\displaystyle\sum_{n=1}^{\infty} \frac{1}{(3n-2)(3n+1)} = \frac{1}{1 \times 4} + \frac{1}{4 \times 7} + \frac{1}{7 \times 10} + \cdots$

9. $\displaystyle\sum_{n=1}^{\infty} \frac{1}{(2n-1)(2n+1)} = \frac{1}{1 \times 3} + \frac{1}{3 \times 5} + \frac{1}{5 \times 7} + \cdots$

10.* $\displaystyle\sum_{n=1}^{\infty} \frac{1}{n(n+1)(n+2)} = \frac{1}{1 \times 2 \times 3} + \frac{1}{2 \times 3 \times 4} + \frac{1}{3 \times 4 \times 5} + \cdots$

11. $\displaystyle\sum_{n=1}^{\infty} \frac{1}{2n-1} = 1 + \frac{1}{3} + \frac{1}{5} + \cdots$

12. $\displaystyle\sum_{n=1}^{\infty} \frac{n}{n+2}$

13. $\displaystyle\sum_{n=1}^{\infty} n^{-1/2}$

14. $\displaystyle\sum_{n=1}^{\infty} \frac{2}{n+1}$

15. Obtain a simple expression for the partial sum s_n of the series $\Sigma_{n=1}^{\infty} (-1)^n$ and use it to show that this series diverges.

16. When dropped, an elastic ball bounces back up to a height three-quarters of that from which it fell. If the ball is dropped from a height of 2 m and allowed to bounce up and down indefinitely, what is the total distance it travels before coming to rest?

17. If a bank account pays 10% simple interest into an account once a year, what is the balance in the account at the end of 8 years if \$1000 is deposited into the account at the beginning of each of the 8 years? (Assume there was no balance in the account initially.)

18.* Prove Theorem 4 of Section 8.2.

19.* Prove Theorem 5 of Section 8.2.

20.* State a theorem analogous to Theorem 5 but for a negative sequence.

In Exercises 21–26 decide whether the given statement is true or false. If it is true, prove it. If it is false, give a counterexample showing the falsehood.

21.* If $a_n = 0$ for every n, then Σa_n converges.

22.* If Σa_n converges, then $\Sigma(1/a_n)$ diverges to infinity.

23.* If Σa_n and Σb_n both diverge, then so does $\Sigma(a_n + b_n)$.

24.* If $a_n \geq c > 0$ for every n, then Σa_n diverges to infinity.

25.* If Σa_n diverges and $\{b_n\}$ is bounded, then $\Sigma a_n b_n$ diverges.

26.* If $a_n > 0$ and Σa_n converges, then $\Sigma(a_n)^2$ converges.

8.3 CONVERGENCE TESTS FOR POSITIVE SERIES

In the previous section we saw a few examples of convergent series (geometric and telescoping series) whose sums could be determined exactly because the partial sums

s_n could be algebraically manipulated into explicit functions of n whose limits as $n \to \infty$ could be easily evaluated. It is not usually possible to do this with a given series, and therefore it is not usually possible to determine exactly the sum of a series. However, there are many techniques for determining whether a given series converges and, if it does, for approximating the sum to any desired degree of accuracy. We shall investigate some of these techniques in the remaining sections of this chapter.

In this section we deal exclusively with (ultimately) *positive series,* that is, series of the form

$$\sum_{n=1}^{\infty} a_n = a_1 + a_2 + a_3 + \cdots$$

where, for some positive integer N, $a_n \geq 0$ for $n = N, N + 1, N + 2, \ldots$. As noted in Theorem 5 of Section 8.2, such a series will converge if its partial sums are bounded above and will diverge to infinity otherwise.

The Comparison Tests

The first test we consider is analogous to the comparison test for improper integrals. (See Theorem 10 of Section 6.9.) It enables us to determine convergence or divergence of one series by comparing it with another series that is known to converge or diverge.

Theorem 7 (*The Comparison Test*) Let $\{a_n\}$ and $\{b_n\}$ be ultimately positive sequences, and suppose there exist a positive integer N and a positive constant K such that

$$0 \leq a_n \leq Kb_n \qquad \text{if } n \geq N.$$

If the series $\sum_{n=1}^{\infty} b_n$ converges, then so does the series $\sum_{n=1}^{\infty} a_n$. Equivalently, if $\sum_{n=1}^{\infty} a_n$ diverges to infinity, then so does $\sum_{n=1}^{\infty} b_n$.

Proof Let $s_n = a_1 + a_2 + \cdots + a_n$ and $S_n = b_1 + b_2 + \cdots + b_n$. If $n \geq N$ we have

$$\begin{aligned} s_n &= a_1 + a_2 + \cdots + a_{N-1} + a_N + a_{N+1} + \cdots + a_n \\ &= s_{N-1} + a_N + a_{N+1} + \cdots + a_n \\ &\leq s_{N-1} + Kb_N + Kb_{N+1} + \cdots + Kb_n \\ &= s_{N-1} + K(S_n - S_{N-1}). \end{aligned}$$

If Σb_n converges, then $\{S_n\}$ is convergent and hence bounded by Theorem 1 of Section 8.1. Hence $\{s_n\}$ is bounded above. By Theorem 5, Σa_n converges. Since the convergence of Σb_n guarantees that of Σa_n, if the latter series diverges to infinity, that is, does not converge, then the former cannot converge either, and so it must diverge to infinity too. □

Example 1 Test for convergence

a) $\displaystyle\sum_{n=1}^{\infty} \frac{1}{2^n + 1}$

b) $\sum_{n=1}^{\infty} \frac{1}{n^2}$

c) $\sum_{n=2}^{\infty} \frac{1}{\ln n}$

Solution a) Since $0 < \frac{1}{2^n + 1} < \frac{1}{2^n}$ for $n = 1, 2, 3, \ldots$, and since $\sum_{n=1}^{\infty} (1/2^n)$ is a convergent geometric series, the series $\sum_{n=1}^{\infty} (1/(2^n + 1))$ also converges by comparison. Evidently in this case,

$$0 < \sum_{n=1}^{\infty} \frac{1}{2^n + 1} < \sum_{n=1}^{\infty} \frac{1}{2^n} = \frac{\frac{1}{2}}{1 - \frac{1}{2}} = 1.$$

b) Since $0 < \frac{1}{n^2} < \frac{1}{(n-1)n}$ for $n = 2, 3, 4, \ldots$, and since

$$\sum_{n=2}^{\infty} \frac{1}{(n-1)n} = \frac{1}{1 \times 2} + \frac{1}{2 \times 3} + \frac{1}{3 \times 4} + \cdots = \sum_{k=1}^{\infty} \frac{1}{k(k+1)} = 1$$

(by Example 3 of Section 8.2), $\sum_{n=1}^{\infty} (1/n^2)$ converges by comparison. In this case we have the estimate

$$\sum_{n=1}^{\infty} \frac{1}{n^2} = 1 + \sum_{n=2}^{\infty} \frac{1}{n^2} < 1 + \sum_{n=2}^{\infty} \frac{1}{(n-1)n} = 1 + 1 = 2.$$

c) For $n = 2, 3, 4, \ldots$ we have $0 < \ln n < n$. Thus $1/(\ln n) > 1/n$. Since $\sum_{n=2}^{\infty}(1/n)$ diverges to infinity (harmonic series), so does $\sum_{n=2}^{\infty} 1/(\ln n)$ by comparison.

The following theorem provides a version of the comparison test that is not quite as general as that of Theorem 7 but is often easier to apply in specific cases.

Theorem 8 (*The Limit Comparison Test*) Suppose that $\{a_n\}$ and $\{b_n\}$ are ultimately positive sequences and that

$$\lim \frac{a_n}{b_n} = L,$$

where L is either a finite number (nonnegative) or $+\infty$.

a) If $L < \infty$ and Σb_n converges, then Σa_n also converges.

b) If $L > 0$ and Σb_n diverges to infinity, then so does Σa_n.

Proof If $L < \infty$, then for n sufficiently large we have $b_n > 0$ and

$$0 \leq \frac{a_n}{b_n} \leq L + 1,$$

so $0 \le a_n \le (L + 1)b_n$. Hence Σa_n converges if Σb_n converges, by Theorem 7. If $L > 0$, then for n sufficiently large

$$\frac{a_n}{b_n} \ge \frac{L}{2},$$

so $0 < b_n \le (2/L)a_n$ and Σa_n diverges to infinity if Σb_n does, by Theorem 7. □

Example 2 Test the following sequences for convergence using the limit comparison test.

a) $\displaystyle\sum_{n=1}^{\infty} \frac{1}{1 + \sqrt{n}}$

b) $\displaystyle\sum_{n=1}^{\infty} \frac{n + 5}{n^3 - 2n + 3}$

c) $\displaystyle\sum_{n=1}^{\infty} \frac{2^n + 1}{3^n - 1}$

Solution a) Since the sequence $\{\sqrt{n}\}$ does not grow as fast as $\{n\}$, we suspect that the series diverges to infinity by comparison with $\Sigma(1/n)$. Observe that

$$L = \lim \frac{\dfrac{1}{1 + \sqrt{n}}}{\dfrac{1}{n}} = \lim \frac{n}{1 + \sqrt{n}} = \infty > 0.$$

Therefore $\Sigma 1/(1 + \sqrt{n})$ does indeed diverge to infinity by comparison with $\Sigma(1/n)$.

b) For large n, the terms behave like n/n^3 so let us compare with the series $\Sigma 1/n^2$, which is known to converge. ($\Sigma 1/n(n + 1)$ would also do.)

$$L = \lim \frac{\dfrac{n + 5}{n^3 - 2n + 3}}{\dfrac{1}{n^2}} = \lim \frac{n^3 + 5n^2}{n^3 - 2n + 3} = 1.$$

Since $L < \infty$, the series $\Sigma_{n=1}^{\infty} ((n + 5)/(n^3 - 2n + 3))$ converges by comparison with $\Sigma(1/n^2)$.

c) In this case the terms behave like $(2/3)^n$ for large n. We have

$$L = \lim \frac{\dfrac{2^n + 1}{3^n - 1}}{\left(\dfrac{2}{3}\right)^n} = \lim \frac{2^n + 1}{2^n} \frac{3^n}{3^n - 1} = \lim \frac{1 + \dfrac{1}{2^n}}{1 - \dfrac{1}{3^n}} = 1.$$

Again $L < \infty$ and the series $\Sigma_{n=1}^{\infty} ((2^n + 1)/(3^n - 1))$ converges by comparison with the convergent geometric series $\Sigma_{n=1}^{\infty} (2/3)^n$.

In order to apply successfully the original version of the comparison test (Theorem 7) it is important to have an intuitive feeling for whether the given series is likely to converge or diverge. The form of the comparison will depend on whether you are trying to prove convergence or divergence. For instance, if you did not know intuitively that $\sum_{n=1}^{\infty} (1/(100n + 20{,}000))$ would have to diverge to infinity, you might try to argue

$$\frac{1}{100n + 20{,}000} < \frac{1}{n} \qquad \text{for } n = 1, 2, 3, \ldots .$$

While true, this doesn't help at all. $\Sigma 1/n$ diverges to infinity, and therefore Theorem 7 yields no information from this comparison. We could, of course, argue instead that

$$\frac{1}{100n + 20{,}000} \geq \frac{1}{200n} \qquad \text{if } n \geq 200,$$

and conclude by Theorem 7 that $\sum_{n=1}^{\infty} (1/(100n + 20{,}000))$ diverges to infinity by comparison with the divergent harmonic series $\Sigma(1/n)$. An easier way to use Theorem 8:

$$L = \lim \frac{\dfrac{1}{100n + 20{,}000}}{\dfrac{1}{n}} = \lim \frac{n}{100n + 20{,}000} = \frac{1}{100} > 0.$$

The limit comparison test has two disadvantages, however:

i) Its use does not lead to any obvious bounds for the sum of a convergent series such as those bounds given in Examples 1(a) and 1(b) above.

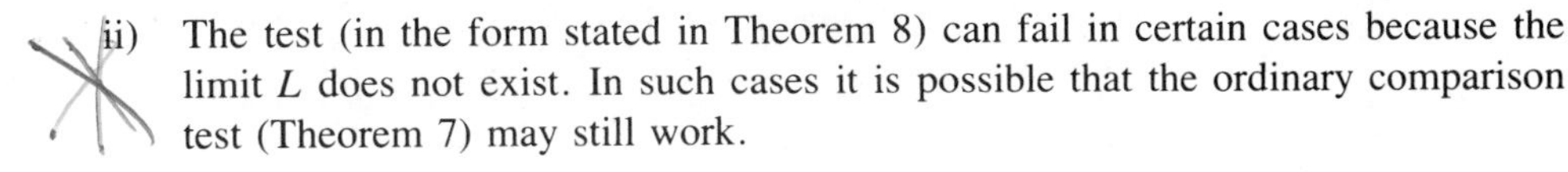

ii) The test (in the form stated in Theorem 8) can fail in certain cases because the limit L does not exist. In such cases it is possible that the ordinary comparison test (Theorem 7) may still work.

Example 3 Test the series $\displaystyle\sum_{n=1}^{\infty} \frac{1 + \sin n}{n^2}$ for convergence.

Solution The terms decrease like those of $\sum_{n=1}^{\infty} (1/n^2)$, so we should use this latter convergent series for comparison. Since

$$\lim \frac{\dfrac{1 + \sin n}{n^2}}{\dfrac{1}{n^2}} = \lim\,(1 + \sin n)$$

does not exist, Theorem 8 gives no information. However, it is evident that

$$0 \leq \frac{1 + \sin n}{n^2} \leq \frac{2}{n^2} \qquad \text{for } n = 1, 2, 3, \ldots,$$

so that the given series does, in fact, converge by comparison with $\Sigma(1/n^2)$, using Theorem 7.

The Integral Test

The integral test provides a means for determining whether an ultimately positive series converges or diverges by comparison with an improper integral that behaves similarly. Example 4 of Section 8.2 was an example of the use of this technique. We formalize the method in the following theorem.

Theorem 9 (*The Integral Test*) Suppose that for $n \geq N$ we have $a_n = f(n)$ where $f(x)$ is positive, continuous, and nonincreasing on $[N, \infty)$. Then

$$\sum_{n=1}^{\infty} a_n \quad \text{and} \quad \int_N^{\infty} f(t)\, dt$$

either both converge or both diverge to infinity.

Proof Let $s_n = a_1 + a_2 + \cdots + a_n$. Then if $n > N$ we have

$$\begin{aligned} s_n &= s_N + a_{N+1} + \cdots a_n \\ &= s_N + f(N+1) + \cdots + f(n) \\ &= s_N + \text{sum of areas of rectangles shaded in Fig. 8.4.} \\ &\leq s_N + \int_N^{\infty} f(t)\, dt. \end{aligned}$$

Thus, if the improper integral $\int_N^\infty f(t)\, dt$ converges, then $\{s_n\}$ is bounded above and Σa_n converges.

Conversely, suppose that $\Sigma_{n=1}^\infty a_n$ converges to the sum s. Then

$$\begin{aligned} \int_N^{\infty} f(t)\, dt &= \text{area under } y = f(t) \text{ above } y = 0 \text{ from } t = N \text{ to } t = \infty \\ &\leq \text{sum of areas of shaded rectangles in Fig. 8.5} \\ &= a_N + a_{N+1} + \cdots \\ &= s - s_{N-1} < \infty, \end{aligned}$$

so the improper integral represents a finite area and is thus convergent. (We omit the

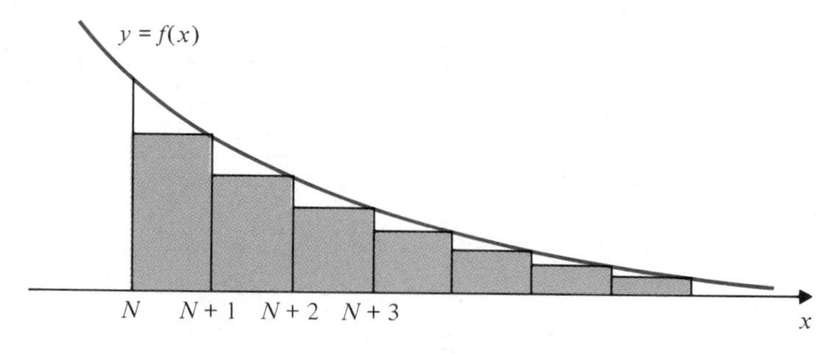

Figure 8.4

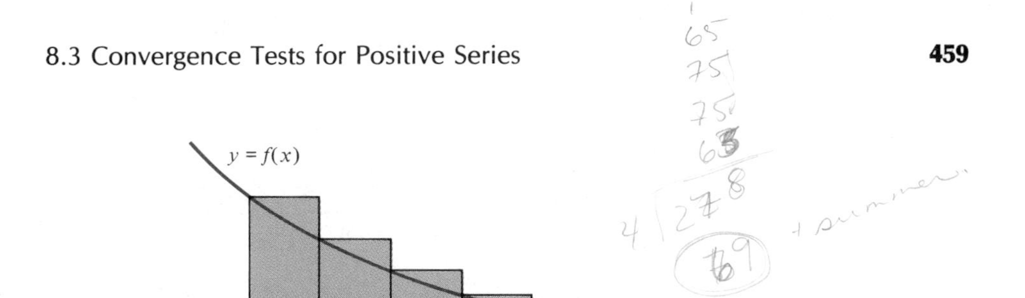

Figure 8.5

remaining details showing that $\lim_{R\to\infty} \int_N^R f(t)\, dt$ exists; like the series case, the argument depends on completeness.) □

The principal use of the integral test is to establish the following two examples. The results of these examples should be memorized; the series involved are very convenient for use with the comparison test.

Example 4 (*p-Series*)

$$\sum_{n=1}^{\infty} \frac{1}{n^p} \begin{cases} \text{converges if } p > 1 \\ \text{diverges to infinity if } p \le 1 \end{cases}$$

Observe that if $p > 0$, then $f(x) = x^{-p}$ is positive, continuous, and decreasing on $[1, \infty)$. Hence the given series converges for $p > 1$ and diverges for $0 < p \le 1$ by comparison with $\int_1^\infty x^{-p}\, dx$. (See Theorem 9 of Section 6.9.) If $p \le 0$, then $\lim (1/n^p) \ne 0$, so the series cannot converge in this case. (See Theorem 3 of Section 8.2.) Being a positive series, it must diverge to infinity.

Example 5

$$\sum_{n=2}^{\infty} \frac{1}{n(\ln n)^p} \begin{cases} \text{converges if } p > 1 \\ \text{diverges to infinity if } p \le 1 \end{cases}$$

Note that the sum starts from $n = 2$ because $\ln 1 = 0$. Once again the result follows by comparison with the improper integral

$$\int_2^\infty \frac{dx}{x(\ln x)^p}.$$

The details are left to the student in Exercise 31 at the end of this section.

Example 5 is a fine-tuning of the result of Example 4. Note that for any real exponents q and r with $q < 1 < r$ we have ultimately for any real p

$$\frac{1}{n^r} < \frac{1}{n(\ln n)^p} < \frac{1}{n^q}.$$

This fine-tuning can be successfully refined even further (see Exercise 32 at the end of this section).

Example 6 Test for convergence

$$\sum_{n=1}^{\infty} \frac{n^3 - n^2 + 5n + 22}{n^4\sqrt{n} + 7n^2 - \sqrt{n} + 3}.$$

Solution Inspection of the highest powers of n occurring in the numerator and denominator leads us to compare the series with $\Sigma_{n=1}^{\infty} 1/(n^{3/2})$, a convergent series by Example 4. Since

$$\lim \frac{n^3 - n^2 + 5n + 22}{n^4\sqrt{n} + 7n^2 - \sqrt{n} + 3} \bigg/ \frac{1}{n^{3/2}} = 1,$$

the given series is also convergent.

The method of comparing the tail of a series with an improper integral can be used to obtain estimates for the error if a positive series is approximated using its partial sums. We look into this further in Section 8.5.

The Ratio Test

Theorem 10 (*The Ratio Test*) Suppose that $a_n > 0$ (ultimately) and that

$$\rho = \lim_{n\to\infty} \frac{a_{n+1}}{a_n}$$

exists or is $+\infty$.

a) If $0 \le \rho < 1$, then $\Sigma_{n=1}^{\infty} a_n$ converges.

b) If $1 < \rho \le \infty$, then $\lim a_n = \infty$ and $\Sigma_{n=1}^{\infty} a_n$ diverges to infinity.

c) If $\rho = 1$, this test gives no information; the series may either converge or diverge to infinity.

Proof a) Suppose $\rho < 1$. Pick a number r such that $\rho < r < 1$. Since $\lim (a_{n+1}/a_n) = \rho$, $a_{n+1}/a_n \le r$ for n sufficiently large; that is, $a_{n+1} \le ra_n$ for $n \ge N$, say. In particular,

$$\begin{aligned} a_{N+1} &\le ra_N \\ a_{N+2} &\le ra_{N+1} \le r^2 a_N \\ a_{N+3} &\le ra_{N+2} \le r^3 a_N \\ &\vdots \\ a_{N+k} &\le r^k a_N \qquad (k = 0, 1, 2, 3, \ldots) \end{aligned}$$

Hence $\Sigma_{n=N}^{\infty} a_n$ converges by comparison with the convergent geometric series $\Sigma_{k=0}^{\infty} r^k$. It follows that $\Sigma_{n=1}^{\infty} a_n = \Sigma_{n=1}^{N-1} a_n + \Sigma_{n=N}^{\infty} a_n$ must converge.

b) Now suppose that $\rho > 1$. Pick a number r such that $1 < r < \rho$. Since $\lim a_{n+1}/a_n = \rho$, $a_{n+1}/a_n \geq r$ for n sufficiently large, say for $n \geq N$. We assume N is chosen large enough that $a_N > 0$. It follows by an argument similar to that used in part (a) that $a_{N+k} \geq r^k a_N$ for $k = 0, 1, 2, \ldots$, and since $r > 1$, $\lim a_n = \infty$.

c) If ρ is computed for each of the series $\Sigma(1/n)$ and $\Sigma(1/n^2)$, we get $\rho = 1$ in each case. Since the first series diverges to infinity and the second converges, it is apparent that the ratio test cannot distinguish between convergence and divergence if $\rho = 1$. □

All p-series (see Example 4) fall into the indecisive category where $\rho = 1$. The ratio test is most useful for series whose terms decrease at least exponentially fast. The presence of factorials in a term also suggests the use of the ratio test.

Example 7 Test for convergence

a) $\displaystyle\sum_{n=1}^{\infty} \frac{99^n}{n!}$ b) $\displaystyle\sum_{n=1}^{\infty} \frac{n^5}{2^n}$

c) $\displaystyle\sum_{n=1}^{\infty} \frac{n!}{n^n}$ d) $\displaystyle\sum_{n=1}^{\infty} \frac{(2n)!}{(n!)^2}$

Solution a)

$$\rho = \lim \frac{99^{n+1}}{(n+1)!} \Big/ \frac{99^n}{n!} = \lim \frac{99}{n+1} = 0 < 1.$$

Thus $\Sigma_{n=1}^{\infty} (99^n/n!)$ converges by the ratio test.

b)

$$\rho = \lim \frac{(n+1)^5}{2^{n+1}} \Big/ \frac{n^5}{2^n} = \lim \frac{1}{2}\left(\frac{n+1}{n}\right)^5 = \frac{1}{2} < 1.$$

Hence $\Sigma_{n=1}^{\infty} (n^5/2^n)$ converges by the ratio test.

c)

$$\rho = \lim \frac{\dfrac{(n+1)!}{(n+1)^{n+1}}}{\dfrac{n!}{n^n}} = \lim \frac{(n+1)!\,n^n}{(n+1)^{n+1}n!} = \lim \left(\frac{n}{n+1}\right)^n$$

$$= \lim \frac{1}{\left(1 + \dfrac{1}{n}\right)^n} = \frac{1}{e} < 1.$$

Thus $\Sigma_{n=1}^{\infty} (n!/n^n)$ converges by the ratio test.

d)

$$\rho = \lim \frac{(2(n+1))!}{((n+1)!)^2} \Big/ \frac{(2n)!}{(n!)^2} = \lim \frac{(2n+2)(2n+1)}{(n+1)^2} = 4 > 1.$$

Thus $\Sigma_{n=1}^{\infty} \dfrac{(2n)!}{(n!)^2}$ diverges to infinity by the ratio test.

EXERCISES

In Exercises 1–30 determine whether the given series converges or diverges by using any appropriate test. The p-series (Example 4 in the text) can be used for comparison, as can geometric series. Be alert for series whose terms do not approach 0.

1. $\displaystyle\sum_{n=1}^{\infty} \frac{1}{n^2+1}$

2. $\displaystyle\sum_{n=1}^{\infty} \frac{n}{n^4-2}$

3. $\displaystyle\sum_{n=1}^{\infty} \frac{n^2+1}{n^3+1}$

4. $\displaystyle\sum_{n=1}^{\infty} \frac{\sqrt{n}}{n^2+n+1}$

5. $\displaystyle\sum_{n=1}^{\infty} \left|\sin \frac{1}{n^2}\right|$

6. $\displaystyle\sum_{n=8}^{\infty} \frac{1}{\pi^n+5}$

7. $\displaystyle\sum_{n=2}^{\infty} \frac{1}{(\ln n)^3}$

8. $\displaystyle\sum_{n=1}^{\infty} \frac{1}{\ln (3n)}$

9. $\displaystyle\sum_{n=1}^{\infty} \frac{1}{\pi^n - n^\pi}$

10. $\displaystyle\sum_{n=0}^{\infty} \frac{1+n}{2+n}$

11. $\displaystyle\sum_{n=2}^{\infty} \frac{1}{\ln (n^3)}$

12. $\displaystyle\sum_{n=1}^{\infty} \frac{1}{ne^n}$

13. $\displaystyle\sum_{n=1}^{\infty} \frac{1+n^{4/3}}{2+n^{5/3}}$

14. $\displaystyle\sum_{n=1}^{\infty} \frac{n^2}{1+n\sqrt{n}}$

15. $\displaystyle\sum_{n=3}^{\infty} \frac{1}{n \ln n \sqrt{\ln \ln n}}$

16. $\displaystyle\sum_{n=2}^{\infty} \frac{1}{n \ln n\,(\ln \ln n)^2}$

17. $\displaystyle\sum_{n=1}^{\infty} \frac{1-(-1)^n}{n^4}$

18. $\displaystyle\sum_{n=1}^{\infty} \frac{1+(-1)^n}{\sqrt{n}}$

19. $\displaystyle\sum_{n=1}^{\infty} \frac{2+\cos n}{n+\ln n}$

20. $\displaystyle\sum_{n=1}^{\infty} \frac{e^n \cos^2 n}{1+\pi^n}$

21. $\displaystyle\sum_{n=1}^{\infty} \frac{1}{2^n(n+1)}$

22. $\displaystyle\sum_{n=1}^{\infty} \frac{n^4}{n!}$

23. $\displaystyle\sum_{n=1}^{\infty} \frac{n!}{n^2 e^n}$

24. $\displaystyle\sum_{n=1}^{\infty} \frac{(2n)!6^n}{(3n)!}$

25. $\displaystyle\sum_{n=2}^{\infty} \frac{\sqrt{n}}{3^n \ln n}$

26. $\displaystyle\sum_{n=0}^{\infty} \frac{n^{100}2^n}{\sqrt{n!}}$

27. $\displaystyle\sum_{n=1}^{\infty} \frac{(2n)!}{(n!)^3}$

28. $\displaystyle\sum_{n=1}^{\infty} \frac{1+n!}{(1+n)!}$

29. $\displaystyle\sum_{n=4}^{\infty} \frac{2^n}{3^n-n^3}$

30. $\displaystyle\sum_{n=1}^{\infty} \frac{n^n}{\pi^n n!}$

31. Complete Example 5 by showing that

$$\int_2^{\infty} \frac{dx}{x(\ln x)^p} \begin{cases} \text{converges if } p > 1 \\ \text{diverges to infinity if } p \le 1. \end{cases}$$

32.* Show that $\sum_{n=3}^{\infty}(1/(n \ln n(\ln \ln n)^p))$ converges if and only if $p > 1$. Generalize this result to series of the form

$$\sum_{n=N}^{\infty} \frac{1}{n(\ln n)(\ln \ln n) \cdots (\ln_j n)(\ln_{j+1} n)^p}$$

where $\ln_j n = \underbrace{\ln \ln \ln \ln \ldots \ln}_{j \text{ ln's}} n.$

33.* (*The Root Test*) Suppose that $a_n > 0$ (ultimately) and that

$$\sigma = \lim_{n\to\infty} (a_n)^{1/n}$$

exists or is $+\infty$. Show that $\Sigma_{n=1}^{\infty} a_n$ converges if $\sigma < 1$ and diverges to infinity if $\sigma > 1$. Show that the series may either converge or diverge if $\sigma = 1$. (*Hint:* Mimic the proof of the ratio test. The root test is actually a little easier.)

34.* Test for convergence $\displaystyle\sum_{n=1}^{\infty} \left(\frac{n}{n+1}\right)^{n^2}$.

35.* Try to use the ratio test to determine whether $\displaystyle\sum_{n=1}^{\infty} \frac{2^{2n}(n!)^2}{(2n)!}$ converges. What happens? Now observe that

$$\frac{2^{2n}(n!)^2}{(2n)!} = \frac{[2n(2n-2)(2n-4)\cdots 6 \times 4 \times 2]^2}{2n(2n-1)(2n-2)\cdots 4 \times 3 \times 2 \times 1}$$

$$= \frac{2n}{2n-1} \times \frac{2n-2}{2n-3} \times \cdots \times \frac{4}{2} \times \frac{2}{1}.$$

Does the given series converge or not? Why?

36.* Try to decide whether the series $\displaystyle\sum_{n=1}^{\infty} \frac{(2n)!}{2^{2n}(n!)^2}$ converges. (*Hint*. Proceed in a manner similar to that suggested in Exercise 35 and show that $a_n \geq \dfrac{1}{2n}$.)

8.4 ABSOLUTE AND CONDITIONAL CONVERGENCE

All of the series $\Sigma_{n=1}^{\infty} a_n$ considered in the previous section were ultimately positive; that is, $a_n \geq 0$ for n sufficiently large. We now drop this restriction and allow arbitrary real terms a_n. We can, however, always obtain a positive series from any given series by replacing all the terms with their absolute values.

Definition 4 The series $\Sigma_{n=1}^{\infty} a_n$ is said to **converge absolutely** (or to be **absolutely convergent**) if $\Sigma_{n=1}^{\infty} |a_n|$ converges.

The series

$$s = \sum_{n=1}^{\infty} \frac{(-1)^n}{n^2} = -1 + \frac{1}{4} - \frac{1}{9} + \frac{1}{16} - \cdots$$

converges absolutely since

$$S = \sum_{n=1}^{\infty} \frac{1}{n^2} = 1 + \frac{1}{4} + \frac{1}{9} + \frac{1}{16} + \cdots$$

converges. It seems intuitively clear that the first series must converge, and its sum s must satisfy $-S \leq s \leq S$. In general, the cancellation that occurs because some terms are negative and others positive makes it ''easier'' for a series to converge than if all the terms were of one sign. We formalize this intuition in the following theorem.

Theorem 11 If a series converges absolutely, then it converges.

Proof Let $\Sigma_{n=1}^{\infty} a_n$ be absolutely convergent. Let

$$s_n = a_1 + a_2 + a_3 + \cdots + a_n$$
$$S_n = |a_1| + |a_2| + |a_3| + \cdots + |a_n|.$$

Then $\lim S_n = S$ exists. Since $\{S_n\}$ is increasing, we have $S_n \leq S$ for every n. We want to show that $\lim s_n$ exists.

For $n = 1, 2, 3, \ldots$ let

$$p_n = \begin{cases} a_n & \text{if } a_n \geq 0 \\ 0 & \text{if } a_n < 0, \end{cases} \qquad q_n = \begin{cases} 0 & \text{if } a_n \geq 0 \\ -a_n & \text{if } a_n < 0. \end{cases}$$

Evidently $a_n = p_n - q_n$ and $|a_n| = p_n + q_n$ for each n. Let

$$P_n = p_1 + p_2 + p_3 + \cdots + p_n \qquad \text{and} \qquad Q_n = q_1 + q_2 + q_3 + \cdots + q_n.$$

Then $s_n = P_n - Q_n$ and $S_n = P_n + Q_n$ for each n.

Since $p_n \geq 0$ and $q_n \geq 0$ for every n, $\{P_n\}$ and $\{Q_n\}$ are increasing sequences. Both of these sequences are bounded above since $P_n \leq S_n \leq S$ and $Q_n \leq S_n \leq S$. Hence $P = \lim P_n$ and $Q = \lim Q_n$ exist by completeness. It follows that $\lim s_n = \lim P_n - \lim Q_n = P - Q$ exists. Therefore $\Sigma_{n=1}^{\infty} a_n$ converges. □

The comparison tests, the integral test and the ratio test, can each be used to test for absolute convergence. They should be applied to the series $\Sigma_{n=1}^{\infty} |a_n|$. For the ratio test we calculate $\rho = \lim |a_{n+1}/a_n|$. If $\rho < 1$, then Σa_n converges absolutely. If $\rho > 1$, then $\lim |a_n| = \infty$, so both $\Sigma|a_n|$ and Σa_n must diverge. If $\rho = 1$ we get no information.

Example 1 Test for absolute convergence

a) $\sum_{n=1}^{\infty} \frac{(-1)^{n-1}}{2n-1}$

b) $\sum_{n=1}^{\infty} \frac{n \cos(n\pi)}{2^n}$

Solution

a) $\lim \frac{\left|\frac{(-1)^{n-1}}{2n-1}\right|}{\frac{1}{n}} = \lim \frac{n}{2n-1} = \frac{1}{2} > 0$

Since $\Sigma_{n=1}^{\infty}(1/n)$ diverges to infinity, $\Sigma_{n=1}^{\infty}((-1)^{n-1}/(2n-1))$ cannot converge absolutely.

b) $\rho = \lim \frac{\left|\frac{(n+1)\cos(n+1)\pi}{2^{n+1}}\right|}{\left|\frac{n\cos(n\pi)}{2^n}\right|} = \lim \frac{n+1}{2n} = \frac{1}{2} < 1.$

Therefore $\Sigma_{n=1}^{\infty}((n\cos(n\pi))/2^n)$ converges absolutely by the ratio test.

The converse of Theorem 11 is false. We show later in this section that, for example, the alternating harmonic series

$$\sum_{n=1}^{\infty} \frac{(-1)^{n-1}}{n} = 1 - \frac{1}{2} + \frac{1}{3} - \frac{1}{4} + \frac{1}{5} - \cdots$$

is convergent, even though it does not converge absolutely (because the harmonic series diverges to infinity):

$$\sum_{n=1}^{\infty} \frac{1}{n} = 1 + \frac{1}{2} + \frac{1}{3} + \frac{1}{4} + \cdots = \infty.$$

Definition 5 If $\Sigma_{n=1}^{\infty} a_n$ is convergent, but not absolutely convergent, then we say that it is **conditionally convergent** or that it **converges conditionally.**

The alternating harmonic series is an example of a conditionally convergent series.

The Alternating Series Test

We cannot use any of the previously developed tests to show that the alternating harmonic series converges; all of those tests apply only to (ultimately) positive series, and so they can test only for absolute convergence. Demonstrating convergence that is not absolute is generally harder to do. We present only one test that can establish such convergence; this test can only be used on a very special kind of series.

Theorem 12 (*The Alternating Series Test*) Suppose that the sequence $\{a_n\}$ is positive, decreasing, and tends to 0, that is, suppose

i) $a_n \geq 0$ for $n = 1, 2, 3, \ldots$;
ii) $a_{n+1} \leq a_n$ for $n = 1, 2, 3, \ldots$;
iii) $\lim a_n = 0$.

Then the alternating series

$$\sum_{n=1}^{\infty} (-1)^{n-1} a_n = a_1 - a_2 + a_3 - a_4 + a_5 - \cdots$$

converges. Moreover, if the sum of the series is s and if s_n denotes the nth partial sum,

$$s_n = \sum_{k=1}^{n} (-1)^{k-1} a_k = a_1 - a_2 + a_3 - \cdots + (-1)^{n-1} a_n,$$

then for every n we have

$$s_{2n} \leq s \leq s_{2n-1},$$
$$s_{2n} \leq s \leq s_{2n+1}.$$

In particular,

$$|s - s_n| \leq a_{n+1}.$$

Thus the error made in using a partial sum to approximate the sum of the series does not exceed the first omitted term in absolute value.

Proof Since the sequence $\{a_n\}$ is decreasing, we have $a_{2n+1} \geq a_{2n+2}$. Therefore $s_{2n+2} = s_{2n} + a_{2n+1} - a_{2n+2} \geq s_{2n}$ for $n = 1, 2, 3, \ldots$; the even partial sums $\{s_{2n}\}$ form an increasing sequence. Similarly $s_{2n+1} = s_{2n-1} - a_{2n} + a_{2n+1} \leq s_{2n-1}$, so the odd partial sums $\{s_{2n-1}\}$ form a decreasing sequence. Since $s_{2n} = s_{2n-1} - a_{2n} \leq s_{2n-1}$, we can say, for any n, that

$$s_2 \leq s_4 \leq s_6 \leq \cdots \leq s_{2n} \leq s_{2n-1} \leq s_{2n-3} \leq \cdots \leq s_5 \leq s_3 \leq s_1.$$

Hence s_2 is a lower bound for the decreasing sequence $\{s_{2n-1}\}$ and s_1 is an upper bound for the increasing sequence $\{s_{2n}\}$. Both of these sequences therefore converge by completeness of the real numbers:

$$\lim_{n\to\infty} s_{2n-1} = s_{\text{odd}}, \qquad \lim_{n\to\infty} s_{2n} = s_{\text{even}}.$$

Since $0 = \lim a_{2n} = \lim (s_{2n-1} - s_{2n}) = s_{\text{odd}} - s_{\text{even}}$, we have $s_{\text{odd}} = s_{\text{even}} = s$, say. Thus $\lim s_n = s$ exists and the series $\Sigma(-1)^{n-1} a_n$ converges to this sum s.

It is evident that every even partial sum is less than or equal to s and every odd partial sum is greater than or equal to s. That is, s lies between s_{2n} and either s_{2n-1} or s_{2n+1}. It follows that

$$|s - s_n| \leq |s_{n+1} - s_n| = a_{n+1}. \square$$

Note that the series $\Sigma_{n=1}^{\infty} (-1)^{n-1} a_n$ begins with a positive term, a_1. The conclusions of Theorem 12 also hold for the series $\Sigma_{n=1}^{\infty} (-1)^n a_n$, which starts with a negative term, $-a_1$. (It is just the negative of the first series.) In this case, however, we have

$$s_{2n-1} \leq s \leq s_{2n} \quad \text{and} \quad s_{2n+1} \leq s \leq s_{2n}.$$

The theorem also remains valid if conditions (i) and (ii) in its statement are replaced by "ultimate" versions:

$$\left.\begin{array}{l} \text{i)}' \ a_n \geq 0 \\ \text{ii)}' \ a_{n+1} \leq a_n \end{array}\right\} \text{for } n = N, N+1, N+2, \ldots.$$

The inequalities bounding s in terms of partial sums s_n are then only valid for $n \geq N$.

Example 2 Each of the series

a) $\displaystyle\sum_{n=1}^{\infty} \frac{(-1)^{n-1}}{n}$,

b) $\displaystyle\sum_{n=2}^{\infty} \frac{\cos(n\pi)}{\ln n}$, and

c) $\displaystyle\sum_{n=1}^{\infty} \frac{(-1)^{n-1}}{n^4}$

satisfies the conditions of the alternating series test, and so converges. Evidently (a) and (b) do not converge absolutely (for (b) note that $1/\ln n > 1/n$), so these are both conditionally convergent series. Series (c), however, is absolutely convergent so we really do not need the alternating series test to show that it converges; Theorem 11 would do as well.

When determining the convergence of a given series it is best to consider first whether the series converges absolutely. If it does not, then there remains the possibility of conditional convergence.

Example 3 For what values of x does the series $\displaystyle\sum_{n=1}^{\infty} \frac{(x-5)^n}{n2^n}$ converge absolutely? Converge conditionally? Diverge?

Solution For such series whose terms involve functions of a variable x, it is usually wisest to begin testing for absolute convergence with the ratio test. We have

$$\rho = \lim \left| \frac{(x-5)^{n+1}}{(n+1)2^{n+1}} \middle/ \frac{(x-5)^n}{n\,2^n} \right| = \lim \frac{n}{n+1} \left| \frac{x-5}{2} \right| = \left| \frac{x-5}{2} \right|.$$

The series converges absolutely if $|(x-5)/2| < 1$, that is, if $|x-5| < 2$, that is, if $3 < x < 7$. If $x < 3$ or $x > 7$, then $|(x-5)/2| > 1$ and the series diverges.

If $x = 3$, the series is $\Sigma_{n=1}^{\infty} ((-1)^n/n)$, which converges conditionally; if $x = 7$, the series is $\Sigma_{n=1}^{\infty} (1/n)$, which diverges to infinity. Hence the given series converges absolutely on the interval (3, 7), converges conditionally at $x = 3$, and diverges everywhere else.

Example 4 For what values of x does the series $\Sigma_{n=0}^{\infty} (n + 1)^2 (x/(x + 2))^n$ converge absolutely? Converge conditionally? Diverge?

Solution Again we begin with the ratio test.

$$\rho = \lim \left| (n + 2)^2 \left(\frac{x}{x + 2}\right)^{n+1} \Big/ (n + 1)^2 \left(\frac{x}{x + 2}\right)^n \right| = \lim \left(\frac{n + 2}{n + 1}\right)^2 \left|\frac{x}{x + 2}\right|$$

$$= \left|\frac{x}{x + 2}\right|$$

The series converges absolutely if $|x/(x + 2)| < 1$, that is, if $|x| < |x + 2|$ or $x > -1$. The series diverges if $|x/(x + 2)| > 1$, that is, if $x < -1$. If $x = -1$, then the series is $\Sigma_{n=0}^{\infty} (-1)^n (n + 1)^2$, which diverges. We conclude that the series converges absolutely for $x > -1$, converges conditionally nowhere, and diverges for $x \leq -1$.

When using the alternating series test it is important to verify (at least mentally) that *all three conditions* (i)–(iii) are satisfied. (As mentioned above, (i) and (ii) need only be satisfied ultimately.)

Example 5 a)

$$\sum_{n=1}^{\infty} (-1)^{n-1} \frac{n + 1}{n}$$

Here $a_n = (n + 1)/n$ is positive and decreases as n increases. However, $\lim a_n = 1 \neq 0$. The test fails. In fact the given series diverges, since its terms do not approach 0.

b) Consider the series

$$1 - \frac{1}{4} + \frac{1}{3} - \frac{1}{16} + \frac{1}{5} - \cdots = \sum_{n=1}^{\infty} (-1)^{n-1} a_n,$$

where

$$a_n = \begin{cases} \dfrac{1}{n} & \text{if } n \text{ is odd,} \\[2ex] \dfrac{1}{n^2} & \text{if } n \text{ is even.} \end{cases}$$

This series alternates, a_n is positive, and $\lim a_n = 0$. However, $\{a_n\}$ is not decreasing (even ultimately). Once again the alternating series test cannot be applied. In fact, since $-1/4 - 1/16 - \cdots - 1/(2n)^2 - \cdots$ converges, and $1 + 1/3 + 1/5 + \cdots + 1/(2n - 1) + \cdots$ diverges to infinity, it is readily seen that the given series diverges to infinity.

The basic difference between absolute and conditional convergence is that when a series Σa_n converges absolutely it does so because its terms $\{a_n\}$ decrease in size fast enough that their sum can be finite even if no cancellation occurs because all the terms are of the same sign. If cancellation is required to make the series converge (because the terms decrease slowly), then the series can only converge conditionally.

Consider the alternating harmonic series

$$1 - \frac{1}{2} + \frac{1}{3} - \frac{1}{4} + \frac{1}{5} - \frac{1}{6} + \cdots .$$

This series converges, but only conditionally. If we take the subseries containing only the positive terms, we get the series

$$1 + \frac{1}{3} + \frac{1}{5} + \frac{1}{7} + \cdots ,$$

which diverges to infinity. Similarly, the subseries of negative terms

$$-\frac{1}{2} - \frac{1}{4} - \frac{1}{6} - \frac{1}{8} - \cdots$$

diverges to negative infinity.

If a series converges absolutely, the subseries consisting of positive terms and the subseries consisting of negative terms must each converge to a finite sum. The following theorem is stated for interest.

Theorem 13

a) If the terms of an absolutely convergent series are rearranged so that the additions occur in a different order, the rearranged series still converges to the same sum as the original series.

b) If a series is conditionally convergent, and L is any real number, then the terms of the series can be rearranged so as to make the series converge (conditionally) to the sum L. It can also be rearranged so as to diverge to ∞ or to $-\infty$. □

We will not attempt any proof. Part (b) shows that conditional convergence is a rather suspect kind of convergence, being dependent on the order in which the terms are added.

EXERCISES

Determine whether the series in Examples 1–12 converge absolutely, converge conditionally, or diverge.

1. $\displaystyle\sum_{n=1}^{\infty} \frac{(-1)^{n-1}}{\sqrt{n}}$

2. $\displaystyle\sum_{n=1}^{\infty} \frac{(-1)^n}{n^2 + \ln n}$

3. $\displaystyle\sum_{n=1}^{\infty} \frac{\cos(n\pi)}{(n+1)\ln(n+1)}$

4. $\displaystyle\sum_{n=1}^{\infty} \frac{(-1)^{2n}}{2^n}$

5. $\displaystyle\sum_{n=0}^{\infty} \frac{(-1)^n(n^2-1)}{n^2+1}$

6. $\displaystyle\sum_{n=1}^{\infty} \frac{(-2)^n}{n!}$

7. $\displaystyle\sum_{n=1}^{\infty} \frac{(-1)^n}{n\pi^n}$

8. $\displaystyle\sum_{n=0}^{\infty} \frac{-n}{n^2+1}$

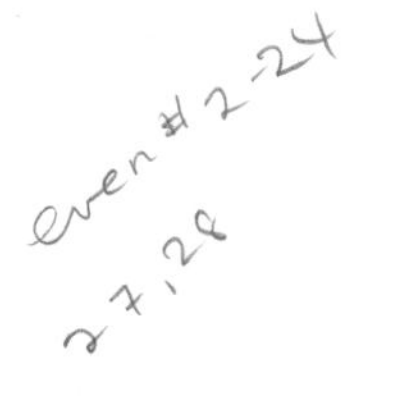

9. $\sum_{n=1}^{\infty} (-1)^n \frac{20n^2 - n - 1}{n^3 + n^2 + 33}$

10. $\sum_{n=1}^{\infty} \frac{100 \cos (n\pi)}{2n + 3}$

11. $\sum_{n=1}^{\infty} \frac{n!}{(-100)^n}$

12. $\sum_{n=10}^{\infty} \frac{\sin (n + 1/2)\pi}{\ln \ln n}$

Determine the values of x for which the series in Examples 13–24 converge absolutely, converge conditionally, or diverge.

13. $\sum_{n=0}^{\infty} \frac{x^n}{\sqrt{n + 1}}$

14. $\sum_{n=1}^{\infty} \frac{(x - 2)^n}{n^2 \, 2^{2n}}$

15. $\sum_{n=0}^{\infty} (-1)^n \frac{(x + 1)^n}{2n + 3}$

16. $\sum_{n=1}^{\infty} \frac{1}{2n - 1} \left(\frac{3x + 2}{-5} \right)^n$

17. $\sum_{n=2}^{\infty} \frac{x^n}{2^n \ln n}$

18. $\sum_{n=1}^{\infty} \frac{(4x + 1)^n}{n^3}$

19. $\sum_{n=1}^{\infty} \frac{(ax + b)^n}{n^{1/3} c^n} \quad (a > 0,\ c > 0)$

20. $\sum_{n=1}^{\infty} \frac{1}{n} \left(1 + \frac{1}{x} \right)^n$

21. $\sum_{n=1}^{\infty} \frac{1}{n^2} \left(1 - \frac{1}{x} \right)^n$

22. $\sum_{n=1}^{\infty} \frac{(x^2 - 1)^n}{\sqrt{n}}$

23. $\sum_{n=0}^{\infty} n(x^2 - x - 1)^n$

24. $\sum_{n=1}^{\infty} \frac{\sin^n x}{n}$

25.* Does the alternating series test apply directly to the series $\sum_{n=1}^{\infty} \frac{\sin (n\pi/2)}{n}$? Determine whether this series converges.

26.* Show that the series $\sum_{n=1}^{\infty} a_n$ converges absolutely if

$$a_n = \begin{cases} \dfrac{10}{n^2} & \text{if } n \text{ is even,} \\[2ex] \dfrac{-1}{10n^3} & \text{if } n \text{ is odd.} \end{cases}$$

27.* Which of the following statements are true and which are false? Justify your assertion of truth, or give a counterexample to show falsehood.

a) If $\sum_{n=1}^{\infty} a_n$ converges, then $\sum_{n=1}^{\infty} (-1)^n a_n$ converges.

b) If $\sum_{n=1}^{\infty} a_n$ converges and $\sum_{n=1}^{\infty} (-1)^n a_n$ converges, then $\sum_{n=1}^{\infty} a_n$ converges absolutely.

c) If $\sum_{n=1}^{\infty} a_n$ converges absolutely, then $\sum_{n=1}^{\infty} (-1)^n a_n$ converges absolutely.

28.* **a)** Use a Riemann sum argument to show that

$$\ln n! \geq \int_1^n \ln t \, dt = n \ln n - n + 1.$$

b) For what values of x does the series $\sum_{n=1}^{\infty} \frac{n! x^n}{n^n}$ converge absolutely? Converge conditionally? Diverge? (*Hint:* First use the ratio test. To test the cases where $\rho = 1$ you will find the inequality in 28(a) useful.)

29.* For what values of x does the series $\sum_{n=1}^{\infty} \frac{(2n)! x^n}{2^{2n} (n!)^2}$ converge absolutely? Converge conditionally? Diverge? (*Hint:* See Example 36 of Section 8.3.)

30.* Devise a procedure for rearranging the terms of the alternating harmonic series so that the rearranged series converges to 8.

8.5 ESTIMATING THE SUM OF A SERIES

So far we have been concerned mainly with the problem of determining whether a given series converges or diverges, that is, whether it has a finite sum. Except for very special series (for example, geometric or telescoping series) we have not actually calculated the sums of the series we have shown to converge. Indeed, there are no elementary techniques for calculating the sums of most series, though we will see how some other series can be summed in the next chapter.

When it is not practical to find the exact sum of a convergent series, we can still approximate that sum by using a partial sum s_n of the series:

$$s = \sum_{k=1}^{\infty} a_k = a_1 + a_2 + a_3 + \cdots$$

$$s \simeq s_n = \sum_{k=1}^{n} a_k = a_1 + a_2 + a_3 + \cdots + a_n.$$

Of course, such an approximation is of little use unless we know how good an approximation it is, that is, unless we have a bound for the size of the error, $|s - s_n|$.

Observe that the error $s - s_n$ is just the *tail* of the given series; it is the sum of the rest of the terms of the series beyond those included in s_n:

$$s - s_n = \sum_{k=n+1}^{\infty} a_k = a_{n+1} + a_{n+2} + a_{n+3} + \cdots.$$

If the given series converges, then the error $s - s_n$ approaches 0 as n increases toward infinity (since $\lim s_n = s$), but we would still like to know how fast that approach is and how large n need be to ensure that the error is within tolerable bounds. Various convergence tests involve, implicitly or explicitly, techniques for finding bounds for the error. We shall investigate error-bounding techniques arising from the integral test, the ratio test, and the alternating series test.

Integral Bounds

Suppose that $a_k = f(k)$ for $k = n + 1, n + 2, n + 3, \ldots$ where f is a positive, continuous function, decreasing on the interval $[n, \infty)$. Evidently,

$$\begin{aligned} s - s_n &= \sum_{k=n+1}^{\infty} f(k) \\ &= \text{sum of areas of rectangles shaded in Fig. 8.6.} \\ &\leq \int_n^{\infty} f(x)\,dx. \end{aligned}$$

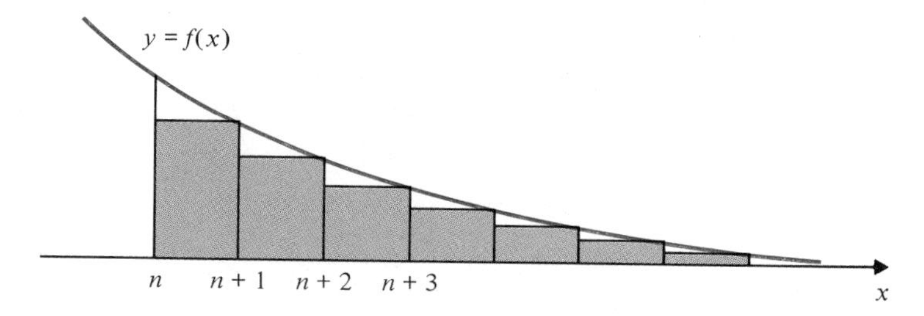

Figure 8.6

Similarly,

$$s - s_n = \text{sum of areas of rectangles in Fig 8.7.}$$
$$\geq \int_{n+1}^{\infty} f(x)\,dx.$$

That is,

$$\int_{n+1}^{\infty} f(x)\,dx \leq s - s_n \leq \int_{n}^{\infty} f(x)\,dx.$$

Let $A_n = \int_{n+1}^{\infty} f(x)\,dx$ and $B_n = \int_{n}^{\infty} f(x)\,dx$. The series $\Sigma_{n=1}^{\infty} a_n$ converges if and only if A_n and B_n are finite; in this case its sum s satisfies

$$s_n + A_n \leq s \leq s_n + B_n.$$

Given any $n = 1, 2, 3, \ldots$, the error in the approximation $s \simeq s_n$ satisfies $0 \leq s - s_n \leq B_n$.

However, since s must lie in the interval $[s_n + A_n, s_n + B_n]$, we can do better by using the approximation

$$s \simeq s_n^* = s_n + \frac{1}{2}(A_n + B_n),$$

that is, the midpoint of the interval $[s_n + A_n, s_n, + B_n]$, as shown in Fig. 8.8.

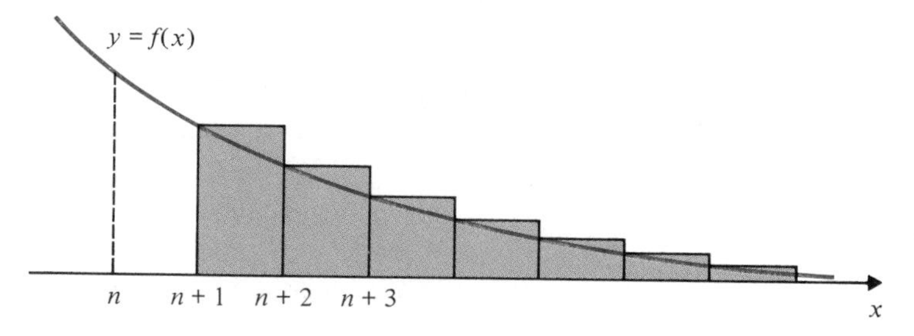

Figure 8.7

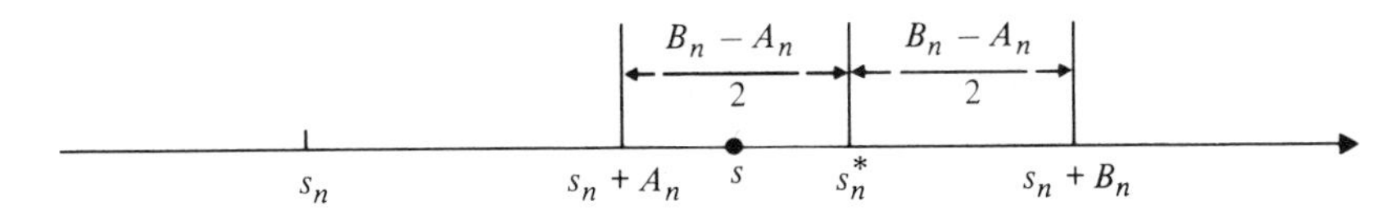

Figure 8.8

Evidently the error in this approximation, $s - s_n^*$, satisfies

$$|s - s_n^*| \leq \frac{1}{2}(B_n - A_n),$$

half the length of the interval containing s.

(Whenever an unknown quantity is known to lie in a certain interval, the midpoint of that interval is the best approximation to the quantity, and the error in that approximation does not exceed half the length of the interval in absolute value.)

Example 1 Find the best approximation to the sum s of the series $\sum_{n=1}^{\infty} (1/n^2)$ making use of the partial sum s_n of the first n terms. How large should n be taken to ensure that the approximation $s \simeq s_n$ has error less than 1/1000? How large should n be taken to ensure that the best approximation using the terms of s_n has error less than 1/1000 in absolute value?

Solution Since $f(x) = 1/x^2$ is positive, continuous, and decreasing on $[1, \infty)$ for any $n = 1, 2, 3, \ldots$, we have

$$A_n = \int_{n+1}^{\infty} \frac{1}{x^2}\,dx = \lim_{R\to\infty} \left(-\frac{1}{x}\right)\Bigg|_{n+1}^{R} = \frac{1}{n+1}$$

and, similarly, $B_n = 1/n$. Therefore, $s_n + 1/(n+1) \leq s \leq s_n + (1/n)$. The best approximation to s (using s_n) is

$$s_n^* = s_n + \frac{1}{2}\left(\frac{1}{n+1} + \frac{1}{n}\right) = s_n + \frac{2n+1}{2n(n+1)}$$

$$= 1 + \frac{1}{4} + \frac{1}{9} + \frac{1}{16} + \cdots + \frac{1}{n^2} + \frac{2n+1}{2n(n+1)}.$$

The error in the approximation $s \simeq s_n$ is $s - s_n$ and satisfies

$$0 \leq s - s_n \leq B_n = \frac{1}{n} < \frac{1}{1000}$$

provided $n > 1000$. The error in the approximation $s \simeq s_n^*$ satisfies

$$|s - s_n^*| \leq \frac{1}{2}(B_n - A_n) = \frac{1}{2}\left(\frac{1}{n} - \frac{1}{n+1}\right) = \frac{1}{2n(n+1)} \leq \frac{1}{1000}$$

provided $n(n + 1) \geq 500$. This latter condition will be met if $n \geq 22$. Taking $n = 22$, we can be sure that

$$s_{22}^* = 1 + \frac{1}{4} + \frac{1}{9} + \cdots + \frac{1}{22^2} + \frac{45}{44 \times 23}$$

approximates $\sum_{n=1}^{\infty} (1/n^2)$ with error not exceeding $1/1000$ in absolute value.

Example 2 Find the sum of the series $\sum_{k=1}^{\infty} (1/k^4)$ with error less than $1/1000$.

Solution Let $f(x) = 1/x^4$, which is positive, continuous, and decreasing on $[1, \infty)$. Let

$$A_n = \int_{n+1}^{\infty} \frac{dx}{x^4} = \lim_{R\to\infty} \left(-\frac{1}{3x^3}\right)\Bigg|_{n+1}^{R} = \frac{1}{3(n + 1)^3},$$

$$B_n = \int_{n}^{\infty} \frac{dx}{x^4} = \frac{1}{3n^3}.$$

We use the approximation

$$s \simeq s_n^* = s_n + \frac{1}{2}\left(\frac{1}{3(n + 1)^3} + \frac{1}{3n^3}\right).$$

The error satisfies

$$|s - s_n^*| \leq \frac{1}{2}\left(\frac{1}{3n^3} - \frac{1}{3(n + 1)^3}\right) = \frac{1}{6}\frac{(n + 1)^3 - n^3}{n^3(n + 1)^3}$$

$$= \frac{1}{6}\frac{3n^2 + 3n + 1}{n^3(n + 1)^3} < \frac{7}{6}\frac{1}{n^4}.$$

We have used $3n^2 + 3n + 1 < 7n^2$ and $n^3(n + 1)^3 > n^6$ to obtain the last inequality. We will have $|s - s_n^*| < 1/1000$ provided

$$\frac{7}{6n^4} < \frac{1}{1000},$$

that is, $n^4 > 7000/6$. Since $6^4 = 1296 > 7000/6$, $n = 6$ will do. Thus

$$\sum_{n=1}^{\infty} \frac{1}{n^4} \simeq s_6^* = 1 + \frac{1}{2^4} + \frac{1}{3^4} + \frac{1}{4^4} + \frac{1}{5^4} + \frac{1}{6^4} + \frac{1}{6}\left(\frac{1}{6^3} + \frac{1}{7^3}\right)$$

$$\simeq 1.08238 \qquad \text{with error less than } \frac{1}{1000}.$$

Observe that only six terms are required to estimate the sum of the series $\sum_{n=1}^{\infty} (1/n^4)$ in Example 2 within the same tolerance for which we needed 22 terms of the series $\sum_{n=1}^{\infty} (1/n^2)$ in Example 1. Evidently the series $\sum_{n=1}^{\infty} (1/n^2)$ *converges more slowly* than does $\sum_{n=1}^{\infty} (1/n^4)$. The use of integrals to bound the tails of series is appropriate for those series for which the integral test would be appropriate to demonstrate convergence. In particular, this class includes series whose terms are rational functions of n.

Geometric Bounds

Suppose that an inequality of the form

$$0 \le a_k \le Kr^k$$

holds for $k = n + 1, n + 2, n + 3, \ldots$, where K and r are constants and $r < 1$. We can then use a geometric series to bound the tail of $\sum_{n=1}^{\infty} a_n$.

$$\begin{aligned} 0 \le s - s_n = \sum_{k=n+1}^{\infty} a_k &\le \sum_{k=n+1}^{\infty} Kr^k \\ &= Kr^{n+1}(1 + r + r^2 + \cdots) \\ &= \frac{Kr^{n+1}}{1 - r}. \end{aligned}$$

Since $r < 1$ the series converges and the error approaches 0 at an exponential rate as n increases.

Example 3 In Chapter 9 we will show that

$$e = \frac{1}{0!} + \frac{1}{1!} + \frac{1}{2!} + \frac{1}{3!} + \cdots = \sum_{n=0}^{\infty} \frac{1}{n!}$$

(recall that $0! = 1$). Estimate the error if the sum s_n of the first n terms of the series is used to approximate e. Find e to three-decimal-place accuracy using the series.

Solution We have

$$\begin{aligned} s_n &= \frac{1}{0!} + \frac{1}{1!} + \frac{1}{2!} + \frac{1}{3!} + \cdots + \frac{1}{(n-1)!} \\ &= 1 + 1 + \frac{1}{2} + \frac{1}{6} + \frac{1}{24} + \cdots + \frac{1}{(n-1)!}. \end{aligned}$$

(Since the series starts with $n = 0$, the nth term is $1/(n - 1)!$.) We can estimate the error in the approximation $s \simeq s_n$ as follows:

$$\begin{aligned} 0 < s - s_n &= \frac{1}{n!} + \frac{1}{(n+1)!} + \frac{1}{(n+2)!} + \frac{1}{(n+3)!} + \cdots \\ &= \frac{1}{n!}\left(1 + \frac{1}{n+1} + \frac{1}{(n+1)(n+2)} + \frac{1}{(n+1)(n+2)(n+3)} + \cdots\right) \\ &< \frac{1}{n!}\left(1 + \frac{1}{n+1} + \frac{1}{(n+1)^2} + \frac{1}{(n+1)^3} + \cdots\right) \end{aligned}$$

since $n + 2 > n + 1$, $n + 3 > n + 1$, and so on. Thus

$$0 < s - s_n < \frac{1}{n!} \frac{1}{1 - \dfrac{1}{n+1}} = \frac{n+1}{n(n!)}.$$

If we wish to evaluate e accurately to three decimal places, then we would like to ensure that the error is less than 5 in the fourth decimal place, that is, that the error is less than 0.0005. Hence we want

$$\frac{n+1}{n}\frac{1}{n!} < 0.0005 = \frac{1}{2000}.$$

Since $7! = 5040$, but $6! = 720$ only, we can use $n = 7$ but no smaller. We have

$$\begin{aligned} e &\simeq 1 + 1 + \frac{1}{2!} + \frac{1}{3!} + \frac{1}{4!} + \frac{1}{5!} + \frac{1}{6!} \\ &= 2 + \frac{1}{2} + \frac{1}{6} + \frac{1}{24} + \frac{1}{120} + \frac{1}{720} \simeq 2.718 \qquad \text{(rounded to three decimals).} \end{aligned}$$

It is appropriate to use geometric series to bound the tails of positive series whose convergence would be demonstrated by the ratio test. Such series converge ultimately faster than any p-series $\Sigma_{n=1}^{\infty} (1/n^p)$, for which the limit ratio is $\rho = 1$.

Alternating Series Bounds

The alternating series test (Theorem 12 of Section 8.4) provides a direct estimate for the tail of any series that satisfies its conditions. Suppose that

i) $a_k \geq 0$ for $k = n, n + 1, n + 2, \ldots,$

ii) $a_{k+1} \leq a_k$ for $k = n, n + 1, n + 2, \ldots,$ and

iii) $\lim a_k = 0$.

The error $s - s_n$ encountered if we use the sum of the first n terms to approximate the sum of the alternating series $s = \Sigma_{k=1}^{\infty} (-1)^{k-1} a_k$ does not exceed the $(n + 1)$st term in absolute value and is in fact of the same sign as that term.

Example 4 How many terms are sufficient to approximate the sum of the alternating harmonic series to within 1/1000?

Solution

$$s = 1 - \frac{1}{2} + \frac{1}{3} - \frac{1}{4} + \cdots = s_n + \frac{(-1)^n}{n+1} + \frac{(-1)^{n+1}}{n+2} + \cdots$$

We have $|s - s_n| \leq 1/(n + 1) < 1/1000$, provided $n + 1 > 1000$. To be sure that the approximation error is less than 1/1000 we need at least 1000 terms. (Evidently there is not much future in computing the sum of this series to any great accuracy; it converges very slowly!)

Example 5 Find an approximate value for $\sin 10° = \sin(\pi/18)$ accurate to four decimal places, using the series

$$\sin x = x - \frac{x^3}{3!} + \frac{x^5}{5!} - \cdots = \sum_{k=0}^{\infty} (-1)^k \frac{x^{2k+1}}{(2k+1)!},$$

which will be established in Chapter 9.

Solution

$$\sin 10° = \sin\frac{\pi}{18} = \frac{\pi}{18} - \frac{(\pi/18)^3}{3!} + \frac{(\pi/18)^5}{5!} - \cdots$$

If we stop with the term $(-1)^{n-1}\left(\frac{\pi}{18}\right)^{2n-1}\frac{1}{(2n-1)!}$, the error E will satisfy

$$|E| \le \left(\frac{\pi}{18}\right)^{2n+1}\frac{1}{(2n+1)!}.$$

Now $\pi/18 < 4/16 = 1/4$, so $|E| < 1/(4^{2n+1}(2n+1)!) < 0.00005 = 1/20{,}000$ provided $4^{2n+1}(2n+1)! > 20{,}000$. This is satisfied even for $n = 2$ ($4^5 5! = 122{,}880$). Hence

$$\sin 10° \simeq \frac{\pi}{18} - \frac{1}{6}\left(\frac{\pi}{18}\right)^3 \simeq 0.1736 \qquad \text{(to four decimal places).}$$

EXERCISES

In Exercises 1–6 let s_n denote the sum of the first n terms of the given series. Using s_n, find the smallest interval that you can be sure contains the sum of the series. If the midpoint of this interval is used to approximate s, how large should n be chosen to ensure that the error is less than 1/1000?

1. $\sum_{k=1}^{\infty}\frac{1}{k^{10}}$

2. $\sum_{k=1}^{\infty}\frac{1}{k^3}$

3. $\sum_{k=1}^{\infty}\frac{1}{k^{3/2}}$

4. $\sum_{k=2}^{\infty}\frac{1}{k(\ln k)^2}$

5. $\sum_{k=1}^{\infty}\frac{1}{e^k}$

6. $\sum_{k=1}^{\infty}\frac{1}{k^2+k}$

For each positive series in Exercises 7–10 find the best upper bound you can for the error $s - s_n$ encountered if the partial sum s_n is used to approximate the sum of the series. How many terms of each series do you need to be sure that the approximation has error less than 1/1000?

7. $\sum_{k=1}^{\infty}\frac{1}{2^k k!}$

8. $\sum_{n=1}^{\infty}\frac{1}{(2n-1)!}$

9. $\sum_{n=0}^{\infty}\frac{2^n}{(2n)!}$

10. $\sum_{n=1}^{\infty}\frac{1}{n^n}$

For the series in Exercises 11–14 determine how many terms are sufficient to ensure that the partial sum s_n approximates the sum s of the series with error less than 1/1000 in absolute value.

11. $\sum_{n=1}^{\infty}(-1)^{n-1}\frac{n}{n^2+1}$

12. $\sum_{n=0}^{\infty}\frac{(-1)^n}{(2n)!}$

13. $\sum_{n=1}^{\infty}(-1)^{n-1}\frac{n}{2^n}$

14. $\sum_{n=0}^{\infty}(-1)^n\frac{3^n}{n!}$

15.* **a)** Show that if $k > 0$ and n is a positive integer, then

$$n < \frac{1}{k}(1+k)^n.$$

b) Use the estimate in (a) with $0 < k < 1$ to obtain an upper bound for the sum of the series $\Sigma_{n=0}^{\infty} (n/2^n)$. For what value of k is this bound least?

c) If we use the sum s_n of the first n terms to approximate the sum s of the series in (b), obtain an upper bound for the error $s - s_n$ using the inequality from (a). For given n, find k to minimize this upper bound.

16.* (*Improving the Convergence of a Series*) We know (from Example 3 of Section 8.2) that $\displaystyle\sum_{n=0}^{\infty} \frac{1}{n(n+1)} = 1$. Since

$$\frac{1}{n^2} = \frac{1}{n(n+1)} + c_n, \qquad \text{where } c_n = \frac{1}{n^2(n+1)},$$

we have

$$\sum_{n=1}^{\infty} \frac{1}{n^2} = 1 + \sum_{n=1}^{\infty} c_n.$$

Evidently, Σc_n converges more rapidly than $\Sigma(1/n^2)$. Hence fewer terms of that series will be needed to compute $\Sigma(1/n^2)$ to any desired degree of accuracy than would be needed if we calculated with $\Sigma(1/n^2)$ directly. Using integral upper and lower bounds, determine a value of n for which the modified partial sum s_n^* for the series $\Sigma_{n=1}^{\infty} c_n$ approximates the sum of that series with error less than 1/1000 in absolute value. Hence determine $\Sigma_{n=1}^{\infty} (1/n^2)$ to within 1/1000 of its true value.

(The technique exhibited in this exercise is known as improving the convergence of a series. It can be applied to estimating the sum Σa_n if we know the sum Σb_n and if $a_n - b_n = c_n$ where $|c_n|$ decreases faster than $|a_n|$ as n tends to infinity.)

9.1 Power Series
9.2 Differentiation and Integration of Power Series
9.3 Taylor and Maclaurin Series
9.4 Applications of Taylor and Maclaurin Series
9.5 Taylor's Theorem

9

Power Series Representations of Functions

The principal reason for studying infinite series in an elementary calculus course is that many of the most important elementary functions $f(x)$ have convenient and useful representations as series whose terms are constant multiples of nonnegative integer powers of x. Such series are called power series; they generalize polynomials that are sums of finitely many such terms. Among the functions studied in calculus, polynomials are the easiest to manipulate, to calculate with, to differentiate, and to integrate. Power series, behaving for some purposes like "polynomials of infinite degree," inherit many of these benign qualities. Therefore, the representation of such transcendental functions as e^x, $\sin x$, Arctan x, and others as power series renders these functions even more useful in the mathematical modeling of concrete problems.

In this final chapter we develop the machinery of power series and establish several power series representations of elementary functions. The climax comes with Taylor's theorem, which provides not only a framework for representing functions as series but also estimates for the tails when partial sums of such series are used to approximate the functions. Taylor's theorem simultaneously generalizes the mean-value theorem, the error formula for the tangent-line approximation, and the fundamental theorem of calculus.

In addition to power series, there are many other useful series representations of functions, in particular trigonometric series (Fourier series), that are beyond the scope of this book. They are encountered in higher-level courses in mathematical analysis and differential equations and, like power series, constitute a basic tool of the mathematician, the engineer, and other mathematically oriented scientists.

9.1 POWER SERIES

Definition 1

> A series of the form
>
> $$\sum_{n=0}^{\infty} a_n(x - c)^n = a_0 + a_1(x - c) + a_2(x - c)^2 + a_3(x - c)^3 + \cdots,$$
>
> where $a_0, a_1, a_2, \ldots$ are constants, is called a **power series in powers of $x - c$** or a **power series about the point $x = c$.** The numbers $a_0, a_1, a_2, \ldots$ are called the coefficients of the power series.

Since the terms of a power series are functions of a variable x, the series may or may not converge for each value of x. For those values of x for which the series does converge, the sum defines a function of x. For example, if $-1 < x < 1$, then

$$1 + x + x^2 + x^3 + \cdots = \frac{1}{1 - x}.$$

The geometric series on the left-hand side is a power series *representation* of the function $1/(1 - x)$ in powers of x (or about the point $x = 0$). Note that the representation is valid only in the open interval $(-1, 1)$ even though $1/(1 - x)$ is well defined for all real x except $x = 1$. For $x = -1$ and for $|x| > 1$ the series does not converge, so it cannot represent $1/(1 - x)$ at these points.

The point c is called the **center of convergence** of the power series $\sum_{n=0}^{\infty} a_n(x - c)^n$. Evidently the series converges (to a_0) at $x = c$. (All the terms except possibly the first are 0.) Theorem 1 shows that if the series converges anywhere else, then it converges on an interval (possibly infinite) centered at $x = c$, and it converges absolutely everywhere in that interval except possibly at one or both of the endpoints if the interval is finite. An example of this situation is the series

$$\sum_{n=1}^{\infty} \frac{1}{2^n n} (x - 5)^n,$$

which we discussed in Example 3 of Section 8.4. We showed that the series converged in the interval $[3, 7)$, an interval with center $x = 5$, and that the convergence was absolute on the interval $(3, 7)$ but only conditional at the endpoint $x = 3$.

Theorem 1 Suppose the series $\sum_{n=0}^{\infty} a_n(x_0 - c)^n$ converges, where $x_0 \neq c$. If x satisfies $|x - c| < |x_0 - c|$, that is, if x is closer to c than x_0 is, then $\sum_{n=0}^{\infty} a_n(x - c)^n$ converges absolutely.

Proof Since $\sum_{n=0}^{\infty} a_n(x_0 - c)^n$ converges, $\lim a_n(x_0 - c)^n = 0$ and so $|a_n(x_0 - c)^n| \leq K$ for all n, where K is some constant. (Theorem 1 of Section 8.1.) If $r = |x - c|/|x_0 - c| < 1$, then

$$\sum_{n=0}^{\infty} |a_n(x - c)^n| = \sum_{n=0}^{\infty} |a_n(x_0 - c)^n| \left|\frac{x - c}{x_0 - c}\right|^n \leq K \sum_{n=0}^{\infty} r^n = \frac{K}{1 - r} < \infty.$$

Thus $\sum_{n=0}^{\infty} a_n(x - c)^n$ converges absolutely. □

Theorem 1 shows that the set of values x for which the power series $\sum_{n=0}^{\infty} a_n(x - c)^n$ converges is an interval centered at $x = c$. We call this interval the **interval of convergence** of the power series. It must be of one of the following forms:

i) the isolated point $x = c$,
ii) $[c - R, c + R]$, or $[c - R, c + R)$, or $(c - R, c + R]$, or $(c - R, c + R)$,
iii) the entire real line.

The number R in (ii) is called the **radius of convergence** of the power series. In case (i) we say the radius of convergence is 0; in case (iii) it is infinite.

The radius of convergence, R, can often be found by using the ratio test on the power series:

$$\frac{1}{R} = \lim_{n \to \infty} \left|\frac{a_{n+1}}{a_n}\right|,$$

provided the limit exists. (If $1/R = 0$, then $R = \infty$; if $1/R = \infty$, then $R = 0$.) To see this, note that the series converges absolutely if

$$\rho = \lim_{n\to\infty} \left| \frac{a_{n+1}(x-c)^{n+1}}{a_n(x-c)^n} \right| = |x-c| \lim_{n\to\infty} \left| \frac{a_{n+1}}{a_n} \right| = \frac{|x-c|}{R} < 1,$$

that is, if $|x - c| < R$. Similarly, it diverges if $|x - c| > R$. Thus R is the radius of convergence.

Example 1 Determine the center, radius, and interval of convergence of

$$\sum_{n=0}^{\infty} \frac{(2x-5)^n}{(n^2+1)3^n}.$$

Solution The series can be rewritten

$$\sum_{n=0}^{\infty} \left(\frac{2}{3}\right)^n \frac{1}{n^2+1} \left(x - \frac{5}{2}\right)^n.$$

The center of convergence is $x = 5/2$. The radius of convergence, R, is given by

$$\frac{1}{R} = \lim \frac{\left(\frac{2}{3}\right)^{n+1} \dfrac{1}{(n+1)^2+1}}{\left(\frac{2}{3}\right)^n \dfrac{1}{n^2+1}} = \lim \frac{2}{3} \frac{n^2+1}{(n+1)^2+1} = \frac{2}{3}.$$

That is, $R = 3/2$. The series converges absolutely on $(5/2 - 3/2, 5/2 + 3/2) = (1, 4)$, and it diverges on $(-\infty, 1)$ and on $(4, \infty)$. At $x = 1$ the series is $\sum_{n=0}^{\infty} (-1)^n/(n^2 + 1)$; at $x = 4$ it is $\sum_{n=0}^{\infty} 1/(n^2 + 1)$. Both series converge (absolutely). The interval of convergence of the given power series is therefore $[1, 4]$.

Example 2 Determine the radii of convergence of the series

a) $\displaystyle\sum_{n=0}^{\infty} \frac{x^n}{n!}$, b) $\displaystyle\sum_{n=0}^{\infty} n!\, x^n$.

Solution a) $\dfrac{1}{R} = \lim \dfrac{n!}{(n+1)!} = \lim \dfrac{1}{n+1} = 0$, so $R = \infty$. The series converges for all x.

b) $\dfrac{1}{R} = \lim \dfrac{(n+1)!}{n!} = \lim (n+1) = \infty$, so $R = 0$. The series converges only at $x = 0$.

Algebraic Operations on Power Series

For purposes of the following discussion, we will consider only power series with $x = 0$ as center of convergence, that is, series of the form

$$\sum_{n=0}^{\infty} a_n x^n = a_0 + a_1 x + a_2 x^2 + a_3 x^3 + \cdots.$$

Any properties we demonstrate for such series extend automatically to more general series of the form $\sum_{n=0}^{\infty} a_n(y - y_0)^n$ via the change of variable $x = y - y_0$.

First we observe that series having the same center of convergence can be added or substracted on whatever interval is common to their intervals of convergence.

Theorem 2

$$\sum_{n=0}^{\infty} (a_n + b_n)x^n = \sum_{n=0}^{\infty} a_n x^n + \sum_{n=0}^{\infty} b_n x^n$$

This series on the left-hand side converges whenever both series on the right-hand side converge. If the two series on the right have radii of convergence R_a and R_b, respectively, then the series on the left-hand side has radius of convergence $R \geq \min(R_a, R_b)$ (that is, at least as large as the smaller of R_a and R_b. Similarly,

$$\sum_{n=0}^{\infty} (a_n - b_n)x^n = \sum_{n=0}^{\infty} a_n x^n - \sum_{n=0}^{\infty} b_n x^n,$$

and the series on the left also has radius of convergence not less than $\min(R_a, R_b)$.

Theorem 3 If $\sum_{n=0}^{\infty} a_n x^n$ has radius of convergence R, then so does $\sum_{n=0}^{\infty} (ca_n)x^n$ for any nonzero constant c. Moreover,

$$\sum_{n=0}^{\infty} (ca_n)x^n = c\sum_{n=0}^{\infty} a_n x^n. \square$$

Theorems 2 and 3 are straightforward consequences of Theorem 6 of Section 8.2. No proofs are required.

The situation regarding multiplication and division of power series is more complicated. We will mention only the results and not attempt any proofs of our assertions. You can refer to a textbook in mathematical analysis for more details.

Long multiplication of the form

$$(a_0 + a_1 x + a_2 x^2 + \cdots)(b_0 + b_1 x + b_2 x^2 + \cdots)$$
$$= a_0 b_0 + (a_0 b_1 + a_1 b_0)x + (a_0 b_2 + a_1 b_1 + a_2 b_0)x^2 + \cdots$$

leads one to conjecture the formula

$$\left(\sum_{n=0}^{\infty} a_n x^n\right)\left(\sum_{n=0}^{\infty} b_n x^n\right) = \sum_{n=0}^{\infty} c_n x^n,$$

where

$$c_n = a_0 b_n + a_1 b_{n-1} + \cdots + a_n b_0 = \sum_{j=0}^{n} a_j b_{n-j}.$$

The series $\sum_{n=0}^{\infty} c_n x^n$ is called the **Cauchy product** of the series $\sum_{n=0}^{\infty} a_n x^n$ and $\sum_{n=0}^{\infty} b_n x^n$. The Cauchy product also has radius of convergence at least equal to the lesser of those of its factor series.

Example 3 Since

$$\frac{1}{1 - x} = 1 + x + x^2 + x^3 + \cdots = \sum_{n=0}^{\infty} x^n$$

holds for $-1 < x < 1$, we can determine a power series representation for $1/(1 - x)^2$ by taking the Cauchy product of this series with itself. Since $a_n = b_n = 1$ for $n = 0, 1, 2, \ldots$ we have

$$c_n = \sum_{j=0}^{n} 1 = n + 1$$

and

$$\frac{1}{(1-x)^2} = 1 + 2x + 3x^2 + 4x^3 + \cdots \sum_{n=0}^{\infty} (n+1)x^n,$$

which must also hold for $-1 < x < 1$. The same series can be obtained by direct long multiplication of the series:

$$\begin{array}{rrrrl}
 & 1 + & x + & x^2 + & x^3 + \cdots \\
\times & 1 + & x + & x^2 + & x^3 + \cdots \\
\hline
 & 1 + & x + & x^2 + & x^3 + \cdots \\
 & & x + & x^2 + & x^3 + \cdots \\
 & & & x^2 + & x^3 + \cdots \\
 & & & & x^3 + \cdots \\
 & & & & \quad\ \ \cdots \\
\hline
 & 1 + & 2x + & 3x^2 + & 4x^3 + \cdots .
\end{array}$$

Power series can also be divided by long division, but there is no simple rule for determining the coefficients of the quotient series. The radius of convergence of the quotient series is the least of the three numbers R_1, R_2, and R_3 where R_1 and R_2 are the radii of convergence of the divisor and dividend series and R_3 is the distance from the center of convergence to the nearest complex number where the divisor series has sum equal to 0.

EXERCISES

Determine the center, radius, and interval of convergence of the power series in Exercises 1–10.

1. $\displaystyle\sum_{n=0}^{\infty} \frac{x^{2n}}{\sqrt{n+1}}$

2. $\displaystyle\sum_{n=0}^{\infty} 3^n(x+1)^n$

3. $\displaystyle\sum_{n=1}^{\infty} \frac{1}{n}\left(\frac{x+2}{2}\right)^n$

4. $\displaystyle\sum_{n=1}^{\infty} \frac{(-1)^n}{n^4 2^{2n}} x^n$

5. $\displaystyle\sum_{n=0}^{\infty} n^3(2x-3)^n$

6. $\displaystyle\sum_{n=1}^{\infty} \frac{e^n}{n^3}(4-x)^n$

7. $\displaystyle\sum_{n=0}^{\infty} \frac{(1+5^n)}{n!} x^n$

8. $\displaystyle\sum_{n=2}^{\infty} \frac{1}{\ln n}\left(\frac{2-5x}{3}\right)^{3n}$

9. $\displaystyle\sum_{n=1}^{\infty} \frac{(4x-1)^n}{n^n}$

10.* $\displaystyle\sum_{n=1}^{\infty} \left(1 + \sin\left(\frac{n\pi}{2}\right)\right)\frac{x^n}{n^{3/2}}$

11.* Use multiplication of series to find a power series representation of $1/(1-x)^3$ valid in the interval $(-1, 1)$.

12.* Determine the Cauchy product of the series $1 + x + x^2 + x^3 + \cdots$ and $1 - x + x^2 - x^3 + \cdots$. On what interval and to what function does the product series converge?

13.* Determine the power series expansion of $1/(1 - x)^2$ by formally dividing $1 - 2x + x^2$ into 1.

9.2 DIFFERENTIATION AND INTEGRATION OF POWER SERIES

If a power series has positive radius of convergence it can be differentiated or integrated term by term, and the resulting series will converge to the appropriate derivative or integral of the sum of the original series everywhere except possibly at the endpoints of the interval of convergence of the original series. This very important fact ensures that, for purposes of calculation, power series behave in an essentially similar way to polynomials, the easiest functions to differentiate and integrate. We formalize the differentiation and integration properties of power series in Theorem 4. The proof of this theorem is both lengthy and difficult so we postpone it to the end of this section. Understanding this proof is not as important for our purposes as understanding what the theorem tells us.

Theorem 4 Suppose that

$$f(x) = \sum_{n=0}^{\infty} a_n x^n = a_0 + a_1 x + a_2 x^2 + a_3 x^3 + \cdots$$

where the series converges for $-R < x < R$ for some $R > 0$. Then

a) f is differentiable on $(-R, R)$, and in this interval we can obtain $f'(x)$ by differentiating the series for $f(x)$ term by term:

$$f'(x) = \sum_{n=1}^{\infty} n a_n x^{n-1} = a_1 + 2a_2 x + 3a_3 x^2 + 4a_4 x^3 + \cdots; \text{ and}$$

b) f is integrable over any closed subinterval of $(-R, R)$ and if $|x| < R$, then

$$\int_0^x f(t)\, dt = \sum_{n=0}^{\infty} \frac{a_n}{n+1} x^{n+1} = a_0 x + \frac{a_1}{2} x^2 + \frac{a_2}{3} x^3 + \cdots. \quad \square$$

Together these results imply that the termwise differentiated or integrated series have the same radius of convergence as the given series. In fact, as the following examples illustrate, the interval of convergence of the differentiated series is the same as that of the original series except for the *possible* loss of an endpoint if the original series converges at an endpoint of its interval of convergence. Similarly, the integrated

* Difficult and/or theoretical problems.

series will converge everywhere on the interval of convergence of the original series and possibly at one or both endpoints of that interval even if the original series does not converge at the endpoints.

Being differentiable on $(-R, R)$, where R is the radius of convergence, the sum $f(x)$ of a power series is necessarily continuous on that open interval. If the series happens to converge at either or both of the endpoints $-R$ and R, then f is also continuous (on one side) up to these endpoints.

Theorem 5 The sum of a power series is a continuous function everywhere on the interval of convergence of that series. In particular, if $\sum_{n=0}^{\infty} a_n R^n$ converges for some $R > 0$, then

$$\lim_{x \to R-} \sum_{n=0}^{\infty} a_n x^n = \sum_{n=0}^{\infty} a_n R^n,$$

and if $\sum_{n=0}^{\infty} a_n(-R)^n$ converges, then

$$\lim_{x \to -R+} \sum_{n=0}^{\infty} a_n x^n = \sum_{n=0}^{\infty} a_n(-R)^n. \ \square$$

The proof of this theorem is also postponed to the end of this section.

Example 1 Recall that

$$\sum_{n=0}^{\infty} x^n = 1 + x + x^2 + x^3 + \cdots = \frac{1}{1-x} \qquad \text{if } -1 < x < 1.$$

Differentiating term by term according to Theorem 4(a) we obtain

$$\sum_{n=1}^{\infty} nx^{n-1} = 1 + 2x + 3x^2 + 4x^3 + \cdots = \frac{1}{(1-x)^2} \qquad \text{if } -1 < x < 1.$$

Differentiating again we get

$$\sum_{n=2}^{\infty} n(n-1)x^{n-2} = 2 + 6x + 12x^2 + \cdots = \frac{2}{(1-x)^3} \qquad \text{if } -1 < x < 1.$$

On the other hand, integrating the original geometric series from 0 to x, where $|x| < 1$, we obtain

$$\sum_{n=0}^{\infty} \frac{x^{n+1}}{n+1} = x + \frac{x^2}{2} + \frac{x^3}{3} + \frac{x^4}{4} + \cdots = -\ln(1-x) \qquad \text{if } -1 < x < 1.$$

Observe, however, that this latter series converges (conditionally) at $x = -1$. Since $\ln(1-x)$ is continuous at $x = -1$, we have, by Theorem 5,

$$\ln(1-x) = -x - \frac{x^2}{2} - \frac{x^3}{3} - \frac{x^4}{4} - \cdots = -\sum_{n=1}^{\infty} \frac{x^n}{n} \qquad \text{if } -1 \le x < 1$$

or, replacing x with $-x$,

$$\ln(1+x) = x - \frac{x^2}{2} + \frac{x^3}{3} - \frac{x^4}{4} + \cdots = \sum_{n=1}^{\infty} \frac{(-1)^{n-1}}{n} x^n \qquad \text{if } -1 < x \le 1.$$

In particular, the alternating harmonic series converges to log 2:

$$\ln 2 = 1 - \frac{1}{2} + \frac{1}{3} - \frac{1}{4} + \frac{1}{5} - \cdots = \sum_{n=1}^{\infty} \frac{(-1)^{n-1}}{n}.$$

This would not, however, be a very useful formula for calculating the value of ln 2. (Why not?)

Example 2 Starting again with the geometric series

$$\frac{1}{1-x} = \sum_{n=0}^{\infty} x^n = 1 + x + x^2 + x^3 + x^4 + \cdots \qquad \text{for } -1 < x < 1,$$

let us substitute $-x^2$ in place of x. Note that $-1 < -x^2 \leq 0$ if $-1 < x < 1$, so

$$\frac{1}{1+x^2} = 1 - x^2 + x^4 - x^6 + x^8 - \cdots \qquad \text{for } -1 < x < 1.$$

Integration leads as follows:

$$\begin{aligned} \operatorname{Arctan} x &= \int_0^x \frac{dt}{1+t^2} = \int_0^x (1 - t^2 + t^4 - t^6 + t^8 - \cdots)\, dt \\ &= x - \frac{x^3}{3} + \frac{x^5}{5} - \frac{x^7}{7} + \frac{x^9}{9} - \cdots \\ &= \sum_{n=0}^{\infty} (-1)^n \frac{x^{2n+1}}{2n+1} \qquad \text{if } -1 < x < 1. \end{aligned}$$

However, note that the series also converges (conditionally) at $x = -1$ and 1. Hence the above series representation for Arctan x also holds for these values, by Theorem 5. Letting $x = 1$ we get an interesting series:

$$\frac{\pi}{4} = 1 - \frac{1}{3} + \frac{1}{5} - \frac{1}{7} + \frac{1}{9} - \cdots.$$

Again, however, this would not be a good formula with which to calculate a numerical value of π. (Why not?)

Example 3 Find the sum of the series

$$\sum_{n=1}^{\infty} n^2 x^n = x + 4x^2 + 9x^3 + 16x^4 + \cdots.$$

Solution First use the ratio test to find the radius of convergence:

$$\rho = \lim \left| \frac{(n+1)^2 x^{n+1}}{n^2 x^n} \right| = |x| \lim \left(\frac{n+1}{n} \right)^2 = |x| < 1 \qquad \text{if } -1 < x < 1.$$

The radius of convergence is $R = 1/\rho = 1$. Evidently the series does not converge at $x = 1$ or -1. Let the sum of the series be $f(x)$ on $(-1, 1)$. We have

$$\begin{aligned} f(x) &= x + 4x^2 + 9x^3 + 16x^4 + \cdots \\ &= x(1 + 4x + 9x^2 + 16x^3 + \cdots) \end{aligned}$$

$$= x\frac{d}{dx}(x + 2x^2 + 3x^3 + 4x^4 + \cdots)$$

$$= x\frac{d}{dx}\left(x(1 + 2x + 3x^2 + 4x^3 + \cdots)\right)$$

$$= x\frac{d}{dx}\left(x\frac{d}{dx}(1 + x + x^2 + x^3 + \cdots)\right)$$

$$= x\frac{d}{dx}\left(x\frac{d}{dx}\frac{1}{1-x}\right)$$

$$= x\frac{d}{dx}\left(x\frac{1}{(1-x)^2}\right) = \frac{x(1+x)}{(1-x)^3} \qquad (\text{for } -1 < x < 1).$$

Observe that twice in the above calculation a series was recognized to be, and was replaced by, the derivative of a simpler series.

Example 4 Find a series representation of $f(x) = 1/(2 + x)$ in powers of $x - 1$. What is the interval of convergence of this series?

Solution

$$f(x) = \frac{1}{3 + (x - 1)} = \frac{1}{3}\frac{1}{1 - \left(-\dfrac{x-1}{3}\right)} = \frac{1}{3}\frac{1}{1-u} \qquad \text{where } u = -\frac{x-1}{3}$$

$$= \frac{1}{3}(1 + u + u^2 + u^3 + \cdots)$$

$$= \frac{1}{3}\left(1 - \frac{x-1}{3} + \frac{(x-1)^2}{9} - \frac{(x-1)^3}{27} + \cdots\right) = \sum_{n=0}^{\infty}\frac{(-1)^n(x-1)^n}{3^{n+1}}.$$

This series has radius of convergence 3, center of convergence 1, and interval of convergence $(-2, 4)$. It does not converge at either endpoint.

Proofs of Theorems 4 and 5

These proofs really belong in a course in elementary mathematical analysis. We present them here because the theorems themselves are very important in applications of power series. These proofs can be omitted, but it is suggested that serious mathematics students should at least read them through because they provide some insights into not only the behavior of series but also the nature of an analytical proof.

Proof of Theorem 4(a) We are given that

$$f(x) = \sum_{n=0}^{\infty} a_n x^n = a_0 + a_1x + a_2x^2 + a_3x^3 + \cdots$$

on the interval $(-R, R)$. We want to show that

i) $\sum_{n=1}^{\infty} na_nx^{n-1}$ converges on $(-R, R)$ and

ii) the sum of the series in (i) is $f'(x)$.

First we verify (i). Pick any x in the interval $(-R, R)$, and let r satisfy $|x| < r < R$. Since $\Sigma_{n=0}^{\infty} a_n r^n$ converges, we know that $\{a_n r^n\}$ is a bounded sequence; that is, $|a_n| r^n \le K$ for some constant K. Now

$$\sum_{n=1}^{\infty} |na_n x^{n-1}| = \sum_{n=1}^{\infty} |a_n r^n| \frac{n}{r^n} |x|^{n-1} \le \frac{K}{r} \sum_{n=1}^{\infty} n \left|\frac{x}{r}\right|^{n-1}.$$

This latter series converges by the ratio test since $|x/r| < 1$. Hence the termwise differentiated series $\Sigma_{n=1}^{\infty} na_n x^{n-1}$ converges (absolutely) on $(-R, R)$ to some sum, say $g(x)$:

$$g(x) = \sum_{n=1}^{\infty} na_n x^{n-1} = a_1 + 2a_2 x + 3a_3 x^2 + \cdots \qquad \text{on } (-R, R).$$

To complete the proof (by verifying (ii)) we must show that $f'(x) = g(x)$ on $(-R, R)$. Note that the series defining $g(x)$ could also be written $\Sigma_{n=0}^{\infty} na_n x^{n-1}$, but there is no reason to start with $n = 0$ since the corresponding term is 0.

For x in $(-R, R)$ and h sufficiently small that $x + h$ also lies in that interval, we have

$$\frac{f(x+h) - f(x)}{h} - g(x) = \sum_{n=0}^{\infty} a_n \frac{(x+h)^n - x^n}{h} - \sum_{n=1}^{\infty} na_n x^{n-1}.$$

Now $(x + h)^n - x^n = 0$ if $n = 0$. For $n \ge 1$ we apply the error formula for the tangent-line approximation to $y = x^n$ at x. (See Theorem 6 of Section 4.6.) We obtain

$$\frac{(x+h)^n - x^n}{h} = nx^{n-1} + \frac{n(n-1)}{2} c_n^{n-2} h,$$

where c_n is some number between x and $x + h$ (a different number for each n, of course). Hence

$$\begin{aligned}\frac{f(x+h) - f(x)}{h} - g(x) &= \sum_{n=1}^{\infty} a_n \left(nx^{n-1} + \frac{n(n-1)}{2} c_n^{n-2} h - nx^{n-1} \right) \\ &= \frac{h}{2} \sum_{n=2}^{\infty} n(n-1) a_n c_n^{n-2}.\end{aligned}$$

If we can show that $|\Sigma_{n=2}^{\infty} n(n-1) a_n c_n^{n-2}| \le M$, where M is a constant independent of h for sufficiently small h, then the equation above implies that

$$f'(x) - g(x) = \lim_{h \to 0} \left(\frac{f(x+h) - f(x)}{h} - g(x) \right) = 0,$$

so that $f'(x) = g(x)$ as required.

Pick numbers r and s such that $|x| < s < r < R$. If h is small enough in absolute value that $|x + h| \le s$, then $|c_n| \le s$ for each n. Hence

$$\begin{aligned}\sum_{n=2}^{\infty} |n(n-1) a_n c_n^{n-2}| &\le \sum_{n=2}^{\infty} |a_n r^n| \frac{n(n-1)}{r^2} \left(\frac{s}{r}\right)^{n-2} \\ &\le \frac{K}{r^2} \sum_{n=2}^{\infty} n(n-1) \left(\frac{s}{r}\right)^{n-2} = M < \infty.\end{aligned}$$

(We have used the estimate $|a_n r^n| \le K$ obtained in the earlier part of the proof.) The last series above converges by the ratio test, since $s/r < 1$. This completes the proof of part (a). □

Proof of Theorem 4(b) Given that

$$f(x) = \sum_{n=0}^{\infty} a_n x^n = a_0 + a_1 x + a_2 x^2 + a_3 x^3 + \cdots$$

holds for x in the interval $(-R, R)$, we want to show that for any such x,

$$\int_0^x f(t)\,dt = \sum_{n=0}^{\infty} \frac{a_n}{n+1} x^{n+1} = a_0 x + \frac{a_1}{2} x^2 + \frac{a_2}{3} x^3 + \cdots.$$

Observe that the coefficients $a_n/(n + 1)$ of the integrated series are smaller in absolute value than those of the original series. Therefore, the integrated series converges (absolutely) to a sum, say $h(x)$, at least on the interval $(-R, R)$. Using part (a) we conclude that

$$h'(x) = \sum_{n=0}^{\infty} a_n x^n = f(x).$$

Since $h(0) = 0$, we have

$$\int_0^x f(t)\,dt = h(t)\Big|_0^x = h(x),$$

as required. This completes the proof of part (b). □

If the original series has radius of convergence R, then the termwise integrated series cannot have radius of convergence any larger than R because part (a) could then be used to show that the original series also had the larger radius of convergence, a contradiction. Hence the original series, the termwise differentiated series, and the termwise integrated series all have the same radius of convergence.

Proof of Theorem 5 Suppose that $\Sigma_{n=0}^{\infty} a_n R^n$ converges for some $R > 0$. Let $f(x)$ be the sum of the series $\Sigma_{n=0}^{\infty} a_n x^n$ wherever it converges (that is, at least on the interval $(-R, R]$). We want to show that

$$\lim_{x \to R-} f(x) = f(R) = \sum_{n=0}^{\infty} a_n R^n.$$

Let $s_n = a_0 + a_1 R + a_2 R^2 + \cdots + a_n R^n$ for $n = 0, 1, 2, \ldots$. Then $s_k - s_{k-1} = a_k R^k$ for $k = 1, 2, 3, \ldots$. Now

$$\sum_{k=0}^{n} a_k x^k = \sum_{k=0}^{n} a_k R^k \left(\frac{x}{R}\right)^k = a_0 + \sum_{k=1}^{n} (s_k - s_{k-1}) \left(\frac{x}{R}\right)^k$$

$$= a_0 + \sum_{k=1}^{n} s_k \left(\frac{x}{R}\right)^k - \sum_{k=1}^{n} s_{k-1} \left(\frac{x}{R}\right)^k \qquad \text{in the second sum replace } k \text{ with } k+1$$

$$= a_0 + \sum_{k=1}^{n} s_k\left(\frac{x}{R}\right)^k - \sum_{k=0}^{n-1} s_k\left(\frac{x}{R}\right)^{k+1}$$

$$= \left(1 - \frac{x}{R}\right)\sum_{k=0}^{n-1} s_k\left(\frac{x}{R}\right)^k + s_n\left(\frac{x}{R}\right)^n.$$

For any x such that $|x| < R$ we can take the limit as $n \to \infty$. Since $\{s_n\}$ converges to $f(R)$ it is bounded, and so the last term above approaches 0. Hence

$$f(x) = \lim_{n\to\infty} \sum_{k=0}^{n} a_k x^k = \left(1 - \frac{x}{R}\right)\sum_{k=0}^{\infty} s_k\left(\frac{x}{R}\right)^k.$$

Now $\Sigma_{k=0}^{\infty} (x/R)^k = 1/(1 - (x/R))$ and so

$$\left(1 - \frac{x}{R}\right)\sum_{k=0}^{\infty}\left(\frac{x}{R}\right)^k = 1.$$

Thus

$$f(x) - f(R) = \left(1 - \frac{x}{R}\right)\sum_{k=0}^{\infty}(s_k - f(R))\left(\frac{x}{R}\right)^k.$$

Observe that for fixed $N > 1$ we can estimate

$$|f(x) - f(R)| \le \left|1 - \frac{x}{R}\right|\left(\sum_{k=0}^{N}|s_k - f(R)|\left|\frac{x}{R}\right|^k + \sum_{k=N+1}^{\infty}|s_k - f(R)|\left|\frac{x}{R}\right|^k\right)$$

$$\le \left|1 - \frac{x}{R}\right|\sum_{k=0}^{N}|s_k - f(R)| + 2\sum_{k=N+1}^{\infty}|s_k - f(R)|\left|\frac{x}{R}\right|^k.$$

Let ϵ be any given positive number. We can ensure that the second sum on the right is less than $\epsilon/2$ by choosing N large enough, because $\{s_k - f(R)\}$ is bounded (it converges to 0) and $\Sigma\,(x/R)^k$ is a convergent geometric series. Having so chosen N, we can ensure that the first term on the right is also less than $\epsilon/2$ by choosing x close enough to R. Hence we can ensure that $|f(x) - f(R)| < \epsilon$ by choosing x less than, but sufficiently close to, R. That is, $\lim_{x\to R-} f(x) = f(R)$, as required. □

EXERCISES

Starting with the power series representation

$$\frac{1}{1-x} = 1 + x + x^2 + x^3 + \cdots \qquad (-1 < x < 1)$$

1 - 13

determine power series representations for the functions indicated in Exercises 1–10. On what interval is each representation valid?

1. $\dfrac{1}{2-x}$ in powers of x

2. $\dfrac{1}{(2-x)^2}$ in powers of x

3. $\dfrac{1}{1+2x}$ in powers of x

4. $\dfrac{x}{(1+2x)^2}$ in powers of x

5. $\ln(2 - x)$ in powers of x

6. $1/x$ in powers of $x - 1$

7. $1/x^2$ in powers of $x - 1$

8. $\ln x$ in powers of $x - 1$

9. $\dfrac{1 - x}{1 + x}$ in powers of x

10. $\dfrac{x}{1 - x^2}$ in powers of x

Determine the interval of convergence and the sum of the series in Exercises 11–15.

11. $1 - 4x + 16x^2 - 64x^3 + \cdots = \sum_{n=0}^{\infty} (-1)^n (4x)^n$

12.* $3 + 4x + 5x^2 + 6x^3 + \cdots = \sum_{n=0}^{\infty} (n + 3)x^n$

13.* $\dfrac{1}{3} + \dfrac{x}{4} + \dfrac{x^2}{5} + \dfrac{x^3}{6} + \cdots = \sum_{n=0}^{\infty} \dfrac{x^n}{n + 3}$

14.* $1 \times 3 - 2 \times 4x + 3 \times 5x^2 - 4 \times 6x^3 + \cdots = \sum_{n=0}^{\infty} (-1)^n (n + 1)(n + 3)x^n$

15.* $2 + 4x^2 + 6x^4 + 8x^6 + 10x^8 + \cdots = \sum_{n=0}^{\infty} 2(n + 1)x^{2n}$

9.3 TAYLOR AND MACLAURIN SERIES

Definition 2

If $f(x)$ is infinitely often differentiable at $x = c$, that is, if $f^{(k)}(c)$ exists for $k = 0, 1, 2, 3, \cdots$ (recall that $f^{(0)} = f$), then the series

$$\sum_{k=0}^{\infty} \frac{f^{(k)}(c)}{k!}(x - c)^k$$

$$= f(c) + f'(c)(x - c) + \frac{f''(c)}{2!}(x - c)^2 + \frac{f^{(3)}(c)}{3!}(x - c)^3 + \cdots$$

is called the **Taylor series of f about $x = c$** (or the *Taylor series of f in powers of $x - c$*). If $c = 0$, the term **Maclaurin series** is usually used in place of Taylor series.

Note that the definition of Taylor series makes no requirement that the series should converge anywhere except at the obvious point $x = c$. The series exists provided all the derivatives of f exist at $x = c$; in practice this means that each derivative must exist in an open interval containing $x = c$. However, the series may have radius of convergence 0, and even if the radius of convergence is greater than 0 the series may converge to some function other than $f(x)$.

Definition 3

If the Taylor series for $f(x)$ in powers of $x - c$ converges to $f(x)$ in an interval with positive radius centered at $x = c$, then we say that f is **analytic at** $x = c$.

Most of the elementary functions encountered in calculus are analytic wherever they are infinitely often differentiable. However, an example of a function that does not have this property is given in Exercise 38 at the end of this section. Thus not every infinitely differentiable function is analytic. On the other hand, whenever a power series has a positive radius of convergence, its sum $f(x)$ is analytic at the center of convergence and the given series is the Taylor series of $f(x)$ about that point.

Theorem 6 Suppose the series

$$f(x) = \sum_{n=0}^{\infty} a_n(x - c)^n = a_0 + a_1(x - c) + a_2(x - c)^2 + \cdots$$

has radius of convergence $R > 0$. Then

$$a_k = \frac{f^{(k)}(c)}{k!} \qquad \text{for } k = 0, 1, 2, 3, \ldots,$$

and so f is analytic at $x = c$. In particular, a given function can have at most one power series representation about any particular point. If

$$f(x) = \sum_{k=0}^{\infty} a_k(x - c)^k = \sum_{k=0}^{\infty} b_k(x - c)^k,$$

then $a_k = b_k$ for $k = 0, 1, 2, \ldots$.

Proof Successive applications of Theorem 4(a) of Section 9.2 (suitably reformulated for powers of $x - c$) lead to

$$f'(x) = \sum_{n=1}^{\infty} na_n(x - c)^{n-1} = a_1 + 2a_2(x - c) + 3a_3(x - c)^2 + \cdots$$

$$f''(x) = \sum_{n=2}^{\infty} n(n - 1)a_n(x - c)^{n-2} = 2a_2 + 6a_3(x - c) + 12a_4(x - c)^2 + \cdots$$

$$\vdots$$

$$f^{(k)}(x) = \sum_{n=k}^{\infty} n(n - 1)(n - 2)\cdots(n - k + 1)a_n(x - c)^{n-k}$$

$$= k!a_k + \frac{(k + 1)!}{1!}a_{k+1}(x - c) + \frac{(k + 2)!}{2!}a_{k+2}(x - c)^2 + \cdots,$$

each series converging absolutely on $(c - R, c + R)$. Setting $x = c$ we obtain $f^{(k)}(c) = k!a_k$, as required. □

Maclaurin Series for the Elementary Functions

In the previous section we obtained the following three Maclaurin series. Each of the corresponding functions is analytic at $x = 0$. (In fact, each is analytic wherever it is defined, though we cannot prove this yet.)

$$\frac{1}{1-x} = \sum_{n=0}^{\infty} x^n = 1 + x + x^2 + x^3 + \cdots \qquad (-1 < x < 1)$$

$$\ln(1+x) = \sum_{n=1}^{\infty} \frac{(-1)^{n-1}}{n} x^n = x - \frac{x^2}{2} + \frac{x^3}{3} - \frac{x^4}{4} + \cdots \qquad (-1 < x \le 1)$$

$$\operatorname{Arctan} x = \sum_{n=1}^{\infty} \frac{(-1)^{n-1}}{2n-1} x^{2n-1} = x - \frac{x^3}{3} + \frac{x^5}{5} - \frac{x^7}{7} + \cdots \qquad (-1 \le x \le 1)$$

We now extend this list by developing Maclaurin series for e^x, e^{-x}, $\cosh x$, $\sinh x$, $\cos x$, and $\sin x$.

Example 1 We show that

$$e^x = \sum_{n=0}^{\infty} \frac{x^n}{n!} = 1 + x + \frac{x^2}{2!} + \frac{x^3}{3!} + \cdots \qquad \text{(for all real } x\text{).}$$

The series on the right-hand side converges for all real numbers x by the ratio test; the series has infinite radius of convergence. Let its sum be $f(x)$:

$$f(x) = \sum_{n=0}^{\infty} \frac{x^n}{n!} = 1 + x + \frac{x^2}{2!} + \frac{x^3}{3!} + \cdots .$$

By Theorem 4(a) of the previous section we have

$$f'(x) = \sum_{n=1}^{\infty} \frac{nx^{n-1}}{n!} = \sum_{n=1}^{\infty} \frac{x^{n-1}}{(n-1)!} = \sum_{k=0}^{\infty} \frac{x^k}{k!} = 1 + x + \frac{x^2}{2} + \cdots = f(x).$$

By Example 6 of Section 3.5 we must have $f(x) = Ce^x$ for all real x. Evidently $C = f(0) = 1 + 0 + 0 + \cdots = 1$. Thus $f(x) = e^x$ as claimed.

Example 2 a) Replacing x with $-x$ in the series for e^x we obtain

$$e^{-x} = \sum_{n=0}^{\infty} (-1)^n \frac{x^n}{n!} = 1 - x + \frac{x^2}{2!} - \frac{x^3}{3!} + \frac{x^4}{4!} - \cdots \qquad \text{(for all real } x\text{).}$$

b) The Maclaurin series for e^x and e^{-x} can be added and the sum divided by 2 to produce the Maclaurin series for $\cosh x$. The $(n+1)$st term is

$$\frac{1+(-1)^n}{2} \frac{x^n}{n!} = \begin{cases} x^n/n! & \text{if } n \text{ is even} \\ 0 & \text{if } n \text{ is odd.} \end{cases}$$

Thus

$$\cosh x = \frac{e^x + e^{-x}}{2} = \sum_{n=0}^{\infty} \frac{x^{2n}}{(2n)!} = 1 + \frac{x^2}{2!} + \frac{x^4}{4!} + \cdots \qquad \text{(for all real } x\text{).}$$

Similarly, subtracting the series for e^x and e^{-x} leads to

$$\sinh x = \frac{e^x - e^{-x}}{2} = \sum_{n=0}^{\infty} \frac{x^{2n+1}}{(2n+1)!} = x + \frac{x^3}{3!} + \frac{x^5}{5!} + \cdots$$

(for all real x).

Example 3 We show that

$$\cos x = \sum_{n=0}^{\infty} (-1)^n \frac{x^{2n}}{(2n)!} = 1 - \frac{x^2}{2!} + \frac{x^4}{4!} - \frac{x^6}{6!} + \cdots \qquad \text{(for all real } x\text{)}$$

$$\sin x = \sum_{n=0}^{\infty} (-1)^n \frac{x^{2n+1}}{(2n+1)!} = x - \frac{x^3}{3!} + \frac{x^5}{5!} - \frac{x^7}{7!} + \cdots \qquad \text{(for all real } x\text{).}$$

Observe that both series converge absolutely for all real x by the ratio test. Let their sums be $f(x)$ and $g(x)$, respectively:

$$f(x) = \sum_{n=0}^{\infty} (-1)^n \frac{x^{2n}}{(2n)!}, \qquad g(x) = \sum_{n=0}^{\infty} (-1)^n \frac{x^{2n+1}}{(2n+1)!}.$$

Differentiating these series we obtain

$$g'(x) = \frac{d}{dx}\left(x - \frac{x^3}{3!} + \frac{x^5}{5!} - \frac{x^7}{7!} + \cdots\right)$$

$$= 1 - \frac{x^2}{2!} + \frac{x^4}{4!} - \frac{x^6}{6!} + \cdots = f(x)$$

$$g''(x) = f'(x) = \frac{d}{dx}\left(1 - \frac{x^2}{2!} + \frac{x^4}{4!} - \frac{x^6}{6!} + \frac{x^8}{8!} - \cdots\right)$$

$$= -x + \frac{x^3}{3!} - \frac{x^5}{5!} + \frac{x^7}{7!} - \cdots = -g(x).$$

Since $g''(x) + g(x) = 0$ for all real x, we can conclude that

$$g(x) = A \cos x + B \sin x.$$

(See the discussion of simple harmonic motion in Section 3.2 and also Exercises 47–51 of that section.) But $0 = g(0) = A$ and $1 = f(0) = g'(0) = B$, so $g(x) = \sin x$, as desired. Hence also $f(x) = g'(x) = \cos x$.

The nine Maclaurin series quoted above are very useful and you should commit them to memory.

Observe the similarity between the series for $\sin x$ and $\sinh x$ and between those for $\cos x$ and $\cosh x$. If we were to allow complex numbers (numbers of the form $z = x + iy$ where $i^2 = -1$ and x and y are real) as arguments for our functions, and if we were to demonstrate that our operations on series could be extended to series of complex numbers, we could see that $\cos x = \cosh(ix)$ and $\sin x = -i\sinh(ix)$. In fact,

$$e^{ix} = \cos x + i\sin x, \qquad e^{-ix} = \cos x - i\sin x$$

and so

$$\cos x = \frac{e^{ix} + e^{-ix}}{2}, \qquad \sin x = \frac{e^{ix} - e^{-ix}}{2i}.$$

Such formulas are established in courses and textbooks on functions of a complex variable; from the complex point of view the trigonometric and exponential functions are just different manifestations of the same basic function, a complex exponential $e^z = e^{x+iy}$. We content ourselves here with having mentioned the interesting relationship above and invite the reader to verify them formally by calculating with series. (Such formal calculations do not, of course, constitute a proof since we have not established the various rules for operating with complex numbers.

Other Maclaurin and Taylor Series

We now show how other Maclaurin and Taylor series can be obtained from the ones we already know by various operations. We have already seen examples above and in the previous section.

Example 4 Obtain Maclaurin series for the following:

a) $\dfrac{\sin(x^2)}{x}$, b) $\sin^2 x$, c) $e^{-x^2/3}$.

Solution a)

$$\frac{\sin(x^2)}{x} = \frac{1}{x}\left(x^2 - \frac{(x^2)^3}{3!} + \frac{(x^2)^5}{5!} - \cdots\right)$$

$$= x - \frac{x^5}{3!} + \frac{x^9}{5!} - \cdots = \sum_{n=0}^{\infty}(-1)^n\frac{x^{4n+1}}{(2n+1)!} \qquad \text{(for all real } x\text{)}.$$

Note that $(\sin(x^2))/x$ is not defined at $x = 0$ but does have a limit (0) as x approaches 0. The series converges to $(\sin(x^2))/x$ for all $x \neq 0$, and to that limit at $x = 0$.

b)

$$\sin^2 x = \frac{1 - \cos 2x}{2} = \frac{1}{2} - \frac{1}{2}\left(1 - \frac{(2x)^2}{2!} + \frac{(2x)^4}{4!} - \cdots\right)$$

$$= \frac{1}{2}\left(\frac{(2x)^2}{2!} - \frac{(2x)^4}{4!} + \frac{(2x)^6}{6!} - \cdots\right)$$

$$= \sum_{n=0}^{\infty}(-1)^n\frac{2^{2n+1}}{(2n+2)!}x^{2n+2} \qquad \text{(for all real } x\text{)}$$

c)

$$e^{-x^2/3} = 1 - \frac{x^2}{3} + \frac{1}{2!}\left(\frac{x^2}{3}\right)^2 - \frac{1}{3!}\left(\frac{x^2}{3}\right)^3 + \cdots$$

$$= \sum_{n=0}^{\infty}(-1)^n\frac{1}{3^n n!}x^{2n} \qquad \text{(for all real } x\text{)}$$

Sometimes it is quite difficult, if not impossible, to obtain all the terms (that is, the general term) of a Maclaurin or Taylor series. In such cases it is usually possible to obtain the first few terms before the calculations get too cumbersome. Had we attempted to do Example 4(b) by multiplying the series for sin x by itself we might have found ourselves in this bind. Other examples occur when it is necessary to substitute one series into another or divide one by another.

Example 5 Obtain the first three nonzero terms of the Maclaurin series for

a) ln cos x, b) tan x.

Solution a)

$$\begin{aligned}
\ln\cos x &= \ln\left(1+\left(-\frac{x^2}{2!}+\frac{x^4}{4!}-\frac{x^6}{6!}+\cdots\right)\right)\\
&= \left(-\frac{x^2}{2!}+\frac{x^4}{4!}-\frac{x^6}{6!}+\cdots\right)-\frac{1}{2}\left(-\frac{x^2}{2!}+\frac{x^4}{4!}-\frac{x^6}{6!}+\cdots\right)^2\\
&\quad +\frac{1}{3}\left(-\frac{x^2}{2!}+\frac{x^4}{4!}-\frac{x^6}{6!}+\cdots\right)^3-\cdots\\
&= -\frac{x^2}{2}+\frac{x^4}{24}-\frac{x^6}{720}+\cdots-\frac{1}{2}\left(\frac{x^4}{4}-\frac{x^6}{24}+\cdots\right)\\
&\quad +\frac{1}{3}\left(-\frac{x^6}{8}+\cdots\right)-\cdots\\
&= -\frac{x^2}{2}-\frac{x^4}{12}-\frac{x^6}{45}-\cdots
\end{aligned}$$

Note that at each stage of the calculation we kept only enough terms to ensure that we could get all the terms up to x^6. Being an even function, ln cos x has only even powers in its Maclaurin series. We cannot find the general term of this series, and only with considerable computational effort can we find many more terms of the series than we have already found. One can also try to calculate terms by using the formula $a_k = f^{(k)}(0)/k!$ but even this is cumbersome after the first few values of k.

b) tan x = (sin x)/(cos x). We can obtain the first three terms of the Maclaurin series for tan x by long division of the series for cos x into that for sin x:

$$\begin{array}{r|l}
 & x + x^3/3 + (2/15)x^5 + \cdots \\ \hline
1-\dfrac{x^2}{2}+\dfrac{x^4}{24}-\cdots & x - x^3/6 + x^5/120 - \cdots \\
 & x - x^3/2 + x^5/24 - \cdots \\ \hline
 & \quad x^3/3 - x^5/30 + \cdots \\
 & \quad x^3/3 - x^5/6 + \cdots \\ \hline
 & \qquad (2/15)x^5 - \cdots \\
 & \qquad (2/15)x^5 - \cdots \\ \hline
 & \qquad\qquad \cdots
\end{array}$$

$$\tan x = x + \frac{1}{3}x^3 + \frac{2}{15}x^5 + \cdots.$$

Again we cannot easily find all the terms of the series. This Maclaurin series for $\tan x$ converges for $|x| < \pi/2$, but we cannot demonstrate this fact by the techniques we have at our disposal now. Note that the series for $\tan x$ could also have been derived from that of $\ln \cos x$ obtained in part (a) since $\tan x = -(d/dx) \ln \cos x$.

The elementary Maclaurin series can also be used to obtain Taylor series about some point other than 0.

Example 6 Find the Taylor series for

a) e^x in powers of $x - c$,

b) $\cos x$ about the point $x = \pi/3$.

Solution a)

$$\begin{aligned} e^x &= e^{x-c+c} = e^c e^{x-c} \\ &= e^c\left(1 + (x - c) + \frac{(x-c)^2}{2!} + \frac{(x-c)^3}{3!} + \cdots\right) \\ &= \sum_{n=0}^{\infty} \frac{e^c}{n!}(x - c)^n \end{aligned}$$

This representation is valid for all x, and e^x is analytic at every point of the real line.

b)

$$\begin{aligned} \cos x &= \cos\left(x - \frac{\pi}{3} + \frac{\pi}{3}\right) = \cos\left(x - \frac{\pi}{3}\right)\cos\frac{\pi}{3} - \sin\left(x - \frac{\pi}{3}\right)\sin\frac{\pi}{3} \\ &= \frac{1}{2}\left(1 - \frac{\left(x - \frac{\pi}{3}\right)^2}{2!} + \frac{\left(x - \frac{\pi}{3}\right)^4}{4!} - \cdots\right) \\ &\quad - \frac{\sqrt{3}}{2}\left(\left(x - \frac{\pi}{3}\right) - \frac{\left(x - \frac{\pi}{3}\right)^3}{3!} + \cdots\right) \\ &= \frac{1}{2} - \frac{\sqrt{3}}{2}\left(x - \frac{\pi}{3}\right) - \frac{1}{2}\frac{1}{2!}\left(x - \frac{\pi}{3}\right)^2 + \frac{\sqrt{3}}{2}\frac{1}{3!}\left(x - \frac{\pi}{3}\right)^3 \\ &\quad + \frac{1}{2}\frac{1}{4!}\left(x - \frac{\pi}{3}\right)^4 - \cdots . \end{aligned}$$

This series representation is valid for all x. A similar calculation would enable us to expand $\cos x$ or $\sin x$ in powers of $x - c$ for any real c, so both functions are analytic at every point of the real line.

EXERCISES

Find Maclaurin series representations for the functions in Exercises 1–15. For what values of x is each representation valid?

1. e^{3x+1} **2.** $\cos(2x^3)$ **3.** $\sin(x - \pi/4)$

4. $\cos(2x - \pi)$

5. $\cosh x - \cos x$

6. $x^2 \sinh(3x)$

7. $\cos^2(x/2)$

8. $\sin x \cos x$

9. $e^{2x} - \sin(3x) - 1$

10. $\text{Arctan}(5x^2)$

11. $\ln(1 - x)$

12. $\dfrac{1 + x^3}{1 + x^2}$

13. $\ln(2 + x^2)$

14. $\ln \dfrac{1 - x}{1 + x}$

15. $(e^{2x^2} - 1)/x^2$

Find the first three nonzero terms in the Maclaurin series for the functions in Exercises 16–18.

16. $\sec x$

17. $\sec x \tan x$

18. $1/(1 + e^x)$

Find the appropriate Taylor series in Exercises 19–25. Where is each series representation valid?

19. for $f(x) = e^{-2x}$ about the point $x = -1$

20. for $f(x) = \sin x$ about the point $x = \pi/2$

21. for $f(x) = \cos x$ in powers of $x - \pi$

22. for $f(x) = \ln x$ in powers of $x - 3$

23. for $f(x) = \sin x - \cos x$ about the point $x = \pi/4$

24. for $f(x) = \cos^2 x$ about the point $x = \pi/8$

25. for $f(x) = 1/x^2$ in powers of $x + 2$

26. Find the first three nonzero terms of the Maclaurin series for $\text{Arctan}(e^x - 1)$.

27. Find the first three nonzero terms of the Maclaurin series for $e^{\text{Arctan}\, x} - 1$.

28. Use the fact that $(\sqrt{1 + x})^2 = 1 + x$ to find the first three nonzero terms of the Maclaurin series for $\sqrt{1 + x}$.

29. Does $\csc x$ have a Maclaurin series? Why? Find the first three nonzero terms of the Taylor series for $\csc x$ about the point $x = \pi/2$.

Find the sums of the series in Exercises 30–33.

30. $1 + x^2 + \dfrac{x^4}{2!} + \dfrac{x^6}{3!} + \dfrac{x^8}{4!} + \cdots$

31. $x^3 - \dfrac{x^9}{3! \times 4} + \dfrac{x^{15}}{5! \times 16} - \dfrac{x^{21}}{7! \times 64} + \dfrac{x^{27}}{9! \times 256} - \cdots$

32. $1 + \dfrac{x^2}{3!} + \dfrac{x^4}{5!} + \dfrac{x^6}{7!} + \dfrac{x^8}{9!} + \cdots$

33. $1 + \dfrac{1}{2 \times 2!} + \dfrac{1}{4 \times 3!} + \dfrac{1}{8 \times 4!} + \cdots$

34. What is the Maclaurin series for the function $P(x) = 1 + x + x^2$?

35. What is the Taylor series for the polynomial $P(x)$ of Exercise 34 about the point $x = 1$?

36.* Verify for direct calculation that $f(x) = 1/x$ is analytic at $x = a$ for every $a \neq 0$.

37.* Verify by direct calculation that $\ln x$ is analytic at $x = a$ for every $a > 0$.

38.* Review Exercise 50 of Section 4.2. It shows that the function

$$f(x) = \begin{cases} e^{-1/x^2} & \text{if } x \neq 0 \\ 0 & \text{if } x = 0 \end{cases}$$

is differentiable infinitely often at every point of the real line, and $f^{(k)}(0) = 0$ for every positive integer k. What is the Maclaurin series for $f(x)$? What is the interval of convergence of this Maclaurin series? On what interval does the series covnerge to $f(x)$? Is f analytic at $x = 0$?

39.* By direct manipulation of the Maclaurin series for e^x and e^y show that $e^x e^y = e^{x+y}$.

9.4 APPLICATIONS OF TAYLOR AND MACLAURIN SERIES

Approximating the Values of Functions

Partial sums of Taylor and Maclaurin series can be used as polynomial approximations to more complicated functions. We illustrate the methods in the following examples. (The procedures whereby electronic calculators and computers evaluate (approximate) the transcendental functions are similar to these.)

Example 1 Use Maclaurin series to find $\sqrt{e}$ correct to four decimal places, that is, with error less than 0.00005 in absolute value.

Solution

$$\sqrt{e} = e^{1/2} = 1 + \frac{1}{2} + \frac{1}{2!}\frac{1}{2^2} + \frac{1}{3!}\frac{1}{2^3} + \cdots$$

$$\simeq 1 + \frac{1}{2} + \frac{1}{2!}\frac{1}{2^2} + \frac{1}{3!}\frac{1}{2^3} + \cdots + \frac{1}{(n-1)!}\frac{1}{2^{n-1}} = s_n$$

The error in this approximation can be bounded geometrically:

$$0 < \sqrt{e} - s_n = \frac{1}{n!}\frac{1}{2^n} + \frac{1}{(n+1)!}\frac{1}{2^{n+1}} + \frac{1}{(n+2)!}\frac{1}{2^{n+2}} + \cdots$$

$$= \frac{1}{n!}\frac{1}{2^n}\left(1 + \frac{1}{2(n+1)} + \frac{1}{2^2}\frac{1}{(n+2)(n+1)} + \cdots\right)$$

$$< \frac{1}{n!}\frac{1}{2^n}\left(1 + \frac{1}{2(n+1)} + \left(\frac{1}{2(n+1)}\right)^2 + \cdots\right)$$

$$= \frac{1}{n!}\frac{1}{2^n}\frac{1}{1 - \dfrac{1}{2(n+1)}} = \frac{2(n+1)}{(2n+1)n!\,2^n}.$$

We want this error less than $0.00005 = 1/20{,}000$. Thus we want

$$\frac{2^n n!(2n+1)}{2n+2} > 20{,}000.$$

We can get by with $n = 6$. Thus

$$\sqrt{e} \simeq 1 + \frac{1}{2} + \frac{1}{8} + \frac{1}{48} + \frac{1}{384} + \frac{1}{3840} \simeq 1.6487,$$ rounded to four decimal places.

Example 2 Find $\cos 43°$ with error less than 1/10,000.

Solution i) We could proceed using Maclaurin series:

$$\cos 43^\circ = \cos\frac{43\pi}{180} = 1 - \frac{1}{2!}\left(\frac{43\pi}{180}\right)^2 + \frac{1}{4!}\left(\frac{43\pi}{180}\right)^4 - \cdots.$$

Now $43\pi/180 = 0.75049\ldots < 1$, so the series above evidently satisfies the conditions for the alternating series test. If we truncate the series after the term

$$(-1)^{n-1}\frac{1}{(2n-2)!}\left(\frac{43\pi}{180}\right)^{2n-2},$$

then the error E will satisfy

$$|E| \le \frac{1}{(2n)!}\left(\frac{43\pi}{180}\right)^{2n} < \frac{1}{(2n)!}.$$

The error will not exceed 1/10,000 if $(2n)! > 10{,}000$, so $n = 4$ will do.

$$\cos 43^\circ \simeq 1 - \frac{1}{2!}\left(\frac{43\pi}{180}\right)^2 + \frac{1}{4!}\left(\frac{43\pi}{180}\right)^4 - \frac{1}{6!}\left(\frac{43\pi}{180}\right)^6 \simeq 0.73135\ldots$$

ii) Since 43° is close to 45° we can do a bit better by using the Taylor series about $x = \pi/4$ instead of the Maclaurin series.

$$\begin{aligned}\cos 43^\circ &= \cos\left(\frac{\pi}{4} - \frac{\pi}{90}\right) = \cos\frac{\pi}{4}\cos\frac{\pi}{90} + \sin\frac{\pi}{4}\sin\frac{\pi}{90} \\ &= \frac{1}{\sqrt{2}}\left(\left(1 - \frac{1}{2!}\left(\frac{\pi}{90}\right)^2 + \frac{1}{4!}\left(\frac{\pi}{90}\right)^4 - \cdots\right) + \left(\frac{\pi}{90} - \frac{1}{3!}\left(\frac{\pi}{90}\right)^3 + \cdots\right)\right)\end{aligned}$$

Since

$$\frac{1}{4!}\left(\frac{\pi}{90}\right)^4 < \frac{1}{3!}\left(\frac{\pi}{90}\right)^3 < \frac{1}{20000},$$

we need only the first two terms of the first series and the first term of the second series:

$$\cos 43^\circ \simeq \frac{1}{\sqrt{2}}\left(1 + \frac{\pi}{90} - \frac{1}{2}\left(\frac{\pi}{90}\right)^2\right) \simeq 0.731358\ldots.$$

(In fact, $\cos 43^\circ = 0.7313537\ldots$.)

When finding approximate values of functions it is best, whenever possible, to use a power series about a point as close as possible to the point where the approximation is desired.

Functions Defined by Integrals

As we saw in Chapter 6, many functions expressible as simple combinations of elementary functions cannot be antidifferentiated by elementary techniques; their antiderivatives are not simple combinations of elementary functions. We can, however, often find the Taylor series for the antiderivatives of such functions and hence approximate definite integrals not otherwise obtainable.

Example 3 Find the Taylor series for

$$E(x) = \int_0^x e^{-t^2}\,dt$$

and use it to evaluate $E(1)$ to three-decimal-place accuracy.

Solution We have

$$\begin{aligned}
E(x) &= \int_0^x \left(1 - t^2 + \frac{t^4}{2!} - \frac{t^6}{3!} + \frac{t^8}{4!} - \cdots\right) dt \\
&= \left(t - \frac{t^3}{3} + \frac{t^5}{5 \times 2!} - \frac{t^7}{7 \times 3!} + \frac{t^9}{9 \times 4!} - \cdots\right)\Bigg|_0^x \\
&= x - \frac{x^3}{3} + \frac{x^5}{5 \times 2!} - \frac{x^7}{7 \times 3!} + \frac{x^9}{9 \times 4!} - \cdots = \sum_{n=0}^{n} (-1)^n \frac{x^{2n+1}}{(2n+1)n!}.
\end{aligned}$$

Now we have

$$\begin{aligned}
E(1) &= 1 - \frac{1}{3} + \frac{1}{5 \times 2!} - \frac{1}{7 \times 3!} + \cdots \\
&\simeq 1 - \frac{1}{3} + \frac{1}{5 \times 2!} - \frac{1}{7 \times 3!} + \cdots + \frac{(-1)^n}{(2n+1)n!}.
\end{aligned}$$

The error in this approximation does not exceed the first omitted term, so it will be less than 0.0005 provided $(2n + 3)(n + 1)! > 2000$. Since $13 \times 6! = 9360$, $n = 5$ will do. Thus

$$E(1) \simeq 1 - \frac{1}{3} + \frac{1}{10} - \frac{1}{42} + \frac{1}{216} - \frac{1}{1320} \simeq 0.747$$

rounded to three decimal places.

Indeterminate Forms

Maclaurin and Taylor series can provide a useful alternative to l'Hôpital's rules for evaluating limits of indeterminate forms.

Example 4 a)

$$\begin{aligned}
&\lim_{x \to 0} \frac{x - \sin x}{x^3} \qquad \left[\frac{0}{0}\right] \\
&= \lim_{x \to 0} \frac{x - \left(x - \dfrac{x^3}{3!} + \dfrac{x^5}{5!} - \cdots\right)}{x^3} \\
&= \lim_{x \to 0} \frac{\dfrac{x^3}{3!} - \dfrac{x^5}{5!} + \cdots}{x^3}
\end{aligned}$$

$$= \lim_{x\to 0}\left(\frac{1}{3!} - \frac{x^2}{5!} + \cdots\right) = \frac{1}{3!} = \frac{1}{6}.$$

b) $$\lim_{x\to 0}\frac{(e^{2x}-1)\ln(1+x^3)}{(1-\cos 3x)^2} \qquad \left[\frac{0}{0}\right]$$

$$= \lim_{x\to 0}\frac{\left(1+(2x)+\frac{(2x)^2}{2!}+\frac{(2x)^3}{3!}+\cdots-1\right)\left(x^3-\frac{x^6}{2}+\cdots\right)}{\left(1-\left(1-\frac{(3x)^2}{2!}+\frac{(3x)^4}{4!}-\cdots\right)\right)^2}$$

$$= \lim_{x\to 0}\frac{2x^4+2x^5+\cdots}{\left(\frac{9}{2}x^2-\frac{1}{4!}3^4x^4+\cdots\right)^2}$$

$$= \lim_{x\to 0}\frac{2+2x+\cdots}{\left(\frac{9}{2}-\frac{1}{4!}3^4x^2+\cdots\right)^2} = \frac{2}{\left(\frac{9}{2}\right)^2} = \frac{8}{81}.$$

The student can check that the latter of these examples will require much more laborious calculation if attempted using l'Hôpital's rule.

EXERCISES

Use Maclaurin or Taylor series to calculate the function values indicated in Exercises 1–12 with error less than 1/20000 in absolute value.

1. $e^{0.2}$ **2.** $1/e$ **3.** $e^{3/2}$

4. $\sin(0.1)$ **5.** $\cos 5°$ **6.** $\ln(6/5)$

7. $\ln(0.9)$ **8.** $\sin 80°$ **9.** $\cos 65°$

10. Arctan (0.2) **11.** $\cosh(1)$ **12.** $\ln(3/2)$

Find Maclaurin series for the functions in Exercises 13–17.

13. $I(x) = \int_0^x \frac{\sin t}{t}\,dt$

14. $J(x) = \int_0^x \frac{e^t - 1}{t}\,dt$

15. $K(x) = \int_1^{1+x} \frac{\ln t}{t-1}\,dt$

16. $L(x) = \int_0^x \cos(t^2)\,dt$

17. $M(x) = \int_0^x \frac{\text{Arctan}(t^2)}{t^2}\,dt$

18. Find $I(1)$ to three-decimal-place accuracy, with I defined as in Exercise 13.

19. Find $L(0.5)$ to three-decimal-place accuracy, with L defined as in Exercise 16.

Evaluate the limits in Exercises 20–25.

20. $\lim_{x\to 0}\frac{\sin(x^2)}{x\sinh x}$

21. $\lim_{x\to 0}\frac{1-\cos(x^2)}{(1-\cos x)^2}$

22. $\lim_{x\to 0}\frac{(e^x-1-x)^2}{x^2-\ln(1+x^2)}$

23. $\lim_{x\to 0}\frac{2\sin 3x - 3\sin 2x}{5x - \text{Arctan}\,5x}$

24. $\lim_{x \to 0} \dfrac{\sin(\sin x) - x}{x(\cos(\sin x) - 1)}$

25. $\lim_{x \to 0} \dfrac{\sinh x - \sin x}{\cosh x - \cos x}$

26.† Series can provide a useful tool for solving differential equations. Substitute the series $y = \sum_{n=0}^{\infty} a_n x^n$ into the differential equation

$$y'' + xy' + y = 0$$

and thereby determine a relationship between a_{n+2} and a_n that enables all the coefficients in the series to be determined in terms of a_0 and a_1. Solve the initial-value problem

$$\begin{cases} y'' + xy' + y = 0 \\ y(0) = 1 \\ y'(0) = 0 \end{cases}$$

27.† Solve the initial-value problem

$$\begin{cases} y'' + xy' + 2y = 0 \\ y(0) = 1 \\ y'(0) = 2 \end{cases}$$

by the method suggested in Exercise 26.

9.5 TAYLOR'S THEOREM

If the function $f(x)$ has derivatives up to and including order n at the point $x - c$, we can form the polynomial

$$P_n(x) = \sum_{k=0}^{n} \frac{f^{(k)}(c)}{k!}(x - c)^k$$
$$= f(c) + f'(c)(x - c) + \frac{f''(c)}{2!}(x - c)^2 + \cdots + \frac{f^{(n)}(c)}{n!}(x - c)^n.$$

$P_n(x)$ is called the ***n*th degree Taylor polynomial** for $f(x)$ about $x = c$. (If $c = 0$ it is called a **Maclaurin polynomial.**)

If f has derivatives of all orders at $x = c$, and therefore has a Taylor series in powers of $x - c$, then $P_n(x)$ is a partial sum of that Taylor series; it is the sum of the first $(n + 1)$ terms of the series.

Among all polynomials of degree at most n, the Taylor polynomial $P_n(x)$ best describes the behavior of $f(x)$ near $x = c$ in the sense that

$$P_n(c) = f(c), P_n'(c) = f'(c), P_n''(c) = f''(c), \ldots, P_n^{(n)}(c) = f^{(n)}(c).$$

Taylor's theorem provides two formulas for the error involved when $P_n(x)$ is used to approximate $f(x)$.

Theorem 7 (*Taylor's Theorem*) Suppose that f has continuous derivatives up to and including $(n + 1)$st order on an interval containing the points x and c. Then $f(x)$ can be expressed as a sum of the Taylor polynomial $P_n(x)$ and a remainder term $R_n(x)$:

† These exercises involve differential equations or initial-value problems.

$$f(x) = P_n(x) + R_n(x) \qquad \text{(Taylor's formula with remainder)}$$
$$= \sum_{k=0}^{n} \frac{f^{(k)}(c)}{k!}(x-c)^k + R_n(x),$$

where the remainder term $R_n(x)$ is given by either of the following equations:

i) $R_n(x) = \dfrac{1}{n!}\displaystyle\int_c^x (x-t)^n f^{(n+1)}(t)\,dt$

ii) $R_n(x) = \dfrac{f^{(n+1)}(X)}{(n+1)!}(x-c)^{n+1}$, where X is some number between x and c.

Note that $R_n(x) = f(x) - P_n(x)$ is the error involved in the approximation $f(x) \approx P_n(x)$. Form (i) is called the **integral form of the remainder** in Taylor's formula. Form (ii) is called the **Lagrange form of the remainder.**

Proof of Taylor's Theorem with Integral Remainder

We start with the fundamental theorem of calculus written in the form

$$f(x) = f(c) + \int_c^x f'(t)\,dt = P_0(x) + R_0(x).$$

Note that the fundamental theorem is just the special case $n = 0$ of Taylor's formula with integral remainder. We now apply integration by parts to the integral, setting

$$U = f'(t), \qquad dV = dt$$
$$dU = f''(t)\,dt, \qquad V = -(x-t).$$

(Note the choice of a constant of integration, $-x$, in V. We make this choice to ensure that V vanishes when $t = x$.) We have

$$f(x) = f(c) - f'(t)(x-t)\Big|_{t=c}^{t=x} + \int_c^x (x-t)f''(t)\,dt$$
$$= f(c) + f'(c)(x-c) + \int_c^x (x-t)f''(t)\,dt$$
$$= P_1(x) + R_1(x).$$

We have thus proved the case $n = 1$ of Taylor's formula with integral remainder.

Let us complete the proof for general n by induction. Suppose that Taylor's formula holds with integral remainder for some $n = k$:

$$f(x) = P_k(x) + R_k(x) = P_k(x) + \frac{1}{k!}\int_c^x (x-t)^k f^{(k+1)}(t)\,dt.$$

Again we integrate by parts. Let

$$U = f^{(k+1)}(t), \qquad dV = (x - t)^k\,dt,$$

$$dU = f^{(k+2)}(t)\,dt, \qquad V = \frac{-1}{k+1}(x - t)^{k+1}.$$

We have

$$\begin{aligned} f(x) &= P_k(x) + \frac{1}{k!}\left(-\frac{f^{(k+1)}(t)(x-t)^{k+1}}{k+1}\Bigg|_{t=c}^{t=x} + \frac{1}{k+1}\int_c^x (x-t)^{k+1} f^{(k+2)}(t)\,dt \right) \\ &= P_k(x) + \frac{f^{(k+1)}(c)}{(k+1)!}(x-c)^{k+1} + \frac{1}{(k+1)!}\int_c^x (x-t)^{k+1} f^{(k+2)}(t)\,dt \\ &= P_{k+1}(x) + R_{k+1}(x). \end{aligned}$$

Thus Taylor's formula with integral remainder is valid for $n = k + 1$ if it is valid for $n = k$. Having been shown to be valid for $n = 0$ (and $n = 1$), it must therefore be valid for every positive integer n. □

Exercises 30 and 32 of Section 4.6 suggest a direct method for proving Taylor's formula with Lagrange remainder by generalizing the argument used to establish the error formula for the tangent-line approximation. That formula, presented in Theorem 6 of Section 4.6, states that

$$f(x) = f(c) + f'(c)(x - c) + \frac{f''(X)}{2!}(x - c)^2$$

for some number X between x and c; that is, it is the case $n = 1$ of Taylor's formula with Lagrange remainder. Case $n = 0$ is just the mean-value theorem:

$$f(x) = f(c) + f'(X)(x - c)$$

for some X between x and c. An induction argument can be constructed to complete the proof for general n. Instead, however, we shall derive the Lagrange form of the remainder from the integral form already established. To accomplish this we need the following result.

Theorem 8 (*A Generalized Mean-Value Theorem for Integrals*) If f and g are continuous on the interval $[a, b]$ and g does not change sign on that interval, then there exists a number X in $[a, b]$ such that

$$\int_a^b f(x)g(x)\,dx = f(X)\int_a^b g(x)\,dx.$$

Proof We can assume that $g(x) \ge 0$ on $[a, b]$ (a similar proof holds if $g(x) \le 0$ on $[a, b]$). We can also assume that $\int_a^b g(x)\,dx > 0$, because otherwise $g(x)$ will be identically 0 on $[a, b]$ and there is nothing to prove. Since f is continuous on $[a, b]$ there exists points r and s in $[a, b]$ such that for every x in that interval

$$f(r) \le f(x) \le f(s).$$

(See Theorem 5 of Section 1.4.) Thus

$$f(r)g(x) \leq f(x)g(x) \leq f(s)g(x)$$

$$f(r)\int_a^b g(x)\,dx \leq \int_a^b f(x)g(x)\,dx \leq f(s)\int_a^b g(x)\,dx$$

$$f(r) \leq \frac{\int_a^b f(x)g(x)\,dx}{\int_a^b g(x)\,dx} \leq f(s).$$

The middle member of the latter inequality is a number between two values of $f(x)$ on $[a, b]$. By the intermediate-value theorem (Theorem 6 of Section 1.5) that member must itself be a value of $f(x)$ at some point in $[a, b]$:

$$\frac{\int_a^b f(x)g(x)\,dx}{\int_a^b g(x)\,dx} = f(X) \text{ for some } X \text{ in } [a, b].$$

Thus

$$\int_a^b f(x)g(x)\,dx = f(X)\int_a^b g(x)\,dx$$

as required. □

Proof of Taylor's Theorem with Lagrange Remainder Since $(x - t)^n$ does not change sign as t ranges over the interval with endpoints x and c, we can apply Theorem 8 to the integral form of the remainder:

$$\begin{aligned} R_n(x) &= \frac{1}{n!}\int_c^x (x - t)^n f^{(n+1)}(t)\,dt \\ &= \frac{1}{n!}f^{(n+1)}(X)\int_c^x (x - t)^n\,dt \qquad \text{for some } X \text{ between } x \text{ and } c \\ &= -\frac{f^{(n+1)}(X)}{(n + 1)!}(x - t)^{n+1}\Bigg|_{t=c}^{t=x} = \frac{f^{(n+1)}(X)}{(n + 1)!}(x - c)^{n+1} \end{aligned}$$

as required. □

Applications of Taylor's Formula

Since the remainder term in Taylor's formula represents the tail of the Taylor (or Maclaurin) series, it can be used to obtain bounds for the tail and thus error estimates for approximations using Taylor polynomials.

Example 1 Estimate the error in the approximation

$$\tan(0.5) \simeq 0.5 + \frac{(0.5)^3}{3}.$$

Solution Since $\tan x$ is an odd function, its Maclaurin polynomials involve only odd powers of x. If $f(x) = \tan x$ we have $f(0) = 0$ and

$$
\begin{aligned}
f'(x) &= \sec^2 x & f'(0) &= 1\\
f''(x) &= 2\sec^2 x \tan x & f''(0) &= 0\\
f^{(3)}(x) &= 2\sec^4 x + 4\sec^2 x \tan^2 x & f^{(3)}(0) &= 2\\
f^{(4)}(x) &= 16\sec^4 x \tan x + 8\sec^2 x \tan^3 x & f^{(4)}(0) &= 0\\
f^{(5)}(x) &= 88\sec^4 x \tan^2 x + 16\sec^6 x + 16\sec^2 x \tan^4 x.
\end{aligned}
$$

Thus the Maclaurin polynomials of degree 3 and 4 coincide (both are polynomials of degree 3 only):

$$P_3(x) = P_4(x) = x + \frac{x^3}{3}.$$

If we use the approximation $\tan(0.5) \simeq P_4(0.5) = 0.5 + (0.5)^3/3$, then the error is

$$R_4(0.5) = \frac{f^{(5)}(X)}{5!}(0.5)^5 \qquad \text{for some } X \text{ between } 0 \text{ and } 0.5.$$

Now for $0 \le X \le 0.5$ we have

$$
\begin{cases}
0 \le \sec X \le \sec(0.5) < \sec\dfrac{\pi}{6} = \dfrac{2}{\sqrt{3}},\\[2ex]
0 \le \tan X \le \tan(0.5) < \tan\dfrac{\pi}{6} = \dfrac{1}{\sqrt{3}}.
\end{cases}
$$

Hence

$$
\begin{aligned}
|f^{(5)}(X)| &\le 88\left(\frac{2}{\sqrt{3}}\right)^4\left(\frac{1}{\sqrt{3}}\right)^2 + 16\left(\frac{2}{\sqrt{3}}\right)^6 + 16\left(\frac{2}{\sqrt{3}}\right)^2\left(\frac{1}{\sqrt{3}}\right)^4\\
&= \frac{1}{27}(88 \times 16 + 16 \times 64 + 16 \times 4) \simeq 92.44 < 93.
\end{aligned}
$$

Thus

$$|R_4(0.5)| < \frac{93}{5!}(0.5)^5 \simeq 0.024$$

and the error in the approximation $\tan(0.5) \simeq 0.5 + (0.5)^3/3$ is less than 0.025 in absolute value.

Example 2 Use Taylor's formula to obtain the Maclaurin series for e^x.

Solution If $f(x) = e^x$, then $f^{(k)}(x) = e^x$ for every positive integer k, and $f^{(k)}(0) = 1$. Hence

$$e^x = 1 + x + \frac{x^2}{2!} + \frac{x^3}{3!} + \cdots + \frac{x^n}{n!} + \frac{e^X x^{n+1}}{(n+1)!}$$

for some X between 0 and x. Evidently $|e^X| \le e^{|X|} \le e^{|x|}$, so

$$\left|\frac{e^X x^{n+1}}{(n+1)!}\right| \le e^{|x|}\frac{|x|^{n+1}}{(n+1)!} \to 0 \text{ as } n \to \infty$$

for every real number x. (See Example 9 of Section 8.1.) Hence

$$e^x = 1 + x + \frac{x^2}{2!} + \frac{x^3}{3!} + \cdots \qquad \text{(for every real } x\text{)}.$$

The other Maclaurin series obtained in Section 9.2 can also be obtained by using Taylor's formula and taking the limit as $n \to \infty$. We have used the Lagrange form of the remainder in both examples above, but that form is not always appropriate for showing that $\lim_{n\to\infty} R_n(x) = 0$. When it fails to yield convergence of the series on the appropriate interval, the integral form can usually be used instead. Such is the case for the series for $\ln(1 + x)$ and the binomial series we consider now.

The Binomial Series

We use Taylor's formula to establish the binomial series: If r is any real number and $-1 < x < 1$, then

$$\begin{aligned}(1+x)^r &= 1 + rx + \frac{r(r-1)}{2!}x^2 + \frac{r(r-1)(r-2)}{3!}x^3 + \cdots \\ &= 1 + \sum_{n=1}^{\infty}\frac{r(r-1)(r-2)\cdots(r-n+1)}{n!}x^n.\end{aligned}$$

Note that if r is a positive integer or 0 the series has only finitely many terms that are not 0, and so it "converges" for every x. If $r = -1$, the series reduces to the standard geometric series for $1/(1 + x)$. If $r > 0$, the series also converges at $x = 1$.

If $f(x) = (1 + x)^r$, then

$$f'(x) = r(1+x)^{r-1}, \qquad f''(x) = r(r-1)(1+x)^{r-2}, \ldots,$$
$$f^{(n)}(x) = r(r-1)(r-2)(r-3)\cdots(r-n+1)(1+x)^{r-n}.$$

Hence $f^{(n)}(0) = r(r-1)(r-2)\cdots(r-n+1)$. The binomial series is therefore the Maclaurin series for $(1 + x)^r$. The ratio test shows that this series converges absolutely on the interval $(-1, 1)$; we want to show that it converges to $(1 + x)^r$. Writing Taylor's formula with integral remainder for $(1 + x)^r$, we obtain

$$(1+x)^r = 1 + \sum_{k=1}^{n}\frac{r(r-1)(r-2)\cdots(r-k+1)}{k!}x^k + R_n(x)$$

where

$$R_n(x) = \frac{r(r-1)(r-2)\cdots(r-n)}{n!}\int_0^x \frac{(x-t)^n}{(1+t)^{n+1-r}}\,dt.$$

We want to show that $\lim_{n\to\infty} R_n(x) = 0$ for each x in the interval $(-1, 1)$.

Select an x such that $-1 < x < 1$. Since the variable of integration, t, in the formula for $R_n(x)$ lies between 0 and x, we have

$$1 \le 1 + t \le 1 + x < 2 \qquad \text{if } 0 \le x < 1,$$
$$0 < 1 + x \le 1 + t \le 1 \qquad \text{if } -1 < x \le 0.$$

In either case we can find a constant K, depending on x and r, but not on n, such that

$$|1 + t|^{r-1} \le K \qquad \text{for all } t \text{ between 0 and } x.$$

Also, if $0 \le x < 1$, then

$$\left|\frac{x - t}{1 + t}\right| = \frac{x - t}{1 + t} \le x - t \le x = |x|.$$

If $-1 < x \le 0$ then

$$\left|\frac{x - t}{1 + t}\right| = \frac{t - x}{1 + t} \le |x|$$

since $(t - x)/(1 + t)$ increases from 0 to $-x = |x|$ as t increases from x to 0. Thus we have

$$|R_n(x)| = \left|\frac{r(r-1)(r-2)\cdots(r-n)}{n!}\right| \left|\int_0^x (1 + t)^{r-1}\left(\frac{x - t}{1 + t}\right)^n dt\right|$$
$$\le K \left|\frac{r(r-1)(r-2)\cdots(r-n)}{n!}\right| |x|^n \left|\int_0^x dt\right|$$
$$= K \left|\frac{r(r-1)(r-2)\cdots(r-n)}{n!}\right| |x|^{n+1}.$$

Now apply the ratio test to the series $\Sigma_{n=1}^{\infty} \dfrac{r(r-1)(r-2)\cdots(r-n)}{n!} x^{n+1}$:

$$\rho = \lim_{n\to\infty} \left| \frac{\dfrac{r(r-1)(r-2)\cdots(r-n)(r-n-1)}{(n+1)!} x^{n+2}}{\dfrac{r(r-1)(r-2)\cdots(r-n)}{n!} x^{n+1}} \right|$$
$$= \lim_{n\to\infty} \left|\frac{r - n - 1}{n + 1}\right| |x| = |x| < 1.$$

Thus the series converges and so its nth term must approach 0:

$$\lim_{n\to\infty} \left|\frac{r(r-1)(r-2)\cdots(r-n)}{n!}\right| |x|^{n+1} = 0.$$

Therefore $\lim_{n\to\infty} R_n(x) = 0$ and the binomial series representation holds at x as claimed.

Example 3 Find the Maclaurin series for $1/\sqrt{1+x}$.

Solution Here $r = -(1/2)$:

$$\begin{aligned}\frac{1}{\sqrt{1+x}} &= (1+x)^{-1/2}\\ &= 1 - \frac{1}{2}x + \frac{\left(-\frac{1}{2}\right)\left(-\frac{3}{2}\right)}{2!}x^2 + \frac{\left(-\frac{1}{2}\right)\left(-\frac{3}{2}\right)\left(-\frac{5}{2}\right)}{3!}x^3 + \cdots\\ &= 1 - \frac{1}{2}x + \frac{1\times 3}{2^2\,2!}x^2 - \frac{1\times 3\times 5}{2^3\,3!}x^3 + \cdots\\ &= 1 + \sum_{n=1}^{\infty}(-1)^n\frac{1\times 3\times 5\times\cdots\times(2n-1)}{2^n\,n!}x^n.\end{aligned}$$

This series converges absolutely for $-1 < x < 1$.

Example 4 Find the Maclaurin series for Arcsin x.

Solution Replace x with $-x^2$ in the series obtained in Example 3 to get

$$\frac{1}{\sqrt{1-x^2}} = 1 + \sum_{n=1}^{\infty}\frac{1\times 3\times 5\times\cdots\times(2n-1)}{2^n\,n!}x^{2n} \quad (-1<x<1).$$

$$\begin{aligned}\text{Arcsin}\, x &= \int_0^x \frac{dt}{\sqrt{1-t^2}} = \int_0^x\left(1 + \sum_{n=1}^{\infty}\frac{1\times 3\times 5\times\cdots\times(2n-1)}{2^n\,n!}t^{2n}\right)dt\\ &= x + \sum_{n=1}^{\infty}\frac{1\times 3\times 5\times\cdots\times(2n-1)}{2^n\,n!(2n+1)}x^{2n+1}\\ &= x + \frac{x^3}{6} + \frac{3}{40}x^5 + \cdots \qquad (-1<x<1).\end{aligned}$$

EXERCISES

Find the indicated Taylor or Maclaurin polynomials of the specified degrees for the functions given in Exercises 1–10.

1. for $f(x) = \sin x$ in powers of x having degree 5
2. for $f(x) = \sin x$ in powers of x having degree 6
3. for $f(x) = \sin x$ in powers of $x - (\pi/4)$ having degree 4
4. for $f(x) = e^{-x}$ in powers of x having degree n
5. for $f(x) = e^{-x}$ in powers of $x + 2$ having degree n
6. for $f(x) = \ln x$ in powers of $x - 2$ having degree n
7. for $f(x) = \sec x$ in powers of x having degree 4
8. for $f(x) = \text{Arctan}\, x$ in powers of $x - 1$ having degree 3

9. for $f(x) = \sqrt{1 + x}$ in powers of x having degree 4

10. for $f(x) = \text{Arcsin } x$ in powers of $x + (1/2)$ having degree 3

Use Taylor's formula to establish the Maclaurin series for the functions in Exercises 11–20.

11. e^x **12.** e^{-x} **13.** $\cosh x$ **14.** $\sinh x$

15. $\sin x$ **16.** $\cos x$ **17.** $\sin^2 x$ **18.** $\dfrac{1}{1 - x}$

19. $\ln (1 + x)$ **20.** $\dfrac{x}{2 + 3x}$

Use Taylor's formula to obtain the Taylor series indicated in Exercises 21–26.

21. for e^x in powers of $x - a$

22. for $\sin x$ in powers of $x - (\pi/6)$

23. for $\cos x$ in powers of $x - (\pi/4)$

24. for $\ln x$ in powers of $x - 1$

25. for $\ln x$ in powers of $x - 2$

26. for $\cosh x$ in powers of $x - 1$

27. Let $f(x) = x^2 + 2x + 3$. Find the following.

a) the Maclaurin polynomial of degree 2 for $f(x)$

b) the Maclaurin polynomial of degree 99 for $f(x)$

c) the Taylor polynomial of degree 2 for $f(x)$ in powers of $x + 2$

d) the Taylor polynomial of degree 99 for $f(x)$ in powers of $x - 3$

28. Find the Taylor polynomial of degree 6 for $f(x) = x^6$ in powers of $x - 1$.

29. For what functions $f(x)$ is it true that $f(x)$ is identically equal to its Maclaurin polynomial of degree n?

30. For what functions $f(x)$ is it true that the Taylor polynomial of degree n in powers of $x - a$ vanishes identically?

31. Estimate the error if the Maclaurin polynomial of degree 2 for $\sec x$ is used to approximate $\sec (0.2)$.

32. Estimate the error if the Maclaurin polynomial of degree 3 for $\ln (\cos x)$ is used to approximate $\ln (\cos 0.1)$.

33. Estimate the error if the Maclaurin polynomial of degree 3 for $e^{-x} \sin (x/2)$ is used to approximate $e^{-1} \sin (0.5)$.

34. Estimate the error if the Taylor polynomial of degree 3 for $\text{Arctan } x$ in powers of $x - 1$ is used to approximate $\text{Arctan } (99/100)$.

35. Estimate the error if the Taylor polynomial of degree 4 for $\ln x$ in powers of $x - 2$ is used to approximate $\ln (1.95)$.

Write Maclaurin series for the functions in Exercises 36–39.

36.* $\sqrt{1 - x}$

37.* $\sqrt{4 + x}$

38.* $\dfrac{1}{\sqrt{4 + x^2}}$

39.* $F(x) = \displaystyle\int_0^x \frac{dt}{\sqrt{4 + t^2}}$

40.* Show that if $r > 0$, then the binomial series for $(1 + x)^r$ converges at the endpoint $x = 1$ as claimed in the discussion.

10.1 Functions of Several Variables

10.2 Partial Derivatives

10.3 The Chain Rule and Differentials

10.4 Gradients and Directional Derivatives

10.5 Extreme Values

Partial Differentiation

10.1 FUNCTIONS OF SEVERAL VARIABLES

We have been using the notation $y = f(x)$ to denote that the variable y is a function of the single real variable x. The domain of such a function f is a set of real numbers. Many quantities can be regarded as depending on more than one real variable, and thus to be functions of more than one variable. For example, the volume of a circular cylinder of radius r and height h is given by $V = \pi r^2 h$; that is, V is a function of the two variables r and h. If we choose to denote this function by f we could write $V = f(r,h)$ where

$$f(r,h) = \pi r^2 h, \quad (r > 0,\ h > 0).$$

Thus f is a function of two variables having domain the set of points in the rh-plane having coordinates (r,h) satisfying $r > 0$ and $h > 0$. Similarly, the relationship $w = f(x,y,z) = x + 2y - 3z$ defines w as a function of the three variables x, y and z, with domain the three dimensional space of coordinates (x,y,z), or, if we state explicitly, some particular subset of that space.

By analogy with Definition 1 of Section 1.2 we define a function of n variables as follows:

Definition 1

> A **function** f of n real variables is a rule which assigns to each n-tuple $(x_1, x_2,\ldots, x_n)$ of real numbers in some set $D(f)$ of the n-dimensional space of coordinates $(x_1, x_2,\ldots, x_n)$, a **unique** real number $f(x_1, x_2,\ldots, x_n)$. The set $D(f)$ is called the **domain** of f; the set of all real numbers $f(x_1, x_2,\ldots, x_n)$ obtained from points in the domain is called the **range** of f.

As for functions of one variable, it is conventional to take the domain of such a function of n variables to be the largest set of points with coordinates $(x_1, x_2,\ldots, x_n)$ for which $f(x_1, x_2,\ldots, x_n)$ makes sense as a real number, unless the domain is explicitly stated to be a smaller set.

Most of the examples we consider in this chapter will be functions of **two** independent variables. We shall usually denote these variables x and y, and shall denote by z the dependent variable giving the value of the function. Thus $z = f(x,y)$. Theorems will also be stated in the two-variable case; the extensions to three or more variables will usually be obvious.

Graphical Representations

The graph of a function of one variable requires a two-dimensional coordinate system, one dimension each for the independent and dependent variables. The graph of a function of two variables, that is, the graph of the equation $z = f(x,y)$, requires a three-dimensional coordinate system with mutually perpendicular x-, y- and z-axes. Most commonly these axes are drawn as in Fig. 10.1, with the x- and y-

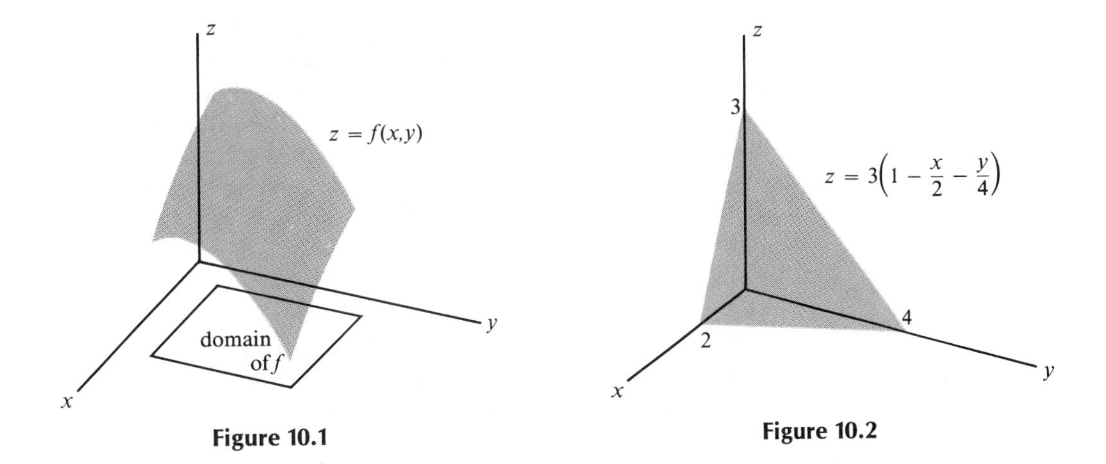

Figure 10.1 Figure 10.2

axes considered horizontal and the z-axis vertical. Such a coordinate system is called right-handed because the thumb, forefinger and middle finger of the right hand can be extended so as to point in the direction of the x-axis, y-axis and z-axis respectively. The graph of $z = f(x,y)$ is, in general, a surface lying above (or below) the region $D(f)$ in the x-y plane.

Example 1

$$f(x,y) = 3\left(1 - \frac{x}{2} - \frac{y}{4}\right), \quad (0 \le x \le 2, 0 \le y \le 4 - 2x).$$

The graph of f is the plane triangular surface with vertices at (2,0,0), (0,4,0) and (0,0,3). See Fig. 10.2. If the domain of f had not been specified explicitly as a subset of the x-y plane, the graph would have been the whole plane through these three points.

Example 2 $f(x,y) = \sqrt{9 - x^2 - y^2}$. Here the domain is the disc $x^2 + y^2 \le 9$ and the graph of f is a hemisphere of radius 3 centered at the origin and lying above the x-y plane. See Fig. 10.3.

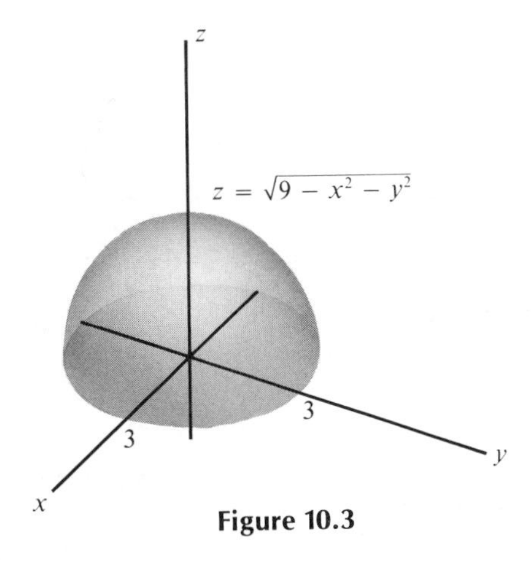

Figure 10.3

Since it is necessary to project the surface $z = f(x,y)$ onto a two-dimensional page, most such graphs are difficult to sketch without considerable artistic talent and training. Another way of representing the function $f(x,y)$ graphically is to produce a two-dimensional "topographic" map of the surface $z = f(x,y)$. Thus we sketch, in the x-y plane, the curves $f(x,y) = C$ for various choices of the constant C. These curves are called **level curves** of f because if we project them vertically onto the surface $z = f(x,y)$ they do not rise or fall, but remain at the same distance above (or below) the x-y plane. The graph and some level curves of the function $f(x,y) = x^2 + y^2$ are shown in Fig. 10.4. The contour curves in the topographic map in Fig. 10.5 show elevations, in 200 ft increments above sea level, on part of Nelson Island on the British Columbia coast.

Example 3 The level curves of the function $f(x,y) = 3\left(1 - \frac{x}{2} - \frac{y}{4}\right)$ of Example 1 are the segments of the straight lines

$$3\left(1 - \frac{x}{2} - \frac{y}{4}\right) = C, \quad \text{or} \quad \frac{x}{2} + \frac{y}{4} = 1 - \frac{C}{3}, \quad (0 \le C \le 3)$$

which lie in the first quadrant: $x \ge 0$, $y \ge 0$. See Fig. 10.6.

Example 4 The level curves of the function $f(x,y) = \sqrt{9 - x^2 - y^2}$ of Example 2 are the concentric circles. See Fig. 10.7.

$$\sqrt{9 - x^2 - y^2} = C, \quad \text{or} \quad x^2 + y^2 = 9 - C^2, \quad (0 \le C \le 3).$$

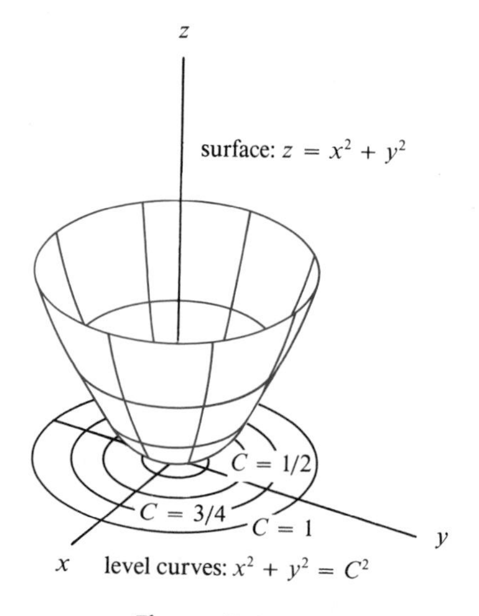

Figure 10.4

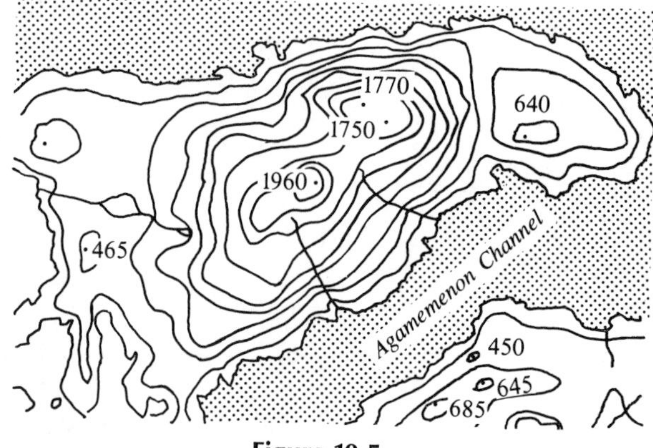

Figure 10.5

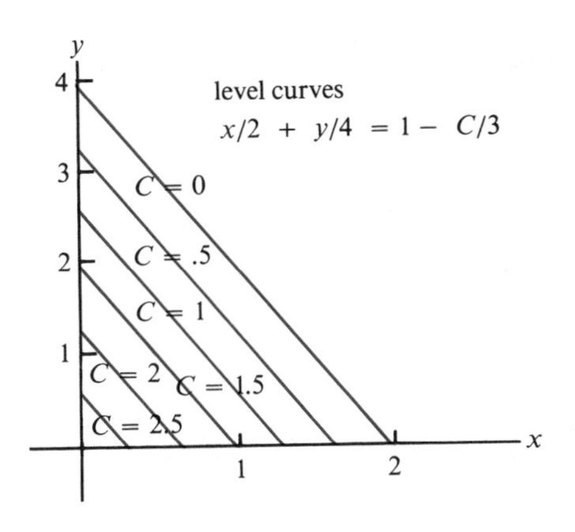

Figure 10.6

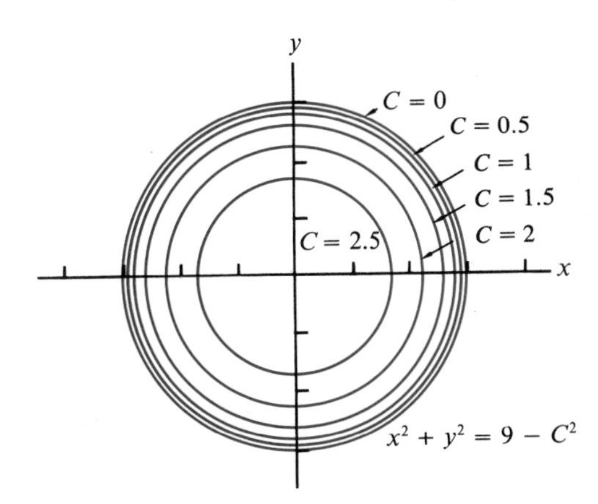

Figure 10.7

Example 5 The level curves of the function $f(x,y) = x^2 - y^2$ are the curves $x^2 - y^2 = C$. For $C = 0$ the level "curve" is the pair of straight lines $x = y$ and $x = -y$. For other values of C they are rectangular hyperbolas with these lines as asymptotes. See Fig. 10.8a. The graph of $z = x^2 - y^2$ is somewhat more difficult to sketch. It is the saddle-like surface of Fig. 10.8b.

Limits and Continuity

The concept of limit of a function of several variables is similar to that for functions of one variable discussed in Section 1.3. For clarity we present the definition for functions of two variables only; the general case is similar.

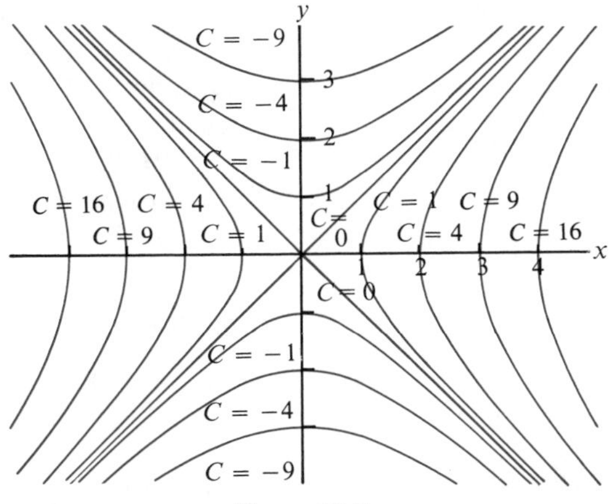

Figure 10.8a

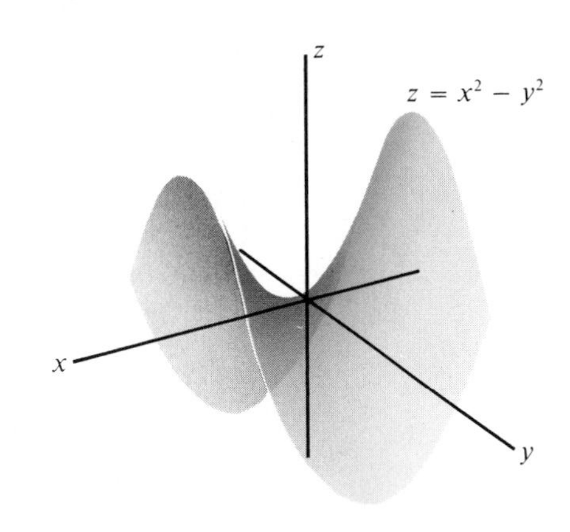

Figure 10.8b

Definition 2

We say that $f(x,y)$ approaches the limit L as the point (x,y) approaches the point (a,b), and we write

$$\lim_{(x,y)\to(a,b)} f(x,y) = L,$$

if all points of a disc of positive radius centered at (a,b) except possibly the point (a,b) itself belong to the domain of f, and if $f(x,y)$ approaches arbitrarily close to L as (x,y) approaches (a,b).

More formally, $\lim_{(x,y)\to(a,b)} f(x,y) = L$ if and only if for every positive number ϵ there exists a positive number $\delta = \delta(\epsilon)$ depending on ϵ such that

$$0 < \sqrt{(x-a)^2 + (y-b)^2} < \delta \text{ implies } |f(x,y) - L| < \epsilon.$$

Just as the existence of $\lim_{x\to a} f(x)$ implies that $f(x)$ approaches the same limit as x approaches a from either the right or the left, so also here the limit cannot exist unless $f(x,y)$ approaches L no matter how (x,y) approaches (a,b). The examples below illustrate this.

All the usual laws of limits (see Theorems 1, 2 and 3 of Section 1.3) extend to limits of functions of several variables in the obvious way.

Example 6

a) $\lim_{(x,y)\to(2,3)} 2x - y^2 = 4 - 9 = -5.$

b) $\lim_{(x,y)\to(a,b)} x^2y = a^2b.$

c) $\lim_{(x,y)\to((\pi/3),2)} y\sin(x/y) = 2\sin(\pi/6) = 1.$

Example 7 Let $f(x,y) = \dfrac{2xy}{x^2 + y^2}$. Note that $f(x,y)$ is defined at all points of the x-y plane except the origin (0,0). We can still ask whether $\lim_{(x,y)\to(0,0)} f(x,y)$ exists or not. If we let (x,y) approach (0,0) along the x-axis we get $f(x,y) = f(x,0) = 0$ so the limit must be 0 if it exists at all. Similarly, at all points of the y-axis we have $f(x,y) = f(0,y) = 0$. However, at points of the line $x = y$, f has a different constant value: $f(x,x) = 1$. Hence the limit of $f(x,y)$ is 1 as the point (x,y) approaches the origin along this line. It follows that $f(x,y)$ cannot have a unique limit at the origin; $\lim_{(x,y)\to(0,0)} f(x,y)$ does not exist. Observe that $f(x,y)$ has constant values on any ray through the origin (on the ray $y = kx$ the value is $2k/(1 + k^2)$), but these values differ on different rays. The level curves of $f(x,y)$ are the rays through the origin. It is very difficult to sketch the graph of f near the origin. See Fig. 10.9.

Example 8 Let $f(x,y) = \dfrac{x^2y}{x^4 + y^2}$. As in the previous example, $f(x,y)$ vanishes identically on the coordinate axes so $\lim(x,y)\to(0,0)$ must be 0 if the limit exists at all. If we examine $f(x,y)$ at points of the ray $y = kx$ we obtain

$$f(x,kx) = \frac{kx^3}{x^4 + k^2x^2} = \frac{kx}{x^2 + k^2} \to 0 \text{ as } x \to 0 \ (k \neq 0).$$

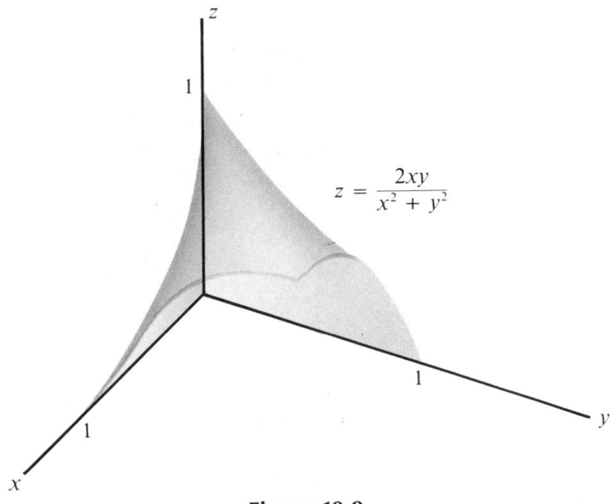

Figure 10.9

Thus $f(x,y) \to 0$ as $(x,y) \to (0,0)$ along any straight line through the origin. We might be led to infer that $\lim_{(x,y)\to(0,0)} f(x,y) = 0$, but this would be incorrect. Observe the behavior of $f(x,y)$ along the parabola $y = x^2$:

$$f(x,x^2) = \frac{x^4}{x^4 + x^4} = 1/2.$$

so that $f(x,y)$ does not approach 0 as $(x,y) \to (0,0)$ along this curve. Thus $\lim_{(x,y)\to(0,0)} f(x,y)$ does not exist. See Fig. 10.10. The level curves of f are parabolas of the form $y = kx^2$.

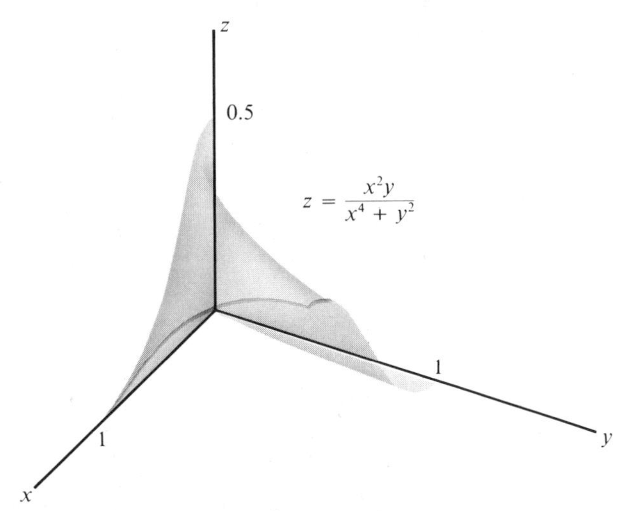

Figure 10.10

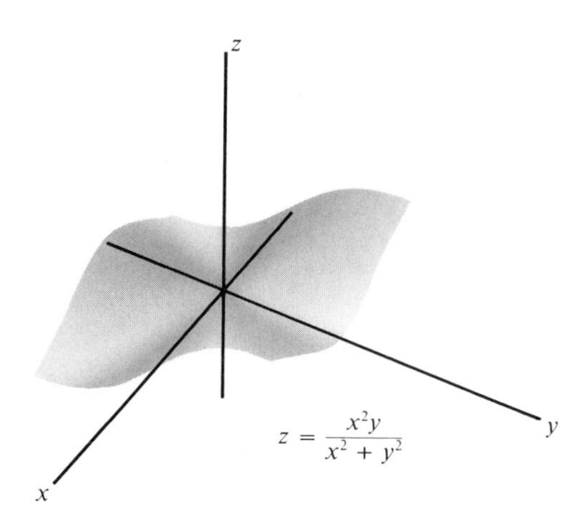

Figure 10.11

Example 9 Let $f(x,y) = \dfrac{x^2y}{x^2 + y^2}$. In this case we actually do have a limit: $\lim_{(x,y)\to(0,0)} f(x,y) = 0$. To see this, observe that since $|x| \leqslant \sqrt{x^2 + y^2}$ and $|y| \leqslant \sqrt{x^2 + y^2}$, therefore $|f(x,y)| \leqslant \sqrt{x^2 + y^2}$ which approaches zero as $(x,y) \to (0,0)$. See Fig. 10.11.

Definition 3

> $f(x,y)$ is continuous at the point (a,b) if $\lim_{(x,y)\to(a,b)} f(x,y)$ exists and equals $f(a,b)$.

As for functions of one variable, suitable combinations of continuous functions are continuous. (See Theorem 4 of Section 1.5.) The functions of Examples 7, 8 and 9 above are continuous wherever they are defined, that is, at all points except the origin. There is no way to define $f(0,0)$ so that the functions $f(x,y)$ of Examples 7 and 8 become continuous at the origin, but the function of Example 9 will be continuous at (0,0) if we define $f(0,0) = 0$.

EXERCISES

Specify the domains of the functions in Exercises 1-10.

1. $f(x,y) = \dfrac{x + y}{x - y}$.

2. $f(x,y) = \sqrt{xy}$.

3. $f(x,y) = \dfrac{x}{x^2 + y^2}$.

4. $f(x,y) = \dfrac{xy}{x^2 - y^2}$.

5. $f(x,y) = \sqrt{4x^2 + 9y^2 - 36}$.

6. $f(x,y) = \dfrac{1}{\sqrt{x^2 - y^2}}$.

7. $f(x,y) = \ln(1 + xy)$.

8. $f(x,y) = \text{Arcsin}\,(x + y)$.

9. $f(x,y,z) = \dfrac{xyz}{x^2 + y^2 + z^2}$.

10. $f(x,y,z) = \dfrac{e^{xyz}}{xy + yz + zx}$.

Sketch the graphs of the functions in Exercises 11-20.

11. $f(x,y) = x, (0 \le x, \le 2, 0 \le y \le 3)$

12. $f(x,y) = 4x^2 + y^2$

13. $f(x,y) = \sqrt{x^2 + y^2}$

14. $f(x,y) = 6 - x - 2y$

15. $f(x,y) = |x| + |y|$

16. $f(x,y) = y^2, (0 \le x \le 1, 0 \le y \le 1)$

17. $f(x,y) = 4 - x^2 - y^2, (x^2 + y^2 \le 4, x \ge 0, y \ge 0)$

18. $f(x,y) = \sin x, (0 \le x \le 2\pi, 0 \le y \le 1)$

19. $f(x,y) = \dfrac{1}{x^2 + 4y^2}$

20. $f(x,y) = c\sqrt{1 + (x/a)^2 + (y/b)^2}$

Sketch some level curves of the functions in Exercises 21-30.

21. $f(x,y) = x - y.$

22. $f(x,y) = x^2 + 2y^2.$

23. $f(x,y) = xy.$

24. $f(x,y) = x^2/y.$

25. $f(x,y) = \dfrac{x - y}{x + y}.$

26. $f(x,y) = \dfrac{y}{x^2 + y^2}.$

27. $f(x,y) = \dfrac{x^2}{x^2 + y^2}.$

28. $f(x,y) = \sin(x + y).$

29. $f(x,y) = \dfrac{1}{x^2 - y^2}.$

30. $f(x,y) = x\,e^{-y}.$

31. Find $f(x,y)$ if each level curve $f(x,y) = C$ is a circle with centre the origin and radius a) C, b) C^2, c) $\sqrt{C}$, d) $\ln(C)$.

In Exercises 32-41 evaluate the indicated limit or explain why it does not exist.

32. $\lim_{(x,y)\to(2,-1)} xy + x^2.$

33. $\lim_{(x,y)\to(0,0)} \dfrac{x}{x^2 + y^2}.$

34. $\lim_{(x,y)\to(0,0)} \dfrac{y^3}{x^2 + y^2}.$

35. $\lim_{(x,y)\to(0,0)} \dfrac{\sin(x - y)}{\cos(x + y)}.$

36. $\lim_{(x,y)\to(0,0)} \dfrac{xy}{x^2 + e^y}.$

37. $\lim_{(x,y)\to(1,2)} \dfrac{2x^2 - xy}{4x^2 - y^2}.$

38. $\lim_{(x,y)\to(0,0)} \dfrac{x^2y^2}{x^2 + y^4}.$

39. $\lim_{(x,y)\to(0,0)} \dfrac{x^2y^2}{2x^4 + y^4}.$

40. $\lim_{(x,y)\to(0,0)} \dfrac{\sin(xy)}{x^2 + y^2}.$

41. $\lim_{(x,y)\to(-1,1)} \dfrac{(xy + 1)^2}{x^2 + y^2 + 2x - 2y + 2}.$

42. How can the function $f(x,y) = \dfrac{x^2 + y^2 - x^3y^3}{x^2 + y^2}$ be defined at the origin so that the resulting function becomes continuous at all points in the x-y plane?

43. What is the domain of $f(x,y) = \dfrac{x - y}{x^2 - y^2}$? Does $f(x,y)$ have a limit as $(x,y) \to (1,1)$? How can the domain of f be extended so that the resulting function is continuous at (1,1)?

44. The definition of limit (Defintion 2) can be modified to the following analogue of the definition of **one-sided limits** given in Section 1.4: we say that $f(x,y)$ tends to the limit L as (x,y) approaches (a,b) if

i) every disc of positive radius about (a,b) contains points of the doman of f distinct from (a,b), and

ii) for every positive number ϵ there exists a positive number $\delta = \delta(\epsilon)$ such that $|f(x,y) - L| < \epsilon$ whenever (x,y) lies in the domain of f and

$$0 < \sqrt{(x - a)^2 + (y - b)^2} < \delta.$$

Show that the function $f(x,y)$ of Exercise 43 does have a limit as $(x,y) \to (1,1)$ according to this definition.

45. Given a function $f(x,y)$ and a point (a,b) in its domain we define single-variable functions g and h as follows:

$$g(x) = f(x,b), \quad h(y) = f(a,y).$$

If g is continuous at $x = a$ and h is continuous at $y = b$ does it follow that f is continuous at (a,b)? Justify your answer.

10.2 PARTIAL DERIVATIVES

In attempting to extend the concepts and techniques of single-variable calculus to functions of more than one variable it is convenient to begin by isolating the dependence of such functions on one variable at a time. Thus we define two **partial derivatives** of a function of two variables—one with respect to each independent variable.

Definition 3

The **first partial derivative** of the function $f(x,y)$ **with respect to the variable** x is the function $\partial f/\partial x$ whose value at (x,y) is given by

$$\frac{\partial f}{\partial x} = \lim_{h \to 0} \frac{f(x + h,y) - f(x,y)}{h}$$

provided the limit exists. Similarly, the **first partial derivative of** f **with respect to** y is the function $\partial f/\partial y$ whose value at (x,y) is given by

$$\frac{\partial f}{\partial y} = \lim_{k \to 0} \frac{f(x,y + k) - f(x,y)}{k}$$

provided the limit exists.

Observe that $\partial f/\partial x$ is just the ordinary first derivative of f considered as a function of x only (regarding y as a constant parameter). At any particular point (a,b), $\partial f/\partial x$ measures the rate of change of $f(x,y)$ with respect to x at $x = a$ while y is held fixed at b. In geometric terms, the surface $z = f(x,y)$ intersects the vertical plane $y = b$ in a curve. If we take horizontal and vertical lines through the point $(0,b,0)$ as coordinate axes in that vertical plane, the curve has equation $z = f(x,b)$, and its slope at $x = a$ is the value of $\partial f/\partial x$ at (a,b). See Fig. 10.12a. Similarly, $\partial f/\partial y$ is the first derivative of $f(x,y)$ considered as a function of y alone, and represents the rate of change of $f(x,y)$ with respect to y as x is held fixed. The surface $z = f(x,y)$ intersects the vertical plane $x = a$ in a curve $z = f(a,y)$ whose slope at $y = b$ is the value of $\partial f/\partial y$ at (a,b). See Fig. 10.12b.

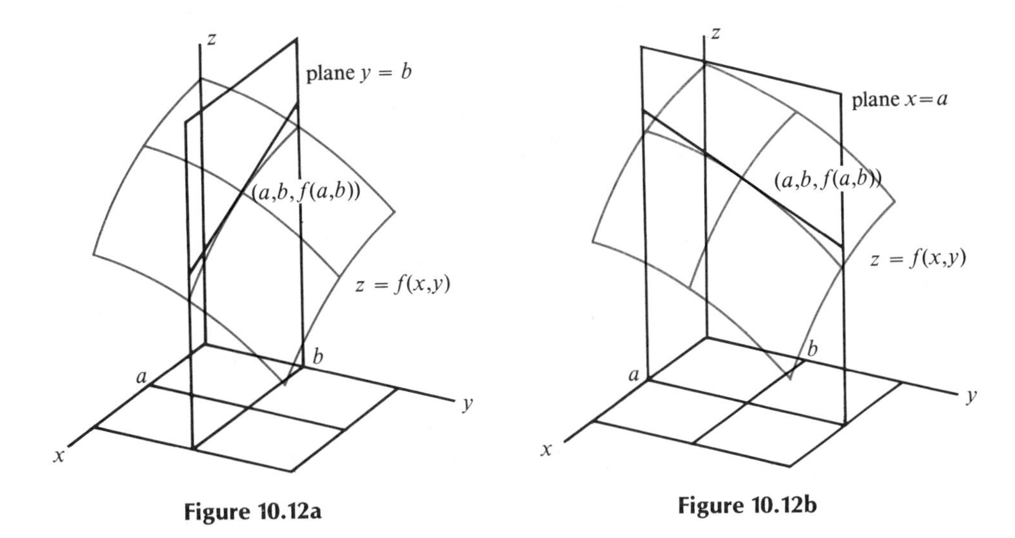

Figure 10.12a **Figure 10.12b**

Various other notations are used to denote the partial derivatives of $z = f(x,y)$ considered as functions of x and y:

$$\frac{\partial z}{\partial x} = \frac{\partial f}{\partial x} = f_1(x,y) = D_1 f(x,y),$$
$$\frac{\partial z}{\partial y} = \frac{\partial f}{\partial y} = f_2(x,y) = D_2 f(x,y).$$

The subscripts "1" and "2" in the latter notations refer to the first and second variables of f. We prefer the notation f_1 to the alternative notation f_x to avoid problems later when we consider compositions of functions (that is, the chain rule). $f_1(x^2,x^3)$ refers to the derivative of $f(u,v)$ with respect to u evaluated at $u = x^2$, $v = x^3$, while $f_x(x^2,x^3)$ ought to refer to the derivative with respect to x of the function $f(x^2,x^3)$, a function of one variable.

Values at a particular point (a,b) are denoted similarly:

$$\left.\frac{\partial z}{\partial x}\right|_{(a,b)} = \left.\frac{\partial f}{\partial x}\right|_{(a,b)} = f_1(a,b) = D_1 f(a,b),$$
$$\left.\frac{\partial z}{\partial y}\right|_{(a,b)} = \left.\frac{\partial f}{\partial y}\right|_{(a,b)} = f_2(a,b) = D_2 f(a,b).$$

Example 1 Find $\partial z/\partial x$ and $\partial z/\partial y$ if $z = x^3y^2 + x^4y + y^4$.

Solution $\partial z/\partial x = 3x^2y^2 + 4x^3y$, $\partial z/\partial y = 2x^3y + x^4 + 4y^3$.

Example 2 Find $f_1(0,\pi)$ if $f(x,y) = e^{xy}\cos(x+y)$.

Solution $f_1(x,y) = y\, e^{xy}\cos(x+y) - e^{xy}\sin(x+y)$,
$f_1(0,\pi) = \pi\, e^0 \cos\pi - e^0 \sin\pi = -\pi$.

Example 3 If f is an everywhere differentiable function of one variable show that $z = f(x/y)$ satisfies the **partial differential equation**

$$x\frac{\partial z}{\partial x} + y\frac{\partial z}{\partial y} = 0.$$

Solution By the chain rule $\frac{\partial z}{\partial x} = f'\left(\frac{x}{y}\right)\left(\frac{1}{y}\right)$ and $\frac{\partial z}{\partial y} = f'\left(\frac{x}{y}\right)\left(\frac{-x}{y^2}\right)$. Hence

$$x\frac{\partial z}{\partial x} + y\frac{\partial z}{\partial y} = f'\left(\frac{x}{y}\right)\left(\frac{x}{y} - \frac{x}{y}\right) = 0.$$

Definition 3 can be extended in the obvious way to cover functions of more than two variables. If f is a function of the n variables $x_1, x_2, \ldots, x_n$ then f has n first partial derivatives, $\partial f/\partial x_1, \partial f/\partial x_2, \ldots, \partial f/\partial x_n$. (The notations $f_1(x_1,\ldots,x_n), \ldots, f_n(x_1,\ldots,x_n)$ are also appropriate.)

Example 4 $$\frac{\partial}{\partial z}\frac{2xy}{zx + yz} = -\frac{2xy}{(xz + yz)^2}(x + y).$$

Tangent Planes

If the graph $z = f(x,y)$ is a smooth surface near the point P with coordinates $(a,b,f(a,b))$ then that graph will have a **tangent plane** at P. This tangent plane will contain the tangent lines to all smooth curves through P which lie in the surface $z = f(x,y)$. In particular, the two straight lines which are tangent at P to the two curves in which the surface $z = f(x,y)$ intersects the vertical planes $y = b$ and $x = a$ will lie in that tangent plane. See Figs. 10.12a and 10.12b, and Fig. 10.13.

The tangent plane to $z = f(x,y)$ at $(a,b,f(a,b))$ will have a linear equation of the form

$$z = f(a,b) + A(x - a) + B(y - b)$$

with certain constants A and B. The intersection of this plane with the vertical plane $y = b$ is the straight line with equations

$$z = f(a,b) + A(x-a), \quad y = b.$$

As observed above, this line is tangent to the curve

$$z = f(x,b), \quad y = b,$$

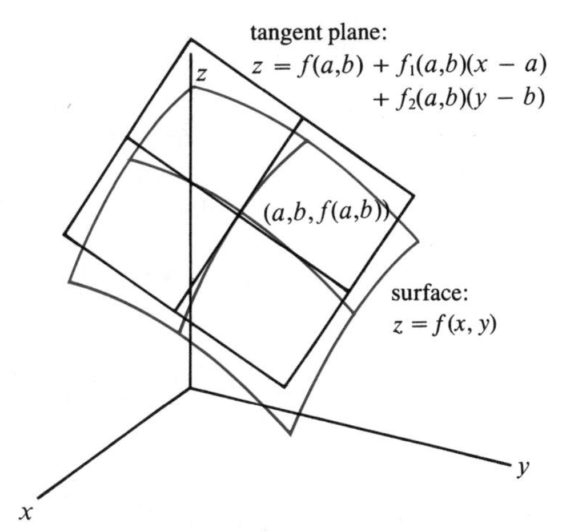

Figure 10.13

so we must have $A = f_1(a,b)$, the slope of that curve at $x = a$. Similarly, the tangent plane meets the vertical plane $x = a$ in the straight line

$$z = f(a,b) + B(y - b), \quad x = a,$$

which is tangent to the curve

$$z = f(a,y), \quad x = a.$$

Hence $B = f_2(a,b)$. Therefore the tangent plane has equation

$$\boxed{z = f(a,b) + f_1(a,b)(x - a) + f_2(a,b)(y - b).}$$

We shall obtain this result by a different method in Section 10.4 when we consider tangency in a more general context.

Example 5 Find an equation of the tangent plane to the graph $z = \sin(xy)$ at the point $(\pi/3, -1)$.

Solution $\partial z/\partial x = y\cos(xy)$, $\partial z/\partial y = x\cos(xy)$.
At $(\pi/3, -1)$ we have $z = -\sqrt{3}/2$, $\partial z/\partial x = -1/2$, and $\partial z/\partial y = \pi/6$.
Thus the tangent plane has equation

$$z = \frac{-\sqrt{3}}{2} - \frac{1}{2}\left(x - \frac{\pi}{3}\right) + \frac{\pi}{6}(y + 1),$$

or,

$$z = -\frac{1}{2}x + \frac{\pi}{6}y - \frac{1}{2}\sqrt{3} + \frac{\pi}{3}.$$

Higher Order Derivatives

Second and higher order partial derivatives are calculated by taking partial derivatives of already calculated partial derivatives. The order in which the differentiations are performed is indicated by the notations in an obvious way. If $z = f(x,y)$ then we can calculate four partial derivatives of second order:

two **pure** second partial derivatives with respect to x or y,

$$\frac{\partial^2 z}{\partial x^2} = \frac{\partial}{\partial x}\frac{\partial z}{\partial x} = f_{11}(x,y), \quad \frac{\partial^2 z}{\partial y^2} = \frac{\partial}{\partial y}\frac{\partial z}{\partial y} = f_{22}(x,y),$$

and two **mixed** second partial derivatives with respect to x and y,

$$\frac{\partial^2 z}{\partial x \partial y} = \frac{\partial}{\partial x}\frac{\partial z}{\partial y} = f_{21}(x,y), \quad \frac{\partial^2 z}{\partial y \partial x} = \frac{\partial}{\partial y}\frac{\partial z}{\partial x} = f_{12}(x,y).$$

similarly, if $w = f(x,y,z)$ then

$$\frac{\partial^5 w}{\partial y \partial x \partial y^2 \partial z} = f_{32212}(x,y,z) = \frac{\partial}{\partial y}\frac{\partial}{\partial x}\frac{\partial}{\partial y}\frac{\partial}{\partial y}\frac{\partial w}{\partial z}.$$

Example 6 Find the four second partial derivatives of $f(x,y) = x^3y^4$.

Solution

$$f_1(x,y) = 3x^2y^4, \qquad f_2(x,y) = 4x^3y^3,$$

$$f_{11}(x,y) = \frac{\partial}{\partial x}(3x^2y^4) = 6xy^4, \qquad f_{21}(x,y) = \frac{\partial}{\partial x}(4x^3y^3) = 12x^2y^3,$$

$$f_{12}(x,y) = \frac{\partial}{\partial y}(3x^2y^4) = 12x^2y^3, \qquad f_{22}(x,y) = \frac{\partial}{\partial y}(4x^3y^3) = 12x^3y^2.$$

Example 7 Calculate $f_{322}(x,y,z)$ and $f_{223}(x,y,z)$ for the function $f(x,y,z) = e^{x-2y+3z}$.

Solution

$$\begin{aligned} f_{322}(x,y,z) &= \frac{\partial}{\partial y}\frac{\partial}{\partial y}\frac{\partial}{\partial z}[e^{x-2y+3z}] = \frac{\partial}{\partial y}\frac{\partial}{\partial y}[3e^{x-2y+3z}] \\ &= \frac{\partial}{\partial y}[-6e^{x-2y+3z}] = 12e^{x-2y+3z}, \\ f_{223}(x,y,z) &= \frac{\partial}{\partial z}\frac{\partial}{\partial y}\frac{\partial}{\partial y}[e^{x-2y+3z}] = \frac{\partial}{\partial z}\frac{\partial}{\partial y}[-2e^{x-2y+3z}] \\ &= \frac{\partial}{\partial z}[4e^{x-2y+3z}] = 12e^{x-2y+3z}. \end{aligned}$$

In both of the examples above observe that the two mixed partial derivatives taken with respect to the same variables but in different orders turned out to be equal. This is not a coincidence. It will occur whenever the mixed partial derivatives involved are **continuous**. The following theorem states formally the simplest case of this important phenomenon.

Theorem 1 (*Equality of mixed partials*) At any point (a,b) where the mixed partial derivatives f_{12} and f_{21} exist and are continuous, we have $f_{12}(a,b) = f_{21}(a,b)$.

Proof If f_{12} and f_{21} are continuous at (a,b) then f, f_1, f_2, f_{12} and f_{21} exist, and f, f_1, and f_2 are continuous, in some disc of positive radius centred at (a,b). Let h and k have absolute value sufficiently small that the point $(a + h,b + k)$ lies in this disc. Then so do all points of the rectangle with these two points as diagonally opposite corners. See Fig. 10.14.

Let $Q = f(a + h,b + k) - f(a + h,b) - f(a,b + k) + f(a,b)$, and define single-variable functions $u(x)$ and $v(y)$ by

$$u(x) = f(x,b + k) - f(x,b), \quad v(y) = f(a + h,y) - f(a,y).$$

Evidently $Q = u(a + h) - u(a)$ and also $Q = v(b + k) - v(b)$. By the mean-value theorem (Theorem 2 of Section 2.8) there is a point $a + \theta_1 h$ between $a + h$ and a (that is, satisfying $0 < \theta_1 < 1$) such that

$$\begin{aligned} Q = u(a + h) - u(a) &= hu'(a + \theta_1 h) \\ &= h[f_1(a + \theta_1 h,b + k) - f_1(a + \theta_1 h,b)]. \end{aligned}$$

Now apply the mean-value theorem again to f_1 considered as a function of its second variable, and obtain another number θ_2 satisfying $0 < \theta_2 < 1$ such that

$$f_1(a + \theta_1 h,b + k) - f_1(a + \theta_1 h,b) = kf_{12}(a + \theta_1 h,b + \theta_2 k).$$

Thus $Q = hkf_{12}(a + \theta_1 h,b + \theta_2 k)$. Two similar applications of the mean-value theorem starting from $Q = v(b + k) - v(b)$ yield two numbers θ_3 and θ_4, each between 0 and 1 such that $Q = khf_{21}(a + \theta_4 h,b + \theta_3 k)$. Thus

$$f_{12}(a + \theta_1 h,b + \theta_2 k) = f_{21}(a + \theta_4 h,b + \theta_3 k).$$

Since f_{12} and f_{21} are continuous at (a,b) we can let $h \to 0$ and $k \to 0$ to obtain $f_{12}(a,b) = f_{21}(a,b)$, as required. □

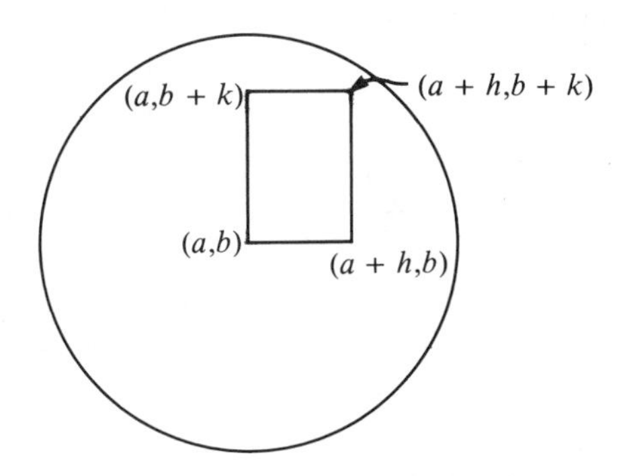

Figure 10.14

Example 8 Show that the function $z = e^{kx}\cos(ky)$ satisfies the partial differential equation

$$\frac{\partial^2 z}{\partial x^2} + \frac{\partial^2 z}{\partial y^2} = 0$$

at every point (x,y).

Solution

$$\frac{\partial z}{\partial x} = k\,e^{kx}\cos(ky), \qquad \frac{\partial z}{\partial y} = -k\,e^{kx}\sin(ky),$$

$$\frac{\partial^2 z}{\partial x^2} = k^2\,e^{kx}\cos(ky), \qquad \frac{\partial^2 z}{\partial y^2} = -k^2\,e^{kx}\cos(ky).$$

Thus $\dfrac{\partial^2 z}{\partial x^2} + \dfrac{\partial^2 z}{\partial y^2} = k^2\,e^{kx}\cos(ky) - k^2\,e^{kx}\cos(ky) = 0.$

Remark The partial differential equation in Example 8 is called the (two-dimensional) **Laplace equation**. A function having continuous second partial derivatives in a region of the plane is said to be **harmonic** there if it satisfies Laplace's equation. Harmonic functions play an important role in many areas of mathematics and its applications. Besides being central to the study of differentiable functions of a **complex variable**, such functions are used to model various physical quantities such as steady-state temperature distributions, fluid flows, and electric and magnetic potential fields. Laplace's equation, and therefore harmonic functions, can be considered in any number of dimensions. (See Exercises 40 and 41.)

Example 9 If f and g are any twice differentiable functions of one variable show that $w = f(x - ct) + g(x + ct)$ satisfies the differential equation

$$\frac{\partial^2 w}{\partial t^2} = c^2\frac{\partial^2 w}{\partial x^2}.$$

Solution

$$\frac{\partial w}{\partial t} = -cf'(x - ct) + cg'(x + ct), \qquad \frac{\partial w}{\partial x} = f'(x - ct) + g'(x + ct),$$

$$\frac{\partial^2 w}{\partial t^2} = c^2f''(x - ct) + c^2g''(x + ct), \qquad \frac{\partial^2 w}{\partial x^2} = f''(x - ct) + g''(x + ct).$$

Thus w satisfies the given differential equation.

Remark The differential equation of Example 8 is called the (one-dimensional) **wave equation**. If t measures time then $f(x - ct)$ represents a waveform travelling to the right along the x-axis with speed c. See Fig. 10.15. Similarly, $g(x + ct)$ represents a waveform travelling to the left with speed c.

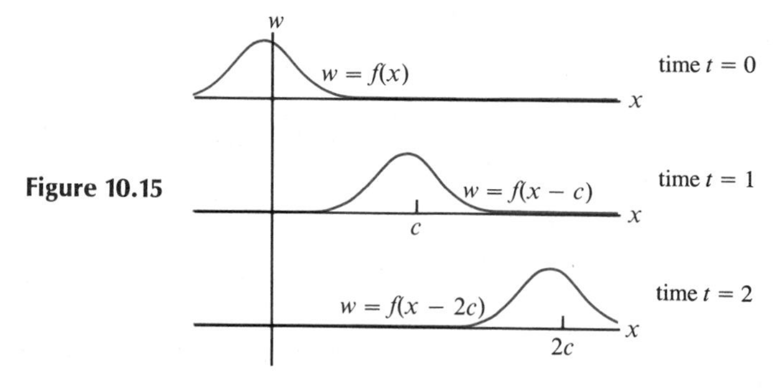

Figure 10.15

EXERCISES

In Exercises 1-9 find all the first partial derivatives of the functions specified. Also evaluate these at the given point.

1. $f(x,y) = x - y + 2$; $(3,2)$

2. $f(x,y) = xy + x^2$; $(2,0)$

3. $f(x,y,z) = x^3y^4z^5$; $(0,-1,-1)$

4. $g(x,y,z) = yz/(x + z)$; $(1,1,1)$

5. $z = \operatorname{Arctan}(y/x)$; $(-1,1)$

6. $w = \ln(1 + e^{xyz})$; $(2,0,-1)$

7. $f(x,y) = \sin(x\sqrt{y})$; $(\pi/3,4)$

8. $f(x,y) = 1/\sqrt{x^2 + y^2}$; $(-3,4)$

9. $w = x^{y \ln z}$; $(e,2,e)$

10. Find $f_1(0,0)$ and $f_2(0,0)$ if $f(x,y) = \begin{cases} \dfrac{x^2 - 2y^2}{x - y} & \text{if } x \neq y \\ 0 & \text{if } x = y \end{cases}$.

In Exercises 11-20 find the equation of the tangent plane to the graph of the given function at the point specified.

11. $f(x,y) = x^2 - y^2$ at $(-2,1)$

12. $f(x,y) = (x - y)/(x + y)$ at $(1,1)$

13. $f(x,y) = \cos(x/y)$ at $(\pi,4)$

14. $f(x,y) = e^{xy}$ at $(2,0)$

15. $f(x,y) = \dfrac{x}{x^2 + y^2}$ at $(1,2)$

16. $f(x,y) = y\, e^{-x^2} \sin z$ at $(0,1,\pi/3)$

17. $f(x,y) = \ln(x^2 + y^2)$ at $(1,-2)$

18. $f(x,y) = \dfrac{2xy}{x^2 + y^2}$ at $(0,2)$

19. $f(x,y) = \operatorname{Arctan}(y/x)$ at $(1,-1)$

20. $f(x,y) = \sqrt{1 + x^3y^2}$ at $(2,1)$

In exercises 21-27 show that the given function satisfies the given partial differential equation.

21. $z = x\, e^y$, $\quad x\dfrac{\partial z}{\partial x} = \dfrac{\partial z}{\partial y}$

22. $z = \dfrac{x + y}{x - y}$, $\quad x\dfrac{\partial z}{\partial x} + y\dfrac{\partial z}{\partial y} = 0$

23. $z = \sqrt{x^2 + y^2}$, $\quad x\dfrac{\partial z}{\partial x} + y\dfrac{\partial z}{\partial y} = z$

24. $w = x^2 + yz$, $\quad x\dfrac{\partial w}{\partial x} + y\dfrac{\partial w}{\partial y} + z\dfrac{\partial w}{\partial z} = 2w$

25. $w = \dfrac{1}{x^2 + y^2 + z^2}$, $\quad x\dfrac{\partial w}{\partial x} + y\dfrac{\partial w}{\partial y} + z\dfrac{\partial w}{\partial z} = -2w$

26. $z = f(x^2 + y^2)$, where f is any differentiable function of 1 variable, $y\dfrac{\partial z}{\partial x} - x\dfrac{\partial z}{\partial y} = 0$

27. $z = f(x^2 - y^2)$, where f is any differentiable function of 1 variable, $y\dfrac{\partial z}{\partial x} + x\dfrac{\partial z}{\partial y} = 0$

In Exercises 28-33 find all the second partial derivatives of the given function.

28. $f(x,y) = x^2 + y^2$

29. $z = x^2(1 + y^2)$

30. $z = \sqrt{3x^2 + y^2}$

31. $w = x^3y^3z^3$

32. $f(x,y) = \ln(1 + \sin(xy))$

33. $z = x\, e^y - y\, e^x$

34. How many mixed partial derivatives of order 3 can a function of three variables have? If they are all continuous how many different values can they have? Find the three mixed partials of order 3 for $f(x,y,z) = x\, e^{xy} \cos(xz)$ which involve two differentiations with respect to z and one with respect to x.

Show that the functions in Exercises 35-39 are harmonic in the plane regions indicated.

35. $f(x,y) = e^{kx} \sin(ky)$ in the whole plane

36. $f(x,y) = A(x^2 - y^2) + Bxy$ in the whole plane (A, B are constants)

37. $f(x,y) = 3x^2y - y^3$ in the whole plane. (Can you think of another polynomial of degree 3 in x and y which is also harmonic?)

38. $f(x,y) = \ln(x^2 + y^2)$ everywhere except at the origin

39. $f(x,y) = \text{Arctan}(y/x)$ except at points on the y-axis

40. Show that $w = e^{3x+4y}\sin(5z)$ is a harmonic function in the whole 3-dimensional space, that is, it satisfies everywhere the 3-dimensional Laplace equation

$$\frac{\partial^2 w}{\partial x^2} + \frac{\partial^2 w}{\partial y^2} + \frac{\partial^2 w}{\partial z^2} = 0.$$

41. Show that if $f(x,y)$ is harmonic in the plane then each of the following functions is harmonic in three dimensional space:
a) $z\,f(x,y)$, b) $x\,f(y,z)$, c) $y\,f(z,x)$.
What condition should the constants a, b and c satisfy to ensure that $f(ax + by,cz)$ is harmonic?

42. Suppose the functions $u(x,y)$ and $v(x,y)$ have continuous second partial derivatives and satisfy the **Cauchy-Riemann equations**

$$\frac{\partial u}{\partial x} = \frac{\partial v}{\partial y}, \quad \frac{\partial v}{\partial x} = -\frac{\partial u}{\partial y}.$$

Show that u and v are both harmonic.

43. Show that $u(x,t) = t^{-\frac{1}{2}}e^{-x^2/4t}$ satisfies the partial differential equation

$$\frac{\partial u}{\partial t} = \frac{\partial^2 u}{\partial x^2}.$$

This equation is called the (one-dimensional) heat equation as it models heat diffusion in an insulated rod (with u representing the temperature at position x at time t) and other similar phenomena.

44. Show that $u(x,y,t) = t^{-1}e^{-(x^2+y^2)/4t}$ satisfies the two-dimensional heat equation

$$\frac{\partial u}{\partial t} = \frac{\partial^2 u}{\partial x^2} + \frac{\partial^2 u}{\partial y^2}.$$

Comparing this result with that of Exercise 43, guess a solution of the three-dimensional heat equation

$$\frac{\partial u}{\partial t} = \frac{\partial^2 u}{\partial x^2} + \frac{\partial^2 u}{\partial y^2} + \frac{\partial^2 u}{\partial z^2}.$$

Verify your guess.

45. Let $f(x,y) = \begin{cases} \dfrac{2xy}{x^2 + y^2} & \text{if } (x,y) \neq (0,0) \\ 0 & \text{if } (x,y) = (0,0) \end{cases}$

As noted in Example 7 of Section 10.1, f is not continuous at (0,0). Therefore its graph is not smooth. Show, however, that $f_1(0,0)$ and $f_2(0,0)$ both exist. (Use Definition 3). Hence the existence of partial derivatives does not imply continuity for functions of several variables. This is in contrast to the single-variable case—see Theorem 1 of Section 2.3.

46. Let $g(x,y) = (x^2 - y^2)f(x,y)$ where f is the function defined in Exercise 45. Calculate $g_1(x,y)$ and $g_2(x,y)$ for $(x,y) \neq (0,0)$. Also calculate $g_1(0,0)$ and $g_2(0,0)$. Finally, show that $g_{21}(0,0) = 2$ but $g_{12}(0,0) = -2$. Does this result contradict Theorem 1? Explain why.

10.3 THE CHAIN RULE AND DIFFERENTIALS

As we did with functions of one variable in Section 2.4, we shall begin this section by stating (some several-variable versions of) the chain rule to illustrate the rule before we attempt to prove the rule formally.

Let us start with a function of two variables, in which each independent variable is itself a function of two other variables:

$$z = f(x,y) \quad \textit{where} \quad x = u(s,t),\ y = v(s,t).$$

We may then form the composite function

$$z = g(s,t) = f(u(s,t),v(s,t)).$$

Let us assume that f, u and v have continuous first partial derivatives with respect to their respective variables. Then so does g, and the first partial derivatives of g are given by

$$g_1(s,t) = f_1(u(s,t),v(s,t))u_1(s,t) + f_2(u(s,t),v(s,t))v_1(s,t),$$
$$g_2(s,t) = f_1(u(s,t),v(s,t))u_2(s,t) + f_2(u(s,t),v(s,t))v_2(s,t),$$

or, expressed more simply using Leibniz notation,

$$\frac{\partial z}{\partial s} = \frac{\partial z}{\partial x}\frac{\partial x}{\partial s} + \frac{\partial z}{\partial y}\frac{\partial y}{\partial s}, \qquad \frac{\partial z}{\partial t} = \frac{\partial z}{\partial x}\frac{\partial x}{\partial t} + \frac{\partial z}{\partial y}\frac{\partial y}{\partial t}.$$

The first term in the expression for $g_1(s,t)$ (or that for $\partial z/\partial s$) represents the contribution to the rate of change of z with respect to s which comes from the dependence of z on s through the first variable of f; the second term is the contribution coming from the second variable of f. Note the significance of the subscripts on the various partial derivatives: the "1" in $g_1(s,t)$ refers to differentiation with respect to s, the first variable on which g depends; by contrast, the "1" in $f_1(u(s,t),v(s,t))$ refers to differentiation with respect to x, the first variable on which f depends.

Example 1 If $z = \sin(x^2y)$ where $x = st^2$ and $y = s^2 + \dfrac{1}{t}$, find $\dfrac{\partial z}{\partial s}$ and $\dfrac{\partial z}{\partial t}$

a) by direct substitution and the single-variable chain rule, and

b) by using the (two-variable) chain rule.

Solution a) by direct substitution and the single-variable chain rule

$$z = \sin\left((st^2)^2\left(s^2 + \frac{1}{t}\right)\right) = \sin(s^4t^4 + s^2t^3)$$

$$\frac{\partial z}{\partial s} = (4s^3t^4 + 2st^3)\cos(s^4t^4 + s^2t^3),$$

$$\frac{\partial z}{\partial t} = (4s^4t^3 + 3s^2t^2)\cos(s^4t^4 + s^2t^3).$$

b) Using the two-variable chain rule

$$\begin{aligned}\frac{\partial z}{\partial s} &= \frac{\partial z}{\partial x}\frac{\partial x}{\partial s} + \frac{\partial z}{\partial y}\frac{\partial y}{\partial s}\\ &= [2xy\cos(x^2y)]t^2 + [x^2\cos(x^2y)]2s\\ &= \left[2st^2\left(s^2 + \frac{1}{t}\right)t^2 + 2s^3t^4\right]\cos(s^4t^4 + s^2t^3)\\ &= (4s^3t^4 + 2st^3)\cos(s^4t^4 + s^2t^3),\end{aligned}$$

$$\begin{aligned}\frac{\partial z}{\partial t} &= \frac{\partial z}{\partial x}\frac{\partial x}{\partial t} + \frac{\partial z}{\partial y}\frac{\partial y}{\partial t}\\ &= [2xy\cos(x^2y)]2st + [x^2\cos(x^2y)]\left(\frac{-1}{t^2}\right)\\ &= \left[2st^2\left(s^2 + \frac{1}{t}\right)2st + s^2t^4\left(\frac{-1}{t^2}\right)\right]\cos(s^4t^4 + s^2t^3)\\ &= (4s^4t^3 + 3s^2t^2)\cos(s^4t^4 + s^2t^3).\end{aligned}$$

Note that we still had to use direct substitution on the derivatives obtained using the chain rule in order to show they were in fact the same as those obtained without using (the two variable form of) the chain rule. (Of course, the one variable form of the chain rule was used even in a).

Now suppose that $z = f(x,y)$ where $x = g(t)$ and $y = h(t)$. Then $z = Q(t) = f(g(t),h(t))$, and z may be regarded as a function of the single variable t. Since z depends on t through both of the variables of f the chain rule for dz/dt still has two terms:

$$\frac{dz}{dt} = \frac{\partial z}{\partial x}\frac{dx}{dt} + \frac{\partial z}{\partial y}\frac{dy}{dt}.$$

Note the use of the straight "d" for derivatives of functions of only one variable: dz/dt means $Q'(t)$ while $\partial z/\partial x$ means $f_1(x,y)$.

In a similar vein, if $z = f(x)$ where $x = g(s,t)$ then we can regard z as being a function of the two variables s and t:

$$z = h(s,t) = f(g(s,t)).$$

Using the chain rule here gives

$$\begin{aligned}\frac{\partial z}{\partial s} &= h_1(s,t) = \frac{dz}{dx}\frac{\partial x}{\partial s} = f'(g(s,t))g_1(s,t),\\ \frac{\partial z}{\partial t} &= h_2(s,t) = \frac{dz}{dx}\frac{\partial x}{\partial t} = f'(g(s,t))g_2(s,t).\end{aligned}$$

In this case we are really only using the single-variable form of the chain rule as developed in Section 2.4.

The following example involves a "hybrid" application of the chain rule to a function depending both directly and indirectly on the variable of differentiation.

Example 2 Find dz/dt where $z = f(x,y,t)$ and $x = g(t)$, $y = h(t)$. (Assume that f, g and h all have continuous derivatives.)

Solution z depends on t through each of the three variables of f so there will be three terms in the appropriate chain rule:

$$\frac{dz}{dt} = \frac{\partial z}{\partial x}\frac{dx}{dt} + \frac{\partial z}{\partial y}\frac{dy}{dt} + \frac{\partial z}{\partial t}$$
$$= f_1(x,y,t)g'(t) + f_2(x,y,t)h'(t) + f_3(x,y,t).$$

Remark In Example 2 we can easily distinguish between the meaning of the symbols dz/dt and $\partial z/\partial t$. If, however, we had been dealing with a function of the form $z = f(x,y,s,t)$ where $x = g(s,t)$ and $y = h(s,t)$ then the meaning of the symbol $\partial z/\partial t$ would be unclear; it could refer to either the simple partial derivative of f with respect to t with x, y and s held constant, or to the composite derivative of $f(g(s,t),h(s,t),s,t)$ with respect to t with only s held constant. We distinguish these notationally by showing which variables are held constant as follows:

$$\left.\frac{\partial z}{\partial t}\right|_{x,y,s} = \frac{\partial}{\partial t} f(x,y,s,t) = f_4(x,y,s,t),$$

$$\left.\frac{\partial z}{\partial t}\right|_{s} = \frac{\partial}{\partial t} f(g(s,t),h(s,t),s,t) = f_1 g_2 + f_2 h_2 + f_4.$$

In applications the variables to be held constant in a given differentiation will usually be clear from the context.

Example 3 Atmospheric temperature depends on position and time. If we denote position by means of three spatial coordinates (x,y,z) and time by t then the temperature T can be represented by a function of four variables, $T(x,y,z,t)$. If a thermometer is attached to a weather balloon and moves through the atmosphere on a path with parametric equations $x = f(t)$, $y = g(t)$, $z = h(t)$, what is the rate of change of the temperature recorded by the thermometer at time t? Find the rate of change of the temperature at time $t = 1$ if

$$T(x,y,z,t) = \frac{e^{-z} \sin t}{5 + x^2 + y^2},$$

and if the balloon moves along the curve

$$x = f(t) = t,\ y = g(t) = 2t,\ z = h(t) = t - t^4.$$

Solution Here the rate must take into account the change of position of the thermometer as well as increasing time. It is given by

$$\frac{dT}{dt} = \frac{\partial T}{\partial x}\frac{dx}{dt} + \frac{\partial T}{\partial y}\frac{dy}{dt} + \frac{\partial T}{\partial z}\frac{dz}{dt} + \frac{\partial T}{\partial t}.$$

Note that the partial derivative $\partial T/\partial t$ refers only to the rate of change of temperature with respect to time at a fixed position in the atmosphere. For the special case we have, at $t = 1$, $x = 1$, $y = 2$, $z = 0$, $\partial x/\partial t = 1$, $\partial y/\partial t = 2$, $\partial z/\partial t = -3$ and

$$\partial T/\partial x = -2x\, e^{-z} \sin t/(5 + x^2 + y^2)^2 = -2 \sin(1)/100 = -\sin(1)/50$$
$$\partial T/\partial y = -2y\, e^{-z} \sin t/(5 + x^2 + y^2)^2 = -4 \sin(1)/100 = -\sin(1)/25$$

$$\partial T/\partial z = -e^{-z}\sin t/(5 + x^2 + y^2) = -\sin(1)/10$$
$$\partial T/\partial t = e^{-z}\cos t/(5 + x^2 + y^2) = \cos(1)/10.$$

Thus

$$\left.\frac{dT}{dt}\right|_{t=1} = -\sin(1)\left(\frac{1}{50} + \frac{2}{25} - \frac{3}{10}\right) + \frac{\cos(1)}{10}$$
$$\approx 0.2223$$

The discussion and examples above show that the chain rule for functions of several variables can take many different forms depending on the nature of the various functions being composed and the numbers of their variables. As a mnemonic device for determining the correct form of the chain rule in a given situation one can construct a simple chart showing the dependencies of the various variables on one another. Fig. 10.16 shows such a chart for the temperature function of Example 3.

The chain rule for dT/dt involves a term for every route from T to t in the chart. The route from T through x to t produces the term $\dfrac{\partial T}{\partial x}\dfrac{dx}{dt}$ and so on.

Example 4 Write the appropriate chain rule for $\partial z/\partial x|_y$, where $z = f(u,v,r)$, $u = g(x,y,r)$, $v = h(x,y,r)$ and $r = k(x,y)$.

Solution With only y held fixed the appropriate chart is shown in Fig. 10.17. There are five routes from z to x:

$$\left.\frac{\partial z}{\partial x}\right|_y = \frac{\partial z}{\partial u}\frac{\partial u}{\partial x} + \frac{\partial z}{\partial u}\frac{\partial u}{\partial r}\frac{\partial r}{\partial x} + \frac{\partial z}{\partial v}\frac{\partial v}{\partial x} + \frac{\partial z}{\partial v}\frac{\partial v}{\partial r}\frac{\partial r}{\partial x} + \frac{\partial z}{\partial r}\frac{\partial r}{\partial x}.$$

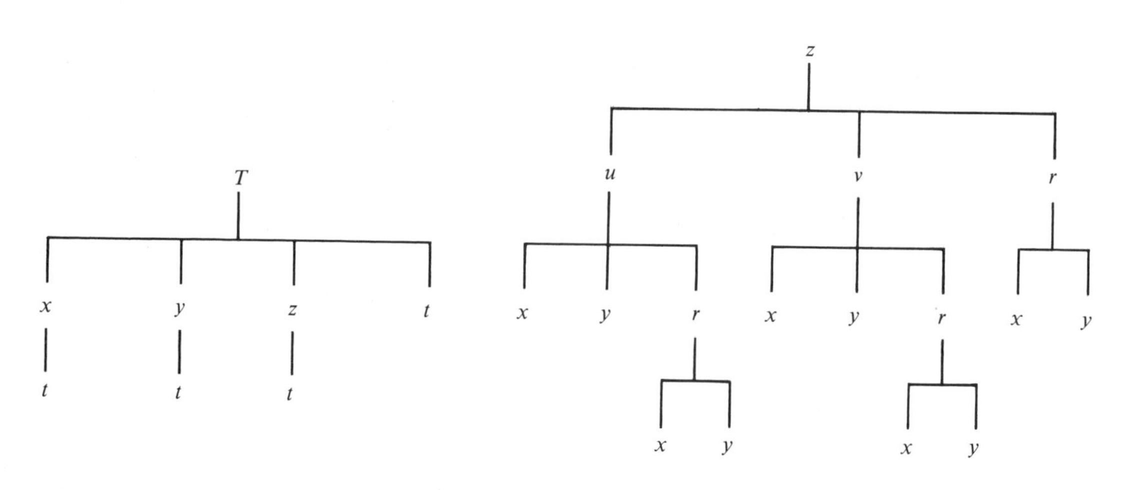

Figure 10.16 **Figure 10.17**

Homogeneous Functions

A function $f(x_1,x_2,\ldots,x_n)$ is said to be **positively homogeneous of degree** k if, for all values of its variables, and any positive real number t we have

$$f(tx_1,tx_2,\ldots tx_n) = t^k f(x_1,x_2,\ldots,x_n).$$

For example,

$$f(x,y) = x^2 + xy - y^2 \text{ is positively homogeneous of degree 2,}$$

$$f(x,y) = \frac{2xy}{x^2 + y^2} \text{ is positively homogeneous of degree 0,}$$

$$f(x,y,z) = \frac{x - y + z}{xy + z^2} \text{ is positively homogeneous of degree } -1.$$

Observe that a positively homogeneous function of degree 0 remains constant along rays from the origin. More generally, along such rays a positively homogeneous function of degree k grows or decays proportionally to the k'th power of distance from the origin. As we shall see in the next section, if f is sufficiently smooth, this observation is equivalent to the following theorem.

Theorem 2 (*Euler's Theorem*) If $f(x_1,\ldots,x_n)$ has continuous first partial derivatives, and is positively homogeneous of degree k then

$$x_1 f_1(x_1,\ldots,x_n) + x_2 f_2(x_1,\ldots,x_n) + \ldots + x_n f_n(x_1,\ldots,x_n) = kf(x_1,\ldots,x_n).$$

Proof Differentiate the equation $f(tx_1,\ldots,tx_n) = t^k f(x_1,\ldots,x_n)$ with respect to t to get

$$x_1 f_1(tx_1,\ldots,tx_n) + x_2 f_2(tx_1,\ldots,tx_n) + \ldots + x_n f_n(tx_1,\ldots,tx_n) = k\, t^{k-1} f(x_1,\ldots,x_n).$$

Now substitute $t = 1$ to get the desired result. □

Note that Exercises 12-15 of Section 10.2 are illustrations of this theorem.

Higher Order Derivatives

Applications of the chain rule can become quite complicated when higher order derivatives are involved. It is important to keep in mind at every stage which variables are independent of one another. The following example shows that a smooth function $v(r,\theta)$ expressed in terms of polar coordinates $x = r\cos\theta$, $y = r\sin\theta$, is harmonic (i.e. satisfies the two dimensional Laplace equation—see Example 7 of Section 10.2) provided

$$\frac{\partial^2 v}{\partial r^2} + \frac{1}{r}\frac{\partial v}{\partial r} + \frac{1}{r^2}\frac{\partial^2 v}{\partial \theta^2} = 0$$

Example 5 If $u(x,y)$ has continuous partial derivatives of second order, and if $v(r,\theta) = u(x,y)$ where $x = r\cos\theta$ and $y = r\sin\theta$, show that

$$\frac{\partial^2 v}{\partial r^2} + \frac{1}{r}\frac{\partial v}{\partial r} + \frac{1}{r^2}\frac{\partial^2 v}{\partial \theta^2} = \frac{\partial^2 u}{\partial x^2} + \frac{\partial^2 u}{\partial y^2}.$$

Solution First note that since $x = r\cos\theta$ and $y = r\sin\theta$,

$$\frac{\partial x}{\partial r} = \cos\theta, \quad \frac{\partial x}{\partial \theta} = -r\sin\theta, \quad \frac{\partial y}{\partial r} = \sin\theta, \quad \frac{\partial y}{\partial \theta} = r\cos\theta.$$

Thus

$$\frac{\partial v}{\partial r} = \frac{\partial u}{\partial x}\frac{\partial x}{\partial r} + \frac{\partial u}{\partial y}\frac{\partial y}{\partial r} = \cos\theta\,\frac{\partial u}{\partial x} + \sin\theta\,\frac{\partial u}{\partial y}.$$

Now differentiate with respect to r again. Remember that r and θ are independent variables, so the factors $\cos\theta$ and $\sin\theta$ can be regarded as constants. However $\partial u/\partial x$ and $\partial u/\partial y$ depend on x and y and therefore on r and θ.

$$\begin{aligned}\frac{\partial^2 v}{\partial r^2} &= \cos\theta\,\frac{\partial}{\partial r}\frac{\partial u}{\partial x} + \sin\theta\,\frac{\partial}{\partial r}\frac{\partial u}{\partial y}\\ &= \cos\theta\left[\cos\theta\,\frac{\partial^2 u}{\partial x^2} + \sin\theta\,\frac{\partial^2 u}{\partial y\partial x}\right] + \sin\theta\left[\cos\theta\,\frac{\partial^2 u}{\partial x\partial y} + \sin\theta\,\frac{\partial^2 u}{\partial y^2}\right]\\ &= \cos^2\theta\,\frac{\partial^2 u}{\partial x^2} + 2\cos\theta\sin\theta\,\frac{\partial^2 u}{\partial x\partial y} + \sin^2\theta\,\frac{\partial^2 u}{\partial y^2}.\end{aligned}$$

We have used the equality of mixed partials to obtain the last line. Similarly,

$$\frac{\partial v}{\partial \theta} = -\,r\sin\theta\,\frac{\partial u}{\partial x} + r\cos\theta\,\frac{\partial u}{\partial y}.$$

When we differentiate with respect to θ a second time we can regard r as constant, but each term is still a product of two functions which depend on θ:

$$\begin{aligned}\frac{\partial^2 v}{\partial \theta^2} &= -r\left[\cos\theta\,\frac{\partial u}{\partial x} + \sin\theta\,\frac{\partial}{\partial \theta}\frac{\partial u}{\partial x}\right] + r\left[-\sin\theta\,\frac{\partial u}{\partial y} + \cos\theta\,\frac{\partial}{\partial \theta}\frac{\partial u}{\partial y}\right]\\ &= -\,r\frac{\partial v}{\partial r} - r\sin\theta\left[-r\sin\theta\,\frac{\partial^2 u}{\partial x^2} + r\cos\theta\,\frac{\partial^2 u}{\partial y\partial x}\right]\\ &\qquad + r\cos\theta\left[-r\sin\theta\,\frac{\partial^2 u}{\partial x\partial y} + r\cos\theta\,\frac{\partial^2 u}{\partial y^2}\right]\\ &= -\,r\frac{\partial v}{\partial r} + r^2\sin^2\theta\,\frac{\partial^2 u}{\partial x^2} - 2r^2\sin\theta\cos\theta\,\frac{\partial^2 u}{\partial x\partial y} + r^2\cos^2\theta\,\frac{\partial^2 u}{\partial y^2}.\end{aligned}$$

Combining these results we obtain.

$$\frac{\partial^2 v}{\partial r^2} + \frac{1}{r}\frac{\partial v}{\partial r} + \frac{1}{r^2}\frac{\partial^2 v}{\partial \theta^2} = \frac{\partial^2 u}{\partial x^2} + \frac{\partial^2 u}{\partial y^2}.$$

Differentials and Differentiability

If the first partial derivatives of a function $f(x_1,x_2,\ldots,x_n)$ exist at a point we may construct a differential of the function in a manner similar to that used for functions of one variable:

$$df = \frac{\partial f}{\partial x_1}\,dx_1 + \frac{\partial f}{\partial x_2}\,dx_2 + \ldots + \frac{\partial f}{\partial x_n}\,dx_n$$

$$= f_1(x_1,x_2,\ldots,x_n)dx_1 + \ldots + f_n(x_1,x_2,\ldots,x_n)dx_n.$$

Here the differential df is considered to be a function of the $2n$ independent variables $x_1, x_2, \ldots, x_n, dx_1, dx_2, \ldots dx_n$.

Differentials can be used to approximate values of functions at points nearby points where their values and those of their derivatives are known. If $\Delta f = f(x + \Delta x, y + \Delta y) - f(x,y)$ then we take $dx = \Delta x$, $dy = \Delta y$ and use df to approximate Δf:

$$f(x + \Delta x, y + \Delta y) - f(x,y) = \Delta f \approx df = f_1(x,y)dx + f_2(x,y)dy.$$

Example 6 Find an approximate value for $f(x,y) = (2x^2 + e^{2y})^{1/2}$ at $x = 2.2$, $y = -0.2$.

Solution $f(2,0) = 3$,
$f_1(x,y) = 2x(2x^2 + e^{2y})^{-1/2}$, $f_1(2,0) = 4/3$,
$f_2(x,y) = e^{2y}(2x^2 + e^{2y})^{-1/2}$, $f_2(2,0) = 1/3$.

$$f(2.2,-0.2) \approx f(2.8) + df = f(2.0) + 0.2f_1(2,0) - 0.2f_2(2,0)$$

$$+ 3 + \frac{0.8}{3} - \frac{0.2}{3} = 3.2.$$

The differential df will be a "good" approximation to Δf if the difference between the two is small compared with the distance between $(x + \Delta x, y + \Delta y)$ and (x,y). We adopt this condition as our definition of **differentiability**.

Definition 4 We shall say that the function $f(x,y)$ is **differentiable** at the point (a,b) if

$$\lim_{h,k \to 0} \frac{f(a + h, b + k) - f(a,b) - hf_1(a,b) - kf_2(a,b)}{\sqrt{h^2 + k^2}} = 0$$

The requirement of differentiability is equivalent to the geometric requirement that the surface $z = f(x,y)$ have a nonvertical tangent plane at (a,b). In using differentials for approximation above, we are really using the tangent plane to approximate the surface. (Compare this with the one-variable situation.) We shall establish a two-variable version of the mean-value theorem and use it to show that functions with continuous first partial derivatives are differentiable.

Theorem 3 (A mean-value theorem) If $f_1(x,y)$ and $f_2(x,y)$ are continuous is a disc of positive radius centred at (a,b) and if h and k have sufficiently small absolute value, then there exist numbers θ_1 and θ_2 between 0 and 1 such that

$$f(a + h, b + k) - f(a,b) = hf_1(a + h, b + \theta_2 k) + kf_2(a + \theta_1 h, b).$$

Proof The proof is very similar to that of Theorem 1 in Section 10.2 so we give only a sketch here. Write

$$f(a + h,b + k) - f(a,b) = [f(a + h,b + k) - f(a + h,b)] + [f(a + h,b) - f(a,b)]$$

Now apply the one-variable mean-value theorem separately to $f(a + h,y)$ on the interval $[b,b + k]$ and to $f(x,b)$ on the interval $[a,a + h]$ to get the desired result. □

Theorem 4 If f_1 and f_2 are continuous nearby the point (a,b) then f is differentiable at (a,b).

Proof Using Theorem 3 and the fact that

$$\left|\frac{h}{\sqrt{h^2 + k^2}}\right| \leq 1 \text{ and } \left|\frac{k}{\sqrt{h^2 + k^2}}\right| \leq 1,$$

we estimate

$$\left|\frac{f(a + h,b + k) - f(a,b) - hf_1(a,b) - kf_2(a,b)}{\sqrt{h^2 + k^2}}\right|$$

$$= \left|\frac{h}{\sqrt{h^2 + k^2}}[f_1(a + h,b + \theta_2 k) - f_1(a,b)] + \frac{k}{\sqrt{h^2 + k^2}}[f_2(a + \theta_1 h,b) - f_2(a,b)\right|$$

$$\leq |f_1(a + h,b + \theta_2 k) - f_1(a,b)| + |f_2(a + \theta_1 h,b - f_2(a,b)|.$$

Since f_1 and f_2 are continuous, each of these latter terms approaches zero as $h, k \to 0$. This is what we wanted to prove.

Let us illustrate differentiability with an example where we can calculate directly the difference between Δf and df.

Example 7 Calculate $\Delta f - df$ at the point (x,y) if $f(x,y) = x^3 + xy^2$.

Solution $f_1(x,y) = 3x^2 + y^2, f_2(x,y) = 2xy$. Thus

$$\begin{aligned}\Delta f - df &= f(x + h,y + k) - f(x,y) - hf_1(x,y) - kf_2(x,y)\\ &= (x + h)^3 + (x + h)(y + k)^2 - x^3 - xy^2 - (3x^2 + y^2)h - 2xyk\\ &= 3xh^2 + h^3 + 2yhk + hk^2 + xk^2\end{aligned}$$

Observe that $\Delta f - df$ is a polynomial in h and k with no term less than second degree in these variables. Evidently this difference approaches zero like the square of the distance from (h,k) to $(0,0)$ as $(h,k) \to (0,0)$. Such is the case for any function f with continuous second partial derivatives. (See Exercise 32 below.)

Proof of the Chain Rule

We conclude this section by stating formally, and proving, a simple but representative case of the chain rule for multivariate functions.

Theorem 5 Let $z = f(x,y)$, where $x = u(s,t)$ and $y = v(s,t)$. Suppose that

i) $u(a,b) = p$ and $v(a,b) = q$,

ii) the first partial derivatives of u and v exist at (a,b).

iii) $f_1(x,y)$ and $f_2(x,y)$ are continuous near the point (p,q).

Then $z = w(s,t) = f(u(s,t),v(s,t))$ has first partial derivatives with respect to s and t at (a,b), and

$$w_1(a,b) = f_1(p,q)u_1(a,b) + f_2(p,q)v_1(a,b)$$
$$w_2(a,b) = f_1(p,q)u_2(a,b) + f_2(p,q)v_2(a,b),$$

that is,

$$\frac{\partial z}{\partial s} = \frac{\partial z}{\partial x}\frac{\partial x}{\partial s} + \frac{\partial z}{\partial y}\frac{\partial y}{\partial s}, \quad \frac{\partial z}{\partial t} = \frac{\partial z}{\partial x}\frac{\partial x}{\partial t} + \frac{\partial z}{\partial y}\frac{\partial y}{\partial t}.$$

Proof We mimic the proof of the one-variable chain rule give in Section 2.4. Let

$$E(0,0) = 0,$$
$$E(h,k) = \frac{f(p + h,q + k) - f(p,q) - hf_1(p,q) - kf_2(p,q)}{\sqrt{h^2 + k^2}} \text{ if } (h,k) \neq (0,0).$$

Since f is differentiable at (p,q) by Theorem 4, $E(h,k)$ is continuous at $(0,0)$. Now

$$f(p + h,q + k) - f(p,q) = hf_1(p,q) + kf_2(p,q) + \sqrt{h^2 + k^2}\, E(h,k).$$

In this formula put $h = u(a + \alpha,b) - u(a,b)$, $k = v(a + \alpha,b) - v(a,b)$ and get

$$\frac{w(a + \alpha,b) - w(a,b)}{\alpha} = \frac{f(u(a + \alpha,b),v(a + \alpha,b)) - f(u(a,b),v(a,b))}{\alpha}$$
$$= f_1(p,q)\frac{h}{\alpha} + f_2(p,q)\frac{k}{\alpha} + \sqrt{(h/\alpha)^2 + (k/\alpha)^2}\, E(h,k)$$

Since $\lim_{\alpha \to 0} h/\alpha = u_1(a,b)$ and $\lim_{\alpha \to 0} k/\alpha = v_1(a,b)$ we have, letting $\alpha \to 0$,

$$w_1(a,b) = f_1(p,q)u_1(a,b) + f_2(p,q)v_1(a,b). \square$$

The proof for w_2 is similar.

EXERCISES

In Exercises 1-6 write appropriate versions of the chain rule for the indicated derivatives.

1. for $\partial w/\partial t$ if $w = f(x,y,z)$ where $x = g(s,t)$, $y = h(s,t)$, $z = k(s,t)$.

2. for $\partial w/\partial t$ if $w = f(x,y,z)$ where $x = g(s)$, $y = h(s,t)$, $z = k(t)$.

3. for $\partial z/\partial u$ if $z = g(x,y)$ where $y = f(x)$, $x = h(u,v)$.

4. for dw/dx if $w = f(x,y,z)$, $y = g(x,z)$, $z = h(x)$.

5. for $\partial w/\partial x|_z$ if $w = f(x,y,z)$, $y = g(x,z)$.

6. for dw/dt if $w = f(x,y)$, $x = g(r,s)$, $y = h(r,t)$, $r = k(s,t)$, $s = m(t)$.

7. If $w = f(x,y,z)$ where $x = g(r,s)$, $y = h(x,r,s)$ and $z = k(x,r,s)$ state appropriate versions of the chain rule for

a) $\left.\dfrac{\partial w}{\partial r}\right|_{x,s}$ b) $\left.\dfrac{\partial w}{\partial r}\right|_{y,s}$ c) $\left.\dfrac{\partial w}{\partial r}\right|_{x,y,s}$ d) $\left.\dfrac{\partial w}{\partial r}\right|_{s}$

8. Find $\partial u/\partial t$ if $u = \sqrt{x^2 + y^2}$, $x = e^{st}$, $y = 1 + s^2 \cos t$.

9. Find $\partial z/\partial t$ if $z = \operatorname{Arctan}(u/v)$, $u = 2x + y$, $v = 3x - y$.

10. Find dz/dt if $z = txy^2$, $x = t + \ln(y + t^2)$, $y = e^t$.

In Exercises 11-14 find the indicated derivatives, assuming that the function $f(x,y)$ has continuous first partial derivatives.

11. $\dfrac{\partial}{\partial v} f(u^2,v^2)$

12. $\dfrac{\partial}{\partial t} f(st^2,s^2 + t)$

13. $\dfrac{\partial}{\partial x} f(f(x,y), f(x,y))$

14. $\dfrac{\partial}{\partial y} f(yf(x,t), f(y,t))$

15. Suppose that the temperature T in a certain liquid varies with depth z and time t according to the formula $T = e^{-t}z$. Find the rate of change of temperature with respect to time at a point which is moving through the liquid of a path whose depth at time t is $f(t)$. What is the rate if $f(t) = e^t$? What is happening in this case?

16. Suppose the strength S of an electric field in space depends on position (x,y,z) and time t according to the formula $S = f(x,y,z,t)$, where f has continuous first partial derivatives. Find the rate of change of the electric field strength with respect to time measured at a point moving through space along the spiral path $x = \sin t$, $y = \cos t$, $z = t$.

In Exercises 17-22 assume $f(x,y)$ has continuous partial derivatives of all orders, and let $z = f(x,y)$.

17. If $x = 2s + 3t$ and $y = 3s - 2t$ find

a) $\dfrac{\partial^2 z}{\partial s^2}$ b) $\dfrac{\partial^2 z}{\partial s \partial t}$ c) $\dfrac{\partial^2 z}{\partial t^2}$ d) $\dfrac{\partial^3 z}{\partial s^3}$.

Guess, and verify by induction, a formula for $\partial^n z/\partial s^n$, $n = 1, 2, 3, \ldots$

18. If $f(x,y)$ is harmonic show that $f(ax + by,bx - ay)$ is also harmonic.

19. If $f(x,y)$ is harmonic show that $f(x^2 - y^2,2xy)$ is also harmonic.

20. If $f(x,y)$ is harmonic show that $f(x/(x^2 + y^2), -y/(x^2 + y^2))$ is also harmonic.

21. If $x = t \sin s$ and $y = s \cos t$ find $\partial^2 z/\partial s \partial t$.

22. Find $\partial^3 z/\partial s \partial t^2$ if $z = f(s^2 - t, s + t^2)$.

23. If $x = e^s \cos t$ and $y = e^s \sin t$ show that $\dfrac{\partial^2 z}{\partial s^2} + \dfrac{\partial^2 z}{\partial t^2} = (x^2 + y^2)\left(\dfrac{\partial^2 z}{\partial x^2} + \dfrac{\partial^2 z}{\partial y^2}\right)$.

24. If $f(x,y)$ is positively homogeneous of degree k and has continuous partial derivatives of second order show that

$$x^2 f_{11}(x,y) + 2xyf_{12}(x,y) + y^2 f_{22}(x,y) = k(k - 1)f(x,y).$$

25. Generalize the result of Exercise 24 to functions $f(x_1,x_2,\ldots,x_n)$.

26. Generalize the result of Exercise 24 to expressions involving n'th partial derivatives of $f(x,y)$.

In Exercises 27-29 use differentials to find an approximate value of the given function at the point indicated.

27. $f(x,y) = \sin(\pi xy + \ln(y))$ at $(0.01,1.05)$.

28. $f(x,y) = \dfrac{24}{x^2 + xy + y^2}$ at $(2.1,1.8)$.

29. $f(x,y,z) = \sqrt{x + 2y + 3z}$ at $(1.9,1.8,1.1)$.

30. By approximately what percentage will $f(x,y,z) = \dfrac{xy^2}{1 + xz^2}$ increase or decrease if $x = 1$ and decreases by 2%, $y = 5$ and increases by 1%, and $z = 3$ and decreases by 3%?

31. Prove the following version of the mean-value theorem: if $f(x,y)$ has first partial derivatives continuous near every point of the straight line segment joining the points $(a + h,b + k)$ and (a,b) then there exists a number 0 satisfying $0 < \theta < 1$ such that

$$f(a + h,b + k) = f(a,b) + hf_1(a + \theta h,b + \theta k) + kf_2(a + \theta h,b + \theta k).$$

Hint: apply the one-variable mean-value theorem to $g(t) = f(a + th,b + tk)$. Why could we not have used this result in place of Theorem 3 to prove Theorem 4 and hence the chain rule?

32. Generalize Exercise 31 as follows: show that if $f(x,y)$ has continuous partial derivatives of second order near the point (a,b) then there exists a number θ satisfying $0 < \theta < 1$ such that for h and k sufficiently small in absolute value,

$$\begin{aligned} f(a + h,b + k) &= f(a,b) + hf_1(a,b) + kf_2(a,b) \\ &\quad + h^2 f_{11}(a + \theta h,b + \theta k) + 2hkf_{12}(a + \theta h,b + \theta k) + k^2 f_{22}(a + \theta h,b + \theta k). \end{aligned}$$

Hence show that there is a constant K such that for sufficiently small h and k,

$$|\Delta f - df| \le \mathrm{K}(h^2 + k^2)$$

where $\Delta x = dx = h$ and $\Delta y = dy = k$.

10.4 GRADIENTS AND DIRECTIONAL DERIVATIVES

It is often useful to combine the first partial derivatives of a function into a single vector called the **gradient**. We will examine carefully the development and interpretation of the gradient vector in two dimensions, and will discuss extensions to three and higher dimensions at the end of the section. The reader may wish to review the material on plane vectors in Section 5.3 before reading this section.

Definition 5

At any point (x,y) where the partial derivatives $\partial f/\partial x$ and $\partial f/\partial y$ of the function $f(x,y)$ exist we define the **gradient vector** ∇f by

$$\nabla f(x,y) = \frac{\partial f}{\partial x}\mathbf{i} + \frac{\partial f}{\partial y}\mathbf{j} = f_1(x,y)\mathbf{i} + f_2(x,y)\mathbf{j}.$$

Recall that **i** and **j** denote the unit basis vectors joining the origin $(0,0)$ to the points $(1,0)$ and $(0,1)$ respectively.

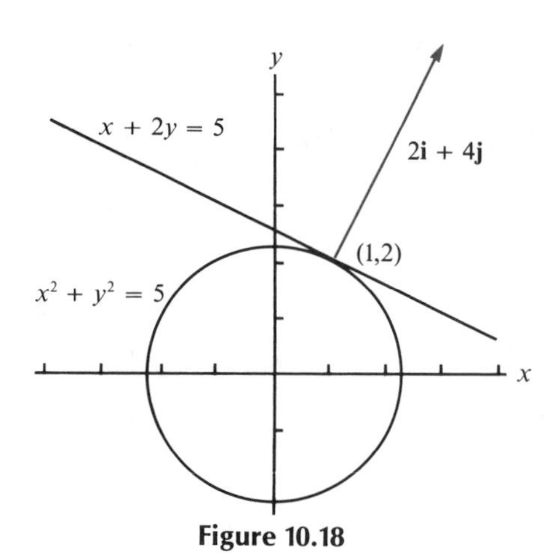

Figure 10.18

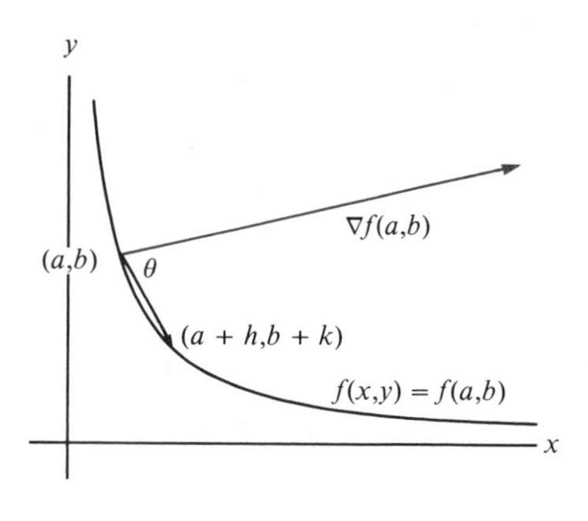

Figure 10.19

Example 1 If $f(x,y) = x^2 + y^2$ then $\nabla f(x,y) = 2x\mathbf{i} + 2y\mathbf{j}$. In particular, $\nabla f(1,2) = 2\mathbf{i} + 4\mathbf{j}$. Observe that this vector is perpendicular to the straight line $x + 2y = 5$ which is tangent to the circle $x^2 + y^2 = 5$, the level curve of $f(x,y)$ which passes through the point (1,2). See Fig. 10.18. This is no accident, as the following theorem shows.

Theorem 6 If $f(x,y)$ is differentiable at (a,b) and $\nabla f(a,b) \neq 0$ then $\nabla f(a,b)$ is a normal vector to the level curve of f which passes through (a,b).

Proof The angle θ between the vector $\nabla f(a,b)$ and the vector $h\mathbf{i} + k\mathbf{j}$ from the point (a,b) to the point $(a + h,b + k)$ (see Fig. 10.19) satisfies

$$\cos\theta = \frac{\nabla f(a,b)\cdot(h\mathbf{i} + k\mathbf{j})}{|\nabla(a,b)|\sqrt{h^2 + k^2}} = \frac{1}{|\nabla f(a,b)|}\,\frac{hf_1(a,b) + kf_2(a,b)}{\sqrt{h^2 + k^2}}$$

If $(a + h,b + k)$ lies on the level curve of f which passes through (a,b) then $f(a + h,b + k) = f(a,b)$, and so, since f is differentiable at (a,b) (see Definition 4 of Section 10.3) we have

$$\lim_{h,k\to 0}\frac{hf_1(a,b) + kf_2(a,b)}{\sqrt{h^2 + k^2}}$$

$$= -\lim_{h,k\to 0}\frac{f(a + h,b + k) - f(a,b) - hf_1(a,b) - kf_2(a,b)}{\sqrt{h^2 + k_2}} = 0$$

Hence $\cos\theta \to 0$ and $\theta \to \pi/2$ as h, $k \to 0$; that is, $\nabla f(a,b)$ is perpendicular in the limit to short secant vectors to the level curve. □

Directional Derivatives

The first partial derivatives $f_1(a,b)$ and $f_2(a,b)$ give the rates of change of $f(x,y)$ at (a,b) measured in the directions of the positive x- and y-axes respectively. If we want to know how fast $f(x,y)$ changes value as we move through (a,b) in some other direction in the plane we require a more general **directional derivative.** If **V** is a nonzero vector the directional derivative of f at (a,b) in the direction of $\mathbf{V}$ is the rate of change of f with respect to distance measured in that direction. It is denoted $D_{\mathbf{V}}f(a,b)$. This derivative depends only on the direction of $\mathbf{V}$, not on its length, and for calculation it is useful to replace $\mathbf{V}$ by a unit vector $\mathbf{U}$ in the same direction:

$$\mathbf{U} = \mathbf{V}/|\mathbf{V}| = u\mathbf{i} + v\mathbf{j},\ u^2 + v^2 = 1.$$

The directional derivative is defined by

$$D_{\mathbf{V}}f(a,b) = D_{\mathbf{U}}f(a,b) = \lim_{h\to0} \frac{f(a + hu,b + hv) - f(a,b)}{h}.$$

The following theorem shows how the gradient can be used to calculate any directional derivative.

Theorem 7 If $f(x,y)$ has continuous first partial derivatives at (a,b) and if $\mathbf{U}$ is a unit vector, then the directional derivative of f in the direction of $\mathbf{U}$ is given by

$$D_{\mathbf{U}}f(a,b) = \mathbf{U}\cdot\nabla f(a,b).$$

Proof For sufficiently small h, the first partial derivatives of f are continuous near all points of the straight line segment from (a,b) to $(a + hu,b + hv)$. If we apply the (one-dimensional) mean-value theorem to the function $g(t) = f(a + thu,b + thv)$ on the interval $[0,1]$ we obtain the formula $g(1) - g(0) = g'(\theta)$ for some θ between 0 and 1, that is,

$$f(a + hu,b + hv) - f(a,b) = huf_1(a + \theta hu,b + \theta hv) + hvf_2(a + \theta hu,b + \theta hv).$$

hence

$$\begin{aligned} D_{\mathbf{U}}f(a,b) &= \lim_{h\to0}[uf_1(a + \theta hu,b + \theta hv) + vf_2(a + \theta hu,b + \theta hv)] \\ &= uf_1(a,b) + vf_2(a,b) = \mathbf{U}\cdot\nabla f(a,b). \end{aligned}$$

Remark The first part of the above proof is essentially the proof of the mean-value theorem referred to in Exercise 31 of Section 10.3.

Example 2 Find the rate of change of $f(x,y) = y^4 + 2xy^3 + x^2y^2$ at $(0,1)$ measured in the directions:

a) $\mathbf{i} + 2\mathbf{j}$

b) $\mathbf{j} - 2\mathbf{i}$

c) $\mathbf{i}$

d) $\mathbf{i} + \mathbf{j}$.

Solution $\nabla f(x,y) = (2y^3 + 2xy^2)\mathbf{i} + (4y^3 + 6xy^2 + 2x^2y)\mathbf{j}$
$\nabla f(0,1) = 2\mathbf{i} + 4\mathbf{j}$.
The four required directional derivatives are:

a) $D_{\mathbf{i}+2\mathbf{j}}f(0,1) = \dfrac{\mathbf{i} + 2\mathbf{j}}{|\mathbf{i} + 2\mathbf{j}|} \cdot (2\mathbf{i} + 4\mathbf{j}) = \dfrac{2 + 8}{\sqrt{5}} = 2\sqrt{5}.$

(Observe that $\mathbf{i} + 2\mathbf{j}$ points in the same direction as $\nabla f(0,1)$ so the directional derivative is positive and equal to the length of $\nabla f(0,1)$.)

b) $D_{\mathbf{j}-2\mathbf{i}}f(0,1) = \dfrac{-2\mathbf{i} + \mathbf{j}}{|\mathbf{j} - 2\mathbf{i}|} \cdot (2\mathbf{i} + 4\mathbf{j}) = \dfrac{-4 + 4}{\sqrt{5}} = 0.$

(Since $\mathbf{j} - 2\mathbf{i}$ is perpendicular to $\nabla f(0,1)$ it is tangent to the level curve of f through $(0,1)$ and so the directional derivative in that direction must be zero.)

c) $D_{\mathbf{i}}f(0,1) = \mathbf{i} \cdot (2\mathbf{i} + 4\mathbf{j}) = 2$. (As noted previously, the directional derivative of f in the direction of the positive x-axis is just $\partial f/\partial x$.

d) $D_{\mathbf{i}+\mathbf{j}}f(0,1) = \dfrac{\mathbf{i} + \mathbf{j}}{|\mathbf{i} + \mathbf{j}|} \cdot (2\mathbf{i} + 4\mathbf{j}) = \dfrac{2 + 4}{\sqrt{2}} = 3\sqrt{2}.$

(If we move along the surface $z = f(x,y)$ through the point $(0,1,1)$ in a direction making horizontal angles of 45 degrees with the positive directions of the x- and y-axes, we would be rising at a rate of $3\sqrt{2}$ vertical units per horizontal unit moved.)

As observed in Example 2, Theorem 7 provides a useful interpretation for the gradient vector. Since $|\mathbf{U}| = 1$ we have

$$D_{\mathbf{U}}f(a,b) = |\nabla f(a,b)| \cos \theta,$$

where θ is the angle between the vectors $\mathbf{U}$ and $\nabla f(a,b)$. For any value of θ we know that $-1 \leq \cos \theta \leq 1$. Thus $D_{\mathbf{U}}f(a,b)$ takes on values between $-|\nabla f(a,b)|$ and $|\nabla f(a,b)|$ for different directions $\mathbf{U}$. We will have $D_{\mathbf{U}}f(a,b) = |\nabla f(a,b)|$ if and only if $\cos \theta = 1$, that is, $\mathbf{U}$ points in the same direction as $\nabla f(a,b)$. Thus $\nabla f(a,b)$ points in the direction in which $f(x,y)$ increases most rapidly at (a,b), and the length of $\nabla f(a,b)$ is equal to this maximum rate of increase:

$$|\nabla f(a,b)| = \max_{|\mathbf{U}| = 1} D_{\mathbf{U}}f(a,b).$$

Similarly, the minimum value of a directional derivative of f at (a,b) is $-|\nabla f(a,b)|$, and this value occurs in the direction opposite to the gradient. For example, the streams shown on the topographic map in Fig. 10.5 (Section 10.1) flow in the direction of steepest descent at all points, that is, in the direction of $-\nabla f$ where f is the function giving the elevation of the land. The streams therefore cross the contours (the level curves of f) at right angles.

The directional derivative is zero when $\theta = \pi/2$, that is, in the direction of the tangent line to the level curve of f at (a,b).

Example 3 In what direction at the point $(2,1)$ does the function $f(x,y) = x^2e^{-y}$ increase most rapidly? What is the rate of increase of f in this direction?

Solution $\nabla f(x,y) = 2xe^{-y}\mathbf{i} - x^2e^{-y}\mathbf{j}$,
$\nabla f(2,1) = (4/e)\mathbf{i} - (4/e)\mathbf{j} = (4/e)\,(\mathbf{i} - \mathbf{j})$.
At $(2,1)$, $f(x,y)$ increases most rapidly in the direction of the vector $\mathbf{i} - \mathbf{j}$. The rate of increase in this direction is $|\nabla f(2,1)| = 4\sqrt{2}/e$.

Example 4 Euler's Theorem (Theorem 2 of Section 10.3) is a statement about the directional derivative of a positively homogeneous function in the direction directly away from the origin. If, for instance, $f(x,y)$ is positively homogeneous of degree k then the directional derivative of $f(x,y)$ in the direction of $x\mathbf{i} + y\mathbf{j}$ is

$$\frac{x\mathbf{i} + y\mathbf{j}}{|x\mathbf{i} + y\mathbf{j}|} \cdot \nabla f(x,y) = \frac{1}{\sqrt{x^2 + y^2}}\left(x\frac{\partial f}{\partial x} + y\frac{\partial f}{\partial y}\right) = \frac{k\,f(x,y)}{\sqrt{x^2 + y^2}},$$

by Euler's Theorem. This is consistent with the interpretation that such a function grows at a rate proportional to the k th power of distance from the origin along rays from the origin.

Tangent Lines to Level Curves

If $f(x,y)$ is differentiable at (a,b) and $\nabla f(a,b) \neq 0$ then the level curve of f passing through (a,b), that is, the curve with equation $f(x,y) = f(a,b)$, is **smooth** at (a,b); since it has a nonzero normal vector, namely the vector $\nabla f(a,b)$, it must also have a tangent line. If (x,y) is any point on the tangent line to the level curve at (a,b) then the vector $(x - a)\mathbf{i} + (y - b)\mathbf{j}$ is perpendicular to $\nabla f(a,b)$. Thus the equation of the tangent line can be written in the form

$$\nabla f(a,b)\cdot[(x - a)\mathbf{i} + (y - b)\mathbf{j}] = 0$$

which simplifies to $Ax + By = C$ where $A = f_1(a,b)$, $B = f_2(a,b)$ and $C = Aa + Bb$.

This equation for the tangent line can also be found by using implicit differentiation as in Section 2.7. Since $\nabla f(a,b)$ is assumed not to vanish, one or other of the first partial derivatives of f is nonzero at (a,b). To be specific, suppose $f_2(a,b) \neq 0$. It can be shown that this implies that the equation $f(x,y) = f(a,b)$ defines y implicitly as a function of x near (a,b). (This is a version of the **implicit function theorem**; we will not attempt to prove it here.) Differentiating the equation of the level curve implicitly with respect to x (using the chain rule) we obtain

$$f_1(x,y) + f_2(x,y)\frac{dy}{dx} = 0.$$

Thus the slope of the tangent line at (a,b) is

$$\left.\frac{dy}{dx}\right|_{x=a,y=b} = -\frac{f_1(a,b)}{f_2(a,b)} = -\frac{A}{B},$$

and once again we obtain $Ax + By = C$ as the equation of the tangent line. The same result would have occurred if we had assumed instead that $f_1(a,b) \neq 0$. In this case we would have regarded $f(x,y) = f(a,b)$ as defining x as a function of y and would have differentiated with respect to y to obtain dx/dy whose reciprocal is the slope of the tangent.

Example 5 Find $\nabla f(1,2)$ if $f(x,y) = x^2y^3$. Find an equation of the tangent line to the curve $x^2y^3 = 8$ at $(1,2)$.

Solution $\nabla f(x,y) = 2xy^3\mathbf{i} + 3x^2y^2\mathbf{j}$, so $\nabla f(1,2) = 16\mathbf{i} + 12\mathbf{j}$. The equation of the tangent line must be

$$0 = [(x - 1)\mathbf{i} + (y - 2)\mathbf{j}] \cdot [16\mathbf{i} + 12\mathbf{j}] = 16(x - 1) + 12(y - 2),$$

or, more simply, $4x + 3y = 10$.

Higher Dimensional Vectors

Section 5.3 contains an introduction to vectors in the plane. Many of the concepts developed there extend in a natural way to vectors in three or more dimensions. For example, an arrow from the point $A = (a,b,c)$ to the point $P = (p,q,r)$ in 3-dimensional space is a vector $\mathbf{V} = \overrightarrow{AP}$ having length

$$|\mathbf{V}| = \sqrt{(p - a)^2 + (q - b)^2 + (r - c)^2}$$

and which can be written in terms of standard basis vectors,

$$\mathbf{V} = (p - a)\mathbf{i} + (q - b)\mathbf{j} + (r - c)\mathbf{k},$$

where $\mathbf{i}$, $\mathbf{j}$ and $\mathbf{k}$ are the mutually perpendicular unit vectors joining the origin (0,0,0) to the points (1,0,0), (0,1,0) and (0,0,1) respectively. See Fig. 10.20. $\mathbf{V}$ has components $(p - a, q - b, r - c)$.

Addition and scalar multiplication can be defined using components in the same way as for plane vectors: if $\mathbf{A} = a\mathbf{i} + b\mathbf{j} + c\mathbf{k}$, $\mathbf{U} = u\mathbf{i} + v\mathbf{j} + w\mathbf{k}$ then

$$\mathbf{A} + \mathbf{U} = (a + u)\mathbf{i} + (b + v)\mathbf{j} + (c + w)\mathbf{k},$$
$$t\mathbf{A} = ta\mathbf{i} + tb\mathbf{j} + tc\mathbf{k}.$$

The dot product is defined similarly:

$$\mathbf{A} \cdot \mathbf{U} = au + bv + wk,$$

and has all the properties mentioned in Section 5.3. In particular, if θ is the angle between A and U then

$$\mathbf{A} \cdot \mathbf{U} = |\mathbf{A}|\ |\mathbf{U}| \cos \theta.$$

Finally, we remark that all these concepts extend to spaces of dimension greater than three in the obvious way, although it is more difficult to visualize vectors of dimension higher than three.

The Gradient in Higher Dimensions

All of the machinery developed earlier for functions of two variables extends in an obvious way to functions of n variables. We shall summarize here the major results as they apply to a function $f(x,y,z)$ of three variables. The reader is invited to formulate the results for functions $f(x_1,x_2,\ldots,x_n)$ of n variables.

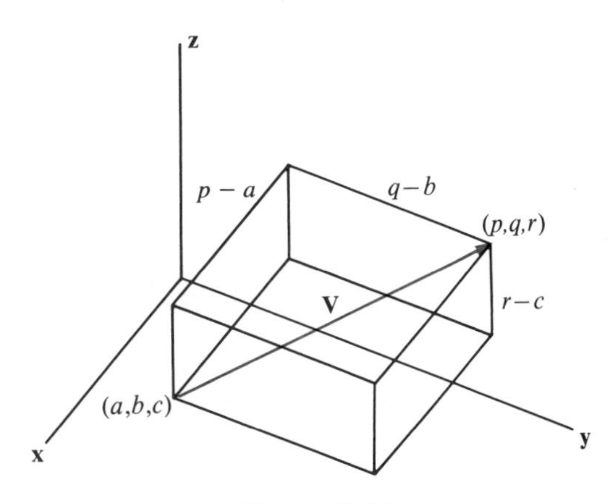

Figure 10.20

The function $f(x,y,z)$ is said to be differentiable at the point (a,b,c) if

$$\lim_{h,k,m\to 0} \frac{f(a+h,b+k,c+m) - f(a,b,c) - hf_1(a,b,c) - kf_2(a,b,c) - mf_3(a,b,c)}{\sqrt{h_2 + k^2 + m^2}} = 0,$$

a condition which will be satisfied if all three first partial derivatives of f are continuous near the point (a,b,c). The gradient of f is the three-dimensional vector

$$\nabla f = \frac{\partial f}{\partial x}\mathbf{i} + \frac{\partial f}{\partial y}\mathbf{j} + \frac{\partial f}{\partial z}\mathbf{k}.$$

If f is differentiable at (a,b,c) and $\nabla f(a,b,c)$ is not the zero vector then it is normal to the level surface of f passing through (a,b,c), that is, to the surface with equation $f(x,y,z) = f(a,b,c)$.

If $\mathbf{U} = u\mathbf{i} + v\mathbf{j} + w\mathbf{k}$ is a vector of unit length (that is, if $u^2 + v^2 + w^2 = 1$) then the directional derivative of f at (a,b,c) in the direction of U is

$$D_{\mathrm{U}}f(a,b,c) = \lim_{h\to 0} \frac{f(a+hu,b+hv,c+hw) - f(a,b,c)}{h}.$$

If f has continuous first partial derivatives at (a,b,c) then the directional derivative is also given by

$$D_{\mathrm{U}}f(a,b,c) = \mathbf{U}\cdot\nabla f(a,b,c).$$

The level surface $f(x,y,z) = f(a,b,c)$ has a nonzero normal vector at (a,b,c) provided f is differentiable there and the gradient vector $\nabla f(a,b,c)$ is nonzero. Therefore the surface must be smooth and must have a tangent plane at (a,b,c), namely the unique plane passing through the point (a,b,c) which is perpendicular to $\nabla f(a,b,c)$. If (x,y,z) is any point on that tangent plane then

$$\nabla f(a,b,c)\cdot[(x-a)\mathbf{i} + (y-b)\mathbf{j} + (z-c)\mathbf{k}] = 0,$$

which is therefore an equation of the tangent plane. This equation can be rewritten in the form $Ax + By + Cz = D$ where $A = f_1(a,b,c)$, $B = f_2(a,b,c)$, $C = f_3(a,b,c)$ and $D = Aa + Bb + Cc$.

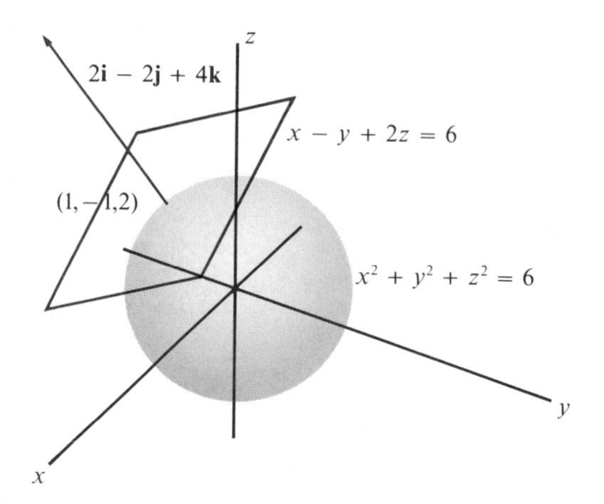

Figure 10.21

Example 6 Find ∇f where $f(x,y,z) = x^2 + y^2 + z^2$ and use it to find an equation of the tangent plane to the sphere $x^2 + y^2 + z^2 = 6$ at the point $(1,-1,2)$. What is the maximum rate of increase of f at $(1,-1,2)$ and in what direction does it occur? What is the rate of change of $f(x,y,z)$ at $(1,-1,2)$ in the direction towards the point $(3,1,1)$?

Solution $\nabla f = 2x\mathbf{i} + 2y\mathbf{j} + 2z\mathbf{k}$ so $\nabla f(1,-1,2) = 2\mathbf{i} - 2\mathbf{j} + 4\mathbf{k}$. The tangent plane has equation $2(x - 1) - 2(y + 1) + 4(z - 2) = 0$, or $x - y + 2z = 6$. See Fig. 10.21. The maximum rate of increase of f at $(1,-1,2)$ is $|\nabla f(1,-1,2)| = \sqrt{24} = 2\sqrt{6}$, and it occurs in the direction of $\nabla f(1,-1,2)$, that is, in the direction of the vector $\mathbf{i} - \mathbf{j} + 2\mathbf{k}$. The rate of change of f in the direction from $(1,-1,2)$ towards $(3,1,1)$, that is, in the direction of the vector $2\mathbf{i} + 2\mathbf{j} - \mathbf{k}$, is given by

$$\frac{2\mathbf{i} + 2\mathbf{j} - \mathbf{k}}{\sqrt{(4 + 4 + 1)}} \cdot (2\mathbf{i} - 2\mathbf{j} + 4\mathbf{k}) = \frac{4 - 4 - 4}{3} = -\frac{4}{3},$$

and f is decreasing at a rate 4/3 in this direction.

Tangents To The Graphs of Functions

In Section 10.2 we derived an equation of the tangent plane to the graph of a function $f(x,y)$ by an argument based on direct interpretation of the partial derivatives. This equation can also be obtained by considering gradients.

The graph of the function $f(x,y)$ is the surface $z = f(x,y)$ in three-dimensional space. This surface can be thought of as the level surface $g(x,y,z) = 0$ of the 3-variable function

$$g(x,y,z) = f(x,y) - z.$$

Since $g_1(x,y,z) = f_1(x,y)$, $g_2(x,y,z) = f_2(x,y)$ and $g_3(x,y,z) = -1$, g will be differentiable at any point (a,b,c) if f is differentiable at (a,b), and will have a nonzero

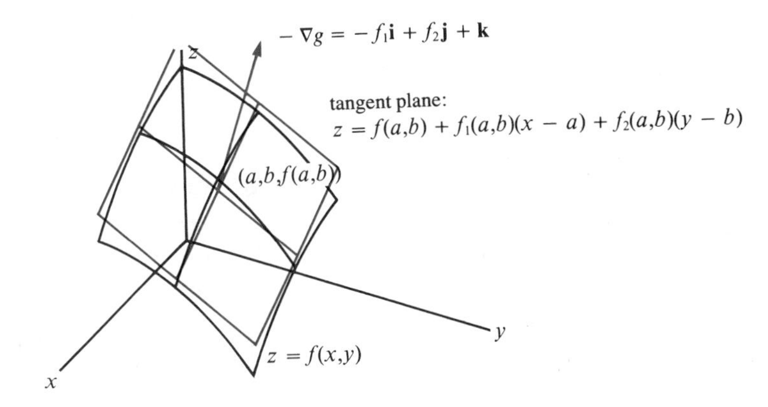

Figure 10.22

gradient vector in this case:

$$\nabla g(a,b,c) = f_1(a,b)\mathbf{i} + f_2(a,b)\mathbf{j} - \mathbf{k}.$$

Therefore, at any point (a,b) where $f(x,y)$ is differentiable the graph $z = f(x,y)$ of this function is smooth and has a tangent plane (see Fig. 10.22) whose equation is given by

$$[f_1(a,b)\mathbf{i} + f_2(a,b)\mathbf{j} - \mathbf{k}] \cdot [(x - a)\mathbf{i} + (y - b)\mathbf{j} + (z - f(a,b))\mathbf{k}] = 0,$$

or, more simply,

$$z = f(a,b) + f_1(a,b)(x - a) + f_2(a,b)(y - b).$$

This is the same equation derived in Section 10.2.

The notion of tangent plane can also be extended to functions of three or more variables, though the situation is more difficult to visualize since the graphs and tangent planes are no longer two-dimensional surfaces in a three dimensional space. The graph of a function $f(x_1,\ldots,x_n)$ of n variables is an n-dimensional "hypersurface" in an $(n + 1)$-dimensional space. Such a hypersurface will have a tangent "hyperplane" (n-dimensional plane) at any point $(a_1,\ldots,a_n)$ where f is differentiable. This tangent hyperplane contains all $(n + 1)$-dimensional vectors with tails at $(a_1,\ldots,a_n)$ which are perpendicular to the $(n + 1)$-dimensional vector

$$f_1(a_1,\ldots,a_n)\mathbf{e}_1 + f_2(a_1,\ldots,a_n)\mathbf{e}_2 + \ldots + f_n(a_1,\ldots a_n)\mathbf{e}_n - \mathbf{e}_{n+1},$$

where $\mathbf{e}_1, \mathbf{e}_2, \ldots, \mathbf{e}_{n+1}$ are the standard basis vectors in $(n + 1)$-dimensional space.

EXERCISES

In Exercises 1-10 find the gradients of the given functions at the points indicated. In 1-5 and 7-8 find the tangent plane to the graph of f at the given point.

1. $f(x,y) = x^2 - y^2$ at $(-2,1)$

2. $f(x,y) = (x - y)/(x + y)$ at $(1,1)$

3. $f(x,y) = \cos(x/y)$ at $(\pi,4)$

4. $f(x,y) = e^{xy}$ at $(2,0)$

5. $f(x,y) = \dfrac{x}{x^2 + y^2}$ at $(1,2)$

6. $f(x,y,z) = y\, e^{-x^2} \sin z$ at $(0,1,\pi/3)$

7. $f(x,y) = \ln(x^2 + y^2)$ at $(1,-2)$

8. $f(x,y) = \dfrac{2xy}{x^2 + y^2}$ at $(0,2)$

9. $f(x,y,z) = x^2y + y^2z + z^2x$ at $(1,-1,1)$

10. $f(x,y,z) = \cos(x + 2y + 3z)$ at (π,π,π)

In Exercises 11-16 find the rate of change of the given function at the given point in the specified direction.

11. $f(x,y) = 3x - 4y$ at $(0,2)$ in the direction of the vector $-\mathbf{i}$

12. $f(x,y) = x^2y$ at $(-1 -1)$ in the direction of the vector $\mathbf{i} + 2\mathbf{j}$

13. $f(x,y) = \dfrac{x}{1 + y}$ at $(0,0)$ in the direction of the vector $\mathbf{i} - \mathbf{j}$

14. $f(x,y) = x^2 + y^2$ at $(1,2)$ in a direction making an angle of $60°$ with the positive x-axis.

15. $f(x,y) = e^{x+2y}$ at the origin in the direction towards the point $(4,-3)$.

16. $f(x,y) = x^2 - y^2$ at $(1,0)$ in the direction towards the point $(2,-1)$.

17. In what directions at the point $(1,1)$ does the function $f(x,y) = xy^2$

a) increase most rapidly?

b) decrease most rapidly?

c) instantaneously neither increase nor decrease?

d) increase at rate 2?

18. Repeat Exercise 17 a), b) and c) for $f(x,y) = \text{Arctan}\,(y/x)$ at (a,b) where $a \neq 0$.

19. Find a vector normal to the surface with equation $x^2 + 2y^2 + 3z^2 = 6$ at the point $(1,1,1)$.

20. Find the rate of change of the function $f(x,y,z) = xy^2z^3$ at the point $(0,-1,1)$ in the direction from this point towards the point $(2,-2,-1)$.

In Exercises 21-28 find an equation of the straight line tangent to the level curve of the function given in the indicated exercise at the point given in that exercise.

21. Exercise 1

22. Exercise 2

23. Exercise 3

24. Exercise 4

25. Exercise 5

26. Exercise 8

27. Exercise 7

28. Exercise 18

29. Find an equation of the tangent plane to the level surface of the function in Exercise 9 at the point specified in that exercise.

30. Repeat Exercise 29 but with reference to Exercise 10.

31. Why do we require $\nabla f(a,b) \neq 0$ in Theorem 6? (Consider the function $f(x,y) = xy$ at $(0,0)$.)

32. If $\nabla f(x,y) = 0$ throughout the disc $x^2 + y^2 < r^2$ prove that $f(x,y) = C$ (constant) throughout that disc.

33. Let $f(x,y)$ have continuous first partial derivatives in a disc of positive radius centred at (a,b). If the directional derivatives of f at (a,b) is zero in each of two non-parallel directions show that $\nabla f(a,b) = 0$.

34. Reprove a version of Theorem 6 for a function of three variables.

35. Reprove a version of Theorem 7 for a function of three variables.

36. Find an equation of the hyperplane tangent to the hypersurface $w = x^2 - y^3 + z^4$ at the point (a,b,c).

10.5 EXTREME VALUES

The subject of maximum and minimum values of functions of several variables is, like the one-dimensional case, very fruitful in applications of mathematics. Unfortunately, it is also much more complicated to present, and, within the context of this introductory chapter, cannot be dealt with in much generality. We shall limit ourselves mostly to the consideration of functions of two variables; most of the discussion can be extended to functions of three or more variables but the form of the extension will not always be obvious.

Recall that a single-variable function $f(x)$ has a local maximum value (or a local minimum value) at a point a in its domain if $f(x) \leq f(a)$ (or $f(x) \geq f(a)$) for all x in the domain of f which are sufficiently close to a. (If the appropriate inequality holds for all x in the domain of f we can call the maximum or minimum value absolute instead of local.) Moreover, local (or absolute) extreme values can occur only at points which are

critical points—where the derivative $f'(x) = 0$,

or singular points—where $f'(x)$ does not exist,

or endpoints of the domain of f.

A similar situation obtains for functions of several variables. We say that $f(x,y)$ has a **local maximum** (or **minimum**) value at the point (a,b) if $f(x,y) \leq f(a,b)$ (or $f(x,y) \geq f(a,b)$) for all points (x,y) in the domain of f which are sufficiently close to the point (a,b). (If the condition holds for all (x,y) in the domain of f we can call the maximum or mimimum value **absolute**.) Also, such local or absolute extreme values can only occur at points which are

critical points—where $\nabla f(x,y) = 0$, i.e. $f_1(x,y) = f_2(x,y) = 0$,

or singular points—where $\nabla f(x,y)$ fails to exist,

or boundary points of the domain of f.

A point P is a **boundary point** of a set S in the plane if every disc of positive radius centred at P contains points in S and points not in S. A boundary point may or may not belong to the set. If all boundary points of S belong to S then we say that S is a **closed set.** Of course, if a function $f(x,y)$ is to have a maximum value at a boundary point of its domain then that boundary point must belong to the domain. Points of a set which are not boundary points are called **interior points** of the set. For example, if S is the set of points in the plane which satisfy $x^2 + y^2 \leq 1$ (S is called a closed disc) then the boundary of S is the circle $x^2 + y^2 = 1$, and the interior of S consists of points satisfying $x^2 + y^2 < 1$.

The trichotomy of possibilities for points where extreme values can occur can be established as follows. If (a,b) is not a boundary point of f, and if $\nabla f(a,b)$ exists and is not the zero vector, then $f(x,y)$ is increasing as we move away from (a,b) in the direction of $\nabla f(a,b)$, and is decreasing if we move in the opposite direction. Thus f cannot have either a maximum or a minimum value at (a,b).

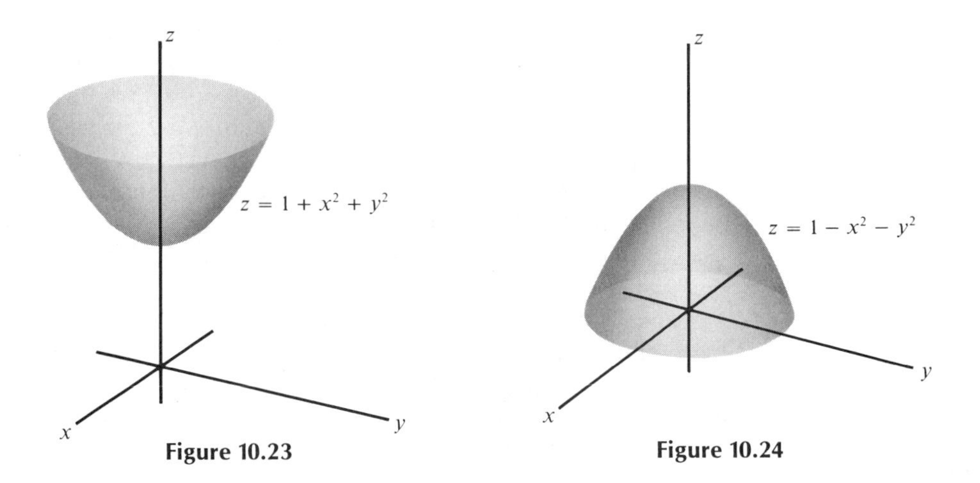

Figure 10.23

Figure 10.24

Example 1 The function $f(x,y) = 1 + x^2 + y^2$ has a critical point at (0,0) since $f_1(x,y) = 2x$ and $f_2(x,y) = 2y$ both vanish at that point. Evidently $f(x,y) > 1 = f(0,0)$ if $(x,y) \neq (0,0)$, so f has an absolute minimum value at (0,0). See Fig. 10.23. Similarly, $g(x,y) = 1 - x^2 - y^2$ has an absolute maximum value at its critical point (0,0). See Fig. 10.24.

Example 2 The function $f(x,y) = 1 + x^2 - y^2$ has a critical point at (0,0) but has neither a local maximum nor a local minimum value at this point: $f(0,0) = 1$ but $f(x,0) > 1$ and $f(0,y) < 1$ for all nonzero values of x and y. In view of the shape of the graph of f (see Fig. 18.25) we say that f has a **saddle point** at (0,0).

In general we shall somewhat loosely call any interior critical point of $f(x,y)$ a saddle point if f does not have a local extreme value there. For example, $f(x,y) = -x^3$ has a whole line of saddle points along the y-axis though its graph (see Fig. 10.26) does not resemble a saddle near them. Saddle points generalize the horizontal inflection points of functions of one variable.

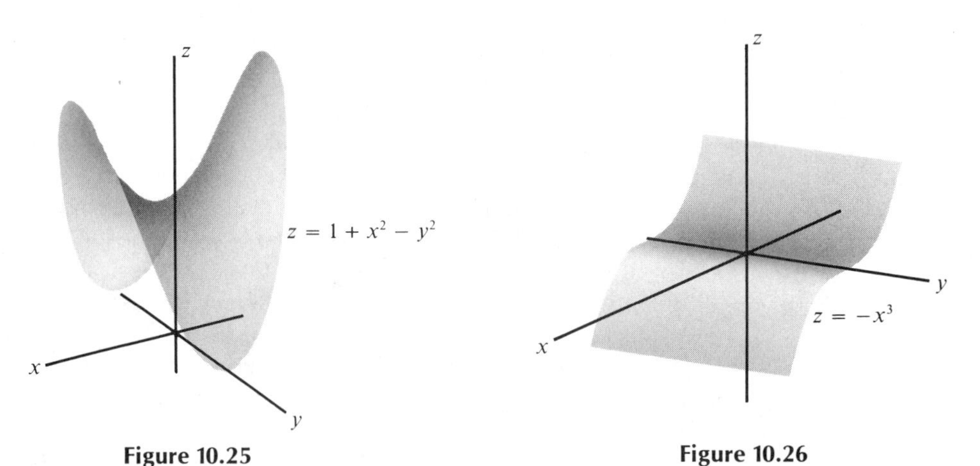

Figure 10.25

Figure 10.26

Example 3 The function $f(x,y) = \sqrt{x^2 + y^2}$ has no critical points, but it has a singular point at (0,0) where it has a local minimum value. The graph of f is a circular cone. See Fig. 10.27.

Example 4 The function $f(x,y) = 1 - x$ has no critical or singular points ($\nabla(x,y) = -\mathbf{i}$ at every point (x,y)) and so has no local extreme values. However, if we restrict the domain of f to consist only of points in the disc $x^2 + y^2 \leq 1$ then f has a local (and absolute) maximum value at the boundary point $(-1,0)$ and a minimum value at the boundary point $(1,0)$. See Fig. 10.28.

The following theorem is a generalization of the second derivative test for extreme values of functions of one variable introduced in Section 4.2. In most cases it enables us to determine whether an interior critical point is a local maximum or local minimum or a saddle point.

Theorem 8 (*A Second Derivative Test*) Suppose that (a,b) is a critical point of $f(x,y)$ and is an interior point of the domain of f. Suppose also that all the second partial derivatives of $f(x,y)$ exist and are continuous near (a,b). Let

$$A = f_{11}(a,b), \qquad B = f_{12}(a,b) = f_{21}(a,b), \qquad C = f_{22}(a,b).$$

a) If $B^2 - AC < 0$ and $A > 0$ then $f(x,y)$ has a local minimum value at (a,b).

b) If $B^2 - AC < 0$ and $A < 0$ then $f(x,y)$ has a local maximum value at (a,b).

c) If $B^2 - AC > 0$ then $f(x,y)$ has a saddle point at (a,b)

The test provides no information if $B^2 - AC = 0$.

Before proving this theorem let us illustrate it with examples.

Example 5 Find and classify the critical points of the functions

a) $f(x,y) = x^3 + y^3 - 3xy$,

b) $g(x,y) = xye^{-(x^2 + y^2)/2}$.

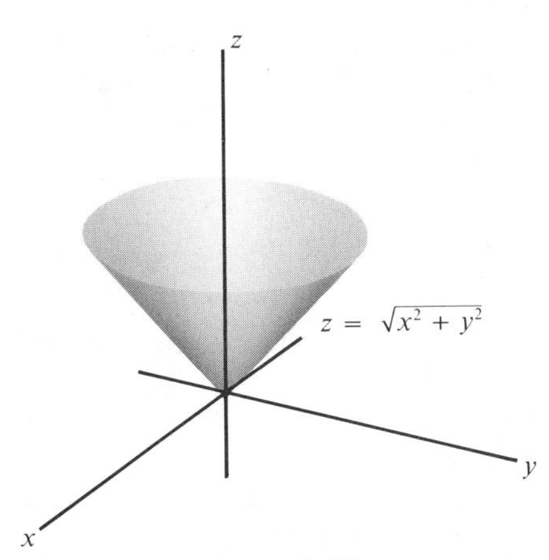

Figure 10.27

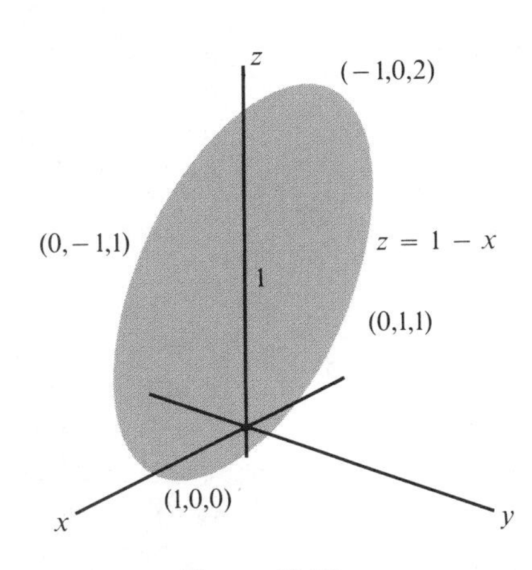

Figure 10.28

Solution a) $f_1(x,y) = 3x^2 - 3y, f_2(x,y) = 3y^2 - 3x$.
For critical points $x^2 = y$ and $y^2 = x$. Thus $x^4 = x$ so $x = 0$ (and $y = 0$) or $x = 1$ (and $y = 1$). The critical points are (0,0) and (1,1).
Now $f_{11}(x,y) = 6x, \quad f_{12}(x,y) = -3, \quad f_{22}(x,y) = 3y$.
At (0,0): $A = 0 = C, B = -3$. Thus $B^2 - AC = 9 > 0$
and f has a saddle point at (0,0).
At (1,1): $A = 6 = C, B = -3$. Thus $B^2 - AC = -27 < 0$
and f has a local minimum at (1,1).

b) $g_1(x,y) = y(1 - x^2)e^{-(x^2+y^2)/2}, g_2(x,y) = x(1 - y^2)e^{-(x^2+y^2)/2}$.
The critical points are (0,0), (1,1), (1,−1), (−1,1) and (−1,−1).

$$g_{11}(x,y) = xy(x^2 - 3)e^{-(x^2+y^2)/2}, g_{22}(x,y) = xy(y^2 - 3)e^{-(x^2+y^2)/2},$$
$$g_{12}(x,y) = (1 - x^2)(1 - y^2)e^{-(x^2+y^2)/2}.$$

At (0,0): $A = C = 0, B = 1$. Thus $B^2 - AC > 0$ and g has a saddle point at (0,0).
At (1,1) and (−1,−1): $A = C = -2e^{-1} < 0, B = 0$. Thus $B^2 - AC < 0$ and g has local maximum values at these points.
At (−1,1) and (1,−1): $A = C = 2e^{-1} > 0, B = 0$. Thus $B^2 - AC < 0$ and g has local minimum values at these points.

Proof of Theorem 8 Let $g(t) = f(a + th, b + tk)$ where h and k are sufficiently small numbers not both zero. By the chain rule and the equality of the mixed partial derivatives f_{12} and f_{21}

$$g'(t) = hf_1(a + th, b + tk) + kf_2(a + th, b + tk),$$
$$g''(t) = h^2f_{11}(a + th, b + tk) + 2hkf_{12}(a + th, b + tk) + k^2f_{22}(a + th, b + tk).$$

Since (a,b) is a critical point of f we have $g'(0) = 0$ and $t = 0$ is a critical point of g. Also, $g''(0) = Ah^2 + 2Bhk + Ck^2$.

Now $f(x,y)$ will have a local minimum (or maximum) value at (a,b) provided $g(t)$ has a local minimum (or maximum) value at $t = 0$ for every choice of the point (h,k) sufficiently near but not equal to (0,0). By the one-variable form of the second derivative test (Theorem 5 of Section 4.2) it is sufficient to show that $g''(0) > 0$ (or $g''(0) < 0$) for all such points (h,k). Likewise, $f(x,y)$ will have a saddle point at (a,b) if in any disc of positive radius centred at (0,0) there are points (h,k) for which $g''(0) < 0$ and other such points for which $g''(0) > 0$. Since $g''(0)$ is a quadratic function in the variables h and k, we can complete the square and write (assuming $A \neq 0$)

$$g''(0) = A\left[\left(h + \frac{B}{A}k\right)^2 + \left(\frac{AC - B^2}{A^2}\right)k^2\right].$$

If $B^2 - AC < 0$ the expression in the large brackets is a sum of squares and so is positive for all $(h, k) \neq (0,0)$. Thus we will have $g''(0) > 0$ if $A > 0$ and $g''(0) < 0$ if $A < 0$. If $B^2 - AC > 0$ then the expression in the large brackets is a difference of squares and can be positive or negative for various values of h and k as close to 0 as we like. This completes the proof of the theorem provided $A \neq 0$. If $A = 0$ but $C \neq 0$ a similar proof works. If $A = C = 0$ but $B \neq 0$ then $g''(0) = 2Bhk$ which can clearly take on positive or negative values and so corresponds to a saddle point of f. □

Extreme Values of Functions Defined on Closed, Bounded Sets

Theorem 5 of Section 1.5 states that if a function $f(x)$ is continuous on a closed, finite interval then it assumes absolute maximum and minimum values at points of that interval. (This is proved in Appendix 2.) A similar result holds for functions of several variables. We state the two-dimensional version; we say that a set S in the plane is **bounded** if S is contained inside some disc of finite radius.

Theorem 9 If $f(x,y)$ is continuous on a closed, bounded set in the plane then f takes on absolute maximum and minimum values on S. That is, there exist points (x_0,y_0) and (x_1,y_1) in S such that for every point (x,y) in S we have

$$f(x_0,y_0) \le f(x,y) \le f(x_1,y_1).$$

We will not attempt to prove this, but note that the points (x_0,y_0) and (x_1,y_1) must be critical points or singular points of f, or boundary points of the domain of f.

Example 6 Find the maximum values of $f(x,y) = xy - y^2$ on the disc $x^2 + y^2 \le 1$.

Solution Since f is continuous and the disc is a closed, bounded set, f must have absolute maximum and minimum values at some points of the disc. The partial derivatives of f are

$$\frac{\partial f}{\partial x} = y, \qquad \frac{\partial f}{\partial y} = x - y.$$

Both of these derivatives are continuous everywhere so f has no singular points. Its only critical point satisfies $y = 0$, $x - y = 0$ so is the origin $(0,0)$, an interior point of the domain of f. We have $f(0,0) = 0$.

We must also look at boundary points of the disc, that is, points of the circle $x^2 + y^2 = 1$. We can express $f(x,y)$ as a function of one variable on this circle by using a convenient parametrization of the circle, say

$$x = \cos(t),\ y = \sin(t),\ 0 \le t \le 2\pi.$$

We have

$$\begin{aligned} f(\cos(t),\sin(t)) &= \cos(t)\sin(t) - \sin^2(t) \\ &= \frac{\sin(2t)}{2} - \frac{1 - \cos(2t)}{2} = g(t). \end{aligned}$$

In order to find any extreme values of $g(t)$ we look at its citical points:

$$\begin{aligned} 0 &= g'(t) = \cos(2t) - \sin(2t), \\ \tan(2t) = 1, \quad 2t &= \pi/4,\ 5\pi/4,\ 9\pi/4,\ 13\pi/4, \\ t &= \pi/8,\ 5\pi/8,\ 9\pi/8,\ 13\pi/8. \end{aligned}$$

Evidently g is periodic with period π so

$$g(\pi/8) = g(9\pi/8) \ = (2 - \sqrt{2})/(2\sqrt{2}) > 0,$$
$$g(5\pi/8) = g(13\pi/8) = -(2 + \sqrt{2})/(2\sqrt{2}) < 0.$$

Because g is periodic, the endpoints $t = 0$ and $t = 2\pi$ will not contribute any local extreme values to f unless they are critical points of g. (Why is this?) Thus the maximum value of $f(x,y)$ over the disc must be $(2 - \sqrt{2})/\ (2\sqrt{2})$, and the minimum value must be $-(2 + \sqrt{2})/(2\sqrt{2})$, each value occurring at two boundary points. The second derivative test shows that, in this example, the interior critical point, (0,0), is a saddle point of f.

Example 7 Find the extreme values of the function $f(x,y) = x^2ye^{-(x+y)}$ on the triangular region $x \geq 0$, $y \geq 0$, $x + y \leq 4$.

Solution First we look for critical points:

$$0 = \frac{\partial f}{\partial x} = (2xy - x^2y)e^{-(x+y)} \Leftrightarrow xy(2-x) = 0 \Leftrightarrow x = 0,\ y = 0 \text{ or } x = 2,$$

$$0 = \frac{\partial f}{\partial y} - (x^2 - x^2y)e^{-(x+y)} \Leftrightarrow x^2(1-y) = 0 \Leftrightarrow x = 0 \text{ or } y = 1.$$

The critical points are $(0,y)$ for any y, and (2,1). Note that (2,1) is interior to the triangular region. See Fig. 10.29. There are no singular points. We have $f(0,y) = 0$, $f(2,1) = 4/e^3 \approx 0.199$. The boundary of the domain of f consists of three straight line segments:

i) $x = 0,\ 0 \leq y \leq 4$,

ii) $y = 0,\ 0 \leq x \leq 4$, and

iii) $x + y = 4,\ 0 \leq x \leq 4$.

Note that $f(x,y) = 0$ at all points of the first two segments, and $f(x,y) > 0$ at all other points in the triangular region. On the third boundary segment we can

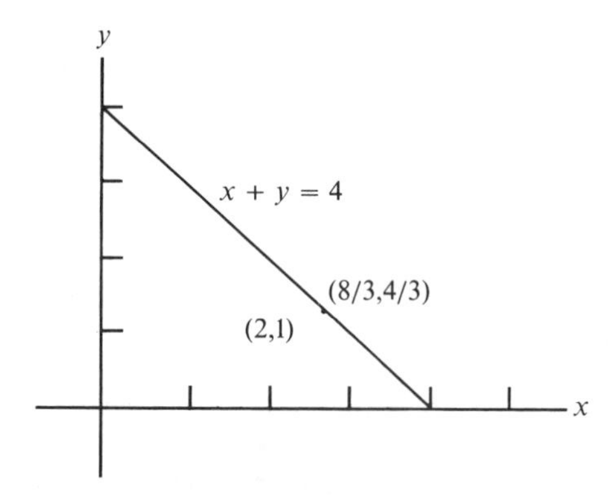

Figure 10.29

express $f(x,y)$ as a function $g(x)$ of the single variable x:

$$g(x) = f(x, 4 - x) = x^2(4 - x)e^{-4}.$$

Clearly $g(0) = g(4) = 0$, and $g(x) > 0$ if $0 < x < 4$. The critical points of g are given by $0 = g'(x) = (8x - 3x^2)e^{-4}$, so they are $x = 0$ and $x = 8/3$. We have $g(8/3) = (256/27)e^{-4} \approx 0.174 < f(2,1)$. We conclude that the maximum value of f over the triangular region is $4/e^3$ and it occurs at the point $(2,1)$. The minimum value of f over the region is 0 and this value occurs at all points of the two perpendicular sides of the triangular boundary. Note that f has neither a local maximum nor a local minimum at the boundary point $(8/3, 4/3)$ although g has a local maximum there.

Extreme Value Problems with Constraints

A constrained extreme value problem is one in which the variables of the function to be maximized or minimized are not independent of one another, but must satisfy one or more constraint equations or inequalities. For example, the problems

$$\textbf{maximize } f(x,y) \textbf{ subject to } g(x,y) = C,$$

and

$$\textbf{minimize } f(x,y,z,w) \textbf{ subject to } g(x,y,z,w) = C_1,\ h(x,y,z,w) = C_2$$

have, respectively, one and two constraint equations, while the problem

$$\textbf{maximize } f(x,y) \textbf{ subject to } g(x,y) \le C$$

has a single constraint inequality.

Generally, inequality constraints may be regarded as restricting the domain of the function to be extremized to a smaller set which still, however, has interior points. Examples 6 and 7 above are illustrations of problems of this sort.

For the rest of this section we shall be dealing with problems with equation constraints. Such constraints usually restrict the domain of the function to be extremized so that it has no interior points, and the methods used above cannot be directly applied. If the constraint equations can be solved for some variables in other terms, then the problem can be reduced by substitution to a free (that is, unconstrained) problem for a function with fewer variables. For example, if the constraint equation $g(x,y) = C$ can be solved for $y = h(x)$ then the problem

$$\textbf{maximize } f(x,y) \textbf{ subject to } g(x,y) = C$$

can be reduced to the unconstrained, single-variable problem

$$\textbf{maximize } F(x) = f(x,h(x)),$$

which is subject to the methods of Section 4.4. However, this approach to equation constraints is not very satisfactory because it is often difficult, if not impossible, to solve the constraint equations.

In order to discover a somewhat better approach to dealing with equation constraints we shall look at a specific concrete example.

Example 8 Find the shortest distance from the origin to the curve $x^2y = 16$.

Solution It is sufficient to minimize the square of the distance from the point (x,y) on the curve $x^2y = 16$ to the origin, that is, to solve the problem

$$\textbf{minimize } f(x,y) = x^2 + y^2 \textbf{ subject to } g(x,y) = x^2y = 16.$$

As noted above, we could solve the constraint equation for $y = 16/x^2$ and substitute into f, thus reducing the problem to one of finding the unconstrained minimum value of $F(x) = f(x,16/x^2) = x^2 + 1/x^4$. Instead let us examine the problem geometrically. Fig. 10.30 shows the graph of the curve C with equation $x^2y = 16$, as well as some level curves of f. (These latter are circles centred at the origin.) Let P be a point on C which is closest to the origin. (By symmetry, there are two such points.) Evidently the level curve of f passing through P is tangent to C at P. Thus the gradient vectors of $f(x,y)$ and $g(x,y)$ at P, which are normal to these two curves at P, must be parallel to one another.

Now $\nabla f(x,y) = 2x\mathbf{i} + 2y\mathbf{j}$ and $\nabla g(x,y) = 2xy\mathbf{i} + x^2\mathbf{j}$. These vectors are parallel if one is a scalar multiple of the other, say

$$\nabla f(x,y) = -\lambda \nabla g(x,y) \text{ for some real number } \lambda,$$

that is,

$$2x = -\lambda 2xy, \; 2y = \lambda x^2.$$

Eliminating λ from these two equations we get $x/y = 2y/x$, or $x^2 = 2y^2$. The coordinates of P must satisfy this equation, and also the constraint equation $x^2y = 16$. Thus $2y^3 = 16$ and $y = 2$. Finally, $x = \pm 2\sqrt{2}$ and the two points on C closest to the origin are $(\pm 2\sqrt{2},2)$. The shortest distance from C to the origin is $\sqrt{8 + 4} = 2\sqrt{3}$ units.

The crux of the above solution was the observation that at the point where the extreme value occurred, the constraint curve and the level curve of f were tangent and so had parallel gradients; $\nabla f = -\lambda \nabla g$, or $\nabla(f + \lambda g) = 0$. Thus the extreme value occurs at critical points of the function $L(x,y) = f(x,y) + \lambda g(x,y)$. This

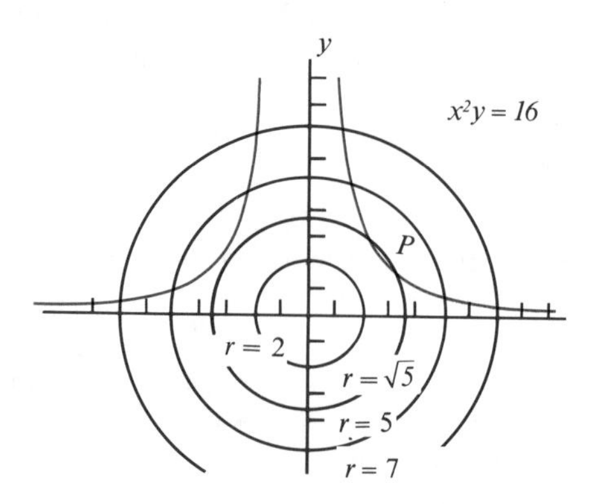

Figure 10.30

function is called the **Lagrangian** for the constrained extreme value problem, the number λ is called a **Lagrange multiplier**, and the method used to solve the problem is called the **method of Lagrange multipliers**.

The method of Lagrange multipliers can be applied to find extreme values of functions of n variables subject to at most $n-1$ equation constraints in those variables. The general procedure involves defining a Lagrangian function by adding constant multiples of the constraint functions to the function to be extremized and then looking for critical or singular points of the Lagrangian as well as endpoints or edgepoints of the constraint set.

Consider, for example, a problem of the form

$$\text{extremize } f(x,y,z) \text{ subject to constraints } g(x,y,z) = C_1,\ h(x,y,z) = C_2.$$

For suitably well-behaved functions f, g and h a geometric argument can still be given to show that extreme values should occur at critical points of the Lagrangian function

$$L(x,y,z) = f(x,y,z) + \lambda g(x,y,z) + \mu h\,(x,y,z).$$

The constraint equations $g(x,y,z) = C_1$ and $h(x,y,z) = C_2$ represent two 2-dimensional surfaces in 3-space, which in general will intersect along a curve C. At an extreme point of f which is not an endpoint of C we would expect C to be tangent to the level surface of f; thus ∇f will be perpendicular to C at such a point. Since ∇g and ∇h are both perpendicular to C (why?) we would expect ∇f to lie in the plane spanned by these two gradients and hence to be a linear combination of them:

$$\nabla f(x,y,z) = -\lambda\,\nabla g(x,y,z) - \mu\,\nabla h\,(x,y,z) \text{ for some constants } \lambda \text{ and } \mu.$$

Thus we should look for points (x,y,z) for which there exist numbers λ and μ such that

$$\begin{aligned} f_1(x,y,z) + \lambda g_1(x,y,z) + \mu h_1(x,y,z) &= 0 \\ f_2(x,y,z) + \lambda g_2(x,y,z) + \mu h_2(x,y,z) &= 0 \\ f_3(x,y,z) + \lambda g_3(x,y,z) + \mu h_3(x,y,z) &= 0 \\ g(x,y,z) &= 0 \\ h(x,y,z) &= 0, \end{aligned}$$

that is, we should look for critical points of the Lagrangian. Solving such a system of equations can, of course, present a considerable obstacle. In electing to use Lagrange multipliers we have traded off the problem of solving two constraint equations for two of their variables as **functions** of the third (which would have reduced the given problem to a single-variable unconstrained problem) for the problem of solving five equations for **numerical** values of the five unknowns x, y, z, λ and μ.

Example 9 Find the maximum and minimum values of the function $f(x,y,z) = xy + 2z$ on the circle which is the intersection of the plane $x + y + z = 0$ and the sphere $x^2 + y^2 + z^2 = 24$.

Solution The circle has no endpoints so we look for critical points of the Lagrangian function $L = xy + 2z + \lambda(x + y + z) + \mu(x^2 + y^2 + z^2)$. Setting the first partial derivatives of L equal to zero we obtain

$$y + \lambda + 2x\mu = 0 \quad (A)$$
$$x + \lambda + 2y\mu = 0 \quad (B)$$
$$2 + \lambda + 2z\mu = 0 \quad (C)$$

These three equations must be combined with the constraint equations

$$x + y + z = 0 \quad (D)$$
$$x^2 + y^2 + z^2 = 0 \quad (E)$$

to locate points where extreme values can occur.

Subtracting (A) from (B) we get $(x - y)(1 - 2\mu) = 0$. Therefore either $\mu = \frac{1}{2}$ or $x = y$. We analyze both possibilities.

If $\mu = \frac{1}{2}$ then we obtain from (B) and (C)

$$x + \lambda + y = 0, 2 + \lambda + z = 0.$$

Thus $x + y = 2 + z$. Combining this equation with (D) we get $z = -1$ and $x + y = 1$. Now by (E), $x^2 + y^2 = 24 - z^2 = 23$. Since $x^2 + y^2 + 2xy = (x + y)^2 = 1$, we have $2xy = 1 - 23 = -22$ and $xy = -11$. Now $(x - y)^2 = x^2 + y^2 - 2xy = 23 + 22 = 45$, so $x - y = \pm 3\sqrt{5}$. Combining this with $x + y = 1$ we find the two critical points arising from $\mu = \frac{1}{2}$, namely

$$\left(\frac{1 + 3\sqrt{5}}{2}, \frac{1 - 3\sqrt{5}}{2}, -1\right) \text{ and } \left(\frac{1 - 3\sqrt{5}}{2}, \frac{1 + 3\sqrt{5}}{2}, -1\right).$$

At both of these points we have $f(x,y,z) = xy + 2z = -11 - 2 = -13$.

If $x = y$ then (B) and (C) become

$$x + 2\mu x + \lambda = 0, 2 + 2\mu z + \lambda = 0.$$

Subtracting these equations we get $x + 2\mu x - 2\mu z - 2 = 0$. From (D) we have $z = -x - y = -2x$ so $x + 2\mu x + 4\mu x - 2 = 0$, that is, $x = y = 2/(1 + 6\mu)$. From (E), $24 = x^2 + y^2 + z^2 = 6x^2 = 24/(1 + 6\mu)^2$. Hence $(1 + 6\mu)^2 = 1$, $1 + 6\mu = \pm 1$, and $\mu = 0$ or $\mu = -1/3$.

If $\mu = 0$ then $x = y = 2$, $z = -4$ and $f(2,2,-4) = 4 - 8 = -4$.

If $\mu = -1/3$ then $x = y = -2$, $z = 4$ and $f(-2, -2,4) = 4 + 8 = 12$.

We conclude that the maximum value of f on the given circle is 12 and the minimum value is -13.

EXERCISES

In Exercises 1-10 find and classify the critical points of the given functions.

1. $f(x,y) = x^2 + 2y^2 - 4x + 4y$

2. $f(x,y) = xy - x + y$

3. $f(x,y) = 2x^3 - 6xy + 3y^2$

4. $f(x,y) = x^4 + y^4 - 4xy$

5. $f(x,y) = \dfrac{x}{y} + \dfrac{8}{x} - y$

6. $f(x,y) = \cos(x + y)$

7. $f(x,y) = x \sin y$

8. $f(x,y) = \cos x + \cos y$

9. $f(x,y) = xye^{-(x^2 + y^2)/2}$

10. $f(x,y) = \dfrac{xy}{2 + x^4 + y^4}$

11. Find the critical points of the function $f(x,y,z) = xyz - x^2 - y^2 - z^2$. Determine whether each critical point is a local maximum, a local minimum, or neither. Hint: Let $\mathbf{U} = u\mathbf{i} + v\mathbf{j} + w\mathbf{k}$ be a unit vector. Calculate the **second directional derivative** of f in the direction of $\mathbf{U}$, that is, $D_{\mathbf{U}}(D_{\mathbf{U}}f(x,y,z))$, and evaluate it at each critical point of f.

12. Repeat Exercise 11 for the function $f(x,y) = xy + x^2z - x^2 - y$.

13. Find the maximum and minimum values of $f(x,y) = x + 2y$ on the disc $x^2 + y^2 \leq 1$.

14. Find the maximum value of $f(x,y) = xy - x^3y^2$ on the square $0 \leq x \leq 1$, $0 \leq y \leq 1$.

15. Find the maximum and minimum values of $f(x,y) = \sin x \cos y$ on the closed triangular region bounded by the coordinate axes and the line $x + y = 2\pi$.

16. Find the maximum and minimum values of the function $f(x,y) = x^2y$ on the disc $x^2 + y^2 \leq 1$.

17. What is meant by "boundary point" of a set in 3-dimensional space? closed set? interior of a set? If S is the set of points in 3-space which form the plane with equation $x + y + z = 0$, what is the boundary of S? the interior of S? Is S a closed set?

18. Find the shortest distance from the point (3,0) to the curve $y = x^2$
   a) by reducing to an unconstrained problem in one variable.
   b) by using the method of Lagrange multipliers.

19. Find the shortest distance from the origin (0,0,0) to the plane with equation $x + 2y + 2z = 3$
   a) by using geometric arguments (vectors, but no calculus).
   b) by reducing the problem to an unconstrained problem in two variables.
   c) by using the method of Lagrange multipliers.

It is suggested that Lagrange multipliers be used in Exercises 20-29.

20. Find the maximum and minimum values of $f(x,y,z) = x + y - z$ over the sphere $x^2 + y^2 + z^2 = 1$.

21. Find the shortest distance from the point $(2,1,-2)$ to the sphere $x^2 + y^2 + z^2 = 1$.

22. Find the shortest distance from the origin to the surface $xyz^2 = 2$.

23. Find the maximum and minimum values of the function $f(x,y,z) = x$ over the curve of intersection of the surfaces $x^2 + 2y^2 + 2z^2 = 8$ and $z = x + y$.

24. Find the maximum and minimum values of $f(x,y,z) = x^2 + y^2 + z^2$ on the ellipse formed by the intersection of the cone $z^2 = x^2 + y^2$ and the plane $x - 2z = 3$.

25. Find the maximum and minimum values of $f(x,y,z) = x^2 + y^2 - z$ on ellipse formed by the intersection of the cylinder $x^2 + y^2 = 4$ and the plane $x + y + z = 1$.

26. Find the maximum and minimum values of $f(x,y,z) = x + y^2z$ subject to the constraints $y^2 + z^2 = 2$ and $z = x$.

27. Find the volume of the largest rectangular box with sides parallel to the coordinate planes which can be inscribed inside the ellipsoid

$$x^2 + \frac{y^2}{4} + \frac{z^2}{4} = 1.$$

28. Show that among all rectangular boxes with specified volume V, the one which is a cube has the least total surface area.

29. If x, y, z are the angles of a triangle, show that

$$\sin\frac{x}{2}\sin\frac{y}{2}\sin\frac{z}{2} \le \frac{1}{8}.$$

For what triangles does equality occur?

30. Suppose that f and g are differentiable functions in the plane, and suppose that the equation $g(x,y) = C$ defines y implicitly as a function of x nearby the point $x = a$. Use the chain rule to show that if $f(x,y)$ has a local extreme value at (a,b) subject to the constraint $g(x,y) = C$ then (a,b) is a critical point of the function $L(x,y) = f(x,y) + \lambda g(x,y)$. This constitutes a more formal justification for the method of Lagrange multipliers in this case.

31. Repeat Exercise 30 for the situation where $f(x,y,z)$ is to be extremized subject to constraints $g(x,y,z) = C_1$ and $h(x,y,z) = C_2$. (Suppose the latter equations define y and z implicitly as differentiable functions of x near $x = a$.)

Appendix I Mathematical Induction

Appendix II The Theoretical Foundations of Calculus

Appendix III The Riemann Integral

Appendices

APPENDIX I: MATHEMATICAL INDUCTION

Mathematical induction is a technique that can be used to prove the validity of statements about the integer n that are suspected to be valid for all integers greater than or equal to some starting integer n_0.

Example 1 Show that for $n = 1, 2, 3, \ldots,$ the sum of the integers from 1 up to n is $n(n + 1)/2$, that is,

$$\sum_{k=1}^{n} k = 1 + 2 + 3 + \cdots + n = \frac{n}{2}(n + 1). \qquad (*)$$

Proof A direct proof of this fact was given in Section 6.1. An alternate inductive proof can be given as follows

First, observe that the formula (*) is true for $n = 1$; in this case the left-hand side is 1 and the right-hand side is $(1/2)(1 + 1) = 1$. Now *assume* that formula (*) is true for some value of $n \geq 1$, say for $n = k$; that is, we are assuming that

$$1 + 2 + 3 + \cdots + k = \frac{k(k + 1)}{2}.$$

Consider the sum of the numbers from 1 up to $k + 1$. Using the assumption above we calculate

$$\begin{aligned}1 + 2 + 3 + \cdots + k + (k + 1) &= (1 + 2 + 3 + \cdots + k) + (k + 1)\\ &= \frac{k}{2}(k + 1) + (k + 1) = \frac{(k + 1)}{2}(k + 2).\end{aligned}$$

This is just what formula (*) says if $n = k + 1$. Thus formula (*) must be true for $n = k + 1$ *if* it is true for $n = k$.

Since (*) is true for $n = 1$, therefore it is true for $n = 2$; since it is true for $n = 2$, therefore it is true for $n = 3$, and so on. We summarize this last sentence by saying formula (*) holds for all $n \geq 1$ *by induction*. □

As suggested in Example 1, mathematical induction is carried out in the following way: If S is a statement about the integer n that is to be proved for $n = n_0, n_0 + 1, n_0 + 2, \ldots$ (that is, for all integers $n \geq n_0$), then we should

i) prove by direct means that S is true for $n = n_0$;

iii) assume that S is true for $n = k$ for some $k \geq n_0$ (or even for all n between n_0 and some such k); this is called the **induction hypothesis;**

iii) using the induction hypothesis, show that S must also be true for $n = k + 1$; this is called the **induction step;**

iv) conclude that S is true for all integers $n \geq n_0$ **by induction.**

The justification for mathematical induction rests on the fact that every nonempty set of positive integers contains a smallest integer. If the statement S to be proved is false for some $n \geq n_0$, then there must be a smallest n, say n_1, for which it is false. $n_1 > n_0$ since S is true for n_0. S is true for $n_1 - 1$ since n_1 is the smallest n for which it is false. But then the induction step shows that S is true for n_1 and we have a contradiction. Since there cannot be any smallest false value of n, there cannot be any $n \geq n_0$ for which S is false.

Students are sometimes bothered by the fact that in an inductive proof we seem to be assuming what we want to prove. This is not so; we merely show that *if* S holds for $n = k$, then it also holds for $n = k + 1$.

Example 2 Show that $(d^n/dx^n)x^n = n!$ (for $n \geq 1$).

Proof The assertion holds for $n = 1$ since $(d/dx)x = 1 = 1!$. Assume that for some $k \geq 1$ we have $(d^k/dx^k)x^k = k!$. Observe that

$$\begin{aligned}\frac{d^{k+1}}{dx^{k+1}}x^{k+1} = \frac{d^k}{dx^k}\frac{d}{dx}x^{k+1} &= \frac{d^k}{dx^k}(k + 1)x^k\\ &= (k + 1)\frac{d^k}{dx^k}x^k = (k + 1)k! = (k + 1)!.\end{aligned}$$

Since the assertion holds for $n = k + 1$ if it holds for $n = k$, it must hold for all $n \geq 1$ by induction. □

We remark that the inductive technique can also be used in definitions. For instance, $n!$ is defined inductively:

$$\begin{aligned}0! &= 1\\ (n + 1)! &= (n + 1)n! \qquad \text{if } n = 0, 1, 2, 3, \ldots.\end{aligned}$$

There is no ready formula for computing $n!$ directly for any n. We must actually multiply $1 \times 2 \times 3 \times \cdots \times n$, that is, we must calculate $1!, 2!, 3!, \ldots$ up to $(n - 1)!$ and finally $n!$ to find the value of $n!$. See Example 1(h) of Section 8.1 for another example of an inductive definition.

EXERCISES

1. Use mathematical induction to prove that $(d/dx)x^n = nx^{n-1}$ for $n = 1, 2, 3, \ldots$.

2. Use induction to prove that

$$\sum_{k=1}^{n} k^2 = 1^2 + 2^2 + 3^2 + \cdots + n^2 = \frac{n(n+1)(2n+1)}{6}.$$

3. Prove that

$$\sum_{k=1}^{n} (2k - 1) = 1 + 3 + 5 + \cdots + (2n - 1) = n^2.$$

4. Use induction to prove the formula for a finite geometric sum:

$$\sum_{k=0}^{n} r^k = 1 + r + r^2 + \cdots + r^n = \frac{1 - r^{n+1}}{1 - r} \qquad (r \neq 1).$$

5. Prove that

$$\sum_{k=1}^{n} k^3 = 1^3 + 2^3 + 3^3 + \cdots + n^3 = \left(\frac{n(n+1)}{2}\right)^2.$$

It is an interesting observation that $\sum_{k=1}^{n} k^3 = (\sum_{k=1}^{n} k)^2$.

6. Prove that

$$\sum_{k=1}^{n} (-1)^{k-1}k^2 = (-1)^{n-1} \sum_{k=1}^{n} k.$$

7. If $r > 0$ prove that $(1 + r)^n \geq 1 + nr$ for $n = 1, 2, 3, \ldots$.

8. Prove that

$$\sum_{k=1}^{n} \frac{k}{2^k} = \frac{1}{2} + \frac{2}{2^2} + \frac{3}{2^3} + \cdots + \frac{n}{2^n} = 2 - \frac{n+2}{2^n}.$$

9. Guess a formula for $f^{(n)}(x)$ where $f(x) = \sqrt{1 + x}$ and prove it by induction.

10. Guess a formula for $(d^n/dx^n)xe^{-x}$ and prove it by induction.

11. Do not use induction in this problem; direct proofs are easier. The results will be needed in Exercises 12 and 13.

For $0 \leq k \leq n$ (k and n are nonnegative integers) we define the *binomial coefficients* $\binom{n}{k}$ by

$$\binom{n}{k} = \frac{n!}{k!(n-k)!}.$$

Thus, for instance, $\binom{5}{2} = 5!/(2!3!) = 120/12 = 10$. Show that

i) $\binom{n}{0} = \binom{n}{n} = 1$ for every n; and

ii) if $0 \leq k \leq n$, then $\binom{n}{k-1} + \binom{n}{k} = \binom{n+1}{k}$.

It follows that for fixed $n \geq 1$ the binomial coefficients $\binom{n}{0}, \binom{n}{1}, \binom{n}{2}, \ldots, \binom{n}{n}$ are the elements of the n row of *Pascal's triangle*

$$\begin{array}{ccccccccccc} & & & & 1 & & 1 & & & & \\ & & & 1 & & 2 & & 1 & & & \\ & & 1 & & 3 & & 3 & & 1 & & \\ & 1 & & 4 & & 6 & & 4 & & 1 & \\ 1 & & 5 & & 10 & & 10 & & 5 & & 1 \\ \cdot & & & & & & & & & & \cdot \\ \cdot & & & & & & & & & & \cdot \\ \cdot & & & & & & & & & & \cdot \end{array}$$

where each element is the sum of the two diagonally above it.

12. Use mathematical induction and the results of Exercise 11 to prove the binomial theorem:

$$(a + b)^n = \sum_{k=0}^{n} \binom{n}{k} a^{n-k}b^k$$
$$= a^n + na^{n-1}b + \binom{n}{2} a^{n-2}b^2 + \cdots + b^n.$$

13. Use mathematical induction, the product rule, and Exercise 11 to verify the *Leibniz rule* for the nth derivative of a product of two functions:

$$(fg)^{(n)} = \sum_{k=0}^{n} \binom{n}{k} f^{(n-k)}g^{(k)}.$$
$$= f^{(n)}g + nf^{(n-1)}g' + \binom{n}{2} f^{(n-2)}g'' + \cdots + fg^{(n)}.$$

14. Show that every positive integer is interesting.

APPENDIX II: THE THEORETICAL FOUNDATIONS OF CALCULUS

As noted in Chapter 1, the development of calculus depends in an essential way on the concept of limit of a function and thereby on properties of the real number system. In Chapter 1 we presented these notions in an intuitive way and did not attempt to prove any of the quoted results on limits or continuity of functions. Indeed, these results seem quite obvious and most students and users of calculus are not bothered by applying them without proof.

Nevertheless, mathematics is a highly logical and rigorous discipline, and any statement, however obvious, that cannot be proved by strictly logical arguments from acceptable assumptions must be considered suspect. In this appendix we present formal proofs of the properties of limits and continuous functions given in Chapter 1. The branch of mathematics that deals with such proofs is called mathematical analysis. This subject is not usually pursued by students in introductory calculus courses but is postponed to higher years and is studied by students in majors or honors programs

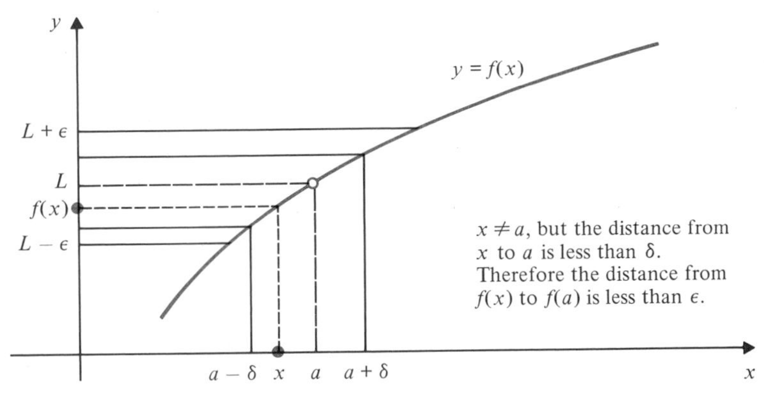

Figure II.1

in mathematics. We include here only the basics of mathematical analysis, enough to get us as far as the proof that a function continuous on a closed, finite interval assumes a maximum and a minimum value and has the intermediate-value property. It is hoped that some of this material will be of value to honors-level calculus courses and individual students with a deeper interest in understanding calculus. The reader should still refer back to Sections 1.3 and 1.5 of Chapter 1 for more discussion of the topics covered in this appendix.

The material is divided into sections dealing with the formal definition and properties of limits, continuous functions, limits of sequences, and finally the properties of continuous functions on closed, finite intervals.

Limits of Functions

At the heart of mathematical analysis is the formal definition of limit, Definition 8a of Section 1.3, which we restate here as follows:

> We say that $\lim_{x\to a} f(x) = L$ if for every positive number ϵ there exists a positive number δ depending on ϵ ($\delta = \delta(\epsilon)$) such that $|f(x) - L| < \epsilon$ whenever $0 < |x - a| < \delta$.

(It is traditional to use the Greek letters ϵ (epsilon) and δ (delta) in this context.) Note in Fig. II.1 that $\lim_{x\to a} f(x) = L$ can exist even if $f(x)$ is not defined at $x = a$, and even if $f(a)$ is defined we need not have $f(a) = L$.

Let us begin by illustrating this definition with a few examples.

Example 1 Prove that $\lim_{x\to 2} (5 - 2x) = 1$.

Proof Let ϵ be any given positive number. Referring to Fig. II.2 we have

$$|(5 - 2x) - 1| = |4 - 2x| = 2|2 - x| = 2|x - 2| < \epsilon,$$

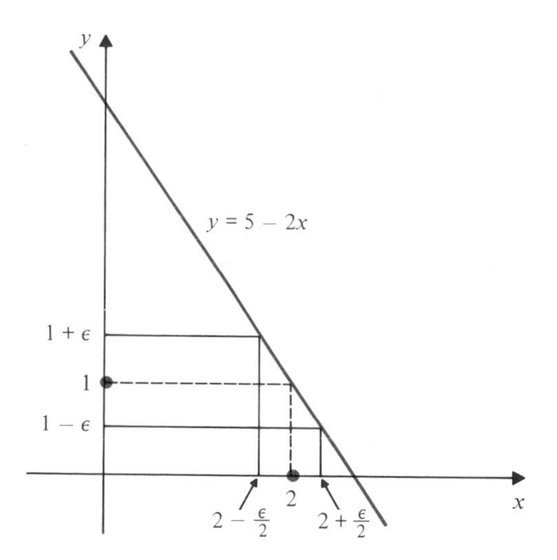

Figure II.2

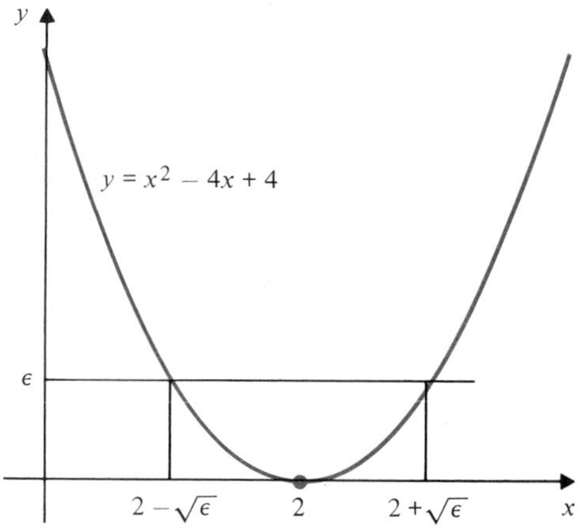

Figure II.3

provided $|x - 2| < \epsilon/2$. Thus we may take $\delta = \epsilon/2$ and ensure that $|(5 - 2x) - 1| < \epsilon$ by taking x to satisfy $0 < |x - 2| < \delta$. (Indeed, in this case we can even allow $|x - 2| = 0$.) Thus $\lim_{x\to 2}(5 - 2x) = 1$. □

Example 2 Show that $\lim_{x\to 2}(x^2 - 4x + 4) = 0$.

Proof Let ϵ be any positive number. We have

$$|(x^2 - 4x + 4) - 0| = |(x - 2)^2| = |x - 2|^2 < \epsilon,$$

provided $0 < |x - 2| < \delta = \sqrt{\epsilon}$. Thus $\lim_{x\to 2}(x^2 - 4x + 4) = 0$. See Fig. II.3. □

Example 3 Show that $\lim_{x\to 4}\sqrt{x} = 2$.

Proof Let ϵ be any positive number. If $|x - 4| < 4$, then $0 < x$ and

$$|\sqrt{x} - 2| = \left|\frac{(\sqrt{x} - 2)(\sqrt{x} + 2)}{\sqrt{x} + 2}\right| = \frac{1}{\sqrt{x} + 2}|x - 4| < \frac{1}{2}|x - 4|.$$

Let $\delta = \min\{4, 2\epsilon\}$, that is, whichever of the numbers 4 and 2ϵ is smaller. If $0 < |x - 4| < \delta$, then $|\sqrt{x} - 2| < \frac{1}{2}2\epsilon = \epsilon$. Thus $\lim_{x\to 4}\sqrt{x} = 2$. □

Example 4 Show that $\lim_{x\to 1}((x^3 - x)/(x - 1)) = 2$.

Proof Let ϵ be any positive number. If $x \neq 1$ we have

$$\left|\frac{x^3 - x}{x - 1} - 2\right| = |x(x + 1) - 2| = |x^2 + x - 2| = |x - 1||x + 2|.$$

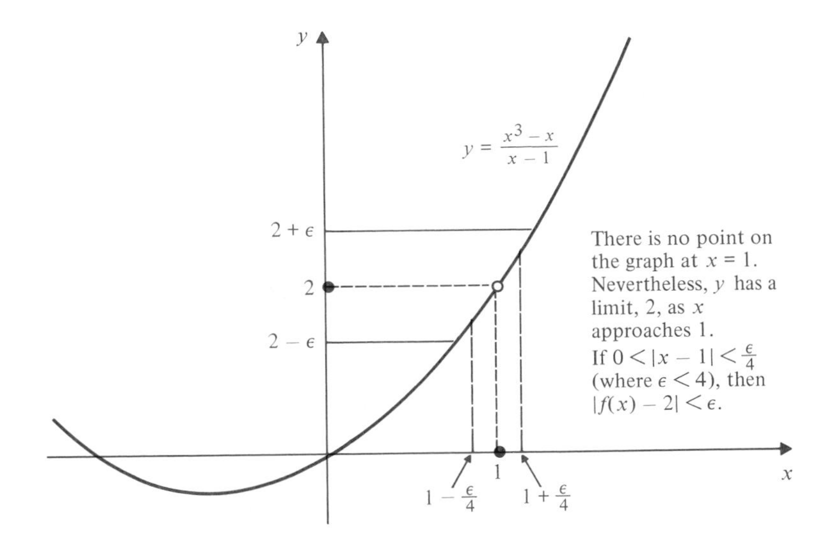

Figure II.4

If $|x - 1| < 1$, then $0 < x < 2$ and $2 < x + 2 < 4$, so $|x + 2| < 4$. Hence if $0 < |x - 1| < \delta = \min\left\{1, \frac{\epsilon}{4}\right\}$ we have $|x + 2| < 4$ and $|x - 1| < \frac{\epsilon}{4}$, so $|(x^3 - x)/(x - 1) - 2| < \epsilon$. Hence $\lim_{x \to 1} ((x^3 - x)/(x - 1)) = 2$. See Fig. II.4. Observe that in this example we really must restrict $|x - 1| > 0$; the fraction is not defined at $x = 1$. □

While the definition of limit provides a means of checking whether a given number is the limit of a particular function at a particular point (as in the four examples above) it provides no means for finding an unknown limit. The limit in each of the examples was intuitively obvious. The main reason for having a formal definition of limit is to prove general theorems about limits such as Theorems 1–3 of Section 1.3, to which we turn our attention now.

Theorem 1 (*Uniqueness of Limits*) If $\lim_{x \to a} f(x) = L$ and $\lim_{x \to a} f(x) = M$, then, necessarily, $L = M$.

Proof It may seem to the student that nothing needs to be proved here; things that are equal to the same thing are necessarily equal to each other. There is, however, the possibility that the definition might have been satisfied for more than one number L. If this were the case we should really have used $\lim_{x \to a} f(x)$ to denote a set of numbers rather than just one number. In proving this theorem we are really justifying our use of the phrase $\lim_{x \to a} f(x) = L$ in the definition of limit.

It suffices for us to prove that for any positive number ϵ we have $|L - M| < \epsilon$. It is then impossible that $L \neq M$. (Why?) Let $\epsilon > 0$ be given. Since $\lim_{x \to a} f(x) = L$, there exists a number δ_1 such that if $0 < |x - a| < \delta_1$ then $|f(x) - L| < \epsilon/2$. Similarly, since $\lim_{x \to a} f(x) = M$, there exists a number δ_2 such that if $0 < |x - a| < \delta_2$ then $|f(x) - M| < \epsilon/2$. Now if $0 < |x - a| < \delta = \min\{\delta_1, \delta_2\}$, then we obtain, using the triangle inequality for real numbers,

$$|L - M| = |(L - f(x)) + (f(x) - M)| \le |f(x) - L| + |f(x) - M| < \frac{\epsilon}{2} + \frac{\epsilon}{2} = \epsilon.$$

Therefore $L = M$. □

Theorem 2 If $\lim_{x\to a} f(x) = L$ and $\lim_{x\to a} g(x) = M$, then

i) $\lim_{x\to a} (f(x) + g(x)) = L + M$

ii) $\lim_{x\to a} (f(x) - g(x)) = L - M$

iii) $\lim_{x\to a} (f(x)g(x)) = LM$

iv) $\lim_{x\to a} \dfrac{f(x)}{g(x)} = \dfrac{L}{M}$, provided $M \neq 0$

v) $\lim_{x\to a} (cf(x)) = cL$ (c = constant)

vi) If $f(x) \le g(x)$ for $0 < |x - a| < \delta$ for some $\delta > 0$, then $L \le M$.

Proof We will prove only parts (i) and (iii) to demonstrate the techniques. The remaining parts are left as exercises. (See Exercises 7–9 below.)

i) Let $\epsilon > 0$ be given. Since $\lim_{x\to a} f(x) = L$, there exists a positive number δ_1 such that if $0 < |x - a| < \delta_1$ then $|f(x) - L| < \epsilon/2$. Similarly, since $\lim_{x\to a} g(x) = M$, there exists a positive number δ_2 such that if $0 < |x - a| < \delta_2$ then $|g(x) - M| < \epsilon/2$. Let $\delta = \min\{\delta_1, \delta_2\}$. If $0 < |x - a| < \delta$, then

$$|(f(x) + g(x)) - (L + M)| = |(f(x) - L) + (g(x) - M)| \le |f(x) - L| + |g(x) - M| < \frac{\epsilon}{2} + \frac{\epsilon}{2} = \epsilon.$$

Thus $\lim_{x\to a} (f(x) + g(x)) = L + M$. □

iii) Let $\epsilon > 0$ be given. Since $\lim_{x\to a} f(x) = L$, there exists a positive number δ_1 such that if $0 < |x - a| < \delta_1$ then $|f(x) - L| < 1$, and hence $|f(x)| = |f(x) - L + L| < 1 + |L|$. (This says that f is bounded near the point a except possibly at a.) Also, there exists a positive number δ_2 such that if $0 < |x - a| < \delta_2$ then $|f(x) - L| < \epsilon/(2(|M| + 1))$. Since $\lim_{x\to a} g(x) = M$, there exists a positive number δ_3 such that if $0 < |x - a| < \delta_3$ then $|g(x) - M| < \epsilon/(2(|L| + 1))$. Let $\delta = \min\{\delta_1, \delta_2, \delta_3\}$. If $0 < |x - a| < \delta$, then

$$\begin{aligned}|f(x)g(x) - LM| &= |f(x)g(x) - f(x)M + f(x)M - LM| \\ &\le |f(x)(g(x) - M)| + |M(f(x) - L)| \\ &= |f(x)||g(x) - M| + |M||f(x) - L| \\ &< (|L| + 1)\frac{\epsilon}{2(|L| + 1)} + |M|\frac{\epsilon}{2(|M| + 1)} < \epsilon.\end{aligned}$$

Hence $\lim_{x\to a} f(x)g(x) = LM$. □

It should be remarked that one does not construct a proof such as this by starting at the top and pulling numbers like $\epsilon/2$ and $\epsilon/(2(|M| + 1))$ out of a hat. One really starts by considering the expressions $|(f(x) + g(x)) - (L + M)|$ and

$|f(x)g(x) - LM|$, which must be made less than ϵ and rewriting them to show their explicit dependence on $|f(x) - L|$ and $|g(x) - M|$. Then one arranges to take x sufficiently close to a to ensure that each of these quantities is sufficiently small that the desired combination is in fact less than ϵ.

Theorem 3 (*The Squeeze Theorem*) If $f(x) \leq g(x) \leq h(x)$ for $0 < |x - a| < k$ (for some $k > 0$) and if

$$\lim_{x \to a} f(x) = L \quad \text{and} \quad \lim_{x \to a} h(x) = L,$$

then $\lim_{x \to a} g(x)$ exists and equals L also. (See Fig. II.5.)

Proof

$$\begin{aligned} |g(x) - L| &= |g(x) - f(x) + f(x) - L| \\ &\leq (g(x) - f(x)) + |f(x) - L| \qquad \text{if } 0 < |x - a| < k \\ &\leq (h(x) - f(x)) + |f(x) - L| \\ &\leq |h(x) - L| + |f(x) - L| + |f(x) - L| \\ &= |h(x) - L| + 2|f(x) - L|. \end{aligned}$$

Let $\epsilon > 0$ be given. Since $\lim_{x \to a} f(x) = \lim_{x \to a} h(x) = L$, there exist positive numbers δ_1 and δ_2 such that if $0 < |x - a| < \delta_1$ then $|f(x - L| < \epsilon/3$ and if $0 < |x - a| < \delta_2$ then $|h(x) - L| < \epsilon/3$. Hence if $0 < |x - a| < \delta = \min\{k, \delta_1, \delta_2\}$, then $|g(x) - L| < \epsilon/3 + 2\epsilon/3 = \epsilon$. Therefore $\lim_{x \to a} g(x) = L$. □

Minor modifications can be made to the definition of limit to allow for one-sided limits, infinite limits, and limits at infinity.

One-Sided Limits

> We say that $\lim_{x \to a+} f(x) = L$ if for every positive number ϵ there exists a positive number $\delta = \delta(\epsilon)$ such that $|f(x) - L| < \epsilon$ holds whenever $a < x < a + \delta$. See Fig. II.6.

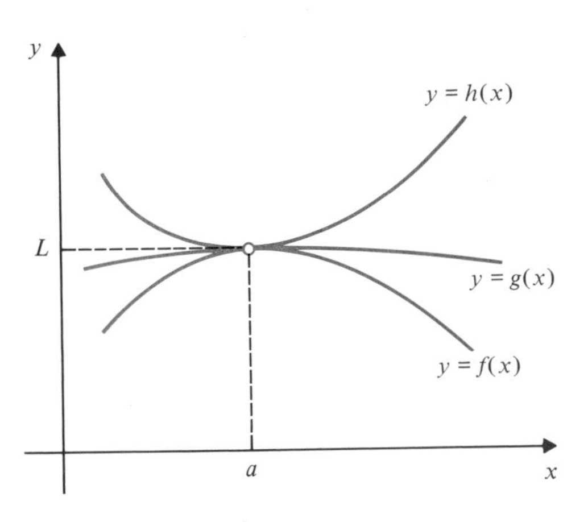

Figure II.5

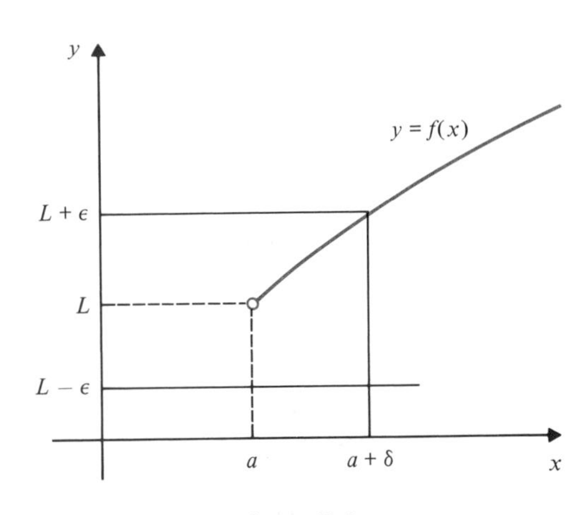

Figure II.6

The definition for $\lim_{x \to a-} f(x) = L$ is similar except that $a < x < a + \delta$ is replaced by $a - \delta < x < a$.

Infinite Limits

> We say that $\lim_{x \to a} f(x) = \infty$ if for every positive number R (no matter how large) there exists a positive number $\delta = \delta(R)$ such that $f(x) > R$ holds whenever $0 < |x - a| < \delta$. See Fig. II.7.

The definition for $\lim_{x \to a} f(x) = -\infty$ is similar except that we replace the condition $f(x) > R$ with $f(x) < -R$.

Limits at Infinity

> We say $\lim_{x \to \infty} f(x) = L$ if for every positive number ϵ there exists a positive number $R = R(\epsilon)$ such that $|f(x) - L| < \epsilon$ holds whenever $x > R$. See Fig. II.8.

The definition for $\lim_{x \to -\infty} f(x) = L$ is similar with $x < -R$ replacing $x > R$.

Appropriate versions of Theorems 1–3 hold for such extensions of the limit concept.

Example 5 Show that $\lim_{x \to 0+} \sqrt{x} = 0$.

Proof Let $\epsilon > 0$ be given. We have

$$|\sqrt{x} - 0| = \sqrt{x} < \epsilon \qquad \text{if } 0 < x < \epsilon^2 = \delta.$$

Hence $\lim_{x \to 0+} \sqrt{x} = 0$. □

Example 6 Show that $\lim_{x \to 2+} x/(x - 2) = \infty$.

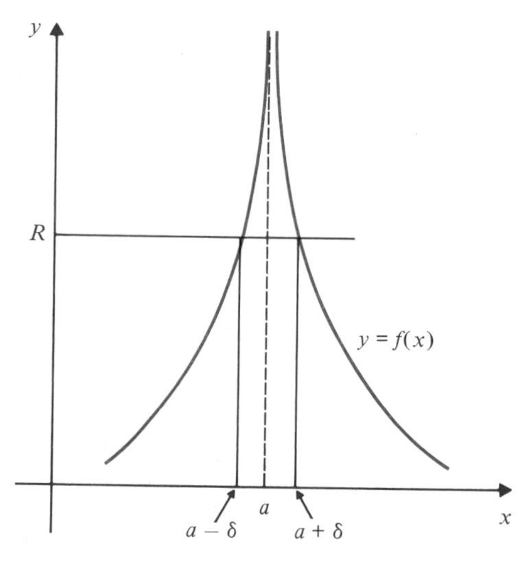

Figure II.7

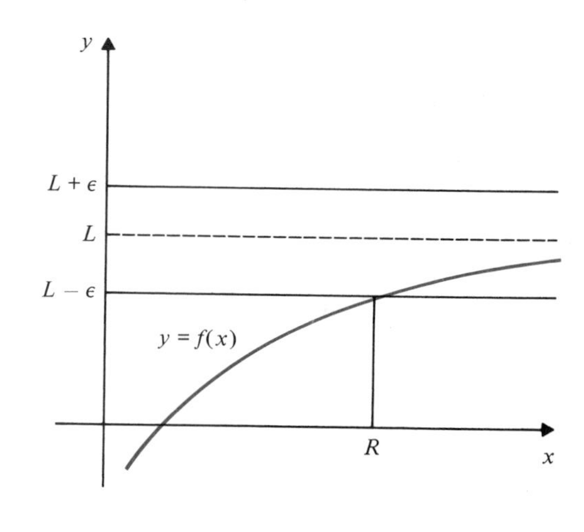

Figure II.8

Proof Let $R > 0$ be given. If $0 < x - 2 < \delta = 2/R$, then $x > 2$ and $x/(x - 2) > 2/(x - 2) > 2\,(R/2) = R$. Hence $\lim_{x\to 2+} x/(x - 2) = \infty$. □

Continuous Functions

A function f on an open interval containing the point a is said to be continuous at the point a if

$$\lim_{x\to a} f(x) = f(a),$$

that is, if for every $\epsilon > 0$ there exists $\delta > 0$ such that if $|x - a| < \delta$ then $|f(x) - f(a)| < \epsilon$.

A function f is continous on an interval if it is continuous at every point of that interval. In the case of an endpoint of a closed interval, f need only be continuous on one side. Thus, f is continuous on the interval $[a, b]$ if

$$\lim_{t\to x} f(t) = f(x)$$

for each x satisfying $a < x < b$, and

$$\lim_{t\to a+} f(t) = f(a) \quad \text{and} \quad \lim_{t\to b-} f(t) = f(b).$$

These concepts are illustrated in Fig. II.9.

Some important results about continuous functions are collected in Theorem 4 of Section 1.5, which we restate here.

Theorem 4 a) If f and g are continuous at the point a, then so are $f + g$, $f - g$, fg, and, if $g(a) \neq 0$, f/g.

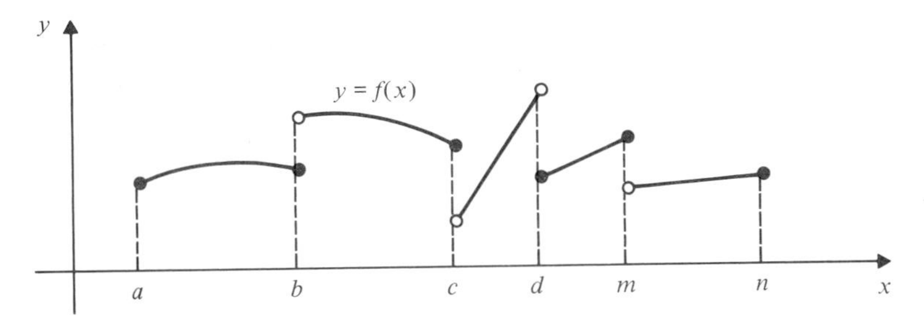

f is continuous on the intervals $[a, b]$, $(b, c]$, (c, d), $[d, m]$, and $(m, n]$

Figure II.9

b) If f is continuous at the point L and if $\lim_{x\to a} g(x) = L$, then $\lim_{x\to a} f(g(x)) = f(L) = f(\lim_{x\to a} g(x))$. In particular, if g is continuous at the point a (so that $L = g(a)$), then $\lim_{x\to a} f(g(x)) = f(g(a))$, that is, $f\circ g(x) = f(g(x))$ is continuous at $x = a$.

c) The function $f(x) = C$ (constant) is continuous on the whole real line.

d) For any rational number r the function $f(x) = x^r$ is continuous at every real number where it is defined.

Proof Part (a) is merely a restatement of various parts of Theorem 2 above; for example,

$$\lim_{x\to a} f(x)g(x) = (\lim_{x\to a} f(x))(\lim_{x\to a} g(x)) = f(a)g(a).$$

Part (b) can be proved as follows. Let $\epsilon > 0$ be given. Since f is continuous at L, there exists $k > 0$ such that $|f(g(x)) - f(L)| < \epsilon$ whenever $|g(x) - L| < k$. Since $\lim_{x\to a} g(x) = L$, there exists $\delta > 0$ such that if $0 < |x - a| < \delta$ then $|g(x) - L| < k$. Hence if $0 < |x - a| < \delta$, then $|f(g(x)) - f(L)| < \epsilon$, and $\lim_{x\to a} f(g(x)) = f(L)$.

The proofs of (c) and (d) are left to the student in Exercises 11–13 at the end of this appendix. □

Completeness and Sequential Limits

A real number u is said to be an **upper bound** for the nonempty set S of real numbers if $x \le u$ for every x in S. The number u^* is called the **least upper bound** of S if u^* is an upper bound for S and $u^* \le u$ for every upper bound u of S.

Similarly, ℓ is a lower bound of S if $\ell \le x$ for every x in S and ℓ^* is the greatest lower bound of S if ℓ^* is a lower bound and $\ell \le \ell^*$ for every lower bound ℓ of S.

Example 7 Set $S_1 = [2, 3]$ and $S_2 = (2, \infty)$. Any number $u \ge 3$ is an upper bound for S_1. S_2 has no upper bound; we say that it is not bounded above. 3 is the least upper bound of S_1. Any real number $\ell \le 2$ is a lower bound for both S_1 and S_2. $\ell^* = 2$ is the greatest lower bound for each set.

We now recall the completeness axiom for the real number system, which we discussed briefly in Section 1.1.

> *Completeness Axiom* A nonempty set of real numbers that has an upper bound must have a least upper bound.

Equivalently, a nonempty set of real numbers having a lower bound must have a greatest lower bound.

We stress that this is an *axiom* to be assumed without proof. It cannot be deduced from the more elementary algebraic and order properties of the real numbers. These other properties are shared by the rational numbers, a set that is not complete. The completeness axiom is essential for the proof of the most important results about continuous functions, in particular those in Theorems 5 and 6 of Section 1.5. Before attempting these proofs, however, we must develop a little more machinery.

In Section 8.1 there is stated a version of the completeness axiom that is pertinent to *sequences* of real numbers; specifically, an increasing sequence that is bounded above converges to a limit. We begin by verifying that this follows from the earlier statement of completeness. (Both statements are, in fact, equivalent.) As noted in Section 8.1, the sequence

$$\{x_n\} = \{x_1, x_2, x_3, \ldots\}$$

is a function on the positive integers, that is, $x_n = x(n)$. We say that the sequence converges to the limit L, and we write $\lim x_n = L$, if the corresponding function $x(t)$ satisfies $\lim_{t\to\infty} x(t) = L$ as defined above. Thus

> $\lim x_n = L$ if for every positive number ϵ there exists a positive number $N = N(\epsilon)$ such that $|x_n - L| < \epsilon$ holds whenever $n > N$.

Theorem 5 If $\{x_n\}$ is an increasing sequence that is bounded above, that is,

$$x_{n+1} \geq x_n \quad \text{and} \quad x_n \leq K \qquad \text{for } n = 1, 2, 3, \ldots,$$

then $\lim x_n = L$ exists. (Equivalently, if $\{x_n\}$ is decreasing and bounded below, then $\lim x_n$ exists.)

Proof Let $\{x_n\}$ be increasing and bounded above. The set of real numbers $S = x_n$ has an upper bound, K, and so has a least upper bound, say L. Thus $x_n \leq L$ for every n, and if $\epsilon > 0$ then there exists a positive integer N such that $x_N > L - \epsilon$. (Otherwise $L - \epsilon$ would be an upper bound for S that is lower than the least upper bound.) If $n \geq N$, then we have $L - \epsilon < x_N \leq x_n \leq L$, so $|x_n - L| < \epsilon$. Thus $\lim x_n = L$. The proof for a decreasing sequence that is bounded below is similar. □

Theorem 6 If $a \leq x_n \leq b$ for each n, and if $\lim x_n = L$, then $a \leq L \leq b$.

Proof Suppose that $L > b$. Let $\epsilon = (L - b)/2$. Since $\lim x_n = L$, there exists n such that $|x_n - L| < \epsilon$. Thus $x_n > L - \epsilon = L - (L - b)/2 = (L + b)/2 > b$, which is a contradiction since we are given that $x_n \leq b$. Thus $L \leq b$. A similar argument shows that $L \geq a$. □

Theorem 7 If f is continuous on $[a, b]$, if $a \leq x_n \leq b$ for each n, and if $\lim x_n = L$, then $\lim f(x_n) = f(L)$. □

The proof is similar to that of Theorem 4(b) and is left as Exercise 16 at the end of this appendix.

Continuous Functions on a Closed, Finite Interval

We are now in a position to prove Theorems 5 and 6 of Section 1.5.

Theorem 8 If f is continuous on $[a, b]$ then f is bounded there; that is, there exists a constant K such that $|f(x)| \leq K$ if $a \leq x \leq b$.

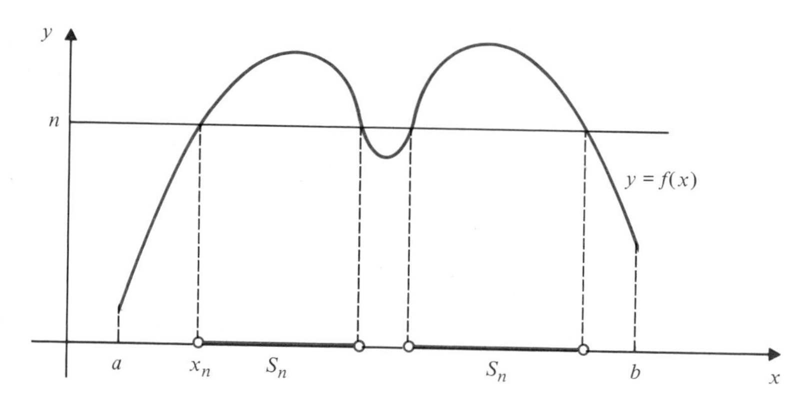

Figure II.10

Proof We show that f is bounded above; a similar proof shows that f is bounded below. For each positive integer n let S_n be the set of points x in $[a, b]$ such that $f(x) > n$:

$$S_n = \{x : a \le x \le b \quad \text{and} \quad f(x) > n\}.$$

We would like to show that S_n is empty for some n. It would then follow that $f(x) \le n$ for all x in $[a, b]$; that is, n would be an upper bound for f on $[a, b]$.

Suppose to the contrary, that S_n is nonempty for every n. We will show that this leads to a contradiction. Since S_n is bounded below (a is a lower bound) by completeness S_n has a greatest lower bound; call it x_n. See Fig. II.10. Evidently $a \le x_n$. Since $f(x) > n$ at some point of $[a, b]$ and f is continuous at that point, $f(x) > n$ on some interval contained in $[a, b]$. Hence $x_n < b$. It follows that $f(x_n) \ge n$. (If $f(x_n) < n$ then by continuity $f(x) < n$ for some distance to the right of x_n, and x_n could not be the greatest lower bound of S_n.)

Since $S_{n+1} \subset S_n$ for each n, $x_{n+1} \ge x_n$ so $\{x_n\}$ is an increasing sequence. Being bounded above (b is an upper bound) this sequence converges, by Theorem 5. Let $\lim x_n = L$. By Theorem 6, $a \le L \le b$. Since f is continuous at L, $\lim f(x_n) = f(L)$ exists by Theorem 7. But since $f(x_n) \ge n$, $\lim f(x_n)$ cannot exist. This is the desired contradiction, and the proof is complete. □

Theorem 9 If f is continuous on $[a, b]$, then there are points v and u in $[a, b]$ such that for any x in $[a, b]$ we have

$$f(v) \le f(x) \le f(u);$$

that is, f assumes a maximum and a minimum value on $[a, b]$.

Proof By Theorem 8 we know that the set $S = \{f(x) : a \le x \le b\}$ has an upper bound and therefore by completeness a least upper bound. Call this least upper bound M. Suppose there exists no point u in $[a, b]$ such that $f(u) = M$. Then by Theorem 4(a), $1/(M - f(x))$ is continuous on $[a, b]$. By Theorem 8, there exists a constant K such that $1/(M - f(x)) \le K$ for all x in $[a, b]$. Thus $f(x) \le M - 1/K$, which contradicts the fact that M is the *least* upper bound for the values of f. Hence there must exist some point u in $[a, b]$ such that $f(u) = M$. Since M is an upper bound for the values of f on $[a, b]$, we have $f(x) \le f(u) = M$ for all x in $[a, b]$.

The proof that there must exist a point v in $[a, b]$ such that $f(x) \geq f(v)$ for all x in $[a, b]$ is similar. □

Theorem 10 (*The Intermediate-Value Theorem*) If f is continuous on $[a, b]$ and s is a real number lying between the numbers $f(a)$ and $f(b)$, then there exists a point c in $[a, b]$ such that $f(c) = s$.

Proof To be specific, we assume that $f(a) < s < f(b)$. (The proof for $f(a) > s > f(b)$ is similar.) Let $S = \{x : a \leq x \leq b \text{ and } f(x) \leq s\}$. S is nonempty (a belongs to S) and bounded above (b is an upper bound) so by completeness S has a least upper bound; call it c.

Suppose that $f(c) > s$. Then $c \neq a$ and, by continuity, $f(x) > s$ on some interval $(c - \delta, c]$ where $\delta > 0$. But this says $c - \delta$ is an upper bound for S lower than the least upper bound, which is impossible. Thus $f(c) \leq s$.

Suppose $f(c) < s$. Then $c \neq b$ and, by continuity, $f(x) < s$ on some interval of the form $[c, c + \delta)$ for some $\delta > 0$. But this says that $[c, c + \delta) \subset S$, which contradicts the fact that c is an upper bound for S. Hence we cannot have $f(c) < s$. Therefore, $f(c) = s$. □

For more discussion of Theorems 8–10 and some applications, see Section 1.5.

EXERCISES

Use the formal definition of limit to verify the limits asserted in Exercises 1–6.

1. $\lim_{x \to 2} (3x - 1) = 5$

2. $\lim_{x \to -2} (x^2 + 3x) = -2$

3. $\lim_{x \to 0} \dfrac{x - 2}{x + 1} = -2$

4. $\lim_{x \to -1} \dfrac{(x + 1)^2}{x^2 - 1} = 0$

5. $\lim_{x \to 1} \dfrac{x^2 + x - 2}{x^2 + 2x - 3} = \dfrac{3}{4}$

6. $\lim_{x \to 2} \sqrt{x^2 + 4} = 2\sqrt{2}$

7. Prove parts (ii) and (v) of Theorem 2.

8. a) If $\lim_{x \to a} g(x) = M$ where $M \neq 0$, prove that there exists $\delta_1 > 0$ such that $|g(x)| > |M|/2$ whenever $0 < |x - a| < \delta_1$.

   b) Prove part (iv) of Theorem 2.

9. Prove part (vi) of Theorem 2 by contradiction; that is, assume that $L > M$ and deduce that $f(x) > g(x)$ for all x sufficiently near a. This contradicts the condition that $f(x) \leq g(x)$ for $0 < |x - a| < \delta$.

10. If $f(x) \leq K$ on the intervals $[a, b)$ and $(b, c]$, and if $\lim_{x \to b} f(x) = L$, prove that $L \leq K$.

11. Prove that $\lim_{x \to 0+} x^r = 0$ for any positive, rational number r.

12. a) Prove that $f(x) = C$ (constant) and $g(x) = x$ are both continuous on the whole real line.

    b) Deduce that any polynomial is continuous on the whole real line.

    c) Deduce that any rational function (quotient of polynomials) is continuous everywhere except at points where its denominator vanishes.

13. a) If n is a positive integer and $a > 0$, prove that $f(x) = x^{1/n}$ is continuous at $x = a$.

b) Deduce that if $r = m/n$ is a rational number, then $g(x) = x^r$ is continuous at every point $a > 0$.

c) If $r = m/n$ where m and n are integers and n is odd, show that $g(x) = x^r$ is continuous at every point $a < 0$. If $r \geq 0$ show that g is continuous at 0 also.

14. Prove that $f(x) = |x|$ is continuous on the whole real line.

15. Use the definitions of $\cos x$, $\sin x$, $\ln x$, and e^x given in Chapter 3 to prove that these functions are continuous on their respective domains.

16. Prove Theorem 7.

17. Suppose that every function that is continuous and bounded on $[a, b]$ must assume a maximum value and a minimum value on that interval. Without using Theorem 8, prove that every function f that is continuous on $[a, b]$ must be bounded on that interval. (*Hint:* Show that $g(t) = t/(1 + |t|)$ is continuous and increasing on $\mathbb{R}$. Then consider $g(f(x))$.)

APPENDIX III: THE RIEMANN INTEGRAL

In Section 6.2 we defined the definite integral $\int_a^b f(x)\,dx$ of a function f continuous on the finite, closed interval $[a, b]$. The integral was defined to be the common limit of lower and upper Riemann sums:

$$\int_a^b f(x)\,dx = \lim_{n\to\infty} L_n(f) = \lim_{n\to\infty} U_n(f),$$

where, for any positive integer n, $L_n(f)$ and $U_n(f)$ were defined by subdividing $[a, b]$ into n subintervals $[x_{j-1}, x_j]$ $(1 \leq j \leq n)$ each of length $\Delta x = (b - a)/n$ and setting

$$L_n(f) = \sum_{j=1}^{n} \Big(\min_{x_{j-1}\leq x\leq x_j} f(x)\Big)\,\Delta x, \text{ and}$$

$$U_n(f) = \sum_{j=1}^{n} \Big(\max_{x_{j-1}\leq x\leq x_j} f(x)\Big)\,\Delta x.$$

Theorem 1 of Section 6.2 asserted that for f continuous on $[a, b]$, the limits $\lim_{n\to\infty} L_n(f)$ and $\lim_{n\to\infty} U_n(f)$ do in fact exist and are equal. We did not prove this theorem in its full generality, but only under the added hypothesis that f was increasing on $[a, b]$.

In this appendix we modify the above definition of definite integral to allow for functions f that are *bounded,* but not necessarily continuous on $[a, b]$, and also allow for subdivisions of $[a, b]$ in which the subintervals need not be of equal length. The new definition will give the same value as the old for the integral of a continuous function over $[a, b]$.

By a **partition** of $[a, b]$ we mean a finite, ordered set of points $\Pi = (x_0, x_1, x_2, \ldots, x_n)$ where $a = x_0 < x_1 < x_2 < \cdots < x_{n-1} < x_n = b$. Such a partition subdivides $[a, b]$ into n subintervals $[x_0, x_1]$, $[x_1, x_2]$, $\ldots$, $[x_{n-1}, x_n]$ where $n = n(\Pi)$ depends on the partition. The length of the jth subinterval $[x_{j-1}, x_j]$ is $\Delta x_j = x_j - x_{j-1}$.

Suppose that the function f is bounded on $[a, b]$. Given any partition Π, the n sets $S_j = \{f(x): x_{j-1} \leq x \leq x_j\}$ have least upper bounds w_j and greatest lower bounds v_j $(1 \leq j \leq n)$, so that

$$v_j \leq f(x) \leq w_j \qquad \text{on } [x_{j-1}, x_j].$$

We define upper and lower sums for f corresponding to the partition Π to be

$$U(\Pi; f) = \sum_{j=1}^{n(\Pi)} w_j \, \Delta x_j, \text{ and}$$

$$L(\Pi; f) = \sum_{j=1}^{n(\Pi)} v_j \, \Delta x_j.$$

(See Fig. III.1) Note that if Π is a partition of $[a, b]$ into n equal subintervals, and if f is continuous on $[a, b]$, then v_j and w_j are in fact the minimum and maximum values of f over $[x_{j-1}, x_j]$ (by Theorem 9 of Appendix II); that is,

$$v_j = f(\ell_j) \quad \text{and} \quad w_j = f(u_j), \quad \text{where}$$
$$f(\ell_j) \leq f(x) \leq f(u_j) \qquad \text{for } x_{j-1} \leq x \leq x_j.$$

Hence $L(\Pi; f) = L_n(f)$ and $U(\Pi; f) = U_n(f)$.

If Π is any partition of $[a, b]$ and we create a new partition Π^* by adding new subdivision points to those of Π, thus subdividing the subintervals of Π into smaller ones, then we call Π^* a **refinement** of Π.

Theorem 1 If Π^* is a refinement of Π, then $L(\Pi^*; f) \geq L(\Pi; f)$ and $U(\Pi^*; f) \leq U(\Pi; f)$.

Proof If S and T are sets of real numbers, and $S \subset T$, then any lower bound (or upper bound) of T is also a lower bound (or upper bound) of S. Hence the greatest lower

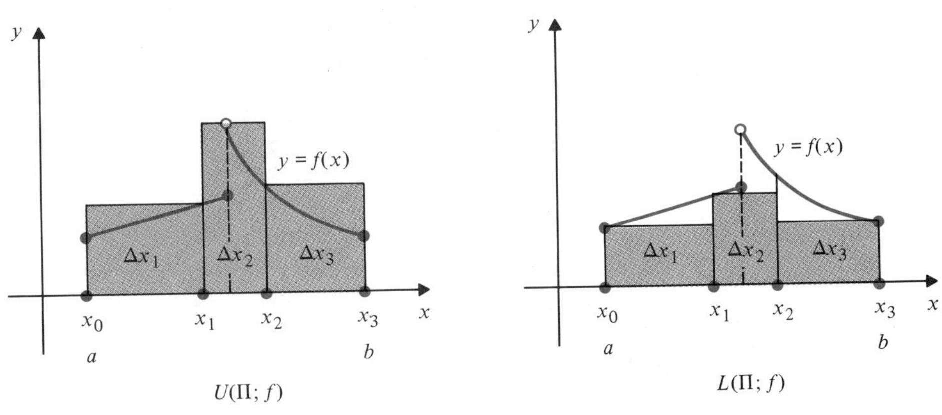

Upper and lower sums corresponding to the partition $\Pi = (x_0, x_1, x_2, x_3)$

Figure III.1

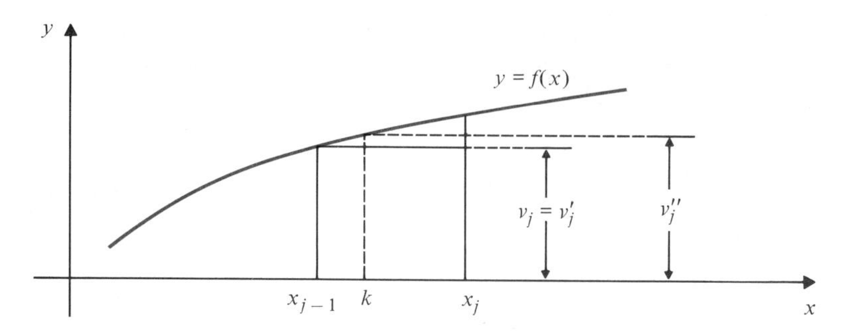

Figure III.2

bound of S is at least as large as that of T, and the least upper bound of S is no greater than that of T.

Let Π be a given partition of $[a, b]$ and form a new partition Π' by adding one subdivision point to those of Π, say the point k dividing the jth subinterval $[x_{j-1}, x_j]$ of Π into two subintervals $[x_{j-1}, k]$ and $[k, x_j]$. See Fig. III.2. Let v_j, v_j', and v_j'' be the greatest lower bounds of the sets of values of $f(x)$ on the intervals $[x_{j-1}, x_j]$, $[x_{j-1}, k]$, and $[k, x_j]$, respectively. Then $v_j \le v_j'$ and $v_j \le v_j''$. Thus $v_j(x_j - x_{j-1}) \le v_j'(k - x_{j-1}) + v_j''(x_j - k)$, and so $L(\Pi; f) \le L(\Pi'; f)$.

If Π^* is a refinement of Π it can be obtained by adding one point at a time to those of Π and thus $L(\Pi; f) \le L(\Pi^*; f)$. We can prove that $U(\Pi; f) \ge U(\Pi^*; f)$ in a similar manner. □

Theorem 2 If Π and Π' are any two partitions of $[a, b]$, then $L(\Pi; f) \le U(\Pi'; f)$.

Proof Combine the subdivisison points of Π and Π' to form a new partition Π^*, which is a refinement of both Π and Π'. See Fig. III.3. Then by Theorem 1,

$$L(\Pi; f) \le L(\Pi^*; f) \le U(\Pi^*; f) \le U(\Pi'; f).\ \square$$

Theorem 2 shows that the set of values of $L(\Pi; f)$ for fixed f and various partitions Π of $[a, b]$ is a bounded set; any upper sum is an upper bound for this set. By completeness, the set has a least upper bound, which we shall denote I_*. Thus $L(\Pi; f) \le I_*$ for any partition Π. Similarly, there exists a greatest lower bound I^* for the set of values of $U(\Pi; f)$ corresponding to different partitions Π. Evidently $I_* \le I^*$.

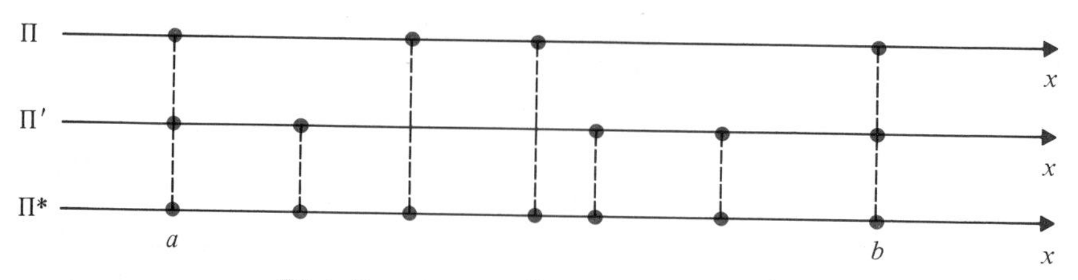

Π^* is the common refinement of Π and Π'.

Figure III.3

Definition

If f is bounded on $[a, b]$ and $I_* = I^*$, then we say that f is (Riemann) integrable on $[a, b]$ and denote by

$$\int_a^b f(x)\, dx = I_* = I^*$$

the Riemann integral of f on $[a, b]$.

We can show (see Exercises 4 and 5 at the end of this appendix) that this definition of integral concides with that given in Section 6.2 if f is continuous on $[a, b]$. That is, such f is integrable by both definitions, and both give the same value for the integral.

The following theorem provides a convenient test for determining whether a given bounded function is integrable.

Theorem 3 The bounded function f is integrable on $[a, b]$ if and only if for every positive number ϵ there exists a partition Π of $[a, b]$ such that $U(\Pi; f) - L(\Pi; f) < \epsilon$.

Proof If for every $\epsilon > 0$ there exists a partition Π such that $U(\Pi; f) - L(\Pi; f) < \epsilon$, then $I^* \le I_*$. Since we already know that $I^* \ge I_*$, $I^* = I_*$ and f is integrable on $[a, b]$.

Conversely, if $I^* = I_*$ and $\epsilon > 0$ are given, we can find a partition Π' such that $L(\Pi'; f) > I_* - \epsilon/2$, and another partition Π'' such that $U(\Pi''; f) < I^* + \epsilon/2$. If Π is a common refinement of Π' and Π'', then by Theorem 1, $U(\Pi; f) - L(\Pi; f) \le U(\Pi''; f) - L(\Pi'; f) < \epsilon/2 + \epsilon/2 = \epsilon$, as required. □

Example 1 Let

$$f(x) = \begin{cases} 0 & \text{if } 0 \le x < 1 \quad \text{or} \quad 1 < x \le 2 \\ 1 & \text{if } x = 1. \end{cases}$$

Show that f is integrable on $[0, 2]$ and find $\int_0^2 f(x)\, dx$.

Solution Let $\epsilon > 0$ be given. Let $\Pi = (0, 1 - \epsilon/3, 1 + \epsilon/3, 2)$. Then $L(\Pi; f) = 0$ since $f(x) = 0$ at points of each of the three subintervals into which Π subdivides $[0, 2]$. See Fig. III.4. Since $f(1) = 1$, we have $U(\Pi; f) = 0(1 - \epsilon/3) + 1(2\epsilon/3) + 0(2 - (1 - \epsilon/3)) = 2\epsilon/3$. Hence $U(\Pi; f) - L(\Pi; f) < \epsilon$ and f is integrable on $[0, 2]$. Evidently $\int_0^2 f(x)\, dx = I_* = 0$.

Example 2 Let $f(x)$ be defined on $[0, 1]$ by

$$f(x) = \begin{cases} 1 & \text{if } x \text{ is rational} \\ 0 & \text{if } x \text{ is irrational.} \end{cases}$$

Show that f is not integrable on $[0, 1]$.

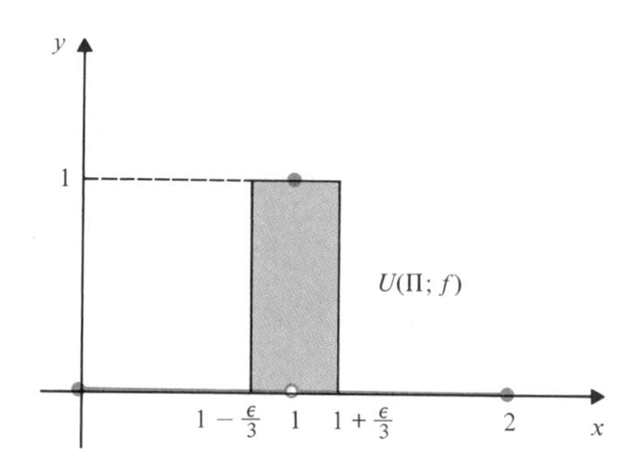

Figure III.4

Proof Every subinterval of $[0, 1]$ having positive length contains both rational and irrational numbers. Hence, for any partition Π of $[0, 1]$ we have $L(\Pi; f) = 0$ and $U(\Pi; f) = 1$. Thus $I_* = 0$ and $I^* = 1$, so f is not integrable on $[0, 1]$. (We could also have used Theorem 3; if $0 < \epsilon < 1$ then there is no partition Π of $[0, 1]$ such that $U(\Pi; f) - L(\Pi; f) < \epsilon$. □

Uniform Continuity

When we assert that a function f is continuous on the interval $[a, b]$ we imply that for every x in that interval, and every $\epsilon > 0$, we can find a positive number δ (depending on *both x and ϵ*) such that $|f(y) - f(x)| < \epsilon$ whenever $|y - x| < \delta$ and y lies in $[a, b]$. In fact, however, it is possible to find a number δ *depending only on* ϵ such that $|f(y - f(x)| < \epsilon$ holds whenever x and y belong to $[a, b]$ and satisfy $|y - x| < \delta$. We describe this phenomenon by saying that f is **uniformly continuous** on the interval $[a, b]$.

Theorem 4 If f is continuous on the closed, finite interval $[a, b]$, then f is uniformly continuous on that interval.

Proof Let $\epsilon > 0$ be given. Define numbers x_n in $[a, b]$ and subsets S_n of $[a, b]$ as follows:

$$x_1 = a$$

$$S_1 = \left\{x : x_1 < x \le b \text{ and } |f(x) - f(x_1)| \ge \frac{\epsilon}{3}\right\}.$$

If S_1 is empty, stop; otherwise let

$$x_2 = \text{the greatest lower bound of } S_1$$

$$S_2 = \left\{x : x_2 < x \le b \text{ and } |f(x) - f(x_2)| \ge \frac{\epsilon}{3}\right\}.$$

If S_2 is empty, stop; otherwise proceed to define x_3 and S_3 analogously. We proceed in this way as long as we can; if x_n and S_n have been defined and S_n is not empty, we define

$$x_{n+1} = \text{the greatest lower bound of } S_n$$
$$S_{n+1} = \left\{ x : x_{n+1} < x \leq b \text{ and } |f(x) - f(x_{n+1})| \geq \frac{\epsilon}{3} \right\}.$$

At any stage where S_n is not empty, the continuity of f at x_n assures us that $x_{n+1} > x_n$ and $|f(x_{n+1}) - f(x_n)| = \epsilon/3$.

We must consider two possibilities for the above procedure: Either S_n is empty for some n, or S_n is nonempty for every n.

Suppose S_n is nonempty for every n. Then we have constructed an infinite, increasing sequence $\{x_n\}$ in $[a, b]$ that, being bounded above (by b), must have a limit by completeness (Theorem 5 of Appendix II). Let $\lim x_n = x^*$. We have $a \leq x^* \leq b$. Since f is continuous at x^* there exists $\delta > 0$ such that $|f(x) - f(x^*)| < \epsilon/8$ whenever $|x - x^*| < \delta$ and x lies in $[a, b]$. Since $\lim x_n = x^*$, there exists a positive integer N such that $|x_n - x^*| < \delta$ whenever $n \geq N$. For such n we have

$$\begin{aligned} \frac{\epsilon}{3} = |f(x_{n+1}) - f(x_n)| &= |f(x_{n+1}) - f(x^*) + f(x^*) - f(x_n)| \\ &\leq |f(x_{n+1}) - f(x^*)| + |f(x_n) - f(x^*)| \\ &< \frac{\epsilon}{8} + \frac{\epsilon}{8} = \frac{\epsilon}{4}, \end{aligned}$$

which is clearly impossible. Thus S_n must, in fact, be empty for some n.

Suppose that S_N is empty. Thus S_n is nonempty for $n < N$, and the procedure for defining x_n stops with x_N. Since S_{N-1} is not empty, $x_N < b$. In this case define $x_{N+1} = b$ and let

$$\delta = \min \{x_2 - x_1, x_3 - x_2, \ldots, x_{N+1} - x_N\}.$$

(The minimum of a finite set of positive numbers is a positive number.) If x lies in $[a, b]$, then x lies in one of the intervals $[x_1, x_2]$, $[x_2, x_3]$, $\ldots$, $[x_N, x_{N+1}]$. Suppose x lies in $[x_k, x_{k+1}]$. If y is in $[a, b]$ and $|y - x| < \delta$, then y lies in either the same subinterval as x or in an adjacent one; that is, y lies in $[x_j, x_{j+1}]$ where $j = k - 1$, k, or $k + 1$. Thus

$$\begin{aligned} |f(y) - f(x)| &= |f(y) - f(x_j) + f(x_j) - f(x_k) + f(x_k) - f(x)| \\ &\leq |f(y) - f(x_j)| + |f(x_j) - f(x_k)| + |f(x_k) - f(x)| \\ &< \frac{\epsilon}{3} + \frac{\epsilon}{3} + \frac{\epsilon}{3} = \epsilon, \end{aligned}$$

which was to be proved. □

We are now in a position to prove that a continuous function is integrable. This completes the proof of Theorem 1 of Section 6.2.

Theorem 5 If f is continuous on $[a, b]$, then f is integrable on $[a, b]$.

Proof By Theorem 4, f is uniformly continuous on $[a, b]$. Let $\epsilon > 0$ be given. Let $\delta > 0$ be such that $|f(x) - f(y)| < \epsilon/(b - a + 1)$ whenever $|x - y| < \delta$ and x and y

belong to $[a, b]$. Choose a partition $\Pi = (x_0, x_1, \ldots, x_n)$ of $[a, b]$ for which each subinterval $[x_{j-1}, x_j]$ has length $\Delta x_j < \delta$. Then the greatest lower bound, v_j, and the least upper bound, w_j, of the set of values of $f(x)$ on $[x_{j-1}, x_j]$ satisfy $w_j - v_j \le \epsilon/(b - a + 1) < \epsilon/(b - a)$. Accordingly,

$$U(\Pi; f) - L(\Pi; f) < \frac{\epsilon}{b - a} \sum_{j=1}^{n(\Pi)} \Delta x_j = \frac{\epsilon}{b - a}(b - a) = \epsilon.$$

Thus f is integrable on $[a, b]$ as asserted. □

EXERCISES

1. Let

$$f(x) = \begin{cases} 1 & \text{if } 0 \le x \le 1 \\ 0 & \text{if } 1 < x \le 2. \end{cases}$$

Show that f is integrable on $[0, 2]$ and find the value of of the integral $\int_0^2 f(x)\, dx$.

2. Let

$$f(x) = \begin{cases} 1 & \text{if } x = 1/n, \quad n = 1, 2, 3, \ldots \\ 0 & \text{for all other values of } x. \end{cases}$$

Show that f is integrable over $[0, 1]$ and find the value of $\int_0^1 f(x)\, dx$.

3. Let

$$f(x) = \begin{cases} 1/n & \text{if } x = m/n \text{ where } m, n \text{ are integers having no common factors} \\ 0 & \text{if } x \text{ is irrational.} \end{cases}$$

Thus $f(1/2) = 1/2$, $f(1/3) = f(2/3) = 1/3$, $f(1/4) = f(3/4) = 1/4, \ldots$. Show that f is integrable on $[0, 1]$ and find $\int_0^1 f(x)\, dx$. (*Hint:* Show that for any $\epsilon > 0$, only finitely many points of the graph of f over $[0, 1]$ lie above the line $y = \epsilon$.)

4. If f is continuous on $[a, b]$ and $\int_a^b f(x)\, dx$ exists in the sense of the definition in Section 6.2, that is, $\lim_{n\to\infty} L_n(f) = \lim_{n\to\infty} U_n(f)$ where L_n and U_n are lower and upper sums on a partition of $[a, b]$ into n equal subintervals, then $\int_a^b f(x)\, dx$ exists in the sense of the definition in this appendix, and both definitions give the same value for the integral. (*Hint:* Show that $\lim_{n\to\infty} L_n(f) \le I_* \le I^* \le \lim_{n\to\infty} U_n(f)$.)

5. If f is continuous on $[a, b]$ and $\int_a^b f(x)\, dx$ exists in the sense of the definition in this appendix (that is, $I_* = I^*$), show that $\int_a^b f(x)\, dx$ also exists in the sense defined in Section 6.2. That is, prove that $\lim_{n\to\infty} L_n(f) = \lim_{n\to\infty} U_n(f) = I_* = I^*$. Together with Exercise 4, this shows that the two definitions of integral are equivalent for continuous functions.

(*Hints:* a) For given $\epsilon > 0$ pick a partition Π for which $L(\Pi; f) > I_* - \frac{\epsilon}{2}$.

b) If $|f(x)| \leq K$ on $[a, b]$ and m is the number of subdivision points of Π, choose n large enough that $(m/n)(b - a)K < \epsilon/4$. Show that $L_n(f) > L(\Pi; f) - \epsilon/2$. Thus $L_n(f) > I_* - \epsilon$.

c) Similarly, show that $U_n(f) < I^* + \epsilon$ for large enough n.)

6. Show directly from the definition of uniform continuity (without using Theorem 8 of Appendix II) that a function f uniformly continuous on a closed, finite interval is necessarily bounded there.

7. Prove parts (c), (d), (e), (f), (g), and (h) of Theorem 2 of Section 6.2 for the Riemann integral. (Note that part (d) would be rather harder to prove if we had not allowed partitions having unequal subinterval lengths.)

8. If f is bounded and integrable on $[a, b]$, prove that $F(x) = \int_a^x f(t)\, dt$ is uniformly continuous on $[a,b]$. (If f were continuous we would have a stronger result; F would be differentiable on (a,b) and $F'(x) = f(x)$ (the fundamental theorem of calculus.)

Table 1 Trigonometric Functions
Table 2 Exponential Functions
Table 3 Natural Logarithms
Table 4 Powers and Roots
Table 5 Table of Integrals

Tables

TABLE 1 TRIGONOMETRIC FUNCTIONS

x					x				
Degree	Radian	**sin x**	**cos x**	**tan x**	Degree	Radian	**sin x**	**cos x**	**tan x**
0°	0.000	0.000	1.000	0.000					
1°	0.017	0.017	1.000	0.017	46°	0.803	0.719	0.695	1.036
2°	0.035	0.035	0.999	0.035	47°	0.820	0.731	0.682	1.072
3°	0.052	0.052	0.999	0.052	48°	0.838	0.743	0.669	1.111
4°	0.070	0.070	0.998	0.070	49°	0.855	0.755	0.656	1.150
5°	0.087	0.087	0.996	0.087	50°	0.873	0.766	0.643	1.192
6°	0.105	0.105	0.995	0.105	51°	0.890	0.777	0.629	1.235
7°	0.122	0.122	0.993	0.123	52°	0.908	0.788	0.616	1.280
8°	0.140	0.139	0.990	0.141	53°	0.925	0.799	0.602	1.327
9°	0.157	0.156	0.988	0.158	54°	0.942	0.809	0.588	1.376
10°	0.175	0.174	0.985	0.176	55°	0.960	0.819	0.574	1.428
11°	0.192	0.191	0.982	0.194	56°	0.977	0.829	0.559	1.483
12°	0.209	0.208	0.978	0.213	57°	0.995	0.839	0.545	1.540
13°	0.227	0.225	0.974	0.231	58°	1.012	0.848	0.530	1.600
14°	0.244	0.242	0.970	0.249	59°	1.030	0.857	0.515	1.664
15°	0.262	0.259	0.966	0.268	60°	1.047	0.866	0.500	1.732

TABLE 1 TRIGONOMETRIC FUNCTIONS (Cont.)

16°	0.279	0.276	0.961	0.287	61°	1.065	0.875	0.485	1.804
17°	0.297	0.292	0.956	0.306	62°	1.082	0.883	0.469	1.881
18°	0.314	0.309	0.951	0.325	63°	1.100	0.891	0.454	1.963
19°	0.332	0.326	0.946	0.344	64°	1.117	0.899	0.438	2.050
20°	0.349	0.342	0.940	0.364	65°	1.134	0.906	0.423	2.145
21°	0.367	0.358	0.934	0.384	66°	1.152	0.914	0.407	2.246
22°	0.384	0.375	0.927	0.404	67°	1.169	0.921	0.391	2.356
23°	0.401	0.391	0.921	0.424	68°	1.187	0.927	0.375	2.475
24°	0.419	0.407	0.914	0.445	69°	1.204	0.934	0.358	2.605
25°	0.436	0.423	0.906	0.466	70°	1.222	0.940	0.342	2.748
26°	0.454	0.438	0.899	0.488	71°	1.239	0.946	0.326	2.904
27°	0.471	0.454	0.891	0.510	72°	1.257	0.951	0.309	3.078
28°	0.489	0.469	0.883	0.532	73°	1.274	0.956	0.292	3.271
29°	0.506	0.485	0.875	0.554	74°	1.292	0.961	0.276	3.487
30°	0.524	0.500	0.866	0.577	75°	1.309	0.966	0.259	3.732
31°	0.541	0.515	0.857	0.601	76°	1.326	0.970	0.242	4.011
32°	0.559	0.530	0.848	0.625	77°	1.344	0.974	0.225	4.332
33°	0.576	0.545	0.839	0.649	78°	1.361	0.978	0.208	4.705
34°	0.593	0.559	0.829	0.675	79°	1.379	0.982	0.191	5.145
35°	0.611	0.574	0.819	0.700	80°	1.396	0.985	0.174	5.671
36°	0.628	0.588	0.809	0.727	81°	1.414	0.988	0.156	6.314
37°	0.646	0.602	0.799	0.754	82°	1.431	0.990	0.139	7.115
38°	0.663	0.616	0.788	0.781	83°	1.449	0.993	0.122	8.144
39°	0.681	0.629	0.777	0.810	84°	1.466	0.995	0.105	9.514
40°	0.698	0.643	0.766	0.839	85°	1.484	0.996	0.087	11.43
41°	0.716	0.656	0.755	0.869	86°	1.501	0.998	0.070	14.30
42°	0.733	0.669	0.743	0.900	87°	1.518	0.999	0.052	19.08
43°	0.750	0.682	0.731	0.933	88°	1.536	0.999	0.035	28.64
44°	0.768	0.695	0.719	0.966	89°	1.553	1.000	0.017	57.29
45°	0.785	0.707	0.707	1.000	90°	1.571	1.000	0.000	

TABLE 2 EXPONENTIAL FUNCTIONS

x	e^x	e^{-x}	x	e^x	e^{-x}
0.00	1.0000	1.0000	2.5	12.182	0.0821
0.05	1.0513	0.9512	2.6	13.464	0.0743
0.10	1.1052	0.9048	2.7	14.880	0.0672
0.15	1.1618	0.8607	2.8	16.445	0.0608
0.20	1.2214	0.8187	2.9	18.174	0.0550
0.25	1.2840	0.7788	3.0	20.086	0.0498
0.30	1.3499	0.7408	3.1	22.198	0.0450
0.35	1.4191	0.7047	3.2	24.533	0.0408
0.40	1.4918	0.6703	3.3	27.113	0.0369
0.45	1.5683	0.6376	3.4	29.964	0.0334
0.50	1.6487	0.6065	3.5	33.115	0.0302
0.55	1.7333	0.5769	3.6	36.598	0.0273
0.60	1.8221	0.5488	3.7	40.447	0.0247
0.65	1.9155	0.5220	3.8	44.701	0.0224
0.70	2.0138	0.4966	3.9	49.402	0.0202
0.75	2.1170	0.4724	4.0	54.598	0.0183
0.80	2.2255	0.4493	4.1	60.340	0.0166
0.85	2.3396	0.4274	4.2	66.686	0.0150
0.90	2.4596	0.4066	4.3	73.700	0.0136
0.95	2.5857	0.3867	4.4	81.451	0.0123
1.0	2.7183	0.3679	4.5	90.017	0.0111
1.1	3.0042	0.3329	4.6	99.484	0.0101
1.2	3.3201	0.3012	4.7	109.95	0.0091
1.3	3.6693	0.2725	4.8	121.51	0.0082
1.4	4.0552	0.2466	4.9	134.29	0.0074
1.5	4.4817	0.2231	5	148.41	0.0067
1.6	4.9530	0.2019	6	403.43	0.0025
1.7	5.4739	0.1827	7	1096.6	0.0009
1.8	6.0496	0.1653	8	2981.0	0.0003
1.9	6.6859	0.1496	9	8103.1	0.0001
2.0	7.3891	0.1353	10	22026.	0.00005
2.1	8.1662	0.1225			
2.2	9.0250	0.1108			
2.3	9.9742	0.1003			
2.4	11.023	0.0907			

TABLE 3 NATURAL LOGARITHMS

x	ln x	x	ln x	x	ln x
0.0		4.5	1.5041	9.0	2.1972
0.1	−2.3026	4.6	1.5261	9.1	2.2083
0.2	−1.6094	4.7	1.5476	9.2	2.2192
0.3	−1.2040	4.8	1.5686	9.3	2.2300
0.4	−0.9163	4.9	1.5892	9.4	2.2407
0.5	−0.6931	5.0	1.6094	9.5	2.2513
0.6	−0.5108	5.1	1.6292	9.6	2.2618
0.7	−0.3567	5.2	1.6487	9.7	2.2721
0.8	−0.2231	5.3	1.6677	9.8	2.2824
0.9	−0.1054	5.4	1.6864	9.9	2.2925
1.0	0.0000	5.5	1.7047	10	2.3026
1.1	0.0953	5.6	1.7228	11	2.3979
1.2	0.1823	5.7	1.7405	12	2.4849
1.3	0.2624	5.8	1.7579	13	2.5649
1.4	0.3365	5.9	1.7750	14	2.6391
1.5	0.4055	6.0	1.7918	15	2.7081
1.6	0.4700	6.1	1.8083	16	2.7726
1.7	0.5306	6.2	1.8245	17	2.8332
1.8	0.5878	6.3	1.8405	18	2.8904
1.9	0.6419	6.4	1.8563	19	2.9444
2.0	0.6931	6.5	1.8718	20	2.9957
2.1	0.7419	6.6	1.8871	25	3.2189
2.2	0.7885	6.7	1.9021	30	3.4012
2.3	0.8329	6.8	1.9169	35	3.5553
2.4	0.8755	6.9	1.9315	40	3.6889
2.5	0.9163	7.0	1.9459	45	3.8067
2.6	0.9555	7.1	1.9601	50	3.9120
2.7	0.9933	7.2	1.9741	55	4.0073
2.8	1.0296	7.3	1.9879	60	4.0943
2.9	1.0647	7.4	2.0015	65	4.1744
3.0	1.0986	7.5	2.0149	70	4.2485
3.1	1.1314	7.6	2.0281	75	4.3175
3.2	1.1632	7.7	2.0412	80	4.3820
3.3	1.1939	7.8	2.0541	85	4.4427
3.4	1.2238	7.9	2.0669	90	4.4998
3.5	1.2528	8.0	2.0794	95	4.5539
3.6	1.2809	8.1	2.0919	100	4.6052
3.7	1.3083	8.2	2.1041		
3.8	1.3350	8.3	2.1163		

TABLE 3 NATURAL LOGARITHMS (Cont.)

3.9	1.3610	8.4	2.1282		
4.0	1.3863	8.5	2.1401		
4.1	1.4110	8.6	2.1518		
4.2	1.4351	8.7	2.1633		
4.3	1.4586	8.8	2.1748		
4.4	1.4816	8.9	2.1861		

TABLE 4 POWERS AND ROOTS

No.	Sq.	Sq. Root	Cube	Cube Root	No.	Sq.	Sq. Root	Cube	Cube Root
1	1	1.000	1	1.000	51	2,601	7.141	132,651	3.708
2	4	1.414	8	1.260	52	2,704	7.211	140,608	3.733
3	9	1.732	27	1.442	53	2,809	7.280	148,877	3.756
4	16	2.000	64	1.587	54	2,916	7.348	157,464	3.780
5	25	2.236	125	1.710	55	3,025	7.416	166,375	3.803
6	36	2.449	216	1.817	56	3,136	7.483	175,616	3.826
7	49	2.646	343	1.913	57	3,249	7.550	185,193	3.849
8	64	2.828	512	2.000	58	3,364	7.616	195,112	3.871
9	81	3.000	729	2.080	59	3,481	7.681	205,379	3.893
10	100	3.162	1,000	2.154	60	3,600	7.746	216,000	3.915
11	121	3.317	1,331	2.224	61	3,721	7.810	226,981	3.936
12	144	3.464	1,728	2.289	62	3,844	7.874	238,328	3.958
13	169	3.606	2,197	2.351	63	3,969	7.937	250,047	3.979
14	196	3.742	2,744	2.410	64	4,096	8.000	262,144	4.000
15	225	3.873	3,375	2.466	65	4,225	8.062	274,625	4.021
16	256	4.000	4,096	2.520	66	4,356	8.124	287,496	4.041
17	289	4.123	4,913	2.571	67	4,489	8.185	300,763	4.062
18	324	4.243	5,832	2.621	68	4,624	8.246	314,432	4.082
19	361	4.359	6,859	2.668	69	4,761	8.307	328,509	4.102
20	400	4.472	8,000	2.714	70	4,900	8.367	343,000	4.121
21	441	4.583	9,261	2.759	71	5,041	8.426	357,911	4.141
22	484	4.690	10,648	2.802	72	5,184	8.485	373,248	4.160
23	529	4.796	12,167	2.844	73	5,329	8.544	389,017	4.179
24	576	4.899	13,824	2.884	74	5,476	8.602	405,224	4.198
25	625	5.000	15,625	2.924	75	5,625	8.660	421,875	4.217
26	676	5.099	17,576	2.962	76	5,776	8.718	438,976	4.236
27	729	5.196	19,683	3.000	77	5,929	8.775	456,533	4.254
28	784	5.292	21,952	3.037	78	6,084	8.832	474,552	4.273
29	841	5.385	24,389	3.072	79	6,241	8.888	493,039	4.291
30	900	5.477	27,000	3.107	80	6,400	8.944	512,000	4.309

TABLE 4 POWERS AND ROOTS (Cont.)

31	961	5.568	29,791	3.141	81	6,561	9.000	531,441	4.327
32	1,024	5.657	32,768	3.175	82	6,724	9.055	551,368	4.344
33	1,089	5.745	35,937	3.208	83	6,889	9.110	571,787	4.362
34	1,156	5.831	39,304	3.240	84	7,056	9.165	592,704	4.380
35	1,225	5.916	42,875	3.271	85	7,225	9.220	614,125	4.397
36	1,296	6.000	46,656	3.302	86	7,396	9.274	636,056	4.414
37	1,369	6.083	50,653	3.332	87	7,569	9.327	658,503	4.431
38	1,444	6.164	54,872	3.362	88	7,744	9.381	681,472	4.448
39	1,521	6.245	59,319	3.391	89	7,921	9.434	704,969	4.465
40	1,600	6.325	64,000	3.420	90	8,100	9.487	729,000	4.481
41	1,681	6.403	68,921	3.448	91	8,281	9.539	753,571	4.498
42	1,764	6.481	74,088	3.476	92	8,464	9.592	778,688	4.514
43	1,849	6.557	79,507	3.503	93	8,649	9.644	804,357	4.531
44	1,936	6.633	85,184	3.530	94	8,836	9.695	830,584	4.547
45	2,025	6.708	91,125	3.557	95	9,025	9.747	857,375	4.563
46	2,116	6.782	97,336	3.583	96	9,216	9.798	884,736	4.579
47	2,209	6.856	103,823	3.609	97	9,409	9.849	912,673	4.595
48	2,304	6.928	110,592	3.634	98	9,604	9.899	941,192	4.610
49	2,401	7.000	117,649	3.659	99	9,801	9.950	970,299	4.626
50	2,500	7.071	125,000	3.684	100	10,000	10.000	1,000,000	4.642

TABLE 5 TABLE OF INTEGRALS

Integration Rules

$$\int (Af(x) + Bg(x))\,dx = A\int f(x)\,dx + B\int g(x)\,dx$$

$$\int f'(g(x))g'(x)\,dx = f(g(x)) + C$$

$$\int U(x)\,dV(x) = U(x)V(x) - \int V(x)\,dU(x)$$

$$\int_a^b f'(x)\,dx = f(b) - f(a)$$

$$\frac{d}{dx}\int_a^x f(t)\,dt = f(x)$$

Elementary Integrals

$$\int x^r\,dx = \frac{1}{r+1}x^{r+1} + C \text{ if } n \neq -1$$

$$\int \frac{dx}{x} = \ln|x| + C$$

$$\int e^x\,dx = e^x + C$$

$$\int a^x\,dx = \frac{a^x}{\ln a} + C$$

$$\int \sin x\,dx = -\cos x + C$$

$$\int \cos x\,dx = \sin x + C$$

$$\int \sec^2 x\, dx = \tan x + C$$

$$\int \sec x \tan x\, dx = \sec x + C$$

$$\int \tan x\, dx = \ln |\sec x| + C$$

$$\int \sec x\, dx = \ln |\sec x + \tan x| + C$$

$$\int \frac{dx}{\sqrt{a^2 - x^2}} = \operatorname{Arcsin} \frac{x}{a} + C$$

$$\int \frac{dx}{a^2 - x^2} = \frac{1}{2a} \ln \left| \frac{x + a}{x - a} \right| + C$$

$$\int \csc^2 x\, dx = -\cot x + C$$

$$\int \csc x \cot x\, dx = -\csc x + C$$

$$\int \cot x\, dx = \ln |\sin x| + C$$

$$\int \csc x\, dx = \ln |\csc x - \cot x| + C$$

$$\int \frac{dx}{a^2 + x^2} = \frac{1}{a} \operatorname{Arctan} \frac{x}{a} + C$$

$$\int \frac{dx}{x\sqrt{x^2 - a^2}} = \frac{1}{a} \operatorname{Arcsec} \left| \frac{x}{a} \right| + C$$

Trigonometric Integrals

$$\int \sin^2 x\, dx = \frac{1}{2} x - \frac{1}{4} \sin 2x + C$$

$$\int \cos^2 x\, dx = \frac{1}{2} x + \frac{1}{4} \sin 2x + C$$

$$\int \tan^2 x\, dx = \tan x - x + C$$

$$\int \cot^2 x\, dx = -\cot x - x + C$$

$$\int \sec^3 x\, dx = \frac{1}{2} \sec x \tan x + \frac{1}{2} \ln |\sec x + \tan x| + C$$

$$\int \csc^3 x\, dx = -\frac{1}{2} \csc x \cot x + \frac{1}{2} \ln |\csc x - \cot x| + C$$

$$\int \sin ax \sin bx\, dx = \frac{\sin (a - b)x}{2(a - b)} - \frac{\sin (a + b)x}{2(a + b)} + C \quad \text{if} \quad a^2 \neq b^2$$

$$\int \cos ax \cos bx\, dx = \frac{\sin (a - b)x}{2(a - b)} + \frac{\sin (a + b)x}{2(a + b)} + C \quad \text{if} \quad a^2 \neq b^2$$

$$\int \sin ax \cos bx\, dx = -\frac{\cos (a - b)x}{2(a - b)} - \frac{\cos (a + b)x}{2(a + b)} + C \quad \text{if} \quad a^2 \neq b^2$$

$$\int \sin^n x\, dx = -\frac{1}{n} \sin^{n-1} x \cos x + \frac{n - 1}{n} \int \sin^{n-2} x\, dx$$

$$\int \cos^n x\, dx = \frac{1}{n} \cos^{n-1} x \sin x + \frac{n - 1}{n} \int \cos^{n-2} x\, dx$$

$$\int \tan^n x\, dx = \frac{1}{n - 1} \tan^{n-1} x - \int \tan^{n-2} x\, dx \text{ if } n \neq 1$$

$$\int \cot^n x\, dx = \frac{-1}{n - 1} \cos^{n-1} x - \int \cot^{n-2} x\, dx \text{ if } n \neq 1$$

$$\int \sec^n x\, dx = \frac{1}{n - 1} \sec^{n-2} x \tan x + \frac{n - 2}{n - 1} \int \sec^{n-2} x\, dx \text{ if } n \neq 1$$

$$\int \csc^n x\, dx = \frac{-1}{n - 1} \csc^{n-2} x \cot x + \frac{n - 2}{n - 1} \int \csc^{n-2} x\, dx \text{ if } n \neq 1$$

$$\int \sin^n x \cos^m x\, dx = -\frac{\sin^{n-1} x \cos^{m+1} x}{n+m} + \frac{n-1}{n+m}\int \sin^{n-2} x \cos^m x\, dx \text{ if } n \neq -m$$

$$\int \sin^n x \cos^m x\, dx = \frac{\sin^{n+1} x \cos^{m-1} x}{n+m} + \frac{m-1}{n+m}\int \sin^{n} x \cos^{m-2} x\, dx \text{ if } m \neq -n$$

$$\int x \sin x\, dx = \sin x - x \cos x + C$$

$$\int x^n \sin x\, dx = -x^n \cos x + n\int x^{n-1} \cos x\, dx$$

$$\int x \cos x\, dx = \cos x + x \sin x + C$$

$$\int x^n \cos x\, dx = x^n \sin x - n\int x^{n-1} \sin x\, dx$$

Integrals Involving $\sqrt{x^2 \pm a^2}$

$$\int \sqrt{x^2 \pm a^2}\, dx = \frac{x}{2}\sqrt{x^2 \pm a^2} \pm \frac{a^2}{2}\ln\left|x + \sqrt{x^2 \pm a^2}\right| + C$$

$$\int \frac{\sqrt{x^2 + a^2}}{x}\, dx = \sqrt{x^2 + a^2} - a\ln\left(\frac{a + \sqrt{x^2 + a^2}}{x}\right) + C$$

$$\int \frac{dx}{\sqrt{x^2 \pm a^2}} = \ln\left|x + \sqrt{x^2 \pm a^2}\right| + C$$

$$\int \frac{\sqrt{x^2 - a^2}}{x}\, dx = \sqrt{x^2 - a^2} - a\sec^{-1}\frac{x}{a} + C$$

$$\int x^2\sqrt{x^2 \pm a^2}\, dx = \frac{x}{8}(2x^2 \pm a^2)\sqrt{x^2 \pm a^2} - \frac{a^4}{8}\ln\left|x + \sqrt{x^2 \pm a^2}\right| + C$$

$$\int \frac{x^2\, dx}{\sqrt{x^2 \pm a^2}} = \frac{x}{2}\sqrt{x^2 \pm a^2} \mp \frac{a^2}{2}\ln\left|x + \sqrt{x^2 \pm a^2}\right| + C$$

$$\int \frac{\sqrt{x^2 \pm a^2}}{x^2}\, dx = -\frac{\sqrt{x^2 \pm a^2}}{x} + \ln\left|x + \sqrt{x^2 \pm a^2}\right| + C$$

$$\int \frac{dx}{x^2\sqrt{x^2 \pm a^2}} = \mp\frac{\sqrt{x^2 \pm a^2}}{a^2 x} + C$$

$$\int \frac{dx}{(x^2 \pm a^2)^{3/2}} = \frac{\pm x}{a^2\sqrt{x^2 \pm a^2}} + C$$

$$\int (x^2 \pm a^2)^{3/2}\, dx = \frac{x}{8}(2x^2 \pm 5a^2)\sqrt{x^2 \pm a^2} + \frac{3a^4}{8}\ln\left|x + \sqrt{x^2 \pm a^2}\right| + C$$

Integrals Involving $\sqrt{a^2 - x^2}$

$$\int \sqrt{a^2 - x^2}\, dx = \frac{x}{2}\sqrt{a^2 - x^2} + \frac{a^2}{2}\text{Arcsin}\frac{x}{a} + C$$

$$\int \frac{\sqrt{a^2 - x^2}}{x}\, dx = \sqrt{a^2 - x^2} - a\ln\left|\frac{a + \sqrt{a^2 - x^2}}{x}\right| + C$$

$$\int \frac{x^2\, dx}{\sqrt{a^2 - x^2}} = -\frac{x}{2}\sqrt{a^2 - x^2} + \frac{a^2}{2}\text{Arcsin}\frac{x}{a} + C$$

$$\int x^2\sqrt{a^2 - x^2}\, dx = \frac{x}{8}(2x^2 - a^2)\sqrt{a^2 - x^2} + \frac{a^4}{8}\text{Arcsin}\frac{x}{a} + C$$

$$\int \frac{dx}{x^2\sqrt{a^2 - x^2}} = -\frac{\sqrt{a^2 - x^2}}{a^2 x} + C$$

$$\int \frac{dx}{x\sqrt{a^2 - x^2}} = -\frac{1}{a}\ln\left|\frac{a + \sqrt{a^2 - x^2}}{x}\right| + C$$

$$\int \frac{\sqrt{a^2 - x^2}}{x^2}\, dx = -\frac{\sqrt{a^2 - x^2}}{x} - \text{Arcsin}\frac{x}{a} + C$$

$$\int \frac{dx}{(a^2 - x^2)^{3/2}} = \frac{x}{a^2\sqrt{a^2 - x^2}} + C$$

$$\int (a^2 - x^2)^{3/2}\, dx = \frac{x}{8}(5a^2 - 2x^2)\sqrt{a^2 - x^2} + \frac{3a^4}{8}\text{Arcsin}\frac{x}{a} + C$$

Exponential and Logarithmic Integrals

$$\int xe^x\,dx = (x - 1)e^x + C$$

$$\int \ln x\,dx = x\ln x - x + C$$

$$\int e^{ax}\sin bx\,dx = \frac{e^{ax}}{a^2 + b^2}(a\sin bx - b\cos bx) + C$$

$$\int e^{ax}\cos bx\,dx = \frac{e^{ax}}{a^2 + b^2}(a\cos bx + b\sin bx) + C$$

$$\int x^n e^x\,dx = x^n e^x - n\int x^{n-1}e^x\,dx$$

$$\int x^n \ln x\,dx = \frac{x^{n+1}}{n+1}\ln x - \frac{x^{n+1}}{(n+1)^2} + C$$

Integrals Involving Inverse Trigonometric Functions

$$\int \text{Arcsin}\,x\,dx = x\,\text{Arcsin}\,x + \sqrt{1 - x^2} + C$$

$$\int \sec^{-1} x\,dx = x\sec^{-1} x - \ln|x + \sqrt{x^2 - 1}| + C$$

$$\int x\tan^{-1} x\,dx = \frac{1}{2}(x^2 + 1)\tan^{-1} x - \frac{x}{2} + C$$

$$\int \text{Arctan}\,x\,dx = x\,\text{Arctan}\,x - \frac{1}{2}\ln(1 + x^2) + C$$

$$\int x\sin^{-1} x\,dx = \frac{1}{4}(2x^2 - 1)\sin^{-1} x + \frac{x}{4}\sqrt{1 - x^2} + C$$

$$\int x\sec^{-1} x\,dx = \frac{x^2}{2}\sec^{-1} x - \frac{1}{2}\sqrt{x^2 - 1} + C$$

$$\int x^n\sin^{-1} x\,dx = \frac{x^{n+1}}{n+1}\sin^{-1} x - \frac{1}{n+1}\int \frac{x^{n+1}}{\sqrt{1 - x^2}}\,dx + C \text{ if } n \neq -1$$

$$\int x^n\,\text{Arctan}\,x\,dx = \frac{x^{n+1}}{n+1}\tan^{-1} x - \frac{1}{n+1}\int \frac{x^{n+1}}{1 + x^2}\,dx + C \text{ if } n \neq -1$$

$$\int x^n\sec^{-1} x\,dx = \frac{x^{n+1}}{n+1}\sec^{-1} x - \frac{1}{n+1}\int \frac{x^n}{\sqrt{x^2 - 1}}\,dx + C \text{ if } n \neq -1$$

Integrals Involving Hyperbolic Functions

$$\int \sinh x\,dx = \cosh x + C$$

$$\int \cosh x\,dx = \sinh x + C$$

$$\int \tanh x\,dx = \ln(\cosh x) + C$$

$$\int \coth x\,dx = \ln|\sinh x| + C$$

$$\int \text{sech}\,x\,dx = \text{Arctan}\,|\sinh x| + C$$

$$\int \text{csch}\,x\,dx = \ln\left|\tanh\frac{x}{2}\right| + C$$

$$\int \sinh^2 x\,dx = \frac{1}{4}\sinh 2x - \frac{x}{2} + C$$

$$\int \cosh^2 x\,dx = \frac{1}{4}\sinh 2x + \frac{x}{2} + C$$

$$\int \tanh^2 x\,dx = x - \tanh x + C$$

$$\int \coth^2 x\,dx = x - \coth x + C$$

$$\int \text{sech}^2 x\,dx = \tanh x + C$$

$$\int \text{csch}^2 x\,dx = -\coth x + C$$

$$\int \text{sech}\,x\tanh x\,dx = -\text{sech}\,x + C$$

$$\int \text{csch}\,x\coth x\,dx = -\text{csch}\,x + C$$

Some Algebraic Integrals

$$\int x(ax+b)^{-1}\,dx = \frac{x}{a} - \frac{b}{a^2}\ln|ax+b| + C$$

$$\int x(ax+b)^{-2}\,dx = \frac{1}{a^2}\left[\ln|ax+b| + \frac{b}{ax+b}\right] + C$$

$$\int x(ax+b)^{n}\,dx = \frac{(ax+b)^{n+1}}{a^2}\left(\frac{ax+b}{n+2} - \frac{b}{n+1}\right) + C \text{ if } n \neq -1, -2$$

$$\int \frac{dx}{(a^2 \pm x^2)^n} = \frac{1}{2a^2(n-1)}\left(\frac{x}{(a^2 \pm x^2)^{n-1}} + (2n-3)\int \frac{dx}{(a^2 \pm x^2)^{n-1}}\right) \text{ if } n \neq 1$$

$$\int x\sqrt{ax+b}\,dx = \frac{2}{15a^2}(3ax-2b)(ax+b)^{3/2} + C$$

$$\int x^n\sqrt{ax+b}\,dx = \frac{2}{a(2n+3)}\left(x^n(ax+b)^{3/2} - nb\int x^{n-1}\sqrt{ax+b}\,dx\right)$$

$$\int \frac{x\,dx}{\sqrt{ax+b}} = \frac{2}{3a^2}(ax-2b)\sqrt{ax+b} + C$$

$$\int \frac{x^n\,dx}{\sqrt{ax+b}} = \frac{2}{a(2n+1)}\left(x^n\sqrt{ax+b} - nb\int \frac{x^{n-1}\,dx}{\sqrt{ax+b}}\right)$$

$$\int \frac{dx}{x\sqrt{ax+b}} = \frac{1}{\sqrt{b}}\ln\left|\frac{\sqrt{ax+b}-\sqrt{b}}{\sqrt{ax+b}+\sqrt{b}}\right| + C \text{ if } b > 0$$

$$\int \frac{dx}{x\sqrt{ax+b}} = \frac{2}{\sqrt{-b}}\operatorname{Arctan}\sqrt{\frac{ax+b}{-b}} + C \text{ if } b < 0$$

$$\int \frac{dx}{x^n\sqrt{ax+b}} = -\frac{\sqrt{ax+b}}{b(n-1)x^{n-1}} - \frac{(2n-3)a}{(2n-2)b}\int \frac{dx}{x^{n-1}\sqrt{ax+b}} \text{ if } n \neq 1$$

$$\int \sqrt{2ax-x^2}\,dx = \frac{x-a}{2}\sqrt{2ax-x^2} + \frac{a^2}{2}\operatorname{Arcsin}\frac{x-a}{a} + C$$

$$\int \frac{dx}{\sqrt{2ax-x^2}} = \operatorname{Arcsin}\frac{x-a}{a} + C$$

$$\int x^n\sqrt{2ax-x^2}\,dx = -\frac{x^{n-1}(2ax-x^2)^{3/2}}{n+2} + \frac{(2n+1)a}{n+2}\int x^{n-1}\sqrt{2ax-x^2}\,dx$$

$$\int \frac{x^n\,dx}{\sqrt{2ax-x^2}} = -\frac{x^{n-1}}{n}\sqrt{2ax-x^2} + \frac{(2n-1)a}{n}\int \frac{x^{n-1}\,dx}{\sqrt{2ax-x^2}}$$

$$\int \frac{\sqrt{2ax-x^2}}{x}\,dx = \sqrt{2ax-x^2} + a\operatorname{Arcsin}\frac{x-a}{a} + C$$

$$\int \frac{\sqrt{2ax-x^2}}{x^n}\,dx = \frac{(2ax-x^2)^{3/2}}{(3-2n)ax^n} + \frac{n-3}{(2n-3)a}\int \frac{\sqrt{2ax-x^2}}{x^{n-1}}\,dx$$

$$\int \frac{dx}{x^n\sqrt{2ax - x^2}} = \frac{\sqrt{2ax - x^2}}{a(1-2n)x^n} + \frac{n-1}{(2n-1)a}\int \frac{dx}{x^{n-1}\sqrt{2ax - x^2}}$$

$$\int (\sqrt{2ax - x^2})^n \, dx = \frac{x-a}{n+1}(2ax - x^2)^{n/2} + \frac{na^2}{n+1}\int (\sqrt{2ax - x^2})^{n-2}\, dx$$

$$\int \frac{dx}{(\sqrt{2ax - x^2})^n} = \frac{x-a}{(n-2)a^2}(\sqrt{2ax - x^2})^{2-n} + \frac{n-3}{(n-2)a^2}\int \frac{dx}{(\sqrt{2ax - x^2})^{n-2}}$$

Definite Integrals

$$\int_0^\infty x^n e^{-x}\, dx = n! \; (n \geq 0)$$

$$\int_0^\infty e^{-ax^2}\, dx = \frac{1}{2}\sqrt{\frac{\pi}{a}} \; (a > 0)$$

$$\int_0^{\pi/2} \sin^n x \, dx = \int_0^{\pi/2} \cos^n x \, dx = \begin{cases} \dfrac{1 \cdot 3 \cdot 5 \cdot \cdots \cdot (n-1)}{2 \cdot 4 \cdot 6 \cdot \cdots \cdot n}\dfrac{\pi}{2} & \text{if } n \text{ is an even integer and } n \geq 2 \\[2ex] \dfrac{2 \cdot 4 \cdot 6 \cdot \cdots \cdot (n-1)}{3 \cdot 5 \cdot 7 \cdot \cdots \cdot n} & \text{if } n \text{ is an odd integer and } n \geq 3 \end{cases}$$

Answers to Odd-Numbered Exercises

CHAPTER 1 FUNCTIONS, LIMITS, AND CONTINUITY

Section 1.1

1. a) $\dfrac{100}{11}$ b) $\dfrac{1589}{495}$ c) $\dfrac{11}{24975}$ d) $\dfrac{2}{7}$

3. $x > 0$

5. $x \geq 1 + \sqrt{2}$ or $x \leq 1 - \sqrt{2}$

7. $x < 5/3$ or $x > 2$

9. $0 < x < 1$

11. $x < -5$ or $-1 < x < 1$

13. $x \leq -(5/2)$

15. $x \geq 1$ or $x \leq -3$

17. $y = -3$

19. $y = 2x - 3$

21. $x + 3y = 0$

23. $y = -(1/3)x + 4/3$

25.

$\frac{x}{4} + y = 1$

Straight line

27.

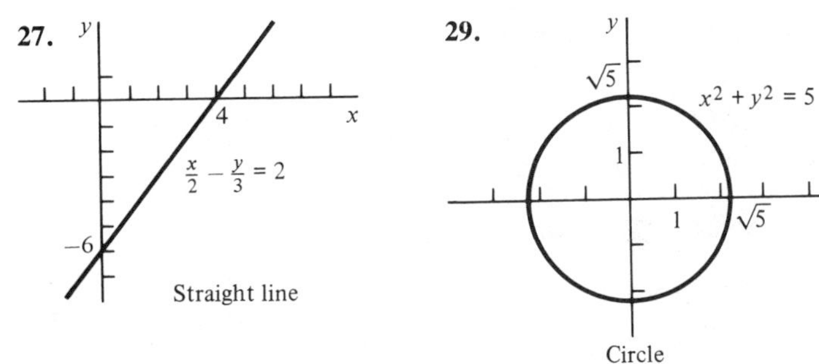

Straight line

29.

Circle

31.

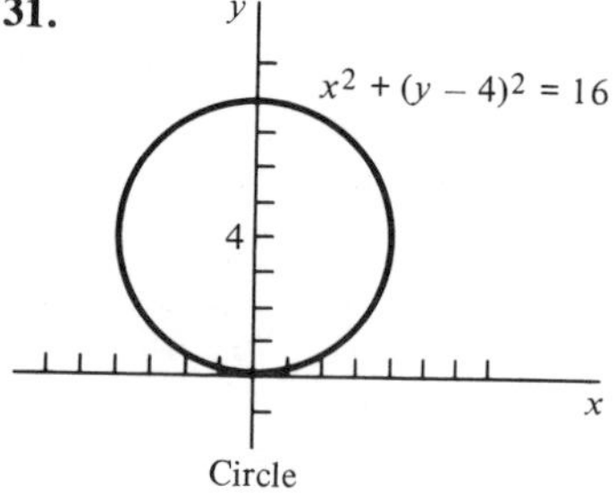

Circle

33.

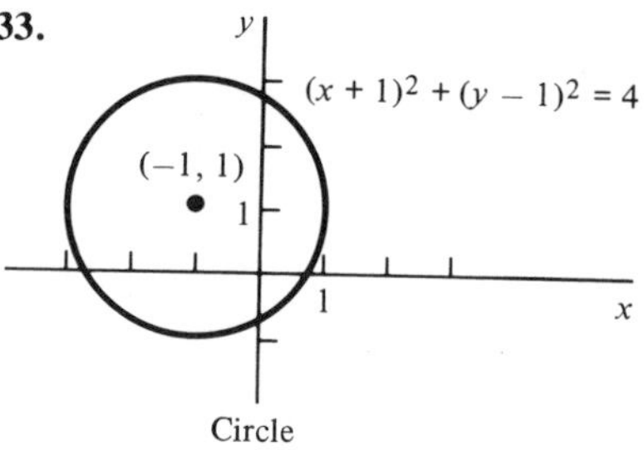

Circle

35.

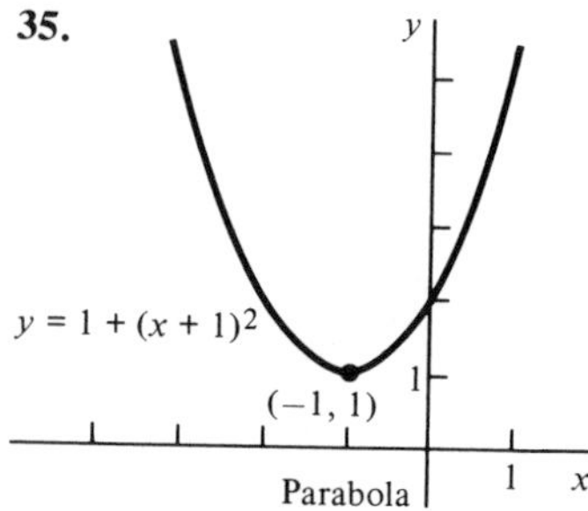

Parabola

37.

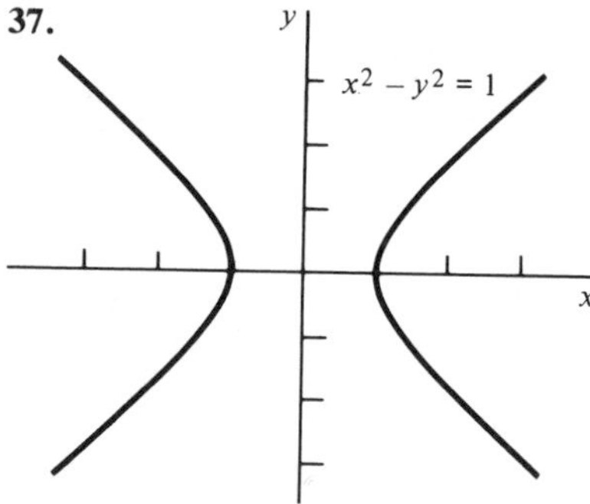

Hyperbola

39.

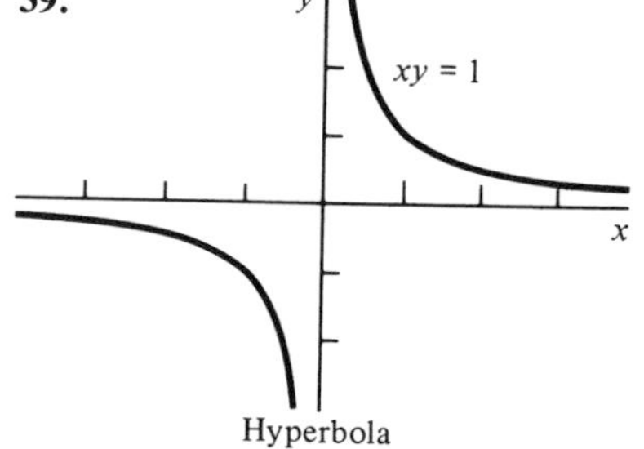

Hyperbola

41.

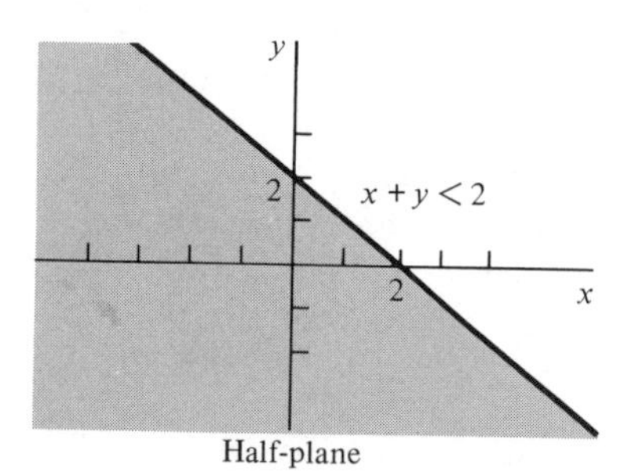

Half-plane

43.

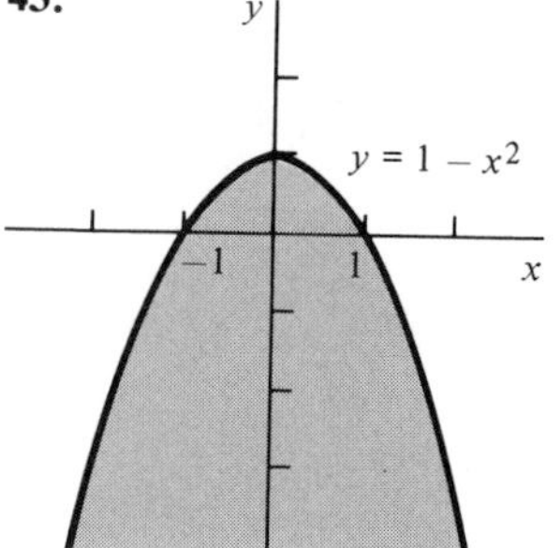

Region on and below parabola

Section 1.2

1. $\mathscr{D}(f) = (-\infty, 0) \cup (0, \infty)$, $\mathscr{R}(f) = (-\infty, 1) \cup (1, \infty)$

3. $\mathscr{D}(F) = (-\infty, 0) \cup (0, 1) \cup (1, \infty)$, $\mathscr{R}(F) = (-\infty, -1) \cup (-1, 0) \cup (0, \infty)$

5. $\mathscr{D}(h) = (-\infty, \infty)$, $\mathscr{R}(h) = [-4, \infty)$

7.

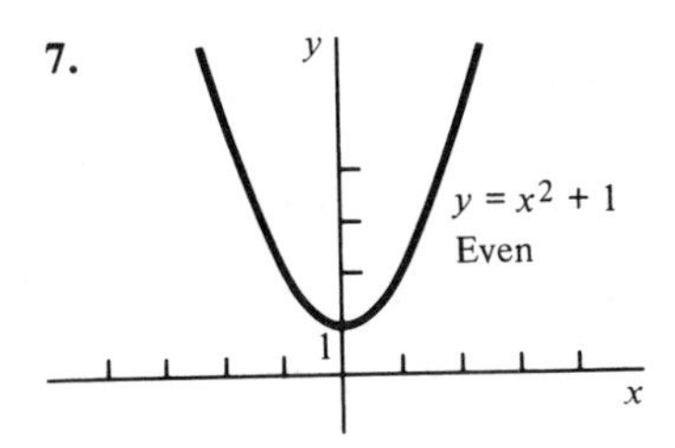

9.

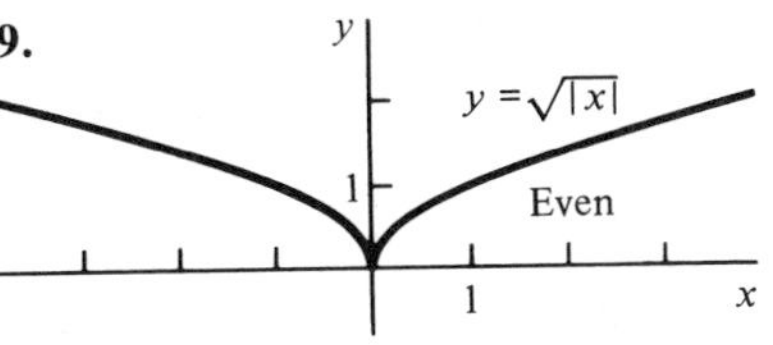

11.

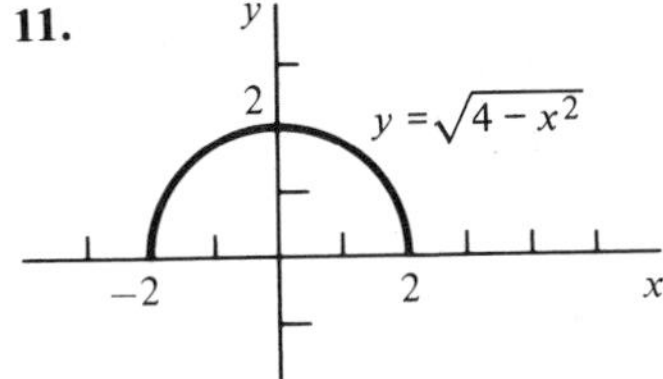

13.

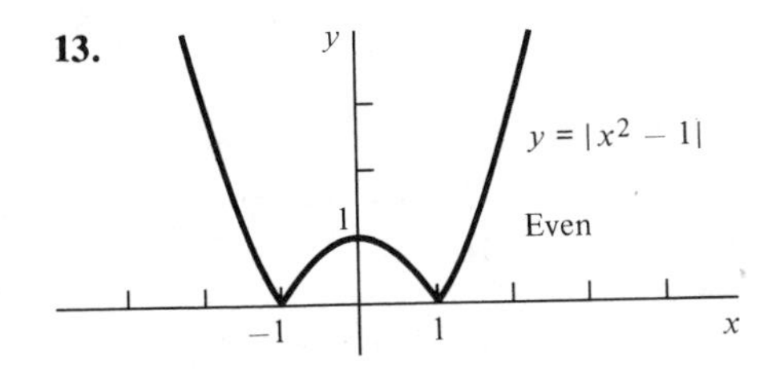

15.

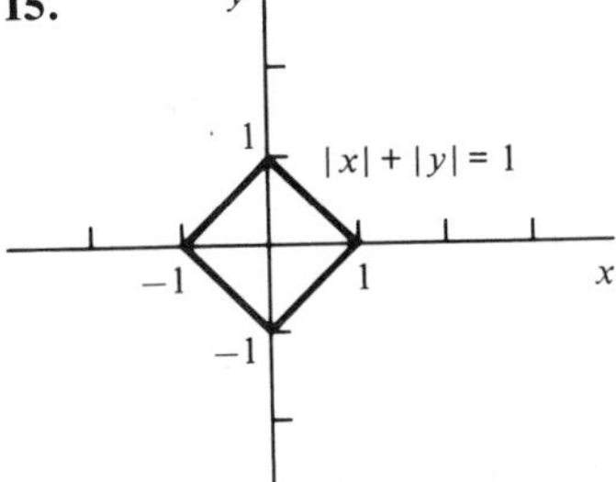

17. $(f+g)(x) = \dfrac{1}{x^2-1} + \dfrac{1}{x}$, $(fg)(x) = \dfrac{1}{x(x^2-1)}$, $\left(\dfrac{f}{g}\right)(x) = \dfrac{x}{x^2-1}$, $\left(\dfrac{g}{f}\right)(x) = \dfrac{x^2-1}{x}$, $f\circ f(x) = \dfrac{(x^2-1)^2}{2x^2-x^4}$, $f\circ g(x) = \dfrac{x^2}{1-x^2}$, $g\circ f(x) = x^2 - 1$, $g\circ g(x) = x$, $\mathcal{D}(f+g) = \mathcal{D}(fg) = \mathcal{D}\left(\dfrac{f}{g}\right) = \mathcal{D}\left(\dfrac{g}{f}\right) = \mathcal{D}(f\circ g) = \{x{:}x \neq 0, x \neq \pm 1\}$, $\mathcal{D}(f\circ f) = \{x{:}x \neq \pm\sqrt{2}, \pm 1, 0\}$, $\mathcal{D}(g\circ f) = \{x{:}x \neq \pm 1\}$, $\mathcal{D}(g\circ g) = \{x{:}x \neq 0\}$

19. $(f+g)(x) = \sqrt{1-x^2} + 2 + x$, $(fg)(x) = \sqrt{1-x^2}(2+x)$, $\left(\dfrac{f}{g}\right)(x) = \dfrac{\sqrt{1-x^2}}{2+x}$, $\left(\dfrac{g}{f}\right)(x) = \dfrac{2+x}{\sqrt{1-x^2}}$, $f\circ f(x) = |x|$, $f\circ g(x) = \sqrt{-x^2-4x-3}$, $g\circ f(x) = 2 + \sqrt{1-x^2}$, $g\circ g(x) = 4 + x$, $\mathcal{D}(f+g) = \mathcal{D}(fg) = \mathcal{D}\left(\dfrac{f}{g}\right) = [-1, 1]$, $\mathcal{D}\left(\dfrac{g}{f}\right) = (-1, 1)$, $\mathcal{D}(f\circ f) = \mathcal{D}(g\circ f) = [-1, 1]$, $\mathcal{D}(f\circ g) = [-3, -1]$, $\mathcal{D}(g\circ g) = (-\infty, \infty)$

21. $f^{-1}(x) = x + 1$, $\mathcal{D}(f^{-1}) = \mathcal{R}(f) = \mathcal{D}(f) = \mathcal{R}(f^{-1}) = (-\infty, \infty)$

23. $f^{-1}(x) = 1 + x^2$, $\mathcal{D}(f^{-1}) = \mathcal{R}(f) = [0, \infty)$, $\mathcal{R}(f^{-1}) = \mathcal{D}(f) = [1, \infty)$

25. $f^{-1}(x) = (x-1)^3$, $\mathcal{D}(f^{-1}) = \mathcal{R}(f) = (-\infty, \infty)$, $\mathcal{R}(f^{-1}) = \mathcal{D}(f) = (-\infty, \infty)$.

27. $f^{-1}(x) = -\sqrt{x}$, $\mathcal{D}(f^{-1}) = \mathcal{R}(f) = [0, \infty)$, $\mathcal{R}(f^{-1}) = \mathcal{D}(f) = (-\infty, 0]$

29. $f^{-1}(x) = \dfrac{x}{\sqrt{1-x^2}}$, $\mathcal{D}(f^{-1}) = \mathcal{R}(f) = (-1, 1)$, $\mathcal{R}(f^{-1}) = \mathcal{D}(f) = (-\infty, \infty)$

31. $f^{-1}(x) = \begin{cases} \sqrt{x-1} & \text{if } x \geq 1 \\ x - 1 & \text{if } x < 1 \end{cases}$ $\mathcal{D}(f^{-1}) = \mathcal{R}(f) = \mathcal{R}(f^{-1}) = \mathcal{D}(f) = (-\infty, \infty)$

33. $f\circ f, f\circ g, g\circ f$ are even, $g\circ g$ is odd.

Section 1.3

1. 7 **3.** 0 **5.** 0
7. 0 **9.** -3 **11.** 3
13. 0 **15.** -1 **17.** does not exist
19. $-(1/2)$ **21.** $-(8/3)$ **23.** 3/16
25. 4 **27.** 1/4 **29.** $1/\sqrt{2}$
31. 1/3 **33.** $\lim_{x\to\pm 1} f(x) = 1$, $\lim_{x\to 0} f(x) = 0$

Section 1.4

1. does not exist **3.** ∞ **5.** ∞ **7.** 0
9. does not exist **11.** does not exist **13.** $-\infty$ **15.** $-\infty$
17. ∞ **19.** 4 **21.** 1 **23.** does not exist
25. $-\infty$ **27.** $-(3/5)$ **29.** $-(\sqrt{2}/4)$ **31.** ∞
33. -3 **35.** 0 **37.** -1

Section 1.5

1. continuous everywhere
3. continuous on $[1/3, \infty)$ ((i) fails)
5. continuous everywhere
7. continuous except at ± 1 ((i) fails at ± 1)
9. continuous except at -2 ((i) fails at -2)
11. continuous everywhere
13. continuous everywhere except ± 1
15. continuous except at 2 ((iii) fails at 2)
17. continuous except at 0 ((ii) fails at 0)
19. largest value is 16
21. f positive on $(-\infty, -3)$ and $(-1, \infty)$, f negative on $(-3, -1)$
23. f positive on $(-2, 0)$ and $(1, \infty)$, f negative on $(-\infty, -2)$ and $(0, 1)$

CHAPTER 2 DIFFERENTIATION—INTERPRETATION AND TECHNIQUES

Section 2.1

1. $y = 8x - 13$, $y = -\dfrac{x}{8} + \dfrac{13}{4}$

3. $y = 3x + 10$, $y = -\dfrac{x}{3} + \dfrac{10}{3}$

5. $y = 12x + 24$, $y = -\dfrac{x}{12} - \dfrac{1}{6}$

7. $y = -\dfrac{2x}{27} + \dfrac{1}{3}$, $y = \dfrac{27}{2}x - \dfrac{727}{18}$

9. $y = \dfrac{x}{4} + \dfrac{5}{4}$, $y = -4x + 14$

11. $y = \dfrac{x}{4} + \dfrac{1}{2}$, $y = -4x + 9$

13. $y = 2x_0x - x_0^2$, $y = x_0^2 - \dfrac{1}{2x_0}(x - x_0)$

15. $y = (2ax_0 + b)x - ax_0^2 + c$, $y = ax_0^2 + bx_0 + c - \dfrac{1}{2ax_0 + b}(x - x_0)$

17. a) $3a^2$ b) $y = 2x \pm \dfrac{4}{3}\sqrt{\dfrac{2}{3}}$

Section 2.2

1. $y' = 2x - 3$

3. $f'(x) = 3x^2$

5. $g'(x) = -\dfrac{4}{(2 + x)^2}$

7. $F'(t) = \dfrac{1}{\sqrt{2t + 1}}$

9. $y' = 1 - \dfrac{1}{x^2}$

11. $\dfrac{dz}{ds} = \dfrac{1}{(1 + s)^2}$

13. $y' = -\dfrac{2}{x^3}$

15. $f'(t) = \dfrac{12t}{(t^2 + 3)^2}$

17. $F'(x) = \dfrac{-x}{(1 + x^2)^{3/2}}$

19.

Δx	Δy	$\dfrac{\Delta y}{\Delta x}$
0.1	0.131	1.31
0.01	0.010301	1.03010
0.001	0.001003	1.00300

$$\frac{d}{dx}(x^3 - 2x)\bigg|_{x=1} = 1$$

21. $f'(x) = \dfrac{1}{3}x^{-2/3}$

23. $y = \dfrac{1}{6}x + \dfrac{5}{2}$

25. $y = -\dfrac{3}{2}t + \dfrac{5}{2}$

27. $\dfrac{dy}{dt} = 22t^{21}$

29. $y = x_0^2x - x_0^3 + \dfrac{1}{x_0}$

31. one line if $b = a^2$, none if $b > a^2$

Section 2.3

1. $y' = 8x^7 - 4x^3$

3. $y' = x^2 - x + 1$

5. $f'(x) = 2Ax + B$

7. $g'(t) = \dfrac{1}{3}t^{-2/3} + \dfrac{1}{2}t^{-3/4} + \dfrac{3}{5}t^{-4/5}$

9. $\dfrac{dz}{ds} = \dfrac{1}{3}s^4 - \dfrac{1}{5}s^2$

11. $F'(x) = 3(1 - 5x) - 5(3x - 2) = 13 - 30x$

13. $g'(t) = (t^2 + 1)4t^3 + 2t(t^4 + 2) = 6t^5 + 4t^3 + 4t$

15. $f'(x) = (1 + 2x)(1 + 3x)(1 + 4x) + 2(1 + x)(1 + 3x)(1 + 4x) + 3(1 + x)(1 + 2x)(1 + 4x) + 4(1 + x)(1 + 2x)(1 + 3x) = 10 + 70x + 150x^2 + 96x^3$

17. $\dfrac{dy}{dx} = -\dfrac{2x + 5}{(x^2 + 5x)^2}$

19. $f'(t) = \dfrac{20}{(2 - 5t)^2}$

21. $\dfrac{dy}{dt} = \dfrac{2t}{7} + \dfrac{4}{7}$

23. $y' = \dfrac{2x}{(a^2 + b^2x^2)(c^2 - d^2x^2)}\left(\dfrac{d^2}{c^2 - d^2x^2} - \dfrac{b^2}{a^2 + b^2x^2}\right)$

25. $y' = \dfrac{4}{(x + 2)^2}$

27. $\dfrac{dz}{dt} = \dfrac{-8t}{(t^2 - 2)^2}$

29. $f'(x) = \dfrac{2x^3 + 3x^2 + 4}{(x + 1)^2}$

31. $F'(t) = \dfrac{-8t^2 + 18t - 1}{(t^2 - t + 1)^2}$

33. $y' = \dfrac{2x+1}{(x^2+x+1)^2}$

37. 16

39. 6

41. $y = -2x + 7,\ y = \dfrac{x}{2} + 2$

43. $y = 0,\ y = 4$

45. $y = -\dfrac{1}{2}x^3$

Section 2.4

1. $y' = 12(2x+3)^5$

3. $f'(x) = -20x(4 - x^2)^9$

5. $y' = Ar(Ax + B)^{r-1}$

7. $y' = \dfrac{12}{(5-4x)^2}$

9. $y' = \dfrac{5}{2\sqrt{5x+3}}$

11. $y' = -\dfrac{3}{2}x^2(2 + x^3)^{-3/2}$

13. $f'(t) = 96(2at + b)(at^2 + bt + c)^{95}$

15. $y' = \dfrac{-x}{2(x+2)^2\sqrt{x+1}}$

17. $y' = \dfrac{-3}{2\sqrt{3x+4}(2+\sqrt{3x+4})^2}$

19. $g'(t) = \left(12t - 3 - \dfrac{11}{2}t^2\right)(2 + t^2)\sqrt{3-t}$

21. $\dfrac{dz}{dx} = -\dfrac{5}{3}\left(u + \dfrac{1}{u-1}\right)^{-8/3}\left(1 - \dfrac{1}{(u-1)^2}\right)$

23. $\dfrac{dy}{dt} = -\dfrac{1}{3}\dfrac{1}{(t+1)^2}$

25. $f'(t) = 3t^2 \operatorname{sgn}(2 + t^3) = \dfrac{3t^2(2+t^3)}{|2+t^3|}$

27. $y' = x^2(2 + |x|^3)^{-2/3} \operatorname{sgn} x = x|x|(2 + |x|^3)^{-2/3}$

29. $\dfrac{dz}{dx} = \dfrac{\operatorname{sgn}(x+4)}{2\sqrt{|x+4|}} = \dfrac{x+4}{2|x+4|^{3/2}}$

31.

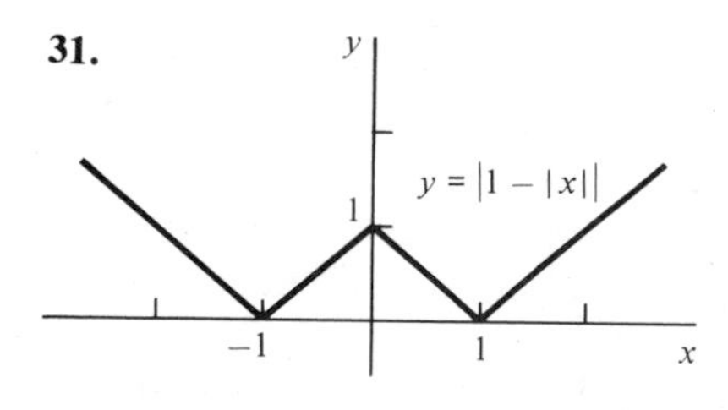

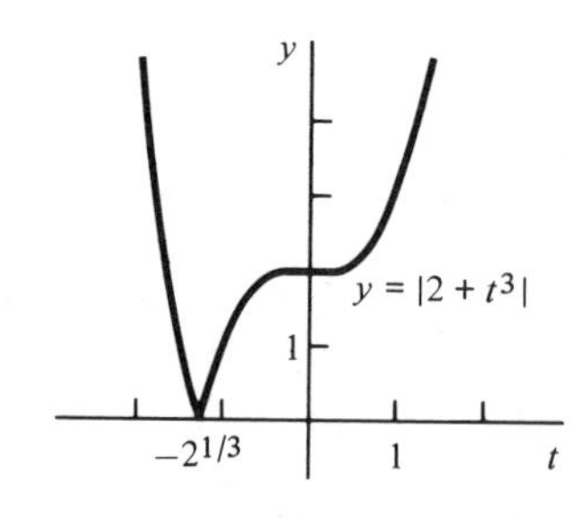

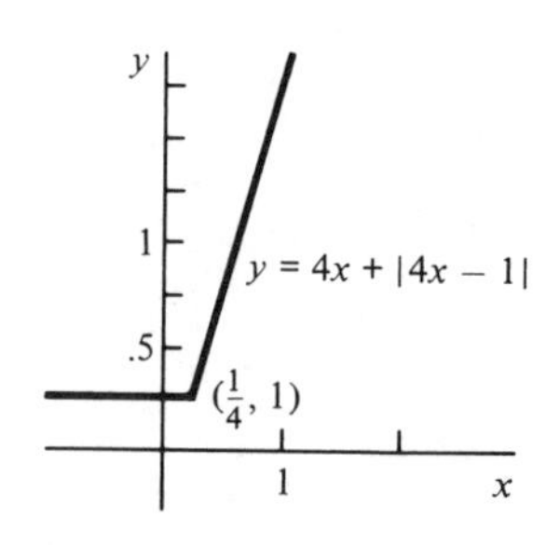

33. $-2xf'(5 - x^2)$

35. $12f'(4t)f'(3f(4t))$

37. $3/(2\sqrt{2})$

39. 102

41. $y = -6\left(x + ((3x)^5 - 2)^{-1/2}\right)^{-7}\left(1 - \dfrac{15}{2}(3x)^4((3x)^5 - 2)^{-3/2}\right)$

43. $24x^2 \cos(5 - 4x^3)\sin(5 - 4x^3)$

45. $\dfrac{b}{(bx+c)\ln(bx+c)}$

47. $-2\cos x \sin(\sin x)\cos(\sin x)\cos[(\cos(\sin x))^2]$

49. $y = \sqrt{2}(x+1)$

51. $y = 5x/162 + 4/81$

53. $1/13$

Section 2.5

1. 10 cm^2/cm **3.** 1600 π cm^3/cm **5.** $\sqrt{\pi}/\sqrt{A}$ cm/cm^2

7. a) $(2, \infty)$ b) $(-\infty, 2)$ c) $t = 2$ d) $a = 2$ e) 0 f) 0

9. a) $(-\infty, 0)$ and $(0, \infty)$ b) none c) $t = 0$ d) $a = 0$ e) 16 f) 12

11. a) $(-1, 1)$ b) $(-\infty, -1)$ and $(1, \infty)$ c) $t = \pm 1$ d) $-(1/2)$ at $t = 1$, $1/2$ at $t = -1$
e) 1/17 f) $-(3/25)$

13. $t \simeq 4.318$ sec, $t \simeq 2.159$ sec **17.** a) $-\$2.00$ b) approx. \$9.11

19. \$32

Section 2.6

1. $y' = 2x + \frac{1}{x^2}, y'' = 2 - \frac{2}{x^3}, y''' = \frac{6}{x^4}$

3. $y' = \frac{2}{(x+1)^2}, y'' = \frac{-4}{(x+1)^3}, y''' = \frac{12}{(x+1)^4}$

5. $y' = 10x^9 + 16x^7, y'' = 90x^8 + 112x^6, y''' = 720x^7 + 672x^5$

7. $f^{(n)}(x) = \frac{(-1)^n n!}{x^{n+1}}$

9. $f'(x) = \frac{1}{2}x^{-1/2}$

$$f^{(n)}(x) = (-1)^{n-1}\frac{1 \cdot 3 \cdot 5 \cdots (2n-3)}{2^n}x^{-(2n-1)/2} \text{ (if } n \geq 2)$$

11. $f'(x) = \frac{2}{3}x^{-1/3}, f^{(n)}(x) = 2(-1)^{n-1}\frac{1 \cdot 4 \cdot 7 \cdots (3n-5)}{3^n}x^{-(3n-2)/3}$ (if $n \geq 2$)

13. $y = 3x - \frac{1}{x}$ **15.** $y = \frac{3}{5}x^3 + \frac{2}{5}x^{-2}$

19. $(fg)^{(3)} = f^{(3)}g + 3f''g' + 3f'g'' + fg^{(3)}$
$(fg)^{(4)} = f^{(4)}g + 4f^{(3)}g' + 6f''g'' + 4f'g^{(3)} + fg^{(4)}$

Section 2.7

1. $\frac{dy}{dx} = \frac{1-y}{2+x}$ **3.** $\frac{dy}{dx} = \frac{2(1-xy^3)}{1+3x^2y^2}$ **5.** $\frac{dy}{dx} = \frac{2xy(x+y)^2 - 2y^3}{x^2(x+y)^2 - 2xy^2}$

7. $2x + 3y = 5$ **9.** $y = x$ **11.** $y'' = \frac{2(y-1)}{(1-x)^2}$

13. $\frac{4xy}{(x-y^2)^3}$ **19.** Calculate $y' = y/x$; however, the given equation is equivalent to $(x + (y/2))^2 + (7/4)y^2 = 0$, and is satisfied by no points (x, y).

Section 2.8

1. Any c in (a, b) will do. **3.** $c = \pm\left(\frac{b^2 + ab + a^2}{3}\right)^{1/2}$ At least one of these lies in (a, b).

5. $c = \pm\frac{2}{\sqrt{3}}$

11. f is inc. on $(-1, \infty)$, dec. on $(-\infty, -1)$. -1 is a C.P., f is locally minimum there.

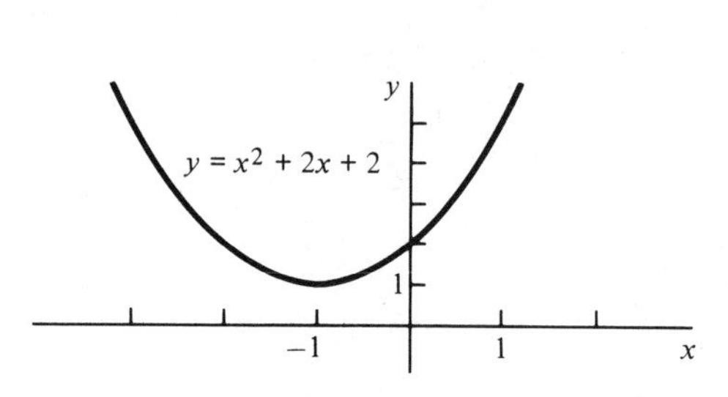

13. f is increasing on the whole real line.

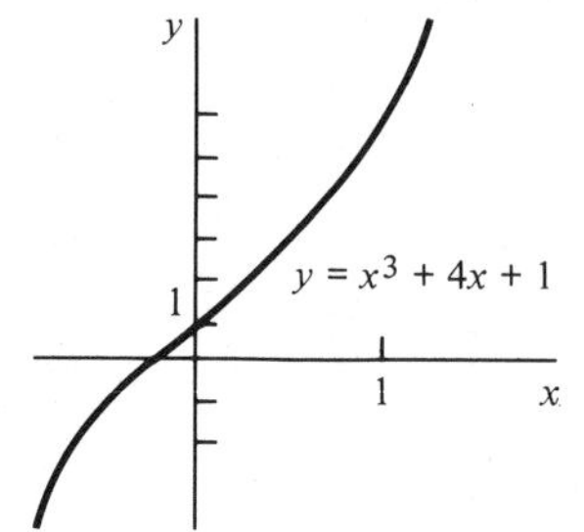

15. f is inc. on $(-\infty, 0)$, f is dec. on $(0, \infty)$, f is locally maximum at C.P. 0.

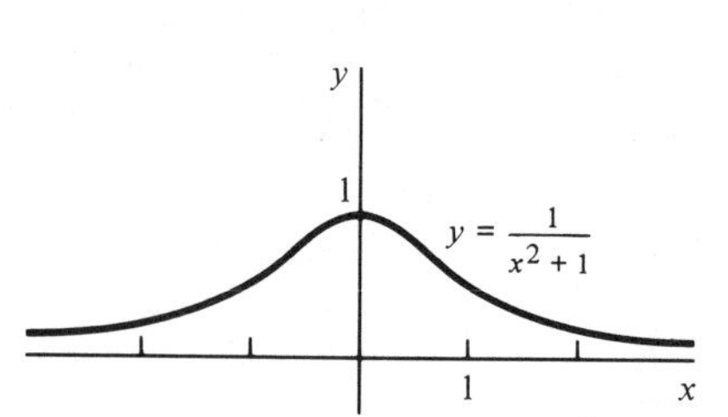

Section 2.9

1. $5x + C$

3. $\frac{2}{3}x^{3/2} + C$

5. $\frac{1}{4}x^4 + C$

7. $a^2x - \frac{1}{3}x^3 + C$

9. $\frac{4}{3}x^{3/2} + \frac{9}{4}x^{4/3} + C$

11. $\frac{1}{12}x^4 - \frac{1}{6}x^3 + \frac{1}{2}x^2 - x + C$

13. $105t + 35t^3 + 21t^5 + 15t^7 + C$

15. $\frac{-1}{1 + x} + C$

19. $y = -\frac{1}{x} + \frac{1}{2x^2} - \frac{3}{2}$, $(-\infty, 0)$

21. $y = \frac{3}{4}x^{4/3} + 5$, $(-\infty < x < \infty)$

23. $y = -\frac{7}{2}x^{-2/7} - \frac{1}{2}$, $(0, \infty)$

25. $y = \frac{1}{6}x^{-2} + \frac{7}{3}x - \frac{3}{2}$, $(0, \infty)$

27. $y = \frac{5}{12}x^4 - 4x^{3/2} + \frac{19}{3}x - \frac{11}{4}$, $(0, \infty)$

Review Exercises for Chapter 2

1. $f'(x) = \frac{3 + 4x - x^2}{(3 + x^2)^2}$

3. $2 - 3/x^2$

5. $-1/(x - 1)^2$

7. $80(2x + 1)^{39}$

9. $\frac{-y}{x + 4y}$

11. $\frac{-8x}{(2 + x^2)^2}$

13. $-(x/y)^2$

15. $\frac{2x - y}{2y + x}$

17. $(x/y)^4$

19. $\frac{1}{2y(2x + 1)^2}$

21. $\frac{3}{\sqrt{2x + 1}(1 - \sqrt{2x + 1})^{5/2}}$

23. y/x

25. $-2\sqrt{y/x} - y/x$

27. $\dfrac{3 - 2x}{(2x + 3)^3}$

29. $-\operatorname{sgn}(x/y) = -(x/y)|y/x|$

31. $\dfrac{x^2 + 2(x + 2)\sqrt{x + 2}}{2\sqrt{x + 2}(x + \sqrt{x + 2})^2}$

33. $y = 2$

35. $4x + 5y = 13$

37. $13x + 12y = 38$

39. $4x - 5y + 1 = 0$

41. $y = -\dfrac{9}{16}x - 2$

43. $f''(x) = \dfrac{4}{(1 + x)^3}$

45. $f'(x) = 2x - 4x^{-5}, f''(x) = 2 + 20x^{-6}, f^{(n)}(x) = (-1)^n \dfrac{(n + 3)!}{6} x^{-(n+4)}$ (if $n \geq 3$)

47. $\dfrac{d^2y}{dx^2} = \dfrac{1}{3y^2 - x}\left(\dfrac{2y - 4x}{3y^2 - x} - \dfrac{6y(y - 2x)^2}{(3y^2 - x)^2} - 2\right)$

49. (1/3)(volume)$^{-2/3}$ units/unit^3

51. 32 cm

55. $y = -(1/x) + 3/2$

57. $y = \begin{cases} 1 + \dfrac{1}{2}x^2 & \text{if } x \geq 0 \\ 1 - \dfrac{1}{2}x^2 & \text{if } x < 0 \end{cases}$

59. $y = \dfrac{1}{6}x^3 - \dfrac{1}{2}x - \dfrac{5}{3}$

CHAPTER 3 THE ELEMENTARY TRANSCENDENTAL FUNCTIONS

Section 3.1

1. $\dfrac{1}{\sqrt{2}}$

3. $\dfrac{\sqrt{3}}{2}$

5. $-\dfrac{1}{\sqrt{2}}$

7. $\dfrac{\sqrt{3} - 1}{2\sqrt{2}}$

9. $\dfrac{\sqrt{3} + 1}{2\sqrt{2}}$

11. $\dfrac{1 - \sqrt{3}}{2\sqrt{2}}$

13. $-\dfrac{\sqrt{3}}{2}$

15. $\dfrac{1 - \sqrt{3}}{2\sqrt{2}}$

17. $-\cos x$

19. $\cos x$

21. $-\sin x$

23. $-\cos x$

25. $\dfrac{1 - \sin x}{\cos x}$

27. $\sin^2 x - \cos^2 x$

33.

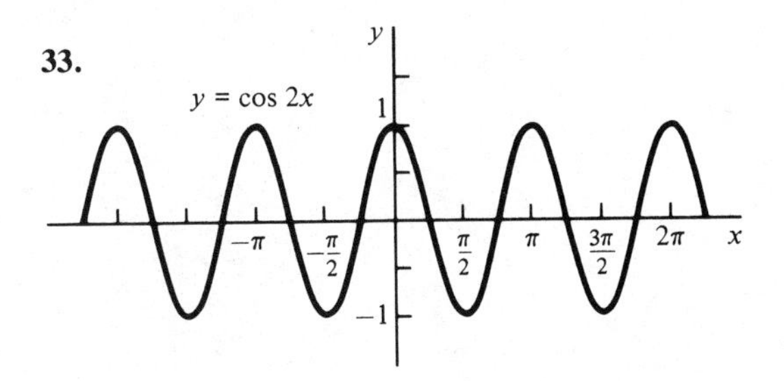

35.

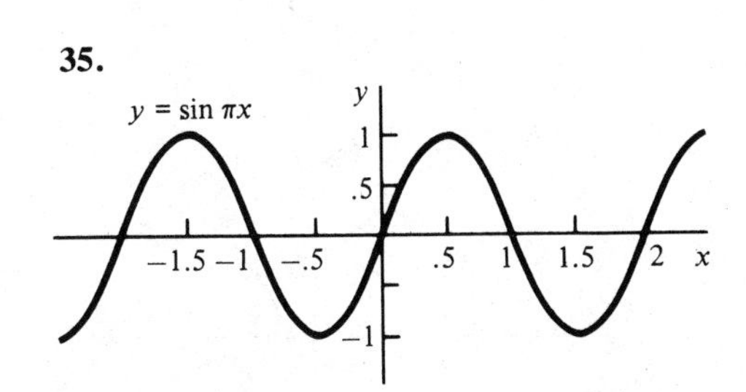

37.

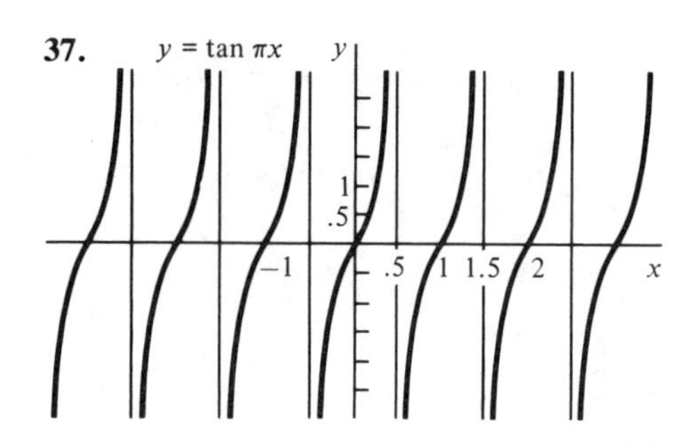

39. $\cos x = -(4/5)$, $\tan x = -(3/4)$

41. $\sin x = -(2\sqrt{2}/3)$, $\tan x = -2\sqrt{2}$

43. $\cos x = -(\sqrt{3}/2)$, $\tan x = 1/\sqrt{3}$

45. $a = 1$, $b = \sqrt{3}$

47. $b = 5/\sqrt{3}$, $c = 10/\sqrt{3}$

49. $a = b \tan A$

51. $a = b \cot B$

53. $c = b \sec A$

55. $\sin A = \sqrt{c^2 - b^2}/c$

57. $\sin B = \dfrac{3}{4\sqrt{2}}$

59. $\sin B = \sqrt{135}/16$

61. $a = 6/(\sqrt{3} + 1)$

63. $b = 4\,\dfrac{\sin 40°}{\sin 70°} \simeq 2.736$

67. $\dfrac{2 \tan 40° \tan 70°}{\tan 40° + \tan 70°} \simeq 1.286$ km

Section 3.2

1. $y' = -3 \sin 3x$

3. $y' = \pi \sec^2 \pi x$

5. $y' = -\dfrac{2}{\pi} \sin \dfrac{2x}{\pi}$

7. $y' = 3 \csc^2 (4 - 3x)$

9. $y' = \dfrac{3}{x^2} \csc \dfrac{3}{x} \cot \dfrac{3}{x}$

11. $f'(x) = r \sin (s - rx)$

13. $F'(t) = a \cos 2at$

15. $\dfrac{du}{dx} = (2x - 4) \sec (x^2 - 4x) \tan (x^2 - 4x)$

17. $\dfrac{dy}{dx} = \dfrac{2}{x} \tan x \sec^2 x - \dfrac{1}{x^2} \tan^2 x$

19. $y' = \sin \dfrac{1}{x} - \dfrac{1}{x} \cos \dfrac{1}{x}$

21. $f'(\theta) = -\dfrac{\sin 2\theta}{\sqrt{\cos 2\theta}}$

23. $y' = \cos t \cos 2t \tan 3t - 2 \sin t \sin 2t \tan 3t + 3 \sin t \cos 2t \sec^2 3t$

25. $f'(t) = \dfrac{\sin t}{\sqrt{1 - \cos t}(1 + \cos t)^{3/2}}$

27. $y' = 2x \cos (x^2 + 1) - 2x(x^2 + 2) \sin (x^2 + 1)$

29. $y' = -\sin (x - 1)$

31. $y = \dfrac{1}{\sqrt{2}} + \dfrac{\pi}{180\sqrt{2}} (x - 45°)$

33. $f'(x) = \dfrac{\sin x \cos x}{|\sin x|} = \cos x \operatorname{sgn} (\sin x)$

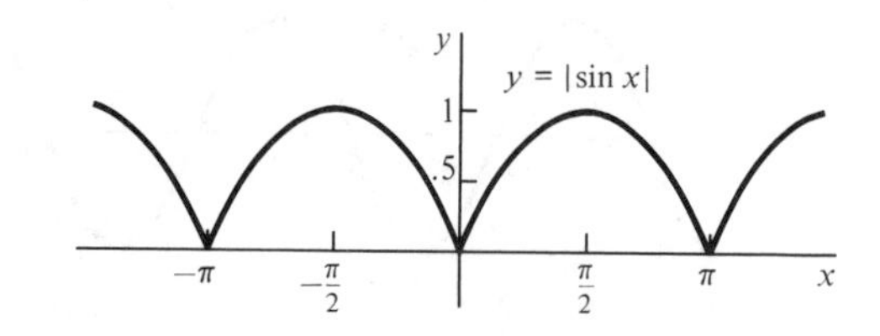

35. $y = 1 - \dfrac{4x - \pi}{4 - \pi}$

37. $y = 1$

43. $\dfrac{d^n}{dx^n}\sin ax = \begin{cases} a^n \cos ax & \text{if } n = 4k+1 \\ -a^n \sin ax & \text{if } n = 4k+2 \\ -a^n \cos ax & \text{if } n = 4k+3 \\ a^n \sin ax & \text{if } n = 4k \end{cases} \quad k = 0, 1, 2, 3, \ldots$

45. $\dfrac{d^n}{dx^n}(x \sin x) = \begin{cases} -n\cos x + x \sin x & \text{if } n = 4k \\ n \sin x + x \cos x & \text{if } n = 4k+1 \\ n \cos x - x \sin x & \text{if } n = 4k+2 \\ -n \sin x - x \cos x & \text{if } n = 4k+3 \end{cases} \quad k = 0, 1, 2, \ldots$

53. $y = \dfrac{3}{10}\sin 10x$

55. $y = 3\cos(x - 2)$

57. $s_0 = \left(\dfrac{4000 \times 9.8}{\sqrt{3}}\right)^{1/2} \simeq 150.44$ m/sec

59. 16 Hz (for 900 g), 48 Hz (for 100 g)

Section 3.3

1. $\pi/3$

3. $-(\pi/4)$

5. $-(\pi/3)$

7. $2\sqrt{2}/3$

9. 200

11. x

13. $\sqrt{1 - x^2}$

15. $1/\sqrt{1 + x^2}$

17. $\sqrt{1 - x^2}/x$

19. $\text{Arccos}(\cos x) = \begin{cases} x - 2k\pi & \text{if } 2k\pi \le x \le (2k+1)\pi \\ (2k+2)\pi - x & \text{if } (2k+1)\pi < x < (2k+2)\pi \end{cases} \quad k = 0, \pm 1, \pm 2, \pm 3, \ldots$

21.

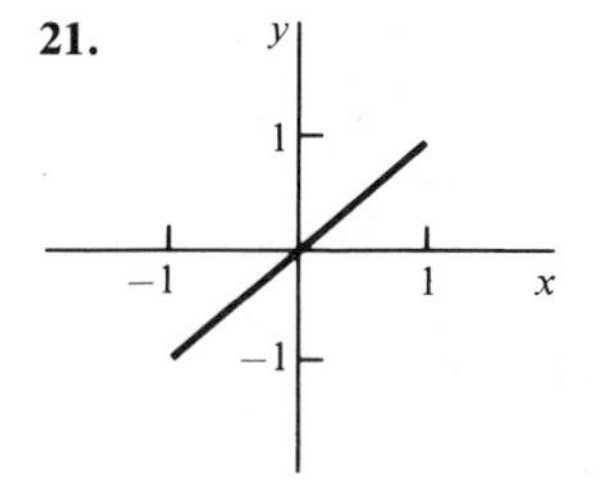

$y = \sin \text{Arcsin}\, x$
$y' = 1$ on $(-1, 1)$

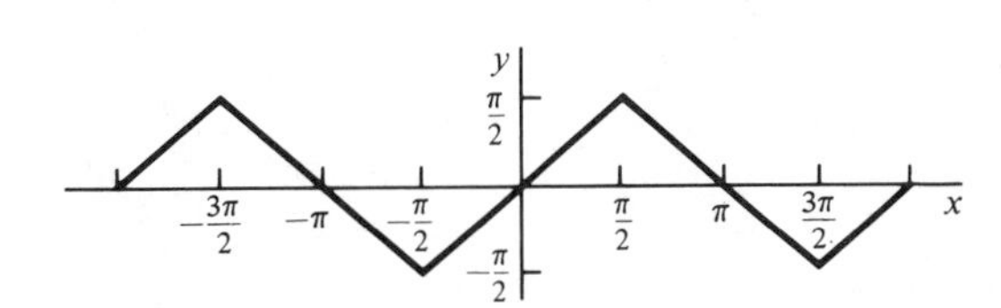

$y = \text{Arcsin} \sin x$

$y' = \begin{cases} 1 & \text{if } \left(2k - \frac{1}{2}\right)\pi < x < \left(2k + \frac{1}{2}\right)\pi \\ -1 & \text{if } \left(2k + \frac{1}{2}\right)\pi < x < \left(2k + \frac{3}{2}\right)\pi \end{cases} \quad k = 0, \pm 1, \pm 2, \ldots$

23. $y' = \dfrac{1}{\sqrt{2 + x - x^2}}$

25. $y' = \dfrac{-1}{\sqrt{a^2 - (x - b)^2}}$

27. $f'(t) = \text{Arctan}\, t + \dfrac{t}{1 + t^2}$

29. $F'(x) = 2x\, \text{Arctan}\, x + 1$

31. $G'(x) = \dfrac{1}{\sqrt{1 - x^2}\,\text{Arcsin}\, 2x} - \dfrac{2\,\text{Arcsin}\, x}{\sqrt{1 - 4x^2}\,(\text{Arcsin}\, 2x)^2}$

33. $f'(x) = \dfrac{x}{\sqrt{1 - x^4}}(\text{Arcsin}\, x^2)^{-1/2}$

35. $y' = \dfrac{1}{2x(x-1)} - \dfrac{\text{Arctan}\sqrt{x-1}}{2(x-1)^{3/2}}$ **37.** $y' = x/\sqrt{2ax - x^2}$ **39.** $C = 3\pi/4$

41. $\text{Csc } x = 1/\sin x$ if $-\pi/2 \le x < 0$ or $0 < x \le \pi/2$, $\text{Arccsc } x = \text{Arcsin } 1/x$,
$\dfrac{d}{dx}\text{Arccsc } x = \dfrac{-1}{|x|\sqrt{x^2-1}}$ $(x < -1$ or $x > 1)$

45. $y = 1 + \text{Arctan } x$ **47.** $y = \text{Arcsin } x + 1 - (\pi/6)$

Section 3.4

1. $\ln \dfrac{64}{81}$ **3.** $\ln x^2(x-2)^5$ **5.** $2 \ln \sin x$

7. $y' = \dfrac{3}{3x-2}$ **9.** $y' = \dfrac{5x^4}{1+x^5}$ **11.** $y' = \dfrac{1}{x \ln x}$

13. $y' = \dfrac{2}{x \ln x}$ **15.** $y' = 2x \ln x$ **17.** $y' = \cot x$

19. $y' = -\csc x$ **21.** $y' = \dfrac{-1}{\sqrt{x^2+a^2}}$ **23.** $y' = \dfrac{-a}{x\sqrt{x^2+a^2}}$

Section 3.5

1. $\sqrt{e}$ **3.** x^5 **5.** $-3x$

7. 5 **9.** $y' = 5e^{5x}$ **11.** $y' = \dfrac{1-x}{e^x}$

13. $y' = \dfrac{e^x - e^{-x}}{2}$ **15.** $y' = e^x e^{(e^x)}$ **17.** $y' = rx^{r-1}e^{ax} + ax^r e^{ax}$

19. $y' = \dfrac{e^x}{(1+e^x)^2}$ **21.** $y' = e^x \sin x + e^x \cos x$

23. $y' = e^x \sin x + xe^x \sin x + xe^x \cos x$ **25.** $y' = e^x \cos e^x$

27. $y' = \dfrac{e^{2x}}{\sqrt{1+e^{2x}}}$ **29.** $f^{(n)}(x) = ne^x + xe^x$

31. $f^{(n)}(x) = na^{n-1}e^{ax} + xa^n e^{ax}$ **33.**

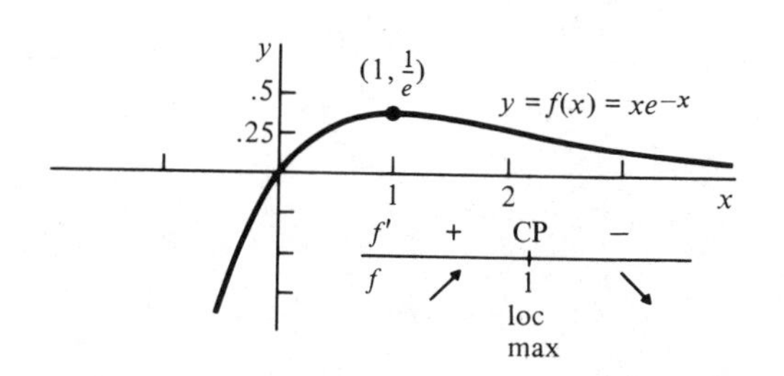

35. $y = 2e^2x - e^2$

37.

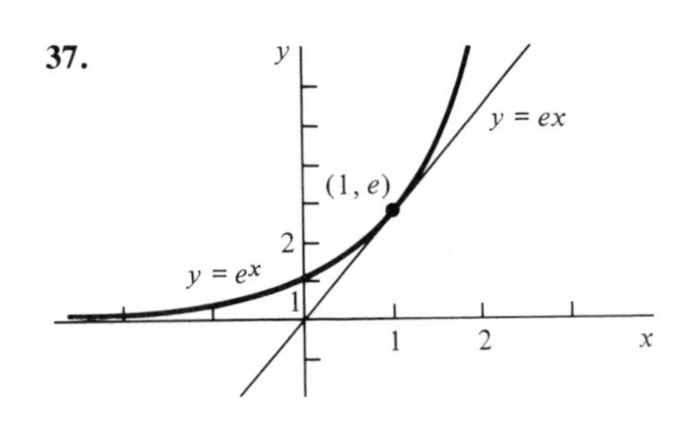

39.

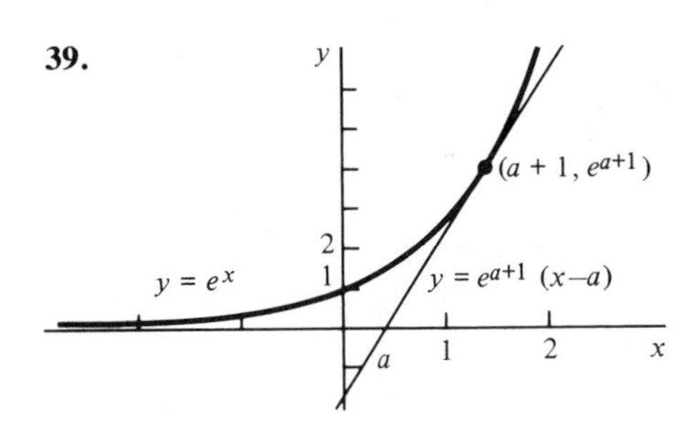

41. $-1/e^2$

45. $\frac{d}{dx}(Ae^{ax}\cos bx + Be^{ax}\sin bx) = (Aa + Bb)\,e^{ax}\cos bx + (Ba - Ab)\,e^{ax}\sin bx$

$$\int e^{ax}\cos bx\,dx = \frac{1}{a^2 + b^2}(ae^{ax}\cos bx + be^{ax}\sin bx) + C$$

$$\int e^{ax}\sin bx\,dx = \frac{1}{a^2 + b^2}(ae^{ax}\sin bx - be^{ax}\cos bx) + C$$

49. $y = 2\,e^{-x}$

51. $y = 3\,e^{-5(x+1)/2}$

59. $y = Ae^{-3x} + Bxe^{-3x}$

61. $y = Ae^{-x}\cos x + Be^{-x}\sin x$

63. $y = 7e^x\cos 2x - (7/2)e^x\sin 2x$

65. $y = e^{2x}\cos 2x - (1/2)e^{2x}\sin 2x$

Section 3.6

1. 1

3. $2^{3/2}$

5. 2

7. $x^{-3/2}$

9. ln 8/ln 7

11. $x = 2/3$

13. $x = -(\ln 75/\ln 15)$

15. $x = 3/2$

17. $3^{\sqrt{2}} = e^{\sqrt{2}\ln 3} \cong 4.7288$, $(\sqrt{2})^{\pi} = e^{(\pi\ln 2)/2} \cong 2.970686$, $\pi^e = e^{e\ln\pi} \cong 22.459$

19. $f'(x) = ba^{bx+c}\ln a$

21. $g'(x) = a^x x^a \ln a + a^{x+1}x^{a-1}$

23. $h'(t) = xt^{x-1} - x^t\ln x$

25. $y' = \dfrac{b}{(bx + c)\ln a}$

27. $y' = \dfrac{2\ln x}{x}x^{\ln x}$

29. $y' = (\sin x)^x(x\cot x + \ln\sin x)$

31. $y' = e^x x^{(e^x)}((1/x) + \ln x)$

33. $f'(x) = (x-1)(x-2)(x-3)(x-4)\left[\frac{1}{x-1} + \frac{1}{x-2} + \frac{1}{x-3} + \frac{1}{x-4}\right]$

35. $F'(0) = -(23/6)$ **37.** $y = 1$ **45.** $y = A|x| + \frac{B}{|x|}\ (x \neq 0)$

47. $y = A\cos\ln|x| + B\sin\ln|x|\ (x \neq 0)$

Section 3.7

1. $1000(2^{5/2}) \cong 5657$ **3.** $\frac{15\ln(0.5)}{\ln(0.7)} \cong 29.15$ years

5. $\frac{40\ln(0.1)}{\ln(0.5)} - 40 \cong 92.88$ min **7.** $20 + 52e^{5\ln(28/52)} \cong 22.35°\text{F}$

9. a) $\frac{5\ln 2}{\ln(3/2)} \cong 8.55$ years, b) $100(e^{[\ln(3/2)]/5} - 1) \cong 8.447\%$

11. $\frac{\ln(0.25)}{\ln(0.91)} \cong 14.7$ years **13.** $y_0 y_2(L - y_1)^2 = y_1^2(L - y_0)(L - y_2),\ L = 45/7$

15. solution valid for $t > t_0 = \frac{1}{kL}\ln\frac{y_0}{y_0 - L}$, and $\to\infty$ as $t \to t_0+$

Section 3.8

1. $\frac{d}{dx}\sinh x = \cosh x,\ \frac{d}{dx}\cosh x = \sinh x,\ \frac{d}{dx}\tanh x = \text{sech}^2 x,$

$\frac{d}{dx}\coth x = -\text{csch}^2 x,\ \frac{d}{dx}\text{sech}\, x = -\text{sech}\, x\tanh x,\ \frac{d}{dx}\text{csch}\, x = -\text{csch}\, x\coth x$

3. $\tanh(x \pm y) = \frac{\tanh x \pm \tanh y}{1 \pm \tanh x \tanh y}$

5. $\frac{d}{dx}\sinh^{-1}x = \frac{1}{\sqrt{1+x^2}},\ \frac{d}{dx}\text{Cosh}^{-1}x = \frac{1}{\sqrt{x^2-1}}\ (x > 1),\ \frac{d}{dx}\tanh^{-1}x = \frac{1}{1-x^2},$

$\int\frac{dx}{\sqrt{1+x^2}} = \sinh^{-1}x + C,\ \int\frac{dx}{\sqrt{x^2-1}} = \text{Cosh}^{-1}x + C,\ \int\frac{dx}{1-x^2} = \tanh^{-1}x + C$

7. a) $\frac{1}{2}\left(x - \frac{1}{x}\right)$, b) $\frac{1}{2}\left(x + \frac{1}{x}\right)$, c) $\frac{x^2-1}{x^2+1}$, d) x^2

11. $f_{A,B} = g_{A+B,\,A-B},\ g_{C,D} = f_{(C+D)/2,\,(C-D)/2}$

13. $y = y_o\cosh k(x-a) + (1/k)v_o\sinh k(x-a) = h_{y_o,\,v_o/k}(x)$

Review Exercises for Chapter 3

1. $5\cos 5x - 3\sin 3x$ **3.** $\tan 2x + 2x\sec^2 2x$ **5.** $\frac{-\sin x}{2\sqrt{2+\cos x}}$

7. $-14\,e^{-2x}$ **9.** 0 **11.** $-\tan x$

13. $\frac{\cos x}{\ln(1+x)} - \frac{\sin x}{(1+x)(\ln(1+x))^2}$ **15.** $3x^2(1-x)\,e^{-3x}$ **17.** $\frac{1}{\sqrt{9-x^2}}$

19. $\frac{2}{(1+4x^2)^2} - \frac{8x}{(1+4x^2)^2}\text{Arctan}\,2x$ **21.** $\frac{\cos x}{1+\sin^2 x}$

23. $-\sin x \operatorname{Arcsin} x + \dfrac{\cos x}{\sqrt{1-x^2}}$

25. $\dfrac{1}{\sqrt{1-x^2}\operatorname{Arctan} x} - \dfrac{\operatorname{Arcsin} x}{(1+x^2)(\operatorname{Arctan} x)^2}$

27. $x2^{x^2}\ln 2$

29. $(2x)^{(3x)}(3 + 3\ln(2x))$

31. $x2^{\sqrt{x}+1} + \dfrac{x^2+1}{\sqrt{x}}2^{\sqrt{x}-1}\ln 2$

33. $-2\dfrac{\sin 2x\cos 2x}{|\cos 2x|}$ $(\cos 2x \neq 0)$

35. $\dfrac{2x(x^2+y^2)-y}{2y(x^2+y^2)-x}$

37. $\dfrac{y^2\sec^2 x}{e^y - 2y\tan x}$

39. $\dfrac{1 - y[\tan(xy) + xy\sec^2(xy)]}{x[\tan(xy) + xy\sec^2(xy)] - 1}$

41. $\tanh x$

43. $\dfrac{-3}{x\sqrt{9+x^2}}$

45. $\dfrac{xe^x\ln x}{(x+1)\sin x}\left(\dfrac{1}{x} + 1 + \dfrac{1}{x\ln x} - \dfrac{1}{x+1} - \cot x\right)$

47. $(1+\sqrt{3})/2\sqrt{2}$

49. $\sqrt{\dfrac{1+\sqrt{2}}{2\sqrt{2}}}$

51. $-(\pi/4)$

53. $1/e^2$

55. 1

57. $\sqrt{2}\left|\sin\dfrac{x}{2}\right|$

59. $\dfrac{2x}{\sqrt{1+4x^2}}$

61. $f(x) = \dfrac{3}{2} - \dfrac{1}{2}\cos 2x$

63. $f(x) = 3\operatorname{Arctan} x - \dfrac{3\pi}{4}$

65. $f(x) = e^{-x} + 3x + 2$

67. $y = 4e^{2x}$

69. $y = 2\cos 2x - \dfrac{1}{2}\sin 2x$

71. $y'' = 2e^x\cos x$

73. $3^{89}\cos 3x$

75. $\dfrac{2(3t^2-1)}{(1+t^2)^3}$

CHAPTER 4 VARIOUS APPLICATIONS OF DIFFERENTIATION

Section 4.1

1. no extreme values

3. abs max 3 at $x = 1$, abs min 1 at $x = -1$

5. no max, abs min 1 at -1

7. no max, abs min -1 at $x = 0$

9. abs max 8 at $x = 3$, abs min -1 at $x = 0$

11. no max, abs min -1 at $x = 0$

13. abs max $f(b)$, abs min $f(a)$

15. no max, abs min $f(a)$

17. abs max 2 at $x = 1$, abs min -2 at $x = -1$

19. abs max 1 at $x = 0$, no min

21. no max, abs min 0 at $x = -2$

23.

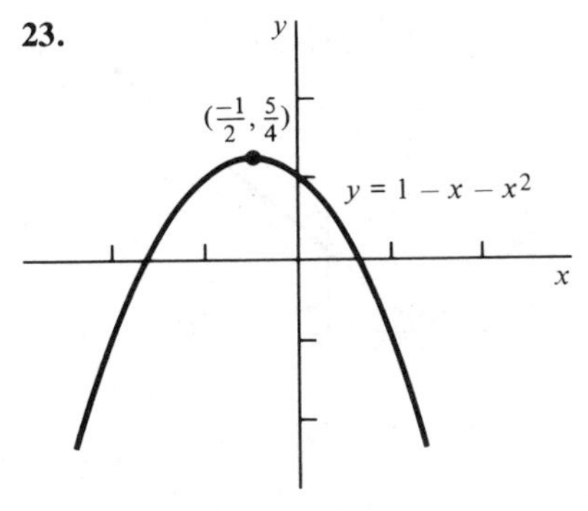

abs max 5/4 at $x = -1/2$

25.

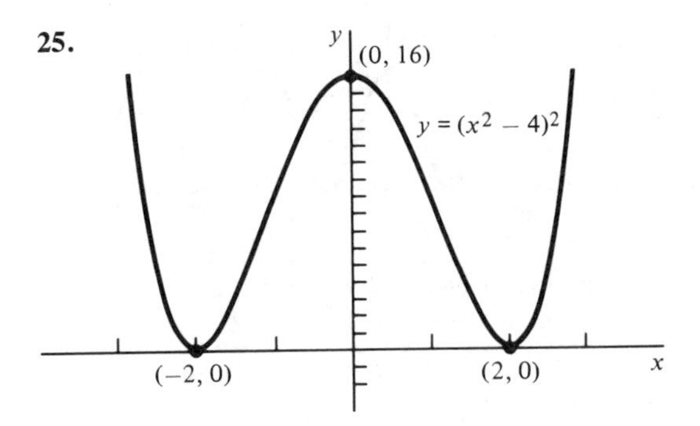

abs min 0 at $x = \pm 2$, loc max 16 at $x = 0$

27.

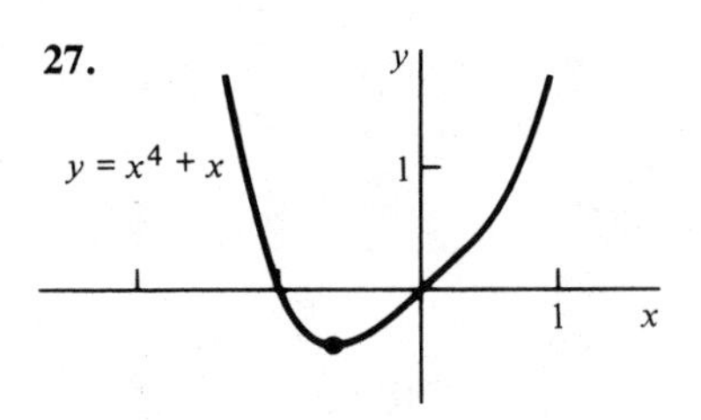

abs min $-\frac{3}{4^{4/3}}$ at $x = -\frac{1}{4^{1/3}}$

29.

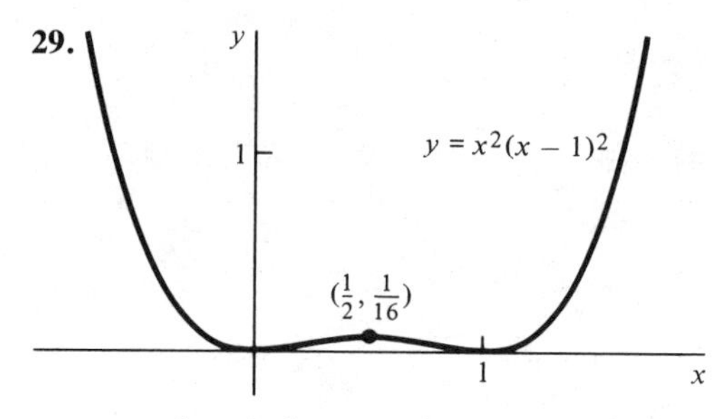

abs min 0 at $x = 0$ and $x = 1$
loc max 1/16 at $x = 1/2$

31.

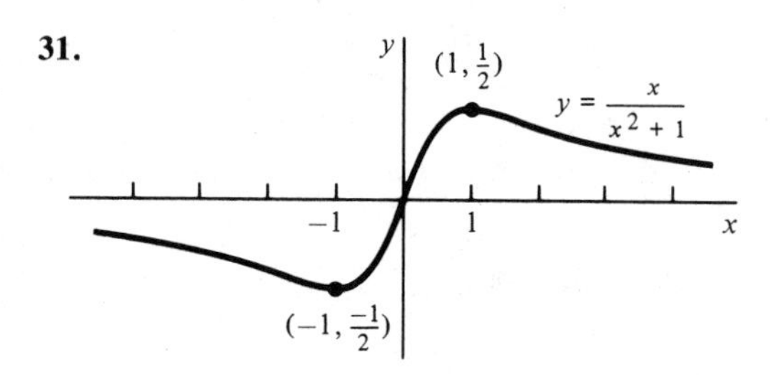

abs max 1/2 at $x = 1$, abs min $-1/2$ at $x = -1$

33.

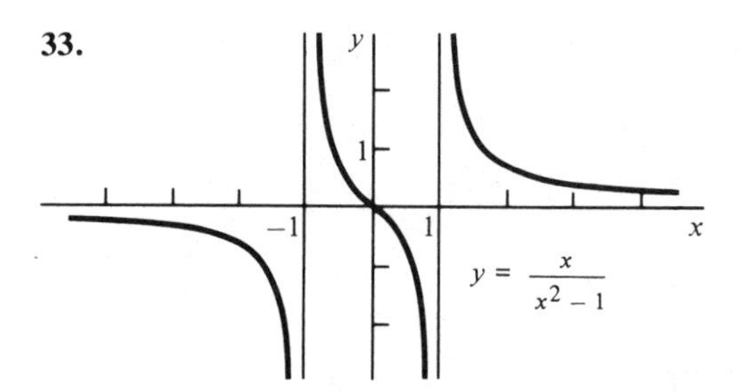

no extreme values

35.

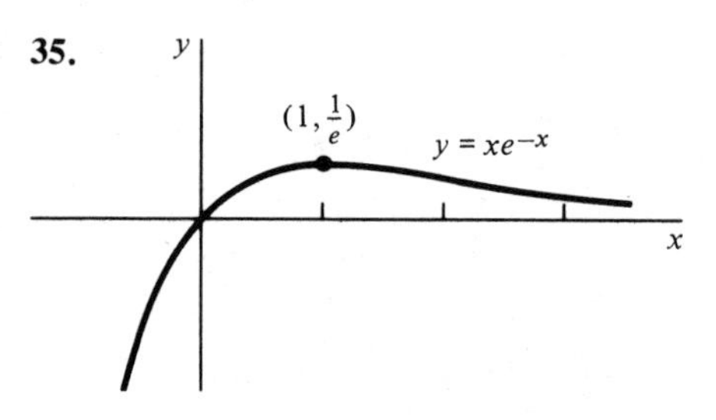

abs max $1/e$ at $x = 1$

37.

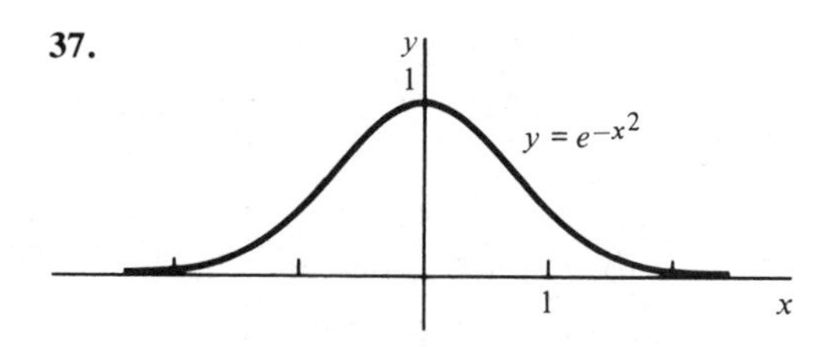

abs max 1 at $x = 0$

39.

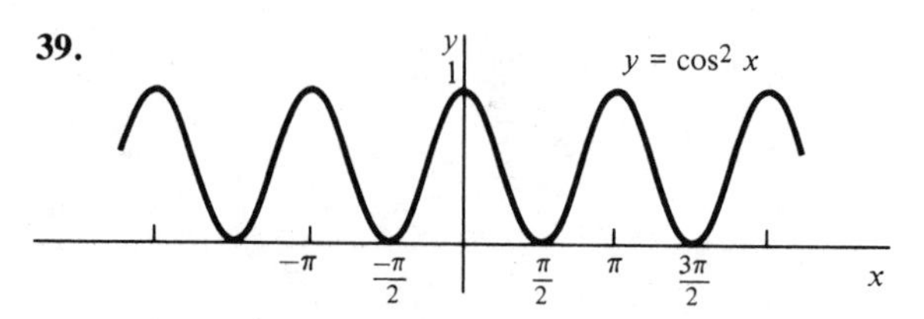

abs max 1 at $x = k\pi$
abs min 0 at $x = (2k + 1)\pi/2$
$k = 0, \pm 1, \pm 2, \ldots$

41.

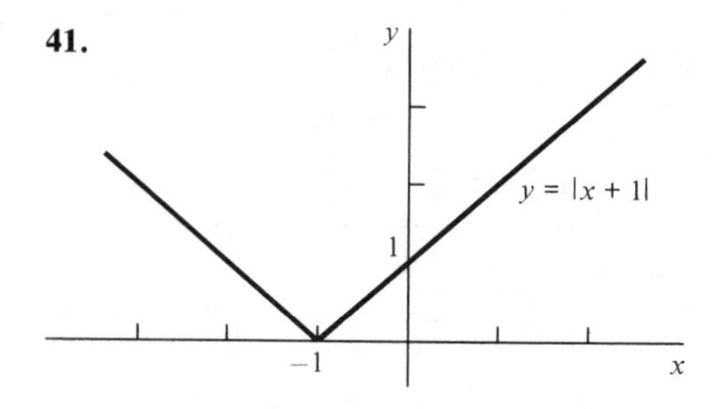

43.

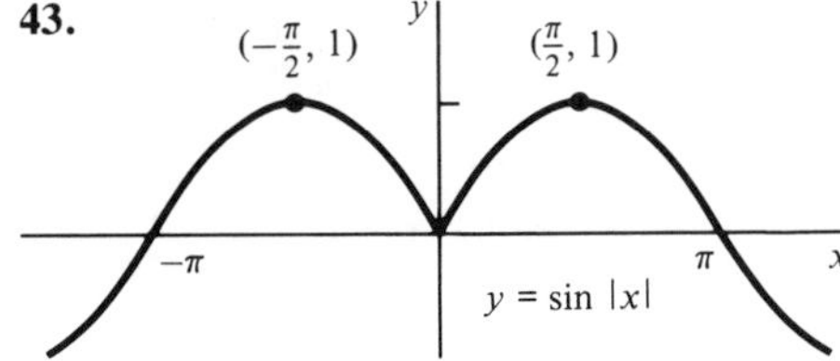

abs max 1 at $x = \pm(4k + 1)\pi/2$
abs min -1 at $x = \pm(4k + 3)\pi/2$
loc min 0 at $x = 0$
$k = 0, 1, 2, \ldots$

45.

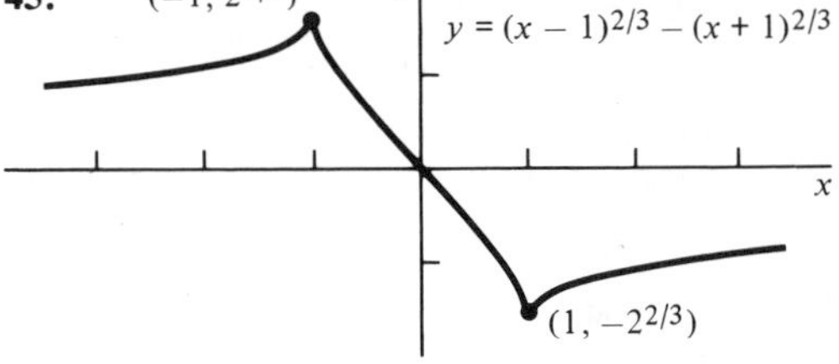

abs max $2^{2/3}$ at $x = -1$
abs min $-2^{2/3}$ at $x = 1$

47.

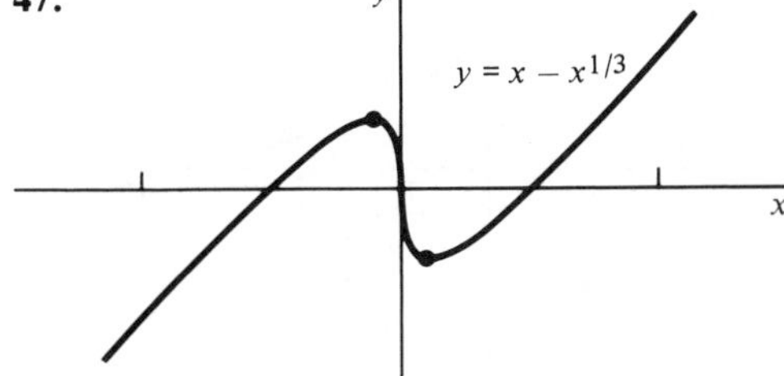

loc max at $-(1/3)^{3/2}$
loc min at $(1/3)^{3/2}$

Section 4.2

1. concave down on $(0, \infty)$

3. concave down on $(-\infty, \infty)$

5. concave up on $(-\infty, \infty)$

7. down on $(-\infty, 0)$, up on $(0, \infty)$, infl $x = 0$

9. concave down on $(-1, 0)$ and $(1, \infty)$, up on $(-\infty, -1)$ and $(0, 1)$, infl $x = 0, \pm 1$

11. concave up on $(0, \infty)$, down on $(-\infty, 0)$, inflection $x = 0$

13. concave down on $(2k\pi, (2k + 1)\pi)$, up on $((2k + 1)\pi, (2k + 2)\pi)$, infl $x = k\pi$, $k = 0, \pm 1, \ldots$

15. concave up on $(-\infty, 0)$, down on $(0, \infty)$, infl $x = 0$

17. concave up on $(-\infty, -\sqrt{3})$ and $(\sqrt{3}, \infty)$, down on $(-\sqrt{3}, \sqrt{3})$, infl at $x = \pm\sqrt{3}$

19. concave up on $(4, \infty)$, down on $(-\infty, 4)$, infl at $x = 4$

21. concave up on $(-\sqrt{3/2}, 0)$ and $(\sqrt{3/2}, \infty)$, down on $(-\infty, -\sqrt{3/2})$ and $(0, \sqrt{3/2})$, infl $0, \pm\sqrt{3/2}$

23. concave up on $\left(\frac{1}{3}, \infty\right)$, down on $\left(-\infty, \frac{1}{3}\right)$, inflection at $x = \frac{1}{3}$

25. concave up on $\left(-\infty, -\frac{1}{\sqrt{3}}\right)$ and $\left(\frac{1}{\sqrt{3}}, \infty\right)$, down on $\left(-\frac{1}{\sqrt{3}}, \frac{1}{\sqrt{3}}\right)$, inflections at $x = \pm\frac{1}{\sqrt{3}}$

27. concave up on $(-1, 0)$ and $(1, \infty)$, down on $(-\infty, -1)$ and $(0, 1)$, infl $x = 0$

29. concave up on $(-1, 1)$, down on $(-\infty, -1)$ and $(1, \infty)$, infl at $x = \pm 1$

31. no concavity, no inflections

33. loc min at $x = 2$, loc max at $x = 2/3$

35. loc min at $x = 1/2$, infl at $x = 2$

37. loc max at $x = -3^{-1/4}$, loc min at $x = 3^{-1/4}$

39. loc min at $x = -1$, loc max at $x = 1$ (both absolute)

41. loc (abs) min at $x = 1/e$

43. loc (abs) min at $x = 0$, inflections $x = \pm 2$

45. loc (abs) min at $x = 0$, loc (abs) max at $x = \pm 1$

Section 4.3

1.

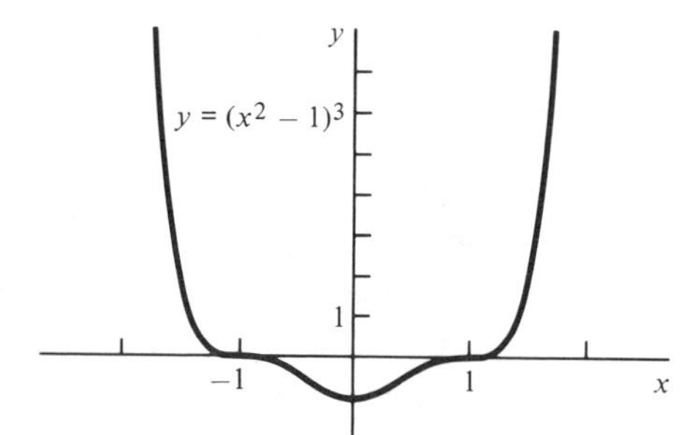

3.

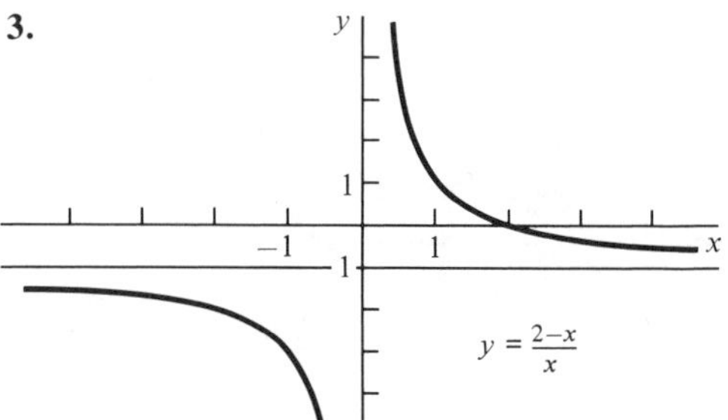

5.

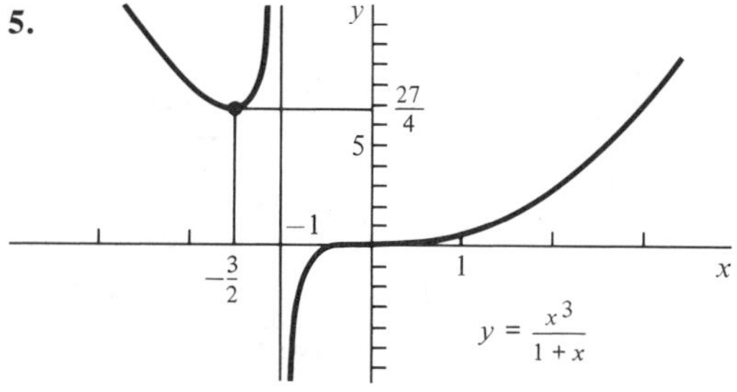

7.

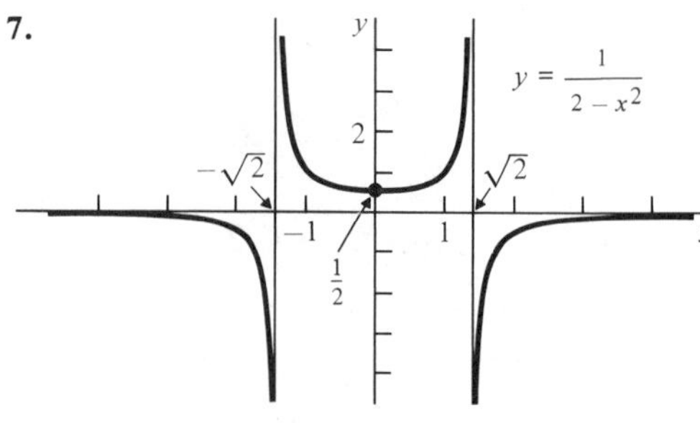

9.

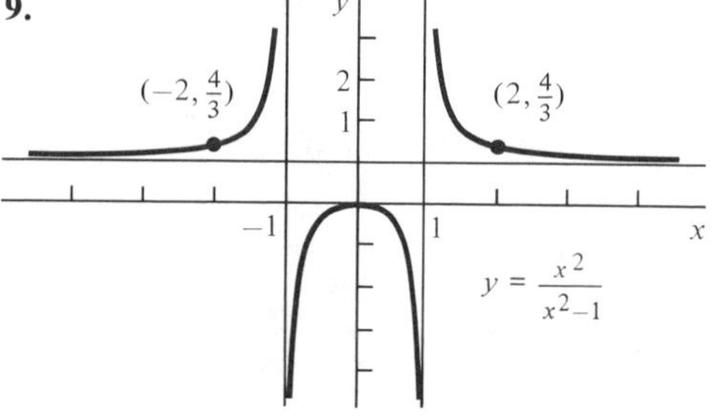

11.

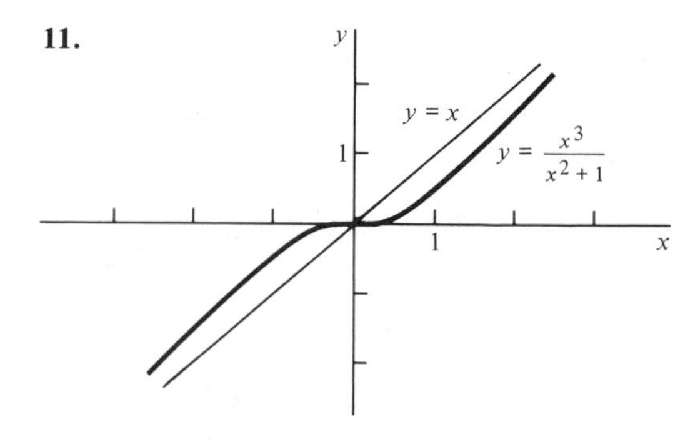

13.

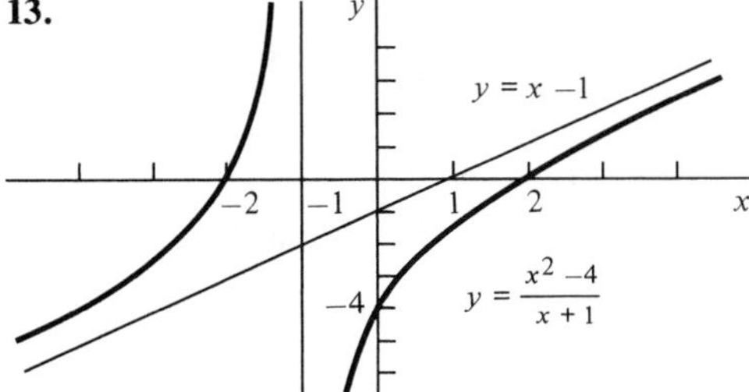

15.

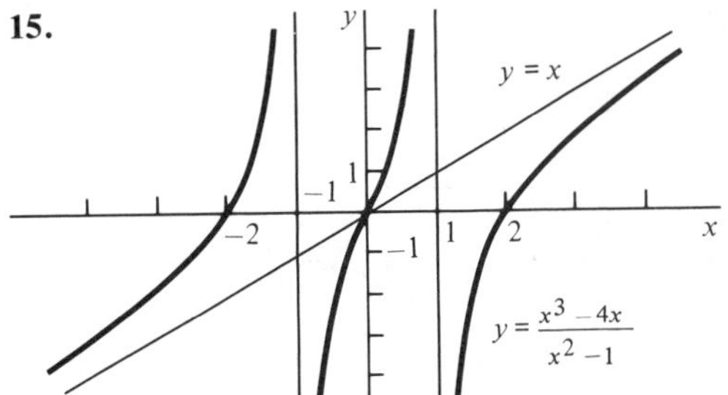

17.

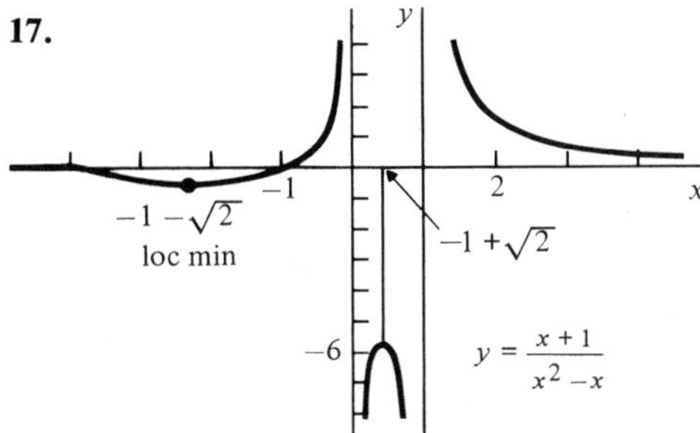

19.

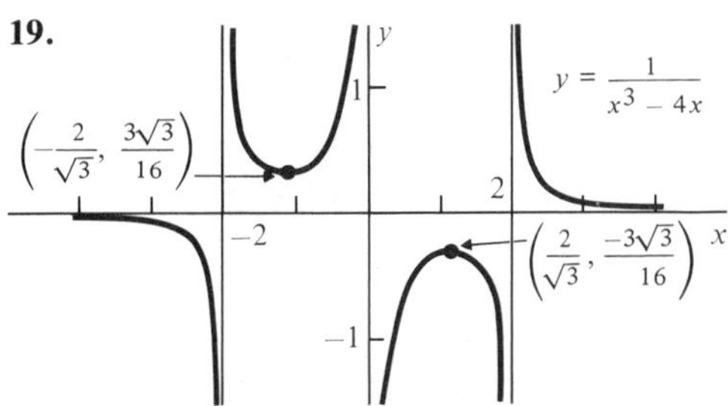

21.

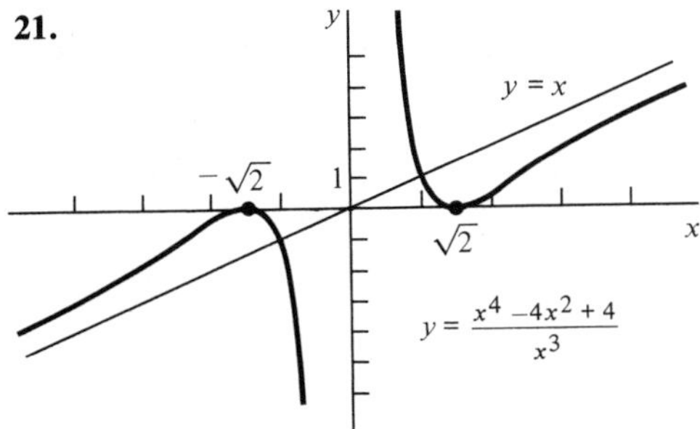

23.

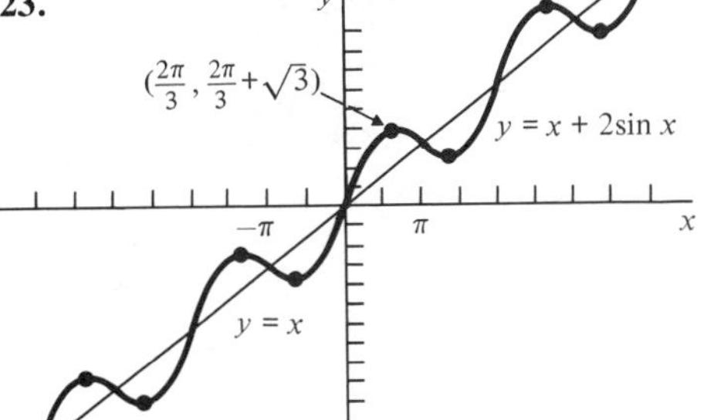

25.

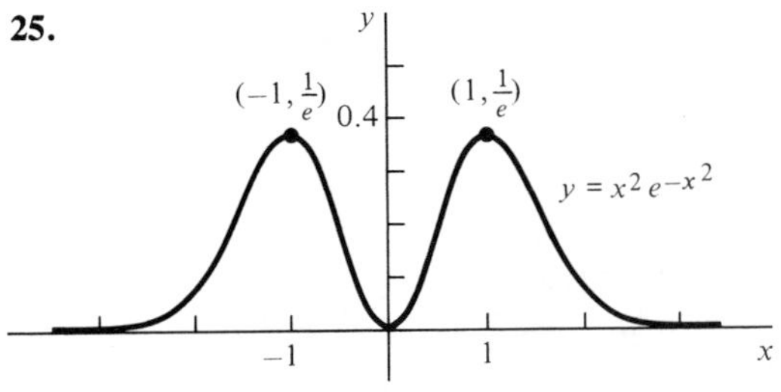

27.

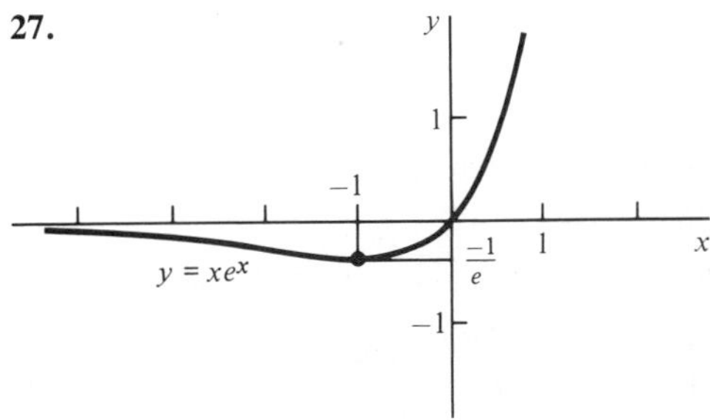

29.

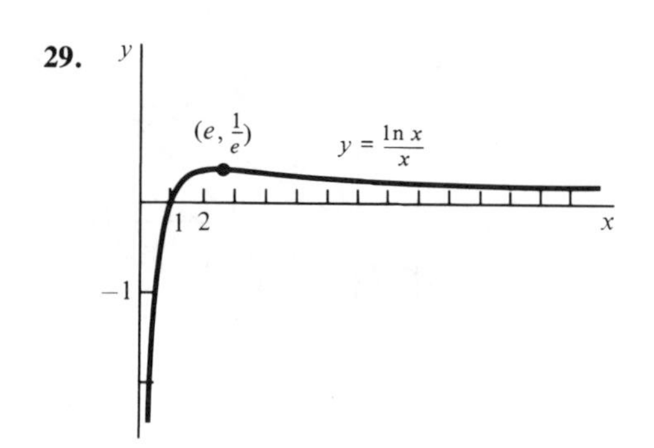

31.

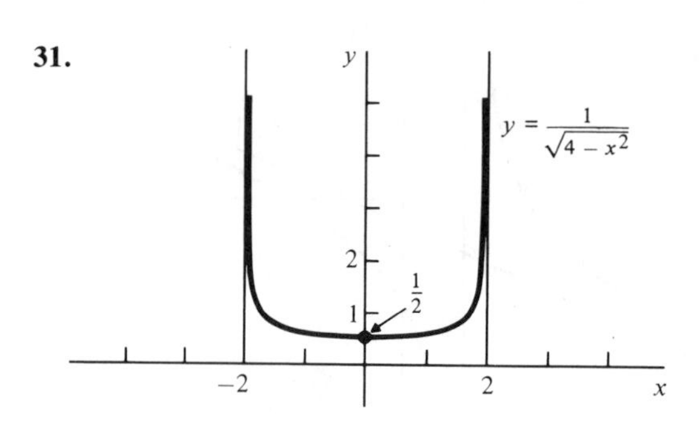

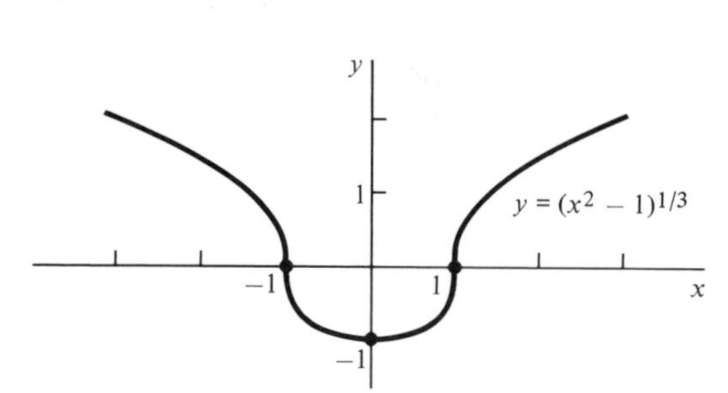

35.

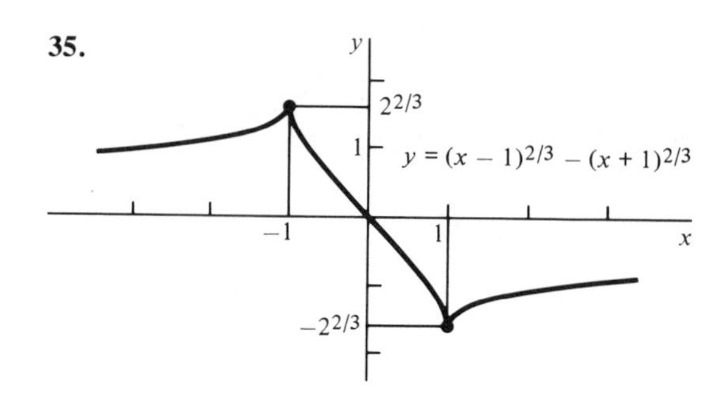

Section 4.4

1. 49/4

3. 20, 40

7. R^2 sq units

9. $2ab$ sq units

11. $8 + 10\sqrt{2}$ m, $4 + 5\sqrt{2}$ m

13. a) 0 m b) $\dfrac{4}{4 + \pi}$ m

15. $(a^{2/3} + b^{2/3})^{3/2}$

17. $|C|/\sqrt{A^2 + B^2}$

19. $2\sqrt{11}$ units

21. $\dfrac{8}{3}\pi R^3$ cu units

23. height = radius

25. radius $= \dfrac{\pi}{2\sqrt{3}} \times$ height

27. width $= \dfrac{20}{4 + \pi}$ m, height $= \dfrac{10}{4 + \pi}$ m

29. $Q = \dfrac{3L}{8}$

31. point 5 km east of A

35. radius $= 2R/3$, height $= H/3$

37. $2\sqrt{6}$ ft

Section 4.5

1. 160π cm²/sec

3. -4 m/sec

5. inc. at 2 cm³/sec

7. inc. at rate 12

9. inc. at rate $2/\sqrt{5}$

11. 4/9 m/sec

13. inc. at $828/\sqrt{1289} \cong 23.06$ mph

15. inc. at $2kV/\pi a^3$ cm/sec

17. inc. at $1/18\pi$ m/min

19. inc. at $36/25000\pi$ m/min, $100^{1/3} \cong 4.64$ m

21. 12π m/min

23. 13/2 m/min

25. $\sqrt{3}/16$ cm/min

27. 25/104 ft/sec

29. dec. at $(2\sqrt{3} - 1)/125 \cong 0.0197$ rad/sec

31. inc. at $4/85 \cong 0.047$ rad/sec

33. $\dfrac{1}{10} \ln 2 \cong 0.0693$ rad/min

Section 4.6

1. $\sqrt{50} \cong 99/14,\ E < 0,\ |E| < 1/2744$

3. $(85)^{1/4} \cong 82/27,\ E < 0,\ |E| < 1/(2 \times 3^6)$

5. $e^{-1/10} \cong 9/10,\ E > 0,\ |E| < 1/200$

7. $\cos 46° \cong \dfrac{180 - \pi}{180\sqrt{2}} \cong 0.694765,\quad E < 0,\quad |E| < \dfrac{1}{2\sqrt{2}}\left(\dfrac{\pi}{180}\right)^2 < 0.00011$

9. $\text{Arctan } 1.05 \cong \dfrac{10\pi + 1}{40} \simeq 0.810398,\quad E < 0,\quad |E| < \dfrac{1}{1600} = 0.000625$

11. $\ln (0.94) \cong -0.06,\quad E < 0,\quad |E| < \dfrac{(0.06)^2}{2(0.94)^2} < 0.0021$

13. 6%

15. 3%

17. 1.259921

19. $0.732051,\ -1,\ -2.732051$

21. 0.636733

23. 0.567143

25. 4.4934095, 7.7252518, 10.904122

31. a) $\sqrt{50} \cong 7 + \dfrac{1}{14} - \dfrac{1}{2744} \cong 7.0710641,\quad E > 0,\quad |E| < \dfrac{1}{268912} < 0.000004$

b) $e^{-1/10} \cong 1 - \dfrac{1}{10} + \dfrac{1}{200} = 0.905,\quad E < 0,\quad |E| < \dfrac{1}{6000}$

Section 4.7

1. 3/4

3. a/b

5. 1

7. 1

9. 0

11. $-(3/2)$

13. ∞

15. $-(1/2)$

17. ∞

19. $(2/\pi)$

21. -2

23. 0

25. a

27. 1

29. $-(1/2)$

31. $1/e^2$

33. 0

35. $f''(x)$

CHAPTER 5 CURVES IN THE PLANE

Section 5.1

1.

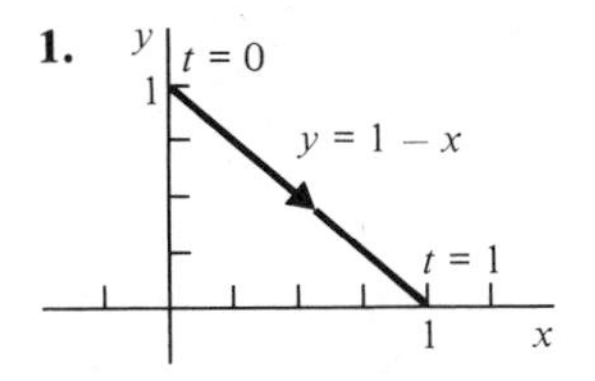

3.

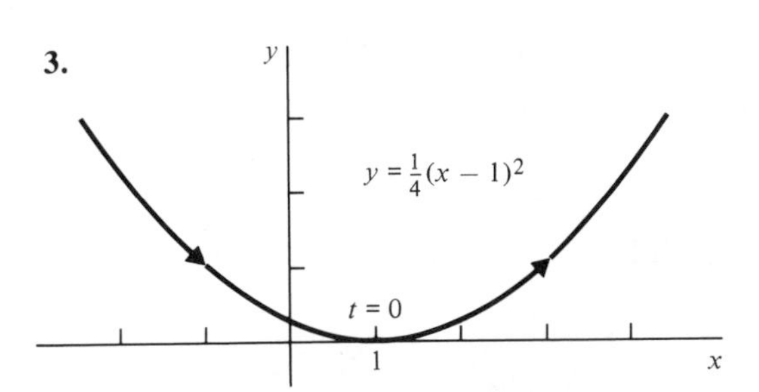

5.

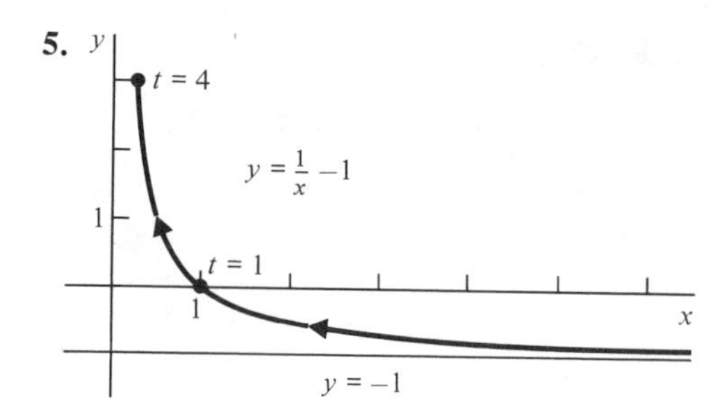

7.

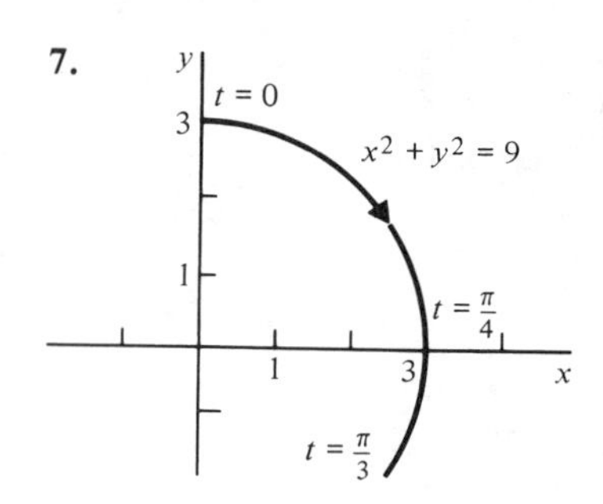

9.

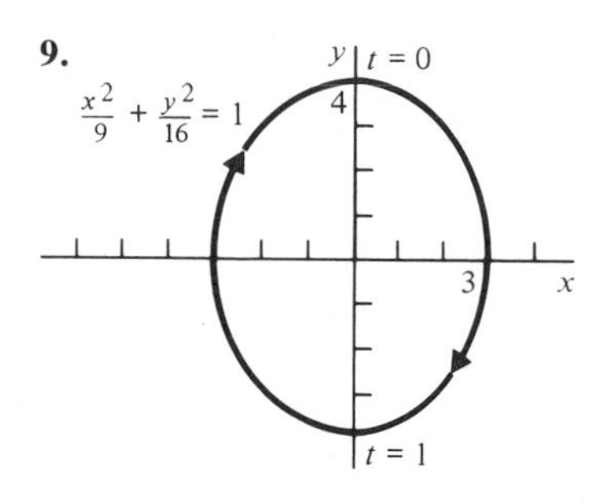

11.

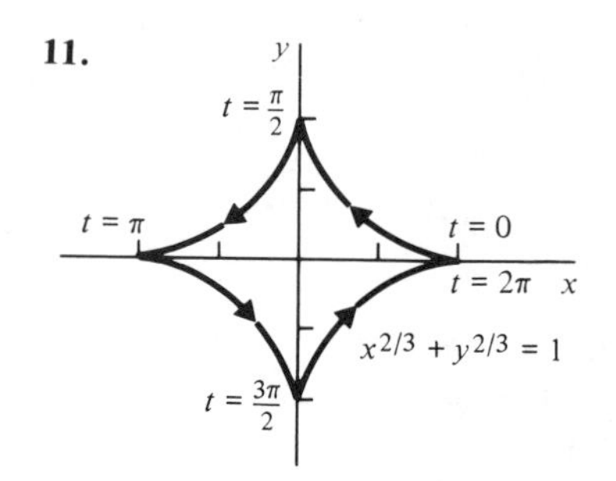

13. $x = m/2$, $y = m^2/4$ (m = slope)

15.

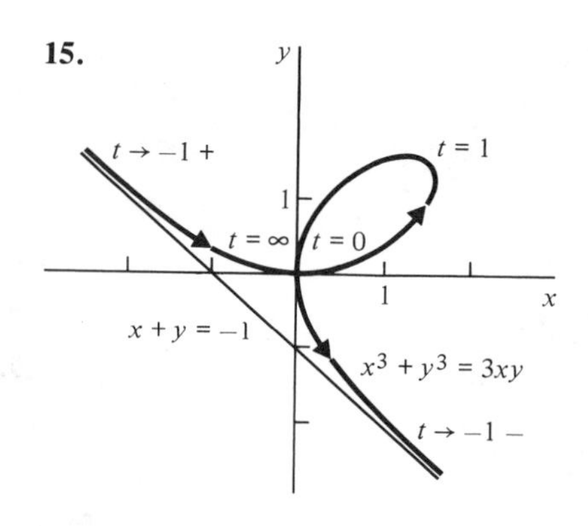

Section 5.2

1.

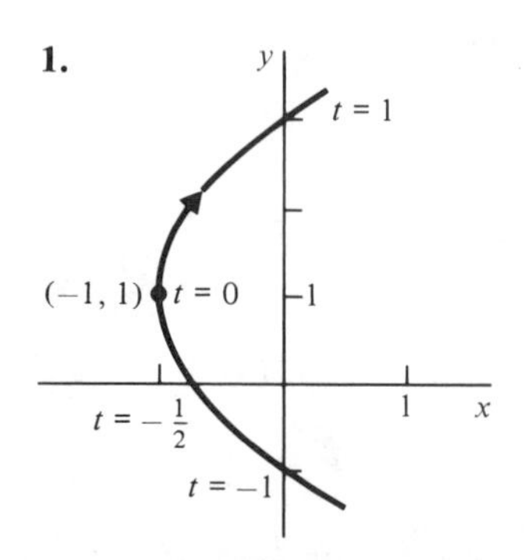

3.

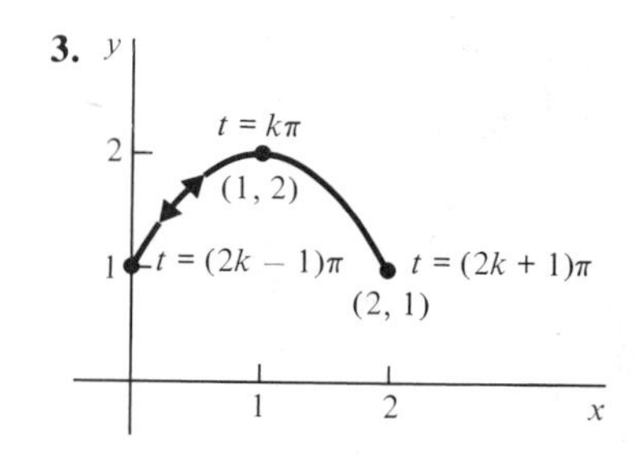

5.

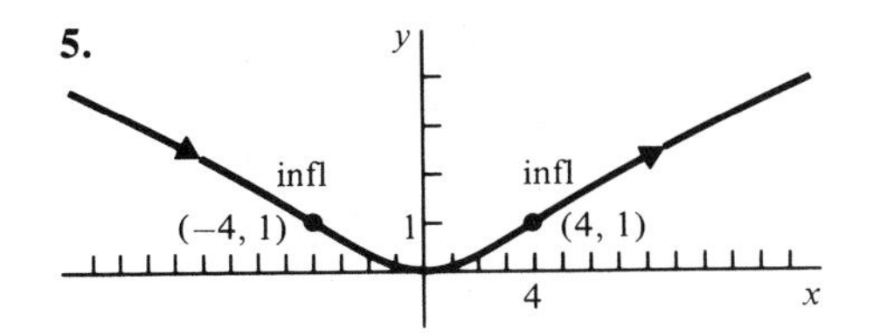

7.

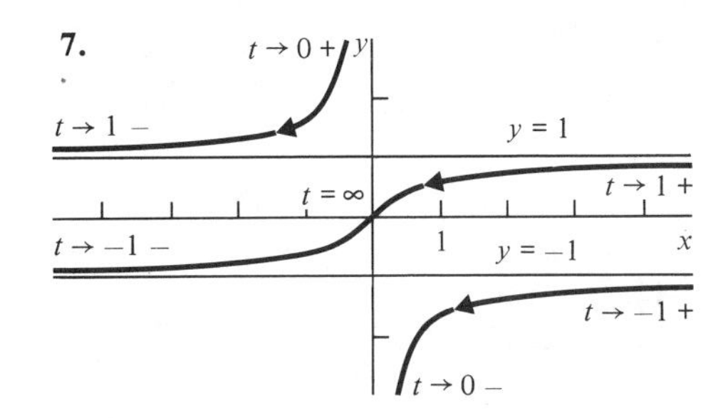

9.

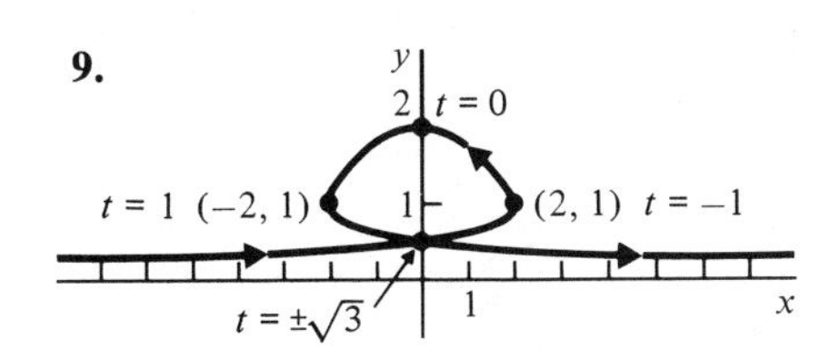

11.

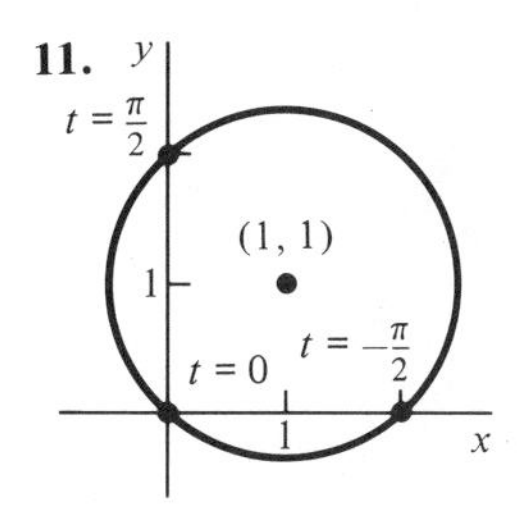

13.

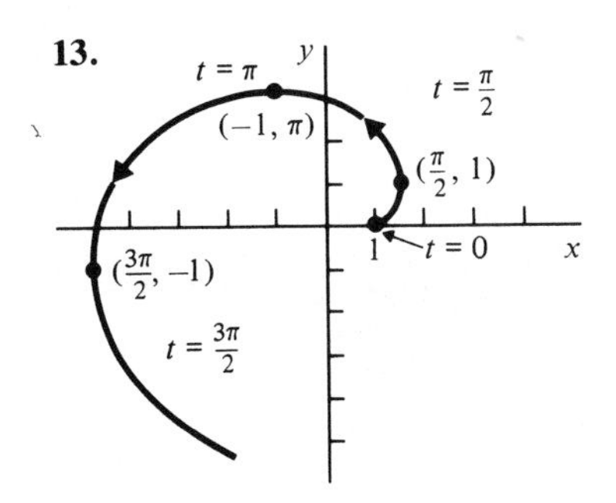

Section 5.3

1. $\overrightarrow{AB} = 3\mathbf{i} - 2\mathbf{j}$, $\overrightarrow{BA} = -3\mathbf{i} + 2\mathbf{j}$, $\overrightarrow{AC} = 2\mathbf{i} - 5\mathbf{j}$
   $\overrightarrow{BD} = -2\mathbf{i} - 4\mathbf{j}$, $\overrightarrow{DA} = -\mathbf{i} + 6\mathbf{j}$, $\overrightarrow{AB} - \overrightarrow{BC} = 4\mathbf{i} + \mathbf{j}$
   $\overrightarrow{AC} - \overrightarrow{2AB} + \overrightarrow{3CD} = -7\mathbf{i} - 4\mathbf{j}$, $\frac{1}{3}(\overrightarrow{AB} + \overrightarrow{AC} + \overrightarrow{AD}) = 2\mathbf{i} - \frac{13}{3}\mathbf{j}$

3. $\mathbf{U} + \mathbf{V} = 2\mathbf{i}$, $\mathbf{U} - \mathbf{V} = 2\mathbf{j}$, $|\mathbf{U}| = |\mathbf{V}| = \sqrt{2}$, $\mathbf{U} \cdot \mathbf{V} = 0$, angle $= \pi/2$

5. $\mathbf{U} + \mathbf{V} = \mathbf{j}$, $\mathbf{U} - \mathbf{V} = 2\mathbf{i} - 5\mathbf{j}$, $|\mathbf{U}| = \sqrt{5}$, $|\mathbf{V}| = \sqrt{10}$, $\mathbf{U} \cdot \mathbf{V} = -7$, angle $= \text{Arccos}\,\dfrac{-7}{5\sqrt{2}} \cong 171.9°$

7. $\mathbf{U} + \mathbf{V} = 3\mathbf{i} + \mathbf{j}$, $\mathbf{U} - \mathbf{V} = -\mathbf{i} + 3\mathbf{j}$, $|\mathbf{U}| = |\mathbf{V}| = \sqrt{5}$, $\mathbf{U} \cdot \mathbf{V} = 0$, angle $= \dfrac{\pi}{2}$

9. $\mathbf{U} + \mathbf{V} = (1 + \sqrt{3})\mathbf{i} - (1 + \sqrt{3})\mathbf{j}$, $\mathbf{U} - \mathbf{V} = (1 - \sqrt{3})\mathbf{i} + (1 - \sqrt{3})\mathbf{j}$, $|\mathbf{U}| = |\mathbf{V}| = 2$, $\mathbf{U} \cdot \mathbf{V} = 2\sqrt{3}$,
   angle $= \text{Arccos}\,\dfrac{2\sqrt{3}}{3} = \dfrac{\pi}{6}$

11. $\mathbf{U} \pm \mathbf{V} = (a \pm c)\mathbf{i} + (b \pm d)\mathbf{j}$, $|\mathbf{U}| = \sqrt{a^2 + b^2}$, $|\mathbf{V}| = \sqrt{c^2 + d^2}$, $\mathbf{U} \cdot \mathbf{V} = ac + bd$,

$$\text{angle} = \text{Arccos}\,\frac{ac + bd}{\sqrt{(a^2 + b^2)(c^2 + d^2)}}$$

17.

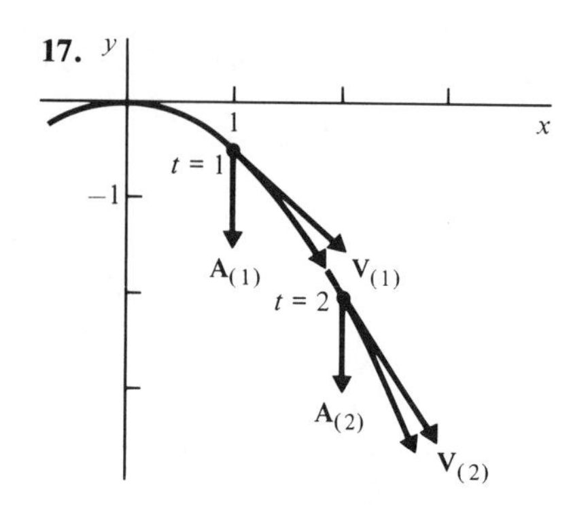

19.

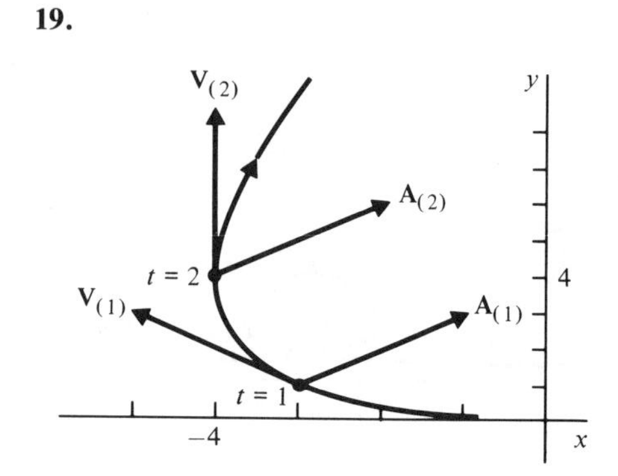

21.

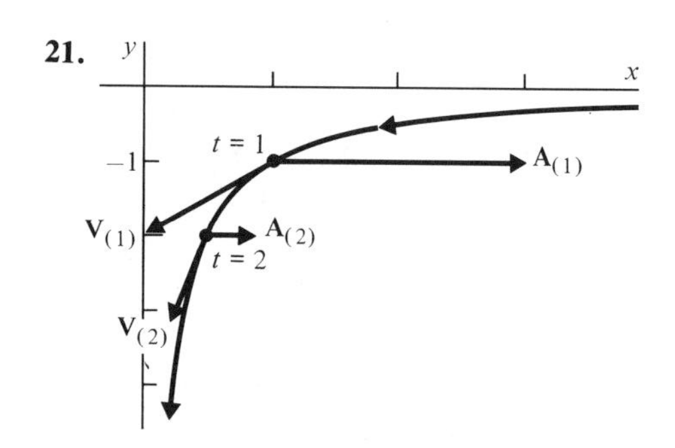

23.

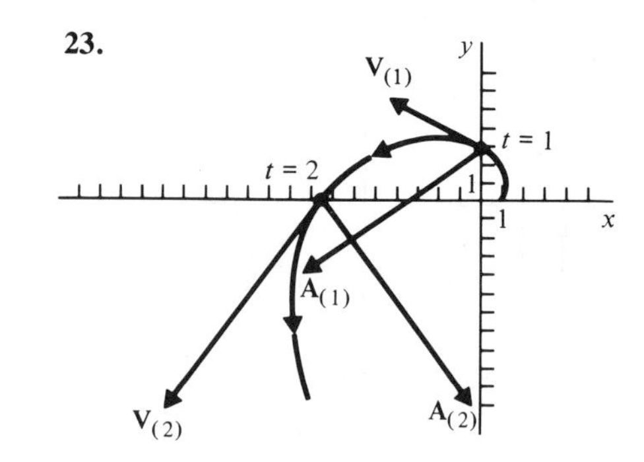

25. $\mathbf{R} = (t^2 + t)\mathbf{i} - \left(\frac{1}{2}t^2 - t + 3\right)\mathbf{j}$

27. $\mathbf{R} = (t^3 + 2t)\mathbf{i} - \frac{1}{2}t^4\mathbf{j}$

29. $\mathbf{A} = a \sin t\, \mathbf{i} + a \cos t\, \mathbf{j}$. $|\mathbf{A}| = |a|$

Section 5.4

1. $x = 3$, vertical line

3. $3y - 4x = 5$, straight line

5. $2xy = 1$, hyperbola

7. $y = x^2 - x$, parabola

9. $y^2 = 2x + 1$, parabola

11. $x^2 - 3y^2 - 8y = 4$, hyperbola

13.

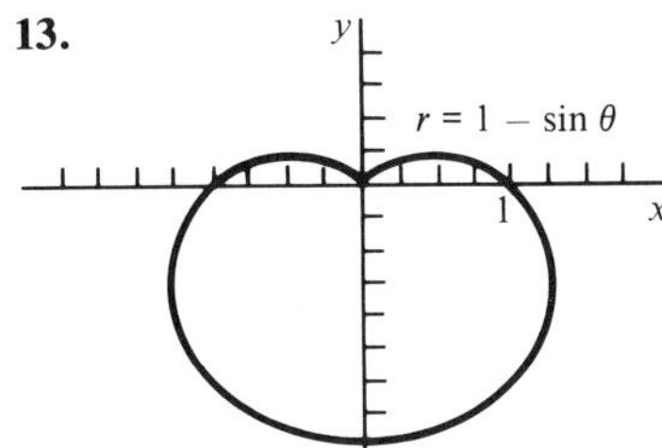

15.

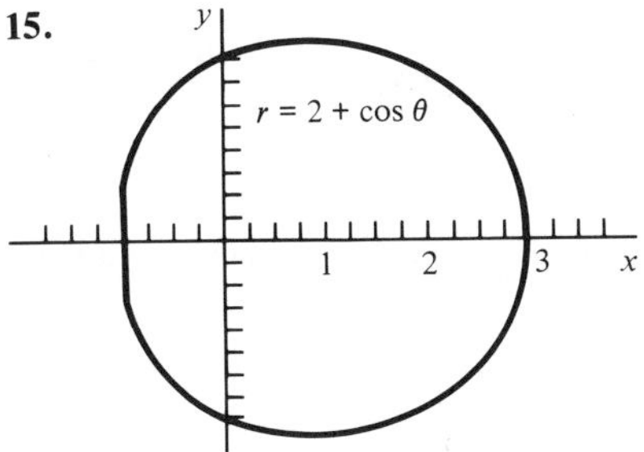

17.

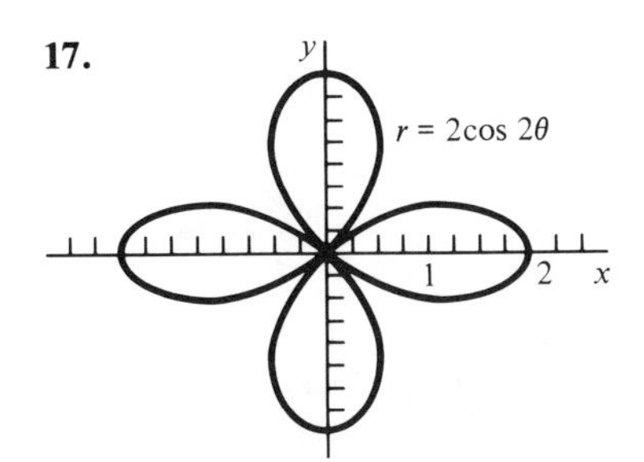

19.

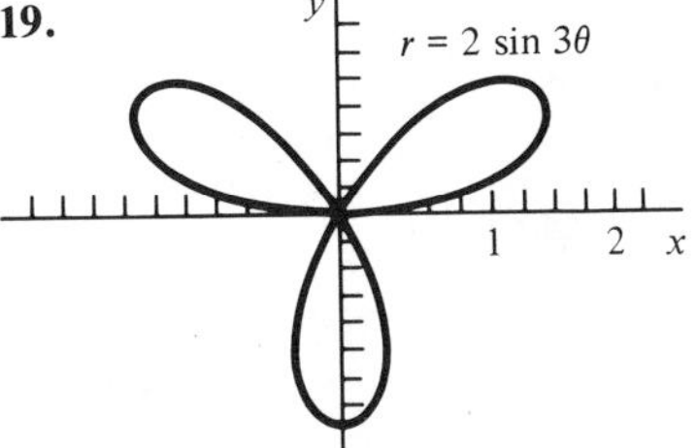

21.

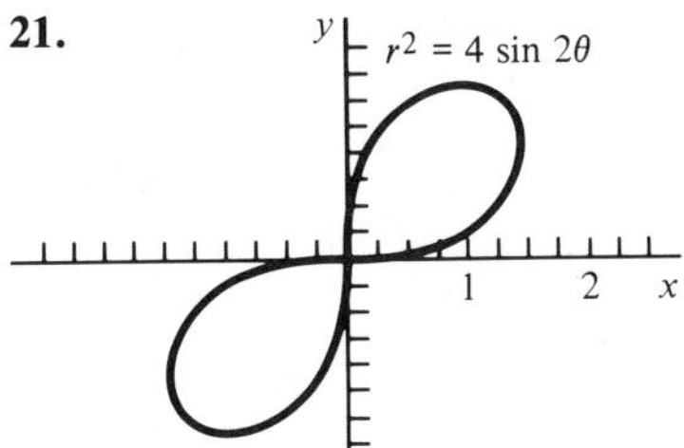

23.

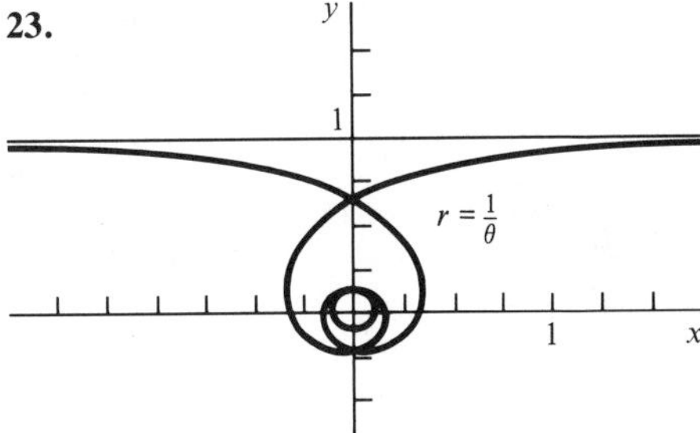

25.

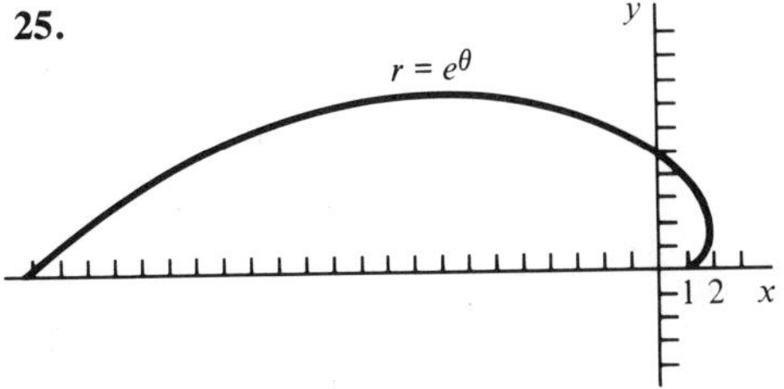

27. horizontal at $\left(\cos\frac{\pi}{8} - \sin\frac{\pi}{8}, -\frac{\pi}{8}\right)$ and $\left(\cos\frac{3\pi}{8} + \sin\frac{3\pi}{8}, \frac{3\pi}{8}\right)$, vertical at $\left(\cos\frac{\pi}{8} + \sin\frac{\pi}{8}, \frac{\pi}{8}\right)$ and $\left(\cos\frac{5\pi}{8} + \sin\frac{5\pi}{8}, \frac{5\pi}{8}\right)$

29. horizontal at $\left(\frac{1}{\sqrt{2}}, \pm\frac{\pi}{6}\right)$ and $\left(\frac{1}{\sqrt{2}}, \pm\frac{5\pi}{6}\right)$, vertical at $(1, 0)$ and $(1, \pi)$

31. horizontal at $\left(e^{k\pi-(\pi/4)}, k\pi - \frac{\pi}{4}\right)$, vertical at $\left(e^{k\pi+(\pi/4)}, k\pi + \frac{\pi}{4}\right)$, $k = 0, \pm 1, \pm 2, \ldots$

33. angle $\pi/4$ at $(\sqrt{2}, \pi/4)$, angles 0 and $\pi/2$ at the origin

CHAPTER 6 INTEGRATION

Section 6.1

1. 3/2 sq units

3. 26/3 sq units

5. $3/(2 \ln 2)$

7. 15 sq units

9. b) $(1/4)a^4$ sq units

11. $\pi/4$

Section 6.2

1. $U_8 = \frac{51}{16}, L_8 = \frac{35}{16}$

3. $U_4 = \frac{e^4 - 1}{e(e-1)}, L_4 = \frac{e^4 - 1}{e^2(e-1)}$

5. $U_6 = \frac{\pi(3 + \sqrt{3})}{6}, L_6 = \frac{\pi(1 + \sqrt{3})}{6}$

7. $\frac{b^3}{3}$

9. $\int_a^b e^x\,dx = e^b - e^a$

11. 8

13. $\pi/2$

15. a^2

17. $\pi/2$

19. $\pi/2$

Section 6.3

1. $\frac{5}{12}$

3. $\frac{1}{\sqrt{2}} - \frac{1}{2}$

5. $\frac{\pi}{3}$

7. $\sqrt{2} - 1$

9. $\frac{32}{3}$

11. $-\ln 2$

13. $\frac{3}{2 \ln 2}$

15. $\frac{1}{5}$ sq units

17. $\frac{1}{12}$ sq units

19. $\frac{9}{2}$ sq units

21. $e^a - 1$ sq units

23. $\frac{1}{3}$ sq units

25. 12 sq units

27. $\frac{1}{12}(e^6 - e^{-6})$

29. $1 + \frac{2}{\pi}$

31. $\frac{2}{3}$

33. $-\frac{2 \sin x^2}{x}$

35. $-2\sqrt{1 + 2t}$

37. $-\csc\theta - \sec\theta$

39. 4

41. $\frac{21}{2}$

43. $\frac{\pi}{4}$

Section 6.4

1. $-\frac{1}{2}e^{5-2x} + C$

3. $\frac{2}{9}(3x+4)^{3/2} + C$

5. $-\frac{2}{9}(x^3+2)^{-3/2} + C$

7. $\frac{-1}{32(4x^2+1)^4} + C$

9. $\cos(\cos x) + C$

11. $\frac{1}{2}e^{x^2} + C$

13. $\frac{1}{2}\text{Arctan}\left(\frac{1}{2}\sin x\right) + C$

15. $2\ln|e^{x/2} - e^{-x/2}| + C = \ln|e^x + e^{-x} - 2| + C$

17. $-\frac{2}{5}\sqrt{4-5s} + C$

19. $\frac{1}{2}\text{Arcsin}\frac{t^2}{2} + C$

21. $-\ln(1+e^{-x}) + C$

23. $-\frac{1}{2}(\ln\cos x)^2 + C$

25. $\ln(t^2+9) + \text{Arctan}\frac{t}{3} + C$

27. $\frac{1}{2}\text{Arctan}\frac{x+3}{2} + C$

29. $\frac{1}{8}\cos^8 x - \frac{1}{6}\cos^6 x + C$

31. $-\frac{1}{3a}\cos^3 ax + C$

33. $\frac{x}{8} - \frac{\sin 4x}{32} + C$

35. $\frac{5x}{16} - \frac{\sin 2x}{4} + \frac{3\sin 4x}{64} + \frac{\sin^3 2x}{48} + C$

37. $\frac{1}{5}\sec^5 x + C$

39. $\frac{2}{3}(\tan x)^{3/2} + \frac{2}{7}(\tan x)^{7/2} + C$

41. $\frac{3}{8}\sin x - \frac{\sin(2\sin x)}{4} + \frac{\sin(4\sin x)}{32} + C$

43. $\frac{1}{3}\tan^3 x + C$

45. $\frac{2}{7}\csc^7 x - \frac{1}{9}\csc^9 x - \frac{1}{5}\csc^5 x + C$

47. $\frac{14\sqrt{17}+2}{3}$

49. $\frac{3\pi}{16}$

51. $\ln 2$

53. a) 2 b) $2(\sqrt{2}-1)$

55. $\frac{\pi}{32}$ sq units

57. $$\int \cos ax \cos bx\,dx = \frac{1}{2}\left(\frac{\sin(a-b)x}{a-b} + \frac{\sin(a+b)x}{a+b}\right) + C$$
$$\int \sin ax \sin bx\,dx = \frac{1}{2}\left(\frac{\sin(a-b)x}{a-b} - \frac{\sin(a+b)x}{a+b}\right) + C$$
$$\int \sin ax \cos bx\,dx = -\frac{1}{2}\left(\frac{\cos(a-b)x}{a-b} + \frac{\cos(a+b)x}{a+b}\right) + C$$

Section 6.5

1. $\frac{1}{2}\text{Arcsin}\,2x + C$

3. $-\frac{1}{4}\sqrt{1-4x^2} + C$

5. $-\sqrt{9-x^2} + \text{Arcsin}\frac{x}{3} + C$

7. $\frac{9}{2}\text{Arcsin}\frac{x}{3} - \frac{x}{2}\sqrt{9-x^2} + C$

9. $\dfrac{x}{a^2\sqrt{a^2 - x^2}} + C$

11. $\dfrac{x}{a^2\sqrt{a^2 + x^2}} + C$

13. $-\dfrac{1}{x}\sqrt{1 - x^2} - \text{Arcsin } x + C$

15. $\dfrac{1}{4}\ln\left|\dfrac{2 + x}{2 - x}\right| + C$

17. $\dfrac{1}{a}\text{Arcsec}\dfrac{x}{a} + C$

19. $\dfrac{1}{3}\text{Arctan}\dfrac{x + 1}{3} + C$

21. $\dfrac{1}{32}\left(\text{Arctan}\dfrac{2x + 1}{2} + \dfrac{4x + 2}{4x^2 + 4x + 5}\right) + C$

23. $-\sqrt{2ax - x^2} + a\,\text{Arcsin}\dfrac{x - a}{a} + C$

25. $\dfrac{1}{4}\dfrac{x + 1}{\sqrt{3 - 2x - x^2}} + C$

27. $\dfrac{3}{8}\text{Arctan } x + \dfrac{5x + 3x^3}{8(1 + x^2)^2} + C$

29. $2\sqrt{x} - 4\ln(2 + \sqrt{x}) + C$

31. $\dfrac{6}{7}x^{7/6} - \dfrac{6}{5}x^{5/6} + \dfrac{3}{2}x^{2/3} + 2x^{1/2} - 3x^{1/3} - 6x^{1/6} + 3\ln|1 + x^{1/3}| + 6\text{ Arctan } x^{1/6} + C$

33. $\dfrac{\pi}{6} - \dfrac{\sqrt{3}}{8}$

35. $\dfrac{\pi}{3}$

37. $\dfrac{2}{9}\sqrt{2} - \dfrac{1}{18}\sqrt{5}$

39. $\dfrac{2}{\sqrt{3}}\text{Arctan}\left(\dfrac{2\tan(\theta/2) + 1}{\sqrt{3}}\right) + C$

41. $\dfrac{2}{\sqrt{5}}\text{Arctan}\left(\dfrac{1}{\sqrt{5}}\tan\dfrac{\theta}{2}\right) + C$

43. $\dfrac{9}{2\sqrt{2}}\text{Arctan}\dfrac{1}{\sqrt{2}} - \dfrac{1}{2}$ sq units

45. $\text{Cosh}^{-1}\dfrac{x}{a} + C = \ln(x + \sqrt{x^2 - a^2}) + C_1,\ (x \geq a),\quad \dfrac{1}{a^2}\tanh\left(\text{Cosh}^{-1}\dfrac{x}{a}\right) + C = \dfrac{1}{a^2}\dfrac{\sqrt{x^2 - a^2}}{x} + C_1$

Section 6.6

1. $-\dfrac{1}{4}\ln|5 - 4x| + C$

3. $\dfrac{1}{6}\ln\left|\dfrac{x - 3}{x + 3}\right| + C$

5. $\dfrac{1}{2a}\ln\left|\dfrac{a + x}{a - x}\right| + C$

7. $x + \dfrac{1}{3}\ln|x - 1| - \dfrac{4}{3}\ln|x + 2| + C$

9. $-2\ln|x| + 3\ln|x + 1| + C$

11. $\dfrac{1}{3(1 - 3x)} + C$

13. $-\dfrac{x}{9} - \dfrac{2}{27}\ln|2 - 3x| + C$

15. $\dfrac{1}{2a^2}\ln\dfrac{|x^2 - a^2|}{x^2} + C$

17. $x + \dfrac{a}{3}\ln|x - a| - \dfrac{a}{6}\ln(x^2 + ax + a^2) - \dfrac{a}{\sqrt{3}}\text{Arctan}\left(\dfrac{2x + a}{\sqrt{3}a}\right) + C$

19. $\dfrac{1}{3}\ln|x| - \dfrac{1}{2}\ln|x - 1| + \dfrac{1}{6}\ln|x - 3| + C$

21. $\dfrac{1}{4}\ln\left|\dfrac{x + 1}{x - 1}\right| - \dfrac{x}{2(x^2 - 1)} + C$

23. $\frac{1}{27}\ln\left|\frac{x-3}{x}\right| + \frac{1}{9x} + \frac{1}{6x^2} + C$

25. $\frac{t-1}{4(t^2+1)} - \frac{1}{4}\ln|t+1| + \frac{1}{8}\ln(t^2+1) + C$

27. $\frac{1}{3}\ln\left|\frac{1-\sqrt{1-x^2}}{x}\right| + \frac{1}{6}\ln\left|\frac{2+\sqrt{1-x^2}}{\sqrt{3+x^2}}\right| + C$

29. $\frac{1}{4}\ln\left|\frac{1+\sin\theta}{1-\sin\theta}\right| - \frac{1}{2(1+\sin\theta)} + C$

Section 6.7

1. $x\sin x + \cos x + C$

3. $\left(\frac{x^2}{k} - \frac{2x}{k^2} + \frac{2}{k^3}\right)e^{kx} + C$

5. $-\frac{1}{2}(x^4 + 2x^2 + 2)e^{-x^2} + C$

7. $\frac{1}{4}x^4\ln x - \frac{1}{16}x^4 + C$

9. $x\,\mathrm{Arctan}\,x - \frac{1}{2}\ln(1+x^2) + C$

11. $\left(\frac{1}{2}x^2 - \frac{1}{4}\right)\mathrm{Arcsin}\,x + \frac{x}{4}\sqrt{1-x^2} + C$

13. $\frac{1}{2}(\mathrm{Arcsin}\,x)^2 + C$

15. $\frac{7}{8}\sqrt{2} + \frac{3}{8}\ln(1+\sqrt{2})$

17. $\frac{2}{13}e^{2x}\sin 3x - \frac{3}{13}e^{2x}\cos 3x + C$

19. $\ln(2+\sqrt{3}) - \frac{\pi}{6}$

21. $x\tan x + \ln|\cos x| + C$

23. $\frac{x}{2}(\cos(\ln x) + \sin(\ln x)) + C$

25. $\ln x\ln(\ln x) - \ln x + C$

27. $x\,\mathrm{Arccos}\,x - \sqrt{1-x^2} + C$

29. $\frac{2\pi}{3} - \ln(2+\sqrt{3})$

31. $\frac{x}{2}\sqrt{9+x^2} + \frac{9}{2}\ln(x+\sqrt{9+x^2}) + C$

33. $I_n = x(\ln x)^n - nI_{n-1}$, $I_4 = x((\ln x)^4 - 4(\ln x)^3 + 12(\ln x)^2 - 24\ln x + 24) + C$

35. $I_n = \frac{1}{2n}\tan x\sec^{2n-1}x + \frac{2n-1}{2n}I_{n-1}$,

$I_3 = \frac{1}{6}\sec^5 x\tan x + \frac{5}{24}\sec^3 x\tan x + \frac{15}{48}\sec x\tan x + \frac{15}{48}\ln|\sec x + \tan x| + C$

Review Exercises on Techniques of Integration

1. $-\frac{1}{6}\ln|2x+1| + \frac{2}{3}\ln|x+2| + C$

3. $\frac{1}{4}\sin^4 x - \frac{1}{6}\sin^6 x + C$

5. $\frac{3}{4}\ln\left|\frac{2x-1}{2x+1}\right| + C$

7. $-\frac{1}{3}\frac{(1-x^2)^{3/2}}{x^3} + C$

9. $\frac{1}{5}(5x^3-2)^{1/3} + C$

11. $\frac{1}{16}\mathrm{Arctan}\,\frac{x}{2} + \frac{x}{8(4+x^2)} + C$

13. $\frac{1}{2\ln 2}(2^x\sqrt{1+4^x} + \ln(2^x + \sqrt{1+4^x})) + C$

15. $\frac{1}{6}\sec^6 x - \frac{1}{4}\sec^4 x + C$

17. $-\frac{2}{5}e^{-x}\cos 2x - \frac{1}{5}e^{-x}\sin 2x + C$

19. $\dfrac{x}{10}\cos(3\ln x) + \dfrac{3x}{10}\sin(3\ln x) + C$

21. $\dfrac{1}{4}(\ln(1 + x^2))^2 + C$

23. $\text{Arcsin}\dfrac{x}{\sqrt{2}} - \dfrac{1}{2}x\sqrt{2 - x^2} + C$

25. $\dfrac{1}{64}\left(\dfrac{-1}{7(4x + 1)^7} + \dfrac{1}{4(4x + 1)^8} - \dfrac{1}{9(4x + 1)^9}\right) + C$

27. $-\dfrac{1}{4}\left(\cos 4x - \dfrac{2}{3}\cos^3 4x + \dfrac{1}{5}\cos^5 4x\right) + C$

29. $-\dfrac{1}{2}\ln(1 + 2e^{-x}) + C$

31. $-4\ln(2 - \sin x) - 2\sin x - \dfrac{1}{2}\sin^2 x + C$

33. $-\dfrac{\sqrt{1 - x^2}}{x} + C$

35. $\dfrac{1}{48}(1 - 4x^2)^{3/2} - \dfrac{1}{16}\sqrt{1 - 4x^2} + C$

37. $\sqrt{x^2 + 1} + \ln(x + \sqrt{1 + x^2}) + C$

39. $x + \dfrac{1}{3}\ln|x| + \dfrac{4}{3}\ln|x - 3| - \dfrac{5}{3}\ln|x + 3| + C$

41. $\dfrac{1}{6}\cos^{12} x - \dfrac{1}{10}\cos^{10} x - \dfrac{1}{14}\cos^{14} x + C$

43. $\dfrac{1}{2}\ln|x^2 + 2x - 1| - \dfrac{1}{2\sqrt{2}}\ln\left|\dfrac{x + 1 - \sqrt{2}}{x + 1 + \sqrt{2}}\right| + C$

45. $\dfrac{1}{3}x^3\,\text{Arcsin}\,2x + \dfrac{1}{24}\sqrt{1 - 4x^2} - \dfrac{1}{72}(1 - 4x^2)^{3/2} + C$

47. $\dfrac{1}{128}\left(3x - \sin 4x + \dfrac{1}{8}\sin 8x\right) + C$

49. $\text{Arctan}\dfrac{\sqrt{x}}{2} + C$

51. $\dfrac{1}{4}\left(2x^2 - 8x + \ln|x| + 15\ln|x + 2| + \dfrac{2}{x}\right) + C$

53. $-\dfrac{1}{2}\cos(2\ln x) + C$

55. $\dfrac{1}{2}e^{2\,\text{Arctan}\,x} + C$

57. $\dfrac{1}{4}(\ln(3 + x^2))^2 + C$

59. $\dfrac{1}{2}\left(\text{Arcsin}\dfrac{x}{2}\right)^2 + C$

61. $\sqrt{x^2 + 6x + 10} - 2\ln(x + 3 + \sqrt{x^2 + 6x + 10}) + C$

63. $\dfrac{2}{5}(x^2 + 2)^{-5/2} - \dfrac{1}{3}(x^2 + 2)^{-3/2} + C$

65. $6\left(\dfrac{1}{7}x^{7/6} - \dfrac{1}{5}x^{5/6} + \dfrac{1}{3}x^{1/2} - x^{1/6} + \text{Arctan}\,x^{1/6}\right) + C$

67. $\dfrac{2}{3}x^{3/2} - x + 4\sqrt{x} - 4\ln(1 + \sqrt{x}) + C$

69. $\dfrac{1}{8 - 2x^2} + C$

71. $\dfrac{1}{3}x^3\,\text{Arctan}\,x - \dfrac{1}{6}x^2 + \dfrac{1}{6}\ln(1 + x^2) + C$

73. $\dfrac{1}{5}\ln\left|3\tan\dfrac{x}{2} - 1\right| - \dfrac{1}{5}\ln\left|\tan\dfrac{x}{2} + 3\right| + C$

75. $\dfrac{1}{4}\ln\left|\dfrac{1 - \cos x}{1 + \cos x}\right| - \dfrac{1}{2 + 2\cos x} + C$

77. $2\sqrt{x} - 2\,\text{Arctan}\sqrt{x} + C$

79. $\frac{1}{2}x^2 + \frac{4}{3}\ln|x - 2| - \frac{2}{3}\ln(x^2 + 2x + 4) + \frac{4}{\sqrt{3}}\text{Arctan}\frac{x + 1}{\sqrt{3}} + C$

Section 6.8

1. TRAP(4) = 4.75, TRAP(8) = 4.6875, SIMP(4) = 4.6666 . . . = SIMP(8) = I, error bounds: $|I - \text{TRAP}(n)| \leq 4/3n^2$, $|I - \text{SIMP}(n)| = 0$ $4/3n^2$, $|I - \text{SIMP}(n)| = 0$

3. TRAP(4) $\cong$ 0.9871158, TRAP(8) $\cong$ 0.9967851, SIMP(4) $\cong$ 1.0001346, SIMP(8) $\cong$ 1.0000083, $I = 1$, $|I - \text{TRAP}(n)| \leq \pi^3/96n^2$, $|I - \text{SIMP}(n)| \leq \pi^5/5760n^4$

5. TRAP(10) $\cong$ 0.7462108, SIMP(4) $\cong$ 0.7468553, $|I - \text{TRAP}(10)| \leq 1/300 = 0.003333\ldots$ $|I - \text{SIMP}(4)| \leq 1/1280 = 0.00078\ldots$

7. SIMP(6) $\cong$ 0.3320533, $|I - \text{SIMP}(6)| \leq (1/90)(1.2)(0.2)^4 \cong 0.0000213$

9. actual error I − SIMP(2) = 1/120, estimated error $|I - \text{SIMP}(2)| \leq 1/120$

Section 6.9

1. 1/2

3. $3(2^{1/3})$

5. 3

7. 1

9. π

11. $\frac{1}{2}$

13. diverges

15. area infinite

19. 2

21. $5 - \sqrt{2} - \sqrt{3}$

23. diverges

25. converges

27. diverges

29. diverges

31. diverges

CHAPTER 7 APPLICATION OF INTEGRATION

Section 7.1

1. 1/6 sq units

3. 64/3 sq units

5. 125/12 sq units

7. 1/2 sq units

9. 5/12 sq units

11. (15/8) − 2 ln 2 sq units

13. $(\pi/2) - (1/3)$ sq units

15. 4/3 sq units

17. $2\sqrt{2}$ sq units

19. a^2 Arccos $(b/a) - b\sqrt{a^2 - b^2}$ sq units

21. $(1/2)(1 + e^{-\pi})$ sq units

23. 1 sq units

25. 4/3 sq units

27. $e/2$ sq units

29. 27/4 sq units

Section 7.2

1. $\frac{\pi}{5}$ cu units

3. $\frac{3\pi}{10}$ cu units

5. a) $\frac{16\pi}{15}$ cu units b) $\frac{8\pi}{3}$ cu units

7. a) $\frac{27\pi}{2}$ cu units b) $\frac{108\pi}{5}$ cu units

9. a) $\frac{15}{4}\pi - \frac{1}{8}\pi^2$ cu units

b) $2\pi - \pi \ln 2$ cu units

11. $\frac{16}{3} r^3$ cu units

13. $V = \frac{\pi}{6} L^3$ cu units

15. $\frac{4}{3} \pi ab^2$ cu units

17. $2\pi^2 abc$ cu units

19. $\frac{a + b}{2} \pi r^2$ cu units

23. $k > \frac{1}{2}$

25. $k = 4\pi$

27. $\frac{\pi}{\sqrt{2}}\left(\frac{57}{8} - 10 \ln 2\right)$ cu units

Section 7.3

1. $\sqrt{1 + a^2}\,|B - A|$ units

3. 6 units

5. $2 \ln 3 - 1$ units

7. $\sqrt{17} + \frac{1}{4} \ln (4 + \sqrt{17})$ units

9. $\ln \frac{\sqrt{2} + 1}{\sqrt{3}}$ units

11. $6a$ units

13. $\frac{2475}{72} \pi$ sq units

15. $\pi a + \frac{\pi}{2} \sinh 2a$ sq units

17. $\pi(e\sqrt{1 + e^2} - \sqrt{2} + \ln (e + \sqrt{1 + e^2}) - \ln (1 + \sqrt{2}))$ sq units

19. $4\pi^2 ab$ sq units

21. a) π cu units

23. $k > -1$

Section 7.4

1. π^2 sq units

3. a^2 sq units

5. $\frac{\pi}{2}$ sq units

7. $2 + \frac{\pi}{4}$ sq units

9. $\frac{\pi}{4}$ sq units

11. $\pi - \frac{3}{2}\sqrt{3}$ sq units

13. $\frac{3}{8} \pi a^2$ sq units

15. $\frac{1}{2} \ln |\sec t + \tan t|$ sq units

17. $\frac{9\pi}{2}$ sq units

19. $5\pi^2 a^3$ cu units

21. $\frac{40\sqrt{40} + 13\sqrt{13} - 16}{27}$ units

23. $2 \ln (1 + \sqrt{2})$ units

25. $\frac{2}{a}\sqrt{1 + a^2} \sinh \pi a$ units

27. $\frac{64}{3} \pi a^2$ sq units

Section 7.5

1. $m = \frac{2L}{\pi} g,\ \bar{x} = \frac{L}{2}$

3. $m = \frac{256k}{15},\ \bar{x} = 0,\ \bar{y} = \frac{16}{7}$

5. $m = \frac{2}{3} a^3 g, \bar{x} = 0, \bar{y}\ \frac{3\pi a}{16}$ cm

7. $m = \frac{k}{3\sqrt{2}} a^3 g,$ center of mass at center of square

9. $m = \frac{8}{3} \pi R^4$ kg, center of mass along perpendicular from center of ball to plane, at distance $(13/6)R$ from the plane.

11. $m = \frac{2}{9} k\pi a^3 b$ g, center of mass on axis at distance $\frac{3b}{20}$ cm above base

Section 7.6

1. $\bar{x} = 0, \bar{y} = \frac{9}{10}\left(\frac{2^{5/2} - 1}{2^{3/2} - 1}\right) - \frac{3}{2} \cong 0.792$

3. $\bar{x} = \bar{y} = \frac{4r}{3\pi}$

5. $\bar{x} = \frac{7}{9}, \bar{y} = \frac{4}{9}$

7. $\bar{x} = 0, \bar{y} = \frac{9\sqrt{3} - 4\pi}{4\pi - \sqrt{3}} \cong 0.279$

9. on axis of symmetry at distance $r/2$ above the base of the hemisphere

11. on axis of symmetry at distance $h/3$ above the base of the cone

13. $\bar{x} = \pi/2, \bar{y} = \pi/8$

15. $\bar{x} = \bar{y} = 0$

17. $5\pi/3$ cu units

19. $\bar{x} = 1, \bar{y} = \frac{1}{5}$

Section 7.7

1. 24,000 kg = 24 tonnes

3. 160 tonnes

5. $\sqrt{3}\left(40 - \frac{8\sqrt{2}}{3}\right) \cong 62.75$ tonnes

7. 72,000 kg-m = 72 tonne-meters

9. $\pi a^3\left(\frac{2h}{3} + \frac{a}{4}\right)$ g-cm

11. 135 kg-m

Section 7.8

1. a) 2/9 b) $\mu = 2, \sigma^2 = 1/2, \sigma = 1/\sqrt{2}$ c) $8/9\sqrt{2} \cong 0.63$

3. a) 3 b) $\mu = 3/4, \sigma^2 = 3/80, \sigma = \sqrt{3/80} \cong 0.194$ c) $6\sqrt{3/80}\left(\frac{9}{16} + \frac{1}{80}\right) \cong 0.668$

5. a) 1/2 b) $\mu = \pi/2, \sigma^2 = (\pi^2 - 8)/4, \sigma = (1/2)\sqrt{\pi^2 - 8} \cong 0.684$ c) $\sin((1/2)\sqrt{\pi^2 - 8}) \cong 0.632$

7. a) 2 b) $\mu = 1/2, \sigma^2 = 3/4, \sigma = \sqrt{3}/2$ c) $1 - e^{-(1+\sqrt{3})^2/4} \cong 0.845$

11. a) 0 b) $e^{-3} \cong 0.05$ c) approx. 0.46

13. 6/1000

Section 7.9

1. $y^2 = Cx$

3. $y^3 = x^3 + C$

5. $Y = Ce^{t^2/2}$

7. $y = \frac{Ce^{2x} - 1}{Ce^{2x} + 1}$

9. $y = -\ln\left(Ce^{-2t} - \frac{1}{2}\right)$

11. $y = x^3 + Cx^2$

13. $y = \frac{1}{2}e^x - Ce^{-x}$

15. $x(t) = \dfrac{a^2kt}{akt + 1}$

17. $v(t) = \dfrac{1}{\omega}\dfrac{e^{2\omega gt} - 1}{e^{2\omega gt} + 1}$ where $\omega = \left(\dfrac{k}{mg}\right)^{1/2}$, $v(\infty) = \left(\dfrac{mg}{k}\right)^{1/2}$

CHAPTER 8 INFINITE SERIES

Section 8.1

1. bounded, positive, increasing, convergent (lim = 1)

3. bounded, positive, convergent (lim = 4)

5. bounded below, positive, increasing, divergent to ∞

7. bounded below, positive, increasing, divergent to ∞

9. divergent

11. bounded, positive, decreasing, convergent (lim = 0)

13. $-2/3$ **15.** 0 **17.** $-(1/3)$ **19.** 1

21. divergent to ∞ **23.** 2 **25.** $1/e^2$ **27.** 0

29. $\lim a_n = 1 + \sqrt{2}$

33. a) true b) false c) true d) false

Section 8.2

1. 1/2 **3.** $\dfrac{1}{(2 + \pi)^{10} - (2 + \pi)^8}$ **5.** 25/4416

7. 3/4 **9.** 1/2 **11.** divergent to ∞ **13.** divergent to ∞

15. $s_n = \begin{cases} 0 \text{ if } n \text{ is even} \\ -1 \text{ if } n \text{ is odd} \end{cases}$

17. \$12,579.48 **21.** true **23.** false **25.** false

Section 8.3

1. convergent **3.** divergent to ∞ **5.** convergent **7.** divergent to ∞

9. convergent **11.** divergent to ∞ **13.** divergent to ∞ **15.** divergent to ∞

17. convergent **19.** divergent to ∞ **21.** convergent **23.** divergent to ∞

25. convergent **27.** convergent **29.** convergent

Section 8.4

1. convergent conditionally **3.** convergent conditionally

5. divergent **7.** convergent absolutely

9. convergent conditionally **11.** divergent

13. convergent absolutely on $(-1, 1)$, conditionally at $x = -1$, divergent elsewhere

15. convergent absolutely on $(-2, 0)$, conditionally at $x = 0$, divergent elsewhere

17. convergent absolutely on $(-2, 2)$, conditionally at $x = -2$, divergent elsewhere

19. convergent absolutely on $\left(-\frac{b+c}{a}, \frac{c-b}{a}\right)$, conditionally at $x = -\frac{b+c}{a}$, divergent elsewhere

21. convergent absolutely if $x \geq 1/2$, divergent for all other x

23. convergent absolutely on $(-1, 0)$ and $(1, 2)$, divergent elsewhere

25. convergent conditionally

27. a) false b) false c) true

29. convergent absolutely on $(-1, 1)$, convergent conditionally at $x = -1$, divergent elsewhere

Section 8.5

1. $s_n + \frac{1}{9(n+1)^9} \leq s \leq s_n + \frac{1}{9n^9}$, 2 terms

3. $s_n + \frac{2}{\sqrt{n+1}} \leq s \leq s_n + \frac{2}{\sqrt{n}}$, 63 terms

5. $s_n + \frac{1}{e^{n+1}} \leq s \leq s_n + \frac{1}{e^n}$, 6 terms

7. $0 < s - s_n < \frac{2n+4}{2n+3} \frac{1}{2^{n+1}(n+1)!}$, 4 terms

9. $0 < s - s_n < \frac{4n^2+6n+2}{4n^2+6n} \frac{2^n}{(2n)!}$, 4 terms

11. 999 terms

13. 13 terms

CHAPTER 9 POWER SERIES REPRESENTATIONS OF FUNCTIONS

Section 9.1

1. $0, 1, (-1, 1)$

3. $-2, 2, [-4, 0)$

5. $3/2, 1/2, (1, 2)$

7. $0, \infty, (-\infty, \infty)$

9. $1/4, \infty, (-\infty, \infty)$

11. $\frac{1}{(1-x)^3} = \frac{1}{2}(2 + 3(2x) + 4(3x) + 5(4x) + \ldots) = \frac{1}{2}\sum_{n=0}^{\infty} (n+2)(n+1)x^n$

13. $\frac{1}{(1-x)^2} = 1 + 2x + 3x^2 + 4x^3 + \ldots = \sum_{n=0}^{\infty} (n+1)x^n$

Section 9.2

1. $\sum_{n=0}^{\infty} \frac{x^n}{2^{n+1}}, (-2, 2)$

3. $\sum_{n=0}^{\infty} (-1)^n(2x)^n, \left(-\frac{1}{2}, \frac{1}{2}\right)$

5. $\ln 2 - \sum_{n=1}^{\infty} \frac{x^n}{2^n n}, [-2, 2)$

7. $\sum_{n=0}^{\infty} (-1)^n(n+1)(x-1)^n, (0, 2)$

9. $1 + 2\sum_{n=1}^{\infty} (-1)^n x^n$, $(-1, 1)$

11. $\left(-\frac{1}{4}, \frac{1}{4}\right), \frac{1}{1 + 4x}$

13. $[-1, 1)$, $-\frac{1}{x^3}\ln(1 - x) - \frac{1}{x^2} - \frac{1}{2x}$

15. $(-1, 1)$, $\frac{2}{(1 - x^2)^2}$

Section 9.3

1. $\sum_{n=0}^{\infty} \frac{e3^n x^n}{n!}$, all x

3. $\sum_{n=0}^{\infty} \frac{(-1)^n}{\sqrt{2}}\left(\frac{x^{2n+1}}{(2n+1)!} - \frac{x^{2n}}{(2n)!}\right)$, all x

5. $2\left(\sum_{n=0}^{\infty} \frac{x^{4n+2}}{(4n+2)!}\right)$, all x

7. $1 + \frac{1}{2}\sum_{n=1}^{\infty} (-1)^n \frac{x^{2n}}{(2n)!}$, all x

9. $-x + \frac{2^2x^2}{2!} + \frac{2^3 + 3^3}{3!}x^3 + \frac{2^4x^4}{4!} + \frac{2^5 - 3^5}{5!}x^5 + \frac{2^6x^6}{6!} + \frac{2^7 + 3^7}{7!}x^7 + \dots$, all x

11. $-\sum_{n=1}^{\infty} \frac{x^n}{n}$, on $[-1, 1)$

13. $\ln 2 + \sum_{n=1}^{\infty} (-1)^{n-1} \frac{x^{2n}}{2^n n}$, on $[-\sqrt{2}, \sqrt{2}]$

15. $\sum_{n=0}^{\infty} \frac{2^{n+1} x^{2n}}{(n+1)!}$, all $x \neq 0$

17. $x + \frac{5}{6}x^3 + \frac{61}{120}x^5$

19. $e^2 \sum_{n=0}^{\infty} (-1)^n \frac{2^n (x+1)^n}{n!}$, all x

21. $-\sum_{n=0}^{\infty} (-1)^n \frac{(x - \pi)^{2n}}{(2n)!}$, all x

23. $\sqrt{2}\sum_{n=0}^{\infty} \frac{(-1)^n}{(2n+1)!}\left(x - \frac{\pi}{4}\right)^{2n+1}$, all x

25. $\frac{1}{4}\sum_{n=0}^{\infty} \frac{(n+1)(x+2)^n}{2^n}$, on $(-4, 0)$

27. $x + \frac{x^2}{2} - \frac{x^3}{6}$

29. no, $1 + \frac{1}{2}\left(x - \frac{\pi}{2}\right)^2 + \frac{5}{24}\left(x - \frac{\pi}{2}\right)^4$

31. $2 \sin \frac{x^3}{2}$

33. $2\sqrt{e} - 2$

35. $3 + 3(x - 1) + (x - 1)^2$

Section 9.4

1. 1.22140

3. 4.481686

5. 0.996194

7. −0.10536

9. 0.422618

11. 1.54308

13. $\sum_{n=0}^{\infty} (-1)^n \frac{x^{2n+1}}{(2n+1)(2n+1)!}$

15. $\sum_{n=1}^{\infty} (-1)^{n-1} \frac{x^n}{n^2}$

17. $\sum_{n=0}^{\infty} (-1)^n \frac{x^{4n+1}}{(4n+1)(2n+1)}$

19. 0.497

21. 2

23. $-(3/25)$

25. 0

27. $y = 1 + 2x - x^2 - x^3 + \frac{1}{3}x^4 + \frac{1}{4}x^5 - \frac{1}{15}x^6 - \frac{1}{24}x^7 + \dots$

$$= 1 + \sum_{n=0}^{\infty} (-1)^n \left(\frac{x^{2n+1}}{2^{n-1}n!} - \frac{x^{2n+2}}{1 \times 3 \times 5 \times \cdots \times (2n+1)}\right)$$

Section 9.5

1. $x - \frac{x^3}{6} + \frac{x^5}{120}$

3. $\frac{1}{\sqrt{2}}\left(1 + \left(x - \frac{\pi}{4}\right) - \frac{1}{2}\left(x - \frac{\pi}{4}\right)^2 - \frac{1}{6}\left(x - \frac{\pi}{4}\right)^3 + \frac{1}{24}\left(x - \frac{\pi}{4}\right)^4\right)$

5. $e^2\left(1 - (x + 2) + \frac{(x+2)^2}{2!} - \cdots + (-1)^n\frac{(x+2)^n}{n!}\right)$

7. $1 + \frac{1}{2}x^2 + \frac{5}{24}x^4$

9. $1 + \frac{1}{2}x - \frac{1}{8}x^2 + \frac{1}{16}x^3 - \frac{5}{128}x^4$

11. $\sum_{n=0}^{\infty} \frac{x^n}{n!}$

13. $\sum_{n=0}^{\infty} \frac{x^{2n}}{(2n)!}$

15. $\sum_{n=0}^{\infty} (-1)^n \frac{x^{2n+1}}{(2n+1)!}$

17. $\sum_{n=1}^{\infty} (-1)^{n-1} \frac{2^{2n-1}x^{2n}}{(2n)!}$

19. $\sum_{n=1}^{\infty} (-1)^{n-1} \frac{x^n}{n}$

21. $e^a \sum_{n=0}^{\infty} \frac{(x-a)^n}{n!}$

23. $\frac{1}{\sqrt{2}} \sum_{n=0}^{\infty} (-1)^n \left(\frac{1}{(2n)!}\left(x - \frac{\pi}{4}\right)^{2n} - \frac{1}{(2n+1)!}\left(x - \frac{\pi}{4}\right)^{2n+1}\right)$

25. $\ln 2 + \sum_{n=1}^{\infty} (-1)^{n-1} \frac{(x-2)^n}{2^n n}$

27. a) $3 + 2x + x^2$ b) $3 + 2x + x^2$ c) $3 - 2(x + 2) + (x + 2)^2$
d) $18 + 8(x - 3) + (x - 3)^2$

29. polynomials of degree at most n

31. $|\text{error}| < (\tan^3(0.2) + 5\tan(0.2))\frac{(0.2)^3}{3!} < (0.2)^3 = 0.008$

33. If $f(x) = e^{-x}\sin\frac{x}{2}$, then $f^{(4)}(x) = -\frac{7}{16}e^{-x}\sin\frac{x}{2} - \frac{3}{2}e^{-x}\cos\frac{x}{2}$

On $[0, 1]$, $|f^{(4)}(x)| < \frac{7}{16}\sin\frac{1}{2} + \frac{3}{2} < \frac{7}{32} + \frac{3}{2} = \frac{55}{32}$

Thus $|\text{error}| < \frac{55}{32} \cdot \frac{1^4}{4!} = \frac{55}{768}$

35. $|\text{error}| < \frac{1}{5 \times (1.95)^5}(0.05)^5 < 2.3 \times 10^{-9}$.

37. $2 + \frac{x}{4} + 2\sum_{n=2}^{\infty} (-1)^{n-1} \frac{1 \times 3 \times 5 \times \cdots \times (2n-3)}{2^{3n}\, n!} x^n$

39. $\frac{x}{2} + \sum_{n=1}^{\infty} (-1)^n \frac{1 \times 3 \times 5 \times \cdots \times (2n-1)}{2^{3n-1} n!(2n+1)} x^{2n+1}$

CHAPTER 10 PARTIAL DIFFERENTIATION

Section 10.1

1. $D(f) = \{(x,y): x \neq y\}$

3. $D(f) = \{(x,y): (x,y) \neq (0,0)\}$

5. $D(f) = \{(x,y): 4x^2 + 9y^2 \geq 36)$

7. $D(f) = \{(x,y): xy > -1\}$

9. $D(f) = \{(x,y,z): (x,y,z) \neq (0,0,0)\}$

11.

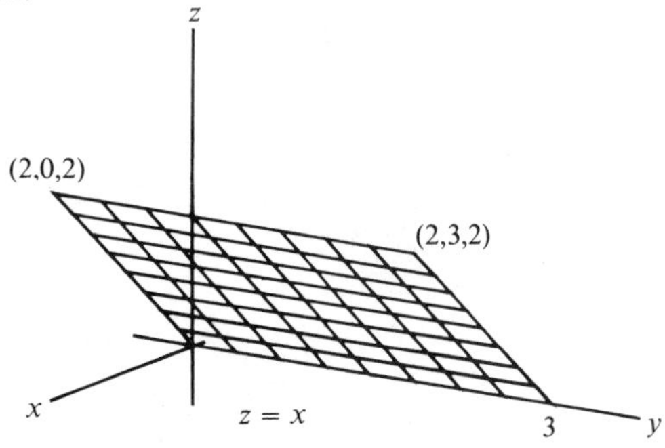

13.

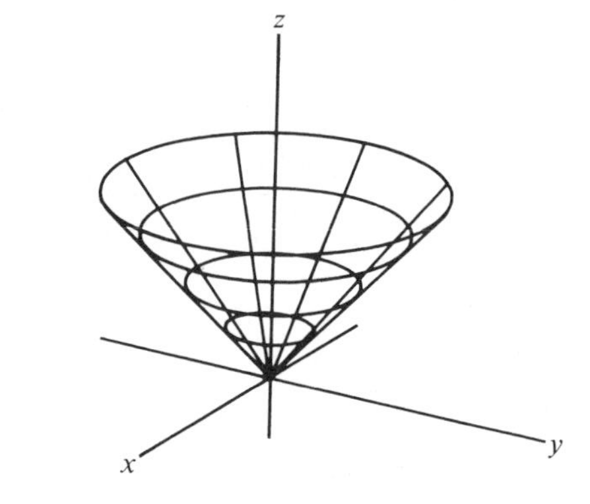

15.

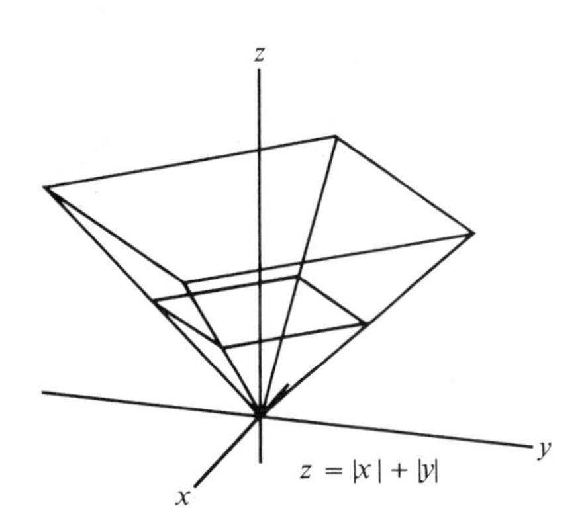

17.

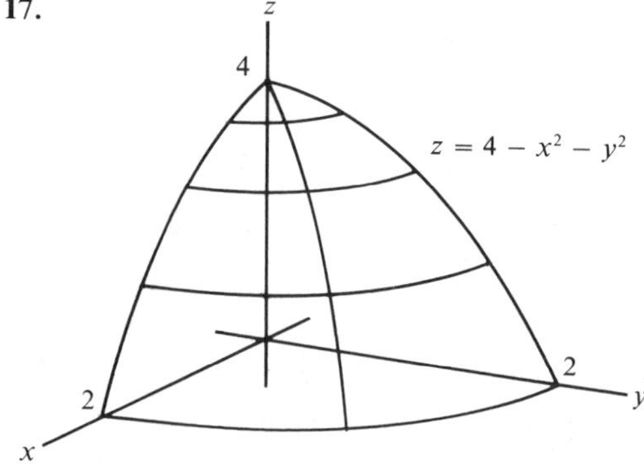

19.

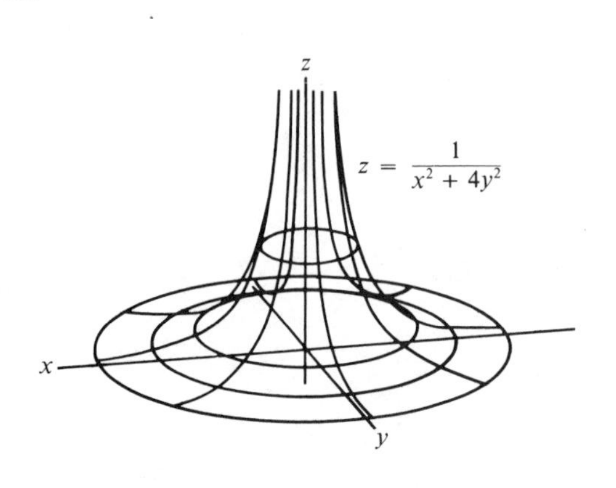

21.

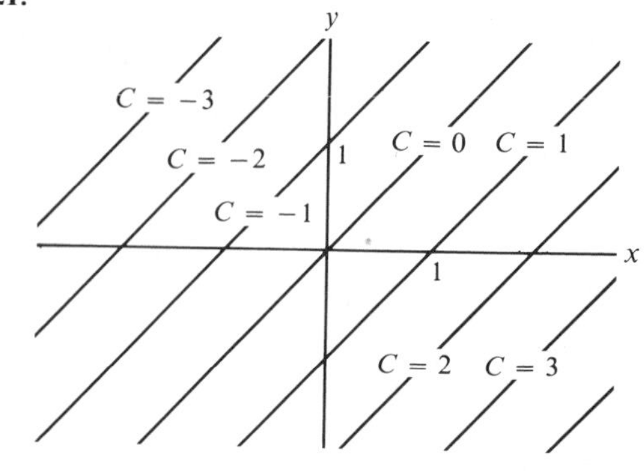

23.

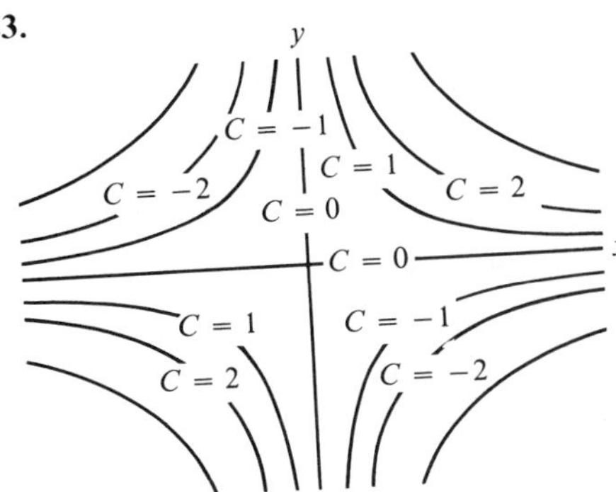

25.

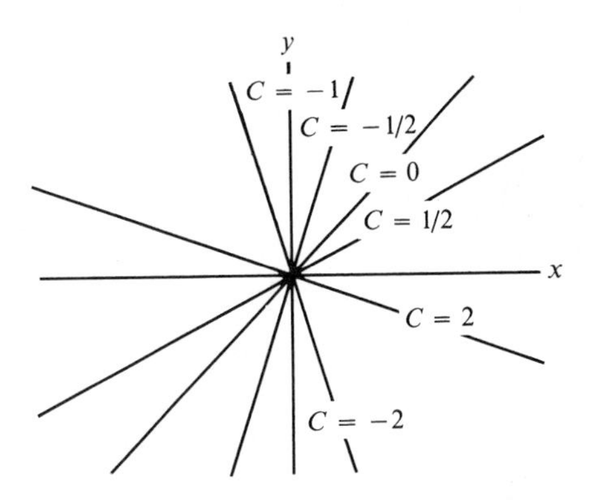

27.

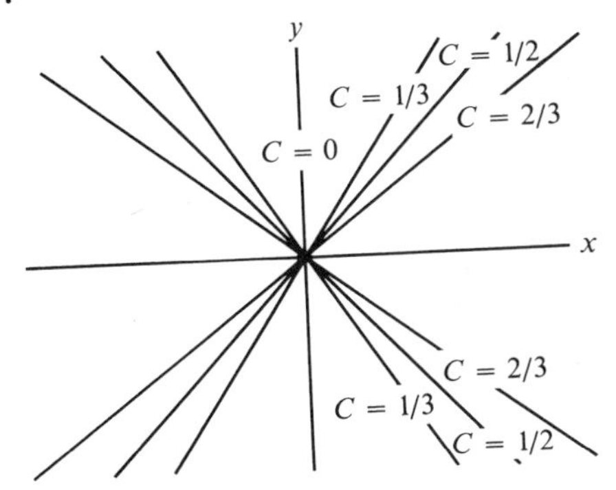

29.

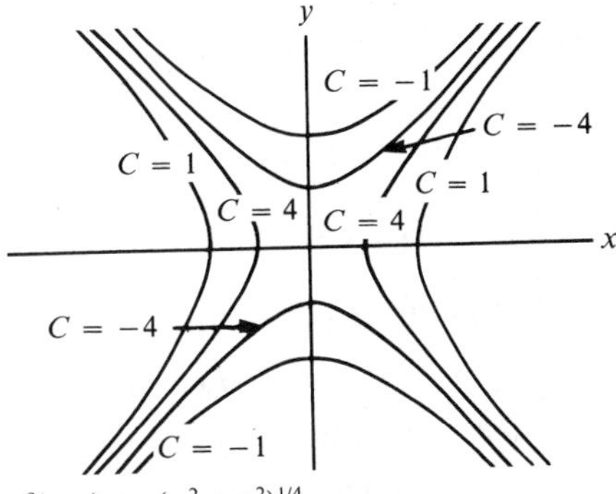

31. **a)** $f(x,y) = (\sqrt{x^2 + y^2}$ **b)** $f(x,y) = (x^2 + y^2)^{1/4}$

c) $f(x,y) = x^2 + y^2$ **d)** $f(x,y) = \exp(\sqrt{x^2 + y^2})$

33. limit does not exist ($f \to \infty$ as $x \to 0+$ along the x-axis)

35. the limit is zero

37. the limit is 1/4

39. the limit does not exist (the limit is zero along the axes but 1/3 along the line $x = y$)

41. the limit does not exist ($f \to \infty$ along the line $y = 0, x = -1 + t$)

43. $D(f) = \{(x,y): x \neq \pm y\}$
At any point of the line $x = y$ except the origin we can define $f(x,x) = 1/(2x)$. So extended, f is continuous at such points.

Section 10.2

1. $f_1(x,y) = f_1(3,2) = 1, f_2(x,y) = f_2(3,2) = -1$

3. $f_1(x,y,z) = 3x^2y^4z^5, f_2(x,y,z) = 4x^3y^3z^5, f_3(x,y,z) = 5x^3y^4z^4$,
$f_1(0,-1,-1) = f_2(0,-1,-1) = f_3(0,-1,-1) = 0$

5. $\partial z/\partial x = -y/(x^2 + y^2), \partial z/\partial y = x/(x^2 + y^2), \partial z/\partial x \Big|_{(-1,1)} = \partial z/\partial y \Big|_{(-1,1)} = -1/2$

7. $f_1(x,y) = \sqrt{y}\cos(x\sqrt{y}),\quad f_2(x,y) = (x/(2\sqrt{y}))\cos(x\sqrt{y}),\quad f_1(\pi/3,4) = -1$,
$f_2(\pi/3,4) = -\pi/24$

9. $\partial w/\partial x = (y\ln z)\,x^{y\ln z - 1},\quad \partial w/\partial y = x^{y\ln z}\ln x\,\ln z$,
$\partial w/\partial z = x^{y\ln z}(y/z)\ln x$,
At $(e,2,e)$: $\partial w/\partial x = 2e, \partial w/\partial y = e^2, \partial w/\partial z = 2e$

11. $4x + 2y + z + 3 = 0$

13. $4x - \pi y + 16\sqrt{2}\,z = 16$

15. $3x - 4y - 25z + 10 = 0$

17. $2x - 4y - 5z - 10 + 5\ln(5) = 0$

19. $2x + 2y - 4z = \pi$

29. $\partial^2 z/\partial x^2 = 2 + 2y^2, \partial^2 z/\partial x\partial y = 4xy = \partial^2 z/\partial y\partial x, \partial^2 z/\partial y^2 = 2x^2$

31. $w_{11} = 6xy^3z^3, w_{22} = 6x^3yz^3, w_{33} = 6x^3y^3z, w_{12} = w_{21} = 9x^2y^2z^3,$
$w_{13} = w_{31} = 9x^2y^3z^2, w_{23} = w_{32} = 9x^3y^2z^2$

33. $z_{11} = -ye^x, z_{12} = z_{21} = e^y - e^x, z_{22} = xe^y$

41. If $f(x,y)$ is harmonic and $c^2 = a^2 + b^2$ then $f(ax + by, cz)$ is harmonic.

Section 10.3

1. $\partial w/\partial t = f_1(x,y,z)g_2(s,t) + f_2(x,y,z)h_2(s,t) + f_3(x,y,z)k_2(s,t)$

3. $\partial z/\partial u = g_1(x,y)h_1(u,v) + g_2(x,y)f'(x)h_1(u,v)$

5. $\partial w/\partial x\,|_z = f_1(x,y,z) + f_2(x,y,z)g_1(x,y)$

7. $\left.\frac{\partial w}{\partial r}\right|_{x,s} = f_2(x,y,z)h_2(x,r,s) + f_3(x,y,z)k_2(x,r,s)$

$\left.\frac{\partial w}{\partial r}\right|_{y,s} = f_1(x,y,z)g_1(r,s) + f_3(x,y,z)[k_2(x,r,s) + k_1(x,r,s)g_1(r,s)]$

$\left.\frac{\partial w}{\partial r}\right|_{x,y,s} = f_3(x,y,z)k_2(x,r,s)$

$\left.\frac{\partial w}{\partial r}\right|_{s} = f_1(x,y,z)g_1(r,s) + f_2(x,y,z)[h_1(x,r,s)g_1(r,s) + h_2(x,r,s)]$
$+ f_3(x,y,z)[k_1(x,r,s)g_1(r,s) + k_2(x,r,s)]$

9. $\partial z/\partial x = -5y/(13x^2 - 2xy + 2y^2)$

11. $2vf_2(u^2,v^2)$

13. $f_1(f(x,y),f(x,y))f_1(x,y) + f_2(f(x,y),f(x,y))f_1(x,y)$

15. $dT/dt = -e^{-t}f(t) + e^{-t}f'(t)$
If $f(t) = e^t$ then $dT/dt = 0$. In this case the decay of temperature with time is exactly compensated by the increase in temperature with depth so the temperature remains constant.

17. $\partial^2 z/\partial s^2 = 4f_{11} + 12f_{12} + 9f_{22}, \partial^2 z/\partial s\partial t = 6f_{11} + 5f_{12} - 6f_{22},$
$\partial^2 z/\partial t^2 = 9f_{11} - 12f_{12} + 4f_{22}, \partial^3 z/\partial s^3 = 8f_{111} + 36f_{112} + 54f_{122} + 27f_{222}$

21. $\partial^2 z/\partial s\partial t = \cos(s)f_1(x,y) + t\sin s\cos s\, f_{11}(x,y)$
$+ \sin s\cos t\, f_{12}(x,y) - \sin t\, f_2(x,y)$
$- st\sin t\cos s\, f_{21}(x,y) - s\sin t\cos t\, f_{22}(x,y)$

25. $\sum_{i=1}^{n} x_i^2 f_{ii}(x_1,x_2,\ldots,x_n) + \sum_{i=1}^{n}\sum_{\substack{j=1\\ j\neq i}}^{n} x_iy_j f_{ij}(x_1,x_2,\ldots,x_n)$
$= k(k-1)f(x_1,x_2,\ldots,x_n)$

27. $f(0.01,1.05) \approx f(0,1) + 0.01f_1(0,1) + 0.05f_2(0,1) \approx 0.0814$

29. $f(1.9,1.8,1.1) \approx f(2,2,1) - 0.1f_1(2,2,1) - 0.2f_2(2,2,1)$
$+ 0.1f_3(2,2,1) \approx 2.9667$

Section 10.4

1. $\nabla f(-2,1) = -4\mathbf{i} - 2\mathbf{j}, 4x + 2y + z + 3 = 0$

3. $\nabla f(\pi,4) = -1/(4\sqrt{2})\mathbf{i} + \pi/(16\sqrt{2})\mathbf{j}, 4x - \pi y + (16\sqrt{2})z = 16$

5. $\nabla f(1,2) = (3/25)\mathbf{i} - (4/25)\mathbf{j}, 3x - 4y - 25z + 10 = 0$

7. $\nabla f(1,-2) = (2/5)\mathbf{i} - (4/5)\mathbf{j}$, $2x - 4y - 5z - 10 + 5\ln(5) = 0$

9. $\nabla f(1,-1,1) = -\mathbf{i} - \mathbf{j} + 3\mathbf{k}$

11. -3

13. $1/(\sqrt{2})$

15. $-2/5$

17. a) $\mathbf{i} + 2\mathbf{j}$ *b*) $-\mathbf{i} - 2\mathbf{j}$ *c*) $\pm(2\mathbf{i} - \mathbf{j})$ *d*) $\mathbf{j}$ or $4\mathbf{i} + 3\mathbf{j}$

19. $i + 2j + 3k$

21. $2x + y + 3 = 0$

23. $4x - \pi y = 0$

25. $3x - 4y + 5 = 0$

27. $x - 2y - 5 = 0$

29. $x + y - 3z + 3 = 0$

31. If $\nabla f(a,b) = 0$ the level curve of f may not be smooth at (a,b).

Section 10.5

1. loc min at $(2,-1)$

3. saddle point at $(-1,1)$

5. loc max at $(-4,2)$

7. saddle points $(0, k\pi)$, k an integer

9. saddle point $(0,0)$, loc max $(1,1)$ and $(-1,-1)$, loc min $(1,-1)$ and $(-1,1)$

11. loc max at $(0,0,0)$, neither max nor min at CPs $(2,2,2)$, $(-2,-2,2)$, $(-2,2,-2)$ and $(2,-2,-2)$

13. max is $\sqrt{5}$ at $(1/(\sqrt{5}), 2/(\sqrt{5}))$, min is $-\sqrt{5}$ at $(-1/(\sqrt{5}), -2/(\sqrt{5}))$

15. max is 1 at $(\pi/2,0)$, min is -1 at $(\pi/2,\pi)$ and $(3\pi/2,0)$

17. P is a boundary point of set S if every ball of positive radius centered at P contains points of S and points not in S. A closed set is one which contains all its boundary points. The interior of S consists of all points of S which are not boundary points of S.
If $S = \{(x,y,z) : x + y + z = 0\}$ then bdry(S) $= S$; interior of S is empty; S is closed.

19. 1 unit

21. 2 units

23. max is 2 at $(2,-1,1)$, min is -2 at $(-2,1,-1)$

25. max is $3 + 2\sqrt{2}$ at $(\sqrt{2}, \sqrt{2}, 1 - 2\sqrt{2})$;
min is $3 - 2\sqrt{2}$ at $(-\sqrt{2}, -\sqrt{2}, 1 + 2\sqrt{2})$

27. $32/(3\sqrt{3})$ cubic units

Index

Absolute convergence, 464
Absolute maximum, 180
Absolute minimum, 180
Absolute value, 6, 60
Acceleration, 80–82, 85, 259, 265–267
Addition formula, 115–119
Addition of vectors, 260
Alternating series test, 466
  bound, 476
Amplitude, 131
Antiderivative, 101, 301
Antidifferentiation, 102
Approximate integration, 342–350
Approximation
  of areas using rectangles, 280–288
  of definite integrals, 342
  of roots, 231–235
  of the sum of a series, 471–477
  of values of functions, 226–229, 500–501, 507
Area, 280–281
  of a circle, 288
  of a region, 291, 362–367, 387–393, 409
  of a surface, 384, 409
Arc length, 380–384, 394–396, 409
Arccos, 140
Arcsec, 140, 319
Arcsin, 134–137, 317
Arctan, 138–139, 318
Asymptote, 36, 197
  horizontal, 36, 98, 197
Asymptote *(continued)*
  oblique, 198
  vertical, 39, 197
Average rate of change, 79
Average value of a function, 297

Basis vector, 261
Binomial coefficients, 565
Binomial series, 509–511
Binomial theorem, 566
Bound
  alternating series, 476
  geometric, 475
  greatest lower, 574
  integral, 471
  least upper, 574
  upper, lower, 574
Boundary point, 551
Bounded function, 44, 356, 555, 575

Cardioid, 271
Cartesian coordinate system, 2
Cauchy-Riemann equation, 530
Center of gravity; *see* Centroid
Centroid, 268, 406–410
Chain rule, 72–78, 531–539
  proof, 538
  with differentials, 531–539
Chord line, 190

Circle, 12, 270–271
  area of, 288
Circular frequency, 131
Circular function, 110–117
Closed curve, 248
Closed interval, 3
Comparison test
  for integrals, 357
  for series, 454–457
Complete elliptic integral, 383
Completeness, 3, 444, 574
Completing the square, 320
Component
  of acceleration, 259
  of a vector, 261
  of velocity, 259
Composition of functions, 19, 72
Concave, 189
Concavity, 189–194, 226
  of a parametric curve, 256
Conditional convergence, 465
Conic, 272–274
Conservation of energy, 416
Conservative force, 416
Constraint, 207
Continuity, 41–48, 517–520, 573–577
  on an interval, 42, 573
  on the left, 42, 573
  at a point, 41, 573
  on the right, 42, 573
  uniform, 582
Continuous random variable, 419
Convergence
  absolute, 464
  conditional, 465
  of improper integrals, 352–359
  of sequences, 442
  of series, 448–469
Coordinate system, 2, 8
  polar, 268–277
  right handed, 515
Cosecant, 118
Cosh, 172
Cosine, 110
  law, 120
  principal value, 140
Cotangent, 118
Critical point, 79, 97, 99, 180–187
  saddle point, 551
Curve
  contour, 516
  level, 516, 519
  parametric, 246
  plane, 50, 249–253
  smooth, 254
Cumulative distribution function, 428
Cusp, 18, 53
Cycloid, 251, 253
Cylinder, 211–213
Cylindrical shell, 375

Decay, 165–169
Definite integral, 290
Density, 318, 411
Derivative, 56–62
  directional, 541–549
  of elementary functions, 60
  higher-order, 84, 535, 537
  partial, 522–528
  second order, 84
  of trigonometric functions, 123–127
Differentiable, 56, 537
Differential, 63, 102, 292, 536–538
  equation, 72, 87, 130, 166, 169, 430–436, 524, 528
  formalism, 64
Differential element
  of arc length, 381
  of area, 363–378
  of surface area, 384
  of volume, 369, 375, 390
Differentiation, 56
  graphical, 50
  implicit, 89–92
  of integrals, 300
  logarithmic, 162
  rules, 66–75
  of series, 485–491
Directional derivatives, 541–549
Directrix, 272
Discrete random variable, 419
Distribution
  cumulative, 428
  exponential, 421, 424
  normal, 426
  uniform, 420–423
Divergence
  of improper integrals, 352–359
  of sequences, 442
  of series, 448–449
Domain, 14, 16, 514
Dot product, 262

Eccentricity, 272
Element; *see* Differential element
Elliptic integral, 383
Endpoint, 180
Energy, 415–416
  conservation, 416
  kinetic, 416
  potential, 415
Error
  in measurement, 229
  relative, 229
  in series approximation, 471–477
  Simpson's rule, 349
  tangent line approximation, 227–228
  trapezoidal rule, 345
Escape velocity, 417
Euler's theorem, 535
Even function, 20, 26, 65, 294
Existence theorem, 92
Expectation, 422
Exponential decay, 165–169

Exponential function, 149–156
  general, 159–160
Exponential growth, 165–169
Exponents
  laws of, 151, 159–160
Extreme values, 551–560

Factorization
  difference of *n*th powers, 61
First-derivative test, 184
Fixed point, 236
Focus, 272
Folium of Descartes, 253
Force, 413
  gravitational, 415
Frequency, 132
Function, 14
  bounded, 356, 575
  circular, 110–117
  composite, 531
  composition, 19
  continuous, 41–48, 573–577
  decreasing, 96
  defined by integrals, 501–502
  differentiable, 56–65, 300
  even, 20, 56, 65
  exponential, 149–160
  greatest integer, 35
  homogeneous, 535
  hyperbolic, 172–174
  identity, 72
  implicit, 89
  implicit function theorem, 545
  increasing, 96
  integrable, 290, 580
  inverse, 21–25
  inverse hyperbolic, 174–176
  inverse trigonometric, 134–142
  Lagrangian, 559, 560
  logarithmic, 143–149
  maximum value of, 44, 180, 182
  minimum value of, 44, 180, 182
  nondecreasing, 96
  nonincreasing, 96
  of *n* real variables, 514
  odd, 20, 26, 65
  one-to-one, 21
  operations on, 19
  piecewise continuous, 356
  self-inverse, 25
  trigonometric, 110–127
  of two variables, 517
Fundamental theorem of calculus, 300

General power rule, 61, 76, 160
Geometric
  bound, 475
  series, 449
  sum, 287, 449
Gradient, 541
  higher dimension, 546–548

Graph
  of an equation, 8
  of a function, 15
  of an inequality, 8
  parametric, 247
  polar, 270
Growth
  exponential, 165–169
  logistic, 169–171

Half-angle formula, 117, 313
Harmonic motion; *see* Simple harmonic motion
Harmonic series, 451
Homogeneous function, 535, 536
Hooke's law, 130, 413
l'Hôpital's rules, 238–243
Hydrostatic pressure, 411
Hyperbola, 273
Hyperbolic functions, 172–174
Hypocycloid, 253–254, 257

Identity
  complementary angle, 113
  Pythagorean, 111
  supplementary angle, 114
Implicit differentiation, 89–92, 213
Improper integral, 304, 351–359
  convergence, divergence, 352
  types I and II, 351
Indefinite integral, 101
Indeterminate form, 237–243, 502
Induction; *see* Mathematical induction
Inequality, 3
  solving, 5–7
  triangle, 6
Infinite series; *see* Series
Infinitesimal, 64
Inflection, 190
Initial-value problem, 72, 87, 102
Integral, 101, 292
  definite, 290, 294
  improper, 304, 351–359
  indefinite, 101, 301, 307–359
  of a piecewise continuous function, 356
  proper, 351
  Riemann, 578–584
Integral remainder, 505
Integral test, 458
  error bound, 471
Integrand, 292
Intercept, 11, 196
Interior point, 551
Intermediate-value theorem, 45, 577
Interval, 3
Inverse function, 21
  derivative of, 75
  hyperbolic, 172–174
  trigonometric, 134–142
Involute of a circle, 250

Kinetic energy, 416

Lagrange
remainder, 237, 505
multiplier, 559
Langrangian, 559
Leibniz notation, 63, 73, 531
Leibniz's rule, 566
Length of a curve, 380–384
l'Hôpital's rules, 238–243
Limit, 26, 567–572
infinite, 38, 571
at infinity, 36, 571
of integration, 292
left-hand, 34, 571
one-sided, 33, 571
properties of, 29
right-hand, 34, 571
several variables, 517
two variables, 517
uniqueness of, 29, 569
Line
equation of, 10
parametric form, 247
point-slope form, 11
slope, y-intercept form, 11
tangent, 50–56, 62, 190
two intercept form, 11
Linear combination, 261
Linear differential equation, 434
Logarithm, 143
common, 161
general, 160
natural, 143
Logarithmic differentiation, 162

Maclaurin polynomial, 504
Maclaurin series, 492–503
Marginal, 82
Mass, 398
center of, 401–405
Mathematical induction, 86, 563–565
Maximum value, 44
absolute, 180, 551
local, 97, 182, 551
Max-min; *see* Optimization
Mean, 422
Mean-value theorem, 92
generalized, 238, 506
for integrals, 296–297, 506
Method of partial fractions, 326–332
Method of substitution, 309–323
Minimum value, 44
absolute, 180, 551
local, 97, 182, 551
Moment, 401–405

Natural logarithm, 143–149
properties of, 146
Natural number, 3
Newton quotient, 51, 264
Newton's method, 231–235
Normal, 54

Odd function, 20, 26, 65, 294
One-to-one function, 21
Open interval, 3
Optimization, 207–218

p-series, 459
Parabola, 12, 129, 247, 253, 273
Parameter, 246
Parametric curve, 129, 246–258, 390–394
Parametrization, 250, 555
Partial derivative, 522–528
pure, 526
mixed, 526
Partial differential equation, 524, 528
Partial fractions, 326–333
Partial integration, 334–339
Partition, 578
Pascal's triangle, 566
Periodic function, 112
Plane curve, 249–253
Plane vector, 259
Polar axis, 260
Polar coordinate system, 268–277
Polar curve, 270–277, 387–390, 395–396
Polynomial, 43, 334
Taylor, 504
Position vector, 263
Positive series, 454
Potential energy, 415
Power rule, 61, 76
Power series, 480–491
Cauchy product, 483
center of convergence, 481
differentiation, integration, 485–491
interval of convergence, 481
radius of convergence, 481
Pressure, 411
Probability, 418–429
density function, 420
Product rule, 67–68
Projectile problem, 127–130
Pythagorean identity, 111
Pythagorean theorem, 8

Quotient rule, 70

Radian, 110
Random variable, 419
Range, 15, 514
Rate of change, 79, 218–225
Ratio test, 460
Rational function, 43, 200, 334
Rational number, 3
Real line, 2
Real number, 2
Reciprocal rule, 69
Reduction formula, 338
Related rates, 218–225
Riemann sum, 289, 305
Rolle's theorem, 100
Root of a function, 231
Root test, 463

Saddle point, 552, 555
Scalar multiplication, 260
Scalar product; *see* Dot product
Secant, 118
Secant line, 51
Second-derivative test, 193
Sector
  area of circular, 111, 173
  area of hyperbolic, 173, 397
Self-inverse function, 25
Separable equation, 430–433
Sequence, 440–446
  bounded, 441
  convergence, divergence, 442
  increasing, decreasing, 441
  limit, 442
  partial sums, 448
Series, 447–477
  convergence, divergence, 448–449
  geometric, 449
  harmonic, 451
  positive, 454
  power, 480–491
  telescoping, 450
Signum, 31
Simple harmonic motion, 130–132
Simpson's rule, 347–350
  error estimate, 349
Sine, 110
  law, 120
  principal value, 134
Singular point, 53, 183, 551, 553, 555, 556
Sinh, 172
Slope
  of a curve, 54
  of a line, 10
  of a parametric curve, 254
  of a polar curve, 275
Smooth curve, 254
Snell's law, 218
Solid of revolution, 372
Speed, 80, 264
Square root, 16
Squeeze theorem, 29
Standard basis, 261
Standard deviation, 423
Stationary, 79
Straight line; *see* Line
Substitution, 309–323
  definite integral, 310
  inverse trigonometric, 317–320
  tan θ/2, 322
  trigonometric, 311–314
Sum
  geometric, 287, 449
  Riemann, 289, 305
  of a series, 448
  upper, lower, 290
Summation, 282
Surface area, 384–385, 396, 409
Symmetry, 21, 112–113, 197

Tail, 471
Tangent, 118
  principal value, 138
Tangent line, 50–56, 62, 190
  approximation, 226–229
  error estimate, 227
  to level curves, 545, 546
Tangent plane, 524, 525
  using gradients, 548, 549
Tanh, 173
Taylor polynomial, 504
Taylor series, 492–493, 498
Taylor's formula, 505
Taylor's theorem, 237, 504
Telescoping series, 450
Test
  first-derivative, 184
  second-derivative, 193
Theorem
  fundamental, 300
  intermediate-value, 45, 577
  mean-value, 92, 238, 297
  Pythagorean, 8
  Rolle's 100
  squeeze, 29
Trapezoidal rule, 343–346
  error estimate, 345
Triangle inequality, 6, 294
Trigonometric functions, 110–127
  inverse, 134–142
Trigonometry, 119–121

Uniform continuity, 582
Uniform distribution, 420–421

Variance, 423
Vector, 259–267
  acceleration, 265–267
  addition, 260
  basis, 261
  gradient, 541
  higher dimensional, 546
  position, 263
  scalar multiplication, 260
  velocity, 264–267
Velocity, 80–82
Volume, 368–378
  of a cone, 370, 373
  by cylindrical shells, 375–378
  for parametric curves, 393
  in polar coordinates, 389
  of a pyramid, 370
  by slicing, 369–374
  of a solid of revolution, 372–378, 409
  of a sphere, 373

Wave equation, 528
Work, 413

Zero of a function, 231
Zero vector, 260